D0842982

THE TOGAVIRUSES
Biology, Structure, Replication

Contributors

Walter E. Brandt
Margo A. Brinton
Dennis T. Brown
Roy W. Chamberlain
Joel Dalrymple
Henrik Garoff
Scott B. Halstead
Ari Helenius
Leevi Kääriäinen
S. Ian T. Kennedy
John Lenard
Frederick A. Murphy
J. S. Porterfield
Leon Rosen
Philip K. Russell
Milton J. Schlesinger
R. Walter Schlesinger
Robert E. Shope
Kai Simons
Victor Stollar
Ellen G. Strauss
James H. Strauss
Gerd Wengler
E. G. Westaway

THE TOGAVIRUSES

Biology, Structure, Replication

EDITED BY
R. Walter Schlesinger

Department of Microbiology
College of Medicine and Dentistry of New Jersey
Rutgers Medical School
Piscataway, New Jersey

ACADEMIC PRESS 1980
A Subsidiary of Harcourt Brace Jovanovich, Publishers
New York London Toronto Sydney San Francisco

ACADEMIC PRESS, INC.
111 Fifth Avenue, New York, New York 10003

United Kingdom Edition published by
ACADEMIC PRESS, INC. (LONDON) LTD.
24/28 Oval Road, London NW1 7DX

Library of Congress Cataloging in Publication Data
Main entry under title:

The Togaviruses.

Includes bibliographies and index.
1. Togaviruses. I. Schlesinger, Robert Walter, 1913– [DNLM: 1. Arboviruses. QW168.5.A7 T645]
QR415.5.T63 576'.6484 79–6783
ISBN 0–12–625380–3

PRINTED IN THE UNITED STATES OF AMERICA

80 81 82 83 9 8 7 6 5 4 3 2 1

Contents

4. Virus–Host Interactions in Natural and Experimental Infections with Alphaviruses and Flaviviruses

R. WALTER SCHLESINGER

5. Immunological Parameters of Togavirus Disease Syndromes

SCOTT B. HALSTEAD

6. Epidemiology of Arthropod-Borne Togaviruses: The Role of Arthropods as Hosts and Vectors and of Vertebrate Hosts in Natural Transmission Cycles

ROY W. CHAMBERLAIN

7. Arthropods as Hosts and Vectors of Alphaviruses and Flaviviruses—Experimental Infections

LEON ROSEN

8. Togavirus Morphology and Morphogenesis

FREDERICK A. MURPHY

9. Alphavirus Proteins

KAI SIMONS, HENRIK GAROFF, AND ARI HELENIUS

10. Lipids of Alphaviruses

JOHN LENARD

11. The Genome of Alphaviruses

S. IAN T. KENNEDY

12. Synthesis of Alphavirus RNA

S. IAN T. KENNEDY

13. Translation and Processing of Alphavirus Proteins

MILTON J. SCHLESINGER AND LEEVI KÄÄRIÄINEN

14. Mutants of Alphaviruses: Genetics and Physiology

ELLEN G. STRAUSS AND JAMES H. STRAUSS

15. Defective Interfering Alphaviruses

VICTOR STOLLAR

16. Effects of Alphaviruses on Host Cell Macromolecular Synthesis

GERD WENGLER

17. The Assembly of Alphaviruses

DENNIS T. BROWN

18. Chemical and Antigenic Structure of Flaviviruses

PHILIP K. RUSSELL, WALTER E. BRANDT, AND JOEL M. DALRYMPLE

19. Replication of Flaviviruses

E. G. WESTAWAY

20. Togaviruses in Cultured Arthropod Cells

VICTOR STOLLAR

21. Non-Arbo Togaviruses

MARGO A. BRINTON

List of Contributors

Numbers in parentheses indicate the pages on which the authors' contributions begin.

Walter E. Brandt (503), Department of Viral Diseases, Walter Reed Army Institute of Research, Washington, D. C. 20012

Margo A. Brinton (623), The Wistar Institute, Philadelphia, Pennsylvania 19104

Dennis T. Brown (473), Cell Research Institute, University of Texas, Austin, Texas 78712

Roy W. Chamberlain (175), Virology Division, Bureau of Laboratories, Center for Disease Control, Atlanta, Georgia 30333

Joel Dalrymple (503), Department of Viral Diseases, Walter Reed Army Institute of Research, Washington, D. C. 20012

Henrik Garoff (317), European Molecular Biology Laboratory, 6900 Heidelberg, West Germany

Scott B. Halstead (107), Department of Tropical Medicine and Medical Microbiology, John A. Burns School of Medicine, University of Hawaii, Honolulu, Hawaii 96816

Ari Helenius (317), European Molecular Biology Laboratory, 6900 Heidelberg, West Germany

Leevi Kääriäinen (371), Department of Virology, University of Helsinki, Haartmaninkatu 3, SF-00290 Helsinki 29, Finland

S. Ian T. Kennedy (343, 351), Department of Biology, University of California, San Diego, La Jolla, California 92093

John Lenard (335), Department of Physiology, College of Medicine and Dentistry of New Jersey, Rutgers Medical School, Piscataway, New Jersey 08854

Frederick A. Murphy (241), College of Veterinary Medicine and Biomedical Sciences, Colorado State University, Fort Collins, Colorado 80525

J. S. Porterfield (13), Sir William Dunn School of Pathology, University of Oxford, Oxford OX1 3RE, England

Leon Rosen (229), Pacific Research Section, Laboratory of Parasitic Diseases, National Institute of Allergy and Infectious Diseases, National Institutes of Health, Honolulu, Hawaii 96806

Philip K. Russell (503), Walter Reed Army Institute of Research, Washington, D. C. 20012

Milton J. Schlesinger (371), Department of Microbiology and Immunology, Washington University School of Medicine, St. Louis, Missouri 63110

R. Walter Schlesinger (1, 83), Department of Microbiology, College of Medicine and Dentistry of New Jersey, Rutgers Medical School, Piscataway, New Jersey 08854

Robert E. Shope (47), Yale Arbovirus Research Unit, Department of Epidemiology and Public Health, Yale University School of Medicine, New Haven, Connecticut 06510

Kai Simons (317), European Molecular Biological Laboratory, 6900 Heidelberg, West Germany

Victor Stollar (427, 583), Department of Microbiology, College of Medicine and Dentistry of New Jersey, Rutgers Medical School, Piscataway, New Jersey 08854

Ellen G. Strauss (393), Division of Biology, California Institute of Technology, Pasadena, California 91125

James H. Strauss (393), Division of Biology, California Institute of Technology, Pasadena, California 91125

Gerd Wengler (459), Institut für Virologie, Justus Liebig-Universität Giessen, 6300 Giessen, West Germany

E. G. Westaway (531), Department of Microbiology, Alfred Hospital, Monash Medical School, Prahran Victoria, Australia 3181

Preface

Frequently during the past year friendly conversations with distinguished colleagues led to exchanges like this: "And what are you up to these days?" "I'm editing a book." "How exciting! What's it about?" "Togaviruses." "Oh, that sounds interesting." "Do you know what they are?" "Well, frankly, no; what are they?" "Oh, you know, alpha- and flaviviruses and some others." "I'm afraid that doesn't help at all!" "Well, Sindbis virus is an example." "Ahh, oh yes, that *is* a fascinating virus!" "They all are." I knew then that this book could fulfill at least one worthwhile mission.

For a long time, interest stimulated by these viruses, specifically those that are transmitted to vertebrates by bloodsucking arthropods (arboviruses), was dominated by a dichotomy reminiscent of nineteenth century biology. E. B. Wilson wrote in 1901 that "a lack of mutual understanding existed between the field naturalist and the laboratory worker which found expression in a somewhat picturesque exchange of compliments, the former receiving the flattering appellation of the 'bug-hunters,' the latter the ignominious title of the 'section-cutters,' which on some irreverent lips was even degraded to that of 'worm-slicers' . . . I daresay there was on both sides justification for these delicate innuendos."*

If we substitute "fraction cutters" and "gel slicers" for their earlier counterparts, we are transported into the more recent past of research on arboviruses. Fortunately, the dichotomy is diminishing rapidly, and therefore it seems a good time to help the process along by publishing this volume now.

In the development of virology as a self-sustaining science, the creation of a rational taxonomic system, based on principles of struc-

*Quoted, with permission, from Garland E. Allen, "Thomas Hunt Morgan, The Man and His Science," Princeton University Press, Princeton, New Jersey, 1978.

ture and molecular function, has played a key role. Sometimes this process has brought together in a single taxon viruses which are strange bedfellows in the sense that their divergent biological or epidemiological behavior patterns would not have suggested fundamental relatedness. The converse is also true: many viruses that are similar in their biological interactions with hosts or cells have turned out to be far apart in their taxonomic properties.

In the case of the viruses that are the subject of this book—the members of the family Togaviridae—both of these paradoxes have posed difficulties. On one hand, no one would have predicted on biological grounds alone that rubella virus and other "non-arbo" viruses would find themselves taxonomically wedded to Saint Louis or equine encephalitis viruses and their many arthropod-transmitted cohorts; on the other hand, one could not have assumed *a priori* that the many viruses which share the singularly complex feature of arthropod–vertebrate transmission cycles (the arboviruses) would prove to be as diverse in basic properties as are the Togaviridae, Bunyaviridae, Rhabdoviridae, and Reoviridae.

Historically, the need for taxonomic order clearly grew out of the profusion of arboviruses which were discovered in the era following World War II. It is to the lasting credit of Jordi Casals and his coworkers at the Rockefeller Foundation Laboratories (more recently at the Yale Arbovirus Research Unit) that they accurately identified, on the basis of painstaking serological analyses, the distinctness of the group A and group B arboviruses (now the Alphavirus and Flavivirus genera of the family Togaviridae) from each other and from the many arboviruses which have now been assigned to other families.

Many of these viruses cause diverse diseases of man and animals. The mode of their transmission involves continuous bidirectional bridging of the phylogenetic gap separating invertebrate from vertebrate hosts and thus stamps them as biologically unique. The ecological demands that this mode of transmission imposes on arboviruses have undoubtedly had a critical impact on the limits of their geographical distribution and on their varying responses to selective pressures for species differentiation. Genus and species differentiation, even within the narrower confines of the family Togaviridae, is such that we are still dealing with an immensely heterogeneous assortment of viruses. This very heterogeneity challenges us to define shared and distinguishing characteristics and to trace, through the use of antigenic, genetic, and biochemical analysis (including nucleic acid and protein sequencing), past and projectable evolutionary relationships.

These are examples of the sources of fascination that the togaviruses (whether or not transmitted by arthropods) hold for physicians and veterinarians, ecologists, entomologists and epidemiologists, cell biologists, immunologists and virologists, physical chemists and biochemists, molecular biologists and geneticists.

The idea to put this book together arose from the realization that the branches emerging from these diverse roots of interests had not previously been gathered and bundled in a single comprehensive volume. Its design is intended to provide (1) a summary of the biological and medical challenges posed by these viruses which might channel the directions of research on their basic properties, and (2) a status report on their structure and biochemistry and on mechanisms of replication which might in turn lead to new perceptions of the capacity to solve biological and epidemiological problems through the concepts and technology of molecular biology.

In trying to meet these aims, some subjects are discussed in great detail, others in summary form, still others not at all in the form of separate chapters. Among the latter are those that are adequately covered in standard textbooks (e.g., clinical details, routine procedures for diagnostic virus isolation and identification and for serological tests) or that are not developed to a point where they would reveal something *specifically* or *uniquely* instructive about togaviruses (e.g., immunological host responses, the role of interferons, antiviral chemotherapy, vaccine development). In general, authors were left free to choose their individual emphases and styles, and editing was kept to a minimum. The field is in flux, especially in research on structural and molecular aspects; therefore the treatment of overlapping subjects inevitably has led to redundancy and sometimes to the presentation of conflicting findings or interpretations. These were permitted to stand because each chapter was intended to be its author's comprehensive and self-sufficient assessment of an area of special competence.

It is a pleasure to acknowledge my gratitude to a group of distinguished authors for their enthusiastic cooperation in this joint venture, for their patience in the face of unavoidable delays, and for their willingness to further improve and update their contributions when the "point of no return" was finally reached in the form of edited copy and page proofs.

R. Walter Schlesinger

1

Introduction

R. WALTER SCHLESINGER

I. HISTORICAL BACKGROUND: THE ARTHROPOD-BORNE VIRUSES (ARBOVIRUSES)

The prefix "toga-" made its first public appearance in 1965 in the proposals and recommendations of a provisional committee for nomenclature of viruses whose task was to fit the sparse physicochemical data then available into a "system of viruses" (Lwoff and Tournier, 1966). The Togavirales were given the exalted status of an order to which the Arboviridae were assigned as a family. Although such subordination of a taxon that was defined in biological terms (arbo = *ar*thropod-*bo*rne) to one that was based on structural features (*toga* = shroud or envelope) was inconsistent with taxonomic principles (see Chapter 2), it reflected the contemporary state of the art: All that was known was that (1) a few prototypes of the viruses that were to be grouped together occupied a position of small to intermediate in the size range of classifiable viruses, (2) they were inactivated by ether or deoxycholate and therefore presumably contained lipid envelopes, and (3) they contained an RNA genome which, in the form of phenol extracts, was infectious. All viruses that had been examined and possessed these three properties were in fact "group-A" or "group-B" arboviruses (Casals, 1957), so designated on the basis of group-specific

THE TOGAVIRUSES

ISBN 0-12-625380-3

antigens and transmission by bloodsucking arthropods to vertebrate hosts.

What was *not* known was (1) whether viruses meeting the same structural specifications would be found among those that were *not* dependent on arthropod transmission, (2) whether other arboviruses—those that were antigenically distinct from group A or B—would fulfill the same three criteria, i.e., size, envelope, and infectious RNA.

As knowledge about the unique features of the structure and replication of group-A and -B arboviruses advanced, these viruses could be clearly delineated from those of other arbovirus groups. The designations in the taxonomic hierarchy were shifted: Within the family Togaviridae, group-A and -B arboviruses became genera and were rechristened, with Linnaean fervor and etymological inconsistency, *alpha-* (for group A) and *flavi-* (for yellow fever) viruses, respectively (Wildy, 1971; Porterfield *et al.*, 1978; see also Chapter 2). This rearrangement allowed the inclusion of non-arthropod-borne viruses in the taxon Togaviridae (see below and Chapter 2), and it eliminated the inappropriate use of a mode of transmission from the taxonomic system.

The process of inching toward a rational classification, while in itself reflecting advancing technology, has in turn contributed greatly to progress in the characterization of those arboviruses which are togaviruses as well as those which are not (notably the family Bunyaviridae). At the same time, these developments affirmed the importance of arboviruses as pathogens for man and diverse animals (see Chapter 3) and promoted interest in the one unique feature by which all arboviruses remain defined: their ability to cross the phylogenetic gap separating invertebrate and vertebrate hosts.

Although the latter property has lost its significance as a taxonomic criterion, it remains the cornerstone of discoveries in the historical sense. Prior to the era of quantitative and molecular virology, many fundamentally important questions arose in investigations of the ecology of arthropod-borne viruses and of their interactions with both invertebrate vectors and vertebrate hosts in natural and experimental infections. These related to epidemiological patterns, viral or host determinants of virulence, viral persistence in nature, antigenic relationships, aspects of pathogenicity and immunity, and many other biological features. The first seven chapters of this book address themselves to these topics, and this chapter briefly summarizes key developments in historical perspective.

The discovery that yellow fever was transmitted to man by bites of the mosquito *Aedes aegypti* (Reed *et al.*, 1900) marked the start of a heroic and successful assault on one of mankind's catastrophic public health hazards (Strode, 1951). Epidemiological and experimental findings established a similar role for mosquitoes (Graham, 1903), specifically *A. aegypti* (Bancroft, 1906) in the transmission of dengue fever and of the sandfly, *Phlebotomus papatasi*, in Mediterranean 3-day fever, now called sandfly fever (Doerr and Russ, 1909). These three—yellow fever, dengue fever, and sandfly fever—were the first specifically *human* afflictions for which *filterable viruses* were demonstrated as etiological agents (Reed and Carroll, 1902; Ashburn and Craig, 1907; Doerr *et al.*, 1909).

During the 1930s, seasonally and geographically limited outbreaks of encephalitis in human populations (e.g., Japanese, St. Louis, and Russian spring-summer—JE, SLE, and RSSE)[1] and in horses (Western, Eastern, and Venezuelan equine—WEE, EEE, and VEE), together with epizootic diseases in other animals (e.g., Louping ill and Rift Valley fever—LI and RVF) focused attention on the medical and veterinary importance of a "new" class of viruses: All of them seemed to share the unusual feature of dependence, for their maintenance in nature, on the ability to multiply in various species of bloodsucking arthropods and in vertebrate animals (Theiler and Downs, 1973). Enormous investigative efforts showed that (1) specific arthropods became infected by a blood meal from an infected vertebrate host, (2) the viruses multiplied in the arthropods, (3) once attaining threshold titers in the arthropod's salivary gland (marking the end of the "extrinsic incubation period"), the viruses could be transmitted, for the rest of the arthropod's life, to new susceptible vertebrates, and (4) in the latter, infection led to virus multiplication and viremia which may or may not be associated with clinically manifest diseases. These characteristics led to the formulation of biological criteria for defining arthropod-borne viruses (arboviruses) (World Health Organization, 1967) as "viruses which are maintained in nature principally, or to an important extent, through biological transmission between susceptible vertebrate hosts by haematophagous arthropods; they multiply and produce viremia in the vertebrates, multiply in the tissues of arthropods, and are passed on to new vertebrates by the bites of arthropods after a period of extrinsic incubation."

[1] Abbreviations cited are those used in the "International Catalogue of Arboviruses" (Berge, 1975).

It is axiomatic that epidemics or epizootics of proved arboviral etiology have been *initiated* only in ecological environments providing for the coexistence of a competent, sufficiently dense vector population and of susceptible vertebrate hosts. This truism has been adduced, along with other arguments, in favor of the suggestion that arboviruses may have their evolutionary origin in arthropods (Schlesinger, 1971).

For the vast majority of arboviruses, alternating arthropod–vertebrate–arthropod transmission cycles seem to be the only means of spreading infection, regardless of whether the virus in question does or does not cause a known disease of man or beast. There are, however, exceptions: For example, certain tick-borne viruses, once established in vertebrate hosts by bites of infected arthropods, can spread by direct vertebrate-to-vertebrate transmission; conversely, tick-borne encephalitis viruses do not "need" the vertebrate link but can be transmitted transovarially from one tick generation to another (see Chapter 6). Similar transstadial transmission of mosquito-borne viruses has been observed in nature (Watts *et al.*, 1973; Coz *et al.*, 1976) and under experimental conditions (Rosen *et al.*, 1978; Aitken *et al.*, 1979). These exceptions do not, however, invalidate the generalization that spread to vertebrates requires, at least initially, biological transmission by competent arthropod vectors.

The number of viral species now accepted as, or suspected of, fulfilling the cited criteria for arboviruses has grown dramatically during the past 30 years and continues to rise. This has come about not merely through investigation of human and animal diseases. Rather, most of the viruses were first isolated as by-products of field investigations connected with yellow fever or encephalitis surveys in various parts of the world (Theiler and Downs, 1973).

Standard procedures used in such surveys involve (1) monitoring of disease incidence in indigenous human and animal populations, (2) testing for virus in blood and tissues of diseased or normal feral and domestic animals, (3) trapping of mosquitoes and other hematophagous arthropods and testing them for virus either by biological transmission to animals or via inoculation of homogenates, (4) posting of cages containing "sentinel" animals in natural environments (e.g., rain forests) where they are subject to bites by bloodsucking arthropods and then testing the animals for viremia. As a matter of routine, test materials for virus isolation are inoculated into mice (or cell cultures) and passaged serially in efforts to produce encephalitis (or cytopathic changes). At the same time, sera of vertebrate donors or recipients of suspect materials are tested for antibodies to known or newly isolated viruses.[2]

Based on techniques used in the propagation and eventual attenuation of yellow fever virus (reviewed in Theiler, 1951) and in the successful isolation of other arboviruses, the *intracerebral inoculation* of *mice* (especially newborn or suckling mice) was for many years the standard procedure for the isolation and identification of new arboviruses (Theiler and Downs, 1973; Berge, 1975). In this host, most arboviruses induce fatal encephalitis, in some cases only after varying numbers of passages required for "adaptation." Despite the fact that *cell cultures* of various origins and intrathoracic inoculation of whole mosquitoes have now partially replaced the mouse as the experimental host system for initial virus isolation, by far the majority of prototype (wild-type) strains of arboviruses in current laboratory use are mouse-adapted neurovirulent derivatives of field isolates. This fact has important implications with respect to the safe handling of these strains and to their genetic or phenotypic relationship to the parental virus (see Chapter 4).

Over the years, the criteria for definitive designation of a given virus as a bona fide arbovirus have been sharpened. The contributions by many investigators have been summarized in the latest "International Catalogue of Arboviruses" (Berge, 1975) and a subsequent supplement (Karabatsos, 1978). Together, these two reports consider a total of 389 viral species and classify them as shown in Table I. Of the 363 viruses designated as definite, probable, or possible arboviruses, the majority (280) have thus far been assigned to only four taxonomic groups, namely, Togaviridae, Bunyaviridae and bunyalike viruses, Rhabdoviridae and Reoviridae (Table II). Thus arthropod–vertebrate–arthropod transmission cycles seem to be particularly characteristic of four families of RNA viruses which, in other respects, differ from each other in readily distinguishable key properties (Table III).

It is tempting to speculate that there must be a rational explanation for the association of viruses representing these four structurally defined groups with the capacity to replicate equally well in two genetically and physiologically disparate phyla of the animal kingdom.[3]

[2] The nature of the searches in a variety of suspect ecological niches, often in tropical or subtropical areas, explains the polyglot names given to these viruses (see Chapter 2): By convention, they are named after the locale in which they were first isolated.

[3] There are, of course, a vast number and variety of viruses which are indigenous and often highly pathogenic to insects and other arthropods; these are not transmitted to

TABLE I

Biological Classification of Viruses Listed in the "International Catalogue of Arboviruses" (Berge, 1975) and Supplement (Karabatsos, 1978)

Classification	Number	Percentage
Definite arboviruses	98	25.2
Probable arboviruses	54	14.0
Possible arboviruses[a]	211	54.2
Probably not arboviruses	9	2.3
Definitely not arboviruses	17	4.3
Total	389	100.0%

[a] "Classification as a *possible* arbovirus was assigned to viruses that were first isolated from an arthropod, or were first isolated from man or a lower vertebrate, under circumstances epidemiologically compatible with arthropod transmission" (Berge, 1975).

They can do so efficiently over a remarkably wide temperature range. Thus, under natural conditions, they must possess a greater inherent adaptability to different organismal and physical environments than is normally required of any other group of animal viruses.

This adaptability extends to the ability of certain arboviruses to adjust their preferences for growth in different arthropod or vertebrate species to the varying fauna available in different ecological habitats (see Chapter 6). Perhaps the limits of this adaptability explain the restricted geographical distribution of individual viral species (Theiler and Downs, 1973); or else, assuming that antigenic relatedness of all species within a given genus betrays evolutionary relatedness, the very differentiation into many species may be an expression of the diversity of selective pressures to which different ecological environments have, over the span of millennia, exposed the branches emerging from their ancestral origin. These questions, as they relate to togaviruses, are further analyzed in Chapters 2 and 6. Because the adaptation to laboratory mice has played such a pivotal role in the discovery and characterization of togaviruses, the expression of neurovirulence in this host species will be discussed in some detail in Chapter 4.

---

vertebrates, nor do they seem to be infectious for higher animals or their cells under experimental conditions (Tinsley and Harrap, 1978). In addition, many *plant* viruses are biologically transmitted by arthropods and thus can cross the barriers even between two different kingdoms.

TABLE II

Taxonomic Designation of Viruses Definitely or Tentatively Identified as Arboviruses

Taxon	Number of definite, probable, or possible arboviruses
Bunyaviridae	90
Bunyalike viruses	55
Togaviridae	
Alphavirus	20
Flavivirus	53
Reoviridae	
Orbivirus	42
Rhabdoviridae	20
	280
Miscellaneous or unclassified	83
Total	363

II. THE TOGAVIRIDAE

Among the arboviruses, those in the family Togaviridae are subdivided into the genera *Alphavirus* and *Flavivirus*, based on their respective group (genus)-specific antigens (Chapter 2). The sharing of common antigenic specificity is paralleled within each of these two genera by common features of structure, composition, and replication. Conversely, the lack of antigenic relationships between alpha- and flaviviruses is only one of several fundamental differences between these two genera (see partial listing in Table IV). These differences, perhaps profound enough to question the placement of both within a single family, will be documented in subsequent chapters.

The guiding principle in assigning family status to certain sets of viruses rests on those structural and chemical characteristics which set them apart from other viruses (see Chapter 2). In accordance with this principle, several viruses which are not arthropod-borne [genera rubivirus (rubella) and pestivirus, several flaviviruses, as well as four thus far unclassified viruses, see Table V] have been included in the family Togaviridae (Porterfield *et al.*, 1978). The relationships between "non-arbo" and "arbo" togaviruses are discussed in detail in Chapters 2, 8, and 21.

TABLE III

Key Properties of Four Virus Families Containing Arboviruses

Family	RNA genome structure					Virion-associated		Capsid symmetry
	Linear	Segmented	Single-stranded	Double-stranded	Polarity	RNA polymerase	Envelope	
Bunyaviridae		X	X		–	Yes	Yes	Helical (closed circles)
Togaviridae	X		X		+	No	Yes	Cubic[a]
Reoviridae		X		X	+/–	Yes	No	Cubic
Rhabdoviridae	X		X		–	Yes	Yes	Helical

[a] See Chapter 8, especially for a consideration of the capsid symmetry of flaviviruses, which remains uncertain.

TABLE IV

Examples of Differences between Alphaviruses and Flaviviruses[a]

Property	Genus *Alphavirus*	Genus *Flavivirus*
Antigenic specificity	Genus (group-A)-specific capsid antigen	Genus (group-B)-specific caspid antigen
Structural features		
Average virion diameter	60–65 nm	48–55 nm
RNA		
Sedimentation coefficient	42–50 S[b]	40–45 S[b]
5′-cap	Type O: $m^7G(5')ppp(5')ApUp...$	Type I: $m^7G(5')ppp(5')AmpNp...$
3′-poly (A)	50–120 Nucleotides	Absent or <12 nucleotides
Polypeptides		
Capsid	30K	13–14K
Envelope	gp 49K–59K[c], gp 47–52K, gp 10K[d]	7–8K (not gp)[e], gp 51–59K
Transcription	Genome size and subgenomic (26 S) mRNAs	Genome size mRNA (40–45 S)
Translation	Precursor polyproteins for nonstructural (from 42–50 S mRNA) and structural (from 26 S mRNA) proteins; posttranslational cleavage	No precursor protein identified; multiple initiation sites proposed
Assembly and release	"Budding" from plasma membrane prominent	Mainly in association with cytoplasmic membranes; "budding" not prominent

[a] Documentation of these and other differences will be presented in appropriate chapters of this volume.
[b] Depending on the method used.
[c] Ranges given for polypeptides represent values reported for different viruses in each genus.
[d] Contains 45% carbohydrate; thus far found only in SFV as separate entity.
[e] Remains associated with the envelope or nucleocapsid, depending on the detergent used for dissociation.

TABLE V

The Family Togaviridae

Genus	Prototype species	Number of species
Alphavirus	Sindbis virus	20[a]
Flavivirus	Yellow fever virus	60[a]
Rubivirus	Rubella virus	1[b]
Pestivirus	Hog cholera virus	3[b]
Unclassified		4[b]

[a] See Chapter 2 for subgroup and species designation.
[b] See Chapter 21 for detailed consideration.

III. PROSPECTUS

In this book all viruses currently assigned to the family Togaviridae will be considered. Alpha- and flaviviruses will be covered most extensively because (1) they include by far the largest number and variety of human pathogens (Chapters 3 and 5); (2) special biological and epidemiological-epizootiological interest attaches to the arthropod–vertebrate–arthropod transmission cycle and its experimental equivalents (Chapters 4, 6, 7, and 20), and (3) much is known about their structure (Chapter 8), composition (Chapters 9–11 and 18), and replication (Chapters 12–17, 19, and 20).

The rubiviruses, pestiviruses and other "non-arbo" togaviruses are also of enormous medical or veterinary importance and offer great challenges in the continuing investigation of their structural and molecular characteristics (Chapters 8, 21).

REFERENCES

Aitken, T. H. G., Tesh, R. B., Beaty, B. J., and Rosen, L. (1979). *Am. J. Trop. Med. Hyg.* **28**, 119–121.

Ashburn, P. M., and Craig, C. F. (1907). *J. Infect. Dis.* **4**, 440–475.

Bancroft, T. L. (1906). *Aust. Med. Gaz.* **25**, 17–18.

Berge, T. O., ed. (1975). "International Catalogue of Arboviruses," 2nd ed., DHEW Publ. No. (CDC) 75-8301. U.S. Dep. Health, Educ. Welfare, Public Health Serv., Washington, D.C.

Casals, J. (1957). *Trans. N.Y. Acad. Sci.* **19**, 219–235.

Coz, J., Valade, M., Cornet, M., and Robin, Y. (1976). *C. R. Acad. Sci.* **283**, 109–122.

Doerr, R., and Russ, V. K. (1909). *Arch. Schiffs- Trop.-Hyg.* **13**, 693–706.

Doerr, R., Franz, K., and Taussig, S. (1909). "Das Pappatacifieber." Franz Deuticke, Leipzig and Vienna.

Graham, H. (1903). *J. Trop. Med.* **6**, 209–214.
Karabatsos, N. (1978). *Am. J. Trop. Med. Hyg.* **27**, 372–440.
Lwoff, A., and Tournier, P. (1966). *Annu. Rev. Microbiol.* **20**, 45–74.
Porterfield, J. S., Casals, J., Chumakov, M. P., Gaidamovich, S. Y., Hannoun, C., Holmes, I. H., Horzinek, M. C., Mussgay, M., Oker-Blom, N., Russell, P. K., and Trent, D. W. (1978). *Intervirology* **9**, 129–148.
Reed, W., and Carroll, J. (1902). *Am. Med.* **3**, 301–305.
Reed, W., Carroll, J., Agramonte, A., and Lanzear, J. W. (1900). *Philadelphia Med. J.* **6**, 790–796.
Rosen, L., Tesch, R. B., Lien, J. C., and Cross, J. H. (1978). *Science* **199**, 909–911.
Schlesinger, R. W. (1971). *Curr. Top. Microbiol. Immunol.* **55**, 241–245.
Strode, G. K., ed. (1951). "Yellow Fever." McGraw-Hill, New York.
Theiler, M. (1951). *In* "Yellow Fever" (G. K. Strode, ed.), pp. 46–136. McGraw-Hill, New York.
Theiler, M., and Downs, W. G. (1973). "The Arthropod-Borne Viruses of Vertebrates." Yale Univ. Press, New Haven, Connecticut.
Tinsley, T., and Harrap, K. A. (1978). *In* "Comprehensive Virology" (H. Frankel-Conrat and R. R. Wagner, eds.), Vol. 12, pp. 1–101. Plenum, New York.
Watts, D. M., Pantuwatana, S., DeFloiart, G. R., Yuill, T. M., and Thompson, W. H. (1973). *Science* **182**, 1140–1141.
Wildy, P. (1971). "Classification and Nomenclature of Viruses," Monographs in Virology, Vol. 5. Karger, Basel.
World Health Organization (1967). Arboviruses and human disease. *W.H.O. Tech. Rep. Ser.* No. 369, p. 9.

2

Antigenic Characteristics and Classification of Togaviridae

J. S. PORTERFIELD

I. INTRODUCTION

A. Classification

Classification is essentially the ordering together of objects into sets on the basis of common characters. In the antigenic classification of

THE TOGAVIRUSES

ISBN 0-12-625380-3

viruses, the common characters are viral antigens detected by serological tests; in the taxonomic classification of viruses the nature of the viral nucleic acid and its mode of replication are the essential criteria. These two approaches overlap to a considerable extent in that viral antigens are determined by viral genes, and viruses which have closely related genomes are likely to have similar or related antigenic properties. Quite minor differences in viral nucleic acids may result in major antigenic differences, such that two viruses could be classified as antigenically distinct. In the reverse direction it is possible, although highly unlikely, for two viruses with different genomes to code for similar amino acid sequences which are recognized as related antigens in serological tests. Determination of the complete nucleotide sequence of the genetic material of a virus is a formidable undertaking and is at present not a realistic basis for classification. Serological techniques, on the other hand, are simple, sensitive, and quite specific and have contributed greatly to our knowledge of interrelationships among viruses.

A further, quite different, criterion for the classification of viruses is based upon the mode of transmission of the virus in nature. Of the various possibilities, some viruses are transmitted between susceptible hosts through the bites of bloodsucking arthropods; arthropod-transmitted viruses may be further subdivided into those in which transmission is purely mechanical and those in which transmission is biological, in the sense that virus replication takes place within the tissues of the arthropod vector. Arthropod-borne viruses that have a biological cycle in their arthropod vectors are known as "arboviruses," a term which is without any taxonomic significance. The biologically defined set of arboviruses contains subsets determined on the basis of antigenic properties and other subsets determined on the basis of chemical and structural criteria which frequently overlap with those defined on antigenic grounds. Representatives from at least four different virus families, the Togaviridae, Bunyaviridae, Reoviridae, and Rhabdoviridae, are found within the biologically defined set of arboviruses, and a single family, such as the Togaviridae, contains some members that are arboviruses in the biological sense and others that are not. Precision in the use of terms is important if ambiguities are to be avoided.

B. Nomenclature and Hierarchy of Subsets

The Linnean terms "family," "genus," and "species" are used in a particular sense in viral nomenclature. Under the rules adopted by the

International Committee on Taxonomy of Viruses (Wildy, 1971; Fenner, 1976), three different levels are recognized, although the criteria adopted are not specifically laid down. Rule 11 states that "for pragmatic purposes the *species* is considered to be a collection of viruses with like characters." Rules 13 and 14 state that "the *genus* is a group of species sharing certain common characters," and "the ending of the name of a viral genus is . . . virus." Rule 16 states that "a *family* is a group of genera with common characters, and the ending of the name of a viral family is . . . viridae."

In the serological hierarchy, the term "group" or "serogroup" is used to cover a set of viruses having common antigenic characters. Subsets of more closely related viruses are placed together in subgroups or complexes of viruses; thus the tick-borne virus subgroup or the flaviviruses complex is a subset of flaviviruses whose members are closely related antigenically (and also share common biological vectors). Below the virus or type it may be possible to recognize different subtypes, and below the subtypes to recognize strain differences.

The ways in which these two hierarchies are related is shown in Table I.

The concept of the serological group is based upon the premise that a virus within a serogroup may be expected to cross-react with other viruses in the same serogroup, but not with viruses in other serogroups. The serogroup is equated with the genus in the taxonomic hierarchy, and group-A arboviruses become members of the genus *Alphavirus;* thus an alphavirus may show cross-reactions with other alphaviruses, but not with viruses in other genera in the same family, i.e., not with flaviviruses, rubiviruses, or pestiviruses.

TABLE I

Relation between Taxonomic and Serological Hierarchies as Applied to Alpha- and Flaviviruses

Linnean taxon	Examples[a]		Serological classification
Family	Togaviridae		
Genus	*Alphavirus* (group A)	*Flavivirus* (group B)	Group or serogroup
Subgenus	WEE	Dengue	Subgroup or complex
Species	SIN	Dengue type 2	Virus or type
Subspecies	?	?	Subtype
Individual	SIN HR strain	Dengue New Guinea B	Strain

[a] For a more complete presentation see Section II,B.

It may be noted in passing that "supergroup" relationships, that is, cross-reactions between viruses in different serogroups, have been demonstrated within the family Bunyaviridae (Whitman and Shope, 1962); but in this family the genus *Bunyavirus* comprises not a single serogroup but at least 10 (Porterfield *et al.*, 1975–1976; Fenner, 1976).

In an earlier review of the nature of serological relationships among arthropod-borne viruses (Porterfield, 1962) it was suggested that the concept of multiple antigenic determinants provided the most reasonable explanation for the facts then known, and recent developments have provided no better interpretation of the observed data. It seems probable that there are at least three different antigenic determinants at the virus surface in alphaviruses and flaviviruses, specifying type-specific, complex (or subgroup)-specific, and group-specific reactivity; on the other hand only two determinants can explain the cross-reactivity within pestiviruses. As pointed out by Gilden *et al.* (1971) in the context of the Oncovirinae,[1] a single virus protein may have multiple antigenic determinants which may be characterized as type- or group-specific; it is questionable whether interspecies determinants such as occur in the Oncovirinae have any equivalent within the Togaviridae. In addition to virion surface antigens, internal nucleocapsid components contribute to the antigenic complexity of the Togaviridae, and the position is further complicated by antigenic overlap between some virion antigens and nonstructural antigens released in the course of viral replication.

A standardized nomenclature following the principles adopted for oncovirus proteins (August *et al.*, 1974) has been proposed for alphavirus polypeptides (Baltimore *et al.*, 1976) and is coming into use in studies on flavivirus polypeptides (Trent, 1977). Viral proteins are designated by a lowercase "p," and glycoproteins by a lowercase "gp," followed by a number representing the molecular weight in thousands. Polypeptides containing nonstructural amino acid sequences are designated "ns," again followed by a number representing the molecular weight. A capital "E" is used to denote envelope proteins. Core or capsid proteins are designated by a capital "C" for alphaviruses (Baltimore *et al.*, 1976) or by a capital "N" for nucleocapsid proteins (Trent, 1977), and a capital "M" is used to denote nonglycosylated membrane protein (Trent, 1977).

In referring to the names of arboviruses, the abbreviations recommended by the American Committee on Arthropod-Borne Viruses (1969) will be followed.

[1] Subfamily of Retroviridae (Fenner, 1976).

II. ANTIGENIC RELATIONSHIPS AMONG THE TOGAVIRIDAE

A. Historical Background

The isolation on different continents of viruses causing encephalitis in man and in animals prompted comparisons among the different isolates, using serological techniques, initially by cross-neutralization and challenge tests and later complement fixation (CF) tests. Webster *et al.* (1935) concluded that St. Louis encephalitis (SLE) virus was distinct from the virus of Japanese encephalitis (JE) and also distinct from louping ill (LI) virus, an agent causing a disease of sheep in Scotland but capable of producing encephalitis in man following accidental infection acquired in the laboratory. Later, however, Webster (1938) reported that convalescent sera from two patients who had recovered from JE neutralized SLE virus in addition to JE virus. The two viruses were clearly not identical, but they did appear to be antigenically related.

Following the isolation in Africa of West Nile (WN) virus (Smithburn *et al.*, 1940) and the realization that the pathology of the neurotropic infection this virus produced in laboratory animals bore striking similarities to the lesions produced by SLE and JE viruses, serological comparisons of these viruses were undertaken, and a complex relationship was revealed (Smithburn, 1942). Some human sera collected in Africa neutralized JE and SLE viruses in addition to WN virus. Antisera prepared in monkeys against JE virus neutralized the homologous virus, but also WN and SLE viruses; antisera prepared against SLE virus neutralized SLE and WN viruses but not JE virus, and antisera against WN virus neutralized only the homologous virus but not JE or SLE virus (Smithburn, 1942).

In Europe, a close relationship had been demonstrated between LI virus and the virus of Russian spring-summer encephalitis (RSSE) (Smorodintseff, 1939–1940), and a definite, but less close relationship was demonstrated between RSSE and JE viruses on the basis of cross-neutralization tests. Smorodintseff regarded JE virus as "simpler" than RSSE virus, which appeared to contain JE antigens plus additional antigens absent from JE virus. Mice immunized against RSSE virus were quite immune to challenge with JE virus, but the converse was not true (Smorodintseff, 1939–1940).

The development of techniques for studying viruses by CF tests greatly facilitated comparisons among different viruses. Casals (1944) found that RSSE and LI viruses were closely related to each other by

CF but were unrelated to JE, WN, SLE, or Western equine encephalitis (WEE) virus by this approach. However, JE, WN, and SLE viruses formed a closely related complex by CF; antiserum against JE virus reacted with all three antigens, SLE antiserum reacted with SLE and JE antigens, and WN antiserum reacted with WN and JE antigens. Western equine encephalitis virus was entirely distinct from the other four viruses studied. Following the isolation of serologically distinct types of dengue viruses which became known as types 1 and 2, Sabin found that monkeys inoculated intracerebrally with either type-1 or type-2 dengue virus developed homologous antidengue antibodies but also developed CF antibodies against yellow fever (YF), WN, and JE viruses, but not against SLE or WEE virus (Sabin, 1950). He also reported that experimental infection of human volunteers with dengue virus was followed by the appearance in their sera of CF antibodies against JE, WN, and YF viruses (Sabin, 1952).

Over the next few years, an increasing number of reports of serological cross-reactions between members of different pairs or sets of viruses appeared in the literature, some detected in the course of experimental studies and others arising out of epidemiological surveys. A new dimension was added to the problem by the development of hemagglutination and hemagglutination inhibition (HI) techniques (Hallauer, 1946; Sabin and Buescher, 1950; Casals and Brown, 1954), which led directly to the concept of antigenic or serological groups of viruses. In an important study of 21 viruses, made up of 19 arboviruses in the biological sense, together with rabies and poliomyelitis viruses which are not arboviruses but are other neurotropic viruses, Casals and Brown (1954) distinguished two major subsets of viruses which they termed group-A and group-B arboviruses. Five viruses, Eastern, Western, and Venezuelan equine encephalitis (EEE, WEE, and VEE) viruses and Sindbis (SIN) and Semliki Forest (SF) viruses, showed extensive cross-reactions on the basis of HI tests and were designated group-A arboviruses. Eleven viruses, dengue types 1 and 2, Ilheus (ILH), JE, LI, Ntaya (NTA), RSSE, SLE, Uganda S (UGS), WN, and YF viruses also showed extensive cross-reactions and were designated group-B arboviruses. Three arboviruses, *Anopheles* A (ANA), California encephalitis (CE), and *Wyeomyia* (WYO) viruses, and two nonarboviruses, rabies and poliovirus, were unrelated to either of the two groups, and there was no interaction between any virus in group A and any antiserum against a group-B virus, or vice versa. Three years later Casals reported that, of 47 arboviruses studied, 7 fell into group A, 17 into group B, and 3 into a newly defined group C; the remaining 20 viruses remained ungrouped (Casals, 1957). These results were based upon three different serological tests, HI, CF, and neutralization tests, of which the first showed the greatest degree of cross-reactivity, the third was the most specific, and the second fell between these two extremes.

Ten years later, a study group of the World Health Organization reported that of the 252 arboviruses known at that time, 19 were placed in group A, 39 in group B, and 138 in 27 other groups which no longer used alphabetic designations but adopted the name of one of the principal viruses in each group, leaving 56 viruses ungrouped (World Health Organization, 1967). By 1975, when the second edition of the "International Catalogue of Arboviruses" was published, the 359 viruses of vertebrates listed therein included 20 group-A arboviruses, alternatively known as alphaviruses, and 57 group-B arboviruses, or flaviviruses (Berge, 1975).

B. The Present State of the Togaviridae

The report of the arbovirus study group (Porterfield *et al.*, 1978) listed 20 alphaviruses, 59 flaviviruses, 1 rubivirus (rubella virus), and 3 pestiviruses (bovine virus diarrhea, hog cholera, and Border disease viruses) as formally recognized members of the family Togaviridae. Equine arteritis virus (EAV) (Bürki, 1970; Zeegers *et al.*, 1976) and lactate dehydrogenase virus (LDV) (Riley, 1974; Rowson and Mahy, 1975; Brinton-Darnell and Plagemann, 1975) were considered by the study group to be additional members of the family, although without formal recognition as of January 1977, and simian hemorrhagic fever virus (SHV) (Wood *et al.*, 1970; Trousdale *et al.*, 1975) and the cell-fusing agent (CFA) (Igarashi *et al.*, 1976) were considered possible additional members. The second edition of the "International Catalogue of Arboviruses" (Berge, 1975) presents comprehensive summaries of the original isolations, characteristics, and relationships of many of the "older" viruses, including 20 alphaviruses and 57 flaviviruses.[2]

[2] Since the publication of the study group's report, at least five additional alphaviruses and three additional flaviviruses have been described. Two alphaviruses were isolated in Colorado–Bijou Bridge virus, which is related to VEE virus, from swallow bugs (*Oeciacus vicarius*) and Fort Morgan virus, which is related to WEE virus, from swallow bugs and from the blood of nestling swallows and sparrows (Hayes *et al.*, 1977; Monath, 1977). Two more alphaviruses, both related to VEE, were isolated in French Guiana, Cabassou virus from *Culex portesi* mosquitoes and Tonate virus from the same mosquito species and from the blood of several bird species (Digoutte and Girault, 1976). Kyzylagach virus, related to Sindbis virus, was isolated in U.S.S.R. from *Culex modestus* mosquitoes (Lvov *et al.*, 1975). (*Continued*)

TABLE II

The Alphaviruses, Their Subgroups or Complexes, and Their Geographic Variants

Subgroup or complex	Virus	Geographic variants
EEE	EEE	At least two
WEE	WEE	At least three
	Fort Morgan	
	SIN	At least three
	Whataroa	
	Aura	
	Middelburg[a]	
	Ndumu[a]	
VEE	VEE	At least five
	Bijou Bridge	
	Everglades	
	Mucambo	
	Pixuna	
	SF	
	Chikungunya	
	O'nyong-nyong	
	Bebaru	
	Getah	
	Ross River	
	Sagiyama	
	Mayaro	
	Una	

[a] Only distantly related to the other members of the subgroup.

Alphaviruses are clearly divisible into 3 subroups or complexes, most clearly separable on the basis of cross-neutralization tests (Porterfield, 1961; Karabatsos, 1975; Chanas *et al.*, 1976) (Table II). The first subgroup contains only EEE virus, which has at least 2 geographic varieties. The second subgroup contains 6 viruses, WEE, SIN, Whataroa, Aura, Middelburg, and Ndumu viruses, the last 2 being much less closely related than the others. The largest subgroup contains the remaining 13 alphaviruses and includes VEE virus and 3

The three new flaviviruses are Rocio, Saumarez Reef and Kedougou. Rocio (ROC) virus (Souza Lopes *et al.*, 1977) was isolated in Brazil from the brain and spinal cord of a patient who died from encephalitis and from the blood of a sparrow, *Zonotrichia capensis*. Saumarez Reef (SRE) virus (St. George *et al.*, 1979) was isolated in Queensland, Australia, from *Ornithodoros capensis* ticks collected from the nest of a sooty tern, *Sterna fuscata;* it appears to be closely related to, but distinguishable from Tyuleniy (TYU) virus, another bird tick isolate. Kedougou virus was isolated in Senegal from *Aedes* (*Aedimorphus*) *minutus* mosquitoes collected on human bait (Robin *et al.*, 1978).

very closely related viruses, as well as SF virus, with its 2 closely related viruses, chikungunya and o'nyong-nyong.

Below the level of the virus, subtypes or geographic variants are recognized for a number of alphaviruses—at least 2 for EEE (Casals, 1964), at least 3 for WEE (Henderson, 1964), 3 for SIN (Casals, 1961), and 5 for VEE (Young and Johnson, 1969).

Subdivision of the flaviviruses presents greater difficulty. Theiler and Downs (1973) and the "International Catalogue of Arboviruses" (Berge, 1975) list 3 subsets: mosquito-borne, tick-borne, and flaviviruses without a known vector. In their cross-neutralization study of 42 flaviviruses, Madrid and Porterfield (1974) placed 36 viruses in one of 7 different subgroups, leaving 6 other viruses which showed completely monospecific reactions. Table III is an attempt to reconcile these two approaches. Subgroup 1 contains 15 viruses, all of which, with the exception of Negishi virus, have been isolated from ticks. Madrid and Porterfield (1974) examined 7 of these viruses and found that 6 of them cross-reacted, the exception being Powassan virus, which gave a monospecific result. Subgroup 2 contains 17 viruses, 16 of which have no known arthropod vector; the exception is Kadam virus, which was isolated from cattle ticks in East Africa and appears to replicate in ticks.

Porterfield and Madrid examined 9 of these viruses and found that 8 cross-reacted, the exception being Montana myotis leukoencephalitis virus, which failed to react with any other virus. The viruses in subgroups 3–7 all are mosquito-borne; it may be justifiable to regard these 4 subgroups as a single large subset, but in their neutralization tests Madrid and Porterfield (1974) recognized four discrete subgroups with only trivial cross-reactions. Subgroup 3 contains 14 viruses, 9 of which cross-reacted in neutralization tests, while 4 more, Bussuquara, ILH, Wesselsbron, and YF gave monospecific reactions; the fourteenth virus, Sepik, was not examined by Madrid and Porterfield but is known to be closely related to Wesselsbron virus. Subgroup 4 contains only 2 viruses, Spondweni and Zika; a third virus, Chuku, was included by Madrid and Porterfield but is generally regarded as being a strain of Spondweni. Subgroup 5 contains 4 viruses, Israel turkey meningoencephalitis, NTA, and Tembusu viruses, which showed extensive cross-reactions, and Bagaza virus, not included in the cross-neutralization study but known to be closely related to NTA virus. Subgroup 6 contains 5 viruses, Banzi, Edge Hill, UGS, Bouboui, and ROC; the last 2 were not included in the cross-neutralization study but are included here because they are known to be closely related to UGS and Banzi viruses. Subgroup 7 contains the four dengue serotypes.

TABLE III

The Flaviviruses and Their Subgroups or Complexes

Subgroup or complex	Virus
Tick-borne viruses	Absettarov[a,b]
	Hanzalova[a,b]
	Hypr[b]
	Karshi[a]
	Kumlinge[a,b]
	Kyasanur Forest disease
	LGT
	LI
	Negishi
	Omsk
	Powassan[c]
	Royal Farm[a]
	RSSE[a]
	SRE[a]
	TY[a]
Viruses with no known vector	Apoi
	Batu Cave[a]
	Bukalasa bat
	Carey Island[a]
	Cowbone Ridge
	Dakar bat
	Entebbe bat
	Jugra[a]
	Jutiapa[a]
	Kadam
	Koutango[a]
	MOD
	Montana myotis leukoencephalitis[c]
	Phnon-Penh bat[a]
	Rio Bravo
	Saboya[a]
	Sokuluk[a]
Mosquito-borne, WN subset	Alfuy
	BUS[c]
	ILH[c]
	JE
	KUN
	Kokobera
	MVE
	SLE
	Sepik[a]
	Stratford
	Usutu
	Wesselsbron[c]

TABLE III (*Continued*)

Subgroup or complex	Virus
	WN
	YF[c]
Mosquito-borne, Spondweni subgroup	Spondweni
	Zika
Mosquito-borne, NTA subgroup	Bagaza[a]
	Israel turkey meningoencephalitis
	NTA
	Tembusu
Mosquito-borne, UGS subgroup	Banzi
	Bouboui[a]
	Edge Hill
	ROC[a]
	UGS
Mosquito-borne, dengue subgroup	Dengue type 1
	Dengue type 2
	Dengue type 3
	Dengue type 4

[a] Not studied by Madrid and Porterfield (1974).
[b] Strains of TBE virus.
[c] Monospecific reaction in cross-neutralization test.

Cross-reactions between the tick-borne virus subgroup 1 and the "no-vector" viruses, subgroup 2, were reported by Madrid and Porterfield (1974) through Langat (LGT) and Negishi antisera reacting with Apoi virus. Subgroups 2 and 3 were linked through JE antiserum, which reacted with Entebbe bat, Bukalasa bat, and Rio Bravo viruses, through Murray Valley encephalitis (MVE) antiserum, which reacted with Rio Bravo virus, and through WN antiserum, which reacted with Entebbe bat virus. There was also a cross-reaction between Spondweni antiserum (subgroup 4) and SLE virus (subgroup 3).

Although given the status of separate viruses in the "International Catalogue of Arboviruses" (Berge, 1975) and in the arbovirus study group report (Porterfield *et al.*, 1979), Absettarov, Hanzalova, Hypr, and Kumlinge should be regarded as strains of tick-borne encephalitis (TBE) virus. Clarke (1964) suggested that two subtypes of TBE virus, central European encephalitis and Far Eastern encephalitis, and two subtypes of Omsk virus, were recognizable. At least two different variants or subtypes of WN virus (Hammam *et al.*, 1965) and two of YF virus (Clarke, 1960) have been described.

The observed antigenic relationships are of more than academic interest and are relevant both from a medical and an evolutionary point of view. Within the flaviviruses, for example, there are viruses with a wide range of pathogenetic potentialities, from nonpathogenic (to humans) to highly and diversely pathogenic members. While the biochemical and genetic bases of these differences are not understood at the present time, it is clearly important to determine whether, and through what specific changes, such differences can be ascribed to the inherent variability of a single viral species (see Chapter 4).

III. METHODS USED TO STUDY INTERRELATIONSHIPS

A. Neutralization Tests

In a virus neutralization test, a preparation of virus is mixed with a sample of immune serum, and after virus–antibody interaction has taken place the residual infectivity of the mixture is determined, with appropriate controls, using either an *in vivo* or an *in vitro* assay method. In its simplest form, a neutralization test may give either a positive or a negative result, indicative of neutralization or no neutralization, but more usually the result is expressed quantitatively. There are two basically different approaches, the α-procedure, which uses constant serum and variable virus dilutions, and the β-procedure, which uses constant virus and variable serum dilutions (Fazekas de St. Groth, 1962).

The first procedure measures how much virus a constant dose of antiserum (usually undiluted) can neutralize, and the result, frequently termed the "neutralization index" (NI) of a serum, represents the difference, often expressed in $\log_{10}$ units, between the infectivity of a given preparation of virus incubated with the test serum and that obtained when the same preparation is incubated with a normal or nonimmune serum. Alternatively, the NI of an antiserum may be expressed as "dex" (Haldane, 1960; the decimal exponent or antilogarithm to base_{10} of a number); i.e., an antiserum with a NI of 3.5 $\log_{10}$ units of difference in infectivity is described as 3.5 dex.

The second procedure measures the antibody titer of a serum and is often expressed as that dilution of serum capable of neutralizing all (or a selected proportion) of a standard dose of virus, frequently 100 mouse LD_{50} or 100 plaque-forming units (PFU) of virus.

Although the "percentage law" (Andrewes and Elford, 1933) relates the relative concentrations of virus and antibody, there is no simple formula for converting the potency of an antiserum as determined by the NI into one expressed as an antibody titer. Extensive use was made in the past of the constant serum–variable virus approach, but this is relatively insensitive and lacking in precision, particularly when the assays are carried out in mice and when serum is used undiluted (Hammon, 1969; Schlesinger, 1977). Kinetic neutralization tests (Lafferty, 1963) may provide better discrimination between closely related viruses.

The constant virus–variable serum method can be appreciably more sensitive and more precise, particularly when it is combined with a satisfactorily reproducible plaque assay. The method is not, however, free of hazards, and results may be distorted by the use of preparations of virus containing a high proportion of antigenically reactive but noninfectious particles. Some sera may combine with virus but may fail to inactivate infectivity, either because of low avidity or because of the formation of infectious virus–antibody complexes (Symington *et al.*, 1977).

Consecutive infections or immunizations with two or more viruses within the same serological group result in a broadening of secondary (group-specific) immune responses, most clearly demonstrated in HI tests but also measurable by neutralization tests (Casals, 1957).

1. The Role of Accessory Factors in Neutralization Tests

Although there is much disagreement on detail, there is substantial evidence for the following points:

1. The neutralizing capacity of a serum for a virus is reduced by heat, by repeated freezing and thawing, and by prolonged storage at 4°C.
2. Part, at least, of the loss of activity produced by heating may be restored by the addition to the serum–virus mixture of fresh, normal serum ("accessory factor").
3. The same serum–virus mixture can give markedly different infectivity end points depending upon the presence or absence of accessory factor(s).

Although initially demonstrated using infectivity titrations carried out in mice (Morgan, 1945; Sabin, 1950; Whitman, 1947), the antibody-potentiating effect of fresh, normal serum can also be demonstrated under the more precise conditions of an *in vitro* plaque assay (Halstead, 1974). Westaway (1968) studied the effect of adding fresh, normal guinea pig serum to different globulin fractions of 7-day rabbit antisera against Kunjin (KUN) virus; all fractions showed significantly more activity in the presence of guinea pig serum than in its absence. In heterologous neutralization tests, the addition of fresh guinea pig serum increased the reactivity of both IgG and IgM fractions of KUN immune serum against WN, MVE, and JE viruses.

Way and Garwes (1970), using a plaque assay in chick embryo cells, found that sera from a range of normal domestic and laboratory animals potentiated the neutralization of SF virus by heated guinea pig antiserum; bovine sera were notable for producing a nonspecific antiviral effect. There was no correlation between the potency of a serum as an antibody-enhancing factor and its complement activity. Equine arteritis virus, a proposed togavirus, has been subjected to detailed study; its neutralization by antiserum is enhanced by guinea pig serum (Hyllseth and Pettersson, 1970). Radwan and Crawford (1974) showed that high concentrations of C4, C2, and C3, with optimal concentrations of C1 were sufficient to neutralize EAV in the presence of excess antibody. The addition of the remaining five components of complement, C5–C9, induced lysis of the previously neutralized virus particle. Immune virolysis of SIN virus has been demonstrated by Stollar (1975).

The simplest interpretation of these findings is that virus–antibody union is enhanced by one or more of the components of the complement system, resulting in progressive coating of active sites on the virion surface with protein layers, thus rendering them unavailable for attachment to cell receptors. The further step of virolysis may follow in some circumstances but does not appear to be an essential requirement for neutralization (Taniguchi and Yoshino, 1965; Daniels *et al.*, 1970; Daniels, 1975; Oldstone, 1975).

2. *Enhancing Effect of Antiglobulin Antibodies*

The addition to a mixture of virus and antiviral antiserum of an antiglobulin prepared against the species in which the antiviral antiserum was prepared can markedly enhance the virus-neutralizing capacity of the serum, whether the assay is carried out by a plaque reduction method (Westaway, 1968), by radioimmune precipitation (Dalrymple *et al.*, 1972), or by any other means. Ozaki *et al.* (1974) showed that infectious complexes of JE virus and IgM against JE virus could be neutralized by anti-rabbit IgG and that this neutralization still took place if the addition of the anti-rabbit globulin followed the adsorption of immune complexes onto susceptible host cells, thus excluding aggregation as the principal effect of the anti-rabbit globulin.

B. Challenge Tests

The degree of protection provided by prior infection (or inoculation) with one virus (X) against subsequent challenge with a second virus (Y) can be used to assess the closeness of the relationship between the two viruses (X and Y). Protection involves whole animals and is usually measured as an increase in the proportion of animals surviving a given viral challenge (Y after X) as compared with deaths in control animals given the same dose of virus (Y only); in some experiments resistance to challenge is measured in terms of a reduced level of viremia (in Y-after-X animals as compared with Y-only animals).

It is well established that infection with one virus frequently induces HI antibodies against other viruses in the same genus and may also induce cross-neutralizing antibodies within the genus (Casals, 1957). It might therefore be supposed that the outcome of challenge experiments would depend upon the presence or absence of such cross-neutralizing antibodies, i.e., that virus X followed by virus Y would result in survival if virus X induced the formation of anti-Y antibodies, but not when virus X failed to induce anti-Y antibodies. However, there are several reports of heterologous challenge resistance in the absence of cross-neutralizing antibodies, both within the alphaviruses (Casals, 1963; Hearn and Rainey, 1963; Cole and McKinney, 1971) and within the flaviviruses (Casals, 1963; Price *et al.*, 1967; Schlesinger, 1977). Conversely, the presence of cross-reactive neutralizing or HI antibodies does not ensure protection against the superinfecting virus (see Chapter 5).

Various interpretations have been placed upon these findings. In their studies with alphaviruses, Hearn and Rainey (1963) suggested that heterologous cross-protection could be due to persistent cellular infection with VEE virus rather than to the effects of humoral antibodies. In studies with inactivated alpha- and flaviviruses, Casals *et al.* (1973) found nonspecific protection on days 1 and 2 and suggested that this was an interferon effect, although they were unable to demonstrate the presence of interferon in mouse serum at any time later than 1 h after vaccination. Eaton (1979) and Igarashi (1979) have described examples of heterologous interference in mosquito cell cultures, which are of special interest because in such cultures neither immunoglobulins nor interferon-like factors appear to play a role (see Chapter 20).

A further example of heterologous protection was reported by Oaten *et al.* (1976) who found that the intracerebral inoculation of avirulent variants of alphaviruses produced significant protection against subsequent intracerebral challenge with a dose of LGT virus (a flavivirus) that killed all normal control mice. When the LGT challenge was given intracerebrally after SF virus or SIN virus given by the intraperitoneal route there were no survivors, but there was prolongation of the average survival time (AST) in both instances. With SIN virus given intraperitoneally this protection was seen only in mice challenged within 1–3 days, during which period the authors were able to demonstrate interferon activity in mouse brains. When SF virus was given intraperitoneally there was an extension of the AST following a LGT challenge given as late as 35 days, a result that could not be attributed to interferon activity. Somewhat surprisingly, moderate neutralizing activity against LGT virus was detected in sera collected 7 and 14 days after SIN virus had been given intracerebrally.

Even in homologous challenge experiments, the nature of the factor(s) in immune serum that provides protection is uncertain. For example, Griffin and Johnson (1977) studied the recovery of mice from encephalitis produced by a neuroadapted strain of SIN virus given intracerebrally and found that mice were protected by passive transfer of SIN immune mouse serum, but not by the passive transfer of SIN immune spleen cells or lymph node cells, or by the transfer of a mouse immune serum prepared against YF virus. The active component in immune mouse serum appeared to be IgG, but when they fractionated SIN mouse immune serum and tested different fractions for their ability to confer passive protection and to neutralize SIN virus *in vitro,* the two functions were clearly separable, suggesting that they involved different classes of immunoglobulins. Levels of virus in the brains of mice that had received immune serum 24 h after virus infection were only slightly lower than those reached in controls without serum, yet significant protection was achieved. The immune response of mice receiving SIN virus plus homologous antiserum was almost indistinguishable from that of mice receiving SIN virus alone.

C. Hemagglutination Inhibition Tests

Much of the foundation of the serological classification of arboviruses is firmly based upon results obtained by HI tests. It is therefore relevant to examine some of the early observations and conclusions in some detail. With alphaviruses, HI antibodies developing in response to a primary infection are first detectable on about day 4 and are substantially monospecific in the early stages (about day 7). Cross-reactions are more likely to be found in serum samples collected later, but these are always substantially lower in magnitude than in the homologous reactions. A second inoculation with the same virus boosts the homologous antibody titer but also broadens the cross-reactivity of the serum. A comparison of HI titers almost always shows that homologous levels are substantially higher than heterologous levels. When an animal immune to one alphavirus is subsequently inoculated with a second, different, alphavirus, antibodies against the initial virus are again boosted, and antibodies against the second virus appear at about the same rate and to about the same titer as they appear in normal animals inoculated in parallel with the same material; however, cross-reactions are now much more extensive, and antibodies may appear against alphaviruses which have not been inoculated, sometimes to titers higher than those produced against viruses that have actually been inoculated (Casals, 1957, 1963).

With primary flavivirus infections, the pattern is in general similar to that seen with primary alphavirus infections. A second or subsequent inoculation with the same flavivirus produces enhanced homologous antibody titers but may so stimulate heterologous antibody levels that they equal the homologous titer. If the second flavivirus is different from the first, the antibody response is likely to be extremely broad, and heterologous titers may well exceed those seen against either of the two viruses inoculated (Casals, 1957, 1963). The important implications of these relationships for hypotheses attempting to provide an immunological basis ("second-infection hypothesis") for dengue hemorrhagic fever and the shock syndrome are discussed in Chapter 5 of this volume.

Envelope glycoproteins are known to be responsible for the hemagglutination observed with alphaviruses, flaviviruses, and rubivirus; no hemagglutination has been reported with pestiviruses, or with the other possible members of the family Togaviridae. Alphaviruses have two or three envelope glycoproteins, of which E1, gp52, is the active hemagglutinin in SIN virus, capable of agglutinating in the purified state (Dalrymple *et al.*, 1976). Anti-E1 antiserum will inhibit agglutination but will not affect the infectivity of SIN virions. Antisera to SIN and WEE viruses will inhibit agglutination with E1 glycoprotein. The hemagglutination of SFV is associated with the second envelope glycoprotein, E2, gp49 (Kennedy and Burke, 1972; Garoff *et al.*, 1975); which of the two glycoproteins is the active hemagglutinin with the other alphaviruses has not been established.

The major envelope glycoprotein of the flaviviruses is the active hemagglutinin in several viruses, gp58 in JE and dengue-2 viruses and gp53 in SLE virus (Trent, 1977). In addition, a 70 S nonstructural antigen extractable from flavivirus-infected cell cultures also has a hemagglutinating capacity (Stollar *et al.*, 1966; Smith *et al.*, 1970; see also Chapters 4 and 18 of this volume).

D. Complement Fixation Tests

Complement fixation tests performed with extracts of virus-infected mouse brains and with antisera prepared against equally crude materials have been immensely valuable in arbovirus studies and have yielded a mass of data on antigenic relationships among Togaviridae (Casals, 1947, 1957). Some investigators have used absorbed antisera (Clarke, 1960, 1964) or hyperimmune ascitic fluids instead of antisera (Sartorelli *et al.*, 1966; Tikasingh *et al.*, 1966), and others have used fractionated antigens (Smith and Holt, 1961; Brandt *et al.*, 1970; Okuno *et al.*, 1968; Trent *et al.*, 1976). Although more is now known about the chemical composition of alphaviruses and flaviviruses, the precise serological specificities of the different virion structural and nonstructural components are still imperfectly understood.

A core or nucleocapsid antigen with broadly group-reactive specificity detectable by CF tests has been described for both alphaviruses (Dalrymple, 1972) and flaviviruses (Trent, 1977). Extracts of flavivirus-infected cells show group cross-reactions by CF tests, but fractions prepared from dengue virus-infected cells by concanavalin A affinity chromatography were dengue virus type-specific and did not cross-react with antisera against WN, SLE, or JE virus (Stohlman *et al.*, 1976).

The classic interpretation placed upon the finding of CF antibody in human sera is that it is indicative of recent antigenic exposure, but there are many instances when this is clearly not the case. Experience with JE virus indicates that as many as 25% of Japanese subjects who have had overt JE virus infection still have detectable CF antibodies 5 years later (Buescher *et al.*, 1959). In follow-up studies on volunteers who took part in the Siler and Simmons dengue infection experiments in the 1920s, eight out of nine subjects were found to have dengue CF antibodies nearly 50 years later, reinfection being extremely unlikely during the intervening years (Halstead, 1974).

Interpretation of the results of CF tests on dengue patients is further complicated by the existence of the soluble complement-fixing (SCF) antigen (ns 39) present in virus-infected mouse brains and cell extracts (Russell and Brandt, 1973). In tests on convalescent sera following primary dengue virus infection only 1/9 sera reacted with SCF antigen, whereas all of 22 convalescent serum samples from secondary dengue virus infections gave positive SCF tests, often to high titer and against two or more dengue virus serotypes (Falkler *et al.*, 1973). Following SLE virus infection 10/10 serum samples contained CF antibodies against SLE gp53 antigen, and 7/10 also gave positive CF tests with SLE nonstructural antigen ns 80 (Trent *et al.*, 1976). Microquantitative CF tests have been used to titrate SIN (Stollar and Stollar, 1970) and dengue virus-specific double-stranded RNA (Lubiniecki *et al.*, 1976).

The conglutinating complement absorption (CCA) test has been applied to detect flavivirus antibodies (Singh, 1970); it appears to be more sensitive and more specific than the CF test.

E. Immunodiffusion Tests

Clarke (1962) showed that it was possible to distinguish between tick-borne flaviviruses by the use of agar gel precipitin techniques using mouse brain antigens and hyperimmune mouse sera; better separation of viruses was obtained through the use of absorbed sera. Chan (1965) claimed that all four dengue serotypes could be distinguished quite simply by agar gel precipitin tests, but Ibrahim and Hammon (1968a) were unable to confirm these findings; in their experience precipitin lines obtained using dengue viruses and dengue antisera and revealed group-specific but not type-specific dengue reactions. Better separation of dengue types was claimed by the use of immunoelectrophoresis (Ibrahim and Hammon, 1968b). Calisher and Maness (1970) included 12 flaviviruses in their extensive series of 75 different arboviruses examined by agar gel precipitin tests. Four viruses, ILH, Rio Bravo, dengue-1, and dengue-2, gave completely monospecific reactions, SLE and WN mouse hyperimmune ascitic fluids reacted fairly broadly, the former with WN, JE, YF, Bussuquara (BUS), and Modoc (MOD) viruses, the latter with SLE, JE, MOD, and MVE viruses. The authors comment that the cross-reactions they encountered were all between viruses known to be antigenically related by other tests.

In the field of the pestiviruses, agar gel diffusion tests have been used to show relationships among swine fever (hog cholera), bovine viral diarrhea (mucosal disease), and Border disease viruses (Darbyshire, 1962; Gutekunst and Malmquist, 1963; Plant *et al.*, 1973).

One major limitation of the studies referred to above is that they were all carried out using extremely crude reactants. It is possible that more specific results might be obtained if purified viral antigens and potent antisera prepared against such materials are applied in immunodiffusion tests.

F. Fluorescent Antibody Tests

Infectivity assays based upon the counting of foci of infected cells stained by an immunofluorescent conjugated antiviral antiserum have been described for YF (Hahon, 1966), VEE (Hahon and Cooke, 1967), Rift Valley fever (Hahon, 1969), chikungunya, and other viruses (Buckley and Clarke, 1970; Hahon and Hankins, 1970; Igarashi and Mantani, 1974). Increased sensitivity can be obtained by allowing adsorption of virus to take place under the influence of centrifugal force. Although there have been many applications of fluorescent antibody (FA) tests for research purposes, diagnostic uses of FA techniques have been of limited value in arbovirus studies, although they have been more useful in the veterinary field (Mengeling and Van der Maaten, 1971). When applied to the diagnosis of JE virus infections in man, the FA technique was less sensitive than the HI test (Kusuda and Takahashi, 1969) and, when applied to human sera following dengue infection, type specificity was poor (Vathanophas *et al.*, 1973). Levitt *et al.* (1975) have described a microprecipitation test for the rapid identification of field isolates of VEE, EEE, and WEE viruses, based upon preliminary incubation of specimens with susceptible duck embryo cells followed by electrophoresis of a mixture of infected culture fluid with potent fluorescent conjugated antisera. Specific fluorescent rings visible under uv light permitted much earlier identification of viruses than was possible by conventional procedures. However, infectivity levels on the order of 9 $\log_{10}$ PFU/ml are required and, while these are attained fairly readily with the three alphaviruses under study, the method is not applicable to viruses with lower infectivity yields.

G. Radioimmunoassays

In recent years there have been several applications of radioimmunoassay methods in the diagnosis of arbovirus infections, and some to the study of cross-reactions between different Togaviridae. As compared with other nonbiological assays, radioimmunoassays offer the advantage of sensitivity which may equal, or possibly even exceed, that provided by sensitive plaque methods. Dalrymple *et al.* (1972) used a radioimmune precipitation (RIP) test to study the relationship among three alphaviruses, EEE, WEE, and SIN. Viruses were grown in chick embryo fibroblast cells and labeled with ^{3}H or ^{14}C isotopes. Infected culture fluids were subjected to ammonium sulfate precipitation followed by sucrose gradient fractionation. The addition of virus-specific antibody, in the form of mouse hyperimmune ascitic fluid, to labeled virus resulted in the precipitation of labeled immune complexes, but the addition of goat anti-mouse IgG or rabbit anti-mouse whole serum to the mixture substantially increased the sensitivity of the system. Homologous titrations established that the concentration of antiglobulin could be varied over a 40-fold range without substantially affecting results except with the lowest concentrations of antiviral antibody, and test conditions were adjusted so that at least 70% RIP was obtained in the standard test. As might be expected, the concentration of virus antigen used had a marked effect upon the 50% RIP titer of an antiviral antibody preparation. To standardize conditions in heterologous reactions, infectivity assays were performed on preparations of the three different viruses, and antigens were standardized to comparable infectivity levels on the assumption that all three viruses had similar ratios of infectivity to antigenic mass. The sensitivity and specificity of the RIP method as compared with that of a conventional plaque-reduction neutralization test are shown in Table IV. Another finding in this study which is of particular relevance in the present context is the biphasic nature of the curve of the interaction between SIN virus and dilutions of WEE antibody. The authors interpret this to mean that WEE antibody preparations contain at least three different components: (1) high-titered WEE-specific antibody,

TABLE IV

Sensitivity and Specificity of Radioimmune Precipitation and Virus Neutralization Tests

Antiserum	Test[a]	Antibody titer against[b]		
		SIN virus	WEE virus	EEE virus
Anti-SIN virus	RIP	40,000	40	<5
	PRN	10,000	100	<5
Anti-WEE virus	RIP	200	300,000	<5
	PRN	100	100,000	<5
Anti-EEE virus	RIP	<5	<5	200,000
	PRN	<5	<5	100,000

[a] RIP, Radioimmune precipitation; PRN, plaque reduction neutralization.
[b] Expressed as reciprocal of dilution exhibiting 50% effect.

(2) a broadly reactive antibody which gave positive results with all three viruses, and (3) a low-titered antibody which reacted with WEE and SIN antigens but not with EEE antigens.

This study was later extended to include VEE virus (Dalrymple, 1972), but RIP detected no significant cross-reactions between VEE and the other three viruses.

A further extension of RIP was carried out using envelope fractions released from SIN, EEE, and WEE viruses by NP-40 treatment, which were reacted with antivirion antisera. The results resembled those obtained with whole-virus antigens, in that cross-reactions were again found between WEE and SIN. Nucleocapsid preparations reacted differently from whole-virus and envelope polypeptide fractions in that type specificity was markedly reduced (Dalrymple, 1972).

More recently, Trent *et al.* (1976) used purified envelope glycoprotein and purified viral nonstructural protein preparations from flavivirus-infected cells and studied homologous and heterologous reactions with SLE, JE, and WN viruses by solid-phase radioimmunoassay (SPRIA). Cross-reactions were found between mouse or human immunoglobulins and viral envelope glycoprotein, which were interpreted as evidence of group- and complex-reactive as well as type-specific determinants. In absolute terms, the SPRIA was slightly more sensitive than a plaque-reduction neutralization test, but cross-reactions were of greater magnitude in the former; both tests were substantially more sensitive than CF and HI tests. The SPRIA tests carried out with 80,000-molecular-weight nonstructural antigens prepared from SLE, JE, and WN virus-infected cells showed a high degree of type specificity.

H. Biochemical Tests

Rosato *et al.* (1974) examined five alphaviruses, three flaviviruses, and five bunyaviruses by velocity sedimentation in sucrose gradients and compared the structural polypeptides of the different viruses by polyacrylamide gel analysis. Major differences among the three sets of viruses were apparent, indicating that biochemical similarities could be useful in placing an uncharacterized virus into its appropriate taxonomic set, but discrimination within sets was difficult or impossible by this approach.

Pedersen and Eddy (1975) have reported that VEE virus strains could be separated into five classes on the basis of electrophoretic differences detected in virion envelope proteins.

In a more detailed study of four different alphaviruses, Wengler *et al.* (1977) studied the base sequence homologies of virion 42 and 26 S RNA and looked for strain differences by RNA–RNA hybridization studies and by RNA fingerprinting. Each of the four viruses gave different fingerprint patterns; strains of the same virus gave similar, but not identical, patterns. The hybridization studies showed very low sequence homology between SF, SIN, and chikungunya viruses, with somewhat more homology between chikungunya and o'nyong-nyong viruses (13%). The authors concluded that amino acid sequence studies of the virion glycoproteins might be a more profitable approach in detecting strain differences than RNA–RNA hybridization studies.

For detailed discussions of these aspects see the appropriate chapters in this volume.

I. Other Methods

Cuadrado and Casals (1967) have described a simple immunoelectrophoresis method for the study of arbovirus antigens. When applied to dengue viruses, the technique gave reactions of identity between dengue-1 and dengue TH-Sman viruses, and between dengue-2 and dengue TH-36 viruses, providing evidence against the separate recognition of these strains as dengue types 6 and 5, respectively. However, when applied to the study of seven strains of SIN virus, no reactions of identity were obtained, suggesting that each of the seven strains had at least one antigenically unique component (Cuadrado and Casals, 1967).

IV. OTHER FACTORS AFFECTING CROSS-PROTECTION

A. Interferon

Most of the evidence that interferon contributes to the survival of an animal infected by a virus is indirect, and it is frequently difficult to assess the relative contributions made by interferon, humoral antibodies, and cell-mediated immunity (CMI). Most attempts to measure interferon in the tissues of animals used in heterologous challenge experiments have given equivocal results. Recently, some experiments on the effects of anti-interferon antibody on the course of viral infections have been more conclusive. Fauconnier (1970, 1971) showed that the administration of sheep anti-mouse interferon to mice infected with SF virus enhanced replication of the virus and increased mortality from low doses of virus. Gresser *et al.* (1976a,b) showed that the early production of interferon in visceral organs was important in limiting the course of several viral infections; they confirmed the somewhat unexpected observation of Fauconnier that, whereas relatively small amounts of interferon were extractable from organs in the course of unmodified infections, more interferon was obtainable from the organs or serum of mice treated with anti-interferon globulin, in which titers of infectious virus were enhanced.

Although no experiments have yet been reported on the possible influence of anti-interferon globulin on early, nonspecific protection such as is seen in some heterologous challenge experiments, such studies should help to unravel the respective contributions made by interferon and other nonantibody mechanisms such as cytotoxic macrophages.

The possible role of defective interfering (DI) virus (or genomes) in regulating persistent infection or replication of superinfecting virus is discussed in Chapters 4 and 15.

B. Cell-Mediated Immunity

Although there are several reports relating to the importance of CMI in recovery from infections with different individual Togaviridae, there are few studies on the relevance of CMI to heterologous cross-protection. Peck *et al.* (1975) showed that mice could be protected against infection with SF virus by passive transfer of spleen cells removed from mice immunized against SIN virus, provided the challenge was given 6 days after the spleen cells; challenge after a 2-day interval gave no protection. By depleting the spleen cell preparation of B or T cells, it was shown that the passive protection was T-cell-dependent. It is unlikely that interferon was responsible for this effect, and the authors failed to recover infectious SIN virus from the spleen cell preparations by direct isolation or by cocultivation attempts. The recipients did not become SIN virus-immune, so it is unlikely that defective interfering particles or SIN virus sequestered in the spleen cells was responsible for the protection. The validity of these results was further strengthened by the demonstration that serial transfer of T-enriched spleen cells from SIN virus-immune mice through three passages in normal mice at 6-day intervals still gave 44% protection against intracerebral challenge with SF virus.

A different manifestation of CMI was described by Rodda and White (1976) in studies of immune cytolysis after SF virus infection of BALB/c mice. In agreement with earlier reports, they found that immune spleen cells collected on day 6 from SF virus-infected mice and inoculated ip into normal mice protected against an immediate challenge with homologous SF virus. Cell-mediated cytotoxicity of effector cells from spleen, peritoneal, and lymph node preparations made from SF virus-infected mice at different time intervals was studied by a complement-dependent radioactive chromium release assay using targets of SF virus-infected mouse plasmacytoma P 815 cells. Day-6 effector cells from spleen and peritoneal cavity, but not lymph node cells, were cytotoxic to SF virus targets cells, but not for normal P 815 cells. However, day-2 effector cells were cytotoxic against both infected and uninfected target cells. When effector cells were prepared from mice infected with the flavivirus KUN, which is in a different genus than SF virus and is completely unrelated to that virus serologically, day-2 cells were as cytotoxic against SF virus targets as were day-2 SF virus effector cells, but day-6 KUN cells were virtually inactive against the SF virus targets. The conclusion from this study is that, while both T cells and complement-dependent antibodies play an important role in recovery from viral infections from day 3 onward, cytotoxic macrophages may be the critical first line of defense which operates before the other two mechanisms are fully mobilized. These early cytotoxic macrophages are not specific for the infecting virus and may explain the early nonspecific cross-protection observed by Hearn and Rainey (1963), by Casals *et al.* (1973), and by others carrying out similar challenge experiments.

V. MEDICAL AND BIOLOGICAL SIGNIFICANCE OF INTERFAMILY RELATIONSHIPS

Some Togaviridae cause important diseases of humans, such as YF, dengue, and TBE (flavivirus infections); EEE, WEE, and VEE (alphavirus infections); and rubella (a rubivirus infection). Others, pestiviruses, cause important diseases of domestic animals. There is very little evidence (other than that discussed in Sections IV,A and B) that viruses in *different* genera interact when they infect the same host. However, there is a mass of evidence, both from the field and from experimental studies, which indicates that prior infection with one virus may substantially alter the response of the host to subsequent infections with other viruses *within the same genus.* The great majority of these interactions are beneficial to the host; that is, they result in a milder disease or in a lower level of viremia than would have been observed in the absence of the initial infection. Some interactions, however, of which sequential infections with different serotypes of dengue virus are the best documented, may be harmful or even lethal to the host (see Halstead, Chapter 5 in this volume).

Under natural conditions, interactions between related viruses can occur only in parts of the world where several viruses in the same genus coexist; these tend to be tropical or subtropical regions. Decades of field studies on YF in Africa and in South and Central America have revealed a number of flaviviruses that are antigenically related to, but are distinct from, YF virus. Does prior infection of man with a different flavivirus modify the course of a subsequent natural infection with YF virus? From a consideration of the results of serological surveys carried out in West Africa, Macnamara *et al.* (1959) suggested that interactions between flaviviruses did occur in nature and resulted in partial or complete suppression of overt YF. However, these conclusions were based on serological findings carried out in the absence of recognized cases of YF. More direct serological studies on confirmed cases of YF in West Africa showed that it was possible to identify primary infections with YF virus and clearly to distinguish these from superinfections (by which is meant YF in an individual with preexisting antibodies against heterologous flaviviruses) on the basis of antibody patterns; however, the clinical severity of the disease was the same in both sets of patients (Monath *et al.*, 1973). In experimentally infected monkeys, there is evidence that prior inoculation with Zika or Wesselsbron virus can protect against YF challenge (Bearcroft, 1957; Henderson *et al.*, 1970), and slight protection has also been reported in monkeys and in mice following UGS virus infection (Macnamara, 1953). The epidemiology of YF in Africa is highly complex, and different factors operate in East and West Africa (Haddow, 1965, 1968). The absence of YF from the Indian subcontinent and from Asia in spite of the presence of abundant *Aedes aegypti* and other potential vectors poses many questions that remain unanswered. It has been suggested that prior infection with dengue virus strains sets up an immunological barrier against the establishment of YF in a population, and the relative resistance of dengue-immune monkeys to challenge with YF virus supports such a suggestion (Theiler and Anderson, 1975). Such protection, if it occurs, is by no means complete, and natural infections with YF virus in individuals having high levels of dengue antibodies have been reported (Spence *et al.*, 1961). However, it is arguable that, even if cross-protection is incomplete, reduced levels of viremia would lessen the possibility of YF virus establishing itself in a population having high levels of antibodies against related flaviviruses.

A highly effective vaccine against YF has been available for over 40 years (Theiler and Smith, 1937), but few other successful flavivirus vaccines have been developed. The question therefore arises: Can YF vaccination of man provide at least partial protection against other flavivirus infections? There is no convincing evidence that YF vaccination of man confers any protection against natural infection with any one of the dengue serotypes (Sabin, 1950; Schlesinger, 1977). However, individuals previously immunized against YF reacted with a broadened anamnestic antibody response when challenged with dengue virus in volunteer experiments (Schlesinger *et al.*, 1956). Many experimental studies of sequential immunization procedures have been published, using different combinations of viruses (Price *et al.*, 1973). When spider monkeys were given live 17D virus intramuscularly, followed 1 month later by live LGT E5 virus intramuscularly, followed 1 month later by live, attenuated dengue-2 virus im, the animals responded by developing high-titered antibodies detected by HI tests and by neutralization tests against 20 different flaviviruses; more significantly, the animals were resistant to subcutaneous challenge with several different flaviviruses (Price *et al.*, 1973). When the order in which the immunizing viruses was changed to dengue, followed by LGT, followed by YF, much less satisfactory protection was obtained.

Experimental studies in primates, such as those just described, encourage the view that it should be possible to protect man against the harmful effects of flaviviruses for which there is as yet no satisfactory vaccine by the sequential use of viruses for which adequately attenuated vaccines are on hand. It is clearly impossible to carry out challenge tests in man, so protection can be assessed only on the basis of the results of neutralization tests. It is extremely difficult to know how to correlate levels of antibodies determined in neutralization tests and real protection against disease, and there are many instances where moderate antibody levels have failed to confer protection and others where there has been clear protection in the absence of demonstrable neutralizing antibodies.

Sabin (1951) and Hammon and Sather (1973) have attempted to estimate how much antibody is required to be given in passive transfer experiments to provide adequate immediate protection against infection with WEE or JE virus. Sabin (1951) suggested that enough antibody should be given to ensure just detectable levels of antibody in undiluted serum at the end of the period of potential exposure. Applying these levels to man, Hammon and Sather (1973) gave doses of IgG of 0.14 to 0.2 ml/kg of body weight using preparations with NI values of about 4 dex. Although these amounts gave lower neutralizing antibody levels than had been expected by extrapolation from mouse data, the procedure appeared to provide some protection, in that laboratory workers who had received accidental exposure to JE virus remained disease-free.

VI. EVOLUTIONARY ASPECTS OF TOGAVIRIDAE

A common evolutionary origin for all alphaviruses seems an entirely reasonable suggestion, as does a common origin for all flaviviruses and a common origin for all pestiviruses. Although the extension of this principle to a common origin for all Togaviridae presents some difficulties, particularly those relating to the apparently different modes of replication of flaviviruses as compared with alphaviruses (Schlesinger *et al.*, 1979; see also Chapter 1), this possibility seems at least as reasonable as the alternative view that all four genera started from separate origins and reached their present degree of similarity by convergent evolution.

Earlier discussions on the possible evolution of arthropod-borne viruses (Baker, 1943; Darlington *et al.*, 1960) favored an arthropod host origin rather than a vertebrate host origin, but these views were developed before the concept of a family containing both arthropod-borne and non-arthropod-borne members was formulated. The non-arthropod-borne flaviviruses (such as MOD virus and a number of viruses isolated from bats) may provide a contemporary link between arthropod-borne flaviviruses (which replicate in both arthropod and nonarthropod hosts) and viruses such as rubivirus, pestiviruses, and the additional members of the family which now replicate solely in vertebrates. These non-arthropod-borne viruses may have had an arthropod stage earlier in their evolutionary history.

The evolutionary stages outlined by Baker (1943) may be used as a framework upon which the following scheme may be constructed: Starting as a subcellular organelle in a primitive arthropod (possibly more related to a tick than to a mosquito, since ticks are regarded as more primitive forms), the primitive togavirus replicated in the tissues of its arthropod host and was vertically transmitted between generations of primitive arthropods. When associations developed between arthropods and vertebrates, some arthropod viruses developed the capacity to replicate in primary vertebrate hosts as well as in primary invertebrate hosts. Tick-borne flaviviruses are present-day examples of this stage of evolution, and it is of interest that some of these viruses still retain the ability to be transmitted between generations in their arthropod hosts by the transovarial route. Once adaptation to a primary vertebrate host had developed, two further evolutionary paths became open, one involving the loss of arthropod dependence and the other involving adaptation to different arthropod hosts such as mosquitoes or other Diptera. Viruses which lost their arthropod dependence became restricted to their vertebrate hosts, in which they were transmitted vertically, horizontally, or both. Nonvector flaviviruses may be regarded as examples of this particular path, pestiviruses, EAV, and LDV being later examples of this course. Viruses which became adapted to different arthropod vectors may have changed relatively little in their antigenic properties and persist today as mosquito-borne flaviviruses, or they may have undergone more radical evolutionary changes to emerge as present-day alphaviruses. The emergence of multiple serological types at the end of each of these paths presents no difficulty, and the existence of a much larger number of discrete viruses in agents that retained the ability to replicate in two classes of host than in those that became restricted to vertebrate hosts only, presents no problem in evolutionary terms.

Looking at present-day Togaviridae from the viewpoint of their evolutionary development, tick-borne flaviviruses may be the most primitive survivors. Russian spring-summer encephalitis virus has retained the ability to be transmitted transovarially through its tick host, in which it produces no disease, and the severity of the illness produced in man could be taken as an indication of fairly recent and rather poor adaptation to man as a vertebrate host. West Nile virus may be a more highly evolved virus, retaining its capacity to replicate in ticks after it had become adapted to mosquitoes as its main arthropod host and developing the ability to replicate in a wide variety of vertebrate hosts, including birds and mammals. Relatively little is known about the natural history of nonvector flaviviruses, although these may provide valuable clues to the evolution of non-arthropod-borne Togaviridae. Lactate dehydrogenase virus may be regarded as the most highly evolved of all the Togaviridae, having lost all arthropod dependence and become almost perfectly adapted to its vertebrate host, the mouse, in which it causes no overt disease but produces lifelong infection with vertical transmission between generations.

In terms of virion structural proteins, which may be expected to be more closely related in viruses having a common evolutionary course, rubella virus seems to be more similar to alphaviruses than to flaviviruses. All the other non-arthropod-borne Togaviridae, pestiviruses, EAV, LDV, and SHV have polypeptide patterns which are more closely related to those of the alpha- or flaviviruses (see Chapter 21). In terms of host pathogenicity, SHV could be regarded as a recently evolved virus, still poorly adapted to its vertebrate host in which it produces severe or fatal disease, whereas LDV, as already noted, is highly adapted to its murine host in which it produces no disease; rubella and pestiviruses come between these two extremes, in that asymptomatic infection and severe disease are both possible. Recent evidence suggests that the natural host of SHV is not the macaque monkey, in which all the known outbreaks of disease have occurred, but the baboon (*Papio papio*) or the African green monkey (*Cercopithecus aethiops*) in both of which species completely silent infections have been found, some persisting for a long as 2 years (London, 1977). Rhesus monkeys appear to be highly susceptible to infection with SHV and acquire their disease either by close contact with carrier animals or by accidental laboratory transfer through injection or tattoo needles used on both species.

Is evolution still taking place within the Togaviridae? This is very difficult to establish given the limited time scale over which observations have been made. It could be argued that the sudden, explosive pandemic of o'nyong-nyong fever in East Africa in the early 1960s is an example of a recent evolutionary shift, o'nyong-nyong virus having developed from chikungunya virus by changes in antigenic structure and in vector potential which gave it new pathogenicity (Haddow *et al.*, 1960). Recombination has never been demonstrated within the Togaviridae, although phenotypic mixing has been reported, and pseudovirions of SIN, an alphavirus, with LDV have been made in the laboratory (Lagwinska *et al.*, 1975). It is unlikely that natural recombinants of Togaviridae play any part in the natural history of these viruses, whereas such a possibility seems much more probable with the Bunyaviridae, which have segmented genomes.

REFERENCES

American Committee on Arthropod-Borne Viruses (1969). *Am. J. Trop. Med. Hyg.* **18**, 731–734.

Andrewes, C. H., and Elford, W. J. (1933). *Br. J. Exp. Pathol.* **23**, 214–220.

August, J. T., Bolognesi, D. P., Fleissner, E., Gilden, R. V., and Nowinski, R. C. (1974). *Virology* **60**, 595–601.

Baker, A. C. (1943). *Am. J. Trop. Med.* **23**, 559–566.

Baltimore, D., Burke, D. C., Horzinek, M. C., Huang, A. S., Kääriäinen, L., Pfefferkorn, E. R., Schlesinger, M. J., Schlesinger, S., Schlesinger, R. W., and Scholtissek, C. (1976). *J. Gen. Virol.* **30**, 273.

Bearcroft, W. G. C. (1957). *J. Pathol. Bacteriol.* **74**, 295–303.
Berge, T. O., ed. (1975). "International Catalogue of Arboviruses," 2nd ed., DHEW Publ. No. (CDC) 75-8301. U.S. Dep. Health, Educ. Welfare, Public Health Serv., Washington, D.C.
Brandt, W. E., Cardiff, R. D., and Russell, P. K. (1970). *J. Virol.* **6**, 500–506.
Brinton-Darnell, M., and Plagemann, P. G. W. (1975). *J. Virol.* **16**, 420–433.
Buckley, S. M., and Clarke, D. H. (1970). *Proc. Soc. Exp. Biol. Med.* **135**, 533–539.
Bürki, F. (1970). *Proc. Int. Conf. Equine Infect. Dis., 2nd, Paris, 1969,* pp. 125–129.
Buescher, E. L., Scherer, W. F., Grossberg, S. E., Chanock, R. M., and Philpot, V. B. (1959). *J. Immunol.* **83**, 582–593.
Calisher, C. H., and Maness, K. S. C. (1970). *Appl. Microbiol.* **19**, 557–564.
Casals, J. (1944). *J. Exp. Med.* **79**, 341–359.
Casals, J. (1947). *J. Immunol.* **56**, 337–341.
Casals, J. (1957). *Trans. N.Y. Acad. Sci.* **19**, 219–235.
Casals, J. (1961). *Pac. Sci. Congr. 10th, Honolulu* Abstr. No. 458.
Casals, J. (1963). *Am. J. Trop. Med. Hyg.* **12**, 587–596.
Casals, J. (1964) . *J. Exp. Med.* **119**, 547–565.
Casals, J., and Brown, L. V. (1954). *J. Exp. Med.* **99**, 429–449.
Casals, J., Buckley, S. M., and Barry, D. W. (1973). *Appl. Microbiol.* **25**, 753–762.
Chan, Y. C. (1965). *Nature (London)* **206**, 116–117.
Chanas, A. C., Johnson, B. K., and Simpson, D. I. H. (1976). *J. Gen. Virol.* **32**, 295–300.
Clarke, D. H. (1960). *J. Exp. Med.* **111**, 1–20.
Clarke, D. H. (1962). *In* "Biology of Viruses of the Tick-Borne Encephalitis Complex" (H. Libiková, ed.), pp. 67–75. Academic Press, New York.
Clarke, D. H. (1964). *Bull. W.H.O.* **31**, 45–56.
Cole, F. E., and McKinney, R. W. (1971). *Infect. Immun.* **4**, 37–43.
Cuadrado, R. R., and Casals, J. (1967). *J. Immunol.* **98**, 314–320.
Dalrymple, J. M. (1972). *In* "Venezuelan Encephalitis," Sci. Publ. No. 243, pp. 56–64. Pan Am. Health Organ., Washington, D.C.
Dalrymple, J. M., Teramoto, A. Y., Cardiff, R. D., and Russell, P. K. (1972). *J. Immunol.* **109**, 426–433.
Dalrymple, J. M., Schlesinger, S., and Russell, P. K. (1976). *Virology* **69**, 93–103.
Daniels, C. A. (1975). *In* "Viral Immunology and Immunopathology" (A. L. Notkins, ed.), pp. 79–97. Academic Press, New York.
Daniels, C. A., Borsos, T., Rapp, H. J., Snyderman, R., and Notkins, A. L. (1970). *Proc. Natl. Acad. Sci. U.S.A.* **65**, 528–535.
Darbyshire, J. H. (1962). *Res. Vet. Sci.* **118**, 125–128.
Darlington, C. D., Mattingley, P. F., and Smith, C. E. G. (1960). *Trans. R. Soc. Trop. Med. Hyg.* **54**, 89–134.
Digoutte, J.-P., and Girault, G. (1976). *Ann. Microbiol. (Paris)* **127**, 429–437.
Eaton, B. T. (1979). *J. Virol.* **30**, 45–55.
Falkler, W. A., Diwan, A. R., and Halstead, S. B. (1973). *J. Immunol.* **111**, 1804–1809.
Fauconnier, B. (1970). *C. R. Acad. Sci., Ser.* **271**, 1464–1466.
Fauconnier, B. (1971). *Pathol. Biol.* **19**, 575–578.
Fazekas de St. Groth, S. (1962). *Adv. Virus Res.* **9**, 1–125.
Fenner, F. (1976). *Intervirology* **7**, 1–115.
Garoff, H., Simons, K., and Renkonen, O. (1975). *Virology* **61**, 493–504.
Gilden, R. V., Oroszlan, S., and Huebner, R. J. (1971). *Nature (London), New Biol.* **231**, 107–108.

Gresser, I., Tovey, M. G., Bandu, M.-T., Maury, C., and Brouty-Boyé, D. (1976a). *J. Exp. Med.* **144,** 1305–1315.
Gresser, I., Tovey, M. G., Maury, C., and Bandu, M.-T. (1976b). *J. Exp. Med.* **144,** 1316–1323.
Griffin, D. E., and Johnson, R. T. (1977). *J. Immunol.* **118,** 1070–1075.
Gutekunst, D. E., and Malmquist, W. A. (1963). *Can. J. Comp. Med. Vet. Sci.* **27,** 121–123.
Haddow, A. J. (1965). *Trans. R. Soc. Trop. Med. Hyg.* **59,** 436–458.
Haddow, A. J. (1968). *Proc. R. Soc. Edinburgh* **70,** 191–227.
Haddow, A. J., Davies, C. W., and Walker, A. J. (1960). *Trans. R. Soc. Trop. Med. Hyg.* **54,** 517–522.
Hahon, N. (1966). *J. Infect. Dis.* **116,** 33–40.
Hahon, N. (1969). *Am. J. Vet. Res.* **30,** 1007–1014.
Hahon, N., and Cooke, K. O. (1967). *J. Virol.* **1,** 317–326.
Hahon, N., and Hankins, W. A. (1970). *Appl. Microbiol.* **19,** 224–231.
Haldane, J. B. S. (1960). *Nature (London)* **187,** 879.
Hallauer, C. (1946). *Schweiz. Z. Pathol. Bakteriol.* **9,** 553–554.
Halstead, S. B. (1974). *Am. J. Trop. Med. Hyg.* **23,** 974–982.
Hammam, H. M., Clarke, D. H., and Price, W. H. (1965). *Am. J. Epidemiol.* **82,** 40–55.
Hammon, W.McD., and Sather, G. E. (1969). *In* "Diagnostic Procedures for Viral and Rickettsial Diseases" (E. H. Lennette and N. J. Schmidt, eds.), 4th ed., Chapter 6, 227–280. Am. Public Health Assoc., New York.
Hammon, W.McD., and Sather, G. E. (1973). *Am. J. Trop. Med. Hyg.* **22,** 524–534.
Hayes, R. O., Francy, D. B., Lazuick, J. S., Smith, G. C., and Gibbs, E. P. J. (1977). *J. Med. Entomol.* **14,** 257–262.
Hearn, H. J., and Rainey, C. T. (1963). *J. Immunol.* **90,** 720–724.
Henderson, B. E., Cheshire, P. P., Kirya, G. B., and Lule, M. (1970). *Am. J. Trop. Med. Hyg.* **19,** 110–118.
Henderson, J. R. (1964). *J. Immunol.* **93,** 452–461.
Hyllseth, B., and Pettersson, V. (1970). *Arch. Gesamte Virusforsch.* **32,** 337–347.
Ibrahim, A. N., and Hammon, W.McD. (1968a). *J. Immunol.* **100,** 86–92.
Ibrahim, A. N., and Hammon, W.McD. (1968b). *J. Immunol.* **100,** 93–98.
Igarashi, A. (1979). *Virology* **98,** 385–392.
Igarashi, A., and Mantani, M. (1974). *Biken J.* **17,** 87–93.
Igarashi, A., Harrap, K. A., Casals, J., and Stollar, V. (1976). *Virology* **74,** 174–187.
Karabatsos, N. (1975). *Am. J. Trop. Med. Hyg.* **24,** 527–532.
Kennedy, S. I. T., and Burke, D. C. (1972). *J. Gen. Virol.* **14,** 87–98.
Kusuda, H., and Takahashi, T. (1969). *Nippon Densembyo Gakkai Zasshi* **43,** 59–63.
Lafferty, K. J. (1963). *Virology* **21,** 61–75.
Lagwinska, E., Stewart, C. C., Adles, C., and Schlesinger, S. (1975). *Virology* **65,** 204–214.
Levitt, N. H., Miller, H. V., Pedersen, C. E., and Eddy, G. A. (1975). *Am. J. Trop. Med. Hyg.* **24,** 127–130.
London, W. T. (1977). *Nature (London)* **268,** 344–345.
Lubiniecki, A. S., Pfancuff, M., and Mitchell, J. (1976). *Int. Arch. Allergy Appl. Immunol.* **50,** 427–435.
Lvov, D. K., Timofeeva, A. A., Gromashevsky, A. L. *et al.* (1975). *Proc. Sci. Sess. Inst. Poliomyelitis Virus Enceph., 18th,* Moscow, U.S.S.R., A.M.S. pp. 322–324.
Macnamara, F. N. (1953). *Br. J. Exp. Pathol.* **34,** 392–399.
Macnamara, F. N., Horne, D. W., and Porterfield, J. S. (1959). *Trans. R. Soc. Trop. Med. Hyg.* **53,** 202–212.

Madrid, A. T. de, and Porterfield, J. S. (1974). *J. Gen. Virol.* **23**, 91–96.
Mengeling, W. L., and Van der Maaten, M. J. (1971). *Am. J. Vet. Res.* **32**, 1825–1833.
Monath, T. P. (1977). Personal communication.
Monath, T. P., Wilson, D. C., and Casals, J. (1973). *Bull. W.H.O.* **49**, 235–244.
Morgan, I. M. (1945). *J. Immunol.* **50**, 359–371.
Oaten, S. W., Webb, H. E., and Bowen, E. T. W. (1976). *J. Gen. Virol.* **33**, 381–388.
Okuno, T. T., Okada, T., Kondo, A., Suzuki, M., Kabayashi, M., and Oya, A. (1968). *Bull. W.H.O.* **38**, 547–563.
Oldstone, M. B. A. (1975). *Prog. Med. Virol.* **19**, 84–119.
Ozaki, Y., Kumagai, K., Kawanishi, M., and Seto, A. (1974). *Arch. Gesamte Virusforsch.* **45**, 7–16.
Peck, R. D., Brown, A., and Wust, C. J. (1975). *J. Immunol.* **114**, 581–584.
Pedersen, C. E., and Eddy, G. A. (1975). *Am. J. Epidemiol.* **101**, 245–252.
Plant, J. W., Littlejohns, I. R., Gardiner, A. C., Vantsis, J. T., and Huck, R. A. (1973). *Vet. Rec.* **92**, 455.
Porterfield, J. S. (1961). *Bull. W.H.O.* **24**, 735–741.
Porterfield, J. S. (1962). *Adv. Virus Res.* **9**, 127–156.
Porterfield, J. S., Casals, J., Chumakov, M. P., Gaidamovich, S. Y., Hannoun, C., Holmes, I. H., Horzinek, M. C., Mussgay, M., Oker-Blom, N., and Russell, P. K. (1975–1976). *Intervirology* **6**, 13–24.
Porterfield, J. S., Casals, J., Chumakov, M. P., Gaidamovich, S. Y., Hannoun, C., Holmes, I. H., Horzinek, M. C., Mussgay, M., Oker-Blom, N., Russell, P. K., and Trent, D. W. (1978). *Intervirology* **9**, 129–148.
Price, W. H., Thind, I. S., O'Leary, W., and El Dadah, A. H. (1967). *Am. J. Epidemiol.* **86**, 11–27.
Price, W. H., Casals, J., Thind, I., and O'Leary, W. (1973). *Am. J. Trop. Med. Hyg.* **22**, 509–523.
Radwan, A. I., and Crawford, T. B. (1974). *J. Gen. Virol.* **25**, 229–237.
Riley, V. (1974). *Prog. Med. Virol.* **18**, 198–213.
Robin, Y., Cornet, M., Le Gonidec, G., Chateau, R., and Heme, G. (1978). *Ann. Microbiol. (Paris)* **129**, 239–244.
Rodda, S. J., and White, D. O. (1976). *J. Immunol.* **117**, 2067–2072.
Rosato, R. R., Dalrymple, J. M., Brandt, W. E., Cardiff, R. D., and Russell, P. K. (1974). *Acta Virol. (Engl. Ed.)* **18**, 25–30.
Rowson, K. E. K., and Mahy, B. W. J. (1975). "Lactic Dehydrogenase Virus." Springer-Verlag, Berlin and New York.
Russell, P. K., and Brandt, W. E. (1973). *Perspect. Virol.* **8**, 263–277.
Sabin, A. B. (1950). *Bacteriol. Rev.* **14**, 225–232.
Sabin, A. B. (1951). *Proc. Soc. Exp. Biol. Med.* **78**, 655–658.
Sabin, A. B. (1952). *In* "Viral and Rickettsial Infections of Man" (T. M. Rivers, ed.), 2nd ed., pp. 556–568. Lippincott, Philadelphia, Pennsylvania.
Sabin, A. B., and Buescher, E. L. (1950). *Proc. Soc. Exp. Biol. Med.* **74**, 222–230.
St. George, T. D., Standfast, H. A., Doherty, R. L., Carley, J. G., Fillipich, C., and Brandsma, J. (1977). *Aust. J. Exp. Biol. Med.* **55**, 493–499.
Sartorelli, A. C., Fischer, D. S., and Downs, W. G. (1966). *J. Immunol.* **96**, 676–682.
Schlesinger, R. W. (1977). "Dengue Viruses," pp. 1–132. Springer-Verlag, Berlin and New York.
Schlesinger, R. W., Gordon, I., Frankel, J. W., Winter, J. W., Patterson, P. R., and Dorrance, W. R. (1956). *J. Immunol.* **77**, 352–364.
Schlesinger, R. W., Stollar, V., Igarashi, A., Guild, G. M., and Cleaves, G. R. (1979). *In*

"Comparative Virology III" (E. Kurstak and K. Maramarosch, eds.), Academic Press, New York.
Singh, G. (1970). *Indian J. Med. Res.* **58,** 1149–1156.
Smith, C. E. G., and Holt, D. (1961). *Bull. W.H.O.* **24,** 749–759.
Smith, T. W., Brandt, W. E., Swanson, J. L., McCown, J. M., and Buescher, E. L. (1970). *J. Virol.* **5,** 524–532.
Smithburn, K. C. (1942). *J. Immunol.* **44,** 25–31.
Smithburn, K. C., Hughes, T. P., Burke, A. W., and Paul, J. H. (1940). *Am. J. Trop. Med.* **20,** 471–493.
Smorodintseff, A. A. (1939–1940). *Arch. Gesamte Virusforsch.* **1,** 468–480.
Souza Lopes, O. de, Coimbra, T. L. M., de Abreu Sachetta, L., and Calisher, C. H. (1978). *Am. J. Epidemiol.* **107,** 444–449.
Spence, L., Downs, W. G., Boyd, C., and Aitken, T. H. (1961). *West. Indian Med. J.* **10,** 54–58.
Stohlman, S. A., Eylar, O. R., and Wisseman, C. L. (1976). *J. Virol.* **18,** 132–140.
Stollar, V. (1975). *Virology* **66,** 620–624.
Stollar, V., and Stollar, B. D. (1970). *Proc. Natl. Acad. Sci. U.S.A.* **65,** 993–1000.
Stollar, V., Stevens, T. M., and Schlesinger, R. W. (1966). *Virology* **30,** 303–312.
Symington, J., McCann, A. K., and Schlesinger, M. J. (1977). *Infect. Immun.* **15,** 720–725.
Taniguchi, S., and Yoshino, K. (1965). *Virology* **26,** 54–60.
Theiler, M., and Anderson, C. R. (1975). *Am. J. Trop. Med. Hyg.* **24,** 115–117.
Theiler, M., and Downs, W. G. (1973). "The Arthropod-Borne Viruses of Vertebrates." Yale Univ. Press, New Haven, Connecticut.
Theiler, M., and Smith, H. H. (1937). *J. Exp. Med.* **65,** 787–800.
Tikasingh, E. S., Spence, L., and Downs, W. G. (1966). *Am. J. Trop. Med. Hyg.* **15,** 219–226.
Trent, D. W. (1977). *J. Virol.* **22,** 608–618.
Trent, D. W., Harvey, C. L., Qureshi, A., and Le Stourgen, D. (1976). *Infect. Immun.* **13,** 1325–1333.
Trousdale, M. D., Trent, D. W., and Shelokov, A. (1975). *Proc. Soc. Exp. Biol. Med.* **150,** 707–711.
Vathanophas, K., Hammon, W.McD., Atchison, R. W., and Sather, G. E. (1973). *Proc. Soc. Exp. Biol. Med.* **142,** 697–702.
Way, H. J., and Garwes, D. J. (1970). *J. Gen. Virol.* **7,** 211–223.
Webster, L. T. (1938). *J. Exp. Med.* **67,** 609–618.
Webster, L. T., Fite, G. L., and Clow, A. D. (1935). *J. Exp. Med.* **62,** 827–847.
Wengler, G., Wengler, G., and Filipe, A. R. (1977). *Virology* **78,** 124–134.
Westaway, E. G. (1968). *J. Immunol.* **100,** 569–580.
Whitman, L. (1947). *J. Immunol.* **56,** 97–108.
Whitman, L., and Shope, R. E. (1962). *Am. J. Trop. Med. Hyg.* **11,** 691–696.
Wildy, P. (1971). *In* "Classification and Nomenclature of Viruses," Monographs in Virology, Vol. 5, pp. 1–81. Karger, Basel.
Wood, O., Tauraso, N., and Leibhaber, H. (1970). *J. Gen. Virol.* **7,** 129–136.
World Health Organization (1967). Arboviruses and human disease. *W.H.O. Tech. Rep. Ser.* No. 369.
Young, N. A., and Johnson, K. M. (1969). *Am. J. Epidemiol.* **89,** 286–307.
Zeegers, J. J. W., van der Zeijst, B. A. M., and Horzinek, M. C. (1976). *Virology* **73,** 200–205.

3

Medical Significance of Togaviruses: An Overview of Diseases Caused by Togaviruses in Man and in Domestic and Wild Vertebrate Animals

ROBERT E. SHOPE

I. INTRODUCTION

In earlier literature togaviruses were known by the diseases they caused—yellow fever, arthropod-borne viral encephalitides, dengue,

THE TOGAVIRUSES

ISBN 0-12-625380-3

German measles, hog cholera—all significant diseases and some of them historic scourges of mankind and domestic animals. We have recently come to recognize that some of these syndromes can be caused by two or more togaviruses: for example, historically, the term "dengue fever" has been applied to diseases which we now recognize as similar but different syndromes caused by chikungunya or dengue virus (Carey, 1971). Even within the classic syndrome caused by dengue virus, one can recognize subtle differences, depending on whether the virus serotype involved is type 1 or type 4 (Halstead, 1974). We have been continuously surprised by the togaviruses.

Dengue, considered a benign disease before 1954, erupted in Manila as hemorrhagic fever with shock, a new clinical syndrome which differed greatly from classic dengue (Hammon *et al.*, 1957). As the ability to isolate viruses has become more sophisticated, so has the recognition of new agents, often nontogaviruses, which cause diseases

TABLE I

Syndromes Caused by Togaviruses

Syndrome	Associated togaviruses
Fever	EEE, WEE, VEE, Mucambo, o'nyong-nyong, chikungunya, Mayaro, Ross River, Sindbis, Semliki Forest, Banzi, Bussuquara, dengue 1–4, Ilheus, Japanese encephalitis, Kunjin, Murray Valley encephalitis, Rocio, Sepik, St. Louis encephalitis, Spondweni, Wesselsbron, West Nile, yellow fever, Zika, tick-borne encephalitis, Kyasanur Forest disease, Omsk, louping ill, Powassan, Rio Bravo, rubella, hog cholera (in hogs)
Fever with rash	Dengue 1–4, West Nile
Rash and polyarthritis	Ross River
Arthralgia or myalgia, fever, and rash	Dengue 1–4, o'nyong-nyong, chikungunya, Mayaro
Hemorrhagic fever and shock syndrome	Dengue 1–4, Kyasanur Forest disease, Omsk, yellow fever, dengue 1–4
Encephalitis	EEE, WEE, VEE, Semliki Forest, Apoi, Ilheus, Japanese encephalitis, Murray Valley encephalitis, Rocio, St. Louis encephalitis, West Nile, tick-borne encephalitis, louping ill, Powassan, Negishi, Rio Bravo
Diabetes (experimental)	VEE
Congenital defects	Rubella
Abortion	Wesselsbron (sheep), Japanese encephalitis (pigs)
Hemorrhagic diarrhea (hogs)	Hog cholera

like hemorrhagic fever. Marburg, Ebola, Machupo, and Crimean hemorrhagic fever viruses have now been characterized; they cause hemorrhagic fever. It is clear that yellow fever, dengue, and Omsk viruses are not the only hemorrhagic fever agents involved in the disease syndrome. Finally, as our diagnostic acumen improved, we recognized additional entirely new clinical manifestations of togavirus infections which we had not previously linked to a togavirus. Thus, congenital defects were associated with rubella virus (Gregg, 1941); a remote disease in Australia, epidemic polyarthritis and rash, was linked serologically to chikungunya infection in Africa (Shope and Anderson, 1960), although the polyarthritis agent was not isolated until several years later (Doherty *et al.*, 1964).

Togaviruses cause an impressive list of clinical syndromes. These are shown in Table I along with representative examples of viruses which have been associated with each clinical category. The togaviruses which cause a defined disease in man or other vertebrate animals are presented in Table II. This list is growing constantly, Rocio virus being the most recent addition. There are scores of known togaviruses which have been incompletely studied, and it may be accepted as an article of faith that some of these cause disease and that there exist others, as yet undiscovered, which will in time be associated with medical or veterinary problems.

II. DESCRIPTION OF MAJOR DISEASE GROUPS

A. Eastern, Western, and Venezuelan Equine Encephalitis

It is an unexplained curiosity that three alphaviruses should cause severe epidemic disease in man and equine animals in the New World and that Africa, Europe, Asia, and Australia, each of which also harbors alphaviruses, do not have recognized alphavirus epidemic encephalitis.

Eastern equine encephalitis (EEE), Western equine encephalitis (WEE), and Venezuelan equine encephalitis (VEE) are viral diseases of man and horses. Mosquitoes transmit the viruses to wild birds or, in the case of VEE (Sudia and Newhouse, 1975), to wild rodents and maintain the natural cycle. Mosquitoes remain infective for life, but the virus has not as yet been demonstrated in mosquito eggs. Thus, it is not known how the viruses overwinter in temperate climates. Man and horses are incidental (except for VEE) hosts of enzootic virus and

TABLE II

Togaviruses Which Cause Disease in Human Beings and Other Vertebrates

Togarvirus	Disease	Diseased hosts	Vector	Distribution
Alphaviruses				
Chikungunya	Fever, arthritis, rash	Man	Mosquito	Africa, Southeast Asia, Philippines
Eastern equine encephalitis	Encephalitis	Man, horse, pheasant	Mosquito	Americas
Everglades	Encephalitis	Man	Mosquito	United States
Mayaro	Fever, arthritis, rash	Man	Mosquito	Americas
Mucambo	Fever	Man	Mosquito	South and Central Americas
O'nyong-nyong	Fever, arthritis, rash	Man	Mosquito	Africa
Ross River	Arthritis, rash	Man	Mosquito	Australia, New Guinea
Semliki Forest	Encephalitis	Man	Laboratory infection	
Sindbis	Fever	Man	Mosquito	Africa, India, Southeast Asia, Philippines, Australia
Venezuelan equine encephalitis	Fever, encephalitis	Man, equines	Mosquito	Americas
Western equine encephalitis	Fever, encephalitis	Man, equines	Mosquito	Americas

Flaviruses				
Apoi	Encephalitis	Man	Unknown	Japan
Banzi	Fever	Man	Mosquito	Africa
Rio Bravo	Encephalitis, aseptic meninigitis	Man	Unknown	Americas
Bussuquara	Fever, hepatitis (monkey)	Man, monkey	Mosquito	South America, Panama
Dengue 1–4	Fever, rash, hemorrhagic fever, shock syndrome	Man	Mosquito	Africa, Asia, Pacific, Australia, Americas
Ilheus	Fever, encephalitis	Man	Mosquito	South America
Japanese encephalitis	Encephalitis	Man, pig, horse	Mosquito	Asia, Pacific islands
Kunjin	Fever	Man	Mosquito	Australia, Sarawak
Kyasanur Forest	Hemorrhagic fever	Man, monkeys	Tick	India
Louping ill	Encephalitis	Sheep, man	Tick	Great Britain
Murray Valley encephalitis	Encephalitis	Man	Mosquito	Australia
Negishi	Encephalitis	Man	Unknown	Japan
Omsk hemorrhagic fever	Hemorrhagic fever	Man	Tick	USSR
Powassan	Encephalitis	Man	Tick	North America, USSR
Rocio	Encephalitis	Man	Unknown	Brazil
Sepik	Fever	Man	Mosquito	New Guinea
Tick-borne encephalitis	Encephalitis	Man	Tick	Europe, Asia
Others				
Rubella	Fever, rash, encephalitis, congenital defects	Man	None	Worldwide
Hog cholera	Fever, hemorrhagic diarrhea	Hog	Unknown	Worldwide

are not essential to maintenance of the cycle. Pheasants are readily infected with EEE virus and suffer high mortality during epizootics. Pheasants transmit EEE virus directly by pecking and develop viremia titers high enough to reinfect mosquitoes. Horses infected with epizootic strains of VEE virus, in addition to becoming sick, develop high levels of virus in the blood and can act as epidemic-amplifying hosts. Thus, vaccination of horses not only protects against disease but prevents spread of virus to mosquitoes.

Eastern equine encephalitis virus is enzootic in eastern North America near large bodies of water including the Atlantic seaboard, the eastern Great Lakes, and the Gulf of Mexico. It is also found throughout the Caribbean and the east and west coasts of South America. The natural cycle in the United States is maintained by *Culiseta melanura* mosquitoes and passerine birds. Since *C. melanura* feed primarily on birds in preference to man, *Aedes* mosquitoes have been invoked to explain the epidemic transmission to man and horses.

Western equine encephalitis virus causes disease over most of the western United States, Central America, and northern South America. It has also been isolated from sick horses in Argentina. The enzootic and epidemic vector of WEE virus in the United States is *Culex tarsalis*. Birds have viremia lasting 1–5 days, and they efficiently maintain the natural cycle.

The Everglades subtype of VEE virus is enzootic in southern Florida where its reservoir consists of *Culex* (*Melanoconion*) species and wild rodents. Encephalitis in persons over 50 years of age has been recorded there on three occasions (Ehrenkranz *et al.*, 1970; Center for Disease Control, 1971). The epidemic form of VEE appears in South and Central America sporadically and unpredictably. In 1971 the virus entered Texas from Mexico and precipitated a major outbreak in horses and man (Zehmer *et al.*, 1974). The virus has not reappeared since, but the risk of introduction again is ever present.

The epidemiology of EEE, WEE, and VEE in Central and South America is similar to that in the United States, except that the viruses may utilize different species of mosquitoes and vertebrate hosts.

In man EEE is a severe disease. Although outbreaks are limited to from 10 to 35 cases, the case fatality rate varies between 50 and 75%. In the epidemic of 1959 in New Jersey, the inapparent/apparent infection ratio was between 8:1 and 50:1 (Goldfield *et al.*, 1968). Older adults and young children are most severely affected. Onset is abrupt, and progression is rapid. The patient may be comatose by the second day of illness. Convulsive seizures are recorded in 75% of patients;

other signs of encephalitis include paralyses and autonomic disturbances such as sialorrhea, irregular breathing, and cyanosis. Edema, either generalized or limited to the periorbital or facial region, is common in children. This may be associated with cerebral edema. Although recovery sometimes occurs starting after 1 week, sequellae such as paralyses, seizures, and mental retardation are severe in up to 30% of survivors (Feemster, 1957).

Outbreaks of EEE may occur in horses alone, as in 1933 in Virginia, Delaware, Maryland, and New Jersey, or in horses followed about 2 weeks later by disease in man. This was the case in 1938 in Massachusetts where 44 persons developed encephalitis and 65% died (Farber *et al.*, 1940). The mortality among 248 horses was 90%. Subsequent small outbreaks occurred in Louisiana, 1947; in the Dominican Republic, 1948–1949 and 1978; in Massachusetts, 1956 (Feemster, 1957); in New Jersey, 1959; and in Jamaica, 1962. Disease in pheasants, man, or horses occurs almost annually in Massachusetts, leading one to believe that there is a persistent focus with overwintering *in situ* rather than introduction by birds or other vertebrate animals.

In contrast to EEE, case fatality rates of WEE infection of man are only about 10% (Reeves and Hammon, 1962). The disease, other than by its epidemiological associations, cannot be distinguished from other forms of encephalitis. It affects mostly children, with a high percentage of infants presenting with convulsions. Sequellae, especially in patients under 1 year of age, are a prominent feature of the disease (Finley and Longshore, 1958; Earnest *et al.*, 1971; Mulder *et al.*, 1952). Sequellae range from epilepsy to minimal brain dysfunction syndromes.

Epizootics of WEE occur in the Pacific Coast region and great plains of the United States and Canada. In 1941, 3000 human cases occurred in Minnesota, North Dakota, and the plains provinces of Canada. In the Central Valley of California, cases of WEE in man occurred virtually every year. There were 375 confirmed cases in 1952, a peak year. In the last decade, however, probably because of exemplary mosquito control efforts, the disease has nearly disappeared from California. In fact, California has demonstrated that, with scientific leadership, popular support, and an enlightened legislature, a focal arboviral disease such as WEE can be controlled. A return to the past may be unfolding with the advent of the cost consciousness of the late 1970s and the threatened cutback of support for encephalitis surveillance and control. Because both the vector, *C. tarsalis*, and WEE virus are still present in California there is every reason to believe human cases of encephalitis may again become the rule rather than the exception.

Venezuelan equine encephalitis is primarily a disease of horses which, as demonstrated in Texas in 1971, can be a major disruptive event. The earliest epizootic occurred in 1938 in Venezuela (Kubes and Rios, 1939); subsequent outbreaks in horses and man in Colombia in 1952 (Sanmartin *et al.*, 1954), in Panama in 1961–1963, and in Colombia and Venezuela in 1962–1964 showed that VEE virus could cause large numbers of human febrile cases and significant numbers of fatal human encephalitides. About 12,000 cases occurred in man concomitantly with thousands of deaths in horses in Colombia and Venezuela (Groot, 1964; Sellers *et al.*, 1965). Again in Cali, Colombia, during 1967 and 1968 both human and equine cases were numerous (Sanmartin *et al.*, 1973); some of the surviving children had residual behavior and learning difficulties (Leon *et al.*, 1975).

During 1969 and 1970, a series of outbreaks in horses and man in Guatemala, Honduras, and Mexico (Borunda-Falcon, 1972) led observers in the United States to predict that epizootic VEE would reach the United States (Reeves, 1972). In spite of the predictions, U.S. authorities were inadequately prepared to stop VEE virus from crossing the southern U.S. border. An excellent live, attenuated VEE vaccine for horses had been stockpiled by the U.S. Army. This was authorized for use by the U.S. Department of Agriculture only after the epizootic was established in Texas. The vaccine was used successfully along with aerial spraying of malathion over a wide area of Texas and 19 other states (Zehmer *et al.*, 1974). In the meantime, thousands of horses died and there were 88 nonfatal VEE cases reported in man (Bowen *et al.*, 1976). The cost of controlling the disease was estimated to be in excess of $20 million (Reeves, 1972). Epizootic VEE has not been recognized in the United States from 1971 to 1980.

B. Chikungunya, O'nyong-nyong, Mayaro, and Epidemic Polyarthritis and Rash

According to Carey (1971), what was called "dengue" in 1823 was actually chikungunya infection; in fact, since as early as 1779, periodic epidemics characterized by fever, joint pain, rash, and residual arthralgias have probably been caused by chikungunya, o'nyong-nyong, or Mayaro virus; or they may have been related to the Australian epidemic polyarthritis-and-rash syndrome caused by Ross River virus. These viruses are closely related serologically, and each is associated with mosquito-borne epidemics.

Chikungunya virus is transmitted by *Aedes aegypti* (Lumsden, 1955; Ross, 1956), o'nyong-nyong by *Anopheles funestes* and *A. gam-*

biae (Haddow *et al.*, 1960), Mayaro by *Haemagogus* (Woodall, 1967), and Ross River by *Aedes vigilax* and *Culex annulirostris* (Doherty *et al.*, 1963; Gard *et al.*, 1973). Major questions remain about the role of vertebrates in the reservoir status of each of these viruses. During chikungunya infection, man develops viremia levels which can easily reinfect *A. aegypti*; it is doubted, however, that man maintains the vertebrate portion of the cycle except during epidemics. A variety of vertebrate animals have antibody; it may be, however, that the monkey is a primary host of chikungunya and, in addition, of o'nyong-nyong and Mayaro viruses; the New Holland mouse may be the host of Ross River virus (Gard *et al.*, 1973). O'nyong-nyong virus, which in 1959 caused one of the largest epidemics of mosquito-borne disease ever recorded, apparently disappeared inexplicably after 1962, leaving doubt of its having a reservoir at all.

Chikungunya was first isolated during the Newala epidemic in southern Tanganyika in 1952 (Ross, 1956). The disease was characterized by sudden onset with high fever and severe joint pains, followed several days later by a maculopapular rash (Robinson, 1955). Children have the typical disease, but less frequently with arthritis and rash than adults (Jadhav *et al.*, 1965). Joint symptoms in adults may persist for several weeks. Hemorrhagic manifestations are recorded (Dasaneyavaja and Pongsupat, 1961), but the syndrome is not as severe as the hemorrhagic fever and shock associated with dengues.

O'nyong-nyong virus caused a very similar illness, but the fever was generally somewhat lower and often absent, and lymphadenitis, especially posterior cervical node involvement, was common (Shore, 1961). Mayaro virus also may cause arthritis, but in most epidemics this has not been a common finding. Mayaro epidemics are usually associated with jungle contact and do not involve urban areas as do o'nyong-nyong and chikungunya.

The clinical picture of Australian epidemic polyarthritis and rash associated with Ross River virus infection differs from that of chikungunya; fever is not prominent, viremia is rare, and therefore, unlike the other viruses in this group, Ross River virus is not usually isolated during the illness. Also unlike chikungunya and o'nyong-nyong infections, the rash of Ross River infections may appear simultaneously with the arthritis (Anderson and French, 1957).

Carey (1971) reviewed "chikungunya" epidemics which had been (probably mistakenly) called dengue. These occurred in 1779 in Batavia and Cairo, in 1823 in Zanzibar, in 1824–1825 in India, in 1827–1828 in the West Indies and southern United States, in 1870 in Zanzibar, in 1901–1902 in Hong Kong, Burma, and Madras, and in

1923 in Calcutta. It is not known which alphavirus caused these epidemics, but Carey's evidence is good that they were not caused by dengue virus. Shope and Anderson (1960) reviewed the early epidemics of polyarthritis and rash which again were defined solely on clinical grounds.

Since 1952 there have been major epidemics of chikungunya in Tanganyika (1952–1953) with an estimated 60,000 cases (Lumsden, 1955), in Thailand in 1962–1964 (Nimmannitya *et al.*, 1969), in Rhodesia in 1962 (McIntosh *et al.*, 1963), in Vellore, India in 1964, in Nigeria in 1969 (Moore *et al.*, 1974), and in American service personnel in Viet Nam in 1966 (Deller and Russell, 1968). A single epidemic of o'nyong-nyong has been recorded, starting in 1959 and lasting 2 years; it involved approximately 2 million persons in Uganda, Kenya, Tanzania, and the Sudan (Williams *et al.*, 1965). Small epidemics of Mayaro were reported by Anderson *et al.* (1957) in 1954 in Trinidad and by Causey and Maroja (1957) in 1955 in the Amazon region of Brazil. Epidemic polyarthritis and rash occurs regularly as outbreaks in the Murray River valley of Australia and eastern Australia. Sporadic cases appear each year in other parts of Australia (Doherty *et al.*, 1971).

C. Yellow Fever

Yellow fever was a major scourge of mankind for over two centuries, and its potential for decimating populations was realized as recently as 1962 in Ethiopia (Serie *et al.*, 1968). Although the 17D yellow fever vaccine is as nearly perfect as can be achieved, its effective use in an epidemic still depends on the early recognition of cases and the availability of a sufficient quantity of vaccine. Should yellow fever break out in Southeast Asia, or even in cities such as Miami, Houston, and New Orleans in the United States, it remains to be proved that an epidemic would be controlled by killing the vector and by vaccination, before substantial mortality occurred (Downs, 1969; Soper, 1969).

Kerr (1951) gives an excellent description of yellow fever from which the following is drawn. Yellow fever is an acute febrile disease with a case fatality of between 2 and 25%. After an incubation of 3–6 days, onset is sudden with fever between 38.5° and 40°C, rarely higher. There is relative bradycardia; back and muscle aches, nausea, headache, and malaise accompany the fever. The skin is hot, and the face flushed. Many cases go on to complete recovery without more severe disease. Others, after a period of remission, develop jaundice and hemorrhagic complications including nosebleeds, vomiting of blood as dark coffee-ground material or as bright red blood; stools may be red or tarry. Renal complications are also common; proteinuria, blood and casts in the urine, oliguria, and azotemia may occur. Patients are usually calm and lucid until shortly before death when hyperexcitability, hallucinations, and aggressive behavior may be manifest. Mild illnesses last 2–3 days, while the most severe last from 5 to 7 days, unless death intervenes on the third or fourth day as in fulminant cases. Recovery from yellow fever is usually uncomplicated, although myocardial failure may occur during convalescence and may result in death (Kirk, 1941).

Patients with yellow fever have leukopenia and an increased sedimentation rate and may develop increased blood urea nitrogen levels. The clotting time is prolonged. Results of liver function tests are not reported in the literature but might be expected to be abnormal since the liver parenchyma undergoes necrotic changes, especially in the midzonal region of the lobules. The kidney undergoes tubular necrosis, accounting for the nephrotic clinical picture. Multiple small hemorrhages of the gut, heart, and brain are reported (Stevenson, 1939).

Jungle yellow fever occurs in the forested tropical regions of South America and Africa. In South America, the infection cycle is believed to be maintained by monkeys and *Haemagogus* mosquitoes in the forest canopy. Man is infected at the forest fringe when mosquitoes descend to the ground to feed. In Africa, the forest cycle is maintained by *Aedes africanus* and monkeys. *Aedes simpsoni* may be responsible for transmitting the virus between monkeys and man. Once established in man in an urban situation, yellow fever is transmitted by *A. aegypti* from man to man. In any given forested area, yellow fever appears only sporadically, usually at several-year intervals; it has been demonstrated by laboratory experiments that the virus is passed transovarially in the mosquito (Aitken *et al.*, 1979), and this may explain how yellow fever virus can be apparently dormant for long periods.

The early epidemics of yellow fever are reviewed by Theiler and Downs (1973) and by Theiler (1948). In the New World, epidemics were recorded in 1667 in Barbados, in 1668 in Yucatan, and in 1695 on military vessels near Martinique; during the eighteenth century, yellow fever epidemics occurred for 36 years, with outbreaks in the United States recorded for 26 of those years. In the nineteenth century, there were epidemics for 56 years, with 37 epidemics recorded in the United States. Between 1793 and 1900, 500,000 cases of yellow fever occurred in the United States; in Cuba, where yellow fever was probably continuously present, 35,900 deaths were attributed to the disease from 1853 to 1900. In the early twentieth century yellow fever urban epidemics terminated, partly because of *A. aegypti* control and partly because of the changes in ships and shipping practices. The last urban epidemic of yellow fever occurred in Rio de Janeiro in 1928.

Yellow fever in Africa, however, continues to be periodically epidemic. A large outbreak in 1940 in the Nuba Mountains of Sudan (Kirk, 1941) was followed in 1960–1962 in Ethiopia by the largest epidemic ever recorded in Africa (Serie *et al.*, 1968). Other outbreaks in the past 20 years have occurred in Senegal, Nigeria, Angola, and in 1978 in Gambia (Center for Disease Control, 1978b).

The potential threat of yellow fever in the New World cannot be emphasized too strongly. Rapid diagnosis will be of extreme importance should an urban outbreak occur. Serological methods are available (Theiler and Casals, 1958); histopathology of the liver, however, may be the initial method of choice, followed by virus isolation in baby mice, insect, and vertebrate tissue culture, or in mosquitoes. Since there is no specific treatment, public health measures such as mosquito control, personal hygiene (bed nets and repellents), and vaccination will be relied upon to control the disease.

D. Dengue

Dengue fever is a disease of man caused by any one of four related flavivirus serotypes transmitted by *Aedes* mosquitoes, usually by *A. aegypti* in the urban setting, or by *Aedes albopictus* in more rural areas of Asia and the Pacific. The basic cycle is mosquito–man–mosquito, although a forest cycle involving monkeys has also been demonstrated in Malaysia (Rudnick, cited in Hammon, 1973). Transovarial transmission in *A. albopictus* mosquitoes has been accomplished in the laboratory (Tesh *et al.*, 1979) but may not be needed as a mechanism to maintain the virus cycle in highly endemic areas such as Nigeria, Hispaniola, and much of Southeast Asia. Epidemics occur in endemic areas primarily in children, since most adults are immune; however, infection with one type confers only temporary immunity to infection and disease caused by another type (Sabin, 1948). In other areas such as the South Pacific islands and many of the Caribbean islands, periodic reintroduction of virus presumably is responsible for epidemics. Under special circumstances, such as during World War II, the introduction of nonimmune persons into an area already seeded with infected mosquitoes resulted in large-scale epidemics. Dengue types 1 and 2 are known to be prevalent in West Africa; types 1–3 in northern South America and the Caribbean; and types 1–4 in Asia.

Classic dengue is characterized by an incubation period of 5–8 days and sudden onset in about one-half of the cases. The patient develops fever up to 40°C lasting 3–6 days, sometimes with a biphasic pattern. Severe headache, nausea, altered taste, malaise, chills, muscle aches and pains, arthralgia, leukopenia, and a maculopapular or scarlatiniform rash are common manifestations. The rash usually appears late in the disease and may be pruritic; there is rarely desquamation. A significant percentage of patients have minor hemorrhagic manifestations such as bleeding gums, epistaxis, or petechial hemorrhage in some outbreaks of otherwise classic dengue (Stewart, 1944). Convalescence is associated with prolonged asthenia and at times mental depression. Excellent detailed clinical descriptions are given by several authors of natural (Sabin, 1948) and of experimental infection of man (Sabin, 1948; Siler *et al.*, 1926; Simmons *et al.*, 1931).

Different outbreaks of the same serotype, and of different serotypes, clearly show minor variations in the clinical picture. Halstead (1974) has compared the human experimental infection studies of Siler *et al.* (1926) and Simmons *et al.* (1931) after determining serologically that they were, respectively, types 4 and 1. Type-1 infections were more severe, with a longer period of fever, more postorbital, limb, and back pains, and more anorexia, vomiting, and altered taste than type-4 dengue. Sabin (1948) reported a shorter and milder clinical course in volunteers infected with type-2 New Guinea dengue virus as compared to that in those infected with type-1 Hawaii virus. In simultaneously occurring outbreaks, Barnes and Rosen (1974) described a higher incidence and greater severity of hemorrhagic manifestations of dengue-2 infection on Niue Island than on Tahiti and other neighboring islands.

Although hemorrhagic disease has been noted frequently in dengue outbreaks, and fatal hemorrhagic dengue has been reported (Wilson, 1904; Goldsmid, 1917; Copanaris, 1928), an apparently new accentuation of the hemorrhagic disease and the dengue shock syndrome appeared in the Philippine Islands in 1953 (Quintos *et al.*, 1954); this syndrome was recognized along with classic cases of dengue during epidemic periods in highly endemic parts of southern Asia and Southeast Asia where multiple dengue serotypes were circulating. Children were almost exclusively at risk, and most cases were in secondary infections, i.e., in patients who had preexisting dengue antibody to a heterologous serotype. The typical case had onset of dengue in the classic picture, but on or after the third day developed hemorrhagic manifestations including petechiae, epistaxis, gastrointestinal bleeding, and hemoptysis. This was followed by a period of shock with return of the temperature to normal, decreased urinary output, increased hematocrit, thrombocytopenia, and markedly diminished pulse pressure; death occurred in about 10% of children (Hammon *et al.*, 1960). Those patients who survived recovered fully.

The hypothesis proposed by Halstead (1970) to explain the dengue shock syndrome involves an immunopathological mechanism which requires sensitization of the host by passive antibody or acquired immunity to a heterologous dengue type during a critical time period not to exceed 5 years (Halstead, 1970) (see Chapter 5).

Hammon (1973), supported by Rosen (1977), suggested that the Southeast Asian strains causing dengue shock syndrome were simply more virulent and represented mutant populations. Neither Hammon nor Rosen has offered evidence for the mode of pathogenesis in the case of the strains of enhanced virulence, and neither has offered plausible reasons why the virulent strains should be of three or four different types arising simultaneously in one part of the world and not elsewhere. Rudnick (cited in Hammon, 1973) suggested that forest monkeys were the source of strains of enhanced virulence for man. It is clear that shock syndrome can occur in primary dengue infections (Scott *et al.*, 1976) and, as suggested by these authors and by Schlesinger (1977), it is still possible that immune complexes can occur in primary dengue infections and thus explain the pathogenesis of the observed shock syndrome.

Much time, effort, and resources have gone into trying to prove the sequential infection hypothesis. It is neither proved nor disproved at this time. The challenge to those who espouse the enhanced virulence hypothesis is clear—show a correlation between enhanced virulence and other biological properties of the virulent virus and explain both the genetics and pathogenesis of the shock syndrome in terms of the virus. The stakes are high, since live virus vaccines are being developed and their use in sequential fashion involves a risk if the sequential infection theory of the shock syndrome is correct (Schlesinger, 1977; see Chapter 5 for a detailed discussion).

Epidemics of dengue have been recognized since 1780 when Benjamin Rush described the disease in Philadelphia (Rush, 1789). Reviews by Carey (1971), Ehrenkranz *et al.* (1971), Hotta (1969), and Schlesinger (1977) detail the distribution and extent of outbreaks. These encompass India, Southeast Asia, northern Australia, the Pacific Islands, much of North, Central, and South America, the

Caribbean, southern Europe, most of Africa, and the Middle East. Major outbreaks have occurred in this century in the southern United States (1922), Western Australia (1925–1926), Greece (1927–1928), Japan (1942–1945), Australia (1954–1955), Viet Nam (1960), Puerto Rico (1963), and Venezuela (1964). Nearly annual outbreaks have occurred in Southeast Asia since 1956, with hundreds of thousands of cases and up to 600 deaths in children in some years. Epidemics in the New World are not associated with death but have been massive; the 1971–1972 outbreak of dengue-2 in Colombia involved 500,000 cases (Morales *et al.*, 1973), and the 1977–1978 outbreak of dengue-1, representing a new type for the New World, was pandemic for the Caribbean and northern South America. As of 1979, dengue-1 infection has not yet become epidemic in the United States, in spite of the high prevalence of *A. aegypti* in presumably receptive southern cities such as Miami, Houston, and New Orleans.

Dengue epidemics in Africa were frequent prior to 1940 but have not been recognized since World War II. Dengue-1 and -2 viruses, however, were isolated frequently from febrile children in Nigeria between 1964 and 1970 (Moore *et al.*, 1975), indicating that the disease is probably hyperendemic in West Africa. Dengue hemorrhagic fever and the shock syndrome have not been reported in Africa.

E. Japanese Encephalitis, Murray Valley Encephalitis, St. Louis Encephalitis, West Nile Encephalitis, and Rocio Encephalitis

Japanese, Murray Valley, St. Louis, West Nile, and Rocio encephalitis viruses cause human central nervous system (CNS) disease, are closely related serologically, and share common epidemiological characteristics. Rocio virus has been only recently discovered; its epidemiology has not yet been completely determined, but the similarity of its clinical and antigenic features to those of St. Louis encephalitis argues for its inclusion in this section.

These viruses are transmitted by *Culex* mosquitoes, have an amplification cycle in birds, and cause epidemic encephalitis in man.

St. Louis encephalitis virus may cause febrile headache, aseptic meningitis, or frank encephalitis. Morbidity and mortality is mostly in the elderly, although people of all ages are infected. Severe cases usually have a sudden onset with fever between 102° and 106°F, nausea, vomiting, and stiffness of the neck. A prodromal phase of fever, myalgia, chills, and lassitude is sometimes described. The illness often includes mental impairment in the form of disorientation, apathy, or drowsiness progressing to coma. Focal neurological signs and abnormal reflexes are common. Convulsions may occur, especially in children.

Fever usually returns to normal within a week, and recovery in about one-third of the cases is rapid and complete. In other instances, weakness and neurasthenia persist; objective neurological deficit remains in less than 10% of St. Louis encephalitis cases.

About one-quarter of St. Louis encephalitis patients have urinary tract symptoms. These include incontinence, frequency, urgency, and retention (Quick *et al.*, 1965). The pathological basis for urinary tract disease is not known. Inappropriate secretion of antidiuretic hormone was also documented in 13 of 52 patients in Dallas during 1966 (White *et al.*, 1969) and may have accounted for the hyponatremia observed in some St. Louis encephalitis cases.

Japanese encephalitis differs from St. Louis encephalitis mainly in its greater severity. Case fatality rates for St. Louis encephalitis vary between 0 and 37%; for Japanese encephalitis they vary between 10 and 90% (Bredeck, 1935). Sequellae of Japanese encephalitis and Murray Valley encephalitis are common, especially in children.

West Nile disease differs from that of the other members of this group; fever, rash, and lymphadenopathy constitute the usual syndrome. Only in elderly persons is encephalitis sometimes seen (Spigland *et al.*, 1958).

Rocio virus is also associated with a slight variation in the clinical syndrome (Tiriba *et al.*, 1976). While encephalitis was prominent as in Japanese encephalitis and in St. Louis encephalitis, there was a notable disturbance of gait and equilibrium accompanied by histological lesions in the cerebellum (Rosemberg, 1977).

In the histopathology of brains of patients dying of St. Louis encephalitis, the necrotic lesions are less marked than those seen in Japanese encephalitis and in Murray Valley encephalitis. All characteristically have perivascular cuffing by round cells, neuronal damage, and congestion with minute hemorrhages. There is nothing pathognomonic among the lesions of any of these infections (McCordock, 1935; Zimmerman, 1946; Robertson, 1952).

Epidemics of St. Louis encephalitis are restricted to North America, although human infection is documented serologically throughout Central America and in Brazil and Argentina. There is some evidence (Mettler and Casals, 1971) that St. Louis encephalitis in South America is a systemic illness and not usually an encephalitic disease.

In the United States, St. Louis encephalitis is endemic in areas where *C. tarsalis* mosquitoes are abundant, such as the Central Valley of California (Reeves and Hammon, 1962) and the high plains area of Texas (Hayes *et al.*, 1967). Urban epidemics cycle at irregular intervals of from 2 to 18 years, often in association with hot weather during June and July (Kokernot *et al.*, 1969), which may accelerate the extrinsic incubation in *Culex pipiens* complex mosquitoes and thus promote more rapid than usual amplification in the urban setting. Dry weather, which leads to stagnation of sewage and increased breeding of *C. pipiens* complex mosquitoes, was a feature of the 1933 outbreak in St. Louis, Missouri (Leake, 1935) and of subsequent epidemics. Table III lists the major St. Louis encephalitis epidemics. Unfortunately, although the disease may persist for 2–3 years in the same city, it has a propensity to find a new site in following years. Thus, anticipation of an epidemic is problematic and, even if a vaccine existed, it would be difficult to know where to use it for disease prevention.

The earlier epidemics of Japanese encephalitis in Japan, as reported by Bredeck (1935), were devastating. The disease was first recognized in 1871, mostly in older persons. Two years later it recurred with a small number of cases but with 90% mortality. In 1901, Japanese encephalitis again affected mostly older persons in large numbers. Other outbreaks occurred in 1903, 1907, 1909, and 1912. In the 1916 epidemic the case fatality rate was 61.5% and the following year only 10%, but the cases were mostly in young adults. In 1919 and again in 1924 major epidemics occurred, the latter affecting nearly 7000 persons with 60% fatality; 76% of cases involved people over 40 years old. After 1924, epidemics occurred annually in Japan until the onset of World War II and are said to have accounted for 27,000 deaths (Olitsky and Casals, 1948).

Major epidemics of Japanese encephalitis continued to occur after 1945, but a drop in the number of cases in children after 1958 was associated with the use of killed vaccine (Kitaoka, 1971). A precipitous drop in the number of cases in all age groups after 1970 was linked to the widespread use of agricultural insecticides, which has diminished the numbers of the vector, *Culex tritaeniorrhynchus*. However, it should be noted that in many prefectures in Japan over 50% of pigs convert serologically each year, indicating widespread viral activity. In pigs, Japanese encephalitis virus affects embryonic tissues, leading to abortion and stillbirth. The rate of human infection is relatively low. If antibody from the vaccination of children with killed vaccine should wane, a large susceptible population would again be at risk.

The post-World War II period, on the other hand, saw the emergence of large epidemics of Japanese encephalitis in Taiwan (Green *et al.*, 1963), Thailand (Grossman *et al.*, 1973), Korea (Chow,

TABLE III

Epidemics of Urban and Suburban St. Louis Encephalitis

Year	Area	Number of reported cases	Attack rate per 100,000	Case fatality ratio (%)	Reference
1932	Paris, Illinois	38	433	37	Leake (1935)
1933	St. Louis, Missouri	1095	100	20	Leake (1935)
1937	St. Louis, Missouri	431		25	Hempelmann (1938)
1954	Lower Rio Grande Valley, Texas	317	164		Luby *et al.* (1969)
1955	Calvert City, Kentucky	13	867	15	Ranzenhofer *et al.* (1957)
1957	Cameron County, Texas	114	155		Luby *et al.* (1969)
1959	Tampa, Florida	68		7	Bond *et al.* (1963)
1961	Tampa, Florida	25		28	Waters *et al.* (1963)
1962	Tampa, Florida	222	23	19	Bond *et al.* (1965)
1964	Houston, Texas	243	20	11	Luby *et al.* (1967)
1964	Camden, New Jersey	117	21	9	Altman and Goldfield (1968)
1966	Corpus Christi, Texas	100		5	Peavy *et al.* (1967)
1966	Dallas, Texas	168	18	12	Luby *et al.* (1969)
1968	Eldorado–Harrisburg, Illinois	27	212	4	Rubin *et al.* (1970)
1974	Memphis, Tennessee (and environs)	23	0.6–4.7	0	Powell and Blakey (1976)
1975	Memphis, Chicago, Greenville and Columbus	2131	1–10.3	8	Creech (1977)
1977	Tampa and other Florida cities	51	0.1–7.9		Beck *et al.* (1977)

1973), and in 1978 in West Bengal and Utar Pradesh, India, where 1869 deaths were recorded (Center for Disease Control, 1978a).

In Australia, epidemics of Murray Valley encephalitis (also called Australian "X" disease, and more recently Australian encephalitis) occur at relatively wide intervals in areas which have experienced high rainfall in the October–December periods of the two preceding years (Forbes, 1975). Epidemics are usually recorded from January to March and occurred in 1917, 1918, 1922, 1925, 1951, 1956, 1974, and 1978. The disease was severe, with mortality rates of 70% in the 1917–1918 outbreaks. Epidemics have been recorded in Victoria and New South Wales, as well as in the Northern Territory and Queensland, usually in association with high population levels of *c. annulirostris*, the presumed vector.

Rocio virus was isolated from cases of a new encephalitic disease in Sao Paulo State, Brazil, which appeared from March to June, 1975, in the coastal zone (Lopes *et al.*, 1978). There were 465 cases with 61 deaths. Antibody with higher titer to Rocio virus than to other flaviviruses was detected in birds. Evidence for a vector is lacking. In September 1975, additional cases were found south of the original epidemic region, indicating perhaps that the area of infection was enlarging. Rocio virus so far has not been isolated outside of Sao Paulo State, Brazil.

F. Wesselsbron

Wesselsbron virus causes fever and epidemic abortion in sheep and death of lambs (Weiss *et al.*, 1956), and an acute febrile illness in man (Smithburn *et al.*, 1957; Justines and Shope, 1969; Heymann *et al.*, 1958), usually in farmers or laboratory workers. The virus can be transmitted by *Aedes circumluteolus* and *A. caballus* mosquitoes, although the reservoir vector and vertebrate host are not known.

The disease in man lasts 2–3 days, with fever, headache, malaise, and pains in the muscles and joints. Hepatic tenderness and enlargement of the liver and spleen have been recorded. The disease is self-limiting in man and without sequellae.

The livers of sheep show necrotic changes, indicating viscerotropism of the virus. There is also a predilection for embryonic tissue. The laboratory model of abortion in rabbits and guinea pigs merits further study at the molecular level. It is not known whether or not Wesselsbron virus causes abortion in human beings.

The infection is widely distributed in man in sub-Saharan Africa, and the virus has been isolated from mosquitoes in Thailand

(Simasathien and Olson, 1973). The closely related Sepik virus has been recovered from mosquitoes in New Guinea (Marshall and Woodroofe, 1975).

G. Tick-Borne Encephalitis

Tick-borne encephalitis is a severe disease known clinically since 1910 in the Asiatic USSR where it primarily affected woodcutters and hunters. At one recognized focus the annual clinical attack rate was 1 in 67, and the case fatality rate was 25% (Smorodintseff, 1940). A milder form of the disease with case fatality rates usually under 5% is recognized in Europe (Chumakov, 1944, cited in Silber and Soloviev, 1946; Blaskovic, 1967). The causative agents of the two forms of tick-borne encephalitis are antigenically closely related (Clarke, 1962) and are called Far Eastern (Russian spring–summer) and central European types, respectively. The virus of louping ill, which causes a disease of sheep, is also closely related. Louping ill virus only rarely produces encephalitis in man. Powassan virus in North America has been associated with severe tick-borne encephalitis in man on 15 occasions; it is a distant relative of the Old World tick-borne flaviviruses and has recently been isolated in the eastern USSR (Lvov *et al.*, 1974).

Far Eastern tick-borne encephalitis is characterized by sudden onset 10–14 days after exposure to a tick bite in the taiga (uncleared virgin forest). The patient complains of severe headache, nausea, photophobia, general hyperesthesia, weakness, fever, and chills. The peak temperature ranges between 39° and 41°C and the fever lasts from 5 to 7 days. Signs of encephalitis appear almost immediately. These may include nuchal rigidity, depressed consciousness or confusion, paresthesias, diplopia, pareses, and convulsions (Silber and Soloviev, 1946).

The Far Eastern type of virus has a tropism for the neurons of the gray matter of the medulla oblongata and the upper cervical cord. This localization leads to a characteristic syndrome affecting both the motor and sensory functions of the upper extremities and neck followed by atrophy and flaccid paralysis. Death may ensue because of bulbar involvement and respiratory paralysis.

Recovery is prolonged in most cases, lasting from 1 to 6 months. The mortality rate is about 30%. Death occurs usually during the acute phase; residua occur in from 30 to 60% of survivors.

Between one-quarter and one-half of the patients with the central European type have a predominant meningitis without localizing neurological signs and with a much more favorable prognosis than patients with the Far Eastern disease. There is a predominance of abortive disease, or meningoencephalitis without serious localizing encephalitic signs. Mortality ranges between 1 and 8%. Interestingly, a small percentage of patients with both the central European and Far Eastern types of disease have progressive chronic encephalitis with remissions and relapses (Silber and Soloviev, 1946).

There are very few descriptions of louping ill and Powassan natural infections of man. Louping ill in eight sheep handlers had a viremic, febrile phase followed by a phase of meningoencephalitis. Characteristics included headache, prostration, stiff neck, arthralgia, ataxia, dysarthria, and pleocytosis of the cerebrospinal fluid (CSF). Some cases were fatal (Williams and Thorburn, 1962).

Louping ill in sheep has a viremic phase which may be accompanied by fever but is usually inapparent. Encephalitis then ensues in a proportion of the infected animals. The infection produces encephalitis with a high mortality. Some sheep remain ill for several weeks with a chronic form of the disease (Gordon, 1934). Many animals exhibit ataxia, and there is destruction of Purkinje cells in the cerebellum, accounting for the characteristic clinical picture (Brownlee and Wilson, 1932).

Powassan encephalitis was first described as a fatal disease by McLean and Donohue (1959) in a 5-year-old Ontario boy with fever, headache, stiff neck, spastic paresis, and CSF pleocytosis. Fourteen subsequent cases have been recognized in Pennsylvania, New York, Quebec, and Ontario; one of these was fatal, and six others had sequellae which included psychosis, headaches, and motor abnormalities (Main, 1980).

The geographic distribution of tick-borne encephalitis follows the distribution of the *Ixodes* ticks which transmit the causative viruses. The Far Eastern type is transmitted by *Ixodes persulcatus* and *I. ricinus* in the USSR. The central European type is transmitted by *I. ricinus* in the European USSR, Hungary, Czechoslovakia, Poland, Bulgaria, Yugoslavia, Austria, Romania, Sweden, Finland, and probably eastern Germany. Louping ill virus also is transmitted by *I. ricinus* in Ireland, northwestern England, and Scotland. Powassan virus is transmitted by *Ixodes cookei* in North America, and possibly by *Haemaphysalis neumanni* in the eastern USSR.

The epidemiology of each of the four tick-borne encephalitides is similar. Cases are first noted in May, peak in June, and then occur as isolated illnesses from July through October. Questing tick populations are greatest about 2 weeks before the peak of the disease. There are large fluctuations from year to year in the number of cases at a given focus. These fluctuations are probably related to the size of the tick populations. The disease is strictly rural and focal; it is more prevalent in males than in females, and more prevalent in the 20–30 year age group than in children. Contact infections do not occur.

Virus has been isolated frequently from naturally infected ticks which remain infected for life. Virus also passes transovarially. Rodents and hares play an important role, both in maintaining the tick populations at taiga foci or in rural forest fringe areas and as probable amplifying hosts.

Virus is also found in the milk of infected goats and is transmitted orally to man in parts of the USSR (Smorodintsev *et al.*, 1954). This disease is called biundulant meningoencephalitis. Milk-borne encephalitis occurs also with the central European virus (Blaskovic, 1967); Powassan virus can be found in milk, although transmission to man has not been demonstrated by this means (Woodall and Roz, 1977).

Cases of encephalitis in the north temperate regions occurring in May and June should arouse a high index of suspicion of tick-borne disease, especially if the patient has been in a forested area. Diagnosis is usually made retrospectively by a hemagglutination inhibition, complement fixation, or neutralization test of acute and convalescent sera. A fourfold or greater rise in antibody is diagnostic of a flavivirus infection. This virus may be isolated postmortem from CNS tissues by inoculation into mice or tissue culture.

The most generally available control measure is personal hygiene, including frequent search of the body for ticks, protective clothing, and repellents. Formalin-inactivated chick embryo vaccine is used in the USSR to protect new workers exposed to Far Eastern virus in the tiaga region (Chumakov *et al.*, 1965). Sheep are protected from louping ill disease by a formalin-killed sheep brain vaccine administered to lambs. This vaccine does not produce serum antibody and is probably effective only in hyperendemic areas where natural immunity gives a high prevalence of maternal antibody (Smith *et al.*, 1964).

H. Omsk Hemorrhagic Fever and Kyasanur Forest Disease

The two flaviviruses causing Omsk and Kyasanur Forest diseases are close serological relatives of tick-borne encephalitis viruses. Why their clinical signs in man should be so distinct from those caused by tick-borne encephalitis virus is not known.

Omsk hemorrhagic fever is a localized disease of the Baraba steppe of the Omsk oblast and the Novosibirsk region of the USSR. Large numbers of cases were seen in Omsk between 1944 and 1946, stimulating an intensive research effort which led to identification of the etiological agent and the associated clinical syndrome and to knowledge of the epidemiology (Chumakov, 1948; Gajdusek, 1953; Smorodintsev *et al.*, 1964). More recently the disease disappeared from Omsk and now is found only in Novosibirsk.

The onset of disease is sudden, with headache, malaise, nausea, muscle pains, fever, flushing of the face, and conjunctivitis. The fever lasts from 4 to 14 days and is accompanied by hemorrhagic signs including hemorrhagic enanthema of the soft palate, hematemesis, epistaxis, and metrorrhagia. In 25% of cases a second peak of fever occurs, and 50% of cases develop bronchopneumonia. Myocarditis with arrhythmia is frequent. Profuse hemorrhage is not the rule, and only 1–2% of patients die.

Leukopenia and thrombocytopenia are characteristic, as are somnolence, bradycardia, and hypotension. Convalescence is prolonged, with weakness and headaches. There is sometimes an encephalitic component, especially in children.

The basic lesion is said to involve dilatation and increased permeability of the capillaries without inflammation or destructive processes. The pathogenesis of the hemorrhage is not well understood.

Isolation of the virus has been made from human blood and from *Dermacentor pictus* ticks, which also transmit the infection experimentally. Persons in all age groups develop the disease, but it is the most common between 11 and 20 years of age in those working in the fields in the steppe region. Peak incidence of disease occurs in May and again in August. This unusual dual seasonality corresponds to the activity of the presumed vector, *D. pictus* (Fediushin, 1948). Shortly after the muskrat was introduced into the Baraba steppe after World War II, the disease was noted in these animals. It is not believed, however, that muskrats are a primary reservoir.

Dermacentor pictus has a much wider distribution than Omsk hemorrhagic fever, leading to the hypothesis that the infection must be geographically limited by the distribution of an as yet undetermined vertebrate reservoir host.

Kyasanur Forest disease was discovered in 1957 in a small, forested area of Mysore (now Karnataka) State, India (Work, 1958). It affects man and monkeys. Since 1957 the enzootic area has become gradually larger and now includes the southeastern and southwestern Shimoga and the North Kanara Districts. During 1977 about 1000 cases were recorded, and in the first 6 months of 1978 there were 600 cases (Hoogstraal, 1980). The virus is transmitted by ticks, primarily *Haemaphysalis spinigera*, although other tick species may be important reservoirs. Transovarial transmission occurs (Singh *et al.*, 1963). The cycle is probably maintained also in small rodents (Boshell, 1969) with monkeys acting as amplifying hosts. Man is infected when entering forested areas and is a dead-end host.

The clinical picture (Work, 1958; Webb and Rao, 1961) is characterized by sudden onset, fever, headache, muscle aches, prostration, and bronchiolar and gastrointestinal symptoms. Gastrointestinal hemorrhage carries a poor prognosis. In some cases after an afebrile period, meningoencephalitis ensues with fever and headache, and increased cells and protein in the CSF. Death occurs in about 5% of those becoming sick (Anderson, 1963). Sequellae have not been reported.

Leukopenia and thrombocytopenia are common; albuminuria and urinary casts are also seen. On postmortem examination there is hemorrhagic pneumonia, necrotic changes in the liver and spleen, and reticuloendothelial cell proliferation (Iyer *et al.*, 1959).

There is no specific treatment for the disease. Vaccination with Far Eastern tick-borne encephalitis killed vaccine did not protect against disease, nor was a formolized Kyasanur Forest disease virus vaccine completely effective in man (Mansharamani *et al.*, 1965; Banerjee *et al.*, 1969).

Basic questions of epidemiology remain unanswered for both Omsk hemorrhagic fever and Kyasanur Forest disease. It is as much a mystery why Omsk hemorrhagic fever disappeared from the Omsk oblast as why Kyasanur Forest disease appeared suddenly in Mysore State in 1957. Studies at the molecular level are also needed to clarify the mechanism of hemorrhagic disease and viscerotropism of these agents.

I. Israel Turkey Meningoencephalitis

During 1958 and 1959 a new disease affecting turkeys appeared in the Shomron area of Israel (Komarov and Kalmar, 1960). The disease appeared as meningoencephalitis; the birds staggered and had drooping wings and greenish diarrhea. The illness progressed to complete paralysis of the legs and wings in some birds. Morbidity approached 50%, and case fatality was 10–12%, although slaughter for market was carried out before many birds died.

Experimental infection by parenteral inoculation was uniformly successful, but attempts to infect turkeys by contact or by intratracheal inoculation failed. Ducks and chickens were not susceptible. Virus could be isolated from brains of infected turkeys in mice and in embryonated hens' eggs.

The seasonality, from August through October, and the lack of contact infection favor an arthropod vector, perhaps a mosquito. The reservoir of the virus is not known, nor has the virus been isolated outside of Israel. Serial passage of the virus 24 times in embryonated hens' eggs resulted in attenuation for turkeys and mice. A vaccine containing such attenuated virus was used successfully until the severe outbreaks of 1970–1971 when many vaccinated birds died.

Virus from the 1971 epidemic was adapted to the Japanese quail which developed persistent infections; and virus was isolated from brain and spleen in the presence of specific antibody (Ianconescu *et al.*, 1974). This quail-adapted virus became attenuated for turkeys and was subsequently used as a live virus vaccine.

J. Rubella

There is no evidence that rubella virus (see also Chapter 21) multiplies in arthropods, and there is no antigenic relationship to known arboviruses or to any other togavirus (Mettler *et al.*, 1968). The virus is distributed worldwide except on some (but not all) isolated islands where its introduction may cause outbreaks (Evans *et al.*, 1974; Gale *et al.*, 1972). It is transmitted from person to person by droplet infection emanating from the respiratory tract. Contagiousness is only moderate; prior to the institution of vaccination, by 5 years of age 20% of children were immune, and by age 20, 80–90% of persons were immune. Epidemics occurred characteristically in the spring at approximately 6- to 9-year intervals.

Man is the only known host of rubella virus. The uncomplicated disease is characterized by minimal prodromal symptoms following an incubation period of 14–23 days. The prodrome consists of mild coryza, fever, and malaise. Virus is found in the oropharyngeal secretions for a few days before and just after the onset. A typical rash follows, appearing first on the head and neck then spreading to the torso and extremities. There are no Koplik's spots. The rash consists of round, slightly raised, discrete macules; the rash fades, leaving areas of desquamation, but may arise in new areas. The soft palate and tonsils may be inflamed; a low-grade fever lasts 2–3 days and usually recedes with the rash. Tender, enlarged occipital and cervical lymph nodes nearly always accompany the rash and may persist for as long as 3 weeks.

The disease is mild in children and is usually uncomplicated; adult disease is complicated in about 25% by joint pains and rarely by full-blown arthritis. Adults may also have encephalitis (0.02%) which appears 3–4 days after onset of the rash. Out of five cases of encephalitis one patient dies (Miller, 1956), but the others usually recover completely. Another rare (0.03%) complication is thrombocytopenic purpura (Wallace, 1963).

Gregg (1941) first reported cataract and cardiac malformations of the fetuses of mothers infected by rubella virus during pregnancy. We now know that congenital infection regularly follows maternal rubella infection, especially during the first 3 months of pregnancy. The risk of fetal abnormalities is between 15 and 20% following maternal infection in the first trimester, and the percentage of abnormalities is even higher if one takes into account effects, especially deafness, which become evident later in childhood (Jackson, 1958).

The infected fetus is smaller in size than normal, reflecting a mitotic arrest of embryonic cells by the rubella virus (Naeye and Blanc, 1965). All organ systems may be involved, especially the heart, lungs, liver (giant cell hepatitis), kidneys, meninges, vascular tissues, eyes, and middle ear. Rubella infection early in pregnancy is more likely to result in fetal wastage than infection late in pregnancy. The organ involved may depend on the stage of pregnancy. The congenital rubella syndrome may have single or multiple organ involvement. Characteristic cardiac lesions include patent ductus arteriosus, ventricular septal defect, and pulmonary stenosis. The ocular lesions are cataracts, glaucoma, microphthalmia, and retinopathy. Deafness secondary to middle ear damage occurs frequently. Encephalitis, aseptic meningitis, microcephaly, and a late manifestation, psychomotor retardation, have been described. Other manifestations include growth retardation, thrombocytopenic purpura, hepatitis, hepatosplenomegaly, and lesions of the long bones.

Clinical rubella may be understood by considering the pathogenesis. The virus is believed to multiply in the upper respiratory tract epithelium and neighboring lymph nodes. Virus is shed from the throat and also seeded throughout the body by means of viremia which may occur as long as a week before the rash (Green *et al.*, 1965). It is not known whether the rash results from direct viral damage or from an antigen–antibody reaction. Fetal infection follows direct seeding of the placenta and fetus during maternal viremia. Marked inflammatory and necrotic lesions are observed in the fetus.

Major U.S. epidemics of rubella occurred in 1935, 1943, 1952, 1958, and 1964. From 200 to 600 cases per 100,000 were recorded. The epidemic of 1964, with significant numbers of resulting congenital rubella cases, stimulated the development of attenuated live virus vaccines for rubella. The HPV_{77} and Cendehill vaccines have been widely used in the United States since 1969 and may be responsible for the gradually diminishing number of reported cases of both rubella and the congenital rubella syndrome. Rubella virus is still transmitted actively among those who are not vaccinated.

K. Hog Cholera, Bovine Viral Diarrhea–Mucosal Disease, and Equine Arteritis

In January 1978, the Secretary of Agriculture declared the United States free of hog cholera (see also Chapter 21) after an eradication campaign which cost $140 million but which the secretary estimated saves the economy $100 million annually (U.S. Department of Agriculture, 1978). This disease still results in more deaths (outside of slaughter) of pigs worldwide than any other single cause. It is prevalent in Europe, Asia, and South America and was previously eradicated from Australia. The risk of reintroduction to the United States demands continued vigilance and strict quarantine procedures.

Hog cholera was first recognized in 1833 in Ohio and as far as is known was a new disease which had not been described in any other part of the world. By 1845, the disease had caused 10 outbreaks in South Carolina, Georgia, Alabama, Florida, Illinois, and Indiana. It then spread rapidly through North America and to all continents where pigs were raised (Shope, 1955).

The clinical features vary with the virulence of the hog cholera virus strain. In epidemics of fully virulent virus, the mortality reaches 90% or greater after an illness characterized by diarrhea, conjunctivitis, erythema of the skin, and high fever. The pigs become weak, do not eat, and may have convulsions prior to death, which occurs from 1 to 3 weeks after onset. In outbreaks of lesser virulence, the mortality rate may be as low as 1%, and runted animals which eat poorly indicate a chronic disease state. The virus may be transmitted *in utero*; this results in baby pigs which carry the virus and can transmit to other susceptibles.

At necropsy there is necrotic enteritis affecting mainly the large intestine and the gastric mucosa. Petechial hemorrhages are found in the bladder, kidneys, lymph nodes, and mucous membranes. There may also be hemorrhagic pneumonia.

The epizootiology of hog cholera is poorly understood. It is clearly transmitted from animal to animal by contact, but the reservoir which accounts for interepizootic maintenance is not known. *Aedes aegypti* mosquitoes transmit the virus mechanically, but attempts to demonstrate biological maintenance have been unsuccessful (Stewart *et al.*, 1975). Persistently infected animals are well-documented and might account for maintenance. The U.S. Department of Agriculture reported decreasing numbers of confirmed cases by year from 1970 as follows: 1231, 418, 76, 163, 2, 0, 18, 0. However, recognition that attenuated virus vaccines can induce persistent infection (especially in breeding sows, leading to abortion or abnormal offspring) has forced suspension of the use of such vaccines in the United States (see Chapter 21).

Bovine viral diarrhea–mucosal disease has been recognized for over 30 years. It is caused by a togavirus serologically related to hog cholera virus. The infection is usually inapparent or recognized only as a mild, self-limiting diarrhea. Serological studies indicate that the infection is ubiquitous in cattle and may also occur in pigs, thus complicating the interpretation of seropositive reactions to the related hog cholera virus.

Rarely, bovine viral diarrhea is a severe disease characterized by anorexia, protracted diarrhea, and erosion of the gut epithelium. The infection can lead either to death or to a chronic debilitating diarrhea which carries the name "mucosal disease." If the infection occurs during pregnancy, abortion may result, or the calf may be born with congenital cerebellar hypoplasia and ocular defects (Kahrs *et al.*, 1970). The disease is prevented by the vaccination of heifers with a modified live virus vaccine (Kahrs *et al.*, 1966; Kahrs, 1968).

Equine viral arteritis is an epidemic disease of horses first recognized in 1954 by Doll *et al.* (1957) who observed an outbreak on a horse farm near Bucyrus, Ohio. The horses developed fever, leukopenia, anorexia, conjunctivitis, coryza, and generalized edema, most marked in the legs and abdomen. A few horses also demonstrated respiratory distress and cough, photophobia, corneal opacity, weakness, and an unsteady gait. The severest disease occurred in pregnant mares; abortion followed in about 50%, 10–33 days later. The principal lesion in natural and experimental infections (Jones, 1969) was necrosis of the muscle cells in the small arteries. This led to inflammation, edema, and hemorrhage, often with infarction of the vessel. Aborted fetuses were edematous and hemorrhagic, but without specific arterial lesions.

Like hog cholera and bovine viral diarrhea viruses, equine arteritis virus is transmitted by contact. The fate of the virus between epizootics is not known. An effective modified live virus vaccine has been developed (McCollum, 1969). The virus is widely distributed throughout the world and is probably a single serotype.

III. LABORATORY INFECTIONS

A fatal laboratory infection ironically had a very direct effect on the author's life. In 1928 the Rockefeller Foundation approached my father to travel to Bahia, Brazil, to carry out research on yellow fever. In deference to my mother's condition (I was to be born in a few months) my father reluctantly rejected the offer, and his mentor, Dr. Paul A. Lewis, took up the challenge. Lewis' death from yellow fever was recorded in history (Strode, 1951) by a single line only. It deserved more. I can only speculate on the influence of that laboratory infection on medical history; it terminated Dr. Lewis' productive career; with "there but for the grace of God. . . ." my father assumed Dr. Lewis' laboratory at the Rockefeller Institute; and I from early childhood was inculcated with the awesome respect due togaviruses in the laboratory.

We no longer fear laboratory infection with yellow fever virus (although we still respect the virus), since the advent of 17D vaccine. There has not been a single laboratory infection with yellow fever virus since the 1930s. Other flaviviruses and, to a lesser extent, alphaviruses continue to pose problems. During his career my father was also infected, inapparently in the laboratory, with Semliki Forest virus, antibody to which was credited later (Clarke, 1961) with aborting an infection with EEE virus. What may have saved one life took another. It was Semliki Forest virus which killed a laboratory scientist in 1978 in Germany following experiments in which the virus was administered by the aerosol route to laboratory animals (Willems *et al.*, 1979).

The togaviruses implicated in overt laboratory infections are Semliki Forest, WEE, EEE, VEE, Mucambo, chikungunya, Mayaro, West Nile, yellow fever, Wesselsbron, St. Louis encephalitis, tick-borne encephalitis, Omsk hemorrhagic fever, Kyasanur Forest disease, louping ill, dengue types 1, 2, and 3, Ilheus, Japanese encephalitis, Kunjin, Spondweni, Zika, Apoi, and Rio Bravo (Berge, 1975; Hammon, 1968;

Hanson *et al.*, 1967; Sulkin *et al.*, 1962). Among the most severe, causing death or requiring hospitalization, are Semliki Forest, WEE, EEE, VEE; yellow fever, St. Louis encephalitis, and the Far Eastern form of tick-borne encephalitis.

The route of infection is usually unknown but may be by needle (Shope, 1972) or by aerosol. In some instances, the person infected was not working directly with the virus (Justines and Shope, 1969). Excreta from laboratory mice may contain large amounts of Kyasanur Forest disease and louping ill viruses; infected newborn chicks excrete EEE and WEE viruses. Therefore togaviruses should be handled in the laboratory with respect, utilizing biological safety cabinets for procedures which might create aerosols. Experience has shown that some togaviruses such as Sindbis and Semliki Forest are relatively safe when used as live antigens on the open bench in the diagnostic laboratory, but even these may conceivably be hazardous when used in high concentrations or when laboratory workers are exposed by the direct aerosol route (Willems *et al.*, 1979).

IV. AN OVERVIEW

It is obvious from the foregoing documentation of the diseases and epidemics resulting from togaviruses that, while there have been major past victories in disease control, there also are huge hurdles ahead. The eradication from the United States of hog cholera, the vaccines for yellow fever and rubella, and the eradication at one time of *A. aegypti* from major cities can be looked upon proudly as significant human accomplishments which took place in the trial-and-error era of virology and epidemiology. We eliminated a disease of hogs by slaughter without knowing whether a nonswine reservoir existed; we made vaccines without understanding why a virus became attenuated; and we controlled mosquitoes without cognizance of their genetics. We still do not know where "new" togavirus diseases (Kyasanur Forest disease and Rocio encephalitis) come from, why given togaviruses attack a specific organ (St. Louis encephalitis virus, the brain; rubella and Wesselsbron viruses, the fetus; and yellow fever virus, the liver) or why a virus such as dengue can be relatively benign in one part of the world and virulent in another.

The answers can be derived by experimental studies at the molecular level. The rewards will not come early or easily. This volume records a substantial amount of molecular information, yet the basis of togavirus attachment, entry, transcription, translation, assembly, and release is not yet well enough understood to lead to an antiviral compound for togaviruses; the genetics of togaviruses are not well enough developed to lead to the engineering of vaccines; attachment or other replication steps have not been well enough studied to explain the tropism or cell specificity of togaviruses; there is as yet no restriction endonuclease for single-stranded RNA. The vectors of togaviruses are only beginning to be approached from the point of view of genetic methods of control; and no one has engineered a virus which will kill specifically a given species of mosquito. Our paucity of understanding has a cure—support for basic studies—since each of the above goals is attainable.

In the meantime, the real world continues to challenge us: steadily increasing numbers of dengue hemorrhagic fever cases in Southeast Asia, a dengue pandemic in 1977–1978 in the Caribbean, the largest epidemic ever of St. Louis encephalitis in 1975 in the United States, the massive Japanese encephalitis epidemic in India in 1978, and potential yellow fever epidemics in the New World. Are we up to the challenge?

REFERENCES

Aitken, T. H. G., Tesh, R. B., Beaty, B. J., and Rosen, L. (1979). *Am. J. Trop. Med. Hyg.* **28**(1), 119–121.

Altman, R., and Goldfield, M. (1968). *Am. J. Epidemiol.* **87**, 470–483.

Anderson, C. R. (1963). *Bull. Natl. Inst. Sci. India* No. 24, 205–216.

Anderson, C. R., Downs, W. G., Wattley, G. H., Ahin, N. W., and Reese, A. A. (1957). *Am. J. Trop. Med. Hyg.* **6**, 1012–1016.

Anderson, S. G., and French, E. L. (1957). *Med. J. Aust.* **2**, 113–117.

Banerjee, K., Dandawate, C. N., Bhatt, P. N., and Rao, T. R. (1969). *Indian J. Med. Res.* **57**, 969–974.

Barnes, W. J. S., and Rosen, L. (1974). *Am. J. Trop. Med. Hyg.* **23**, 495–506.

Beck, E., Buff, E., Janowski, H., Lewis, A., Rogers, A., Schneider, N., Wellings, F., and Yeller, M. (1977). *Cent. Dis. Control: Morbid. Mortal. Wkly. Rep.* **26**, 370.

Berge, T. O., ed. (1975). "International Catalogue of Arboviruses," 2nd ed., DHEW Publ. No. (CDC) 75-8301. U.S. Dep. Health, Educ. Welfare, Public Health Serv., Washington, D.C.

Blaskovic, D. (1967). *Bull. W. H. O.* **36**, Suppl. 1, 5–13.

Bond, J. O., Ballard, W. C., Markush, R. E., and Alexander, E. R. (1963). "The 1959 Outbreak of St. Louis Encephalitis in Florida," Monograph Series, No. 5, p. 2. Fl. State Board Health, Jacksonville.

Bond, J. O., Quick, D. T., Witte, J. J., and Oard, H. C. (1965). *Am. J. Epidemiol.* **81**, 392–404.

Borunda-Falcon, O. (1972). *Salud Publica Mex.* **14**, 329–351.

Boshell, J. M. (1969). *Am. J. Trop. Med. Hyg.* **18**, 67–80.

Bowen, G. S., Fashinell, T. R., Dean, P. B., and Gregg, M. B. (1976). *Bull. Pan Am. Health Organ.* **10,** 46–57.
Bredeck, J. F. (1935). *In* "Report on the St. Louis Outbreak of Encephalitis," Public Health Bull. No. 214, pp. 7–15. U.S. Gov. Print. Off., Washington, D.C.
Brownlee, A., and Wilson, D. R. (1932). *J. Comp. Pathol. Ther.* **14,** 67–92.
Carey, D. E. (1971). *J. Hist. Med. Allied Sci.* **26,** 243–262.
Causey, O. R., and Maroja, O. M. (1957). *Am. J. Trop. Med. Hyg.* **6,** 1017–1023.
Center for Disease Control (1971). *Morbid. Mortal. Wkly. Rep.* **20,** 411–412.
Center for Disease Control (1978a). *Morbid. Mortal. Wkly. Rep.* **27,** 464–465.
Center for Disease Control (1978b). *Morbid. Mortal. Wkly. Rep.* **27,** 520–521.
Chow, C. Y. (1973). *Korean J. Entomol.* **3,** 31–54.
Chumakov, M. P. (1948). *Vestn. Akad. Med. Nauk SSSR* **2,** 19–26. (Engl. transl.)
Chumakov, M. P., Lvov, D. K., Gagarina, A. V., Vil'ner, L. M., Rodin, I. M., Zaklinskaja, V. A., Gol'dfarb, L. G., and Chanina, M. K. (1965). *Vopr. Virusol.* **10,** 168.
Clarke, D. H. (1961). *Am. J. Trop. Med. Hyg.* **10,** 67–70.
Clarke, D. H. (1962). *In* "Biology of Viruses of the Tick-Bone Encephalitis Complex" (H. Libikova, ed.), p. 67. Academic Press, New York.
Copanaris, P. (1928). *Bull. Off. Int. Hyg. Publique* **20,** 1590–1601.
Creech, W. B. (1977). *J. Infect. Dis.* **135,** 1014–1016.
Dasaneyavaja, A. and Pongsupat, S. (1961). *J. Trop. Med. Hyg.* **64,** 310–314.
Deller, J. J., and Russell, P. K. (1968). *Am. J. Trop. Med. Hyg.* **17,** 107–111.
Doherty, R. L., Whitehead, R. H., Gorman, B. M., and O'Gower, A. K. (1963). *Aust. J. Sci.* **26,** 183–189.
Doherty, R. L., Gorman, B. M., Whithead, R. H., and Carley, J. G. (1964). *Aust. Ann. Med.* **13,** 322–327.
Doherty, R. L., Barrett, E. J., Gorman, B. M., and Whithead, R. H. (1971). *Med. J. Aust.* **1,** 5–8.
Doll, E. R., Bryans, J. T., McCollum, W. H., and Crowe, M. E. W. (1957). *Cornell Vet.* **47,** 3–41.
Downs, W. G. (1969). *Am. J. Trop. Med. Hyg.* **18,** 482.
Earnest, M. P., Goolishian, J. A., Calverley, J. R., Hayes, R. O., and Hill, H. R. (1971). *Neurology* **21,** 969–974.
Ehrenkranz, N. J., Sinclair, M. C., Buff, E., and Lyman, D. O. (1970). *N. Engl. J. Med.* **282,** 298–302.
Ehrenkranz, N. J., Ventura, A. K., Cuadrado, R. R., Pond, W. L., and Porter, J. E. (1971). *N. Engl. J. Med.* **285,** 1460–1469.
Evans, A. S., Cox, F., Nankervis, G., Opton, E. M., Shope, R. E., Wells, A. V., and West, B. A. (1974). *Int. J. Epidemiol.* **3,** 167–175.
Farber, S., Hill, A., Connerly, M. L., and Dingle, J. H. (1940). *Am. Med. Assoc.* **114,** 1725–1731.
Fediushin, A. V. (1948). *Tr. Omsk. Med. Inst.* **13**(1), 45–58.
Feemster, R. F. (1957). *N. Engl. J. Med.* **257,** 701–704.
Finlay, K. H., and Longshore, W. A., Jr. (1958). *Trans. Am. Neurol. Assoc.* **83,** 45–49.
Forbes, J. A. (1975). *Abstr., Microbiol. Immunol., ANZAAS Congr., 46th, Canberra, Aust.* Sect. 14, p. 19.
Gajdusek, D. C. (1953). "Acute Infectious Hemorrhagic Fevers and Mycotoxicoses in the Union of Soviet Socialist Republics," Walter Reed Army Med. Cent., Med. Sci. Publ. No. 2, pp. 19–35. U.S. Gov. Print. Off., Washington, D.C.
Gale, J. L., Grayston, J. T., Beasley, R. P., Detels, R., and Kim, K. S. W. (1972). *Int. J. Epidemiol.* **1,** 253–260.

Gard, G., Marshall, I. D., and Woodroofe, G. M. (1973). *Am. J. Trop. Med. Hyg.* **22**, 551–560.
Goldfield, M., Welsh, J. N., and Taylor, B. F. (1968). *Am. J. Epidemiol.* **87**, 32–38.
Goldsmid, J. A. (1917). *Med. J. Aust.* **1**, 7–9.
Gordon, W. S. (1934). *Brit. Med. J.* **1**, 885–888.
Green, I. J., Wang, S.-P., Yen, C.-H., and Hung, S.-C. (1963). *Am. J. Trop. Med. Hyg.* **12**, 668–674.
Green, R. H., Balsamo, M. R., Giles, J. P., Krugman, S., and Mirick, G. S. (1965). *Am. J. Dis. Child.* **110**, 348–365.
Gregg, N. M. (1941). *Trans. Ophthalmol. Soc. Aust.* **3**, 35–46.
Groot, H. (1964). *Rev. Acad. Colomb. Cienc. Exactas, Fis. Nat.* **12**, 1–23.
Grossman, R. A., Edelman, R., Chiewanich, P., Voodhikul, P., and Siriwan, C. (1973). *Am. J. Epidemiol.* **98**, 121–132.
Haddow, A. J., Davies, C. W., and Walker, A. J. (1960). *Trans. R. Soc. Trop. Med. Hyg.* **54**, 517–522.
Halstead, S. B. (1970). *Yale J. Biol. Med.* **42**, 350–362.
Halstead, S. B. (1974). *Am. J. Trop. Med. Hyg.* **23**, 974–982.
Hammon, W. McD. (1968). *J. Am. Med. Assoc.* **203**, 647–648.
Hammon, W. McD. (1973). *Am. J. Trop. Med. Hyg.* **22**, 82–91.
Hammon, W. McD., Rudnick, A., Sather, G. E., Rogers, K. O., Chan, V., Dizon, J. J., and Basaca-Sevilla, V. (1957). *Proc. Pac. Sci. Congr., 9th, Bangkok* pp. 67–72.
Hammon, W. McD., Rudnick, A., and Sather, G. E. (1960). *Science* **131**, 1102–1103.
Hanson, R. P., Sulkin, S. E., Buescher, E. L., Hammon, W. McD., McKinney, R. W., and Work, T. H. (1967). *Science* **158**, 1283–1286.
Hayes, R. O., LaMotte, L. C., and Holden, P. (1967). *Am. J. Trop. Med. Hyg.* **16**, 675–687.
Hempelmann, T. C. (1938). *J. Pediatr.* **13**, 724–729.
Heymann, C. S., Kokernot, R. H., and deMeillon, B. (1958). *S. Afr. J. Med. Sci.* **32**, 543–545.
Hoogstraal, H. (1980). *Proc. FEMS Symp. Arboviruses Mediterr. Ctries, 6th, Brac, Yugo.* Zbl. Bak. Suppl. 9, in press.
Hotta, S. (1969). "Dengue and Related Hemorrhagic Diseases," pp. 3–35. Green, St. Louis, Missouri.
Ianconescu, A., Aharonovici, A., and Samberg, Y. (1974). *Refu. Vet.* **31**, 100–108.
Iyer, C. G. S., Rao, R. L., Work, T. H., and Murthy, O. P. N. (1959). *Indian J. Med. Sci.* **13**, 1011–1022.
Jackson, A. D. M. (1958). *Lancet* **ii**, 1241–1244.
Jadhav, M., Namboodripad, M., Carman, R. H., Carey, D. E., and Myers, R. M. (1965). *Ind. Jour. Med. Res.* **53**, 764–776.
Jones, T. C. (1969). *J. Am. Vet. Med. Assoc.* **155**, 315–317.
Justines, G. A., and Shope, R. E. (1969). *Health Lab. Sci.* **6**, 46–49.
Kahrs, R. F. (1968). *J. Am. Vet. Med. Assoc.* **153**, 1652–1655.
Kahrs, R. F., Robson, D. S., and Baker, J. A. (1966). *Proc. U.S. Livestock Sanit. Assoc.* **70**, 145–153.
Kahrs, R. F., Scott, F. W., and de Lahunta, A. (1970). *J. Am. Vet. Med. Assoc.* **156**, 1443–1450.
Kerr, J. A. (1951). *In* "Yellow Fever" (G. K. Strode, ed.), Chap. 7. McGraw-Hill, New York.
Kirk, R. (1941). *Ann. Trop. Med.* **35**, 67–112.

Kitaoka, M. (1971). *In* "Immunization for Japanese Encephalitis" (W. McD. Hammon, M. Kitaoka, and W. G. Downs, eds.), pp. 287–291. Igaku Shoin, Tokyo.
Kokernot, R. H., Hayes, J., Will, R. L., Tempelis, C. H., Chan, D. H. M., and Radivojevic, B. (1969). *Am. J. Trop. Med. Hyg.* **18,** 750–761.
Komarov, A., and Kalmar, E. (1960). *Vet. Rec.* **72,** 257–261.
Kubes, V., and Rios, F. A. (1939). *Science* **90,** 20–21.
Leake, J. P. (1935). *In* "Report on the St. Louis Outbreak of Encephalitis," Public Health Bull. No. 214, pp. 16–28. U.S. Gov. Print. Off., Washington, D.C.
Leon, C. A., Jaramillo, R., Martinez, S., Fernandez, F., Tellez, H., Lasso, B., and Gugman, R. (1975). *Int. J. Epidemiol.* **4,** 131–140.
Lopes, O. S., Sacchetta, L. A., Coimbra, T. L. M., Pinto, G. H., and Glasser, C. M. (1978). *Am. J. Epidemiol.* **108,** 394–401.
Luby, J. P., Miller, G., Gardner, P., Pigford, C. A., Henderson, B. E., and Eddins, D. (1967). *Am. J. Epidemiol.* **86,** 584–596.
Luby, J. P., Sulkin, S. E., and Sanford, J. P. (1969). *Annu. Rev. Med.* **20,** 329–350.
Lumsden, W. H. R. (1955). *Trans. R. Soc. Trop. Med. Hyg.* **49,** 33–57.
Lvov, D. K., Leonova, G. N., Gromashevsky, V. L., Belikova, N. P., Berezina, L. K., Safronov, A. V., Veselovskaya, O. V., Gofman, Y. P., and Klimento, S. M. (1974). *Vopr. Virusol.* **5,** 538–541. (Engl. transl.)
McCollum, W. H. (1969). *J. Am. Vet. Med. Assoc.* **155,** 318–407.
McCordock, H. A. (1935). *In* "Report on the St. Louis Outbreak of Encephalitis," Public Health Bull. No. 214, pp. 40–53. U.S. Gov. Print. Off., Washington, D.C.
McIntosh, B. M., Harwin, R. M., Paterson, H. E., and Westwater, M. L. (1963). *Cent. Afr. J. Med.* **9,** 351–359.
McLean, D. M., and Donohue, W. L. (1959). *Can. Med. Assoc. J.* **80,** 708–711.
Main, A. J. (1980). *In* "Diseases Transmitted from Animals to Man" (W. T. Hubbert, W. F. McCulloch, and P. R. Schnurrenberger, eds.), 7th ed. Thomas, Springfield, Illinois, in press.
Mansharamani, H. J., Dandawate, C. N., Krishnamurthy, B. G., Nanavoti, A. N. D., and Jhala, H. I. (1965). *Indian J. Pathol. Bacteriol.* **8,** 159–177.
Marshall, I., and Woodroofe, G. M. (1975). *In* "International Catalogue of Arboviruses" (T. O. Berge, ed.), 2nd ed., DHEW Publ. No. (CDC) 75-8301, pp. 642–643. U.S. Dep. Health, Educ. Welfare, Public Health Serv., Washington, D.C.
Mettler, N. E., and Casals, J. (1971). *Acta Virol.* **15,** 148–154.
Mettler, N. E., Petrelli, R. L., and Casals, J. (1968). *Virology* **36,** 503–504.
Miller, H. (1956). *Proc. R. Soc. Med.* **49,** 139–148.
Moore, D. L., Reddy, S., Akinkugbe, F. M., Lee, V. H., David-West, T. S., Causey, O. R., and Carey, D. E. (1974). *Ann. Trop. Med. Parasitol.* **68,** 59–68.
Moore, D. L., Causey, O. R., Carey, D. E., Reedy, S., Cooke, A. R., and Akinkugbe, F. M. (1975). *Ann. Trop. Med. Parasitol.* **69,** 49–64.
Morales, A., Groot, H., Russell, P. K., and McCown, J. M. (1973). *Am. J. Trop. Med. Hyg.* **22,** 785–787.
Mulder, D. W., Parrot, M., and Thaler, M. (1952). *Neurology* **1,** 318.
Naeye, R. L., and Blanc, W. (1965). *J. Am. Med. Assoc.* **194,** 1277–1283.
Nimmannitya, S., Halstead, S. B., Cohen, S. N., and Margiotta, M. R. (1969). *Am. J. Trop. Med. Hyg.* **18,** 954–972.
Olitsky, P. K., and Casals, J. (1948). *In* "Viral and Rickettsial Infections of Man" (T. M. Rivers, ed.), pp. 163–212. Lippincott, Philadelphia, Pennsylvania.
Peavy, J. E., Dewlett, H. J., Metzger, W. R., and Bagby, J. (1967). *Am. J. Public Health* **57,** 2111–2116.

Powell, K. E., and Blakey, D. L. (1976). *South. Med. J.* **69,** 1121–1125.
Quick, D. T., Thompson, J. M., and Bond, J. O. (1965). *Am. J. Epidemiol.* **81,** 415–427.
Quintos, F. N., Lim, L., Juliano, L., Reyes, A., and Lacson, P. (1954). *Phillip. J. Pediatr.* **3,** 1–19.
Ranzenhofer, E. R., Alexander, E. R., Beadle, L. D., Bernstein, A., and Pickard, R. C. (1957). *Am. J. Hyg.* **65,** 147–161.
Reeves, W. C. (1972). *Am. J. Trop. Med. Hyg.* **21,** 251–259.
Reeves, W. C., and Hammon, W. McD. (1962). "Epidemiology of the Arthropod-Borne Viral Encephalitides in Kern County, California 1943–1953." Univ. of California Press, Berkeley.
Robertson, E. G. (1952). *Med. J. Aust.* **1,** 107–110.
Robinson, C. M. (1955). *Trans. R. Soc. Trop. Med. Hyg.* **49,** 28–32.
Rosemberg, S. (1977). *Rev. Inst. Med. Trop. Sao Paulo* **19,** 280–282.
Rosen, L. (1977). *Am. J. Trop. Med. Hyg.* **26,** 337–343.
Ross, R. W. (1956). *J. Hyg.* **54,** 177–191.
Rubin, R. H., Glisk, T. H., and Rose, N. J. (1970). *J. Infect. Dis.* **122,** 347–353.
Rush, B. (1789). "Medical Inquiries and Observations," pp. 104–121. Prichard & Hall, Philadelphia, Pennsylvania.
Sabin, A. (1948). *In* "Viral and Rickettsial Infections of Man" (T. M. Rivers, ed.), 1st ed., Chap. 30. Lippincott, Philadelphia, Pennsylvania.
Sanmartin, C., Groot, H., and Osorno-Mesa, E. (1954). *Am. J. Trop. Med. Hyg.* **3,** 283–293.
Sanmartin, C., MacKenzie, R. B., Trapido, H., Barreto, P., Mullenox, C. H., Gutierrez, E., and Lesmes, C. (1973). *Bol. Of. Sanit. Panam.* **74,** 108–137.
Schlesinger, R. W. (1977). "Dengue Viruses," Virology Monographs, Vol. 16. Springer-Verlag, Berlin and New York.
Scott, R. M., Nimmannitya, S., Bancroft, W. H., and Mansuwan, P. (1976). *Am. J. Trop. Med. Hyg.* **25,** 866–874.
Sellers, R. F., Bergold, G. H., Suarez, O. M., and Morales, A. (1965). *Am. J. Trop. Med. Hyg.* **14,** 460–469.
Serie, C., Andral, L., Poirer, A., Lindrec, A., and Neri, P. (1968). *Bull. W.H.O.* **38,** 879–884.
Shope, R. E. (1955). *Adv. Vet. Sci.* **11,** 1–46.
Shope, R. E. (1972). *In* "Venezuelan Encephalitis," Sci. Publ. No. 243, p. 95. Pan Am. Health Organ., Washington, D.C.
Shope, R. E., and Anderson, S. G. (1960). *Med. J. Aust.* **1,** 156–158.
Shore, H. (1961). *Trans. R. Soc. Trop. Med. Hyg.* **55,** 361–373.
Silber, L. A., and Soloviev, V. D. (1946). *In* "American Review of Soviet Medicine," Spec. Suppl., pp. 1–80. Am.–Sov. Med. Soc., New York.
Siler, J. F., Hall, M. W., and Hitchens, A. P. (1926). "Dengue, its History, Epidemiology, Mechanisms of Transmission, Etiology, Clinical Manifestations, Immunity, and Prevention," Monogr. No. 20. Bur. Sci., Manila.
Simasathien, P., and Olson, L. C. (1973). *J. Med. Entomol.* **10,** 587–590.
Simmons, J. S., St. John, J. H., and Reynolds, F. H. K. (1931). "Experimental Studies of Dengue," Monogr. Bur. Sci., Manila.
Singh, K. R. P., Pavri, K., and Anderson, C. R. (1963). *Nature (London)* **199,** 513.
Smith, C. E. G., McMahon, D. A., O'Reilly, K. J., Wilson, A. L., and Robertson, J. M. (1964). *J. Hyg.* **63,** 53–68.
Smithburn, K. C., Kokernot, R. H., Weinbren, M. P., and deMeillon, B. (1957). *S. Afr. J. Med. Sci.* **22,** 113–120.

Smorodintseff, A. A. (1940). *Arch. Gesamte Virusforsch.* **4,** 468–480.

Smorodintsev, A. A., Drobyshevskaya, A. I., Il'yenko, V. I., Alekseyer, B. P., Gulamova, V. P., and Fedorchuk, L. V. (1954). *In* "Neurotropic Virus Infections" (A. A. Smorodintsev, ed.), pp. 4–37. Medgiz, Leningrad. (Engl. transl.).

Soper, F. L. (1969). *Am. J. Trop. Med. Hyg.* **18,** 482–484.

Spigland, I., Jasinska-Klingberg, W., Hofshi, E., and Goldblum, N. (1958). *Harefuah* **54,** 275–281.

Stevenson, L. D. (1939). *Arch. Pathol.* **27,** 249–266.

Stewart, F. H. (1944). *U.S. Nav. Med. Bull.* **42,** 1233–1240.

Stewart, W. C., Carbrey, E. A., Jenny, E. W., Kresse, J. I., Snyder, M. L., and Wessman, S. J. (1975). *Am. J. Vet. Res.* **36,** 611–614.

Strode, G. E., ed. (1951). "Yellow Fever," p. 633. McGraw-Hill, New York.

Sudia, W. D., and Newhouse, V. F. (1975). *Am. J. Epidemiol.* **101,** 1–13.

Sulkin, S. E., Burns, K. F., Shelton, D. F., and Wallis, C. (1962). *Tex. Rep. Biol. Med.* **20,** 113–127.

Tesh, R. B., Rosen, L., Beaty, B. J., and Aitken, T. H. G. (1979). *In* "Dengue in the Caribbean, 1977," Sci. Publ. No. 375, pp. 179–182. Pan Am. Health Organ., Washington, D.C.

Theiler, M. (1948). *In* "Viral and Rickettsial Infections of Man" (T. M. Rivers, ed.), Chap. 28. Lippincott, Philadelphia, Pennsylvania.

Theiler, M., and Casals, J. (1958). *Am. J. Trop. Med. Hyg.* **7,** 585–594.

Theiler, M., and Downs, W. G. (1973). "The Arthropod-Borne Viruses of Vertebrates," Chap. 21, Yale Univ. Press, New Haven, Connecticut.

Tiriba, A. C., Miziara, A. M., and Lorenco, R. (1976). *Rev. Assoc. Med. Bras.* **22,** 415–420.

U.S. Department of Agriculture (1978). "Foreign Animal Diseases Report," No. 929009. U.S. Gov. Print. Off., Washington, D.C.

Wallace, S. J. (1963). *Lancet* **1,** 139–141.

Waters, C. M., Quick, D. T., Oard, H. C., and Alan, F. K. (1963). "The 1961 Outbreak of St. Louis Encephalitis in Pinellas, Manatee and Sarasota Counties," Monograph Series, No. 5, p. 19. Fl. State Board Health, Jacksonville.

Webb, H. E., and Rao, T. R. (1961). *Trans. R. Soc. Trop. Med. Hyg.* **55,** 284–298.

Weiss, K. E., Haig, D. A., and Alexander, R. A. (1956). *Onderstepoort J. Vet. Res.* **27,** 183–195.

White, M. G., Carter, D. W., Rector, F. C., and Seldin, D. W. (1969). *Ann. Intern. Med.* **71,** 691–702.

Willems, W. R., Kaluza, G., Boschek, C. B., and Bauer, H. (1979). *Science* **203,** 1127–1129.

Williams, H., and Thorburn, H. (1962). *Scott. Med. J.* **7,** 353.

Williams, M. C., Woodall, J. P., and Gillett, J. D. (1965). *Trans. R. Soc. Trop. Med. Hyg.* **59,** 186–197.

Wilson, G. W. (1904). *Public Health Rep.* **19,** 67–70.

Woodall, J. P. (1967). *Atas Simp. Biota Amazon. Patol.* **6,** 31–63.

Woodall, J. P., and Roz, A. (1977). *Am. J. Trop. Med. Hyg.* **26,** 190–192.

Work, T. H. (1958). *Prog. Med. Virol.* **1,** 248–277.

Zehmer, R. B., Dean, P. B., Sudia, W. D., Calisher, C. H., Sather, G. E., and Parker, R. L. (1974). *Health Serv. Rep.* **89,** 278–282.

Zimmerman, H. M. (1946). *Am. J. Pathol.* **22,** 965–991.

4

Virus–Host Interactions in Natural and Experimental Infections with Alphaviruses and Flaviviruses

R. WALTER SCHLESINGER

I. COMPLEXITIES OF "VIRULENCE"

Of all major viral families, the Togaviridae are probably responsible for the largest variety of serious or life-threatening epidemic diseases of man and domestic animals (see Chapters 3 and 21). An important corollary is the frequent occurrence of laboratory infections: At least 7 alphaviruses and 22 flaviviruses have been incriminated in varying numbers of such accidents (estimated from Berge, 1975). Some of these infections have led to fatal or serious and prolonged illnesses (see Chapter 3); some were caused by supposedly "safe" model viruses used with abandon in many laboratories, often in highly concentrated form, for physical and chemical studies.

THE TOGAVIRUSES

ISBN 0-12-625380-3

It seems logical to assume that the diversity of diseases caused by arthropod-borne togaviruses reflects the strong selective pressures the unusual mode of their transmission has imposed on them. The high degree of variability of individual viral species is evident in varying degrees of virulence for any given natural or experimental host species.

The expression of virulence is as variable as the systems in which natural or experimental infections with a given agent can be studied. Its use as a descriptive parameter becomes confusing or misleading mainly when conclusions are based on extrapolation from one host to another. According to Fenner (Fenner *et al.*, 1974):

> Long usage has associated the term "virulence" with those properties of a virus which lead to severe symptoms, of whatever type [are] characteristic for infections with that virus, and what is called high "virulence" for a particular virus obviously depends to a large extent on the pathogenesis of the disease associated with it. . . . With viruses that cause disease by multiplication in vulnerable target organs such as the heart, CNS, or liver, a major factor in our assessment of virulence is whether the virus gains access to susceptible cells, as well as the degree of multiplication and cell damage caused when (it does) reach these cells. . . . Since virulence is a compound of the reaction of host and virus, comparative statements on the level of virulence of a particular strain of virus can be made only if all other factors are kept constant: the species and age of the host, the route of inoculation, the dose, and so on. . . .

At the very least, then, the analysis of virulence of a given virus in a given host requires (1) definition of *vulnerable target cells or organs*, (2) separation of disease expressions due to *primary cell injury* from those due to *secondary* (e.g., *immunological*) *host responses*, (3) *understanding of genetic or nongenetic host determinants* that may affect the operation of (1) or (2). The matter is relatively more straightforward in cell culture systems, where conventional immunological responses can be eliminated. However, unless the cultured cells used happen to be the functionally differentiated equivalents of a particular host's target cells, extrapolations from *in vitro* systems to whole organisms are of dubious value.

II. TARGET ORGANS IN ALPHA- AND FLAVIVIRUS INFECTIONS

A. Natural Infections

For many natural togavirus-related diseases of man, no unequivocal cause-and-effect relationship has been established between virus multiplication in specific organ systems or primary virus-induced cell injury and specific disease manifestations. On the contrary, the protean and systemic nature of such signs or symptoms as fever, rash, hemorrhage, arthralgia, myalgia, etc., as well as the timing of their appearance (see Chapter 3), suggests that in their development immunopathological mechanisms may exceed in importance the direct cytocidal effects associated with viral multiplication (see Chapter 5). An exception is yellow fever virus for which the liver parenchyma is recognized as a target organ (see Chapter 3). Mononuclear phagocytes have recently been implicated as target cells for dengue viruses (see Chapter 5).

It is, however, the central nervous system (CNS) which is the single target organ most directly affected by several alpha- and flaviviruses—those that induce encephalitis. Even so, natural infections with these viruses lead to clinically manifest encephalitis only in a minute proportion of infected (human) individuals, the majority having either no clinical manifestations or transitory mild systemic reactions. All infections are followed by immunity, as evidenced by antiviral antibodies. In those who do develop encephalitis, it is characterized by the presence of virus in the CNS and by degeneration and necrosis of neurons and glial proliferation associated with focal and perivascular inflammatory response. Here, as in paralytic poliomyelitis, we have probably the clearest correlation between severity of a natural disease syndrome and the extent of primary virus-induced injury to specific target cells.

On the other hand, the encephalitis-producing viruses illustrate dramatically the relationship between *expressed* virulence and access to vulnerable target cells: In the vast majority of infected individuals—those who do *not* contract encephalitis—virus replicates to high levels in various extraneural tissues without causing any clinically apparent illness that can be traced to direct cellular injury. Presumably in these individuals the virus does not gain access to the CNS, or does not spread to a critical number of target cells within the CNS, prior to the intervention of protective immune mechanisms. A key question is whether success or failure of a virus in such a contest—either in individual infections or as expressed in terms of greater or lesser severity of epidemics—reflects strain-specific differences in identifiable viral markers, variable host factors, or some fortuitous combination of both. There is no solid evidence bearing on this question from the study of natural infections.

A great deal of work has, however, been done in attempts to dissect the roles of virus and host in the determination of experimental neurovirulence, and in the following sections we shall concentrate on some aspects of these studies.

B. Neurovirulence in Experimental Animals

1. *General Remarks*

Intuitively, one might expect that viruses which display a neurovirulent tendency in nature would do so in suitable experimental animals as well, and that others would not. This is not so; the experimental animal of choice for virtually all alpha- and flaviviruses is the mouse which, after intracerebral inoculation, contracts fatal encephalitis. Sometimes several serial brain-to-brain blind passages are required to bring neurovirulence to full expression; sometimes a given virus is initially pathogenic only for newborn (suckling) mice and acquires encephalitogenic capacity for weanling or adult mice only after multiple passages in sucklings (see below); sometimes "adaptation" to mouse brain is facilitated by prior replication in one kind of cell culture or another. As commonly used, the experimentally infected laboratory mouse is not a valid model for natural alpha- or flavivirus infections: (1) It is questionable whether the degree of neurovirulence for mice, especially after intracerebral inoculation, parallels variations in apparent intrinsic virulence among strains of a virus known to produce occasional encephalitis in natural hosts; (2) it is well established that continued mouse-to-mouse passage generally leads to increased neurovirulence as measured by replication at increasing rates to increasingly higher titers and by reduced average survival time (see below); (3) these changes occur with equal predictability in viruses which, in their natural hosts, show no significant neuroinvasive tendency; (4) conversely, such viruses [e.g., yellow fever (see Strode, 1951) and dengue viruses (see Schlesinger, 1977a)] lose their "normal" pathogenicity for the natural host (man) as the evolution of neurovirulence for experimental animals progresses.

As already mentioned, most laboratory strains of alpha- and flaviviruses in common use are in fact mouse-pathogenic derivatives of original isolates. Therefore it is of special interest to consider viral genetic or phenotypic determinants which may relate to neurovirulence, as well as host factors which may influence the outcome of infections, i.e., the *expression* of virulence.

2. *Alphaviruses*

An early example of the changes which accompany sustained intracerebral passages in mice of an alphavirus was reported by Olitsky *et al.* (1945) in connection with studies on vaccine development for Western equine encephalitis (WEE). These experiments, done with four strains differing in their passage histories, showed unequivocally that the RI strain which had been passed most frequently in mice attained the highest titer in brain tissue and killed mice twice as rapidly as the other, less thoroughly mouse-adapted strains. Subsequently, detailed comparison (Schlesinger, 1949a) of one of the latter strains (Kelser) with the more virulent RI strain revealed a marked difference in overall growth rate, total yield, and onset of encephalitic signs after intracerebral inoculation of relatively small doses (Fig. 1); this was associated with an almost two-fold difference in the latent period in "one-step growth" curves (Fig. 2) and independent of the amounts of virus of each strain inoculated (Fig. 3). The latter finding argues against a major role of autointerference by defective particles in the "slow" Kelser strain. Moreover, after 40 additional intracerebral passages in mice, this strain replicated and killed mice at the same rate as the RI strain. No antigenic differences between the RI and the "slow" or "fast" Kelser strain were detected by quantitative neutralization tests (Schlesinger, 1949a). These findings led to recognition of the decisive role of a rapid viral growth rate in determining the effectiveness of a viral strain in overcoming the host's immune response (Schlesinger, 1949b), thus amplifying the *expression*of inherent virulence (see below).

Neither these early experiments nor subsequent comparisons of several other alphaviruses displaying different levels of neurovirulence (e.g., Taylor and Marshall, 1975a,b) have unequivocally pointed

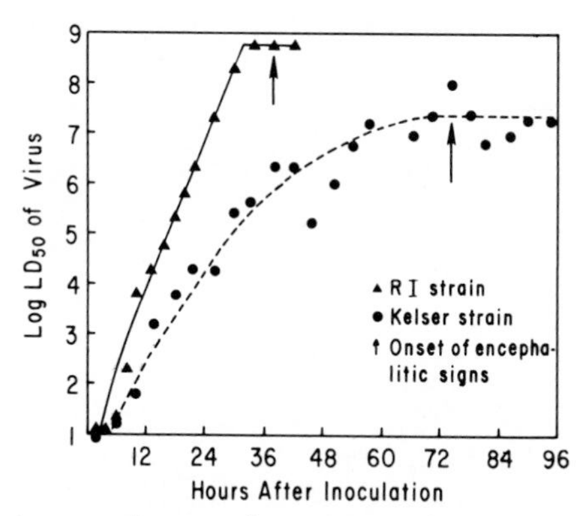

Fig. 1. Comparative growth rates of two strains of WEE virus in mouse brain after intracerebral inoculation. RI strain inoculum $10^{2.6}$ LD_{50}; Kelser strain inoculum $10^{3.4}$ LD_{50}. Each symbol represents a pool of two brains harvested at the indicated time after inoculation. (From Schlesinger, 1949a.)

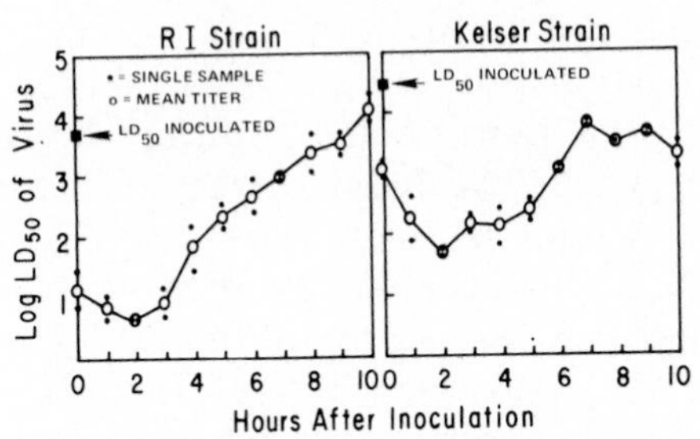

Fig. 2. One-step growth curves of two strains of WEE virus in mouse brain after intracerebral inoculation. (From Schlesinger, 1949a.)

to covariant viral functions other than growth rate and titer attained. The most comprehensive studies have been done with variants of Semliki Forest virus (SFV) with high, intermediate, and low neurovirulence, which were initially derived from several wild-type strains (Bradish *et al.*, 1971). These variants have comparable efficiencies of plating (EOP) in chick embryo cell cultures and are equally lethal for suckling mice after intracerebral or intraperitoneal (ip) inoculation. Marked strain-specific differences in relative virulence [expressed as plaque-forming units (PFU)/LD_{50}] are manifest in mice more than 15–20 days of age inoculated either intracerebrally or intraperitoneally, as well as in rabbits and guinea pigs infected by various routes (Bradish *et al.*, 1971). Careful and complex experiments by Bradish *et al.* (1971, 1972; Bradish and Allner, 1972) indicate that all viral populations studied are intrinsically heterogeneous with regard to neurovirulence for adult mice. Patterns of virus distribution in brain and blood are comparable during the first few hours after infection of

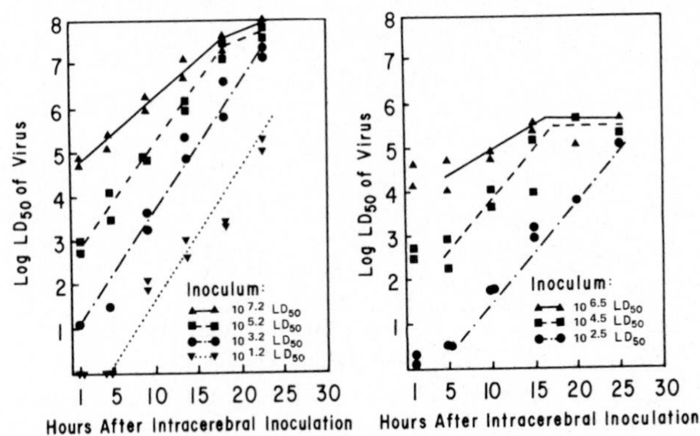

Fig. 3. Comparative growth rates of two strains of WEE virus in mouse brain after intracerebral inoculation of different amounts. (Modified from Schlesinger, 1949a.)

mice with all strains by the intraperitoneal route; however, starting at approximately 48 h after infection, "virulent" variants display higher levels and greater rates of growth of virus in brain than "avirulent" ones (Bradish and Allner, 1972; Fleming, 1977; see also Pusztai *et al.*, 1971). However, if the latter are subjected to only two to three intracerebral passages in baby hamsters, the PFU/LD_{50} ratio decreases from about 10^5 to between 10^1 and 10^2 (Bradish *et al.*, 1972). Pattyn *et al.* (1975) carried studies with these variants further by using organ cultures of muscle, spleen, and brain taken from mice of different ages. In brain cultures, an avirulent variant replicated less well than a virulent one, and replication was also reduced as a function of advancing age of the mice from which the tissue was derived (see also Fleming, 1977). No such differences were found in cultures of spleen or muscle ("the main site of virus replication" *in vivo*). Similar studies with small plaque and large plaque variants of Middelburg virus in brain and muscle organ cultures from newborn and 21-day-old mice showed an age-related decrease of replication in both tissues; no viral replication was demonstrated in macrophage cultures (De Vleeschauver and Pattyn, 1974).

Among the attempts to correlate the differences between virulent and avirulent SFV strains with intrinsic phenotypic or genetic viral properties have been the following.

a. **Plaque Size Phenotype and Viral Surface Properties.** Reports on the possible relationship of plaque size and *in vivo* virulence are reviewed in Chapter 14. The situation is confused in that the viruses and experimental systems investigated by different workers are not always comparable: Sometimes virulence is defined in terms of the natural behavior of the virus under study (e.g., a virulent epizootic versus an avirulent enzootic strain), and sometimes in terms of the neurovirulence of derivative strains.

Interpretations are further complicated by the evidence, also presented in Chapter 14, that "plaque size appears to be related in some cases to the surface charge on the particle." Surface charge, in turn, can be affected by the host system from which a viral population is derived or by the conditions under which it is assayed. Aside from the examples cited in Chapter 14, the following studies may be illuminating. Fleming (1973, 1977) observed a variety of effects on viral phenotype resulting from passage of Bradish's variants of SFV (see above) in chick embryo fibroblasts or L cells in the presence or absence of certain carbohydrates. These manifested themselves in terms of marked differences in (1) elution from hydroxylapatite columns, (2) enhancement of plaque formation under agar by DEAE-dextran, (3) inhibition by certain hexosamines, especially mannosamine, and (4) neurovirulence after intraperitoneal inoculation. His findings led Fleming to conclude that the carbohydrate moiety of the viral envelope glycoproteins may play an important role in determining the outcome of *in vivo* infections and that, for a given strain, low plaque enhancement by polycations was the only viral property of those tested that was positively correlated with high neurovirulence. This was interpreted as being possibly related to surface charge-dependent transfer of virus to the bloodstream and thence to the brain.

Possibly related observations on Getah virus were reported by Kimura and Ueba (1978): A small plaque variant of low neurovirulence for mice, which remained "stable" after eight passages in suckling mice (dose not stated), showed maximal hemagglutination at pH 6.2–6.4 and was effectively neutralized by antiviral serum in the absence of complement. In contrast, a highly virulent large plaque variant failed to hemagglutinate at any pH in the range 6.0–6.8, and its plating efficiency increased in the presence of high concentrations of antiserum without complement.

Although the last example suggests genetic stability of the viral surface characteristics, a *primary* cause-and-effect relationship of such phenotypic markers to viral gene functions responsible for different grades of neurovirulence has not been established in this or in other systems. Rather, genetically determined differences in the rate of virus replication, in viral yield, and in inherent cytopathogenicity are likely to affect differentially the integrity of cellular functions involved in the processing and assembly of viral components. These functions include the activities of sugar transferases involved in the glycosylation of viral envelope polypeptides (Fleming, 1973). Theoretical implications of the "phenotypic polymorphism" which may result from such mechanisms are discussed in Section IV.

b. Temperature Sensitivity. Systematic tests for temperature sensitivity as a correlate of varying *in vivo* neurovirulence of alphaviruses have not been reported. Nor have, conversely, the chemically induced and functionally characterized *ts* mutants of Sindbis virus (SV), SFV, or WEE virus (see Chapter 14) been critically examined for virulence in mice. The point has been made by Bradish *et al.* (1971) that their own virulent and avirulent strains of SFV (see above) probably represent mixtures of "at least two distinct types of genetic material and are distinct from ts^- mutants in which lesions and functional deficiencies are imposed on a single parental genome randomly selected from a population which remains to be defined." Some SV or SFV stocks of *ts* phenotype released from persistently infected mosquito cell cultures (see Chapter 20 for detailed discussion) show reduced virulence for mice (Peleg, 1971, 1975), but it is not known whether they are single-step mutants [a recent report by Maeda *et al.* (1979) on *ts* mutants of WEE virus isolated from persistently infected *Aedes albopictus* cell cultures suggests that *late* isolates (like those of SV and SFV) are multiple-site mutants].

c. **Defective-Interfering Virus.** Woodward and Smith (1975) reported initially that an avirulent strain of SFV, in contrast to a virulent one, generated defective-interfering (DI) particles in brains of mice, but subsequent studies by the same authors (Woodward *et al.*, 1977) did not confirm this impression. It has, however, been reported that uv-irradiated stocks of SFV containing a high proportion of DI particles (but not similar doses of uv-irradiated standard virus) can provide some protection against the lethal effects of 10 LD_{50} of standard SFV administered by the same route (intranasal) (Dimmock and Kennedy, 1978). This suggests that DI virions or genomes, if generated in *in vivo* infections, may have a modulating effect on the disease in mice (see Chapter 15; for general discussion, see Huang and Baltimore, 1977).

3. *Flaviviruses*

Study of the neurovirulence of flaviviruses has been dominated by the model of yellow fever which first demonstrated, in the pace-setting studies of Theiler and his colleagues (see review in Theiler, 1951), the attenuation of human pathogenicity associated with adaptation to the mouse. In this work it was also shown that neurovirulence, once "fixed" in mice, expressed itself in other experimental hosts as well. The successful use of 17D yellow fever vaccine, a moderately neurovirulent derivative, set the precedent for attempts to develop similarly attenuated variants as vaccine candidates for other viruses, notably dengue and Japanese encephalitis. On the other hand, rigorous genetic analysis of flaviviruses has lagged behind that of alphaviruses: There are no banks of well-characterized *ts* mutants, and the generation of DI virus has not been studied systematically.

Biological changes in virus associated with the evolution of neurovirulence have been investigated especially in the case of dengue viruses. For example, type-1 dengue (DV-1) virus, once fixed with regard to neurovirulence in mice, also causes encephalitis in monkeys (Sabin, 1955) and hamsters (Meiklejohn *et al.*, 1952) and multiplies preferentially or exclusively in the CNS tissue of chick embryos (Schlesinger, 1951). Most freshly isolated strains of dengue viruses readily produce encephalitis in suckling mice, but many passages in sucklings or adults may be required before the encephalitogenic capacity is also fully expressed in adult mice (Sabin and Schlesinger, 1945; cf. Hotta, 1965; Schlesinger and Frankel, 1952; Schlesinger, 1977a). This change appears to reside in the viral genome, for RNAs extracted from CNS tissue of early and late suckling mouse passages of DV-2 show the same patterns of age-related pathogenicity as the corresponding whole-virus samples (Schulze, 1962; see also Schlesinger, 1977a). Moreover, virus or RNA derived from early suckling mouse passages shows the phenomenon of autointerference in adult mice in that encephalitis occurs more frequently in those inoculated with high than with low dilutions of the test material (Fig. 4). This paradoxical response, noted earlier by Theiler with yellow fever virus (Theiler, 1951), predicts heterogeneity of such

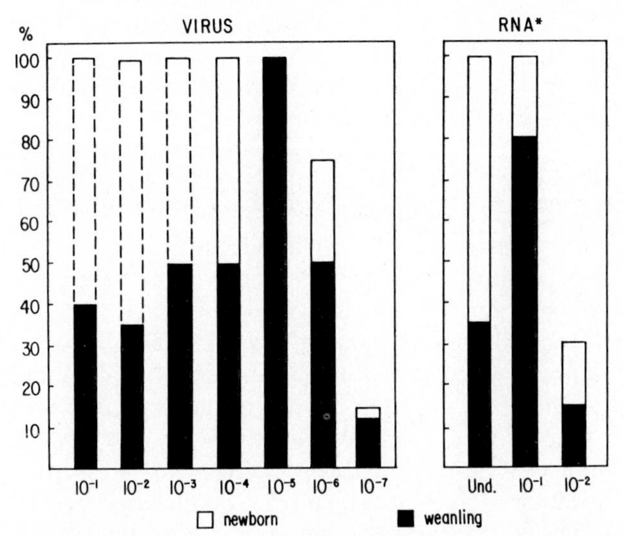

Fig. 4. Dengue-2 virus, New Guinea B strain. Mortality rates in suckling (open areas) and weanling (solid areas) mice inoculated intracerebrally with indicated dilutions of virus derived from suckling mouse brain passage 10 or of RNA isolated therefrom by three sequential cold phenol extractions. Broken lines in "newborn" columns at 10^{-1}–10^{-3} indicate that these dilutions were not included in this experiment but have been extrapolated from consistent results of numerous other experiments. Asterisk indicates no infectivity recovered after treatment with RNase (5 μg/ml) for 20 min at room temperature. Similar treatment had no effect on the infectivity of whole virus. Undiluted (Und.) RNA was equivalent v/v to 20% brain extract. (From Schlesinger, 1977a.)

virus populations: The predominant particles that are avirulent for adult mice replicate in their brains and, given the advantage of sufficiently high multiplicity of infection or time, interfere with the pathological expression of the virulent ones. Reconstruction experiments, employing mixed infections with early-passage virus or RNA and late-passage virus, have borne out this interpretation (Schulze, 1962; see also unpublished data reviewed in Schlesinger, 1977a). In this system, even the fully adapted virus replicates to much higher titers in suckling than in adult mouse brain.

These properties are consistent with genetic heterogeneity of the parental (uncloned) viral population, the suckling mouse serving as the host offering the greatest selective advantage to subpopulations with the greatest neurovirulent potential. Whether the interfering capacity of the early-passage virus is associated either with relative temperature sensitivity or with greater likelihood of inducing the formation of DI particles is not known. Neither of these possible covariant functions has been critically examined for dengue or other flaviviruses at different stages of adaptation to mice. However, Banerjee (1969) and Cole and Wisseman (1969) have presented evidence relating high neurovirulence of mouse-adapted dengue strains to their capacity to grow at high temperatures.

Intrinsic heterogeneity in the expression of several markers, including neurovirulence, of a freshly isolated strain (PR-159) of dengue-2 virus was investigated by Eckels *et al.* (1976). The original isolate, in the form of acute-phase serum from a patient, produced a mixture of large (L) and small (S) plaques on the LLC-MK2 (monkey kidney) cell line. Six serial passages in primary green monkey kidney (PGMK) cell cultures preserved the mixed plaque size. Clones derived from L plaques continued to segregate L and S. It was, however, possible to obtain S clones which appeared to be stable when passaged serially in PGMK cells at a low multiplicity of infection (MOI). However, these clones again yielded mixed L and S plaque progeny when passaged at a high MOI. Covariant with the S phenotype were (1) low neurovirulence for suckling mice and (2) significantly greater temperature sensitivity of replication, both relative to the L phenotype.

In a subsequent paper, the same authors (Harrison *et al.*, 1977) reported that S virus populations, inoculated subcutaneously into primates, failed to induce (or induced minimal) viremia, while parental or L plaque virus caused high-titered and prolonged viremia. Despite the striking difference in the extent to which circulating virus could be recovered, both L and S appeared to be capable of replicating, because both were comparably effective as immunogens, as measured by antibody production. On the basis of these observations, the S variant has been proposed as a suitable candidate for the possible development of a vaccine for human use (Harrison *et al.*, 1977).

These studies illustrate the theoretical considerations governing possible approaches to vaccine development: (1) The stability of the S (low-neurovirulence) phenotype after single passages in primates may not be real if it is capable of interfering with the replication of L; the limitations of the plaque assay would make the detection of L phenotype difficult if it were present at a frequency of less than 0.1–1%. (2) As pointed out by the authors, although viremias in human and simian hosts are in many ways comparable, the relationship of demonstrable viremia to human *virulence* has not been documented: In contrast to humans, monkeys infected even with unmodified dengue viruses do not become sick. (3) In view of the effective immunogenicity of the S variant in primates it is conceivable that viremia does in fact occur but that the virus has been so modified that it no longer produces even small plaques on PGMK cells; in this case, alternate methods would be required to detect virus of viral antigens.

Such questions, raised only in the practical context of the possible use of the S variant for vaccination, do not detract from the fact that these studies by Eckels and Harrison and their colleagues thus far represent the most carefully documented correlation of neurovirulence with other functions of a flavivirus. Together with earlier observations on flaviviruses with attenuated neurovirulence for mice (e.g., Banerjee, 1969; Cole and Wisseman, 1969; Price *et al.*, 1973, 1974; Tarr and Lubiniecki, 1976), they underline the need for intensive genetic studies on flaviviruses and for identification of markers that are stably covariant with defined levels of virulence.

4. *Host Factors Determining the Expression of Neurovirulence*

a. Autosomal Inheritance of Susceptibility and Resistance in Mice. Variations in vulnerability to infection were recognized long ago when it was found that yellow fever virus was highly lethal for certain inbred strains of mice and less so for others (Sawyer and Lloyd, 1931). Lynch and Hughes (1936) established, by cross-breeding experiments, that in this system resistance was an inherited trait. Webster and associates (Webster and Clow, 1936; Webster, 1937; Webster and Johnson, 1941) carried out extensive studies with St. Louis encephalitis (SLE) virus in resistant and susceptible segregants bred out of the RI strain of Swiss mice by continued brother–sister mating. Resistance expressed itself as a >1000-fold reduced threshold susceptibility to lethal infection, reduced viral replication, and the persistence of lesions in the CNS. Webster and Johnson (1941) confirmed the contrasting behavior by infecting brain fragment explant cultures from susceptible and resistant animals.

These early observations were followed up by Sabin (1952a,b) who extended the study of inherited resistance to *flavi*viruses other than SLE and demonstrated that resistant mice retained undiminished susceptibility to *alpha*viruses. He also showed that resistance was a dominant autosomal trait, inherited in Mendelian fashion, which expressed itself as reduced "vulnerability" rather than as an intrinsic inability to support at least some viral replication.

Sabin's conclusions were confirmed in extensive studies by Koprowski and co-workers (Theis *et al.*, 1959; Groschel and Koprowski, 1965) in congenic C3H/RV (resistant, *RR*) and C3H/He (susceptible, *rr*) mice. They showed that resistance to West Nile virus was not associated with a greater capacity for producing neutralizing antibody or interferon (Goodman and Koprowski, 1962; Vainio *et al.*, 1961; Hanson and Koprowski, 1969); a connection first thought to exist between the better ability of *RR* than of *rr* mice to regulate their body temperature (Lagerspetz *et al.*, 1973) and their resistance to infection could not be confirmed subsequently by Darnell and Koprowski (1974a). However, these authors described preliminary evidence strongly suggesting that infection of cell cultures derived from *RR* mice led to significantly greater generation of noninfectious (DI?) virus particles than that of similar cultures from *rr* mice (Darnell and Koprowski, 1974b; Darnell *et al.*, 1974). Although the defective nature of these interfering entities has not been confirmed by studies on their genomes or by analysis of cells for possible subgenomic viral RNA species (see Chapter 15), this work has contributed the most convincing implication to date of nonreplicating particles in determining the outcome of flavivirus infections.

The contrasting behavior of these resistant and susceptible mice, and of cells derived from them, to alpha- and flaviviruses is of special interest. It adds a further dimension to the fundamental differences which separate these two genera with respect to structure, composition, mode of replication, and morphogenesis (see Chapters 1 and 8–20).

In this connection, the recent identification of the major histocompatibility antigens HLA-A and HLA-B on human cells and H2-K and H2-D on mouse cells with receptor activity for SFV (Helenius *et al.*, 1978) is of special interest; in four other virus systems, nonresponsiveness of cytotoxic lymphocytes of specific allotypes is reported to be dominant over responsiveness (Zinkernagel *et al.*, 1978; Doherty *et al.*, 1978).

b. Immune Response. This is not the place to analyze in detail the complex mechanisms by which the host's immune response—both humoral and cellular—can affect the outcome of viral CNS infections. Alpha- and flaviviruses have contributed greatly to this field of investigation, as one would expect in light of their prominent role as producers of encephalitis in natural as well as experimental infections. The general subject has been reviewed by Nathanson and Cole (1970).

Most simply stated, the *expression* of neurovirulence depends on the successful competition of virus-induced cell injury with the evolving defense mechanisms of the host. Therefore, if intrinsic high virulence is associated with an enhanced rate of virus replication, its expression is likely to be amplified by the failure of immune clearance to intervene in time. Conversely, in infection with a relatively avirulent, slowly multiplying strain, an effective immune response is more likely to provide protection. This type of differential response was demonstrated in mice inoculated with the strains of WEE virus discussed above (see Figs. 1–3). When infection by the direct intracerebral route was preceded by immunization with minimally effective doses of inactivated vaccine, mice challenged with the slow strain could be protected, while those challenged with the fast RI strain could not. In the former, survival could be correlated with a marked and, in mice doubly immunized against WEE *and* Eastern equine encephalitis (EEE), *specific* secondary *local* antibody response in the CNS (as measured by a great increase in neutralizing and complement-fixing antibody titer of perfused brain extract relative to that of serum) (Schlesinger, 1949b).

More valid than such intracerebral challenges of previously immunized animals is the demonstration of a differential effect of the *primary* immune response on the outcome of infection after extraneural (peripheral) infection. Morgan (1941), in studies on the age-dependent resistance of mice to encephalitis following intraperitoneal inoculation of EEE virus, sought to correlate the greater susceptibility of infant mice with the immaturity of their antibody-producing capacity. Such a correlation was not confirmed in studies with Sindbis virus by Griffin (1976) and Griffin and Johnson (1977) and with SFV by Fleming (1977). The latter author, in a further elaboration of the results obtained earlier by Bradish *et al.* (1971, 1972; Bradish and Allner, 1972), emphasized the primacy of host age-

and viral strain-dependent variations in susceptibility and of rate of viral replication in determining whether an infected mouse survived long enough to permit protection by antibody. More important in explaining the *mechanism* of successful immunological intervention was his observation that, in mice infected intraperitonally with avirulent SFV, clearance of virus from the brain coincided with the appearance of plaque-neutralizing activity in the brain. This activity was associated with a *selective* accumulation of the IgG_3 subclass of anti-SFV antibody in brain tissue (relative to its proportion in anti-SFV serum Ig). Fleming (1977) assumed that this accumulation represented selective transfer of IgG_3 rather than local production, an interpretation which would require the demonstration that IgG_3 of other specificities underwent a parallel increase in the brain/serum ratio. In the related studies on WEE in doubly immunized mice referred to above, the increase was shown to be specific for antibodies to the challenge virus and therefore interpreted as due to local production (Schlesinger, 1949b). Presumably, infiltrating lymphocytes, sensitized by the immunogenic stimulus of extraneurally replicating virus, could account for such a selective response to massive production of antigen in the CNS (cf. the extraordinarily high cerebrospinal fluid measles antibody titers in patients with subacute sclerosing panencephalitis).

An interesting further elaboration of this general theme was reported by Jagelman *et al.* (1978), who compared intraperitoneally induced infections with the avirulent SFV variant in athymic "nude" (*nu/nu*) mice, their heterozygous littermates (*nu/+*), and conventional Swiss mice. In the *nu/nu* mice, virus was incompletely cleared from the brain, and a secondary phase of viral replication occurred after about 3 weeks, associated with falling titers of neutralizing serum antibody. Earlier it had been reported by Zlotnik *et al.* (1972) that conventional mice that had survived an avirulent SFV infection developed severe CNS disease characterized by hydrocephalus, spongy degeneration, and astrocytosis 2 years later. It will be interesting to see if such a late disease will also occur in the athymic nude mice.

Finally, it should be mentioned that no clear-cut correlation has been observed between expressed levels of virulence of SFV and *interferon* production (Bradish and Allner, 1972; Oaten *et al.*, 1976; Fleming, 1977).

A general inference which can be drawn from the large literature dealing with possible immunological factors associated with high or low neurovirulence is that avirulent viral variants are not intrinsically more effective inducers of antiviral antibodies, cellular immune mechanisms, or interferon than virulent ones.

III. PHYLOGENETIC HOST DETERMINANTS OF VIRULENCE—VERTEBRATE VERSUS ARTHROPOD HOST SYSTEMS

Arthropod-borne togaviruses offer a built-in opportunity for studying extremes of phylogenetically determined vulnerability to virus-induced cell injury: Arthropod vectors, as a rule, show no pathological effects of infection, while naturally and experimentally infected vertebrates often do. The contrast tends to be exaggerated in cell culture systems derived from representatives of these two phyla, especially those infected with alphaviruses: Permissiveness of vertebrate cells for the latter is generally associated with acute cytocidal infection, while permissive arthropod cells display no detectable abnormality; instead they become the progenitors of persistently infected cultures (see Chapter 20 for detailed review).

Briefly, many studies on the behavior of alphaviruses in established cell lines derived from *Aedes albopictus* or *A. aegypti* demonstrate that (1) persistent infection is an inevitable end result, (2) the virus populations from persistently infected cultures score progressively, and finally exclusively, as temperature-sensitive (*ts*) and small plaque (*sp*) phenotypes [the two markers are, however, not covariant (Shenk *et al.*, 1974)], (3) by using these and other differentiating markers, it has been shown that persistently infected cultures are resistant to superinfection with homologous standard virus, (4) mosquito cells are somewhat less efficient in the generation of DI virus than cell cultures of vertebrate origin, (5) in mosquito cells, DI particles derived from vertebrate cells do not interfere with, or replicate in the presence of, helper virus (see details in Chapters 15 and 20).

In considering the possible relationship of the *ts* and *sp* phenotypes to virulence, we have previously raised the theoretical question of whether the cell culture may be a model for events occurring in naturally (or experimentally) infected whole mosquitoes and what the epidemiological implications of such an analogy might be (Schlesinger, 1975, 1977b; Schlesinger *et al.*, 1978). Specifically, we have asked whether the prolonged persistence of virus in overwintering arthropods or during vertical transmission may lead to similar predominance of the *ts* and/or *sp* phenotypes and, conversely, whether the transfer of such virus to vertebrate hosts may select against such phenotypes.

With regard to the latter possibility, Shenk *et al.* (1974) showed that the *ts* character of five RNA$^+$ Sindbis virus clones recovered from persistently infected *A. albopictus* cell cultures displayed very low or undetectable spontaneous reversion rates to *ts*⁺ (6.7×10^{-5} to $<10^{-6}$). However, after four "undiluted serial passages" in BHK-21 cells, maintained at 34°C, particles of non-*ts* phenotype had accumulated to the extent of 8.5×10^{-2} to 1.0×10^{-3}. The reduced proportion of the *ts* phenotype was not associated with reversion of the *sp* character. These results suggested that at least one seemingly stable marker (*ts*) of virus derived from persistently infected mosquito cells tended to be at a selective disadvantage in BHK-21 cells.

Similarly, long-term intrathoracic passages of Sindbis virus in *Drosophila melanogaster* led to the predominance of *ts*, *sp* viral variants (Bras-Herreng, 1976). In our laboratory, the possible evolution of Sindbis virus of *ts* phenotype was studied in *Aedes triseriatus* infected by intrathoracic inoculation and maintained at 20°C (R. L. Regnery, to be published). The results of a large experiment of this type can be summarized as follows: clonal analysis showed that, after a total of 138 days in two consecutive groups of mosquitoes, the proportion of *ts* plaque formers increased from <1 to 29%. The virus contained in 82-day mosquitoes (8% *ts*) was transferred to 1-day-old chicks. These were bled 48 h later. None (<1%) of the virus isolated from the blood scored as temperature-sensitive. The chick blood was inoculated back into *A. triseriatus* which were kept until day 39; at that time 6 of 103 clones tested (6%) were again of *ts* phenotype.

While these percentages are far less impressive than those found in persistently infected *Aedes* cell cultures (Igarashi *et al.*, 1977), they showed a trend in the same direction. This quantitative difference may be explained as follows: In Igarashi's experiments persistently infected *A. albopictus* cell cultures were subjected to weekly subcultures. Each of these led to a great burst of cell division and virus synthesis (see Chapter 20). This process undoubtedly contributed immense selective pressure in favor of the *ts* phenotype ultimately destined to predominate in such cultures. Yet, even under these conditions, it took 20 weeks to reduce non-*ts* virus to an undetectable level (Igarashi *et al.*, 1977). In contrast, cell division in adult mosquitoes such as those used in the inoculation experiments is either rare or absent (W. Trager, personal communication), and the level of virus recovered from persistently infected mosquitoes is remarkably constant [approximately 10^5 PFU per mosquito (R. L. Regnery, to be published)].

For a virus transmitted transovarially and persisting through all developmental stages of the arthropod vector, the situation, in terms of cell division activity, might be more analogous to the cell culture system. If, under such conditions, viral particles of *ts* phenotype or of some other function covariant with low virulence predominated at the beginning of the biting season, one would predict that cases of disease caused by a pathogenic virus early in an epidemic would be clinically milder than later ones. The latter would reflect, in their severity, short-term transmission cycles of the non-*ts* phenotype accumulating in warm-blooded hosts.

In this context, it is of more than historical interest to recount the following translated passage from a paper by Doerr and Russ (1909), dealing with natural and experimental transmission of sandfly fever virus (a bunyalike arbovirus) by *Phlebotomus papataci* in Herzegovina, Yugoslavia:

> . . . the overwintering of the virus of papataci fever remains unexplained . . . overwintering of winged insects has been excluded. . . . in winter or spring, neither recurrences nor new cases have been seen. In view of these observations, the only remaining possibility of connecting the annual summer epidemics is the hereditary transmission by infected *Phlebotomus* females to their progeny. This assumption is all the less unreasonable as Marchoux and Simond [1906] as well as the British Commission were able to produce yellow fever by bites of hereditarily infected *Stegomyia*. Remarkably, the experimental infections transmitted by the progeny of yellow fever-infected mosquitoes were always strikingly mild. Similarly, we observe in the case of papataci fever (as also noted by Mennella in Northern Italy) that the first appearance of sandflies coincides with the occurrence of very mild abortive human cases. These are invariably followed by cases of progressively more severe illnesses. Thus it appears that, in sandfly fever as well as in yellow fever, the virulence of the virus becomes attenuated by hereditary passage from one (insect) to another and increases substantially by transfer to another host, i.e., the presence in the human organism.

The report by Marchoux and Simond (1906) and the concept of transovarial transmission of yellow fever were discounted by subsequent investigators (see Whitman, 1951). However, in light of more recent observations (see Chapter 6), including the transovarial transmission of Japanese encephalitis and yellow fever viruses by experimentally infected *Aedes albopictus* and *A. togoi* (Rosen *et al.*, 1978; Aitken *et al.*, 1979), this question merits intensive reinvestigation with methods designed to uncover *ts* or nonpathogenic viral variants. This will require an assay system other than inoculation into mice or other warm-blooded hosts. For example, Reeves *et al.* (1958) reported on the recovery of WEE virus from naturally infected *Culex tarsalis* collected in Kern County, California, throughout the year. Virus could be isolated by mouse inoculation from specimens caught in all months except December. Strains isolated from January through March were nonpathogenic for mice and poorly immunogenic for chickens. Clearly, these properties would be compatible with temperature sensitivity of the virus samples.

As discussed more fully in Chapter 20, nothing is known at present about the molecular or physiological basis of the differences between mosquito cells and vertebrate cells in their responses to infection with Sindbis and other togaviruses. The recent isolation of *A. albopictus* cell clones which, in contrast to the parental cells, undergo massive cytopathic effects (CPE) after infection with Sindbis virus (Sarver and Stollar, 1977) has introduced a valuable system for elucidating cellular determinants of susceptibility or resistance to virus-induced injury (see Chapter 20).

IV. VIRAL VIRULENCE MARKERS: THE QUESTION OF GENOTYPIC VERSUS PHENOTYPIC POLYMORPHISM

Fundamental to a search for markers that might be covariant with different levels of virulence is the evidence indicating that alpha- and flavivirus populations, in the form of freshly isolated wild-type or laboratory-bred uncloned strains, are heterogeneous in several functions. Examples illustrating this point were mentioned in the preceding sections of this chapter.

Although selective procedures, such as continued passage or cloning in a single host system, usually lead to the predominance or apparent exclusiveness of a given phenotype, segregation of an opposing phenotype is common and greatly enhanced by transfer of the virus to a different set of growth conditions. The predictability of the direction in which some of these changes go, as well as the rapid rate at which they occur, suggests that they are expressions of viral polymorphism, with diverse potential phenotypes being carried as majority or minority subsets in all viral populations. This situation is, of course, not unique to togaviruses; a well-documented example of di- or polymorphism is that presented by influenza A virus (Kilbourne, 1978). To a viral species that maintains itself in nature by continuous alternating transfer between members of one phylum (arthropods) and another (vertebrates), such polymorphism offers the opportunity of ensuring itself of the necessary adaptability to two greatly differing organismal environments.

Conventional wisdom would ascribe such adaptive changes to mutations in the viral gene coding for a product that is associated with a function which we happen to recognize as changed. For example, the surface glycoproteins are responsible for hemagglutination, attachment to cells, surface charge, antigenic characteristics, and other biological activities which may be associated with one or another aspect of virulence. An intriguing and fundamentally important question concerns the real relationship between such recognizable phenotypic expressions and viral gene products which are, in fact, *primarily* involved in determining virulence. At present we know nothing about these relationships in the case of alpha- and flaviviruses.

At the risk of belaboring the (generally unspoken) obvious or the trivial (which it is not if it is true and obvious), one might consider the following speculative thoughts: Suppose a virulent viral mutant A produces a putative "virulence gene" product in a more active form, at a more rapid rate, or in a significantly greater amount than an avirulent mutant B infecting the same kind of cell. The virulence genes presumably code for factors that inhibit or modulate cellular macromolecular syntheses, interfere with structural or functional integrity of cellular membranes and transport mechanisms, and destabilize or otherwise modify cellular enzymes such as sugar transferases or proteases, methylases, and kinases. All these and many other *cellular* functions [including presumably cellular components(s) of RNA polymerase(s) (see the example of *tdCE* mutants of vesicular stomatitis virus described in Pringle, 1978)] play key roles in togavirus replication and/or assembly (see appropriate chapters of this book).

It follows that the finished progeny of mutant A (which disrupts the cell) may be quite different phenotypically from that of mutant B (which does not). Moreover, A virions assembled late in a single growth cycle, when the cell is falling apart, may differ phenotypically from those completed early (there is a hint of this sort of time-dependent change in an experiment described in Fleming, 1973). Such a phenotypic difference may in no way reflect *directly* the expression of the viral gene functions primarily involved in determining the intensity of cellular injury (virulence).

It is important here to separate this hypothetical type of secondary phenotypic polymorphism from mechanisms operating in cell mutants defective in one key function or another (see Chapter 16) or with viruses bearing demonstrable nonstructural or structural gene mutations or defects (see Chapter 14). Obviously, the complexity of the situation would be compounded if we compared mutants A and B in different cell types which respond differently to the putative virulence gene products.

Extrapolating to virus replication in a complex organism of a higher animal or of an invertebrate vector, we can conceive of a virus population emerging that reflects in its phenotypic heterogeneity and predominance the entire spectrum of diverse responses of differentiated cell types to infection with a virus that may be genetically quite homogeneous. Theoretically, such phenotypic polymorphism could, for example, greatly modulate all viral activities that reside in the envelope glycoproteins, specifically in the carbohydrate moieties. Under these circumstances, apparent stability of a phenotypic marker could be enhanced by subjecting the virus to repeated passage in the same host cell system; apparent reversion could be amplified by transfer to another host cell system that responds differently to infection.

The idea of phenotypic polymorphism is advanced here as a possible alternative to a purely genetic interpretation of the apparent variability of certain viral markers. None of the viral stocks thus far employed in studies on virulence in whole animals have been subjected to the rigorous genetic analysis basic to the characterization of chemically induced alphavirus mutants (see Chapter 14). Until this is done, it is essential to consider all options, especially if *apparent* virulence markers are to be used as guides in vaccine development.

V. CONCLUSIONS

Collectively, the alpha-, flavi-, and "non-arbo" togaviruses play a highly significant role as causes of natural diseases or silent infections of man and diverse other vertebrate species. From a general biological viewpoint the arthropod-transmitted members of this family, like other arboviruses, pose uniquely challenging problems with regard to questions of viral evolution, adaptation to phylogenetically disparate organisms, and genetic and/or phenotypic responses to selective pressures encountered in natural transmission cycles. These challenges form a natural rationale for experimental approaches discussed in detail in subsequent chapters of this book.

In this chapter, prominence has been given to the elusive problem of virulence as expressed in natural or experimental infections of whole animals with alpha- and flaviviruses. The reasons for this emphasis are complex: (1) More than in the simpler situation of cell culture systems, virulence for the intact vertebrate host is a compound expression of intrinsic viral properties and host response; (2) one goal of research is the identification of markers related to virulence and of those viral properties and host reactions that would have to be blocked or promoted in effective prophylaxis or therapy; (3) for some of the viruses discussed (e.g., yellow fever and dengue), a critical step often used in making them amenable to laboratory investigation is the adaptation to mice with its attendant acquisition of neurovirulence, and the fact that this step can lead to attenuation of their natural pathogenicity has important implications for an understanding of disease mechanisms and for vaccine development; (4) it is likely that the expression of virulence for various cell culture systems (permissiveness, CPE, plaque production) is even more remote than neurovirulence in mice from virulence measured in terms of injury to target cells or organs in natural hosts; (5) much of the experimental evidence summarized above for alpha- and flaviviruses points to marked heterogeneity of viral stocks (even cloned ones) with respect to several phenotypic markers. An understanding of the essence of such heterogeneity (polymorphism) requires detailed functional and structural analyses of the viruses, especially of chemically induced single-step mutants. Such analyses will lead to clearer definitions of viral as distinct from cellular determinants and of genetic as distinct from nongenetic modifying processes.

In this chapter, alpha- and flaviviruses have been treated as essentially comparable with regard to the problems they pose in nature and in the laboratory. Yet these two genera are far apart in features of their structure and replication (see Chapter 1), a fact which may ultimately force us to explore their interactions with complex biological systems in different ways. This diversity is amplified by the inclusion of "non-arbo" species in the family Togaviridae.

REFERENCES

Aitken, T. H. G., Tesh, R. B., Beaty, B. J., and Rosen, L. (1979). *Am. J. Trop. Med. Hyg.* **28**, 119–121.

Banerjee, K. (1969). *Indian J. Med. Res.* **57**, 1165–1180.

Berge, T. O., ed. (1975). "International Catalogue of Arboviruses," 2nd ed., DHEW Publ. No. (CDC) 75-8301. U.S. Dep. Health, Educ. Welfare, Public Health Serv., Washington, D.C.

Bradish, C. J., and Allner, K. (1972). *J. Gen. Virol.* **15**, 205–218.

Bradish, C. J., Allner, K., and Maber, H. B. (1971). *J. Gen. Virol.* **12**, 141–160.

Bradish, C. J., Allner, K., and Maber, H. B. (1972). *J. Gen. Virol.* **16**, 359–372.

Bras-Herreng, F. (1976). *Ann. Microbiol. (Paris)* **127b**, 541–565.

Cole, G. A., and Wisseman, C. L., Jr. (1969). *Proc. Soc. Exp. Biol. Med.* **130**, 359–363.

Darnell, M. B., and Koprowski, H. (1974a). *J. Infect. Dis.* **129**, 248–256.

Darnell, M. B., and Koprowski, H. (1974b). *In* "Mechanisms of Virus Disease" (W. S. Robinson and C. F. Fox, eds.), pp. 147–158. Benjamin, New York.

Darnell, M. B., Koprowski, H., and Lagerspetz, K. (1974). *J. Infect. Dis.* **129**, 240–247.

De Vleesschauver, L., and Pattyn, S. R. (1974). *Arch. Gesamte Virusforsch.* **45**, 78–85.

Dimmock, N. J., and Kennedy, S. I. T. (1978). *J. Gen. Virol.* **39**, 231–242.

Doerr, R., and Russ, V. K. (1909). *Arch. Schiffs- Trop.-Hyg.* **13**, 693–706.

Doherty, P. C., Biddison, W. E., Bennink, J. R., and Knowles, B. B. (1978). *J. Exp. Med.* **148**, 534–543.
Eckels, K. H., Brandt, W. E., Harrison, V. R., McCown, J. M., and Russell, P. K. (1976). *Infect. Immun.* **14**, 1221–1227.
Fenner, F., McAuslan, B. R., Mims, C. A., Sambrook, J., and White, D. O. (1974). "Biology of Animal Viruses," 2nd ed. Academic Press, New York.
Fleming, P. (1973). *J. Gen. Virol.* **19**, 353–367.
Fleming, P. (1977). *J. Gen. Virol.* **37**, 93–105.
Goodman, G. T., and Koprowski, H. (1962). *J. Cell. Comp. Physiol.* **59**, 333–373.
Griffin, D. E. (1976). *J. Infect. Dis.* **133**, 456–464.
Griffin, D. E., and Johnson, R. T. (1977). *J. Immunol.* **118**, 1070–1075.
Groschel, D., and Koprowski, H. (1965). *Arch. Gesamte Virusforsch.* **18**, 379–391.
Hanson, B., and Koprowski, H. (1969). *Microbios* **1B**, 51–68.
Harrison, V. R., Eckles, K. H., Sagartz, J. W., and Russell, P. K. (1977). *Infect. Immun.* **18**, 151–156.
Helenius, A., Morein, B., Fries, E., Simons, K., Robinson, P., Schirrmacher, V., Terhorst, C., and Strominger, J. L. (1978). *Proc. Natl. Acad. Sci. U.S.A.* **75**, 3846–3850.
Hotta, S. (1965). *In* "Applied Virology" (M. Sanders and E. H. Lennette, eds.), pp. 228–256. Warren H. Green, Inc., St. Louis, Missouri.
Huang, A. S., and Baltimore, D. (1977). *In* "Comprehensive Virology" (H. Frankel-Conrat and R. R. Wagner, eds.,), Vol. 10, pp. 73–116. Plenum, New York.
Igarashi, A., Koo, R., and Stollar, V. (1977). *Virology* **82**, 69–83.
Jagelman, S., Suckling, A. J., Webb, H. E., and Bowen, E. T. W. (1978). *J. Gen. Virol.* **41**, 599–607.
Kilbourne, E. D. (1978). *In* "Viruses and Environment" (E. Kurstak and K. Maramorosch, eds.), pp. 339–350. Academic Press, New York.
Kimura, T., and Ueba, N. (1978). *Arch. Virol.* **57**, 221–229.
Lagerspetz, K. Y. H., Koprowski, H., Darnell, M., and Tarkkonen, H. (1973). *Am. J. Physiol.* **225**, 532–537.
Lynch, C. J., and Hughes, T. P. (1936). *Genetics* **21**, 104–112.
Maeda, S., Hashimoto, K., and Simizu, B. (1979). *Virology* **92**, 532–541.
Marchoux, E., and Simond, P.-L. (1906). *Ann. Inst. Pasteur, Paris* **20**, 16–40.
Meiklejohn, G., England, B., and Lennette, E. H. (1952). *Am. J. Trop. Med. Hyg.* **1**, 59–65.
Morgan, I. M. (1941). *J. Exp. Med.* **74**, 115–132.
Nathanson, N., and Cole, G. A. (1970). *Adv. Virus Res.* **16**, 397–448.
Oaten, S. W., Webb, H. E., and Bowen, E. T. W. (1976). *J. Gen. Virol.* **33**, 381–388.
Olitsky, P. K., Morgan, I. M., and Schlesinger, R. W. (1945). *Proc. Soc. Exp. Biol. Med.* **59**, 93–97.
Pattyn, S. R., De Vlesschauver, L., and Vander Groen, G. (1975). *Arch. Virol.* **49**, 33–37.
Peleg, J. (1971). *Curr. Top. Microbiol. Immunol.* **55**, 155–161.
Peleg, J. (1975). *Ann. N.Y. Acad. Sci.* **266**, 204–213.
Price, W. H., Casals, J., Thind, I., and O'Leary, W. (1973). *Am. J. Trop. Med. Hyg.* **22**, 509–523.
Price, W. H., Casals, J., and O'Leary, W. (1974). *Am. J. Trop. Med. Hyg.* **23**, 118–130.
Pringle, C. R. (1978). *Cell* **15**, 597–606.
Pusztai, R., Gould, E. A., and Smith, H. (1971). *Br. J. Exp. Pathol.* **52**, 669–677.
Reeves, W. C., Bellamy, R. E., and Scrivani, R. P. (1958). *Am. J. Hyg.* **67**, 78–89.
Rosen, L., Tesh, R. B., Lien, J. C., and Cross, J. H. (1978). *Science* **199**, 909–911.

Sabin, A. B. (1952a). *Proc. Natl. Acad. Sci. U.S.A.* **38**, 540–546.
Sabin, A. B. (1952b). *Ann. N.Y. Acad. Sci.* **54**, 936–944.
Sabin, A. B. (1955). *Am. J. Trop. Med. Hyg.* **4**, 198–207.
Sabin, A. B., and Schlesinger, R. W. (1945). *Science* **101**, 640–642.
Sarver, N., and Stollar, V. (1977). *Virology* **80**, 390–400.
Sawyer, W. A., and Lloyd, W. (1931). *J. Exp. Med.* **59**, 533–555.
Schlesinger, R. W. (1949a). *J. Exp. Med.* **89**, 491–505.
Schlesinger, R. W. (1949b). *J. Exp. Med.* **89**, 507–527.
Schlesinger, R. W. (1951). *Proc. Soc. Exp. Biol. Med.* **76**, 817–823.
Schlesinger, R. W. (1975). *Med. Biol.* **53**, 295–301.
Schlesinger, R. W. (1977a). "The Dengue Viruses," Virology Monographs, Vol. 16. Springer-Verlag, Vienna and New York.
Schlesinger, R. W. (1977b). *Med. Microbiol. Immunol.* **164**, 77–85.
Schlesinger, R. W., and Frankel, J. W. (1952). *Am. J. Trop. Med. Hyg.* **1**, 66–67.
Schlesinger, R. W., Stollar, V., Igaraschi, A., Guild, G. M., and Cleaves, G. R. (1978). *In* "Viruses and Environment" (E. Kurstak and K. Maramorosch, eds.), pp. 281–298. Academic Press, New York.
Schulze, I. T. (1962). Ph.D Thesis, Saint Louis Univ., St. Louis, Missouri.
Shenk, T. E., Koshelnyk, K. A., and Stollar, V. (1974). *J. Virol.* **13**, 337–344.
Stollar, V., Peleg, J., and Shenk, T. E. (1974). *Intervirology* **2**, 337–344.
Strode, G. K., ed. (1951). "Yellow Fever." McGraw-Hill, New York.
Tarr, G. C., and Lubiniecki, A. S. (1976). *Infect. Immun.* **13**, 688–695.
Taylor, W. T., and Marshall, I. D. (1975a). *J. Gen. Virol.* **28**, 59–72.
Taylor, W. T., and Marshall, I. D. (1975b). *J. Gen. Virol.* **28**, 73–83.
Theiler, M. (1951). *In* "Yellow Fever" (G. K. Strode, ed.), pp. 46–136. McGraw-Hill, New York.
Theis, G. A., Billingham, R. E., Silvers, W. K., and Koprowski, H. (1959). *Virology* **8**, 264–265.
Vainio, R., Gawatkin, R., and Koprowski, H. (1961). *Virology* **14**, 385–387.
Webster, L. T. (1937). *J. Exp. Med.* **65**, 261–286.
Webster, L. T., and Clow, A. D. (1936). *J. Exp. Med.* **63**, 827–845.
Webster, L. T., and Johnson, M. S. (1941). *J. Exp. Med.* **74**, 489–494.
Whitman, L. (1951). *In* "Yellow Fever" (F. K. Strode, ed.), pp. 237–298. McGraw-Hill, New York.
Woodward, C. G., and Smith, H. (1975). *Br. J. Exp. Pathol.* **56**, 363–372.
Woodward, C. G., Marshall, I. D., and Smith, H. (1977). *Br. J. Exp. Pathol.* **58**, 616–624.
Zinkernagel, R. M., Althage, A., Copper, S., Kreeb, G., Klein, P. A., Sefton, B., Flaherty, L., Stimpfling, J., Shreffler, D., and Klein, J. (1978). *J. Exp. Med.* **148**, 592–606.
Zlotnik, I., Grant, D. P., and Batter-Hatton, D. (1972). *Br. J. Exp. Pathol.* **53**, 125–129.

5

Immunological Parameters of Togavirus Disease Syndromes

SCOTT B. HALSTEAD

I. INTRODUCTION

It is obvious that viral illnesses have a molecular basis and equally obvious that a proportionate contribution toward molecular disturbance is made both by virus and host. It can be argued that each virus disease falls somewhere on a spectrum between those caused by direct viral damage to host cells and those resulting from the immune response directed against viral antigens. A brief reflection on pathophysiology makes it evident that a myriad of mediators and

THE TOGAVIRUSES

ISBN 0-12-625380-3

biochemical amplification systems are involved in producing the localized and generalized signs and symptoms of viral illness. Many of these are immunological. In this sense, all togavirus infections may be said to have an immunological basis.

In addition to the pathological implications of the immune response, viral immunopathology also includes the ability of viruses to interfere with normal function in the immune system. Virus infection may depress delayed hypersensitivity reactions (Von Pirquet, 1908; Woodruff and Woodruff, 1975), prolong the rejection of allografts (Howard *et al.*, 1969b), decrease *in vitro* lymphocyte reactivity to specific antigen (Smithwick and Berkovich, 1966; Zweiman, 1971), or decrease responses to mitogen stimulation (Fireman *et al.*, 1969) or mixed lymphocyte reactions (Twomey, 1974). Correspondingly, viral infection may affect the humoral immune response, producing elevations or depressions in levels of serum immunoglobulins (Notkins *et al.*, 1966; Woodruff and Woodruff, 1975), enhanced or suppressed specific antibody production (Porter and Larsen, 1974; McFarland, 1974; Woodruff and Woodruff, 1975), or suppression of the induction of high zone tolerance (Mergenhagan *et al.*, 1967). While the mechanisms by which viruses cause these and other kinds of immune dysfunction are not fully understood, possibilities include impaired function of infected macrophages, direct or indirect destruction of lymphocytes, inhibition of lymphocyte activation or proliferation, alteration of lymphocyte traffic, or stimulation of suppressor cells (Woodruff and Woodruff, 1975). Some of these phenomena have been described in togavirus infections.

Perhaps the most thoroughly studied is rubella. In congenital rubella, levels of IgM are often elevated, while those of IgG and IgA are reduced (Rawls, 1974). In postnatal rubella transient suppression of tuberculin reactions has been observed (Lamb, 1969; Midulla *et al.*, 1972; Kauffman *et al.*, 1974). Immunopathological aspects of rubella infection were reviewed by Rawls (1974) and will not receive further discussion here.

Venezuelan equine encephalitis (VEE) infections in man are sometimes accompanied by leukopenia (Tigertt *et al.*, 1962). This virus is able to replicate in bone marrow cells in several vertebrate species (Smith *et al.*, 1964). It produces lytic infections of lymphoid tissues in mice, guinea pigs, and hamsters (Gleiser *et al.*, 1962) and, when used as an oncolytic agent, remissions in some human lymphoid tumors (Tigertt *et al.*, 1962). In mice and guinea pigs, when VEE infection was initiated up to 11 days before the inoculation of an unrelated antigen, there was enhanced primary antibody response to that antigen (Howard *et al.*, 1969b; Hruskova *et al.*, 1972a,b). Infection initiated shortly after an antigenic stimulus, however, depressed the antibody response (Hruskova *et al.*, 1972b). It has been suggested that enhanced antibody production might be due to VEE stimulation of the proliferation and phagocytic activity of mononuclear phagocytes (Howard *et al.*, 1969a; Staab *et al.*,1970).

Attenuated yellow fever (YF) virus, has been observed to replicate in human monocytes and in phytohemagglutinin (PHA)-treated lymphocytes (Wheelock and Edelman, 1969). In mice, competent macrophages and lymphocytes are required to assist the host in containing a lethal YF infection (Zisman *et al.*, 1971). Could it be that in man the failure of one or both of these defense mechanisms converts otherwise mild YF infections to fully virulent disease? Questions of this kind have not yet been approached experimentally in man. The only known human immunological idiosyncracy in YF is the prolonged persistence of IgM antibodies following vaccination with the 17D strain (Monath, 1971).

IgM antibodies have also been found many months after clinical Japanese encephalitis (JE) infections. The persistence of antibodies of this immunoglobulin class has been correlated with the severity of the preceding illness (Edelman *et al.*, 1976). Japanese encephalitis may be unique among human systemic infections. Many years after infection, protective antibodies appear to wane, and previously immune individuals apparently again become susceptible to infection (Ishii *et al.*, 1968). Severe JE among the elderly is an important public health problem in Japan (Matsuda, 1962). As yet little is known of immunopathogenetic mechanisms in this intriguing problem.

Although there have been studies on the togavirus antigenic determinants expressed on cell surfaces, as well as attempts in animal systems to elucidate mechanisms of protection or recovery from togavirus infection, the cellular, molecular, and immunological events which contribute to togavirus disease are relatively unstudied.There is one notable exception—dengue virus infections of man. This chapter will consider the evidence for immunopathology in dengue disease and describe the novel mechanisms which may be involved, emphasizing challenges for future researchers.

II. IMMUNOLOGICAL ASPECTS OF DENGUE INFECTION

A. Definitions

Dengue fever is an acute febrile illness syndrome caused by several arthropod-borne viruses and characterized by biphasic fever, myalgia or arthralgia, rash, leukopenia, and lymphadenopathy. Here we will consider only dengue fever caused by dengue viruses, of which there are generally acknowledged to be four types. Dengue hemorrhagic fever (DHF) is a severe febrile disease characterized by abnormalities of hemostasis and increased vascular permeability, which in many instances results in a hypovolemic shock syndrome, dengue shock syndrome (DSS). The degree of vascular permeability and the nature of hemostatic defects in DHF and DSS are precisely defined in a WHO Technical Guide (Technical Guides, 1975).

B. Historical Background

Although "fatal hemorrhagic dengue" has been a sporadic clinical diagnosis for more than 60 years (Goldsmid, 1917), during the first half of the twentieth century dengue was widely regarded more as nuisance than peril. In 1953, the term "hemorrhage fever" was used to describe a disease characterized by shock and hemorrhage—because of similarities to the then well-known Korean hemorrhage fever (Quintos *et al.*, 1954). One early observer recognized differences between the Korean and Philippine syndromes and proposed the name "infectious acute thrombocytopenic purpura" (Stransky and Lim, 1956). In 1956, at least 1000 children with this entity were hospitalized in Manila. From some of these cases Hammon *et al.* (1960a) recovered dengue virus types 2–4. Modern research on DHF had begun.

Because the evolution of symptoms and signs in DSS is distinctive, it is possible to scan the previrological medical literature and with some confidence identify earlier occurrences of this syndrome.

The first outbreak of DSS may have occurred in 1897 in Queensland. Hare (1898) obtained records of 60 fatal cases, 30 in children. The severe illness in children was most characteristic. These cases were

> amongst the most startling that occur in medical practise. In nearly all of these [children] death must be, I think, attributed to the intensity of uncomplicated disease. . . . All . . . were previously healthy children, their ages ranged between 3 and 14 years. . . . The manner of death was in the majority almost identical, vary rapid heart failure and collapse [which] occurred at the crisis on the fifth day of fever, and death ensued from two to 48 hours later. . . . The patient exhibits all the signs of acute hemorrhage, a frightful restlessness, jactitation, extreme irritability of temper . . . terminating [sometimes] in a state exactly resembling the [shock] stage of cholera.

Hare went on to note that many affected communities had been previously attacked by dengue in 1895 and that "large numbers [of persons] who had had the disease two years previously were again attacked equally as badly or even more severely"—evidence that two different etiological agents were involved. This was not an isolated event. Dengue was a major listed cause of death in northern Australia until the beginning of World War I (Lumley and Taylor, 1943).

In refugee-swollen Athens and Piraeus in 1927, there was a barely perceived mild dengue fever outbreak. This was succeeded in 1928 by the most explosive and virulent dengue epidemic ever recorded (Anonymous, 1928). In August and September alone, there were over 650,000 cases with 1060 deaths (Copanaris, 1928). Early retrospective serological studies undertaken before the dengue plaque neutralization test was available suggested that only a single dengue virus (type 1) caused the 1927–1928 epidemics (Theiler *et al.*, 1960; Pavlatos and Smith, 1964). This conclusion was widely accepted despite a report by Cardamatis (1929) of sequential cases of dengue fever in two families at an interval of 7 months. In a recent examination of sera from survivors, Papaevangelou and Halstead (1977) found dengue type-1 and -2 monospecific antibodies and heterotypic antibody in the majority of individuals who had lived in Athens during the outbreak. In unpublished studies *both* type-1 and -2 antibodies were found in persons born in 1928, while no one born in 1929 or subsequently was dengue-immune. Our tentative reconstruction of the Greek epidemic indicates that type-1 transmission began in 1927 and that type 2 was introduced in 1928. After that, dengue was eradicated from Greece.

Japanese physicians described a few cases resembling DSS during a 1931 dengue outbreak in Formosa (Nomura and Akashi, 1931).

From a review of hospital records it was shown that children with DSS-like syndromes were admitted to Siriraj Hospital in Bangkok, Thailand, during every rainy season from 1950 to 1965 (Halstead and Yamarat, 1965). A similar examination of 572 charts of pediatric admissions for the period July and August, 1932–1942, failed to uncover any DSS-like cases.

These early descriptions of DSS-like outbreaks are significant because they suggest rather clearly that this syndrome is not the expression of a unique virulence property of dengue virus. Earlier occurrences were in epidemiological situations in which sequential infection was possible. Finally, *endemic* DHF–DSS is a post-World War II phenomenon which appears to coincide with the endemic establishment of multiple dengue types in the large cities of tropical Asia (Halstead, 1966).

C. The Public Health Problem

The full public health impact of the DHF–DSS problem is difficult to estimate. Sizable epidemics of DHF–DSS have involved every country in tropical Asia except Bangladesh and Sri Lanka. A dengue-3 outbreak in Dacca (Bangladesh) in 1963 resulted in hundreds of hospitalized cases; some patients had mild to severe hemorrhages (Aziz *et al.*, 1967). A fatal case in an adult with meningeal signs and severe obtundation was reported. But clinical descriptions in the published accounts do not suggest that DSS accompanied this outbreak. In Sri Lanka, accounts of sporadic DSS are clinically convincing (Hamza *et al.*, 1966).

Epidemics of DHF–DSS can be devastating, causing the near paralysis of clinical facilities. In 1973, for example, there were 14,320 DHF–DSS hospitalizations in South Vietnam, 10,274 in Indonesia, and 2450 in Singapore and Malaysia, while in 1972 in Thailand there were 23,786 and in 1977 over 38,000. In 1977 in Thailand, DHF–DSS was the leading cause of hospitalization among children and of deaths due to infectious disease at any age.

Dengue viruses of at least two types are endemically established in West Africa and in the Caribbean basin and have recently been the cause of pan-Pacific epidemics (Carey *et al.*, 1971; Ventura and Ehrenkranz, 1976; Dengue Newsletter, 1976). In 1977–1978 a large dengue-1 epidemic swept through Cuba, Haiti, the Caribbean islands, and Colombia. Many of these countries have had recent dengue-2 and -3 activity. With the present resurgence of dengue in the American and African tropics, it seems only a matter of time before DHF–DSS becomes a global problem.

D. Immunity to Dengue Viruses

There are generally considered to be four distinct types of dengue viruses plus a subtype of dengue-3 proposed by Russell and McCown (1972). Dengue type-1 and -2 homotypic and cross-immunity patterns were studied in humans (Sabin, 1952). Six months or more following infection with one of these dengue serotypes, humans were fully susceptible to infection with a different type (Sabin, 1952). Such an infection is referred to as a *secondary infection*. There is no evidence that homotypic reinfection occurs; in fact, in the absence of antigenic stimulation, dengue antibody levels have remained essentially unchanged up to 50 years (Halstead, 1974). Although serial infection with three different dengue virus types has been reported (Myers and

Varkey, 1971), epidemiological observations (Halstead *et al.*, 1969b) suggest that clinical illness in man rarely follows challenge with a third or fourth virus type. Experimental studies on monkeys (Halstead *et al.*, 1973a; Scherer *et al.*, 1972; Whitehead *et al.*, 1970) also showed that infection with two different dengue types frequently conferred resistance to a third viremic infection. Secondary dengue infections stimulate flavivirus group-specific IgG antibody response; it is unusual to find type-specific IgM directed toward neoantigens in the second infecting virus (Russell *et al.*, 1967, 1969; Scott *et al.*, 1972).

E. Clinical Considerations

Since the evolution of symptoms and signs in dengue infections must be the expression of sequential pathogenetic phenomena, insights into pathogenesis necessarily start with a study of disease.

1. Dengue Fever

Clinical findings in mild, self-limiting dengue infections are age-dependent; in young children dengue infections are either undifferentiated febrile illnesses or mild upper respiratory infections (Halstead *et al.*, 1969a; Carey *et al.*, 1968). In adolescents and adults, dengue infections are typically accompanied by the dengue fever syndrome (Siler *et al.*, 1926; Simmons *et al.*, 1931; Sabin, 1952).

It has been suggested that a distinctive mean incubation period, duration of illness, and spectrum of clinical findings may characterize infections with each different dengue type (Halstead, 1974). Unfortunately, there are as yet insufficient studies to test this hypothesis. When dengue is taken as a composite, after an incubation period of 2–7 days there is the sudden onset of fever which rapidly rises to 103°–106°F. This is usually accompanied by frontal or retroorbital headache (Fig. 1). Viremia generally coincides with the febrile period (Simmons *et al.*, 1931). Occasionally, back pain precedes the fever. A transient macular, generalized rash which blanches under pressure may be seen during the first 24–48 h of fever. Myalgia or bone pain occurs soon after onset and increases in severity. During the second to sixth day of fever, nausea and vomiting are apt to occur, and during this phase generalized lymphadenopathy, cutaneous hyperaesthesia or hyperalgesia, taste aberrations, and pronounced anorexia may develop.

Coincident with or 1 or 2 days after defervescence, a generalized morbilliform, maculopapular rash appears, which spares the palms and soles. It disappears in 1–5 days. About the time of the second rash,

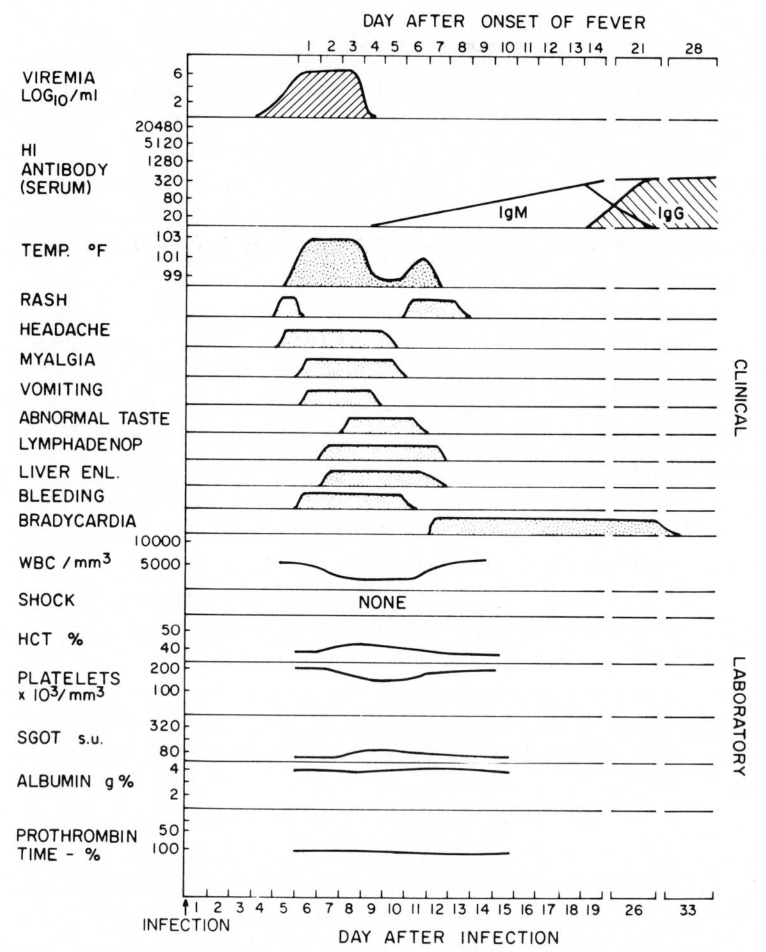

Fig. 1. Schema of clinical and laboratory findings in typical dengue fever syndrome in susceptible adults. The kinetics of infection phenomena represent mean values from published data for several dengue virus types (Siler *et al.*, 1926; Simmons *et al.*, 1931; Halstead *et al.*, 1969b; Kuberski *et al.*, 1977).

the body temperature may rise from previous values, resulting in a biphasic temperature curve.

Epistaxis, petechiae, and purpuric lesions, although uncommon, may occur at any stage of disease. The frequency of these minor hemorrhagic phenomena varies from outbreak to outbreak (Rice, 1923; Sabin, 1952). Swallowed blood from epistaxis may be vomited or passed per rectum and misinterpreted as gastrointestinal bleeding. Gastrointestinal bleeding, menorrhagia, and bleeding from other organs have been described, particularly when outbreaks involve adults (Rice, 1923; Scott, 1923; Kuberski *et al.*, 1977).

The hematopoietic system, heart, and brain are affected by dengue infections by mechanisms which remain to be elucidated. Nelson and Bierman (1964) and Bierman and Nelson (1965), studying predominantly secondary-infection DHF–DSS cases, noted a marked depression of all bone marrow cellular elements during the acute stage of severe dengue illnesses. One of their cases, a 29-year-old American Peace Corps volunteer with classic dengue fever, developed a hypocellular bone marrow during the acute and the early convalescent phases of his disease. Because serological studies were not published with this case, it has not yet been established that bone marrow depression is an intrinsic phenomenon of primary dengue infections. Bone marrow depression could be related to the rapid disappearance of mature polymorphonuclear leukocytes (PMNs), a nearly universal feature of dengue infection in immunological virgins (Vedder, 1907; Siler *et al.*, 1926; Simmons *et al.*, 1931).

Hyman was the first to document T-wave changes during acute dengue fever (Hyman, 1943). Pronounced bradycardia often lasting for weeks is a common but unexplained feature of dengue (Simmons *et al.*, 1931). Psychomotor depression may occur during the convalescent stage of dengue (Rowan, 1956; Kisner and Lisansky, 1944). This was sufficiently pronounced in the 1780 Philadelphia outbreak described by Rush (1789) that he proposed for the disease the name "breakheart fever."

The maculopapular rash and its crucial timing as the *terminal* clinical event in the dengue fever syndrome have been described above. Less well known, but of great interest pathogenetically is the tuberculin-like lesion which appears within 3–5 days after subcutaneous inoculation of viremic human blood or infected mouse brain into susceptible human beings (Sabin, 1952; Schlesinger, 1971). This lesion, which is visible for 2–3 days, could not be transferred by serum (Sabin, 1952). On biopsy, infiltration of mononuclear leukocytes was observed. During the late generalized maculopapular rash in these individuals the site of the initial lesion was conspicuously spared.

2. *Dengue Hemorrhagic Fever–Dengue Shock Syndrome*

The incubation period of DHF is unknown but presumed to be that of dengue fever. In children, the progression of illness is characteristic (Nimmannitya *et al.*, 1969; Johnson *et al.*, 1967; Technical Guides, 1975). A relatively mild first phase with an abrupt onset of fever, malaise, vomiting, headache, anorexia, and cough is succeeded 2–5 days later by rapid deterioration and, sometimes, physical collapse (Fig. 2). In Thailand during 1962–1964, the median day of hospital admission after the onset of fever was day 4 (Nimmannitya *et al.*, 1969). In this second phase the patient usually manifests cold, clammy extremities, a warm trunk, a flushed face, and diaphoresis; is restless

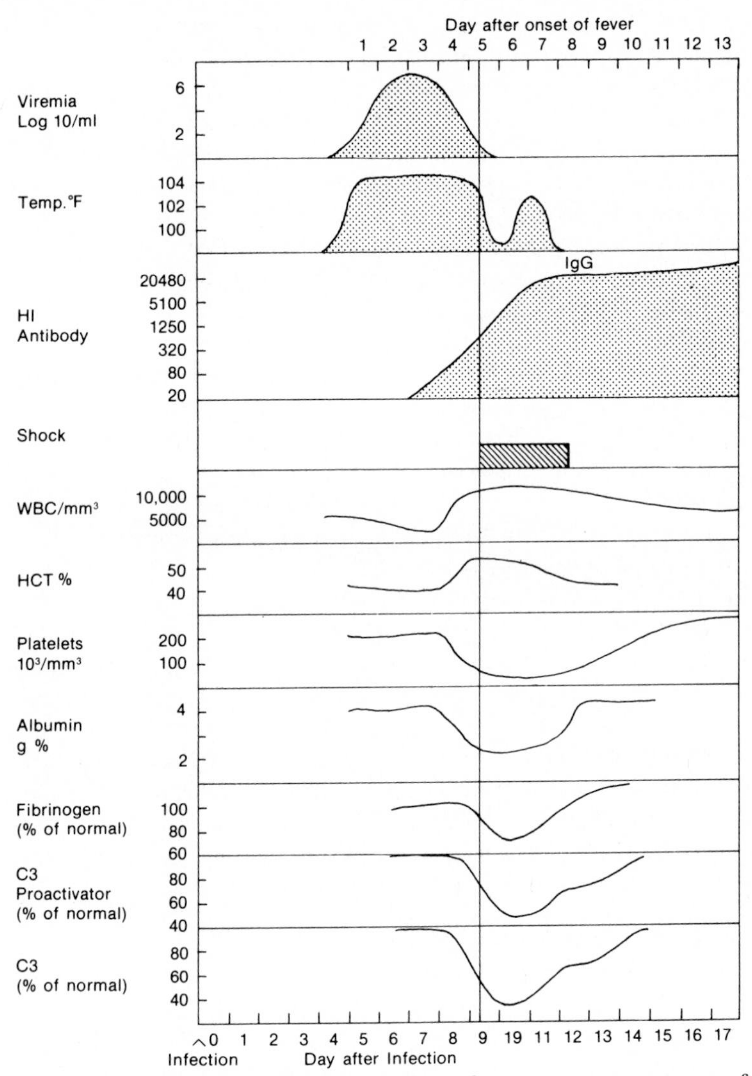

Fig. 2. Schema of clinical laboratory findings for a representative case of DSS (Nimmannitya *et al.*, 1969; Cohen and Halstead, 1966; Bokisch *et al.*, 1973).

and irritable; and complains of midepigastric pain. Frequently, there are scattered petechiae on the forehead and extremities and spontaneous ecchymoses, easy bruisability, and bleeding at sites of venipuncture. There may be circumoral and peripheral cyanosis. Respirations are rapid and often labored. The pulse is weak, rapid, and thready. The child has laboratory abnormalities as described below. If the child stabilizes at this point and does not become hypotensive or develop a narrow pulse pressure, the case is considered to be DHF without shock.

If the pulse pressure is narrow (20 mm Hg or less) or the systolic or diastolic pressure is low or unobtainable, the case is considered to be DHF with shock or DSS. Children without abnormal blood pressure may nonetheless be clinically hypovolemic and are often very ill. The significance of shock is that it identifies a child with a poor prognosis. Since there tends to be a fairly fixed ratio between severely ill children without shock and those with shock, the number of shock cases can be used epidemiologically as a means of comparing dengue epidemics and survival rates among countries and from year to year. Case fatality rates are best expressed using shock as the denominator. Death occurs in 10–40% of shock cases.

Several days after onset of fever the liver may become palpable two or three finger breadths below the costal margin and is usually firm and nontender. Approximately 10% of patients manifest gross ecchymosis or gastrointestinal bleeding. After a 24- to 36-h period of crisis, convalescence is fairly rapid in children who survive. The temperature may return to normal before or during the shock stage.

F. Laboratory Findings

Laboratory findings by severity of disease, age, and type of antibody response in Bangkok children are given in Tables I–III. The number and degree of laboratory abnormalities generally increase with the severity of illness. Elevations in SGOT and SGPT, evidence of metabolic acidosis, hypovolemia, and hypoproteinemia are common in shock cases. At least one-half of children have pleural effusions which can be detected on x ray (Nelson, 1960). The most common hematological abnormalities observed in children in clinical shock are prolonged bleeding and silicone clotting times, moderately prolonged prothrombin time with deficiencies in factors 5, 7, 9, and 10, thrombocytopenia, and mild leukocytosis (Weiss and Halstead, 1965). During shock, blood levels of C1q, C3 and C4, C5–8, and C3 proactivator

TABLE I

Selected Laboratory Findings in the Acute Febrile Stage of Dengue Virus Infection[a]

Finding	Shock, 23 cases (%)	Nonshock, 100 cases (%)	Control, 26 cases (%)[b]
Sodium <135 mEq/liter	65.5	16.0	15.4
CO_2 ≤15 mmol/liter with acid urine	62.0	18.9	11.5
SUN ≥20 mg/100 ml	62.0	22.0	11.5
SGOT ≥150 S.F. units	56.5	16.0	3.9
SGPT ≥100 S.F. units	30.5	8.1	0.0
Proteinuria ≥2+	8.6	8.0	3.8
Hemoconcentration[c]	95.5 (20)[d]	31.8 (63)[d]	11.8 (17)[d]
Total serum protein ≤5.5 g/100 ml	78.5	31.0	7.7

[a] From Johnson *et al.* (1967).
[b] Febrile patients of nondengue etiology included in study.
[c] (Highest hct − recovery hct)/recovery hct ≥ 0.2.
[d] Figure in parentheses shows number examined.

are depressed, and C3 catabolic rates are elevated (Memoranda, 1973; Bokisch *et al.*, 1973). During the acute phase of DSS, Sobel *et al.* (1975) have detected circulating immune complexes using the C1q deviation test. The blood clotting and fibrinolytic systems are mildly activated (Memoranda, 1973; Bokisch *et al.*, 1973). The degree of complement depletion and hypofibrinogenemia correlates directly

TABLE II

Hematological Findings during the Acute Phase of Illness (3–7 Days) in 27 Children Hospitalized with Serologically Confirmed Dengue Virus Infections, Bangkok, Thailand, 1962[a]

Hematological finding	Shock, 10 cases	Nonshock, 17 cases
Tourniquet test positive	5/10	8/15
Bleeding time >7 min	7/9	8/17
Silicone clotting time >45 min	7/10	7/15
Lowest platelet count		
100–200,000	1/10	2/17
50–100,000	3/10	5/17
<50,000	6/10	10/17
Prothrombin time >14.0 s	4/10	1/16
Factor II, 40–80%	4/4	1/1
Factor V, 40–80%	4/4	1/1
Factors VII and X, 40–80%	4/4	1/1
Fibrinogen 150–200 mg%	5/5	5/10

[a] From Johnson *et al.* (1967).

TABLE III

Hematological and Other Laboratory Abnormalities in Hemorrhagic Fever Patients by Etiology, Immunological Response, and Age Group[a]

Finding[b]	Not dengue, 26 cases (%)[d]	Dengue		
		Primary, >1 year, 12 cases (%)[d]	Primary, <1 year, 8 cases (%)[d]	Secondary, 103 cases (%)[d]
CRC study				
Adm. sodium <135 mEq/liter	15.4	0	50.0	26.2
Adm. CO_2 ≤15 mmol/liter	11.5	0	50.0	27.7 (101)
SGOT ≥150 S.F. units	3.9	16.6	37.5	23.3
SGPT ≥100 S.F. units	0.0	16.6	25.0	10.7
Adm. SUN >20 mg%	11.5	0	25.0	33.0
Total protein ≤5.5 g%	7.7	25.0	50.0	40.8
Hemoconcentration[c]	11.8 (17)	12.5 (8)	33.3 (6)	52.2
CH study				
Bleeding time >7 min		0 (3)	66.6 (3)	60.0 (20)
Tourniquet test positive		0 (3)	100.0 (3)	52.6 (19)
Platelet count 100,000/mm^3		33.3 (3)	100.0 (3)	95.2 (21)
Silicone clotting time >45 min		0 (3)	66.6 (3)	63.2 (19)
Prothrombin time >14.0 s		0 (8)	0 (3)	25.0 (20)
Fibrinogen <180 mg%		0 (3)	100.0 (2)	27.3 (11)

[a] Halstead *et al.* (1970).
[b] Adm., Admission.
[c] 20% higher hematocrit on admission than after recovery.
[d] Number in parentheses is number tested if different from that at head of column.

with the onset of shock and with disease severity (Figs. 3 and 4, Table IV). However, the kinin system apparently is not involved (Edelman *et al.*, 1975).

From the clinical and laboratory features described, the World Health Organization has proposed simplified diagnostic criteria for DHF–DSS (Technical Guides, 1975):

"*Dengue hemorrhagic fever* is a dengue illness characterized by fever of 2–7 days duration, hemorrhagic manifestations including at least a positive tourniquet test and, in most cases, enlargement of the liver, thrombocytopenia and hemoconcentration.

"*Dengue shock syndrome* includes the above plus narrow pulse pressure (< 20 mm Hg) or hypotension."

G. Pathology

The general experience with mouse and tissue culture isolation systems has been that dengue virus cannot be recovered from tissues of children dying of DSS. A few strains have been recovered from liver, lymph node, bone marrow, and lung (Dasaneyavaja and Pongsupat, 1961; Nisalak *et al.*, 1970), but the frequency of isolation, as illustrated in Table V is extremely low. Fluorescent antibody studies on autopsy tissues also fail to reveal dengue antigen

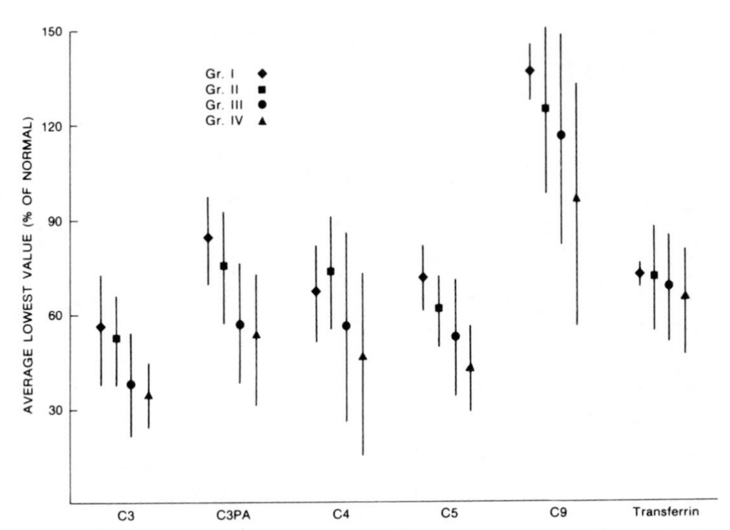

Fig. 3. Average lowest values (±1 S.D.) of various serum complement proteins and transferrin (control) from 55 patients with different grades of disease. (From Bokisch *et al.*, 1973, with permission.)

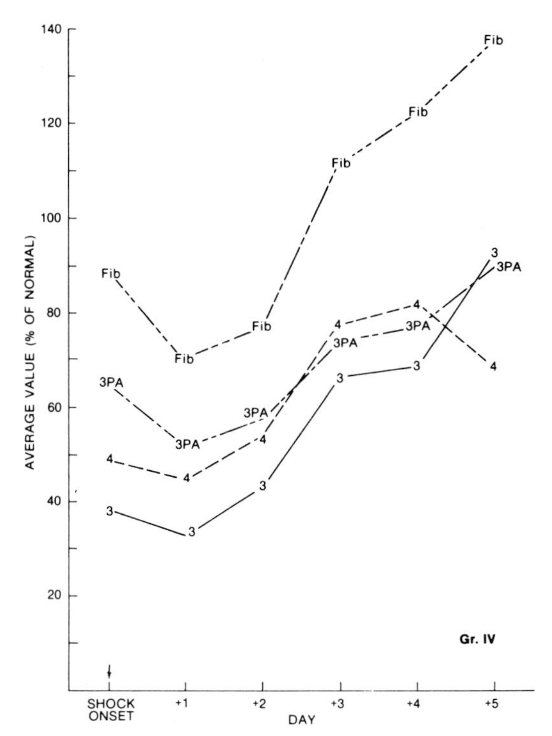

Fig. 4. Mean values of C3, C4, C3 proactivator, and fibrinogen from 13 patients with grade IV DSS. (From Memoranda, 1973.)

(Bhamarapravati and Boonyapaknavik, 1966), although reported studies are far from exhaustive. The low recovery of virus from fatal DSS could be related to the observation of Nisalak *et al.* (1970) that eluates of tissue suspensions neutralize one or more dengue viruses. The mosquito inoculation technique introduced by Rosen and Gubler (1974) is capable of recovering virus from antibody–virus complexes. Recovery of the infecting virus types from a significant number of fatal cases could provide significant insights into the virological antecedents of shock and death in dengue.

The usual observation in fatal dengue is the absence of sufficiently severe gross or microscopic lesions to explain the death of the individual. This is consistent with the study of Cohen and Halstead (1966) who observed terminal hyperkalemia in three children who died with irreversible shock, hypoxia, and metabolic acidosis. In rare instances, death may be due to gastrointestinal or intracranial hemorrhages (Bhamarapravati *et al.*, 1967). Minimal to moderate hemorrhages are

TABLE IV

Evidence for Activation of Blood Coagulation in 52 Patients[a]

		Fibrin and fibrinogen split products		Platelet count	
Grade[b]	Fibrinogen (% normal)[c]	Percentage of positive samples	Average amount (mg/100 ml)	Average	Average lowest
I	87.8	37	1.7	108,000	59,000
II	69.8	41	3.2	82,000	49,000
III	56.9	51	2.9	45,000	24,000
IV	50.0	53	5.1	60,000	20,000

[a] Bokisch *et al.* (1973).

[b] Key to grade of disease severity (Nimmannitya *et al.*, 1969): Grade I—Fever accompanied by nonspecific constitutional symptoms. The only hemorrhagic manifestation is a positive tourniquet test. Grade II—Fever and skin hemorrhage or other bleeding such as epistaxis or gum bleeding. Grade III—Circulatory failure manifested by rapid, weak pulse with narrowing of pulse pressure (<20 mm Hg) or hypotension. Grade IV—Moribund, with undetectable blood pressure and pulse.

[c] Normal level (average of five Thai controls) = 172 mg/100 ml.

usual in the upper gastrointestinal tract, and petechial hemorrhages are frequent in the intraventricular septum of the heart, on the pericardium, and on the subserosal surfaces of major viscera. Focal hemorrhages are occasionally seen in the lungs, liver, adrenals, and subarachnoid spaces. The liver is usually enlarged, often with fatty changes. Yellow, watery, at times blood-tinged, effusions are present in serous cavities in about three-fourths of patients. Retroperitoneal tissues are markedly edematous. Microscopically, there is perivascular edema in soft tissues and widespread diapedesis of red blood cells. Bone marrow megakaryocytes show maturation arrest, and increased numbers of them are seen in lung capillaries, renal glomeruli, and the sinusoids of the liver and spleen. Proliferation of lymphocytoid and plasmacytoid cells, lymphocytolysis and lymphophagocytosis, and germinal center necrosis occur in the spleen and lymph nodes. There is hyalin necrosis of Kupffer cells and reticulum cells in the spleen. In the liver there are varying degrees of fatty metamorphosis, focal midzonal necrosis, and hyperplasia of the Kupffer cells. In the sinusoids are seen nonnucleated cells with vacuolated acidophilic cytoplasm, resembling Councilman bodies. There is a mild, proliferative glomerulonephritis presumably due to the deposition of complexes of dengue antigen, antibody, and complement (S. Boonpucknavig *et al.*,

TABLE V

Specimens from Patients Dying of Clinical Hemorrhagic Fever and Tested for Dengue Virus, Bangkok, 1962–1964[a]

Specimen	1962	1963	1964	Total
Heart blood, serum plasma	41[b]	2	60[c]	103
Liver	15	20	37	72
Liver biopsy			20	20
Spleen	14	19	32	65
Kidney	13	15	31	59
Heart	8	13	32	53
Lung	6	12	30	48
Brain	12	4	24	40
Adrenal	2		22	24
Pancreas		3	3	6
Stomach		1	6	7
Small intestine			19	19
Thymus		3	15	18
Lymph node			22[d]	22
Bone marrow			21[e]	21
Thyroid			13	13
Muscle			21	21
Skin			15	15

[a] Nisalak *et al.* (1970).
[b] One dengue-1 virus recovered.
[c] Three dengue-2 viruses recovered; one untyped dengue virus recovered.
[d] One dengue-2 virus recovered.
[e] One dengue-2 virus recovered in a patient with dengue-2 virus recovered in heart blood.

1976). Biopsies of the skin rash reveal swelling and minimal necrosis of endothelial cells and subcutaneous deposits of fibrinogen. Dengue antigen is found in extravascular mononuclear cells (S. Boonpucknavig *et al.*, 1979). Dengue virionlike particles have been visualized in a macrophage in a glomerulus (S. Boonpucknavig *et al.*, 1976).

H. Immunopathology of DHF–DSS: The Evidence

From the first recognition that dengue could be lethal, explanations have abounded. Many hypotheses have focused on changes in the virulence of the virus (Hammon *et al.*, 1960b; Hammon, 1969, 1973; Rosen, 1977). As a possible selection mechanism, Rudnick (1967) proposed that virus virulence might vary with passage in different arthropod species. Pavri (1976) has made the novel suggestion that DHF may be the result of synergism between dengue infection and infestation with helminths.

The so-called "two-infection" hypothesis of DHF–DSS was proposed by Halstead *et al.* (1967). These workers found secondary-type dengue antibody responses more frequently associated with DSS than with less severe dengue syndromes. This observation was promptly confirmed by two prospective studies in Thailand (Winter *et al.*, 1968; Russell, *et al.*, 1968; Winter *et al.*, 1969). It became necessary to modify the two-infection hypothesis to what was termed the "immunopathological" hypothesis when it was established that infants less than 1 year old could develop DSS during a primary infection (Halstead *et al.*, 1970). Experimental studies have elucidated a single mechanism which now explains "primary- and secondary-infection DHF–DSS." The evidence for this is more fully developed in Section III.

It is well established that DSS has not been detected in every population exposed to multiple dengue types either endemically or sequentially. This enigma is more fully documented and discussed in Sections II,I, 4 and 5. To explain the association of DSS with some but not all secondary infections, there may be crucial variables involving sequential infections which control the severity of disease. These are more fully discussed in Section III.

At this point, it is necessary to review the considerable data associating immune status and disease severity in Southeast Asia.

1. Severity of Disease and Immune Status

Prospective studies of DHF–DSS outbreaks in Bangkok in 1962–1964 and in Koh Samui in 1966 and 1967 provide the data base upon which the immunopathological hypothesis has been constructed (Halstead *et al.*, 1967, 1969b; Winter *et al.*, 1968, 1969; Russell *et al.*, 1968; Halstead and Simasthien, 1970). Two of these studies included DHF or DSS cases within a prospectively studied population (Russell *et al.*, 1968; Halstead *et al.*, 1969b).

The Bangkok study on 19 randomly selected census tracts included 44,408 persons (Halstead *et al.*, 1969b). Each census tract was normalized to approximately 360 households. All family units were identified, and each resident in the study area assigned a number. Areas were visited monthly to ascertain the occurrence of DHF cases. From a 10% random sample of families, 4150 preepidemic blood samples were collected. That the study population was demographically representative of the entire population of Bangkok was extensively documented (Halstead *et al.*, 1969b). After the 1962 rainy season, 1878 previously sampled individuals were rebled; 1253 were children or young adults, aged less than 1–19 years. In this group, there were 205 children who converted from no dengue-1 hemagglutination inhibition (HI) antibody before the epidemic to detectable HI antibody in postepidemic sera. No hospitalizations for DHF–DSS occurred in this group.

Secondary dengue infections were not measured but can be estimated. In the sample, in preepidemic sera, 655 children had dengue HI antibody. During the epidemic, the nonnormalized dengue infection rate in susceptibles was 34.3%. Multiplying this figure by 655 gives a total of 224 *possible* secondary dengue infections. This number considerably overestimates secondary infections.

Some children (an unknown fraction) will be immune to three or four virus types, hence not susceptible to clinically apparent infection. If chance determines the bites of infected mosquitoes, some children immune to a single virus will be bitten by a mosquito carrying that type and will not become infected. In 1962, dengue types 1–3 were the predominant viruses causing human infection. These same types had been recovered in 1960 (Dasaneyavaja and Charansiri, 1961; Hammon and Sather, 1964). In the years 1962–1964, dengue types 1–3 each caused about one-third of all dengue infections (Halstead *et al.*, 1970). If this is assumed to have been a stable feature of dengue transmission in Bangkok, then, in 1962, of 224 immune children bitten by a dengue-infected mosquito, one-third (74) would have been homotypically immune and 150 might have been secondarily infected. In this group, 5 children were hospitalized with a clinical diagnosis of DHF (the incidence of shock in these cases is not known). Thus, there was 1 DHF case for 30 secondary dengue infections, or an attack rate of 33.3 per 1000 secondary dengue infections.

In the 1966 outbreak in Ang Tong, 336 children, aged 2–12 years, were bled in February 1966 and again in February 1967 (Winter *et al.*, 1968; Russell *et al.*, 1968). In the interim these children were observed clinically. Shock syndrome cases among primary and secondary dengue infections were 0/26 and 3/83, respectively, or an attack rate of 36.1 DDS per 1000 secondary dengue infections.

As these two studies illustrate, DHF–DSS comprised about 3% of secondary dengue infections. In the event that a large proportion of a population under study was solidly immune or if dengue infection rates were relatively low, a very large population would have to be bled before and after a dengue outbreak to ensure inclusion of DHF–DSS cases. Time, personnel, and financial constraints make such studies impractical. A more realistic approach is to measure primary and secondary dengue infection rates in a random sample of age groups at risk and to compare these rates with those observed in hospitalized children.

Extrapolating from their serological sample to all children aged <1–15 in Koh Samui, Russell *et al.* (1968) estimated a ratio between shock and total secondary dengue infection of 1:180, or 5.5 per 1000 secondary dengue infections. No shock cases occurred in primarily infected children.

The attack rates of DHF and DSS in primary and secondary dengue infections can be estimated for children 1 year and older for the entire city of Bangkok in 1962. The data and calculations are given in the Appendix (Section V). Our data suggest that there were 1.8 DHF and 0.07 DSS cases per 1000 primary dengue infections and 20.1 DHF and 11.4 DSS cases per 1000 secondary dengue infections. From these calculations it is estimated that shock occurs 160 times more frequently during secondary than during primary dengue infections (excluding infants less than 1 year of age). Thus, four independent estimates of the rate of DHF and DSS per 1000 secondary dengue infections are within two- to threefold of the median value.

The expected proportion of primary and secondary clinically overt dengue can also be derived by measuring the proportion of primary and secondary dengue infections in the whole population. In the 1966 Ang Tong study, when seroconversion data were weighted to be representative of the age group <1–15 years, Russell *et al.* (1968) estimated that secondary infections comprised 58.7% of total dengue infections. If illness severity is independent of antibody response, this should be the proportion of secondary dengue antibody responses among all children hospitalized for dengue. As shown in Table VI, the observed distributions of antibody response by disease syndrome differ significantly from the "expected." Note the parallel increase in disease severity and the proportion of secondary-infection responses.

The full history of dengue virus transmission in a study population may not be known. When this is the case, it may be difficult to make an accurate estimate of the number of secondary infections by studying pre- and postepidemic sera. For example, HI antibody responses following secondary-type infection decline rapidly to preinfection levels, possibly obscuring evidence of a fourfold rise when sera are collected at a long interval (Carey *et al.*, 1965). A crude solution to this problem is to use data from an age-stratified serological survey made prior to an epidemic. The proportion of persons with dengue antibody is taken as the expected proportion of secondary dengue infections.

TABLE VI

Observed and Expected Distributions of Primary and Secondary Dengue Infections by Clinical Syndrome, Using Weighted Estimate of Proportion of Secondary Dengue Infections in Koh Samui Outbreak, 1966[a]

	Number of cases by antibody response[c]				
	Primary		Secondary		
Syndrome[b]	Obs.	Exp.	Obs. (%)	Exp. (%)	Total
UF	7	16	32 (82)[d]	23 (59)	39
DF	4	11	23 (85)	16 (59)	27
HF	2	8	18 (90)	12 (60)	20
SS	0	6	14 (100)	8 (57)	14

[a] Analysis of data from Russell *et al.* (1968).

[b] UF, Undifferentiated fever; DF, fever, petecheal or macular rash; HF, fever, positive tourniquet test, plus one or more of: extensive petechiae, purpura, ecchymosis, epistaxis, hematemesis, hematuria, melena; SS, fever with pulse pressure <20 mm Hg or systolic pressure <90 mm Hg. (From Winter *et al.*, 1968.)

[c] Observed versus expected for all observations = $\chi^2_{[7]} = 34.1 = p < .001$.

[d] Numbers in parentheses indicate percent of total.

This method of estimating fractional secondary infection will be quite accurate if the population is immune to a single dengue virus and a heterologous virus is introduced. When multiple types are endemic, this method will *overestimate* the number of secondary dengue infections. Nonetheless, when this estimate is used, the distribution of secondary antibody in children with severe dengue syndromes differs significantly from the expected values for the 1962 Bangkok and the 1966 and 1967 Koh Samui outbreaks (Table VII).

Russell *et al.* (1968) noted that attack rates of mildly or moderately ill children were higher among children with secondary than with primary infections in the same outbreak. Similarly, in the Bangkok Children's Hospital study the proportion of secondary dengue infections among outpatients is higher than would be expected from the infection experience in the population as a whole (Table VIII). The proportion of secondary dengue infections increases dramatically with the severity of the illness. When primary dengue infections in infants less than 1 year old are removed, 99.9% of DSS cases in the Children's Hospital–Clinical Research Center series are found to have involved a secondary dengue infection.

There is much collateral evidence, all showing a very strong association between secondary-type responses and DSS. As noted by

TABLE VII

Observed Distributions of Secondary Dengue Infections in Hemorrhagic Fever Patients (Aged 1–15 Years) with and without Shock Compared with Expected Values Calculated from the Distribution of Antibody in Preepidemic Serological Samples

	Observed and expected secondary infections by outbreak					
	Bangkok, 1962 (Nimmannitya *et al.*, 1969)[a]		Koh Samui, 1966 (Russell *et al.*, 1968)		Koh Samui, 1967 (Winter *et al.*, 1969)	
Syndrome	Obs. (%)[b]	Exp. (%)	Obs. (%)[c]	Exp. (%)	Obs. (%)[d]	Exp. (%)
DHF (grades I and II)	46 (71.8)[e]	32 (50)	18 (90)	16 (80)	10 (100)	7 (70)
DSS (grades III and IV)	55 (96.4)	29 (50.8)	14 (100)	11 (78)	5 (100)	4 (80)
Preepidemic antibody prevalence	1127/2230 = 50.5% (Halstead *et al.*, 1969c)		515/646 = 79.7% (Winter *et al.*, 1968)		203/278 = 73.0% (Winter *et al.*, 1969)	

[a] Of 59 shock cases studied, two either had no evidence of dengue infection or had specimens inadequate for serological diagnosis; of 57 DSS cases, two had primary-type infections (S. B. Halstead, unpublished data).

[b] Observed (DHF plus DSS) differs from expected; $p < .001$.

[c] Observed (DHF plus DSS) differs from expected; $p < .02$.

[d] Observed (DHF plus DSS) differs from expected; $p < .02$.

[e] Numbers in parentheses are percent of total.

TABLE VIII

Distribution of Primary and Secondary Dengue Infections by Clinical Syndrome, Children's Hospital, CRC Studies, Bangkok, 1962–1964

	Number of cases by antibody response			
Clinical syndrome	Primary	Secondary	Total[a]	Ref.[b]
FUO (outpatient)	33	61 (64.9%)	94 } A	(1)
FUO (inpatient)	13	23 (63.9%)	36 } A	(2)
DHF, grades I and II	65	262 (80.1%)	327 B	(3)
DHF, grades III and IV	6	190 (96.9%)	196 C	(3)

[a] Chi square tests: A versus B: $\chi^2_{[1]} = 12.16$; $p < .001$. A versus C: $\chi^2_{[1]} = 60.89$; $p < .001$. B versus C: $\chi^2_{[1]} = 29.54$; $p < .001$.

[b] Key to references: (1) Halstead *et al.* (1969a); (2) Nimmannitya *et al.* (1969); (3) Halstead *et al.* (1970).

Fischer and Halstead (1970), in Bangkok, where DHF–DSS had been continuously endemic since 1950, an association between severe disease and secondary infections may be surmised from an analysis of the age-specific hospitalization rates. It has been demonstrated that the largest proportion of susceptibles (no antibody) was found in children aged 12–24 months (Halstead *et al.*, 1969c). If it is assumed that dengue viruses are inherently virulent, it follows that the largest number of DSS cases should occur in this age group. Further, of all primary DSS cases, the largest percentage should occur in this age group. The next largest group should be 2-year-olds, etc. In fact, 1-year-olds have the *lowest* rate of hospitalization for DHF–DSS, and no primary infection DSS has yet been reported in 1-year-olds.

2. *Dengue Hemorrhagic Fever–Dengue Shock Syndrome in Infants*

For many years it was recorded that the age distribution of children hospitalized for "Thai hemorrhagic fever" was bimodal (Jatanasen *et al.*, 1961; Halstead *et al.*, 1969c). One peak occurred among infants less than 1 year old, and the other, broader age group peaked at 3 or 4 years old. As shown in Fig. 5, few infants were hospitalized during the first 3 months of life; attack rates then rose steeply, peaking at age 7 or 8 months and then dropping to near the baseline during the eleventh month. The age-specific DHF hospitalization rate for 8-month-old infants was almost twice that for 3-year-old children. Information on this paradox was provided when 30 infants in this age group with serological dengue were included in the 1962–1964 DHF–DSS pros-

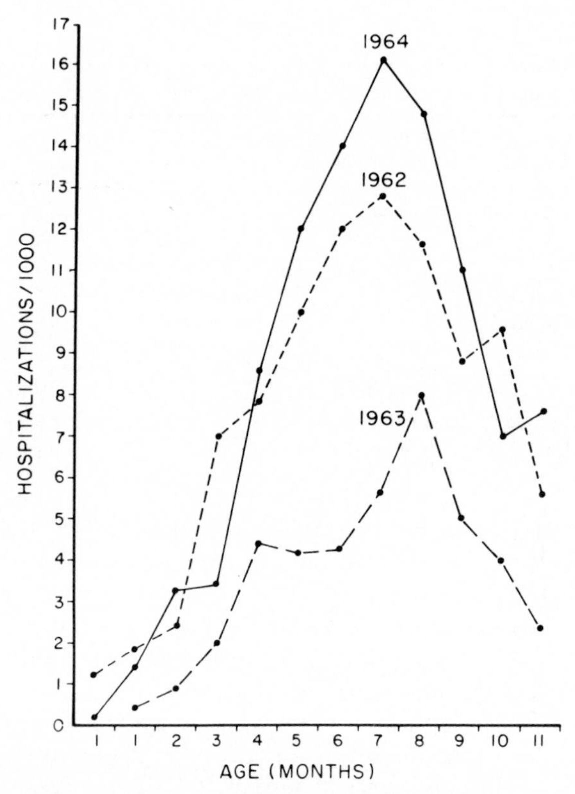

Fig. 5. Monthly age-specific hospitalizations for hemorrhagic fever in 1962–1964 in residents of Bangkok and Thonburi, Thailand, who were less than 1 year old. (From Halstead *et al.*, 1969c.)

pective study. It was found that 28 had a primary-type antibody response. Primary-infection DSS has been carefully studied in a few infants; complement activation patterns (Memoranda, 1973) and the physiological and hematological responses closely resemble those in children with secondary dengue but differ from responses in older children with primary dengue infection or infections of unknown etiology (Tables I–III). Reported studies on mothers of infants with severe primary dengue infections are not numerous, but those studied were dengue-immune (Memoranda, 1973). It must be surmised that this is the case for nearly all infants in Bangkok, because of the high prevalence of dengue antibody in young adults (Halstead *et al.*, 1969c). The pathogenetic significance of passively acquired antibody is discussed in Section III,A,1.

I. Other Factors Affecting the Severity of Dengue Disease

1. Sex

Piyaratn (1961) and Burke (1968) reported an unusually high prevalence of females among DSS autopsies. A predominance of females was also noted among children hospitalized with DHF and DSS in the Koh Samui study (Winter *et al.*, 1968). The relationship between disease severity and sex was carefully studied by Halstead *et al.* (1970). Surprisingly, sex differences were age-dependent. As shown in Table IX, more males than females were hospitalized with primary dengue infections at all ages. For children aged 1–3 years with secondary dengue infections sex ratios were nearly equal. However, from the fourth year of life, a dramatic increase in the hospitalization of girls can be noted. In this age group, twice as many girls had shock during a secondary dengue infection as boys.

2. Nutritional Status

In the Bangkok area study an attempt was made to measure DHF–DSS attack rates in different socioeconomic groups (Halstead *et al.*, 1969c). One of the purposes of this study was to assess the possible effect of nutritional status on the risk of developing DHF. There were no significant differences in hospitalization rates corrected for infection. During 1962–1964, it was the general observation that children

TABLE IX

Etiology, Immune Response, Disease Severity, and Sex in Dengue Hemorrhagic Fever and Controls, Bangkok, 1962–1964[a]

Category	Male	Female	Total
Control			
Chikungunya	18	11	29
No diagnosis	42	34	76
Dengue			
Primary, <1 year	15	13	28
Primary, ≥1 year	25	18	43
Secondary, nonshock, <1–3 years	39	35	74
Secondary, shock, <1–3 years	24	29	53
Secondary, nonshock, ≥4 years	84	104	188
Secondary, shock, ≥4 years	45	92	137
Total	292	336	628

[a] Halstead *et al.* (1970).

in Bangkok were relatively well nourished regardless of socioeconomic status. During the period 1964–1970, independent and unsolicited personal communications from pediatricians in the Philippines, South Vietnam, Thailand, Indonesia, Burma, and Sri Lanka revealed a common experience that children with severe or fatal DSS were almost always well nourished. Obversely, DHF–DSS is rare in children with clinically apparent protein–calorie malnutrition. A well-designed cohort study on Burmese children has shown that DSS patients are somewhat better nourished than controls (U Tin U, personal communication). These observations have considerable theoretical importance, suggesting a sparing effect on the shock syndrome of a nutritionally induced immunosuppression. The relationship between nutrition and the severity of dengue disease should be carefully studied for its pathogenetic as well as therapeutic implications.

3. *Interval between First and Second Dengue Infections*

A high frequency of DHF/DSS occurs during secondary dengue infections; the interval between infections must be examined as a variable affecting the severity of illness. In adult volunteers Sabin (1952) showed that dengue-1 and -2 viruses produced protection against heterologous challenge up to 3 months after infection. In the 1962–1964 Bangkok study, a few 5- to 7-month-old infants were hospitalized with secondary-infection DHF (Halstead *et al.*, 1970). These data suggest that sequential infections at an interval of $\frac{1}{2}$ year or more may be associated with enhanced disease response.

The possibility that two dengue infections acquired at an interval of greater than 5 years may *not* result in DSS was predicted from a mathematical model of dengue infections in Bangkok (Fischer and Halstead, 1970). The significant observation was that hospitalizations for DSS virtually disappeared by age 15. If a sequential infection at any time interval could result in DSS, it was predicted that cases would occur over a broader age range. At realistic average annual dengue infection rates, second dengue infections would be expected to occur in some individuals well into the third decade of life.

4. *Sequence of Infection*

In the 1962–1964 Bangkok study, Halstead *et al.* (1970) recovered dengue type 2 proportionately more frequently from secondarily infected DHF–DSS cases than from wild-caught mosquitoes or primarily infected persons. These data are shown in Tables X and XI. One explanation for this high rate of recovery of dengue-2 from DSS is that

TABLE X

Identification of 108 Dengue Viruses Recovered from Mosquitoes and Patients with Primary Dengue Infections, Bangkok, 1962–1964

Group	Dengue virus type					Ref.[a]
	1	2	3	4	Unidentified	
Nonhospitalized patients[b]	4	5	6	1	2	(1)
Hospitalized patients						
No shock[c]	13	14	10		1	(2)
Shock	3					(2)
Caucasian patients	8	9	9			(3)
Mosquitoes	6	8	8	2		(4)
Total	34	36	33	3	3	

[a] Key to references: (1) Halstead *et al.* (1969a); (2) Nimmannitya *et al.* (1969); (3) Halstead *et al.* (1969b); (4) Halstead *et al.* (1969c).

[b] Children with virologically confirmed dengue infection studied in Children's Hospital Out-Patient Department, Bangkok, 1962–1964.

[c] Children with virologically confirmed dengue infection admitted to Children's Hospital with diagnosis of hemorrhagic fever and other syndromes, Bangkok, 1962–1964.

sequences of infection, dengue-1 and then -2, dengue-3 and then -2 and dengue-4 and then -2, might be more virulent than infections acquired in other sequences. Dengue type 2 has been recovered from patients in nearly every studied outbreak of DHF–DSS (Halstead, 1966). The 1967 Koh Samui outbreak appears to be an exception, since only dengue-4 virus was recovered (Winter *et al.*, 1969). However, dengue type 2 had been present on the island a year earlier. The limited scope of the 1967 study does not preclude the occurrence of dengue-2 infections in a small proportion of patients.

It may be pertinent that enhanced viremias observed in sequentially infected rhesus monkeys were found only in sequences in which type 2 was the challenge virus (Halstead *et al.*, 1973d) (see Section II,K,1).

Unfortunately, comparative long-term prospective studies on dengue virus ecology in DHF endemic and nonendemic areas which might have produced data pertinent to this hypothesis have not been conducted. During 14 years of continuous research on dengue in Vellore, Carey and Myers recovered all four dengue viruses from patients (Carey *et al.*, 1966; Myers *et al.*, 1970). Their studies were largely confined to dengue-susceptible adult hospital personnel residing on the compound of the Christian Medical College. While surveillance

TABLE XI

Identification of 84 Dengue Viruses Recovered from Patients with Secondary Antibody Response by Disease Severity, Bangkok, 1962–1964

	Dengue virus type					
Patient group	1	2	3	4	Unidentified	Ref.[a]
Nonhospitalized[b]	3	11	5			(1)
Hospitalized						
No shock[c]	5	22	14	1	1	(2)
Shock	2	15	3	1	1	(2)
Total	10	48	22	2	2	

[a] Key to references: (1) Halstead *et al.* (1969a); (2) Nimmannitya *et al.* (1969).
[b] Children with virologically confirmed dengue infections studied at the Children's Hospital Out-Patient Department, Bangkok, Thailand, 1962–1964.
[c] Children with virologically confirmed dengue infections admitted to Children's Hospital with diagnoses of hemorrhagic fever or other syndromes, Bangkok, 1962–1964.

among the indigenous population of Vellore uncovered sporadic DSS cases (Myers and Varkey, 1970), infection rates by virus type in indigenous children and hospitalization rates for dengue infections were not established. Without this information not enough is known about primary and secondary dengue infection rates in southern India, the sequences of virus infections to which indigenous children were exposed, or the interval between dengue infections to make comparisons with the Bangkok data or to attempt to explain the rarity of DSS in India. Even less is known about the epidemiology of dengue infections in Nigeria, where type-1 and -2 viruses caused human infections during the period 1967–1970 (Moore *et al.*, 1975).

Table XII summarizes a number of recent episodes of simultaneous or sequential transmission of two or more dengue virus types which were *not* accompanied by epidemic DHF–DSS. Dengue-2 transmission followed a dengue-1 or -3 epidemic by an interval of 5 years or more in Puerto Rico, Jamaica, Samoa, and Tonga. Only in Tahiti in 1971–1973 was dengue-2 introduced at an interval of less than 5 years, following dengue-3 in 1969. Since the 1969 outbreak was small (Saugrain *et al.*, 1970), it must be surmised that relatively few secondary dengue-2 infections occurred in recent dengue-3-immunes. Dengue-1, which followed dengue-2 by 2–4 years in Fiji, quite clearly did *not* cause epidemic DSS (Kuberski *et al.*, 1977). Regretfully, only rarely have primary- and secondary-infection rates been measured, preventing the assimilation of these data into pathogenetic analyses.

TABLE XII

Simultaneous or Sequential Epidemics Involving Two or More Dengue Virus Types but Not Associated with Epidemic Dengue Shock Syndrome[a]

Locale	Date	Virus	Disease	Ref.[b]
Puerto Rico	1963	D3	D	1
	1969–1977	D2	D	2
	1977	D1, D2, D3	D	3
Colombia	1971–1972	D2	D	4
	1975–1977	D3	D	5
Dominican Republic	1972–1973	D2, D3		6
Jamaica	1963	D3?	D	7
	1968–1979	D2, D3	D	7
	1977	D1	D	3
Panama	1941–1954?	D2, D3		8
Tahiti	1963–1969	D3	D	9
	1971–1973	D2	D, H, S?	10
	1975	D1	D, H	11
	1979	D4	D	12
Fiji	1971–1973	D2	D	11
	1974–1975	D1	D, H	11,13
Samoa	1940s	D1	D	14
	1972	D2	D	11
Tonga	1930	D1	D	15
	1972–1974	D2	D, H	15
	1974–1975	D1	D, H, S?	15
Nigeria	1964–1977	D2	D	16
	1967–1970	D1, D2	D	16

[a] Data are representative, not exhaustive. D1, Dengue-1; D2, dengue-2; D3, dengue-3; D, dengue fever; H, hemorrhagic phenomena described—DHF not established; S?, small number of deaths described—DSS not established.

[b] Key to references: (1) Neff *et al.* (1967); (2) Likosky *et al.* (1973); (3) J. H. Woodall (personal communication); (4) Morales *et al.* (1973); (5) Groot (1976); (6) Ventura *et al.* (1975); (7) Ventura and Hewitt (1970); (8) Rosen (1974); (9) Saugrain *et al.* (1970); (10) Moreau *et al.* (1973); (11) Miles (1978); (12) Laigret *et al.* (1979); (13) Kuberski *et al.* (1977); (14) Rosen (1967); (15) Gubler *et al.* (1978); (16) Moore *et al.* (1975).

5. *Inherent Virus Virulence*

It has been proposed that "virulent" serotypes 3–6, recovered from the Philippine and Thai outbreaks of 1956, 1958, and 1960, were responsible for DHF–DSS (Hammon *et al.*, 1960b). This hypothesis was weakened when it was noted that persons with short-term residence in areas with DHF developed dengue fever but not hemorrhagic fever (Halstead and Yamarat, 1965). Subsequently, it was agreed that proposed types 5 and 6 did not differ antigenically from types 2 and 1, respectively. Since 1954, type-3 dengue, or its biological subtype, is known to have been associated with primary-infection dengue fever in Australia, southern India, Pakistan, Tahiti, and the Caribbean without causing DHF–DSS (Doherty *et al.*, 1967; Myers *et al.*, 1970; Aziz *et al.*, 1967; Saugrain *et al.*, 1970; Neff *et al.*, 1967). Dengue type 4 is known to have caused classic dengue fever without DHF in southern India (Carey *et al.*, 1966) and in the 1922 experimental studies of Siler *et al.* in the Philippines (Halstead, 1974).

However, the issue of the inherent virulence of dengue virus is far from resolved. Barnes and Rosen (1974), Rosen (1977), Kuberski *et al.* (1977), and Gubler *et al.* (1978) have suggested that dengue-1 and -2 viruses acquire enhanced virulence properties, producing DSS and death. To test the hypothesis that virulent dengue virus strains cause DSS it is necessary to establish that the clinical response studied in fact is DSS and that the frequency of occurrence of the syndrome approaches the attack rates for children in Southeast Asia. Studies on South Pacific outbreaks fail to satisfy either of these requirements.

On Niue Island (Barnes and Rosen, 1974), "shock" and deaths occurred in an epidemiological setting which suggested that dengue type-2 virus had been introduced into a population in which individuals younger than 25 were without prior flavivirus experience. Serological studies on severe cases were undertaken 8 months after the end of the epidemic. Many of the 23 described cases had signs and symptoms not usually seen in DSS in Southeast Asia, such as extensive ecchymoses, ulcerative skin lesions, hematuria, pleocytosis, and marked leukocytosis. If dengue-2 had acquired the pathogenetic properties seen in Southeast Asia, attack rates of shock cases should have been higher than reported. Barnes and Rosen (1974) estimated 4140 dengue type-2 infections in the Niue outbreak. In Thailand, DSS comprises 1–3% of secondary dengue infections. If these data are applied to Niue, 40–120 shock cases would have been expected; 14 were reported. Finally, much emphasis was given by the authors to the high frequency of hemorrhagic phenomena. The history of hemorrhage, including gastrointestinal hemorrhage, was elicited retrospectively by household interview of nonhospitalized denguelike cases. In Thailand, gastrointestinal hemorrhage is usually observed after the onset of shock, implies a grave prognosis, and is extremely rare in ambulatory patients (Nimmannitya *et al.*, 1969). Hemorrhage, regardless of site and severity, when not accompanied by thrombocytopenia and hypovolemia, does not satisfy diagnostic criteria for the DHF syndrome (see Section II,F).

Gubler *et al.* (1978) studied sequential dengue type-2 and -1 outbreaks on Tongatapu in 1974 and 1975, respectively. In each year a single strain of dengue virus was isolated from young adult women with severe illness. These cases were labeled DSS, although none satisfied acceptable diagnostic criteria. One patient had gastrointestinal hemorrhage without shock, hemoconcentration, or thrombocytopenia; and another had a fatal diarrhea syndrome, apparently without hemoconcentration. In the 1975 outbreak involving thousands of persons, only 11 paired sera were obtained, 5 of which had secondary-type antibody responses. Despite the lack of seroepidemiological studies, the authors concluded that "differences in (severity) of dengue 2 and 1 outbreaks could not be attributed to . . . prior immune status" and that "viral virulence was the most likely explanation."

A majority of severe illnesses and deaths in Pacific island outbreaks have been described in adults (Barnes and Rosen, 1974; Kuberski *et al.*, 1977; Gubler *et al.*, 1978). While these may represent new dengue pathophysiological syndromes distinct from DHF–DSS in children, in the absence of careful clinical studies it is equally possible that many of these patients had nondengue etiological contributions to their disease. For further discussion see Halstead (1978).

Scott *et al.* (1976), Rosen (1977), and Gubler *et al.* (1978) have claimed that individuals dying of DSS from whom a dengue virus is isolated but who do not have detectable dengue antibody at the time of death have had a primary-type infection. This seeming tautology is not necessarily correct. Halstead *et al.* (1970) showed that 8% of patients who survived DHF–DSS and had a secondary-type antibody response to dengue antigens had *no detectable* dengue-1 HI antibody in their acute-phase specimen. Nisalak *et al.* (1970) found that 3 of 101 patients dying of DSS-like disease, aged 1–21 years, were without dengue antibody detectable either in heart blood or spleen suspension. It is reasonable to assume that some of these individuals may have died during the negative phase of their secondary antibody response. A negative-phase antibody response has been observed in a rhesus monkey immune to dengue-4 and challenged with dengue-2 3 months later (Halstead *et al.*, 1973d). This infection is instructive, and its significant serological features are reproduced in Table XIII. Note that during the viremic period the broadly reactive secondary-response antibody which had appeared on day 2 completely disap-

TABLE XIII

Hemagglutination Inhibition Antibody Responses in Monkey 13177 to Sequential Infection with Dengue-4 and -2 Viruses[a,b]

	Reciprocal HI antibody titer[c]																	
	Days after first infection with dengue-4						Days after second infection with dengue-2											
Antigen	0	8	12	21	42	89	0	1	2	3	4	5	7	8	9	11	21	42
d1	0	0	0	0	0	0	0	0	0	40	0	0	0	10	80	320	320	40
d2	0	0	0	0	0	0	0	0	0	320	0	0	10	80	320	>1280	>1280	640
d3	0	0	0	0	0	0	0	0	0	40	0	0	0	20	80	320	320	40
d4	0	0	0	20	40	40	40	40	40	160	40	40	40	80	640	>1280	>1280	320

[a] Halstead *et al.* (1973d).

[b] The monkey had viremia only during secondary infection with dengue-2 virus. The amounts were: day 2 after inoculation, 13 PFU/ml; day 4, 15,000 PFU/ml; day 5, 11,000 PFU/ml.

[c] 0 indicates <1:10.

peared, leaving only the monospecific dengue-4 HI antibody. If a dengue-1 hemagglutination antigen had been used to test acute-phase sera, the animal would have been seronegative. If a dengue-4 HI or neutralization test had been performed and the animal studied on days 4–7 after infection, it would have *appeared* to be in the early phase of a primary dengue infection. The matter could be resolved by separating Ig's. In future studies on fatal putative primary dengue infections it will be important to authenticate cases with greater care. In this connection, a phenomenon described by Marchette *et al.* (1975) may prove useful. These workers noted that leukocytes obtained during the acute phase of secondary dengue infections produced large amounts of antidengue IgG *in vitro*. This was unique to secondary dengue. Similar diagnostic cultures could be established from patients with a fatal outcome.

Despite the lack of convincing evidence for differences in inherent virulence as a cause of DSS, there is some evidence that secondary dengue type-2 infections are more virulent than primary dengue-2 (see Section II,I,4). The relationship between the biological attributes and antigenic structure of dengue viruses and the outcome of human infection is discussed below (see Section III,A,1).

J. Pathogenesis of Dengue Infection in Man and Monkeys

1. *Man*

An understanding of pathogenesis usually begins with the identification of specific organs or cells supporting viral replication *in vivo*. In contrast to many acute fatal human viral infections in which virus can readily be recovered from tissues or viral antigen localized to individual cells, in the usual case of fatal dengue neither virus nor viral antigen can be detected (Bhamarapravati and Boonyapaknavik, 1966; Nisalak *et al.*, 1970). A summary of extensive experience with suckling mouse brain and tissue culture isolation systems is presented in Table V. In addition to the dengue-2 viruses recovered from bone marrow, lymph node, and lung, dengue-4 isolation has been reported from liver (Dasaneyavaja and Pongsupat, 1961). The rate of virus isolation from the blood is inversely related to disease severity (Table XIV). Dengue viral antigen has been visualized by fluorescence microscopy on the surface of human B lymphocytes during the acute stage of DHF–DSS (V. Boonpucknavig *et al.*, 1976). This observation could simply reflect the attachment of circulating complexes to Fc

TABLE XIV

Rate of Dengue Virus Isolation from Blood in Secondary Infections Correlated with Disease Severity[a]

Secondary infections	Isolations/attempts, days 1–7	Virus isolation rate (%)
DHF, no shock	59/352	16.8
DSS, shock	22/179	12.3
DSS, death	4/110	3.6

[a] Halstead *et al.* (1970).

receptors of B lymphocytes and does not imply that this cell is a site of virus replication *in vivo*. Of unknown biological relevance is the demonstration by two groups of workers of the ability of dengue-2 virus to grow in human B lymphoblastoid cells and in mitogen-treated human lymphocytes (Sung *et al.*, 1975; Theofilopoulos *et al.*, 1976). More significant are the reports by S. Boonpucknavig *et al.* (1979) of dengue antigen in skin mononuclear leukocytes and S. Boonpucknavig *et al.* (1976) of dengue virionlike structures in macrophages in glomeruli from DSS patients obtained by percutaneous needle biopsy. Data suggesting mononuclear phagocytes as sites of dengue infection are presented below (see Section II,K,2).

The histopathological findings in fatal dengue (see Section II,G) are consistent with an active anamnestic antibody response and host responses to hypovolemia, heart failure, and severe infection. From pathological study, clues to the sites of dengue virus replication are not obvious.

Three points require emphasis: (1) The absence of recoverable virus or detectable virus antigen implies that immunological elimination of intracellular infection is a preterminal event; (2) the more complete the virus destruction, the sicker the individual; (3) findings in humans, sparse as they are, are consistent with the hypothesis that dengue virus replicates in leukocytes.

2. *Monkeys*

It is well established that subhuman primates can be infected regularly by dengue viruses of all types by peripheral inoculation (Scherer *et al.*, 1972; Whitehead *et al.*, 1970; Halstead *et al.*, 1973a). After a variable incubation period, extracellular virus circulates in the blood. Viremia is followed by a humoral antibody response which is similar qualitatively and quantitatively to that seen in humans (Halstead *et al.*, 1973a).

Marchette *et al.* (1973) studied the distribution of virus in monkeys by two methods: sacrifice of animals at various intervals following infection, and repeated biopsy of selected tissues. A composite reconstruction of a dengue infection in monkeys is shown in Fig. 6. (1)

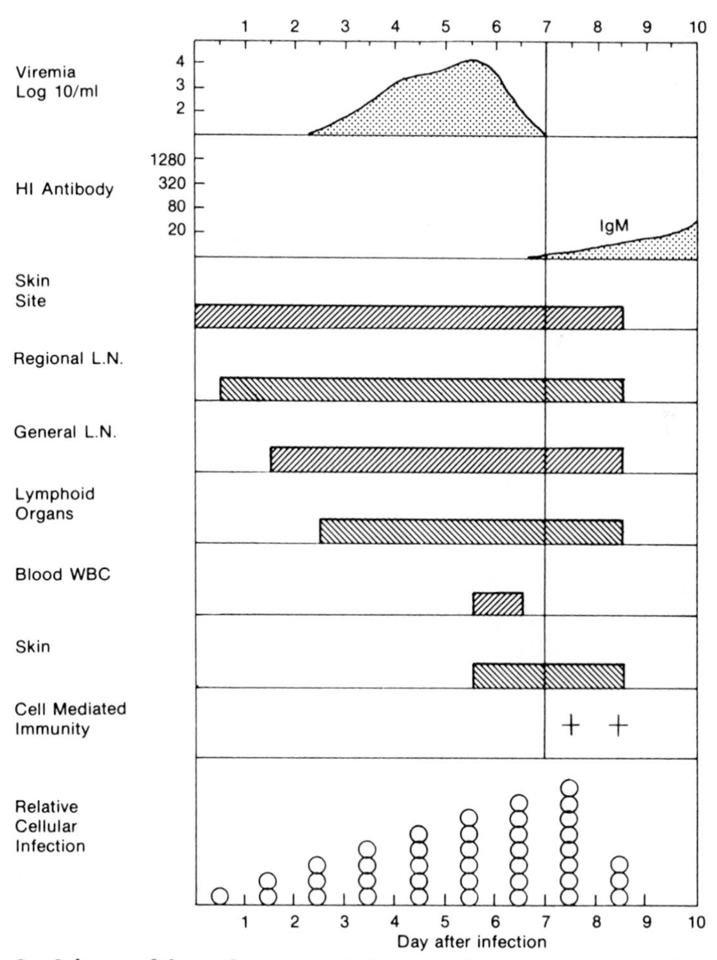

Fig. 6. Schema of the pathogenesis of dengue infection in rhesus monkeys. Data are a composite of results from 31 animals. (Adapted from Marchette *et al.*, 1973.)

Previremia—Following subcutaneous inoculation virus was found in the regional (axillary) lymph nodes within 24 h; 1 day later virus was found in other lymph nodes, such as thoracic, abdominal, and inguinal. (2) *During viremia*—Initially virus was found in lymph nodes and then later in various organs rich in lymphatic cells: spleen, lung, gastrointestinal tract. (3) *Postviremia*—Surprisingly, virus was consistently recovered from tissues after circulating virus was no longer detectable. Virus was present at the inoculation site and in the regional lymph nodes, other lymph nodes, spleen, and various viscera. Either on the last day of viremia or on the first day after the cessation of viremia, virus was recovered from circulating washed peripheral blood leukocytes (PBLs). More recently, this phenomenon was observed in a dengue-2 infection in a human being (Table XV). At this writing, the identity of infected blood leukocytes is unknown. Infected blood leukocytes have been recovered from the acute illness stage in DHF cases. In preliminary studies, infection was associated with the glass-adherent fraction (R. McN. Scott, personal communication).

At about the time that virus is found in circulating monkey leukocytes it can be recovered from skin all over the body. Could it be that virus infection in the skin derives from in-migrating infected leukocytes?

Based on an overview it is readily apparent that the number of dengue-infected cells increases progressively during the course of infection (Fig. 6). Therefore, when intracellular infection is terminated, the number of infected cells is at its acme. The kinetics of intracellular infection are quite different from those of extracellular virus. In monkeys, viremia ends, and 1–3 days later intracellular infection is terminated. It is not known how this happens. In preliminary experiments we have found circulating killer cell(s) which appear variously between days 5 and 9 following the initiation of dengue-2 infection in the rhesus monkey. These cells kill autologous dengue-infected monocytes, while armed K cells and antidengue IgG plus complement do not. The elimination of intracellular flavivirus infection by killer T lymphocytes is consistent with the experimental results obtained by Hirsch and Murphy (1968) is immunosuppressed mice. It seems significant that rash and shock are observed *after* the end of viremia, or just when peak intracellular infection and elimination of infected cells are occurring.

Finally, some but not all secondarily infected monkeys had relatively large amounts of virus in selected tissues. In these animals, the distribution of virus did not differ from that in primarily infected animals (Marchette *et al.*, 1973).

TABLE XV

Accidental Infection with Dengue-2 (16681 Strain) in an Adult Male Caucasian (Yellow Fever-Immune)[a]

Day after infection	Day of disease	Fever	Viremia (PFU/ml)	Infected PBLs	Mononuclear PBL/mm^3	Rash	Dengue-2 HI titer
0					2100		1:20
1					2400		1:20
2							1:20
3			33				1:20
4	Onset	+	5330				1:40
5	1	+	3200		750		1:40
6	2	+	16000		1700		1:40
7	3	+	5800	+	400		1:40
8	4	+	<17		870	±	1:40
9	5		<17		1200	+	1:320
10	6		<17		1100	+	1:1280
11	7		<17		1700	±	≥1:10,240
12	8		<17		1200		≥1:10,240
13	9		<17		2100		≥1:10,240
19	15		<17		3200		≥10,240

[a] N. J. Marchette, unpublished data.

Single, isolated, antigen-bearing cells have been seen in infected monkey skin and lymph node by fluorescent antibody microscopy. Cells in skin have the appearance of histiocytes, while those in lymph nodes are mononuclear cells with abundant cytoplasm (Marchette *et al.*, 1973).

K. Immunopathogenetic Studies

1. Enhanced Infection in Vivo

Several groups of workers have infected susceptible subhuman primates serially with different dengue viruses. Three groups examined animals for evidence of DHF–DSS either clinically (Whitehead *et al.*, 1970) or physiologically (Halstead *et al.*, 1973d; Scherer *et al.*, 1978). Whitehead *et al.* (1970) observed hemorrhage and recovered virus from the liver, lung, and lymph node of a gibbon dying on the ninth day of a secondary dengue-1 infection. The animal had widespread lymphomatous leukemia. Two other gibbons were inoculated with *Bacillus pertussis* vaccine and then challenged with type-1 dengue 1 day later. They had viremias which the authors described as "markedly" longer than those of infected controls.

Halstead *et al.* (1973d) sequentially inoculated the four dengue virus types into groups of rhesus monkeys at intervals of 2, 6, 12, and 26 weeks. After primary and secondary infections, monkeys were bled daily for 10 days to measure viremia, blood platelets, prothrombin time, serum proteins, hematocrit, and in some cases total serum complement. A single animal had a physiological response similar to that seen in DHF. Unexpectedly, secondarily infected animals had mean peak dengue-2 viremias higher than those observed in primary dengue-2 infections (Fig. 7). The mean duration of secondary viremia (3.4 days) was slightly shorter than that of primary viremia (4.0 days). Immunologically enhanced infection has been pivotal in all subsequent studies on the pathogenesis of dengue. It is important that these observations were made *in vivo* and that they antedated the discovery of *in vitro* immunological enhancement of infection in leukocytes.

More recently, the enhancement of dengue viremias has been demonstrated in rhesus monkeys by passive transfer of small amounts of dengue antibody (Halstead, 1979). Five pairs of susceptible rhesus monkeys were given either normal or dengue-immune human cord blood serum to effect a final dilution of 1:300. The immune serum pool had a dengue-2 plaque reduction neutralization titer of 1:140 and a

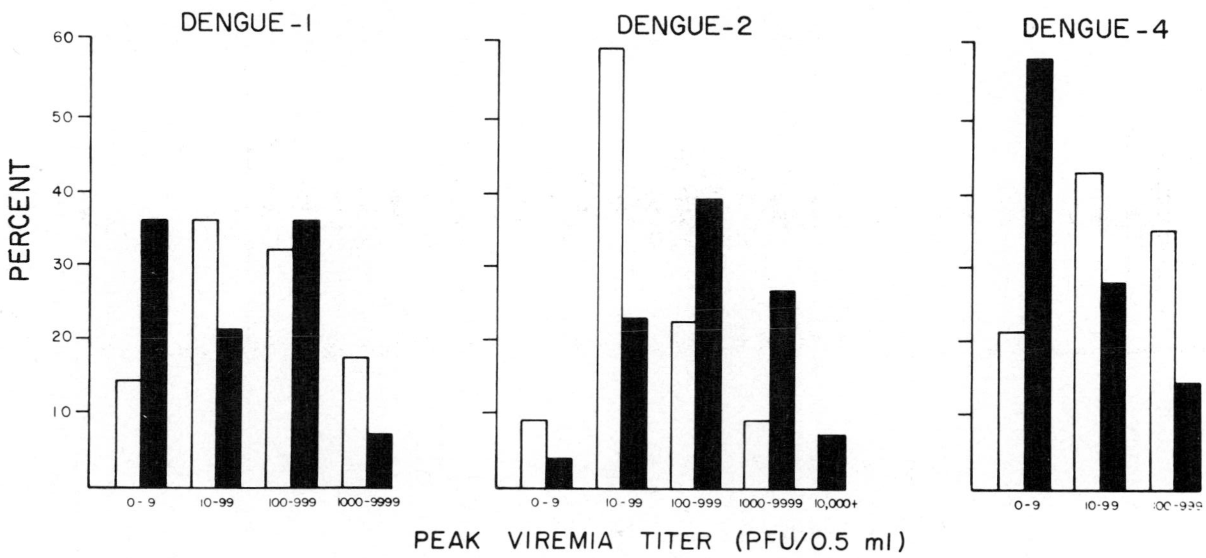

Fig. 7. Distribution of peak titers of viremia in monkeys with primary (open areas) and secondary (solid areas) infections with dengue-1 (strain 16007), dengue-2 (strain 16681), and dengue-4 (strain 4328S). The number of animals with primary dengue-1, -2, and -4 infections was 19, 24 and 37, respectively, and with secondary dengue-1, -2, and -4 infections, 14, 44, and 7, respectively. All primary and secondary infections were initiated using the same dosage of virus and route of infection. (From Halstead *et al.*, 1973a.)

human monocyte infection enhancement titer of greater than 1:1,000,000 (see below). Fifteen minutes after the inoculation of serum, animals were infected. In each pair of monkeys, animals receiving dengue antibody had higher viremias than animals receiving nonimmune cord blood serum. Viremia enhancement varied by 2.7- to 51.4-fold. Results for two monkey pairs are given in Fig. 8.

2. *Immunological Enhancement of Dengue Infection in Mononuclear Phagocytes in Vitro*

While studying cell-mediated immune responses to dengue antigens it was noted that at multiplicities of infection of ≥0.01 all dengue virus types could replicate in cultured PBLs prepared from dengue-immune monkeys but not in cells obtained from susceptible animals (Halstead *et al.*, 1973b) (Fig. 9). A similar result has been observed in man (Halstead *et al.*, 1976). In initial experiments, monkeys were studied just before *in vivo* infection and again several weeks later. At

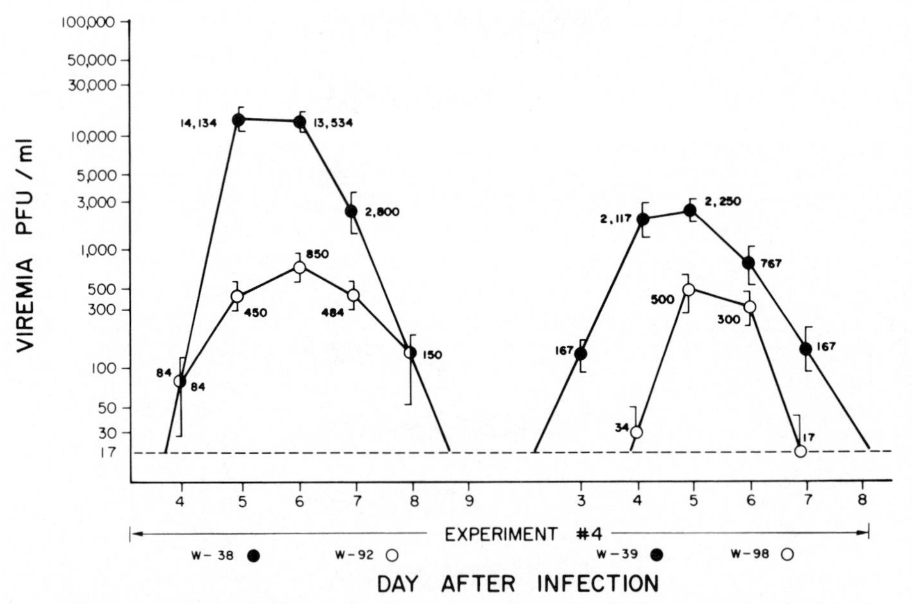

Fig. 8. Daily viremia in five pairs of susceptible juvenile rhesus monkeys infected with dengue-2, strain 16681. Fifteen minutes prior to subcutaneous inoculation of 1000–10,000 PFU of virus, dengue-immune (solid circles) or nonimmune (open circles) human cord blood serum was injected intravenously to effect a final dilution of 1:300 *in vivo*. Animals were bled daily through 12 days after infection. Paired experiments were assayed simultaneously under code. Mean titers are shown ±S.D. as indicated by brackets. (From Halstead, 1979.)

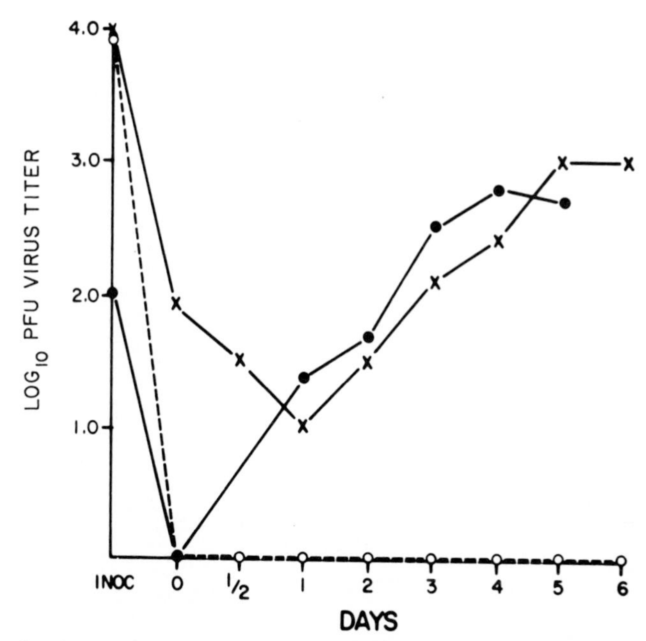

Fig. 9. *In vitro* replication of dengue-2 and -4 viruses in PBLs from monkey H-127 studied before and 1 month after *in vivo* dengue-4 infection. Dengue viruses were added to PBL cultures at a multiplicity of infection of 0.01. After an incubation of 1 h at 37°C, cells were washed and suspended in RPMI-1640 containing 10% fetal calf serum, and extracellular virus assayed after 0–6 days in culture at 37°C. Open circles, normal with dengue-2 virus added; solid circles, dengue-4-immune with dengue-4 virus added; X, dengue-4-immune with dengue-2 virus added. Values are plotted as PFU per milliliter. (From Halstead *et al.*, 1973c.)

this time, the addition of live tissue culture-grown dengue virus stimulated the incorporation of [^{3}H]thymidine by PBLs (Halstead *et al.*, 1973b). Three months later blast transformation could no longer be detected by the [^{3}H]thymidine uptake method. However, PBLs still supported dengue virus replication (Marchette *et al.*, 1976).

Cultured PBLs from susceptible rhesus monkeys and human beings are readily infected when virus and heterotypic antibody are added together to the culture medium (Halstead and O'Rourke, 1977a). If only virus is added, usually little or no replication occurs. Representative experimental results are shown in Fig. 10. This experiment is reproducible when PBLs are separated from erythrocytes using dextran (Alexander and Spriggs, 1960) or by the Ficoll–Hypaque density flotation method (Boyum, 1968). The former method provides a suspension which contains PMNs as well as mononuclear leukocytes.

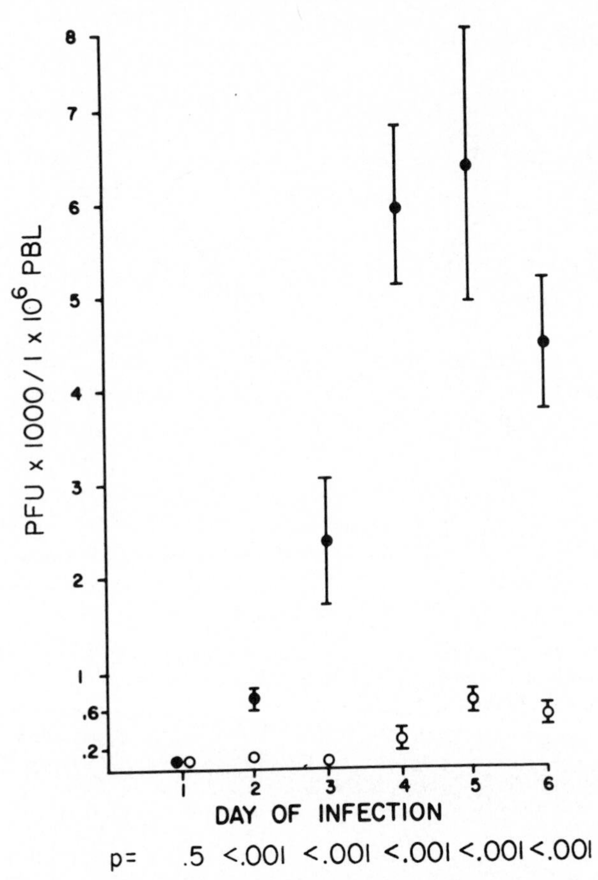

Fig. 10. Dengue-2 virus titers in cultures of normal rhesus PBLs in 20 paired comparisons with (solid circles) and without (open circles) anti-dengue-4. Dengue-2 virus was added to all cultures at a multiplicity of infection between 0.2 and 0.05. Virus was not removed. To one-half of the cultures a single lot of anti-dengue-4 was added at a final dilution of 1:30. Control cultures were incubated with 1:30 autologous serum or pooled normal monkey serum. Mean PFUs were calculated from three 1-oz LLC-MK2 bottles; averages of means are plotted ± S.E. of mean. The *t*-test for paired data was performed using the square roots of mean plaque counts. (From Halstead and O'Rourke, 1977a.)

Theofilopoulos *et al.* (1976) have shown that dengue virus does not replicate in human PMNs, and we have similar unpublished results for both monkey and human PMNs. Cultures are routinely standardized to 1.0 or 1.5 × 10^6 mononuclear leukocytes per milliliter of culture medium (RPMI-1640 with 10% heat-inactivated fetal calf serum). The separation of mononuclear leukocytes using the Ficoll–

Hypaque method results in a dilution of the donor serum of at least 1:40,000.

Homotypic antibody or antibodies from multiply infected humans or monkeys neutralize dengue virus at low dilutions, but at dilutions above the neutralization end point enhancement of infection occurs (Fig. 11) (Halstead and O'Rourke, 1977a,b). All sera were heat-inactivated, and the addition of fresh human or monkey complement to the culture medium did not increase the enhancing effect. As shown in Table XVI, enhancing antidengue is of the IgG class. The nonneutralizing virus–antibody complex requires an intact Fc terminus, which implies that the complex is attached to a Fc receptor prior to initiating infection. The kinetics of attachment and internalization of complexes has been studied; attachment occurs in 10 min, and virus is internalized within 30–60 min (Halstead and O'Rourke, 1977b).

Enhancing antidengue IgG appears relatively early following infection in monkeys and man. In two rhesus monkeys, serum obtained 14 days after infection was partitioned by density gradient ultracentrifugation (Tables XVII–XIX). The IgM fraction contained all the neutralizing and HI antibody activity but did not enhance. The IgG fraction, with little or no antibody activity, had an enhancement titer at a serum dilution equivalent of 1:10,000. Similar results were obtained in a human with a dengue-2 infection. Weak enhancement could be demonstrated in serum fractions containing IgM when fresh comple-

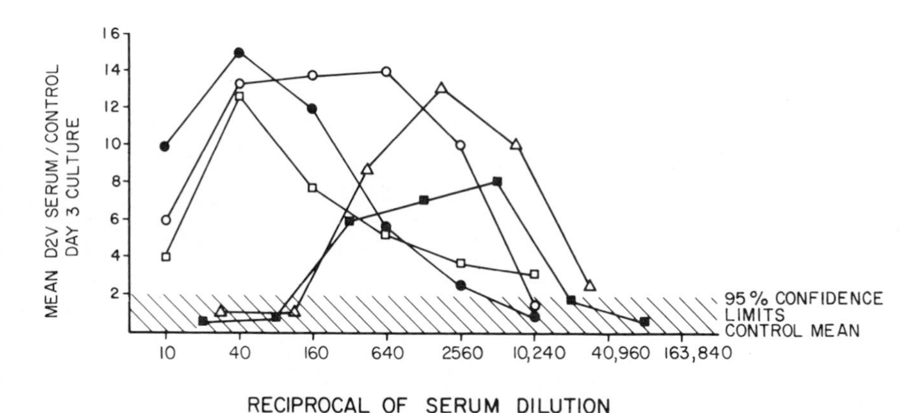

Fig. 11. Fold increase in dengue-2 virus replication in human PBL cultures at varying dilutions of dengue antisera. Dengue-2 plaque reduction neutralization titers of tested sera were: mouse anti-dengue-1, 1:15 (open squares); rabbit anti-dengue-2, 1:320 (solid squares); rabbit anti-dengue-3, 1:40 (open circles); monkey anti-dengue-4, 1:20 (solid circles); human anti-dengue 1-4, 1:717 (triangles). D2V, Dengue-2 virus.

TABLE XVI

Immunoglobulin Concentrations, Anti-Dengue-2, and Replication Enhancement Titers in Serum, IgG, and $F(ab)_2$ Prepared from a Dengue-1–4-Immune Human[a]

Specimen	Lowry (mg/100 ml)	IgG (mg/100 ml)	Reciprocal titer		Dengue-2 virus replication enhancement titer	
			HI[b]	$PRNT_{50}$[c]	Expt. 1	Expt. 2
Serum (SBH)[d]	4,750	1,440	88 ± 8	717	25,600	12,800
IgG (SBH)	170	114	8 ± 0	76	1,280	640
$F(ab)_2$ (SBH)[e]	130		5 ± 0.7	78	<1	<1

[a] Halstead and O'Rourke (1977a).
[b] Mean ± S.D. of 10 replicate titrations.
[c] Fifty percent plaque reduction neutralization titer. Mean of two determinations.
[d] Initials of serum donor.
[e] Versus anti-human F(ab) = 80 mg/100 ml.

TABLE XVII

Temporal Development of Hemagglutination Inhibition and Neutralizing Antibody to Dengue-2 Virus after Primary Infection of Two Rhesus Monkeys[a]

Rhesus 111			Rhesus 112		
Day of infection	Reciprocal titer: HI	Reciprocal titer: $PRNT_{50}$[b]	Day of infection	Reciprocal titer: HI	Reciprocal titer: $PRNT_{50}$
0	<10	<10	0	<10	<10
7	<10		7	<10	
10	<10		10	10	
14	40	4800	14	80	5000
21	40		21	80	
28	40		28	80	
45	40		45	80	

[a] Halstead and O'Rourke (1977a).
[b] Fifty percent plaque reduction neutralization titer.

TABLE XVIII

Immunoglobulin and Antibody Concentrations in Dialyzed Fractions Collected for Centrifugation of Day-14 Sera on a Sucrose Gradient at 100,000 *g* for 16 h

Rhesus 111					Rhesus 112				
			Reciprocal titer					Reciprocal titer	
Fraction	IgM (mg/100 ml)[a]	IgG (mg/100 ml)[a]	HI	$PRNT_{50}$[b]	Fraction	IgM (mg/100 ml)	IgG (mg/100 ml)	HI	$PRNT_{50}$
1	0	0	<1		1	0	0	<1	
2	0	0	1		2	0	0	2	
3	<25 (+)	0	4	320	3	40	0	4	320
4	<25 (+)	0	2		4	+	0	1	
5	0	+	<1		5	0	+	<1	
6	0	+	<1		6	0	+	2	
7	0	305	<1	<10	7	0	481	2	<10
8	0	+	<1		8	0	+	<1	
Whole serum	68	1250	40	4800	Whole serum	172	1960	80	5000

[a] Fractions tested by Ouchterlony versus anti-IgM or anti-IgG. Fractions containing the highest concentration of IgM or IgG were quantitated by radial immunoprecipitation.

[b] Fifty percent plaque reduction neutralization titer.

TABLE XIX

Dengue-2 Virus Replication Enhancement Activity in Rhesus Monkey Sera and Sucrose Gradient Fractions Obtained 14 Days after Primary Infection

Rhesus 111						Rhesus 112					
Serum		Fraction 3		Fraction 7		Serum		Fraction 3		Fraction 7	
Dilution	Enhancement[a]	Dilution[b]	Enhancement	Dilution[b]	Enhancement	Dilution	Enhancement	Dilution[b]	Enhancement	Dilution[b]	Enhancement
10	<1	200	<1	40	7.0	10	<1	400	<1	40	8.0
40	<1	800	<1	160	13.0	40	1.1	1,600	<1	160	12.0
160	2.9	3,200	<1	640	11.0	160	<1	6,400	<1	640	8.0
640	1.7	12,800	<1	2,560	4.4	640	1.3	25,600	<1	2,560	5.4
2,560	2.5	51,200	<1	10,240	2.1	2,560	3.2	102,400	<1	10,240	3.0
10,240	<1			40,960	<1	10,240	2.3			40,960	<1

[a] Day-3 PFUs: test sample/control.
[b] Serum equivalent.

ment was added. Enhancing IgG was first detected in serum collected 10 days after the onset of illness.

Eight dengue-immune human cord blood sera studied by Marchette *et al.* (1979) had infection-enhancing titers between 1:100,000 and 1:1,000,000.

The growth of dengue virus in mitogen-treated monkey PBLs, in cells which were weakly but not strongly adherent to glass surfaces (Halstead *et al.*, 1973c), and in human B lymphoblast cell lines (Sung *et al.*, 1975) favored a hypothesis that dengue virus could stimulate the transformation of committed B lymphocytes which in turn served as cellular hosts for dengue replication (Halstead *et al.*, 1973b).

Further studies on the identity of human and monkey dengue-permissive PBLs have conclusively demonstrated that this is certainly not a major mechanism in leukocyte infection.

When monkey bone marrow, spleen, thymus, and lymph node leukocytes from nonimmune animals were infected with immune complexes, bone marrow and, to a lesser extent, spleen cells supported dengue virus replication (Table XX) (Halstead *et al.*, 1977). Similar results have been obtained with immune monkeys (unpublished). Thymus and lymph node leukocytes are nonpermissive. In peripheral blood it has been concluded that only mononuclear phagocytes support dengue virus replication. The evidence for this is that permissive cells are radioresistant and are removed by passage through nylon wool columns, and that dengue infection is ablated by incubating infected leukocytes with 100 μg/ml of particulate silica (Halstead *et al.*, 1977) (Table XXI). We have shown that particulate silica is a highly specific toxin for mononuclear phagocytes; it has little or no functional effect on other identifiable PBLs (O'Rourke *et al.*, 1978). Silica also reduced D2V replication in bone marrow and spleen suspensions by more than 90% (Halstead *et al.*, 1977).

It was further shown that human PBLs held in culture for more than 48 h became progressively less permissive to infection by dengue virus–nonneutralizing antibody complexes (Halstead and O'Rourke, 1977b). Monkey bone marrow cultures held for a similar period and then infected were fully permissive. Monocyte progenitor cells are found in the bone marrow, and monocytes are produced in bone marrow cultures (Goud *et al.*, 1975). Blood monocytes do not divide but progressively differentiate into mature, active macrophages (Van Furth and Cohn, 1968). Our observations, when evaluated with these known biological attributes of mononuclear phagocytes, suggest that dengue virus replicates in immature monocytes. *A priori* the most likely candidate for a dengue virus-permissive monocyte is a cell

TABLE XX

Replication of Dengue-2 Virus in Nonimmune Rhesus Tissue Leukocyte Cultures in the Presence and Absence of Anti-Dengue-4[a]

Monkey	Day infected	Mean PFU × 1000/10^6 mononuclear leukocytes on days 3–5[b]							
		Bone marrow[c]		Spleen		Thymus		Lymph node	
		aD4	FCS	aD4	FCS	aD4	FCS	aD4	FCS
Rh 113	2	1.3 ± 0.3	0.3	0.2	0.09	ND	ND	0.17	0.1
Rh A	2	0.9 ± 0.5	0.1	0.04	0.04	0.05	0.03	0.07	0.04
Rh B	2	1.3 ± 0.6	0.2	0.06	0.06	0.2	0.05	0.6	<0.02
Rh C	4	67.1 ± 6.0	1.3 ± 0.7	0.15	0.01	0.1	0.07	0.06	<0.01
Rh D	4	3.7 ± 0.8	0.5	0.1	0.08	ND	ND	<0.01	<0.01

[a] Halstead and O'Rourke (1977b).
[b] Virus antibody added after leukocytes were held for 48 or 96 h at 37°C. FCS, fetal calf serum; aD4, anti-dengue-4; ND, no data.
[c] S.E. of mean shown for values >1.0.

TABLE XXI

Effect of Particulate Silica on Human Mononuclear Phagocytes and Dengue-2 Virus Replication in Peripheral Blood Leukocyte Cultures[a]

Silica (μg/ml)	Number of experiments	Mononuclear phagocytes $\times 10^4$, mean $\pm$ S.D.				PFU $\times$ 1000/ml, mean	
		Day 0	Day 1	Day 2	Day 3	Day 2	Day 3
None	14	8.9 ± 0.8	9.3 ± 1.1	10.8 ± 0.9	10.6 ± 0.8	3.1 ± 0.8	1.7 ± 0.4
0.1	4		8.1 ± 1.7	12.1 ± 2.2	14.6 ± 3.2	3.7 ± 0.4	2.8 ± 0.4
1.0	4		8.0 ± 1.4	10.8 ± 2.3	10.2 ± 2.1	3.5 ± 0.5	2.0 ± 0.8
10.0	4		3.7 ± 1.3	4.1 ± 1.2	3.3 ± 1.5	1.7 ± 0.2	0.5 ± 0.1
25.0	1		7.4	5.2	5.3	0.03 ± 0.0	0.2 ± 0.1
50.0	1		0.7	2.2	3.8	0.04 ± 0.0	0.04 ± 0.0
100.0	10		2.9 ± 0.8	1.3 ± 0.4	1.0 ± 0.5	0.07 ± 0.0	0.04 ± 0.0

[a] Halstead and O'Rourke (1977b).

which retains the property of immune phagocytosis but lacks a mature lysosomal apparatus. Replication in promonocytes is consistent with our data that not more than 1% of monocytes are infected *in vitro* (Halstead *et al.*, 1977). This is approximately the fractional proportion of promonocytes in blood (Van Furth and Cohn, 1968).

While it is clear that monocytes can be infected by immune complexes, the mechanism by which washed monocytes from immune hosts are infected is not established. In unpublished studies we have demonstrated that the addition of enhancing antibody to populations of immune monocytes results in additive infection, suggesting the existence of two populations of cells. Simple washing of immune PBLs to dilutions of original serum of over 1:1,000,000 does not reduce permissiveness. Nor has permissiveness been destroyed by the incubation of permissive cells at 37°C followed by washing.

Of great interest, human cord blood PBLs from infants born to dengue-immune mothers also are permissive to dengue infection (Table XXII). Permissive PBLs from passively immune infants or actively immune monkeys or human beings ultimately disappear from circulation. In infants this is a matter of a few months; in *actively immunized* older children and monkeys, it is a matter of years (Marchette *et al.*, 1976).

TABLE XXII

Dengue-2 Virus (Strain 16681) Replication in Cord Blood Leukocytes Obtained from Dengue-Immune and Nonimmune Oriental Women[a]

Maternal immune status	Number studied	Mean PFU/ml/10^6 CBLs: No addition	Autologous plasma	Monkey anti-dengue-4
Dengue-immune[b]	8	794	1660	22,387
Dengue-nonimmune	24	7	5	407

[a] Cord blood leukocytes (CBLs) were obtained from heparinized blood and separated from erythrocytes using dextran T-250. They were washed three times in phosphate-buffered saline and suspended at 1.5×10^6 mononuclear leukocytes per milliliter of RPMI-1640 with 10% fetal calf serum (approximately 1:50,000 dilution). To separate replicate cultures a 1:50 final dilution of autologous cord blood plasma or monkey anti-dengue-4 serum was added. Virus content was assayed on days 3–5 of culture.

[b] In unpublished studies on dengue-immune infants in Thailand, *washed* PBLs were permissive at 3 months but not at 6 months. Addition of infant plasma enhanced infection at 6 months but not at 9 months in 15 individuals studied.

III. THE IMMUNE ENHANCEMENT HYPOTHESIS

In the immunopathological hypothesis of DSS (Halstead, 1970) it was not possible to find a unitary hypersensitivity mechanism which could explain all clinical observations. But it has been noted that, "if delayed hypersensitivity reactions can be altered in the direction of greater virulence by passively or endogenously acquired dengue antibody heterologous to the infecting virus, this might be [the] pathogenetic mechanism."

For several years after this was written, efforts focused on a mechanism proposed by Russell (1970). In this hypothesis, complexing between dengue viral antigens and antidengue IgG interacts with complement to generate anaphylatoxins. These, in turn, are expected to degranulate mast cells and liberate histamine, causing acute vascular permeability. Evidence for C3a elaboration in DSS is provided by the discovery of reduced levels of serum carboxypeptidase B, a C3a inactivator (Corbin *et al.*, 1976). It has also been speculated that an activated complement system might interact with the blood-clotting and fibrinolytic systems (Bokisch *et al.*, 1973). From two comprehensive clinical studies of DHF–DSS there can be no doubt that C3 is catabolized, that complement is activated both by the classic and alternative pathways, and that there is evidence of disseminated intravascular coagulation (Memoranda, 1973; Bokisch *et al.*, 1973). Further, the degree of complement activation is directly related to the severity of illness in the patient. However, it is not known to what extent antigen–antibody complexing contributes to complement activation. Low blood levels of complement components were found in a 6-month-old infant who had no detectable dengue antibody at the time of study (Bokisch *et al.*, 1973). To explain complement activation in this case it was postulated that immune complexes were formed with residual maternal antibody. But, as shown by Scott *et al.* (1976) and Kuberski *et al.* (1977), complement activation can occur during primary dengue infections in 5- to 7-year-old children and in adults who do not have residual maternal IgG.

It must be concluded that complement can be activated by mechanism(s) other than immune complexes. The immune enhancement hypothesis allows for this possibility.

A. Afferent Mechanisms

The central biological property of dengue virus is its ability to survive engulfment by mononuclear phagocytes. Dengue is an im-

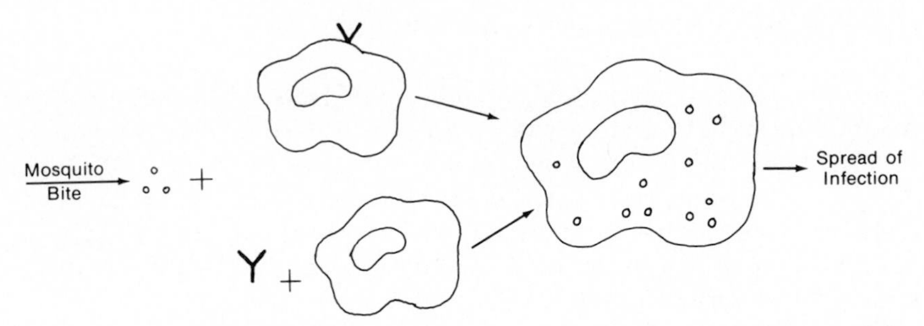

Fig. 12. Immune phagocytosis of dengue virus mediated by humoral or cytophilic enhancing antibody results in the infection of mononuclear phagocytes.

munopathological disease because of the paradoxical ability of antibody to promote infection in mononuclear phagocytes (Fig. 12). To relate to disease, a corollary hypothesis is required—that severity is directly related to the number of infected cells. Thus, dengue fever and DHF/DSS differ in only one essential parameter, the number of cells infected. This hypothesis and sequential events in dengue fever and DHF/DSS are schematically illustrated in Figs. 13 and 14.

From considerations of mononuclear phagocyte biology, the number of dengue-infected cells might be regulated by one or more mechanisms (Fig. 15a–c): (1) the concentration of infection-enhancing antibody, (2) the size of the subpopulation of mononuclear phagocytes permissive to dengue infection, (3) the rate of immune phagocytosis by mononuclear phagocytes, or (4) the yield of infectious virions per infected cell.

1. Optimal Concentration of Enhancing Antibody

It is tentatively assumed that interactions with flavivirus group-antigenic determinants are responsible for infection enhancement, while interactions with type-specific determinants are more likely to result in the destruction of virus. This surely is an oversimplification. Enhancing antibody may either be of endogenous origin or passively acquired. It is easier to predict the behavior of passively transmitted antibody. Such antibodies in an infant, if derived from a mother with broad dengue group immunity, may provide several months of protection from clinical illness but, as the protective threshold is reached, passively acquired antidengue IgG is able to enhance infection *in vivo* by mechanisms which have been described *in vitro* (Fig. 16). The curve of degradation of enhancing antibody conforms with the extremely high titers found in cord blood and provides an explanation of the unusual age-specific hospitalization rate curve. Thus, in the

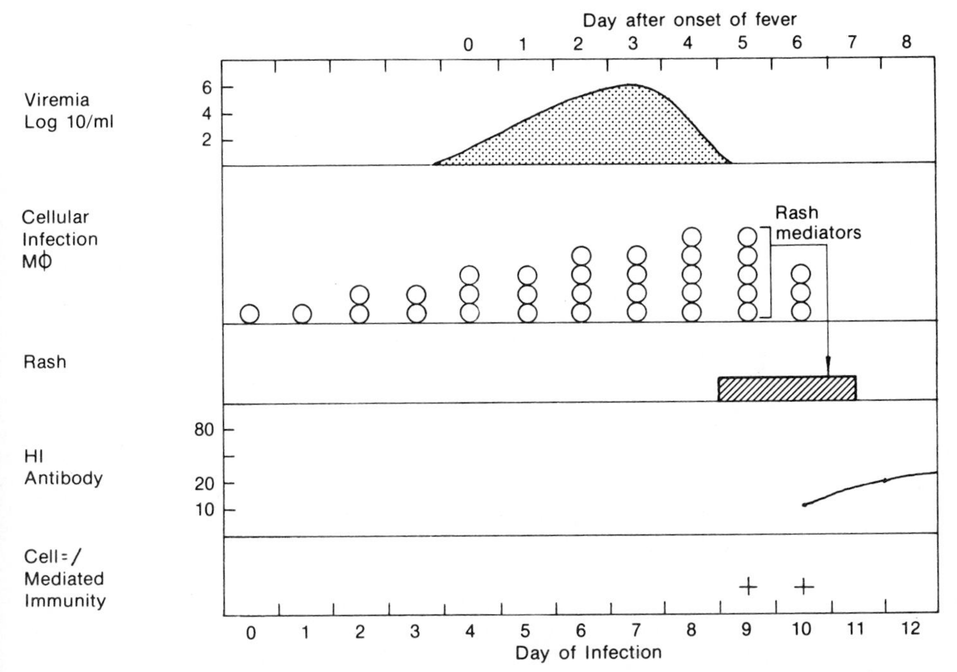

Fig. 13. Schema of the pathogenesis of DHF. It is proposed that the number of cells infected by virus increases progressively until the onset of elimination of infected cells. The appearance of circulating killer PBLs correlates with the defervescence of fever, onset of rash, and disappearance of infected cells.

case of passively acquired enhancing antibody, antibody concentration may regulate cellular infection (Fig. 15a).

The association between DSS outbreaks and secondary dengue infections is notoriously capricious. As noted in Section II,I, DSS has not accompanied secondary dengue infections in the Caribbean, the South Pacific islands, West Africa, or South India. Further, in Southeast Asia, the occurrence of DSS is unpredictable. Although serological studies suggest that dengue virus infections occur year after year, in some countries DSS epidemics are widely spaced.

It is possible to pose two testable hypotheses which independently or together could explain variability in the occurrence of DSS in relation to concentrations of enhancing antibody: The concentration of enhancing antibody is related (1) to the number of antigenic determinants shared by sequentially infecting viruses or (2) to time after infection.

In relation to the first hypothesis it follows that, if two sequential

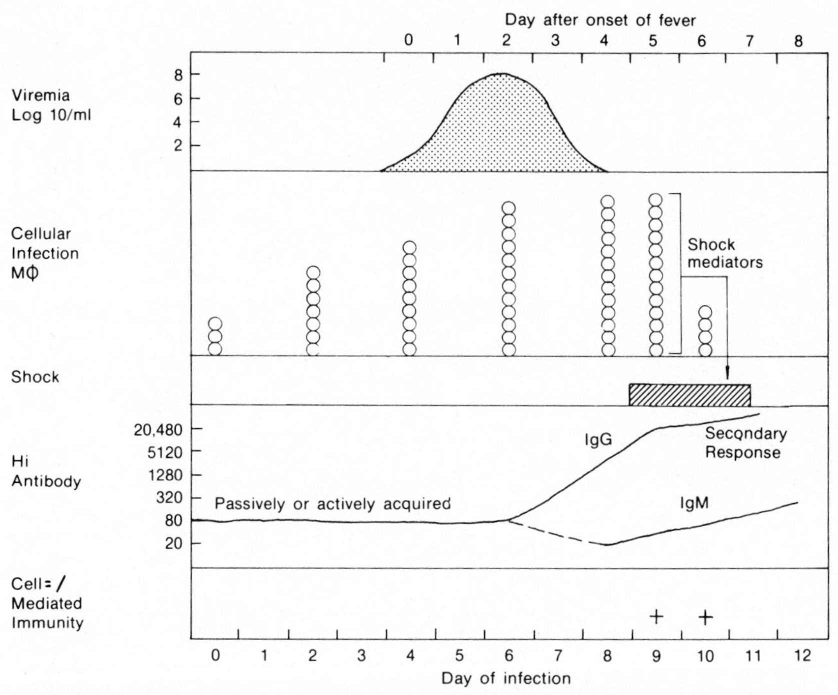

Fig. 14. Schema of the pathogenesis of DSS. It is proposed that the number of infected cells are greater at every stage of infection in DSS compared with DHF. Circulating killer PBLs correlate with the defervescence of fever, elimination of intracellular infection, and onset of shock. In DHF, these events result in rash; in DSS, in shock.

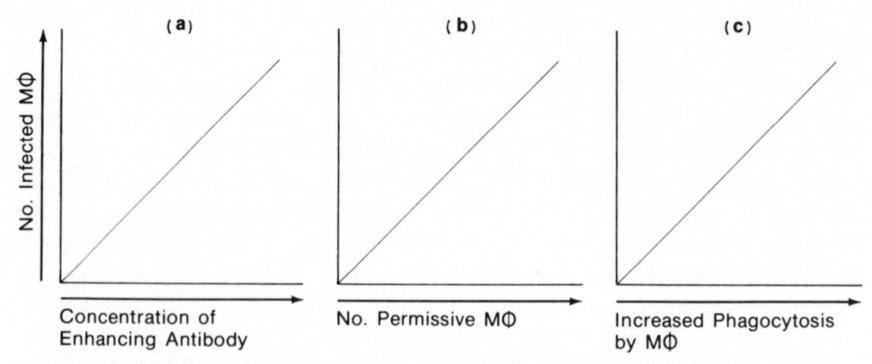

Fig. 15. Hypothesized relationships among the concentration of enhancing antibody, production of permissive mononuclear phagocytes, and rate of immune phagocytosis on the number of dengue-infected mononuclear phagocytes.

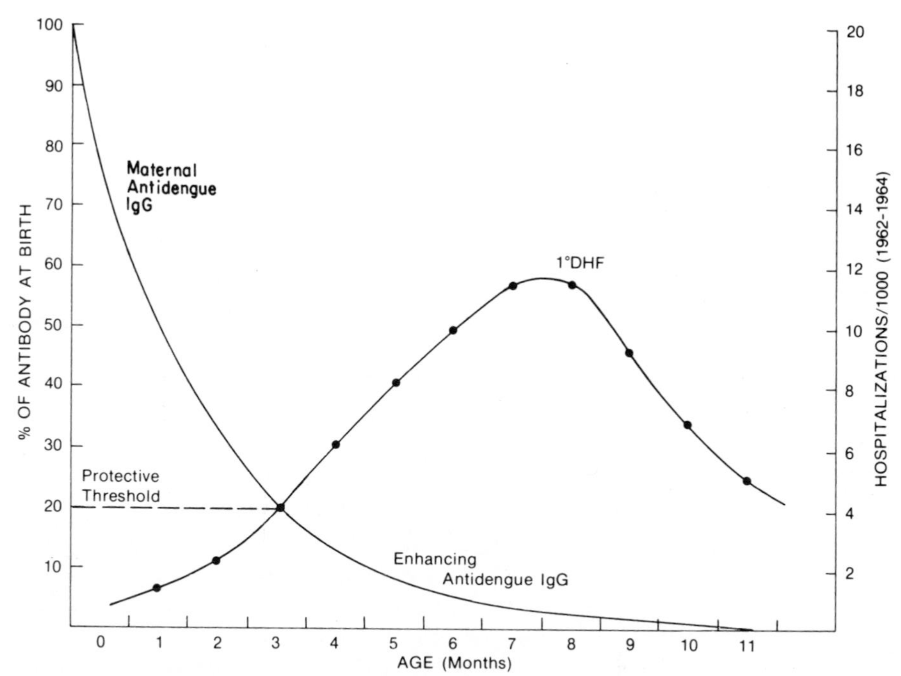

Fig. 16. Schema showing the protective and infection-enhancing role of maternal antidengue IgG. An infant born to a heterotypically immune mother who is bitten by an infected mosquito in the first few months of life will be protected. After maternal antidengue has been catabolized to concentrations below the protective threshold, antibody may enhance dengue infection. As enhancing antibody is eliminated, the risk of enhanced infection wanes (9–12 months). Age-specific hospitalization rates for DHF are for Bangkok, 1962–1964. (From Halstead, 1970.)

infecting viruses share two infection-enhancing antigenic determinants, the concentration of enhancing antibody in the host may be twice that occurring if a virus pair shares only a single determinant. An examination of the antigenic relatedness hypothesis might be approached through a careful antigenic analysis of a collection of viruses isolated from DSS-endemic and nonendemic areas. Virus strains should include those isolated from mosquitoes and mild primary and secondary dengue virus infections as well as DSS cases. Such an analysis might show greater antigenic sharing among viruses associated with endemic DSS than among those recovered from nonendemic areas. Alternatively, viruses isolated from DSS cases might have a significantly larger number of cross-reactive antigenic determinants than control strains. This phenomenon has already been demonstrated in a group of dengue-2 strains isolated from DHF–DSS cases and analyzed by Halstead and Simasthien (1970).

The second hypothesis may relate either to humoral or to cytophilic enhancing antibody. As discussed above, primary DHF during the first year of life correlates with the appearance and disappearance of humoral enhancing antibody. Our group has noted that, with time after dengue infection, permissive monkey leukocytes disappear from circulation (Marchette *et al.*, 1976). In unpublished studies we have found that leukocytes from several adults with remote dengue infections consistently fail to support dengue replication *in vitro*, while PBLs obtained within a few weeks or months of human dengue infection are uniformly permissive (Marchette *et al.*, 1975). Experimentally, the most definitive assessment of the effect of time on the severity of sequential infections can be made by prospective epidemiological studies. This might be accomplished by monitoring annual dengue infections serologically. An alternative approach might be to measure the titer of enhancing antibody in the acute-phase sera of patients with secondary dengue infections using virus isolated during the illness. The experimental hypothesis being tested is that enhancing antibody titers are higher in patients with a severe outcome than in those with milder disease. It will be necessary to determine the fate of enhancing antibody during secondary infections in an experimental animal model before attempting studies on man.

2. *Increase in Permissive Cells*

Dengue-permissive mononuclear phagocytes are not yet sufficiently well characterized to permit educated guesses about the regulation of this population. The limiting variable in determining the number of cells infected by dengue virus *in vivo* could be the size of the dengue-permissive cell population as diagrammed in Fig. 15. Permissiveness resulting from nonspecific stimuli might be either antibody-dependent or antibody-independent. If the promonocyte hypothesis is correct, any stimulus which increases the production of mononuclear phagocytes would increase promonocytes. This could include a transient phenomenon such as acute bacterial infection, or a stimulus of longer duration such as chronic infection with mycobacteria or fungus or a lymphoproliferative disorder. The observations by Whitehead *et al.* (1970) of enhanced dengue disease in a gibbon with lymphomatous leukemia and enhanced viremia following the injection of *B. pertussis* vaccine are exciting and require further study. Individual genetic differences and drugs could affect the production or function of mononuclear phagocytes.

To assess the role of antecedent or concomitant factors on the severity of dengue illnesses, it will be particularly important to make careful clinical studies on individuals who develop DSS during primary dengue infections after the first year of life.

3. *Increase in Immune Phagocytosis*

The uptake of dengue virus, with or without antibody, may increase without a corresponding expansion in the number of mononuclear phagocytes (Fig. 15). In mice, the administration of several steroid hormones increased phagocytosis by macrophages (Vernon-Roberts, 1969). In unpublished studies in this laboratory, progesterone and other steroid hormones markedly increased dengue-2 infection in human monocytes *in vitro*. If steroid hormones in the sexually mature human female more effectively stimulated phagocytosis than did those in males, this might explain the sex difference in the severity of dengue disease which is first seen at about age 4 years and increases until puberty.

4. *Increase in Virus Replication*

The data which suggest that sequential infections ending with dengue-2 are more virulent than infections which occur in other combinations have been reviewed above. It was suggested that this phenomenon might reflect the antigenic relatedness between infecting virus pairs. Another reasonable explanation is that dengue viruses may be heterogeneous with respect to their ability to replicate in mononuclear phagocytes. Viruses which produce more infective progeny would enhance the kinetics of sequential cellular infection. Thus, one or more dengue types in DSS endemic areas might have enhanced replicative properties as compared with viruses recovered from non-DSS areas.

This hypothesis is also testable. A suitable collection of dengue viruses from endemic and nonendemic areas could be systematically studied for replicative kinetics in human monocytes in the presence and absence of enhancing antiserum.

B. Efferent Mechanisms

From clinical considerations, it is predicted that the onset of vascular permeability correlates with a major interaction between dengue-infected cells and immune elimination mechanism(s). The competence of the immune response and the number of infected cells are critical factors in the dengue immunopathogenetic equation.

It has been noted that there is a body of anecdotal evidence that DSS is more severe in the well-nourished than in the clinically undernourished child. The mild immunosuppression provided by protein–calorie malnutrition could depress either the rate of immune elimination and/or the accompanying inflammatory response(s), converting an enhanced dengue infection from a potentially life-threatening process to a mild disease. It is generally conceded that females are more immunocompetent than males, as evidenced by higher rates of autoimmune diseases in females than in males. The observed sex differences in DSS could be a reflection of a sex-related effect on efferent immune mechanisms.

As shown experimentally, antibody has afferent control of dengue virus infection. If large amounts of viral antigen are produced in the presence of IgG antibody, immune complexing and complement activation should result. A more satisfactory unitary hypothesis, one which explains both primary- and secondary-infection DSS, is that complement activation and the generation of other vascular permeability mediators result directly from dengue infection in mononuclear phagocytes. Macrophages are capable of activating both the complement and the blood-clotting systems. C3a and C3b can elicit the release of lysosomal enzymes which in turn cleave C3 to C3b (Schorlemmer *et al.*, 1976; Schorlemmer and Allison, 1976; Dourmashkin and Patterson, 1975). Activated C3b with factor B functions as a proteinase, cleaving further C3 to C3b (Nicholson *et al.*, 1975). Thus, through positive-feedback mechanisms, macrophages are potentially powerful activators of complement via the alternative pathway. Although dengue infection has not been proved to activate complement, viral infections in other cell systems result in the release of factors which activate C3 and C5 (Brier *et al.*, 1970; Ward *et al.*, 1972). Recent studies have shown that activated macrophages release thromboplastin (Prydz *et al.*, 1977). Effector mechanisms in rash and shock are schematically shown in Figs. 17 and 18. It has been emphasized repeatedly that shock and rash are cotemporal events in dengue. Pathogenetically, rash and shock may be identical, differing only quantitatively and in the major site of activity.

A key question requiring experimental study is the identity of the stimulus which elicits the release of lysosomal enzymes, thromboplastin, or other inflammatory mediators from mononuclear phagocytes. For reasons discussed above, it seems probable that an immune response directed against dengue-infected mononuclear phagocytes is a powerful eliciting event. It may be possible to study this phenomenon *in vitro*. Dengue-infected mononuclear phagocytes can be mixed with

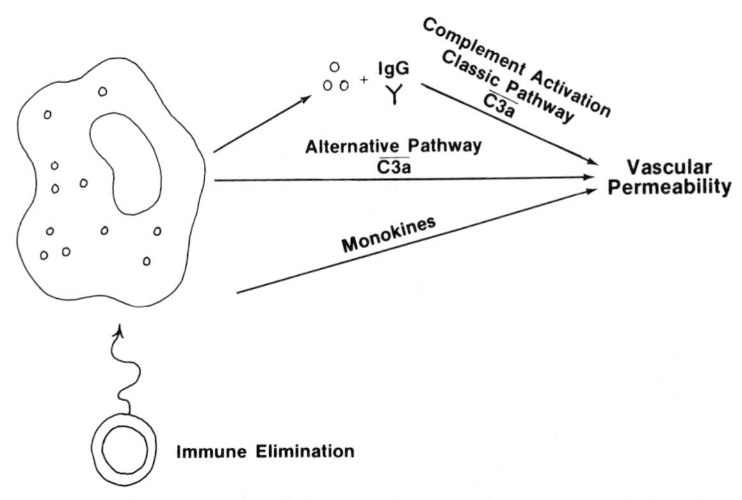

Fig. 17. Proposed effector mechanisms of rash in dengue fever. Infected mononuclear monocytes migrate to the skin. After attack by an effector lymphocyte, infected histiocyte releases either virus which complexes with antibody, activating complement by the classic pathway, or lysosomal enzymes which activate complement by the alternative pathway or by vasoactive lymphokine (either by histiocyte or an effector cell).

identified antidengue humoral or cellular components and examined for the generation of mediators. If this approach is successful, a new era in research on viral pathogenesis will have emerged.

IV. COMMENT

Although the general hypothesis of the pathogenesis of DSS is complex, it is in conformance with established biological attributes of each of its components—virus, antibody, mononuclear phagocytes, and the immune response. Further, it predicts integrative relationships between virus structure and pathogenesis at the molecular level. The roles of antibody and immune elimination response in regulating the severity of dengue disease provide afferent and efferent limbs which can be segregated and studied separately. Finally, it is evident that antibody may be the most efficient mechanism for enhancing cellular infection, but not the only mechanism. Immunological and nonimmunological influences on mononuclear phagocytes may be additive or antagonistic. Careful clinical studies on unusual dengue infections may clarify the mechanisms involved.

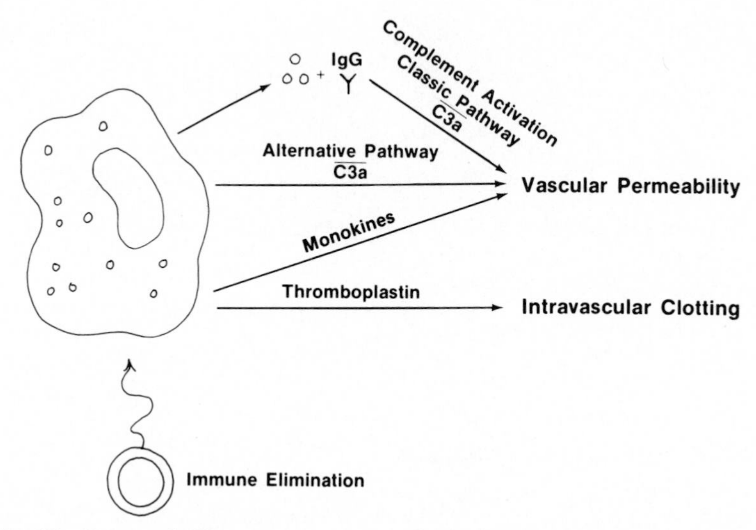

Fig. 18. Proposed effector mechanisms of DSS. Infected mononuclear phagocytes are widely distributed in tissues, especially the liver. After attack by an effector lymphocyte, macrophages release either virus particles which complex with antibody, activating complement by the classic pathway, or lysosomal enzymes which activate complement via the alternative pathway or vasoactive lymphokines (either by macrophages or effector lymphocytes).

The hypothesis has important implications for the control of DHF. For example, if the interval between dengue infections controls the severity of disease, a logical goal in prevention would be to interrupt closely spaced transmission of different dengue viruses. If secondary dengue-2 infection accounts for the vast majority of fatal dengue, a monovalent dengue-2 vaccine might effectively prevent most severe cases. Such a vaccine would be easier to produce and standardize than a polyvalent vaccine. Evidence that the interval or infection sequence is important can only come from longitudinal epidemiological studies. Several comparative studies should be established in areas endemic for DSS and in areas in which dengue transmission occurs but DSS does not. It is important that virological and clinical uniformity be maintained by the use of standardized techniques and interpretive criteria. The classification of primary and secondary dengue antibody responses should be based on Ig separation.

Attempts have been made in this chapter to derive estimates for rates of occurrence of DHF and DSS in primary and secondary dengue infections. To provide quantitative estimates of virulence, further measurements of these rates should be made.

V. APPENDIX

Attack Rates of DHF and DSS in Primary and Secondary Dengue Infections in Bangkok–Thonburi in 1962

Etiology of hospitalized hemorrhagic fever patients	Cases	Percentage of total	Ref.[a]
1. Studied virologically (adequate specimens)	150		(1)
2. Dengue infection	121	80.7	(1)
3. Primary dengue infection, all ages	20	13.3	(1)
4. Secondary dengue infection, all ages	101	67.3	(1)
5. Primary dengue infection, <1 year old	8	5.3[b]	(2)
6. Primary dengue infection, ≥1 year old	12	8.0[b]	(2)
7. Shock, primary dengue infection, ≥1 year old	<1	0.32[b]	(2)
8. Shock, secondary dengue infection	57	38.0	(1)
9. Shock, no etiology	1.3	0.64[b]	(2)

Epidemiological data	Data	Ref.
1. Population, Bangkok and Thonburi, ages <1–14 (1960)	930,451	(3)
2. Population, Bangkok and Thonburi, <1 year old (1960)	88,000	(3)
3. Population, Bangkok and Thonburi, ages 1–14 (1960)	842,451	(3)
4. DHF hospitalizations, Bangkok and Thonburi, ages <1–14 (1962)	3,757	(4)
5. Serological survey, no dengue antibody, ages 1–14	1,006	(4)
6. Serological survey with dengue antibody, ages 1–14	1,091	(4)
7. Percent susceptibles, pre-1962 epidemic, ages 1–14	48.0%	(4)
8. Serological survey, no dengue antibody, <1 year old	97	(4)
9. Serological survey with dengue antibody, <1 year old	36	(4)
10. Percent susceptibles, pre-1962 epidemic, <1 year old	73.5%	(4)
11. Dengue infection rate in Bangkok (1962) (weighted)	41.0%	(4)
12. Dengue types infecting humans in Bangkok (1962) were types 1–3 predominantly		

DHF–DSS attack rates in children	Data
<1 year old	
1. DHF in infants <1 year old = 0.053[c] × 3757	199
2. Total infants without dengue antibody, preepidemic (susceptibles) = 88,000 × 0.735	63,680
3. Total dengue infections during epidemic = susceptibles × 0.41 = 64,680 × 0.41	26,519
4. DHF per 1000 dengue infections = 199/26,519	7.5/1000
1–14 years old	
1. Without antibody, pre-1962 epidemic = 842,451 × 0.48	404,376
2. With antibody = 842,451 − 404,376	438,075
3. Primary dengue infection during epidemic = 404,376 × 0.41	165,794
4. Secondary dengue infections	
a. Total infection exposures = 438,075 × 0.41 = 179,611	

(Continued)

APPENDIX (*Continued*)

DHF–DSS attack rates in children	Data
b. Estimated immunes = 179,611 × 0.3 = 53,883	
c. Estimated infections = 179,611 − 53,883	125,728
5. Primary-infection DHF = 0.08 × 3757	297
6. DHF per 1000 primary dengue infections = 297/165,794	1.8/1000
7. Primary-infection DSS = 0.0032 × 3757	12
8. DSS per 1000 primary dengue infections = 12/165,794	0.07/1000
9. Secondary-infection DHF = 0.673 × 3757	2528
10. DHF per 1000 secondary dengue infections = 2528/125,728	20.1/1000
11. Secondary-infection DSS = 0.38 × 3757	1428
12. DSS per 1000 secondary dengue infections	11.4/1000

[a] Key to references: (1) Nimmannitya *et al.* (1969); (2) Halstead (1970); (3) Thailand Population Census (1960); (4) Halstead *et al.* (1969c).

[b] Based upon combined 1962–1964 data.

[c] Almost certainly an underestimate because of the selective bias against including infants in a virological study. There were 407 infants <1 year old hospitalized for DHF in 1962 and, assuming 80% (326) had a dengue etiology, 12.3 hospitalizations per 1000 infections.

REFERENCES

Alexander, R. F., and Spriggs, A. I. (1960). *J. Clin. Pathol.* **13**, 414–424.

Anonymous (1928). *League Nations Mon. Epidemiol. Rep.* **7**, 334–338.

Aziz, M. A., Gorham, J. R., and Gregg, M. B. (1967). *Pak. J. Med. Res.* **6**, 83–92.

Barnes, W. J. S., and Rosen, L. (1974). *Am. J. Trop. Med. Hyg.* **23**, 495–506.

Bhamarapravati, N., Tuchinda, P., and Boonyapaknavik, V. (1967). *Ann. Trop. Med. Parasitol.* **61**, 500–510.

Bierman, H. R., and Nelson, E. R. (1965). *Ann. Intern. Med.* **62**, 867–884.

Bokisch, V. A., Top, F. H., Russell, P. K., Dixon, F. J., and Müller-Eberhard, H. J. (1973). *N. Engl. J. Med.* **289**, 996–1000.

Boonpucknavig, S., Bhamarapravati, N., Nimmannitya, S., Phalavadhtana, A., and Siripont, J. (1976). *Am. J. Pathol.* **85**, 37–48.

Boonpucknavig, S., Boonpucknavig, V., Bhamarapravati, N., and Nimmannitya, S. (1979). *Arch. Pathol. Lab. Med.* **103**, 463–466.

Boonpucknavig, V., Bhamarapravati, N., Boonpucknavig, S., Futrakul, P., and Tanpaichitr, P. (1976). *Arch. Pathol. Lab. Med.* **100**, 206–212.

Boyum, A. (1968). *Scand. J. Clin. Lab. Invest.* **21**, Suppl., 77–89.

Brier, A. M., Snyderman, R., Mergenhagen, S. E., and Notkins, A. L. (1970). *Science* **170**, 1104–1106.

Burke, T. (1968). *Trans. R. Soc. Trop. Med. Hyg.* **62**, 682–692.

Cardamatis, J. P. (1929). *Bull. Soc. Pathol. Exot.* **22**, 272–296.

Carey, D. E., Myers, R. M., and Rodrigues, F. M. (1965). *Am. J. Trop. Med. Hyg.* **14**, 448–450.

Carey, D. E., Myers, R. M., Reuben, R., and Rodrigues, F. M. (1966). *Am. J. Trop. Med. Hyg.* **15**, 580–587.
Carey, D. E., Rodrigues, F. M., Myers, R. M., and Webb, J. K. G. (1968). *Indian Pediatr.* **5**, 285–296.
Carey, D. E., Causey, O. R., Reddy, S., and Cooke, A. R. (1971). *Lancet* **1**, 105–106.
Cohen, S. N., and Halstead, S. B. (1966). *J. Pediatr.* **68**, 448–456.
Copanaris, P. (1928). *Off. Int. Hyg. Publique Bull.* **20**, 1590–1601.
Corbin, N. C., Hugli, T. E., and Müller-Eberhard, H. J. (1976). *Anal. Biochem.* **73**, 41–51.
Dasaneyavaja, A., and Charansri, U. (1961). *SEATO Med. Res. Monogr.* No. 2, p. 61.
Dasaneyavaja, A., and Pongsupat, S. (1961). *J. Trop. Med. Hyg.* **64**, 310–314.
Dengue Newsletter (1976). *South Pac. Comm.* No. 4. Noumea, New Caledonia.
Doherty, R. L., Westaway, E. G., and Whitehead, R. H. (1967). *Med. J. Aust.* **2**, 1078–1080.
Dourmashkin, R. R., and Patterson, S. (1975). *Inflammation* **1**, 155–161.
Edelman, R., Nimmannitya, S., Colman, R. W., Talamo, R. C., and Top, F. H., Jr. (1975). *J. Lab. Clin. Med.* **86**, 410–421.
Edelman, R., Schneider, R. J., Vejjajwa, A., Pornpibul, R., and Voodhikul, P. (1976). *Am. J. Trop. Med. Hyg.* **25**, 733–738.
Fireman, P., Friday, G., and Kumante, J. (1969). *Pediatrics* **43**, 265–272.
Fischer, D. B., and Halstead, S. B. (1970). *Yale J. Biol. Med.* **42**, 329–349.
Gleiser, C. A., Goschenour, W. S., Jr., Berge, T. O., and Tigertt, W. D. (1962). *J. Infect. Dis.* **110**, 80–97.
Goldsmid, J. A. (1917). *Med. J. Aust.* **1**, 7–8.
Goud, T. J. L. M., Schotte, C., and Van Furth, R. (1975). *J. Exp. Med.* **142**, 1180–1199.
Groot, H. (1976). *Newsl. Dengue, Yellow Fever Aedes aegypti Am.* **5**, No. 2. Pan Am. Health Org., Washington, D.C.
Gubler, D. J., Reed, D., Rosen, L., and Hitchcock, J. C., Jr. (1978). *Am. J. Trop. Med. Hyg.* **27**, 581–589.
Halstead, S. B. (1966). *Bull. W.H.O.* **35**, 3–15.
Halstead, S. B. (1970). *Yale J. Biol. Med.* **42**, 350–362.
Halstead, S. B. (1974). *Am. J. Trop. Med. Hyg.* **23**, 974–982.
Halstead, S. B. (1978). *Asian J. Infect. Dis.* **2**, 59–66.
Halstead, S. B. (1979). *J. Infect. Dis.* **140**, 527–533.
Halstead, S. B., and O'Rourke, E. J. (1977a). *Nature (London)* **265**, 739–741.
Halstead, S. B., and O'Rourke, E. J. (1977b). *J. Exp. Med.* **146**, 201–217.
Halstead, S. B., and Simasthien, P. (1970). *Yale J. Biol. Med.* **42**, 276–292.
Halstead, S. B., and Yamarat, C. (1965). *Am. J. Public Health* **55**, 1386–1395.
Halstead, S. B., Nimmannitya, S., Yamarat, C., and Russell, P. K. (1967). *Jpn. J. Med. Sci. Biol.* **20**, Suppl., 96–102.
Halstead, S. B., Nimmannitya, S., and Margiotta, M. R. (1969a). *Am. J. Trop. Med. Hyg.* **18**, 972–983.
Halstead, S. B., Udomsakdi, S., Singharaj, P., and Nisalak, A. (1969b). *Am. J. Trop. Med. Hyg.* **18**, 984–996.
Halstead, S. B., Scanlon, J. E., Umpaivit, P., and Udomsakdi, S. (1969c). *Am. J. Trop. Med. Hyg.* **18**, 977–1021.
Halstead, S. B., Nimmannitya, S., and Cohen, S. N. (1970). *Yale J. Biol. Med.* **42**, 311–328.
Halstead, S. B., Casals, J., Shotwell, H., and Palumbo, N. (1973a). *Am. J. Trop. Med. Hyg.* **22**, 365–374.

Halstead, S. B., Chow, J. S., and Marchette, N. J. (1973b). *Nature (London), New Biol.* **243**, 24–25.
Halstead, S. B., Marchette, N. J., and Chow, J. S. S. (1973c). *Adv. Biosci.* **12**, 401–416.
Halstead, S. B., Shotwell, H., and Casals, J. (1973d). *J. Infect. Dis.* **128**, 15–22.
Halstead, S. B., Marchette, N. J., Chow, J. S. S., and Lolekha, S. (1976). *Proc. Soc. Exp. Biol. Med.* **151**, 136–139.
Halstead, S. B., O'Rourke, E. J., and Allison, A. C. (1977). *J. Exp. Med.* **146**, 218–229.
Hammon, W. McD. (1969). *Am. J. Trop. Med. Hyg.* **18**, 159–165.
Hammon, W. McD. (1973). *Am. J. Trop. Med. Hyg.* **22**, 82–91.
Hammon, W. McD., and Sather, G. E. (1964). *Mil. Med.* **129**, 130–135.
Hammon, W. McD., Rudnick, A., and Sather, G. E. (1960a). *Science* **131**, 1102–1103.
Hammon, W. McD., Rudnick, A., Sather, G., Rogers, K. D., and Morse, L. J. (1960b). *Trans. Assoc. Am. Physicians* **73**, 140–155.
Hamza, M. H. M., Panabokke, R. G., and Balasubramanium, C. D. (1966). *J. Trop. Med. Hyg.* **69**, 162–164.
Hare, F. E. (1898). *Australas. Med. Gaz.* **17**, 98–107.
Hirsch, M. S., and Murphy, F. A. (1968). *Lancet* **ii**, 37–40.
Howard, R. J., Craig, C. P., Trevino, G. S., Dougherty, S. F., and Mergenhagen, S. E. (1969a). *J. Immunol.* **103**, 699–707.
Howard, R. J., Notkins, A. L., and Mergenhagen, S. E. (1969b). *Nature (London)* **221**, 873–874.
Hruskova, J., Rychterova, V., and Kliment, V. (1972a). *Acta Virol. (Engl. Ed.)* **16**, 115–124.
Hruskova, J., Rychterova, V., and Kliment, V. (1972b). *Acta Virol. (Engl. Ed.)* **16**, 125–132.
Hyman, A. S. (1943). *War Med.* **4**, 497–501.
Ishii, K., Matsunaga, Y., and Kono, R. (1968). *J. Immunol.* **101**, 770–775.
Jatanasen, J., Sakuntanaga, P., and Dhanasiri, C. (1961). *SEATO Med. Res. Monogr.* No. 2, 6–21.
Johnson, K. M., Halstead, S. B., and Cohen, S. N. (1967). *Prog. Med. Virol.* **9**, 105–158.
Kauffman, C. A., Phair, J. P., Linnemann, C. C., Jr., and Schiff, G. M. (1974). *Infect. Immun.* **10**, 212–1215.
Kisner, P., and Lisansky, E. T. (1944). *Ann. Intern. Med.* **20**, 40–51.
Kuberski, T., Rosen, L., and Reed, D., and Mataika, J. (1977). *Am. J. Trop. Med. Hyg.* **26**, 775–783.
Laigret, J., Parc, F., and Rosen, L. (1979). *Morbid. Mortal. Wkly. Rep.* **28**, 194–196.
Lamb, G. A. (1969). *Am. J. Dis. Child.* **118**, 261.
Likosky, W. H., Calisher, C. H., Michelson, A. L., Correa-Coronas, R., Henderson, B. E., and Feldman, R. A. (1973). *Am. J. Trop. Med. Hyg.* **97**, 264–275.
Lumley, G. F., and Taylor, F. H. (1943). "Dengue," Sch. Public Health Trop. Med., Serv. Publ. No. 3. Australas. Med. Publ. Co., Sydney.
McFarland, H. F. (1974). *J. Immunol.* **113**, 173–180.
Marchette, N. J., Halstead, S. B., Falkler, W. A., Stenhouse, A., and Nash, D. (1973). *J. Infect. Dis.* **128**, 23–30.
Marchette, N. J., Sung, J. S., Halstead, S. B., and Lolekha, S. (1975). *Southeast Asian J. Trop. Med. Public Health* **6**, 316–321.
Marchette, N. J., Halstead, S. B., and Chow, J. S. (1976). *J. Infect. Dis.* **133**, 274–282.
Marchette, N. J., O'Rourke, T., and Halstead, S. B. (1979). *Infect. Immun.* **24**, 47–50.
Matsuda, S. (1962). *Koshu Eiseiin Kenkyu Hokoku* **11**, 173–190.
Memoranda (1973). *Bull. W.H.O.* **48**, 117–132.

Mergenhagen, S. E., Notkins, A. L., and Dougherty, S. F. (1967). *J. Immunol.* **99,** 567–581.
Midulla, M., Businco, L., and Moschini, L. (1972). *Acta Paediatr. Scand.* **61,** 609–611.
Miles, J. A. R. (1978). *Asian J. Infect. Dis.* **2,** 1–5.
Monath, T. P. C. (1971). *Am. J. Epidemiol.* **93,** 122–129.
Moore, D. L., Causey, D. R., Carey, D. E., Reddy, S., Cooke, A. R., Akinkugbe, F. M., David-West, T. S., and Kemp, G. E. (1975). *Ann. Trop. Med. Parasitol.* **69,** 49–64.
Morales, A., Groot, H., Russell, P. K., and McCown, J. M. (1973). *Am. J. Trop. Med. Hyg.* **22,** 785–787.
Moreau, J. P., Rosen, L., Saugrain, J., and Lagraulet, J. (1973). *Am. J. Trop. Med. Hyg.* **22,** 237–241.
Myers, R. M., and Varkey, M. J. (1970). *Indian J. Med. Res.* **58,** 1301–1306.
Myers, R. M., Varkey, M. J., Reuben, R., and Jesudass, E. S. (1970). *Indian J. Med. Res.* **58,** 24–30.
Neff, J. M., Morris, L., Gonzalez-Alcover, R., Coleman, P. H., Lyss, S. B., and Negron, H. (1967). *Am. J. Epidemiol.* **86,** 162–184.
Nelson, E. R. (1960). *J. Pediatr.* **56,** 101–108.
Nelson, E. R., and Bierman, H. R. (1964). *J. Am. Med. Assoc.* **190,** 99–103.
Nicholson, A., Brade, V., Schorlemmer, H. U., Burger, R., Bitter-Suermann, C., and Hadding, U. (1975). *J. Immunol.* **115,** 1108–1113.
Nimmannitya, S., Halstead, S. B., Cohen, S. N., and Margiotta, M. R. (1969). *Am. J. Trop. Med. Hyg.* **18,** 954–971.
Nisalak, A., Halstead, S. B., Singharaj, P., Udomsakdi, S., Nye, S. W., and Vinijchaikul, K. (1970). *Yale J. Biol. Med.* **42,** 293–310.
Nomura, S., and Akashi, K. (1931). *Taiwan Igakkai Zasshi* **30,** 1154–1157.
Notkins, A. L., Mergenhagen, S., Rizzo, A., Scheele, C., and Waldmann, T. (1966). *J. Exp. Med.* **123,** 347–364.
O'Rourke, E. J., Halstead, S. B., Allison, A. C., and Platts-Mills, T. A. E. (1978). *J. Immunol. Methods* **19,** 137–151.
Papaevangelou, G., and Halstead, S. B. (1977). *J. Trop. Med. Hyg.* **80,** 46–51.
Pavlatos, M., and Smith, C. E. G. (1964). *Trans. R. Soc. Trop. Med. Hyg.* **58,** 422–424.
Pavri, K. M. (1976). *Indian J. Med. Res.* **64,** 713–729.
Piyaratn, P. (1961). *Am. J. Trop. Med. Hyg.* **10,** 767–772.
Porter, D. D., and Larsen, A. E. (1974). *Prog. Med. Virol.* **18,** 32–47.
Prydz, H., Allison, A. C., and Schorlemmer, H. U. (1977). *Nature (London)* **270,** 173–174.
Quintos, F. N., Lim, L. E., Juliano, L., Reyes, A., and Lacson, P. (1954). *Philipp. J. Pediatr.* **3,** 1–19.
Rawls, W. E. (1974). *Prog. Med. Virol.* **18,** 273–288.
Rice, L. (1923). *Am. J. Trop. Med.* **3,** 73–90.
Rosen, L. (1967). *Jpn. J. Med. Sci. Biol.* **20,** Suppl., 67–69.
Rosen, L. (1974). *Am. J. Trop. Med. Hyg.* **23,** 1205–1206.
Rosen, L. (1977). *Am. J. Trop. Med. Hyg.* **26,** 337–343.
Rosen, L., and Gubler, D. (1974). *Am. J. Trop. Med. Hyg.* **23,** 1153–1160.
Rowan, L. C. (1956). *Med. J. Aust.* **43,** 651–655.
Rudnick, A. (1967). *Bull. W.H.O.* **36,** 528–532.
Rush, B. (1789). "Medical Inquiries and Reflections," pp. 104–121. Prichard & Hall, Philadelphia, Pennsylvania.
Russell, P. K. (1970). *In* "Immunopathology" (P. A. Miescher, ed.), pp. 426–435. Schwabe, Basel.

Russell, P. K., and McCown, J. M. (1972). *Am. J. Trop. Med. Hyg.* **21**, 97–99.
Russell, P. K., Udomsakdi, S., and Halstead, S. B. (1967). *Jpn. J. Med. Sci. Biol.* **20**, Suppl., 103–108.
Russell, P. K., Yuill, T. M., Nisalak, A., Udomsakdi, S., Gould, D. J., and Winter, P. E. (1968). *Am. J. Trop. Med. Hyg.* **17**, 600–608.
Russell, P. K., Intavivat, A., and Kanchanapilant, S. (1969). *J. Immunol.* **102**, 412–420.
Sabin, A. B. (1952). *Am. J. Trop. Med. Hyg.* **1**, 30–50.
Saugrain, J., Rosen, L., Outin-Fabre, D., and Moreau, J. P. (1970). *Bull. Soc. Pathol. Exot.* **63**, 636–642.
Scherer, W. F., Breakenridge, F. A., and Dickerman, R. W. (1972). *Am. J. Epidemiol.* **95**, 67–79.
Scherer, W. F., Russell, P., Rosen, L., Casals, J., and Dickerman, R. (1978). *Am. J. Trop. Med. Hyg.* **27**, 590–599.
Schlesinger, R. W. (1971). *Proc. Conf. Pathogenesis Arboviral Infect., Honolulu* pp. 95–109.
Schorlemmer, H. U., and Allison, A. C. (1976). *Immunology* **31**, 781–788.
Schorlemmer, H. U., Davies, P., and Allison, A. C. (1976). *Nature (London)* **261**, 48–49.
Scott, L. C. (1923). *J. Am. Med. Assoc.* **80**, 387–393.
Scott, R. M., McCown, J. M., and Russell, P. K. (1972). *Infect. Immun.* **6**, 277–281.
Scott, R. M., Nimmannitya, S., Bancroft, W. H., and Mansuwan, P. (1976). *Am. J. Trop. Med. Hyg.* **25**, 866–874.
Siler, J. F., Hall, M. W., and Hitchens, A. P. (1926). *Philipp. J. Sci.* **29**, 1–211.
Simmons, J. S., St. John, J. H., and Reynolds, F. H. K. (1931). *Philipp. J. Sci.* **44**, 1–251.
Smith, T. J., McKinney, R. W., and Sawyer, W. D. (1964). *Proc. Soc. Exp. Biol. Med.* **117**, 271–275.
Smithwick, E. M., and Berkovich, S. (1966). *Proc. Soc. Exp. Biol. Med.* **123**, 276–278.
Sobel, A. T., Bokisch, V. A., and Müller-Eberhard, H. J. (1975). *J. Exp. Med.* **142**, 139–150.
Staab, E. V., Normann, S. J., and Craig, C. P. (1970). *RES J. Reticuloendothel. Soc.* **8**, 342–348.
Stransky, E., and Lim, L. E. (1956). *Ann. Paediatr.* **187**, 309–320.
Sung, J. S., Diwan, A. R., Falkler, W. A., Jr., Yang, H. Y., and Halstead, S. B. (1975). *Intervirology* **5**, 137–149.
Technical Guides for Diagnosis, Treatment, Surveillance, Prevention and Control of Dengue Hemorrhagic Fever (1975). World Health Organ., Geneva.
Thailand Population Census (1960). Cent. Stat. Off., Natl. Econ. Dev. Board, Bangkok.
Theiler, M., Casals, J., and Moutousses, C. (1960). *Proc. Soc. Exp. Biol. Med.* **103**, 244–246.
Theofilopoulos, A. N., Brandt, W. E., Russell, P. K., and Dixon, F. J. (1976). *J. Immunol.* **117**, 953–961.
Tigertt, W. D., Crosby, W. H., Berge, T. O., Howie, D. L., Kress, S., Dangerfield, H. G., Bass, J. W., and Frank, W. (1962). *Cancer (Philadelphia)* **15**, 628–632.
Twomey, J. J. (1974). *J. Immunol.* **112**, 2278–2281.
Van Furth, R., and Cohn, Z. A. (1968). *J. Exp. Med.* **128**, 415–435.
Vedder, E. B. (1907). *N.Y. Med. J.* **86**, 203–206.
Ventura, A. K., and Ehrenkranz, M. J. (1976). *J. Inf. Dis.* **134**, 436–441.
Ventura, A. K., and Hewitt, C. M. (1970). *Am. J. Trop. Med. Hyg.* **19**, 712–715.
Ventura, A. K., Ehrenkranz, M. J., and Rosenthal, D. (1975). *J. Infect. Dis.* **131**, Suppl., 62–68.
Vernon-Roberts, B. (1969). *Int. Rev. Cytol.* **25**, 131–159.

Von Pirquet, C. (1908). *Dtsch. Med. Wochenschr.* **34,** 1297–1300.
Ward, P. A., Cohen, S., and Flanagan, T. D. (1972). *J. Exp. Med.* **135,** 1095–1103.
Weiss, H. J., and Halstead, S. B. (1965). *J. Pediatr.* **66,** 918–926.
Wheelock, F., and Edelman, R. (1969). *J. Immunol.* **103,** 429–436.
Whitehead, R. H., Chaicumpa, V., Olson, L. C., and Russell, P. K. (1970). *Am. J. Trop. Med. Hyg.* **19,** 94–102.
Winter, P. E., Yuill, T. M., Udomsakdi, S., Gould, D., Nantapanich, S., and Russell, P. K. (1968). *Am. J. Trop. Med. Hyg.* **17,** 590–599.
Winter, P. E., Nantapanich, S., Nisalak, A., Udomsakdi, S., Dewey, R. W., and Russell, P. K. (1969). *Am. J. Trop. Med. Hyg.* **18,** 573–579.
Woodruff, J. F., and Woodruff, J. J. (1975). *In* "Viral Immunology and Immunopathology" (A. L. Notkins, ed.), pp. 393–418. Academic Press, New York.
Zisman, B., Wheelock, E. F., and Allison, A. C. (1971). *J. Immunol.* **107,** 236–243.
Zweiman, B. (1971). *J. Immunol.* **106,** 1154–1158.

Epidemiology of Arthropod-Borne Togaviruses: The Role of Arthropods as Hosts and Vectors and of Vertebrate Hosts in Natural Transmission Cycles

ROY W. CHAMBERLAIN

I. INTRODUCTION

The genera *Alphavirus* and *Flavivirus*, which correspond to Casals' antigenic groups A and B, are the "typical" arboviruses, that is, the ones with which we are most familiar. The viruses included in these genera cause a large share of known human and animal arboviral disease and are generally transmitted by mosquitoes or ticks. Superficially, their transmission cycles appear to be quite simple; however, even after many years of study, there are still many unknowns which continue to fascinate and challenge ecologists, entomologists, and epidemiologists alike. These viruses are also numerous and widely distributed. Of the nearly 400 viruses presently registered in the "International Catalogue of Arboviruses" (Berge, 1975; Karabatsos,

THE TOGAVIRUSES

ISBN 0-12-625380-3

1978), they account for 78 (20 alphaviruses, 58 flaviviruses) and have representatives in all tropical and temperate regions of the world.

Alphaviruses are almost exclusively mosquito-borne. Flaviviruses, on the other hand, fall into three separate subgroups according to vector—the mosquito-borne, the tick-borne, and those with no known vector at all. These vector subgroups of flaviviruses generally show distinct serological differences, possibly as a consequence of adaptation to vector type. Mosquito-borne flaviviruses and alphaviruses share some of the same vectors; the members of the tick-borne flavivirus subgroup are, however, quite distinct in this respect and generally will not infect mosquitoes, even experimentally. Non-vector-borne flavivirus types include a number of bat and rodent viruses that serologically appear to be about equally related to the mosquito-borne and tick-borne forms (Berge, 1975) and seem to get along quite well without any arthropod vectors whatsoever. As one can appreciate, non-vector-borne flaviviruses have been frustrating for arbovirus taxonomists of the old school—how should they classify black sheep such as these, i.e., "arthropod-borne" viruses which are obviously not arthropod-borne? Nonetheless, speculation as to how such virus–host relations came about has proved to be an interesting pastime. The general consensus is that these viruses probably split off from vector-borne ancestral stock at some time in the distant past, for the simple reason that direct transmission between their vertebrate hosts had become so completely adequate in ensuring virus survival that there was no longer the need for an arthropod vector. Without the usual alternation of virus between vertebrate and arthropod to ensure bilateral selection for infectivity, arthropod infectivity was thereupon lost. For those speculators who disagree with this view, an opposite one can be almost as convincingly argued, namely, that none of the original flaviviruses had vectors but some later gained them.

At the outset, certain basic parameters of vector and host infection that the alphaviruses and flaviviruses have in common with other arthropod-borne viruses should be mentioned, since knowledge of them will enable the reader to better understand vector–virus–host interactions as they occur in the field. Notably, a threshold phenomenon for virus infection exists in the gut of the mosquito (Chamberlain *et al.*, 1954) and, presumably, in the gut of other kinds of arthropod vectors as well (Burgdorfer and Varma, 1967). Briefly stated, a critical concentration of virus in the blood meal is required to establish infection in a significant proportion of the vectors. This requirement varies with different vector species and different viruses. For example, one species of mosquito may require blood with a virus titer of only 3 logs/ml to infect it, while another may require 7 logs/ml; also, a mosquito species may have a very low threshold for one kind of virus but a very high threshold for a different kind of virus. It is clear that, other factors being equal, the more susceptible vector species has an edge. Once the threshold "barrier" in the gut is overcome, however, the ingested virus appears to develop about equally as well in a high-threshold species as in a low-threshold species, so that a high-threshold vector can assume importance if associated with high-viremia vertebrate hosts. This has strong epidemiological implications.

Obviously, infection threshold cannot be considered alone, but must be viewed together with host viremia and the degree of association between vector and host. The viremia produced in the natural host must be adequate to infect an effective proportion of the associated vector species. This situation does not appear to develop by accident but rather as the result of a continuing selection, with every virus cycle, for virus strains producing viremias of sufficient level (and infectivity) to infect the primary vector species (Chamberlain *et al.*, 1959). Virus strains which do not meet this criterion fail to be further transmitted and presumably die out.

After the vector takes an infective meal, an incubation period of several days to two or more weeks (depending on ambient temperature) is required for virus to develop and infect the salivary glands. Thereafter, the vector generally remains infected for life and is capable of transmitting several times if it lives long enough and feeds on susceptible hosts.

The infected vertebrate host also requires an incubation period before it becomes viremic, but this period is generally quite short, from a few hours to a few days. In warm-blooded hosts detectable viremia usually lasts only 2–6 days and is at the peak levels required for vector infection for only part of this time. Therefore, close association between host and vector is essential to continued virus propagation. As the viremia subsides, neutralizing antibodies develop and for some viruses may persist for the life of the animal.

Those Togaviridae of the most obvious disease importance have attracted the most attention and have therefore been the most intensively investigated. As a result, we know quite a bit about the vector relations of certain viruses, such as those which cause eastern equine encephalomyelitis (EEE), Japanese encephalitis (JE), and yellow fever (YF), but know essentially nothing of some others of lesser disease importance. This spotty information on vectors and hosts makes generalization difficult. In an attempt to fit the different viruses into defined patterns of vector–host interaction, it will often be necessary in the following sections to make some rather broad assumptions, based on the fuller knowledge of the members that have been studied more extensively.

II. THE ALPHAVIRUSES

A. General Biological Features

As mentioned above, alphaviruses are, with rare exception, mosquito-borne. Although occasionally one virus or another has been isolated from other field-collected arthropods besides mosquitoes and experimental infection of a few such arthropods has been demonstrated in the laboratory, there is little evidence that they play significant roles in naturally occurring transmission cycles. Arthropods in question include such diverse groups as blackflies (Anderson *et al.*, 1961; Trapido, 1972), *Culicoides* gnats (Karstad *et al.*, 1957), chicken lice (Howitt *et al.*, 1948), bird mites (Howitt *et al.*, 1948; Sulkin *et al.*, 1955; Reeves *et al.*, 1955; Miles *et al.*, 1951; Chamberlain and Sikes, 1955; Theiler and Downs, 1973, p. 144), rodent mites (Theiler and Downs, 1973, p. 133), kissing bugs (Kitselman and Grundman, 1940), and ticks (Burgdorfer and Varma, 1967). A possible exception (to be considered later) is the bird bug, *Oeciacus vicarius,* which appears to be important in the transmission of both a newly recognized western equine encephalomyelitis (WEE)-related virus and a new Venezuelan equine encephalitis (VEE) group member (Monath, 1977).

The basic enzootic (maintenance) cycle of most alphaviruses appears to involve, in a particular geographic zone, only a single vector species and a single vertebrate host species (or several closely affiliated host species), restricted to or abundant in a common ecological niche favoring vector–host interaction. The level of virus that circulates in the blood of the infected vertebrate is adequate to infect the vector species; and the vector, in turn, after an appropriate incubation period, is able to transmit infection to other members of the host group efficiently enough to ensure continual virus propagation throughout the transmission season. Neither the vector mosquito nor the host vertebrate in this basic virus focus suffers appreciable ill effects as a result of its involvement; if either did, the chance of virus survival could be reduced. And rarely, if ever, does the host population become totally immune; as the transmission season progresses, a balance is eventually reached between vector infection and host immunity. Three major factors are involved: host turnover, which determines the supply of new susceptibles; vector breeding and longevity, which directly affect the size and parity of the vector population; and the increasing level of host herd immunity as the transmission season progresses. The increased herd immunity dampens transmitting efficiency, because vectors already infected may feed on immune hosts instead of susceptible ones. It also interferes with the acquisition of virus by vectors not yet infected, because they, too, may feed on immune hosts instead of on those with a viremia.

Continuous transmission of virus by mosquitoes in enzootic cycles as described above—rapid in some years and slower in others, according to the variable abundance of vectors and hosts—could possibly occur in tropical and subtropical zones, but as an only means of ensuring virus perpetuation, it appears to be tenuous. Therefore, it has been presupposed by many workers in this field (Reeves, 1974) that there must be some other mechanism or mechanisms for ensuring virus persistence through inclement or otherwise critical periods besides continuous transmission: delayed, chronic, or recrudescent infections in the basic vertebrate hosts; involvement of still unknown accessory hosts or vectors; persistence of virus in aestivating or hibernating mosquitoes; and, perhaps, transovarial transmission in mosquitoes. Where winged vertebrate hosts may be involved, seasonal introduction of virus into temperate regions in the spring and its return to the tropics in the fall, as an extension of a basic tropical cycle, has been hypothesized and has received considerable support.

However, none of these suggested survival mechanisms has been proved. This is not surprising, considering the extreme difficulty of accumulating the quantitative field and laboratory data required for such proof. Other than demonstration of extended viremias of a few representative alphaviruses in cold-blooded animals (Karstad, 1961; Thomas *et al.*, 1959; Thomas and Eklund, 1960; Bowen, 1977), isolation of WEE virus from snakes and frogs (Gebhardt *et al.*, 1964; Burton *et al.*, 1966), and rare isolation of WEE virus from organs of immune birds (Reeves *et al.*, 1958), there is little evidence that vertebrates might serve in a reservoir capacity. Nor is there solid evidence that transovarial transmission in vector mosquitoes either occurs at all in nature or is significant. A few clues along the way, however, suggest that transovarial transmission should be studied further: isolation of WEE virus on one occasion from field-collected male *Culiseta melanura* in Alabama (Stamm *et al.*, 1962); isolation of EEE virus from *Cs. melanura* larvae in Massachusetts (Daniels, 1960); precise seasonal (early summer to midsummer) reappearance of both EEE and WEE viruses, year after year, in *Cs. melanura* (with nearly identical infection rates) in the same swamp habitat of the southeastern United States, long after peaks of bird influx and reproduction had passed (Chamberlain *et al*., 1964–1965). In the tropics and subtropics, aestivation of infected adult mosquitoes also deserves further consideration; continuous transmission of alphaviruses in a tropical environment may indeed be possible if it is aided by such a delaying mechanism in carrying the virus over periods of low mosquito activity (Davies, 1972). The cessation of transmitting activity during vector aestivation could also give the associated host population time to reproduce and dilute out its immunes.

The basic vertebrate hosts for most alphaviruses are either wild birds or various species of wild rodents. There are, of course, some exceptions, e.g., chikungunya virus is apparently maintained in man without the need for bird or rodent hosts and also is known to infect other primates. The basic, enzootic vectors for alphaviruses are culicine mosquitoes—predominantly banded-legged members of *Culex* (*Culex*) for the bird viruses, *Culex* (*Melanoconion*) spp. for the rodent viruses, and *Aedes* spp. for chikungunya. Again, exceptions can and do occur; *Cs. melanura*, for example, is the primary vector for both EEE and WEE in the eastern United States (Chamberlain, 1958); and *Anopheles funestus* and *An. gambiae* were the known vectors of o'nyong-nyong virus, a close relative of chikungunya, when it swept through East Africa during 1959–1962 and infected an estimated 2 million people (Williams *et al*., 1965a,b).

In general, pest species of *Aedes* and *Psorophora* do not appear to be primary vectors in basic enzootic cycles of any of the well-studied bird or rodent viruses, although they undoubtedly do occasionally play significant roles in the epidemic spread of some of these viruses beyond their enzootic ecotopes to affect man, domestic animals, and other vertebrates.

B. Vector–Host Relations of Representative Alphaviruses

The known alphaviruses are listed in Table I, divided into subgroups on the basis of cross-reactivity as indicated by the serum neutralization test. The main known or suspected vectors and hosts of each alphavirus are also given. In some instances, vector status has been quite solidly established by extensive field and laboratory investigations; in others, however, it is based on no firmer grounds than one or a few virus isolations from field-collected specimens or an epidemiological association. No attempt has been made to list all suspected vectors of the more widely distributed viruses.

1. The Venezuelan Equine Encephalitis Subgroup

Most of the members of the VEE subgroup appear to have a simple enzootic cycle, mostly involving various species of wild rats and *Culex* mosquitoes of the subgenus *Melanoconion*. The basic habitats are tropical to subtropical, wooded, freshwater swamplands which permit continuous breeding of both the rodent hosts and closely associated mosquito vectors. This type of habitat comprises a multitude of geographically discontinuous sites, from northern South America to southern Florida, that can serve as foci of infection and that have apparently provided the isolation required for the various subgroup members to diverge from a common ancestral stock.

On the basis of kinetic hemagglutination inhibition (HI) test results, Young and Johnson (1969) have divided the VEE subgroup into four main types (I, II, III, and IV). Type I is further divided into five subtypes, A–E. Types IA, IB, and IC are quite closely related, are indigenous only to the northern hump of South America, and are highly virulent for equine animals and man. Studies show that IA, IB, and IC are capable of producing very high viremias in equines (Kissling *et al.*, 1956; Henderson *et al.*, 1971; Walton *et al.*, 1973), high enough to infect practically any kind of mosquito—*Aedes*, *Psorophora*, *Mansonia*, *Deinocerites*—a fact which favors rapid epidemic spread (Sudia and Newhouse, 1975). In special situations, other biting flies may play an accessory role; in Colombia, for example, an outbreak occurred which apparently was transmitted by *Simulium* (Trapido, 1972). However, the basic maintenance cycle of these epidemic VEE subtypes is still unknown. On the other hand, ID, known from Colombia and the Canal Zone, and IE, found in Almirante on the Atlantic coast of Panama, Vera Cruz, Mexico, and probably in much of Central America (Young and Johnson, 1969; Scherer *et al.*, 1970), have not been involved epidemically except on a limited scale. They have been isolated from wild rodents (spiny rat, *Proechimys semispinosus*, and cotton rat, *Sigmodon hispidus*), birds, and *Culex* (*Melanoconion*) mosquitoes (*C. taeniopus* and *C. vomifer* in the Panama area) (Galindo, 1963; Grayson and Galindo, 1968). Experimentally, both ID and IE produce only a low-level viremia in horses (Walton *et al.*, 1973), which undoubtedly restricts their epidemic potential. Type II (Everglades virus) is serologically the most distinct of the four types and is known only from hammocks of the southern Florida Everglades (Chamberlain *et al.*, 1969; Young and

TABLE I

Cataloged Alphaviruses

Subgroup	Virus	Known or suspected vectors	Hosts	Geographic region
VEE	Bijou Bridge[a]	*Oeciacus vicarius* (swallow bug)	Birds	Colorado
	Everglades	*Culex* (*Melanoconion*) spp.	Rodents	Southern Florida
	Mucambo	*Culex* (*Melanoconion*) *portesi*, *Aedes* spp., *Mansonia* spp.	Rodents	Brazil, Trinidad, French Guiana
	Pixuna	*Anopheles nimbus*, *Trichoprosopon digitatum*	Rodents	Brazil
	VEE	*Culex* (*Melanoconion*) spp., *Aedes* spp., *Psorophora* spp.	Rodents, equines	Mexico, Central and South America
EEE	EEE	*Culiseta melanura*, *Aedes* spp., *Psorophora* spp.	Birds	Eastern United States, Central and South America
Middelburg	Middelburg	*Aedes caballus*, *Aedes* spp.	Sheep, goats, cattle?	South Africa, Cameroon
Ndumu	Ndumu	*Mansonia uniformis*, *Aedes circumluteolus*, *A. abnormalis*	?	South Africa, Central Africa
WEE	Aura	*Culex* (*Melanoconion*) spp., *Aedes serratus*	Rodents?	Brazil, Argentina
	Fort Morgan	*Oeciacus vicarius*	Birds	Colorado
	Sindbis	*Culex univittatus*, *C. antennatus*, *Mansonia fuscopennata*	Birds	Egypt, South Africa, Uganda
		Culex annulirostris, *C. pseudovishnui*	Birds	Australia, Borneo

		Culex tritaeniorhynchus, C. bitaeniorhynchus	Birds	Malaya, Philippines, India
	WEE	*Culex tarsalis, Culiseta melanura*	Birds	Western and eastern United States
	Whataroa	*Culex pervigilans, Culiseta tonnoiri*	Birds	New Zealand
	Y6233[a]	*Aedes cantans, A. cinereus*	Birds?	USSR
Semliki Forest	Bebaru	*Culex* (*Lophoceratomyia*) spp., *Aedes butleri*	Monkeys, humans?	Malaya
	Chikungunya	*Aedes aegypti, A. africanus, Aedes* spp.	Humans, monkeys?	Southeast Asia, India, Africa
	Getah	*Culex bitaeniorhynchus, C. tritaeniorhynchus, C. gelidus, Aedes vexans*	Humans, pigs?	Australia, Malaya, Cambodia, Japan
	Mayaro	*Haemagogus, Mansonia venezuelensis, Psorophora* spp.	Monkeys, rodents, humans?	Trinidad, Brazil, Colombia
	Ross River	*Aedes vigilax, Culex annulirostris*	Birds, marsupials	Australia
	O'nyong-nyong	*Anopheles funestus, An. gambiae*	Humans	Uganda, Kenya
	Sagiyama	*Culex tritaeniorhynchus, Aedes vexans*	Pigs?	Japan
	Semliki Forest	*Aedes abnormalis, Aedes* spp.	Birds, rodents?	Uganda, East Africa
	Una	*Aedes serratus, Mansonia arribalzagai, Psorophora ferox, P. albipes*	Rodents?	Trinidad, Brazil, Colombia

[a] Not cataloged.

Johnson, 1969); there the cotton rat and the cotton mouse (*Peromyscus gossypinus*) serve as natural hosts, and *Culex* (*Melanoconion*) *cedecei*, a species very close to *C. opisthopus* of more tropical areas (Stone and Hair, 1968), serves as the basic vector. *Aedes taeniorhynchus* collected at the focus of infection has on occasion yielded isolations of Everglades virus but does not appear to be a necessary vector. Experimentally, viremias in horses are minimal (Henderson *et al.*, 1971). Type III (Mucambo) has been isolated on many occasions from *Culex* (*Melanoconion*) *portesi* in Trinidad and Brazil and from forest rats, particularly *Oryzomys* and *Proechimys* (Theiler and Downs, 1973, pp. 126, 130–131). A few cases of febrile illness in man due to Mucambo infection are on record, but equine disease has not been reported.

Type IV (Pixuna virus) is another relatively nonpathogenic form, which also has been isolated from *Proechimys* (Theiler and Downs, 1973, p. 134). *Anopheles nimbus* and *Trichoprosopon digitatum* have been found naturally infected; whether or not these are basic vectors or represent fortuitously infected species is still not known.

Cross-protection studies have also indicated close antigenic interrelationship of the VEE subgroup viruses, closer than is generally the case in the other alphavirus subgroups. It has been experimentally demonstrated, for example, that infection of horses with Everglades virus from southern Florida will confer full protection against challenge with a highly virulent epidemic strain (Henderson *et al.*, 1971). Cross-protection between Pixuna and Mucambo viruses has also been demonstrated (Shope *et al.*, 1964; Karabatsos, 1975). With cross-reactivity so clearly evident, the factors presumed to have permitted divergence of strains bear further inspection. The clues appear to lie in the special vector–host relations involved, as well as in the geographic isolation provided by the discontinuous swamplike habitats. The restricted feeding habits of the enzootic vector mosquitoes and the limited movement of their associated host rodents tend to keep each virus strain within its habitat boundaries. Other animals of the habitat which may incidentally become infected probably play no active role in virus spread beyond the habitat margins unless their viremias are high enough to infect mosquitoes other than the basic vectors; more likely, the infection of various nonrodent vertebrate species builds up a halo of immunity around the infection focus that serves both to deter escape of a virulent strain, should one emerge through mutation, and to prevent invasion of the focus by another member of the VEE complex from the outside, either by encroachment or by direct introduction by a viremic bird or bat.

It is clear that epidemic strains of VEE virus differ from less virulent forms, such as Mucambo and Everglades, by their production of very high viremias in equine animals that permits their extension beyond enzootic boundaries by vectors other than swamp-bound *Culex* (*Melanoconion*) species. Because virtually any kind of mosquito can become infected from such a virus source, rapid spread over hundreds, even thousands, of miles can then occur for as long a time as there are susceptible equines to serve as epidemic hosts and mosquitoes to feed upon them.

During the great 1969–1971 excursion of epidemic VEE from Ecuador to Texas, 22 species of mosquitoes of 7 genera were implicated as vectors on the basis of virus isolation from field-collected specimens (Sudia and Newhouse, 1975). Some, such as *Aedes sollicitans* and *Psorophora confinnis*, definitely played a major role. During this epidemic, areas of Central America and Mexico where less virulent members of the VEE subgroup resided were literally overrun. But the indigenous virus types were not displaced (Scherer *et al.*, 1976a,b). Quite to the contrary, the epidemic virus seemed to pass around the areas where the indigenous types were ensconced as though they were, indeed, protected by a local immune barrier.

Too little is known of the recently recognized Bijou Bridge virus (Monath, 1977) to allow more than brief mention here. Serologically, it belongs in the VEE subgroup. It is of exceptional interest because it is one of the first alphaviruses that appears to have a nonmosquito as a basic vector. This virus has been isolated only from *O. vicarius* (swallow bug), taken from abandoned mud nests of cliff swallows in eastern Colorado, and from house sparrows that have taken over these nests. Even then, the virus has been found only in association with a new WEE-related virus (Fort Morgan virus) in the same bug pool or bird blood sample. One may speculate that this virus represents a unique adaptation of a VEE virus type to a nonculicine vector. Whether its concurrent association with Fort Morgan virus is fortuitous or dependent has not yet been established. The results of further studies now in progress, which may shed light on heretofore unsuspected aspects of virus–vector–host interrelations, are awaited with considerable interest.

2. *Eastern Equine Encephalitis, Middelburg, and Ndumu Subgroups*

Results of neutralization tests (Karabatsos, 1975) indicate that EEE, Middelburg, and Ndumu viruses are sufficiently unique to be considered apart from each other and from the three major alphavirus subgroups (VEE, WEE, and Semliki Forest). So little is known of Middelburg and Ndumu (Berge, 1975) that there is essentially nothing to add to the data given in Table I. However, EEE virus has been studied extensively.

In the United States EEE virus occurs in the Atlantic and Gulf of Mexico coastal areas from Massachusetts to Texas, with an extension up the Mississippi River valley. Its characteristic habitat is freshwater swamps; the primary hosts are various species of wild birds; and the enzootic vector is *Cs. melanura*, which breeds in swampy environs and feeds almost exclusively on birds. Epidemic spread from the basic swamp habitat occasionally occurs, associated with the seasonal movement of various bird species and the fortuitous presence of large populations of aedine mosquitoes that can transmit infection from viremic birds to equines and man. Viremias in horses are generally too low to be infectious for mosquitoes. However, experimental studies (Kissling *et al.*, 1954; Sudia *et al.*, 1956) have shown that about 1 horse in 20 can develop a viremia high enough to infect some *Aedes* species, which could be important in fanning an epidemic already under way.

On the East Coast *Aedes sollicitans* is undoubtedly the main vector for epidemic spread (Crans, 1977), although some other species, such as *A. vexans*, *A. cantator*, and *Psorophora* spp., have also been suspected on epidemiological grounds. On the Gulf Coast, *A. sollicitans* apparently shares this task with abundant species of that area, primarily *A. taeniorhynchus* and *P. confinnis*.

The ecology of EEE virus has generally been less studied in areas where it has occurred outside the United States: Jamaica, Dominican Republic, Panama, Trinidad, and South America. The apparent vector in outbreaks has usually been *A. taeniorhynchus*. Basic maintenance cycles are still unknown; however, evidence from Trinidad and Brazil (Theiler and Downs, 1973, pp. 118–125) suggests that birds are the principal hosts, although an isolation was also made from a *Proechimys* rat near Belem. The virus has been isolated from *Culex* (*Melanoconion*) *taeniopus* and *C.* (*Culex*) *nigripalpus* in Trinidad and from *C. taeniopus*, several other *Melanoconion* spp., and *A. taeniorhynchus* in Brazil. These findings indicate that *Culex* (*Melanoconion*) mosquitoes may serve the same basic vector role in the tropics as *Cs. melanura* does in the United States, but that rodents may also be involved to some extent, as well as birds.

Southward transport of considerable amounts of EEE virus to the tropics from the United States by birds during their fall migration has been fairly well documented (Stamm and Newman, 1963; Lord and Calisher, 1970); however, Casals (1964) has found antigenic differences between North American and South American strains which suggest that such periodic movement of virus from the north may not be necessary for virus perpetuation in the tropics. Evidence of northward transport of virus by birds in the spring is less firmly established; only two isolations of EEE virus have been made from birds entering the United States via the Gulf of Mexico (Calisher *et al.*, 1971). These strains were more akin to South American strains than to the type indigenous to the United States.

Evidence such as this weakens the prevalent theory that annual recurrence of EEE activity in the United States is dependent upon timely reintroduction of the virus from the tropics by migrating birds. To be considered along with strain differences and the paucity of virus in incoming birds are the late initiation of EEE virus activity in the United States habitats (generally well after the peak of the bird nesting season), recurrence of virus activity at the same sites year after year, and recurrence of almost identical infection rates year after year in the swamp-bound vector—these facts all suggest a locally based mechanism of virus persistence, largely if not entirely independent of annual introduction by migrant birds. However, birds undoubtedly are involved in moving the virus about later in the season, after virus transmission at the infected foci is already in progress. And it is tenable that late-season activity in the Caribbean islands is related to fall migration of birds from the United States (Lord and Calisher, 1970).

There is some evidence from New Jersey that rodents may be involved in a winter cycle of EEE virus. Goldfield and Sussman (1964) have reported isolations of the virus from a white-footed mouse collected during the winter on a game farm where infection is known to occur regularly in *Cs. melanura* and wild birds during the summer and fall months. However, no winter vector has been defined; obviously, the climate of New Jersey precludes active involvement of mosquitoes during that season.

The possibility that EEE virus persists from year to year through transovarial transmission in a mosquito (such as *Cs. melanura,* which overwinters in the larval stage) should be seriously reconsidered. Clues mentioned earlier, i.e., isolation of EEE virus from a pool of *Cs. melanura* larvae in Massachusetts and isolation of WEE virus from a pool of male *Cs. melanura* in south Alabama, indicate either that laboratory contamination occurred (which is possible) or that transovarial transmission can occur and might be significant. The habits of *Cs. melanura* are compatible with this last view: There is a surge of egg laying in the fall, following a peak of virus activity in the birds and mosquitoes that could ensure a high infection rate in egg-laying vectors. Certainly, transovarial transmission could rationally explain EEE overwintering in temperate regions, much as it has clarified the means of persistence of two viruses of the Bunyaviridae family, La Crosse in Wisconsin (Watts *et al.*, 1974) and Keystone in the Delaware–Maryland–Virginia peninsula (Le Duc *et al.*, 1975). In the tropics, however, delayed transmission, possibly aided by aestivating infected adult vectors, may be a more likely mechanism for carrying the virus through difficult climatic periods.

3. *Western Equine Encephalitis Subgroup*

Five recognized members of the WEE subgroup are listed in Table I. All, with the possible exception of Aura and Fort Morgan viruses, appear basically to be bird viruses transmitted mainly by *Culex* (*Culex*) species. Essentially nothing is known of the ecology of Y62-33, except that it has been isolated from *Aedes cantans* and *A. cinereus* mosquitoes in the USSR (Theiler and Downs, 1973, p. 144); but its close relationship to WEE virus (Karabatsos, 1975) suggests that its cycle may be similar to that of WEE.

Western equine encephalomyelitis virus will be considered first because it has been studied more intensively than the other members of the group. It has been "reported" from Europe and the Orient (Berge, 1975) but is probably strictly a New World form and is best known in the western United States and Canada. It is also present in the eastern United States in close association with EEE virus and has been found in Mexico, British Guiana, Brazil, Uruguay, and Argentina as well.

West of the Mississippi River in the United States and in western Canada, WEE virus closely follows the distribution of *Culex* (*Culex*) *tarsalis*. Throughout its principal range, this mosquito can serve as an all-purpose vector, transmitting the virus enzootically to various species of wild birds and, on occasion, epidemically involving horses and man. The habits of this mosquito appear to fit it well for its dual vector role. During late spring and throughout the summer it feeds preferentially on birds, but will also feed opportunely on mammals, including horses and man. In early fall, blood feeding drops off almost entirely, and carbohydrate meals are taken instead, preparatory to developing fat body for overwintering as hibernating adults (Bennington *et al.*, 1958).

Other mosquitoes undoubtedly serve as accessory vectors in epidemic transmission during the summer months. *Aedes dorsalis*, a ubiquitous mosquito of the western plains, has been taken naturally infected in the field and has been shown experimentally to be a capable transmitter. Also, in California and other western states, various other species of mosquitoes, such as *Aedes nigromaculis* and *Culiseta inornata,* probably have been involved. Their role is not the free-wheeling type as with epidemic VEE, however. The viremias are lower in infected horses than in infected birds, and *Aedes* and other mosquitoes involved are less susceptible than *C. tarsalis*; therefore, it is assumed that transmission by the accessory vectors is more likely to be from bird to horse than from horse to horse, and that horses are generally dead-end hosts. Man is even more certainly a dead-end host.

One of the most interesting aspects of WEE virus ecology is the striking shift in vector type east of the Mississippi River, concurrent with an almost total absence of WEE disease in horses and man. *Culex tarsalis* is rarely found east of the Mississippi; there *Cs. melanura,* the enzootic vector of EEE, takes its place as the WEE enzootic vector. In the process, the virus is fitted into a completely different ecological niche than it had in the West, one of freshwater swamps rather than one of ditches, slow-moving streams, and water catchments in the open countryside. In these new environs, WEE virus has an enzootic cycle apparently identical to that of EEE, with *Cs. melanura* transmitting it among the same kinds of wild birds. However, the occasional epidemic excursions involving horses and man that mark WEE in the West do not occur, probably because *Cs. melanura* is so restricted in its habitat and feeding habits that it lacks the ability to play *C. tarsalis'* double-vector role. In addition, it is apparent that the accessory vectors which permit occasional epidemic spread of EEE in the east are not as effective in this regard for WEE.

It is possible, of course, that the reduced disease rate east of the Mississippi River is due to lesser virulence of the WEE virus strains that occur there. Comparative serological studies by Henderson (1964) have indicated detectable antigenic differences between the eastern and western strains of WEE virus. The prototype he selected for the eastern form has been called Highlands J virus, which was isolated from a Florida jay at Highlands, Florida, to the north of Lake Okechobee (Henderson *et al.*, 1962). (This locality just about marks the southern range of *Cs. melanura*; south of Lake Okechobee *Cs. melanura* becomes increasingly scarce in Florida, and so do EEE and WEE viruses.) Some workers look upon Highlands J as different enough from typical WEE virus to be a separate entity; others have judged it merely to be an eastern variant of the western form. The virulence of the eastern form for experimental animals is essentially the same as that of the western form, which does not support the assumption that its virulence for man and horses is less. Rather, differences in its vector–host relations may well account for its reduced association with disease. This view is supported by very low antibody rates in equine animals and man, even in areas where the virus is known to be active in birds and *Cs. melanura* year after year.

The antigenic differences detectable between the eastern and western forms may have come about as a result of the virus' adaptation to the new vector species. The fact that such differentiation did occur also suggests that geographic isolation has been pronounced—that little reciprocal traffic of WEE virus by birds, west to east and vice versa, has occurred. This is generally in keeping with the known migrating habits and seasonal shifts of birds in the United States. Bird traffic is mostly on the north–south axis; lateral movement is much more restricted.

Cohabitation of the same eastern swamp areas by EEE and WEE viruses, and their sharing the same vector and the same bird hosts, probably are possible because there is virtually no cross-protection between the two viruses. If there were, a high antibody rate in the hosts to one of the viruses could greatly reduce host availability for the other. Eventually the more efficient of the two viruses could win out and claim the territory. Such a takeover apparently is not occurring. Cross-protection is quite weak, expressed at best as slightly depressed and shortened viremias in birds already immune to the heterologous virus. However, alternating years of ascendancy have been noted for the two viruses; this alternating pattern seems to be, on the one hand, a reflection of mild cross-protection and, on the other, the effect of carryover of immunes from the preceding year's predominant infection.

The means of overwintering in temperate regions is no better understood for WEE virus than for EEE virus. The same theories concerning the role of migrating birds prevail, and have the same shortcomings. For the most part, evidence appears to be stronger for perennial persistence of the virus at local infected foci than for its annual reintroduction by birds from the tropics. However, as with EEE virus, there is little doubt that WEE virus is moved southward in the fall by many viremic migrants (Stamm and Newman, 1963) and that these migrants reinforce infection in some parts of southern United States during this season. Whether or not the southward migration plays any significant role in maintaining tropical infections is still unknown.

As with EEE virus, there is no clear evidence of transovarial transmission of WEE virus by mosquitoes, although, as earlier noted, it has been isolated on one occasion from a pool of male *Cs. melanura*. Furthermore, the finding that *C. tarsalis*, the primary vector of WEE in the western United States, requires carbohydrate rather than blood for fat body development weakens the theory that the virus overwinters in the hibernating adult. It appears that female *C. tarsalis* that take blood meals before hibernating (and therefore have an opportunity to become infected) have little chance of surviving through the hibernation period.

There are some other findings, however, that may have a bearing on an overwintering mechanism. In New Jersey (Goldfield and Sussman, 1964), WEE virus has been isolated from nonavian vertebrates during parts of the year when no bird infection could be detected. In Utah (Gebhardt *et al.*, 1964), WEE virus has been isolated from the blood of snakes captured early in the spring; in Canada (Burton *et al.*, 1966), it was isolated from leopard frogs; and in New Jersey, it was isolated from *Culex territans*, a mosquito presumed to feed only on cold-blooded vertebrates (Chamberlain and Sudia, 1960). Obviously, there are many aspects possibly related to overwintering that could be further investigated; and it should be borne in mind that overwintering mechanisms for a particular virus need not be the same (or even necessary) in all of its ecological niches.

Sindbis virus is the Old World counterpart of WEE virus. It is the most widely distributed member of the WEE subgroup and has been recorded in the Near East, Africa, Southeast Asia, India, Borneo, Australia, the USSR, and Czechoslovakia (Berge, 1975). By the kinetic HI test, there appear to be three more or less distinct subtypes: African, Oriental, and Australian (Casals, 1961).

In its ecology Sindbis appears to be much like WEE, but less virulent. Infection of domestic animals does not pose a veterinary problem, and human cases, when they occur, are generally mild. Many of the basic investigations of vector–host relations of Sindbis virus were carried out in Egypt and South Africa. Early work by Taylor *et al.* (1955) in the Nile Delta indicated the importance of *Culex* (*Culex*) *univittatus* (primarily a bird feeder) and *Culex* (*Culex*) *attenatus*; a strain of the virus was also isolated from a hooded crow. They found that antibodies were common in a variety of birds, as well as in domestic animals. In South Africa (Theiler and Downs, 1973, p. 143) several isolations were also made from *C. univittatus*. However, in addition, the virus was isolated from *Aedes circumluteolus* and *Culex theileri*, both of which are believed to feed mainly on animals other than birds; and human cases were also reported in the same general area. This is reminiscent of WEE in the western United States (and EEE in the eastern United States) where accessory vectors may aid in spreading virus beyond a basic bird cycle.

In Uganda, Sindbis virus has been isolated from *Mansonia fuscopennata,* and human cases have also been reported (Theiler and Downs, 1973, p. 143). It may be worthy of note that *Mansonia* spp. usually have broad feeding habits that include both birds and large mammals as blood sources.

Banded-legged *Culex* (*Culex*) spp. appear to be the leading Sindbis virus vectors in the Philippine Islands and India (*Culex bitaeniorhynchus*), Malaya (*C. tritaeniorhynchus*), Borneo (*C. pseudovishnui*), and Australia (*C. annulirostris*) (Berge, 1975). In India, isolations have also been made from wild birds, a pool of mixed *Culex* spp., and the tropical fowl mite *Bdellonyssus bursa* (Shah *et al.*, 1960). The bird isolations are in keeping with the general pattern but, without further evidence of the importance of bird mites as vectors, the mite isolation should probably be considered in the same light as those of WEE virus made from *Bdellonyssus bursa* and *B. sylviarum* in the United States—merely an indication of recent feeding on viremic hosts (Reeves *et al.*, 1955).

At this time no special mechanisms which ensure survival of Sindbis virus through difficult periods are known. However, since much of its range includes tropical and subtropical areas, continuous cycling in bird hosts (faster in favorable periods, slower in unfavorable periods), coupled with local movement and long-distance migrations of birds, probably serves in large part to ensure persistence. Its relationship to WEE virus suggests that in the northern part of its range, at least, the two viruses may have similar survival mechanisms.

Whataroa virus is known only from New Zealand, Fiji, and Western Samoa (Berge, 1975) and is apparently an offshoot of Sindbis that came about because of geographic isolation. The virus has been isolated from *Culex* (*Culex*) *pervigilans* and *Culiseta tonnoiri* in New Zealand, and neutralizing antibodies have been detected in man in New Zealand and Fiji. Antibodies have also been found in an opossum, *Trichosurus vulpecula*, in New Zealand and in a variety of passerine, water, and shore birds in New Zealand and Western Samoa. Birds appear more likely to be important in the transmission cycle than opossums, which experimentally developed only a minimal viremia (Ross *et al.*, 1964; Maguire *et al.*, 1967). It may be significant that the antibody rate in passerine birds in the open countryside in New Zealand was considerably higher than in passerines from the bush.

Aura virus has been isolated thus far only from *Culex* (*Melanocon-*

ion) spp. and *Aedes serratus* in Brazil and from *A. serratus* in Argentina. In the Amazon region, according to Causey *et al.* (1963), HI antibodies have been rarely encountered in humans, marsupials, rodents, and horses, and not at all in monkeys, bats, edentates, lizards, rabbits, carnivores, birds, frogs, cows, and sheep. Such a paucity of information on this virus precludes fitting it into a known vector–host pattern with any degree of certainty; however, the isolation from *Culex* (*Melanoconion*) spp. and the presence of antibodies in wild rodents suggest that an enzootic VEE-like cycle may be possible. If so, *A. serratus* could be more involved as an accessory vector for epizootic spread than as a necessary part of the basic maintenance team.

Fort Morgan virus, like Bijou Bridge virus of the VEE subgroup, has been isolated only from the swallow bug *O. vicarius,* in mud nests of cliff swallows in Colorado, and from the swallows and house sparrows inhabiting these nests. It is serologically related to WEE virus but, unlike WEE virus, it does not kill infant mice. It can survive the winter in the bird bug, and experimentally has been shown not to be infective for *C. tarsalis,* either by the oral route or by intrathoracic inoculation; nor is it infectious for *Aedes albopictus* cell cultures. This is in contrast to Bijou Bridge virus which is infectious for mosquitoes, both orally and by injection (Monath, 1977). As mentioned earlier, arbovirologists are following current studies on both Fort Morgan and Bijou Bridge viruses with considerable interest, and speculations are many. One suggestion is that these viruses are aberrant offshoots from typical WEE and VEE virus stocks and are of little significance other than indicating the extremes of vector–virus–host association that can occur in restricted niches. Another possibility is that a hitherto unknown basic maintenance mechanism is at work which may broadly apply throughout the alphavirus group.

4. *The Semliki Forest Virus Subgroup*

This alphavirus subgroup is of particular interest because it contains more members than any of the other alphavirus subgroups and because two of its members, chikungunya and o'nyong-nyong, can be transmitted by mosquitoes from man to man directly, without need for another vertebrate host. In this man-mosquito-man transmission, and also in the clinical disease produced in man, chikungunya and o'nyong-nyong mimic the dengue viruses of the flavivirus group. Epidemic vector–host combinations apparently have been so successful for these two viruses (and for the dengue viruses) that basic enzootic cycles have been lost altogether in some parts of their range or have assumed minor importance. Ross River virus, which causes epidemic polyarthritis in man in Australia, is also a notable member of the Semliki Forest subgroup.

The type virus of the subgroup, Semliki Forest, is highly successful and widely distributed; it is known throughout much of Africa by isolations from sentinel mice, wild birds, and a number of species of mosquitoes (Berge, 1975). *Aedes* spp. appear to be most involved, although the virus also has been isolated from *Eretmopodites grahami* in the Cameroons and *An. funestus* in Kenya. Infected *A. vexans* and *Culex pipiens* have also been found in the far eastern area of the USSR (Gaidamovich *et al.*, 1975). An even wider distribution is apparent on the basis of antibody studies. Antibodies have been detected in a high proportion of chimpanzees in the Congo, and in about 5% of the wild rodents and domestic animals tested in South Africa. In man, neutralizing antibodies have been detected in Africa in 46%; in India, 1.5%; in Malaya and Borneo, up to 37%; in Vietnam and Thailand, up to 34%; and in the Philippines, 16% (Berge, 1975).

There is essentially no information to indicate clearly the natural vector–host cycle of Semliki Forest virus. However, sentinel mice exposed in sylvan environs in Nigeria became infected, and virus or antibodies have been detected in a variety of animals, including birds. These are findings which suggest the possibility of an enzootic cycle involving small wild animals, including birds, and because of the generally tropical distribution of the virus, subhuman primates and man might also be of importance. The means by which Semliki Forest virus persists so successfully from year to year is still unknown but, since most of its distribution is tropical, continuous transmission appears to be the most likely mechanism. However, in areas where *Aedes* spp. appear to be the main vectors, transovarial transmission can be suspect as well. The possibility of chronic infections and long-lasting viremias in some vertebrate hosts might also bear consideration. The general lack of pathogenicity for vertebrate species with which it is associated is in keeping with a long history of adaptation between the virus and these vertebrates.

Although little is known for certain about the basic vector–host relations of Semliki Forest virus, even less is known about some other members of the subgroup. This is particularly true of Bebaru, Getah, and Sagiyama viruses (Berge, 1975). Bebaru virus, known only in Malaya, has been isolated from *Culex* (*Lophoceratomyia*) spp. and *Aedes butleri,* and neutralizing antibodies have been found in approximately one-third of the humans sampled. Whether or not man serves as a principal host and critical source of virus for mosquitoes is a moot point. The role of monkeys and other animals is also vague. Getah virus appears to be more widely distributed than Bebaru, with isolations having been made from *Culex* (*Culex*) *gelidus* in Malaya, *C. tritaeniorhynchus* in Malaya and Cambodia, *C. bitaeniorhynchus* and *Anopheles amictus* in Australia, and *A. vexans* in Japan. Getah virus has also been isolated from a pig in Japan. As with Bebaru virus, neutralizing antibodies have been detected in the blood of about one-third of the humans sampled in Malaya.

Sagiyama virus (Scherer *et al.*, 1962) has thus far been found only in Japan and Okinawa. A number of strains have been isolated from *C. tritaeniorhynchus* and *A. vexans* in the Tokyo area. In the vicinity of Tokyo, neutralization testing showed that about 70% of the swine, 10% of the people, 3% of herons and egrets, and two of two horses had antibodies to this virus. In view of the known broad feeding habits of *C. tritaeniorhynchus* and *A. vexans,* with domestic pigs among their favored blood sources, it appears likely that pigs may play a significant role in maintaining the virus in this area. Involvement with domestic pigs (which have a very rapid turnover rate) may have supplanted any preexisting natural maintenance cycle with wild hosts and may actually have fostered divergence of the virus originally from Getah-like ancestral stock. If so, Sagiyama could be a relatively new virus, because it is only in the last century that rapid expansion of the pig-rearing industry in Japan has made a predominant pig–mosquito cycle in that area appear likely.

Una and Mayaro viruses (Theiler and Downs, 1973, pp. 129, 131–133), both of which are known to occur in Trinidad and the northern hump of South America, have been isolated on numerous occasions from a variety of mosquitoes: Una from *Psorophora, Aedes, Culex* (*Melanoconion*) spp., and *Mansonia*; and Mayaro primarily from *Haemagogus*, but also from *Culex* spp., *Mansonia*, and *Psorophora*. Mayaro virus was also isolated from *Gigantolaelaps* mites collected from an *Oryzomys* rat. On the basis of antibody incidence, monkeys and man appear to be predominant hosts for Mayaro virus, although rodents are also involved. The isolations from *Haemagogus*, a canopy mosquito, are in keeping with the monkey infections. Evidence also indicates some exposure of rodents to Una virus.

Since the mid 1960s, Ross River virus, known only from parts of Australia, has commanded considerable interest because of its association with epidemic polyarthritis in man (Doherty *et al.*, 1963a,b, 1964). The suspected vectors are *C. annulirostris* and *Aedes vigilax*, both of which have been found infected in the field. It has also been isolated from birds (Whitehead *et al.*, 1968) and wallabies (Doherty *et al.*, 1971). Antibody occurs in a high percentage of wallabies and kangaroos and has also been commonly detected in various large mammals (man, cattle, horses) (Doherty *et al.*, 1966). Antibody incidence in wild birds, however, has been quite low, which suggests that birds may not be significant hosts. From the evidence at hand, it appears that Ross River virus can be maintained basically in marsupials but be spread tangentially to domestic animals and man as well. Whether mosquito transmission between domestic animals and man can occur during periods of epidemic activity is still unknown.

Chikungunya virus is widely distributed throughout Africa, southeast Asia, and India (Berge, 1975). The disease it causes in man is virtually indistinguishable from dengue (cf. Chapter 3) and can occur together with dengue in the same human populations. Where such concurrent infections occur, its prevalence has undoubtedly been grossly underestimated.

Man is the primary host of chikungunya virus throughout much of its range, and *Aedes aegypti* (the primary vector of dengue) is the main transmitter. *Aedes aegypti* rapidly spreads the infection, frequently in communities where one or more of the four known dengue virus types is already active.

Maintenance of chikungunya virus by continuous transmission in man appears likely in many of the highly populated areas of southeast Asia. In some areas where the virus is found, however, particularly in Africa, monkeys and various other animals can also be involved. The virus has been isolated from bats (East African Virus Research Institute, 1962–1963) and an African green monkey in Senegal (Institute Pasteur, Dakar, 1972) and from a bird and sentinel mice in Nigeria (Berge, 1975). High antibody rates have been found in chimpanzees in Liberia and the Congo, and in monkeys of various kinds in Ethiopia, Rhodesia, South Africa, and Thailand. Infected mosquitoes other than *A. aegypti* have also been taken in the field: *Aedes africanus* in Uganda and Bangui; other *Aedes* in Nigeria, Senegal, and India; *Mansonia* spp. in Uganda and Nigeria; and *Culex fatigans* in Tanganyika and Thailand (Berge, 1975).

With involvement of hosts other than man and of mosquitoes other than the highly domesticated *A. aegypti*, it is possible that natural maintenance cycles without man exist at selected foci. As pointed out by Theiler and Downs (1973, p. 140), the association of chikungunya virus with *A. aegypti*, *A. africanus*, and monkeys in Africa is reminiscent of sylvan YF. The more apparent epidemic transmission of this virus in man could in some areas be obscuring a less evident natural maintenance system which probably predates it. The virus appears to be one which can involve primates in general, with man having gradually assumed the leading host role as his population has increased and he has altered the environment to favor breeding of *A. aegypti.*

O'nyong-nyong virus is closely related to chikungunya virus and causes the same type of disease in man. It differs from chikungunya mainly in its vectors—*An. funestus* and *An. gambiae* rather than *A. aegypti* or *A. africanus.* It appeared suddenly in East Africa in the late 1950s, and in a 3-year period swept through a large area of East Africa and infected an estimated 2–5 million people (Haddow *et al.*, 1960; Williams *et al.*, 1965a,b; East African Virus Research Institute, 1959–1963). Then it disappeared.

Whether o'nyong-nyong virus was a mutant form of chikungunya virus infectious for *Anopheles* which appeared suddenly or was an already established virus which merely extended epidemically beyond some still-unknown focus remains undecided. Because of cross-reactions between o'nyong-nyong and chikungunya viruses, serological definition based on field-collected sera is uncertain. In its short rise, however, o'nyong-nyong virus earned recognition on two counts: It was the first (and thus far the only) arboviral disease agent to be spread epidemically in man by species of *Anopheles*; and it was the cause of one of the most intense and extensive epidemics of an arboviral disease on record.

III. THE FLAVIVIRUSES

A. General Biological Features

The *Flavivirus* genus comprises the largest antigenic group of the Togaviridae family and is by far the most important of all arbovirus groups in terms of human infection. Essentially half of its members are known to cause human diseases, including such virulent and epidemically significant ones as YF, dengue, JE, St. Louis encephalitis (SLE), and Russian spring-summer encephalitis (RSSE).

As mentioned earlier, flaviviruses have a greater variety of arthropod vectors than alphaviruses—mosquitoes and ticks—or no vector at all. It is also of interest that, although ticks can be experimentally infected with some mosquito-borne flaviviruses, the reverse is rare. Perhaps this indicates that the long-standing virus infections occurring in ticks lead to greater adaptation to (and dependence on) tick tissues.

Most flaviviruses which do not have an arthropod vector apparently grow poorly or not at all in arthropods, suggesting that some key factor required for infection may have been lost upon their adaptation to direct vertebrate-to-vertebrate transmission. And as would be expected in a closed transmission system, such viruses are highly restricted as to their vertebrate hosts. It should be emphasized, however, that the vector status of some "nonvectored" viruses is still far from being clearly defined. In many instances the basis for this classification is merely negative data—no more than a consistent failure to isolate virus from field-collected mosquitoes or ticks. Some of the viruses which have been relegated to this category may, indeed, depend to a variable degree upon some kind of arthropod vector for their maintenance.

In general, the biological features of mosquito-borne flaviviruses are similar to those of alphaviruses. According to the kind of virus, a variety of vertebrate host types may be involved—birds for some, and rodents, primates, hoofed animals, or marsupials for others. Usually a single mosquito species accounts for most of the enzootic transmission of a particular virus in a particular ecological zone, although other species may serve in an accessory capacity to a greater or lesser degree. The epidemic vector or vectors may be different from the enzootic vector and may have a wider geographic and host range. With flaviviruses, as with alphaviruses, successful vector–host combinations at enzootic foci call for a close association of vector and host.

There is very little evidence that arthropods other than mosquitoes play significant vector roles for the mosquito-borne flaviviruses, notably a few isolations of SLE virus from bird mites in the United States and from rodent mites in Brazil as already mentioned, and a few of West Nile (WN) virus from ticks in Egypt (Berge, 1975). For the most part, epidemiological patterns preclude nonmosquitoes as being important in the active spread of infection. In all fairness, however, it should be pointed out that in most ecological studies involving this group it has been rare, indeed, that arthropods other than mosquitoes or ticks have been assiduously collected or tested.

The means by which they survive winters in temperate regions or dry seasons in the tropics remain as uncertain for mosquito-borne flaviviruses as for alphaviruses, but it is reasonable to assume that mechanisms of virus survival appropriate to the particular ecological situation have selectively emerged, or else the virus in question would not still persist. Virus survival could depend upon successful operation of one or more of the following mechanisms: continuous transmission in favorable tropical areas; carryover through dry seasons in aestivating mosquitoes; shuttling of virus back and forth seasonally between temperate and tropical zones by migrating aerial hosts; transovarial transmission of virus by selected mosquito vectors; overwintering of virus in adult mosquitoes; delayed, chronic, or recrudescent infections in vertebrates and persistent or recurring viremias; and persistence of virus in arthropods other than mosquitoes. Despite up to a half-century of research on some of these viruses, however, none of the above mechanisms has been shown unequivocally to be necessary for the survival of a single one.

Because of the entirely different life-styles of ticks and mosquitoes, the biological features of tick-borne flaviviruses are in many respects quite apart from those of mosquito-borne viruses. The great mystery of the mosquito-borne agents, namely, how they manage to survive from year to year, appears to be amply solved for the tick-borne forms. Survival is ensured by at least one and possibly all of the following three mechanisms: transovarial passage by infected female ticks to their eggs; transstadial passage through the larval and/or nymphal stage to the adult tick; and carryover in extremely long-lived immature or adult ticks. For some tick species, the life cycle may take as long as 3 years.

It appears that all the mobile stages of a tick vector (larva, nymph, adult) are susceptible to infection with a particular virus if they feed upon a suitably viremic host. It also appears that they can, after an adequate incubation period, transmit the virus to normal, susceptible vertebrates. The usual variables which affect infection and transmission efficiency of mosquito vectors also apply to ticks: Some kinds of ticks are more susceptible to particular viruses than others; some are better transmitters to their hosts; some are more efficient in passing the virus from one stage to another or to their eggs; some vertebrate hosts are more susceptible to infection than others; and some vertebrate hosts produce viremias that are higher and of longer duration.

The hardy nature of ticks permits certain tick-borne flaviviruses to flourish in many parts of the world, even in the harsh climate of Siberia. The type of tick, its life cycle, and its feeding habits all directly affect the biological features of the virus. Soft ticks (family Argasidae), which are intermittent feeders and commonly inhabit nests or burrows of certain birds (largely seabirds) and rodents, tend to remain in or near these sites from year to year and infect the young of each generation. As would be expected from such close association with a particular host type, the host range is quite limited, with chance for spread to other kinds of vertebrates occurring only when ticks become extremely numerous or when the infected nidus is invaded. However, such species-restricted viruses may be spread over large geographic areas according to the host's migration pattern.

I know of no studies on transovarial transmission of flaviviruses by argasid ticks but, since infected nymphs and adults can survive in nest material from one year to the next, hungrily awaiting the next crop of hosts, the need for efficient virus transfer through eggs to ensure virus perpetuation does not appear to be crucial. This mechanism would seem to be much more important to hard ticks (Ixodidae), which feed but once per molt; without transovarial transmission of virus, infections acquired by feeding in the adult stage would be dead-end.

Hard ticks promote a different pattern of infection than soft ticks because of their different life cycle and feeding behavior. Those known to be vectors of flaviviruses are generally three-host ticks which feed but once per stage—on small vertebrate species as larvae and on different (generally larger) vertebrates as nymphs and adults. Since with this system host selection is largely a matter of chance, the host spectrum tends to be broad. Small rodents or ground-inhabiting birds usually comprise the hosts of the larvae and may be of greater importance as virus sources than the larger animals more commonly fed upon by the nymphs and adults.

B. Vector–Host Relations of Representative Flaviviruses

The cataloged mosquito-borne flaviviruses are listed in Table II, together with their known or suspected vectors, hosts, and general geographic distribution. All these viruses are related antigenically and commonly cross-react to a considerable extent in the broadly reactive HI and complement fixation tests. Attempts to subgroup them by the more definitive neutralization test has shown most of them to fall into one or another of three natural subdivisions—JE-related viruses, which infect mostly birds; dengues, which infect man and monkeys; and YF-related viruses, some of which also infect man and monkeys. However, there is still disagreement as to the position of some of the individual viruses. (See Chapter 2 for a more detailed discussion.) The subgroupings used in Table II were set up in part for convenience of discussion and are not based entirely upon demonstrated cross-reactivity by neutralization. Therefore, the degree of relationship between some of the included viruses may not be truly expressed. (Compare with Table III, Chapter 2.)

The tick-borne flaviviruses are listed in Table III. The flaviviruses for which there are no known arthropod vectors are given in Table IV;

however, since the latter are outside the main thesis of this chapter, they will be considered only briefly.

1. The Mosquito-Borne Flaviviruses

a. Japanese Encephalitis Virus Complex. Japanese encephalitis virus occurs in eastern Asia from Japan and the southeastern USSR to Indonesia and India (Berge, 1975). The basic, enzootic cycle clearly involves various species of birds, particularly ardeids, and *Culex* mosquitoes (Buescher and Scherer, 1959; Scherer *et al.*, 1959a,b). However, in the northern part of the range, particularly in Japan and Korea, young pigs-of-the-year play a significant host role in conjunction with *C. tritaeniorhynchus* as the main vector. This banded-legged mosquito has broad feeding habits, with its blood sources including pigs, other domestic and wild mammals, man and birds. Although birds are judged to be necessary in the initiation and maintenance of infection in the area, epidemics apparently require a buildup of the virus pool in pigs high enough to permit spillover into the human population. *Culex tritaeniorhynchus* is the primary intermediary, but it is possible that other common mosquito species also assist during the peak of virus activity. Total human exposure in some areas is exceptionally great; antibody rates in older age groups may approach 100%. Fortunately, as is common throughout the JE complex of viruses, subclinical infections in man far outnumber the clinically recognized ones, as much as 200–1000 to 1.

Farther south, the pig–*C. tritaeniorhynchus* association weakens and human infections are sparser. Other species of mosquitoes (*Culex annulus, C. gelidus, C. pipiens, C. vishnui*) may assume the major vector role according to local ecology, and, overall, birds appear to be the major hosts. However, high levels of infection locally in pigs and *C. tritaeniorhynchus* have been reported as far south as Java (Van Peenen *et al.*, 1974), and in such situations human cases of JE are more likely.

Whether or not JE virus survives from year to year in the northern part of its range or must depend upon annual reintroduction from the south by migrating birds is still not resolved. There is some evidence that transovarial transmission of virus by *Aedes* mosquitoes may occur; early Japanese, Russian, and Chinese workers all reported isolation of the virus from adult mosquitoes reared from field-collected larvae, but their findings were later discounted when they could not be confirmed (cited in Rosen *et al.*, 1978). However, a recent experimental demonstration by Rosen *et al.* (1978) of transovarial passage of JE virus by *Aedes albopictus* and *A. togoi* has renewed interest in this

TABLE II

Cataloged Mosquito-Borne Flaviviruses

Subgroup[a]	Virus	Known or suspected vectors	Hosts	Geographic region
JE	JE	*Culex tritaeniorhynchus, C. gelidus, C. annulus, C. vishnui*	Birds, pigs	Northern and Southeast Asia, India
	SLE	*Culex tarsalis, C. pipiens pipiens, C. p. quinquefasciatus, C. nigripalpus*	Birds	North, Central, and South America, Carribean
	MVE	*Culex annulirostris, C. bitaeniorhynchus, Aedes normanensis*	Birds	Australia, New Guinea
	WN	*Culex univittatus, C. antennatus, C. modestus, C. vishnui*	Birds	Africa, Asia, Europe
	Usutu	*Culex univittatus, C. perfuscus, Aedes aegypti, Mansonia aurites*	Birds	Africa
	Koutango	?	Rodents	Senegal, Central African Empire
	Rocio	*Aedes serratus?, A. scapularis?, Culex taeniopis?*	Birds, humans?	Sao Paulo State, Brazil
	Bussuquara	*Culex taeniopus, C. vomifer, Mansonia titillans; Culex (Melanoconion) crybda, Trichoprosopon* sp.	Rodents	Brazil, Colombia, Panama
	Ilheus	*Psorophora, Aedes, Culex, Sabethes, Haemagogus*	Birds, humans?	Central and South America
	Jugra	*Aedes (Cancraedes)* spp., *Uranotaenia*	Bats?	Malaysia
	Alfuy	*Aedeomyia catastica*	Birds	Australia
	Kunjin	*Culex annulirostris, C. squamosus, C. pseudovishnui*	Birds	Australia, Sarawak
	Kokobera	*Culex annulirostris, Aedes vigilax*	Marsupials	Australia, New Guinea
	Stratford	*Aedes vigilax*	Cattle?	Australia
Dengue	Dengue-1	*Aedes aegypti, A. albopictus*	Humans, monkeys?	Southeast Asia, India, New Guinea, Nigeria, Carribean
	Dengue-2	*Aedes aegypti, A. albopictus*	Humans, monkeys?	Southeast Asia, India, New Guinea, Nigeria, Car-

			monkeys?	Guinea, Nigeria, Carribean
	Dengue-4	*Aedes aegypti*, *A. albopictus*	Humans, monkeys?	Southeast Asia, India, Philippines
Yellow fever	Yellow fever	*Aedes aegypti*, *A. africanus*, *A. simpsoni; Haemagogus* spp., *Sabethes chloropterus*	Monkeys, humans	Africa, South America
	Uganda S	*Aedes longipalpis–A. ingrami–A. natronius*	Birds, humans?	Uganda, Nigeria, Central African Empire
	Wesselsbron	*Aedes caballus*, *H. circumluteolus*, *A. lineatopennis*, *A. mediolineatus*	Sheep	East and Central Africa, Thailand
	Banzi	*Culex rubinotus*, *C. nakuruensis*, *Mansonia africana*	Humans?, cattle	South and East Africa
	Bouboui	*Anopheles paludis*, *Aedes africanus*, *A. fulcifer*	Humans?	Central African Empire, Senegal, Cameroon
	Spondweni	*Aedes circumluteolus*, *Mansonia africana*, *M. uniformis*	Cattle, sheep, goats	South and West Africa
	Zika	*Aedes africanus*, *A. luteocephalus*, *A. aegypti*	Monkeys, humans?	East, Central, and West Africa; Malaya
	Edge Hill	*Aedes vigilax*, *Culex annulirostris*, *Anopheles meraukensis*	Marsupials	Australia
	Sepik	*Mansonia septempunctata*, *Ficalbia* spp., *Armigeres* spp.	?	New Guinea
Ntaya	Ntaya	*Culex guiarti*, *Culex* spp.	Humans?	East and Central Africa
	Bagaza	*Culex perfuscus*, *C. guiarti*, *C. ingrami*, *C. thalassius*	?	East and South Africa
	Israel turkey meningoencephalitis	?	Birds?	Israel
	Tembusu	*Culex tritaeniorhynchus*, *C. gelidus*, *C. annulus*	Birds?	Malaya, Sarawak, Thailand

[a] The subgroupings as used here are for convenience of discussion and may not truly express the degree of relationship among the various viruses.

TABLE III

Cataloged Tick-Borne Flaviviruses

Subgroup[a]	Virus[a]	Known or suspected vectors	Hosts	Geographic region
RSSE (TBE)	RSSE	*Ixodes persulcatus*, *I. ricinus*	Rodents, birds	USSR
	CEE[b]	*Ixodes ricinus*	Rodents, birds, goats	Western USSR, Western and Central Europe
	Kumlinge	*Ixodes ricinus*	Rodents, hares, birds, cattle(?)	Finland
	Louping ill	*Ixodes ricinus*	Rodents, sheep, birds(?)	British Isles
	KFD	*Haemaphysalis spinigera*, *H. turturis*, *Haemaphysalis* spp., *Ixodes* spp., *Ornithodoros*	Rodents, monkeys, birds	India
	Langat	*Ixodes granulatus*, *Haemaphysalis papuana*	Rodents	Malaya, Thailand
	OHF	*Dermacentor pictus*, *D. marginatus*, *D. silvarum*, *Ixodes persulcatus*	Muskrat, mice, frogs	Western and Southwestern Siberia
Powassan	Powassan	*Ixodes cookei*, *I. marxii*, *I. spinipalpus*, *Dermacentor andersoni*, *Haemaphysalis neumanni*	Marmots, squirrels, chipmunks, mice	Canada, United States; South Primorye, USSR
	Royal Farm	*Argas hermanni*	Pigeons	Afghanistan
Tyuleniy	Tyuleniy	*Ixodes putus*, *I. uriae*	Seabirds	Eastern USSR, Oregon
	Saumarez Reef	*Ixodes eudyptidis*, *Ornithodoros capensis*	Seabirds	Queensland, Tasmania
Other[c]	Kadam	*Rhipicephalus pravus*	Cattle	Uganda
	Karshi	*Ornithodoros papillipes*	Great gerbil	Uzbekistan, USSR

[a] TBE, Tick-borne encephalitis; RSSE, Russian spring-summer encephalitis; CEE, central European encephalitis; KFD, Kyasanur Forest disease; OHF, Omsk hemorrhagic fever.

[b] Absetterov, Hanzelova, and Hypr are cataloged viruses which are similar if not identical to CEE virus serologically, in distribution, and in range of hosts and vectors.

[c] Kadam shows a relationship by neutralization test to Apoi and Entebbe bat viruses of the no-vector "arboviruses"; Karshi is close to WN virus of the mosquito-borne flaviviruses.

TABLE IV

Cataloged Flaviviruses, No Vector Demonstrated

Host	Virus	Geographic region
Bats	Batu Cave	Peninsular Malaysia
	Carey Island	Peninsular Malaysia
	Dakar bat	Senegal, Nigeria, Central African Empire, Uganda
	Entebbe bat	Uganda
	Montana myotis leukoencephalitis	Montana
	Phnom-Penh	Cambodia
	Rio Bravo	California, New Mexico, Texas; Sonora, Mexico
	Sokuluk	USSR
Rodents	Apoi	Japan
	Cowbone Ridge	Florida
	Jutiapa	Guatemala
	Modoc	Western United States
	Saboya[a]	Senegal
Unknown	Negishi[b]	Japan

[a] HI antibodies (primary infection pattern) found in a variety of animals besides rodents, which suggests the existence of an arthropod vector.
[b] Serologically quite close to RSSE, which suggests it may be tick-borne.

area. Successful operation of such a mechanism in the northern part of the JE range could rationally account for the consistent reappearance of the virus year after year in the same localities.

St. Louis encephalitis virus is the New World counterpart of JE virus and is similarly a bird virus transmitted by various *Culex* mosquitoes. Also like JE, it has been most studied in the northern part of its range where the largest epidemics have occurred. Unlike JE, however, neither pigs nor any other mammalian species appear to play a critical host role. Human infections, when they occur, are believed to be transmitted from viremic avian hosts to man by the vector mosquitoes.

St. Louis encephalitis virus is active as far north as Canada and as far south as Brazil, Argentina, and Ecuador. In the western United States and Canada, the primary vector is *C. tarsalis*, the same banded-legged mosquito that serves as the primary vector of WEE virus throughout that geographic area, although SLE infections occur generally later in the season than WEE. Human cases of SLE disease are scattered, as befits the rural habits of this vector species. In the south-central and eastern part of the United States, however, *C. tarsalis* is rare or absent, and the vector role is assumed by the southern house mosquito (*C. pipiens quinquefasciatus*) in the southern half of the country, and by the northern house mosquito (*C. pipiens pipiens*), commencing at about the level of Tennessee and continuing northward. Their habits are quite different from those of *C. tarsalis*, and as a result the virus–host relations are also different. Since they are mosquitoes adapted to breeding in polluted waters in urban environs, their host association is mainly with peridomestic birds such as house sparrows, pigeons, and jays, but they will also feed on man. Therefore, when SLE virus is introduced into a suitable community, presumably by an infected bird, it literally becomes active in the urban backyard, and its association with man is much closer than in the West. As a result large urban outbreaks can occur (Chamberlain, 1978; Monath, 1980). Urban SLE, transmitted by these domesticated vectors, is by far the most important arboviral disease in the United States. As with JE, the subclinical/clinical ratio is generally high, about 200 or more to 1, so that an epidemic with 300–400 hospitalized cases may represent up to 80,000 actual infections. Human antibody rates after an outbreak may range in some communities from 25 to 40%. Up to 40% of the "pest" birds have antibodies at the conclusion of an epidemic episode (Lord *et al.*, 1974).

Subtropical Florida has a different vector pattern. There, *C. nigripalpus*, a common species in tropical areas, breeds in large numbers in essentially any standing fresh water, in rural and urban areas alike. It feeds largely on birds but also on humans and other animals as opportunity affords. Infection occurs late in the season, generally not before August, and may extend up to November. This pattern is compatible with introduction of virus into the area with the first southward migration of birds from the north in late summer. Infections may occur both as urban outbreaks and as scattered rural cases. The epidemic of over 200 proved cases which occurred in the Tampa–St. Petersburg area in 1962 is an example of the urban type of activity (Bond *et al.*, 1965); the more than 50 scattered cases in several counties, from central to southern Florida, which occurred in the late summer and fall of 1977, also exemplify the rural type of transmission (Morbidity and Mortality Weekly Report, 1977). The role of *C. nigripalpus* as vector in both these situations has been amply documented by repeated isolations of virus from field-collected specimens.

Culex nigripalpus is also an apparent vector of SLE in the Caribbean islands and tropical America; virus has been isolated from this species in Jamaica (Belle *et al.*, 1964; Sudia *et al.*, 1966), Trinidad (Aitken *et al.*, 1964), Guatemala (Sudia, 1970), and Ecuador (Francy, 1976). However, other species also appear to be important in the tropics. St. Louis encephalitis virus has been isolated from several *Culex* (*Melanoconion*) spp. in Trinidad (suggesting possible involvement of rodents), and also from *Psorophora ferox, Aedes scapularis,* and *A. serratus* (Aitken *et al.*, 1964). In Panama it has been found in *Sabethes, Trichoprosopon, Wyeomyia,* and *Deinocerites* (Berge, 1975); and there is recent evidence suggesting that *Mansonia indubitans* may also serve in a vector capacity in a newly impounded lake area in eastern Panama (Galindo, 1977).

Thus far there is no proof that there is an effective overwintering mechanism for SLE virus in temperate regions. Some experimental findings suggest that low-level transmission by surviving female *C. tarsalis* could possibly occur throughout the winter in California, but this does not seem very likely (Bellamy *et al.*, 1968). Even in the true tropics, studies have been inadequate to prove the occurrence of continual transmission throughout the year. Nevertheless, there is some reason to believe that a means of virus carryover in temperate areas does indeed exist. The Tampa–St. Petersburg epidemic of 1962 was preceded by 3 years of low-level SLE virus activity (Bond *et al.*, 1965), an indication that the virus had somehow managed to survive in the same area from 1959 to 1962, to then burn itself out in a final great flare-up. This seems to be a more reasonable explanation than that birds selectively brought the virus back to the same area 3 years in a row and then suddenly stopped doing so. A somewhat similar but reversed situation was apparent in McLeansboro, Illinois. That area suffered an epidemic in 1964; for the next 3 years virus was isolated there each summer from *C. pipiens pipiens* (Kokernot *et al.*, 1969).

Studies by Eldridge (1966) have indicated that blood-fed females of *C. pipiens pipiens* exposed to a critical combination of reduced ambient temperature and light day may prepare physiologically for hibernation. This is in contrast to the findings of Bennington *et al.* (1958) that *C. tarsalis* requires a carbohydrate meal prior to hibernation and could mean that *C. pipiens pipiens* females which have had an opportunity to become infected by blood feeding may be able to hibernate and thus carry virus through the winter. In fact, isolations of SLE virus from overwintering *C. pipiens pipiens* in January and February in Maryland and Pennsylvania have recently been reported (Bailey *et al.*, 1978). Moreover, the recent report of Rosen *et al.* (1978) that JE virus can pass through the eggs of *Aedes* mosquitoes suggests the advisability of looking further into transovarial transmission of SLE. Thus far the only deliberate efforts to demonstrate experimental transovarial transmission of SLE virus by mosquitoes were made with *C. pipiens quinquefasciatus* (Chamberlain *et al.*, 1964), whose eggs hatch soon after they are laid. Results were not particularly convincing. This was probably a poor choice of species for this type of experiment. With hindsight, it now seems more logical that virus adaptation to transovarial transmission would occur in mosquitoes like *Aedes*, which overwinter in the egg stage. This could present a distinct selective advantage, namely, survival of the adapted virus strain into the next year.

In Texas, virus was isolated from bats for as long as 2 years after the 1966 outbreaks in Dallas and Corpus Christi (Sulkin *et al.*, 1966; Allen *et al.*, 1970), which suggests chronic infections. Any role of arthropod vectors in these instances was not elucidated. Transovarial transmission of SLE virus by bird mites has been seriously considered (Smith *et al.*, 1945; Reeves *et al.*, 1955; Sulkin *et al.*, 1955). However, this theory was later discounted as unlikely, and isolations of virus made from field-collected bird mites were attributed to recent engorgement upon viremic blood rather than to true biological infection.

Murray Valley encephalitis (MVE) virus is an SLE- and JE-like virus in Australia and New Guinea. *Culex annulirostris* serves as the main vector and apparently transmits the virus between birds and from birds to man, much as *C. tarsalis* transmits SLE virus in the western United States. In parts of Victoria and New South Wales, where most of the human cases have occurred, up to half of the water birds and domestic fowl and a fourth of the land birds may develop antibodies, indicating that in the epidemic areas birds are strongly involved (Anderson, 1954). However, in northern Australia and New Guinea, human cases have been few despite high antibody rates, and experimental host studies in northern Australia have indicated that the *C. annulirostris* found there prefer to feed on calf, dog, feral pig, grey kangaroo, man and domestic fowls, in that order. Under natural conditions, dogs were much more severely attacked than man (Doherty, 1976). In northern Australia isolations of MVE virus have been made from *Aedes normanensis* and *C. bitaeniorhynchus*, as well as *C. annulirostris* (Doherty *et al.*, 1963a).

Obviously, there is still much to be learned about the maintenance and epidemic cycles of MVE virus, but in view of the close relationship of this virus to both SLE and JE viruses, findings relating to them may well apply in principle to it also.

West Nile virus is the most widespread of any of the JE subgroup and has been reported from much of Africa, the Near East, the USSR, parts of Europe, India, and Indonesia. Wild birds are the primary hosts, but high antibody rates in a wide variety of other animals, including man, domestic fowl, large domestic animals, monkeys, and chimpanzees, indicate a broad infection spectrum. As would be expected from its large geographic range, various kinds of mosquitoes, according to area, serve as principal vectors. In Egypt, where much of the early study on this virus was conducted (Taylor *et al.*, 1956), *C. univittatus* and *C. antennatus* were found to be most involved. In South Africa *C. univittatus* is also the apparent vector (Berge, 1975). Other vector associations indicated by the isolation of virus from field-collected specimens include: Central African Empire, *Culex weschei*; Uganda, *Mansonia metallica*; India, *C. vishnui*, *C. fatigans*, *Anopheles subpictus*; Israel, *C. univittatus*, *C. molestus*, *An. coustani*; France, *C. modestus*; Czechoslovakia, *A. cantans*. Isolations have also been reported from *Argas hermanni* ticks in Egypt and *Hyalomma plumbeum* in USSR, but are of unknown significance.

Although WN infection in man is widespread, the morbidity rate is relatively low since it is of considerably less virulence than JE, SLE, or MVE. Severe cases of encephalitis occur occasionally, but the subclinical/clinical ratio is very high. In parts of Africa, up to 70% of the human population may possess antibodies.

In the more tropical parts of its range, it is possible that WN virus could be maintained by continual transmission, particularly if some of the great variety of nonavian animal species known to be susceptible actually enter into the transmission chain. Its mechanism of annual recurrence in the more temperate regions is still unproved, but introduction by migrating birds has been suggested. In Egypt, Hurlbut (1956) showed experimentally that WN virus could propagate in *C. pipiens* at temperatures prevailing there during the winter months and that the mosquitoes could subsequently transmit by bite.

Usutu virus is closely related antigenically to WN, occurs in parts of Africa where WN virus also occurs, and is apparently transmitted by similar mosquito species (Berge, 1975). It has been isolated from *C. univittatus* in South Africa, *Mansonia aurites* in Uganda, *Culex perfuscus* in the Central African Empire and the Congo, and *A. aegypti* in Cameroon. It has also been isolated from birds. It seems to be able to cohabit the same ecological niches with WN virus, sharing some of the same vector and host species. Probably this is possible because it is only weakly neutralized by WN antibodies, whereas WN virus is strongly neutralized by Usutu antibodies. This cross-reaction pattern permits Usutu virus to infect and multiply in WN-immune hosts but prevents WN virus from growing well in Usutu-immune hosts. Such a one-way cross may serve to temper the strong competition for host space that a highly successful virus such as WN must be capable of exerting. Furthermore, it is theoretically possible that the growth of Usutu virus in the presence of WN antibody allows neutralization of variants which do not resist effects of WN antibody, thus ensuring purity of the Usutu virus strain and its ability to compete. The converse—an occasional breakthrough of WN virus in an Usutu-immune host—would do little to affect the total WN gene pool.

Koutango virus is related serologically to both WN and Usutu viruses and has been isolated from rodents in Senegal (*Tatera kempi* and *Mastomys* sp.) and the Central African Empire (*Mastomys* sp. and *Lemnyscomys* sp.) (Berge, 1975). It had been relegated to the "no-vector-demonstrated" category for the lack of a field isolation from a bloodsucking arthropod. However, Coz *et al.* (1975) have experimentally shown that it can infect *A. aegypti* and be transmitted by bite. A complete transmission cycle, suckling mouse–mosquito–suckling mouse, was accomplished with this species. However, the true vector species in nature is still unknown.

Rocio virus is the newest member of the JE virus complex, and one which may well become recognized as being of major medical importance. It is an SLE-like virus of considerable virulence which caused a sizable outbreak on the coastal plain of Sao Paulo State, Brazil, in 1975 (de Sousa Lopez *et al.*, 1978). Antibody studies and a virus isolation from a species of sparrow have indicated that birds may have been the main host (Karabatsos, 1978), and circumstantial evidence suggests that mosquitoes were the vectors. Predominant mosquito species in the affected areas were *Aedes scapularis*, *A. serratus*, and *Culex* (*Melanoconion*) spp., including *C. taeniopus* (Forattini, 1977). Where the virus came from and whether it will return is still unknown.

Bussuquara and Ilheus are two viruses of the American tropics which for convenience I have included in the JE virus complex, although de Madrid and Porterfield (1974) failed to demonstrate cross-reactions with JE by the neutralization test. Broad crossing by other test methods suggests that this subgrouping is as appropriate as any. Bussuquara has been reported in Brazil, Colombia, and Panama and has been isolated on many occasions in Para State, Brazil, from the rodent *Proechimys guyenensis* (Shope, 1963). Birds do not seem to be involved as hosts. *Culex* (*Melanoconion*) spp., including *C. taeniopus*, which feed largely on rodents, have also yielded many isolations of the virus in Para State, and antibody rates as high as 75% have been recorded in *Proechimys* in this same locality. Human disease is rare. No HI or neutralization test antibodies were found in humans in Brazil, but were present in 46% of 383 human blood specimens sampled in central Panama (Srihongse and Johnson, 1971). Whether this difference in human exposure was due to difference in vectors is not known; however, in Panama the virus was isolated from *Culex crybda* and *Trichoprosopon* sp. (Galindo *et al.*, 1966).

Ilheus virus is much more widespread than Bussuquara and has been isolated in Brazil, Trinidad, Colombia, Panama, Honduras, Guatemala, and Argentina (Berge, 1975). Disease caused by Ilheus is uncommon in man although antibody rates in some areas reach 36% and a number of isolations have been made from man. In contrast to Bussuquara, Ilheus appears to have birds as its primary hosts rather than rodents. It has been isolated on a number of occasions from birds in Trinidad and Panama, and antibody has been found in up to 15% of bird bloods sampled. However, the presence of antibodies in a variety of other vertebrates in Brazil, Trinidad, Colombia, Honduras, and Guatemala attests to the broad feeding habits of the vector species: rodents, 10%; marsupials, 15%; sloth, 4%; cattle, 10%; and equines, 35%. In addition, 8 of 287 monkeys had antibodies in Venezuela and 58 of 89 monkeys in Peru. *Psorophora* spp. have been the mosquitoes found most frequently infected, followed by *Aedes, Culex, Sabethes, Haemagogus, Coquillettidia, Wyeomyia,* and *Trichoprosopon.* With such a variety of possible vectors, we appear to be viewing intermediaries of epizootic spread rather than of a restricted enzootic cycle.

Jugra virus from the Malaysian peninsula (Berge, 1975) was isolated by Rudnick and co-workers from *Aedes* (*Cancraedes*) spp. collected at ground level and in crab holes, and also from *Uranotaenia* spp. and from the blood of a bat, *Cynopterus brachyotis.* A low level of serological crossing by neutralization test occurs between Jugra virus and Dakar bat, Rio Bravo, Langat, and Japanese encephalitis (JE) antisera. Little else is yet known. Possibly this virus could be of interest as a link between the mosquito-borne viruses (JE), tick-borne viruses (Langat), and non-vector-borne bat viruses (Dakar bat and Rio Bravo).

Alfuy, Kokobera, Kunjin, and Stratford are all viruses from north Queensland, Australia, which are closely related to each other and to MVE. Alfuy and Kunjin have both been isolated from birds in essentially the same locality, at or near the Mitchell River mission, despite their being cross-protective to a considerable extent. Possibly a difference in vectors may account for their apparent cohabitation; Alfuy was isolated from *Aedeomyia catastica,* whereas Kunjin was isolated from *Culex annulirostris* and *C. squamosus* (Doherty *et al.*, 1968). Kunjin virus has also been isolated from the *Culex pseudovishnui* group in Sarawak (Bowen *et al.*, 1970).

Kokobera virus has been isolated from *A. vigilax* and *C. annuliros-*

tris at the Mitchell River mission; and Stratford has also been isolated from *A. vigilax*, but at Cairns, on the opposite side of Cape York (Doherty *et al.*, 1963a, 1971). Stratford shows a very strong one-way cross-reaction with Kokobera virus (Stratford antiserum neutralizes Kokobera virus, but not the reverse), which should give a survival edge to Stratford should these two viruses have occasion to use the same host species and compete in the same geographic area. Their types of hosts are still uncertain, but suggestive serological evidence indicates wallabies and kangaroos for Kokobera and cattle for Stratford.

b. The Dengues. The dengue fevers are among the most medically important of all the arbovirus diseases, with literally millions of new cases each year (see Chapter 5 for a detailed discussion). The four recognized types of dengue virus (types 1, 2, 3, and 4) are obviously very similar entities which produce the same kind of disease, broadly cross-react by the complement fixation and HI tests, and have the same vectors and hosts. However, they are clearly distinguishable by the neutralization test and provide little cross-protection. This last characteristic undoubtedly accounts for their concurrent or successive activity in the same human populations, where antibody rates in older age groups may approach 100%. It is likely that, if the four types were appreciably cross-protective, only the most successful form would survive in a particular territory. Since the different dengue viruses are poorly cross-protective, a dengue virus of one type can circulate in the blood of a patient who already has antibody to another type from a past infection. Perhaps such natural cross-absorptions nurture the antigenic independence of the different types, as may be required for their cohabitation of the same ecological niche.

All the dengue viruses are or have been widely distributed throughout the world where *A. aegypti* is abundant. Types 1, 2, and 3 are common throughout Southeast Asia, India, New Guinea, parts of Africa, and the Caribbean. Type 4 has been a bit more restricted but has been isolated from essentially all countries in Southeast Asia, India, and the Philippine Islands. In most of the entire area of dengue distribution, *Aedes* (*Stegomyia*) *aegypti* is the sole known vector, and man is essentially the only host. Antibodies in monkeys in some parts of the range indicate that these animals can be infected but do not prove that they are a necessary part of the maintenance cycle. Again, as seems to be the case with chikungunya virus, an epidemic cycle is apparently so self-sustaining as to make restricted enzootic cycles unnecessary to ensure virus perpetuation. In Indonesia *A. albopictus*, another *Stegomyia,* also takes part in dengue transmission in areas where it is in close association with dense human populations. *Aedes (Stegomyia) scutellaris* and *A. (Stegomyia) polynesiensis* may also serve as transmitters in some South Pacific Islands (Clarke and Casals, 1965).

Recent findings by Rudnick (1976) and co-workers in peninsular Malaysia have demonstrated that a jungle dengue cycle involving monkeys exists. They have isolated dengue-4 virus from *Aedes (Finlaya) "niveus"* mosquitoes collected in a sentinal monkey trap in the high canopy of the Gunong Besout Forest Reserve. This is the first isolation of a dengue virus from a jungle mosquito, and the first from a non-*Stegomyia.* Their studies have also shown that monkeys in the high canopy are infected with at least three of the four known types of dengue, and that the canopy vector, *A. "niveus,"* will descend to the ground level to feed on man. The infections may then be further spread by *A. albopictus* and *A. aegypti.*

Part of the great significance of the various dengue viruses lies in their constant disease potential. With modern global transportation, virtually any *A. aegypti*-infested area in the world is subject to epidemic threat.

c. The Yellow Fever Subgroup. Just making the decision as to which viruses would be discussed under this heading was a problem. Theiler and Downs (1973, p. 147) have included Uganda S, Wesselsbron, and Banzi as part of the YF complex. However, de Madrid and Porterfield (1974) failed to show a relationship between these viruses and YF virus by neutralization tests.

Serological cross-reaction of YF virus with Bouboui (Berge, 1975) and Wesselsbron viruses (Casals, 1957) and of Sepik virus with Wesselsbron (Berge, 1975) has been reported. Banzi and Edge Hill are brought in through crossing with Uganda S (de Madrid and Porterfield, 1974), assuming that Uganda S and YF are related. Zika, according to Casals (in Berge, 1975), seems to be closer to Uganda S and YF than to several other group-B viruses, and Henderson *et al.* (1970) have demonstrated a reduced viremia in Zika-immune vervet monkeys challenged with a virulent strain of YF virus. Spondweni shows a relationship to Zika by a neutralization test but not directly to YF; and to add to the complexity of the interrelationships, Spondweni antiserum also partially neutralizes SLE virus of the JE subgroup (de Madrid and Porterfield, 1974).

The voluminous literature on YF printed since the turn of the century precludes more than the most cursory coverage here. The reader is referred to the book "Yellow Fever," edited by Strode (1951), for history and details concerning development of our present knowledge of vector–virus–host interrelations of this important disease.

Yellow fever is endemic in the rain forests of Africa and South America, usually within 12° north and south of the equator. At intervals, epidemic extension may occur, both in Africa and South and Central America. It is conspicuously absent throughout Asia, Indonesia, and Australia.

Until the early 1930s YF was considered to be essentially an urban disease and *A. aegypti* to be its only vector. The great success achieved in suppressing the disease in South and Central America by control measures directed against *A. aegypti* strengthened this concept. Then a rural outbreak in Brazil in the absence of *A. aegypti* led to the recognition of jungle YF, transmitted among monkeys high in the tropical forests by a canopy-inhabiting mosquito, *Haemagogus spegazzinii*. It was found that woodcutters could be exposed to bites of infected *Haemagogus* when the tall trees were felled. It was also found that, if the villages of these workers were infested with *A. aegypti*, local urban-type transmission could then ensue, stemming from the forest-acquired cases. Subsequent studies in East Africa in the late 1930s and early 1940s confirmed the existence of jungle YF there also, but in a somewhat different transmission pattern. *Aedes africanus* was found to be the transmitter among monkeys, with peak feeding activity just after sunset when the monkeys were settled in the treetops for the night. Monkeys (some infected ones included) commonly raid cultivated areas outside the forest and come in contact with a more domesticated mosquito, *Aedes simpsoni*, which becomes infected and transmits to man.

It has been observed that *A. africanus* survives the dry season in the adult stage and thus probably serves to carry virus through periods of low activity. Efforts to find contributing hosts other than monkeys and man have been unsuccessful, although a considerable variety of animals possess antibody as evidence of exposure to the virus.

Quite possibly in some other parts of Africa other mosquitoes besides *A. africanus* play significant roles in the transmission of YF virus among monkeys. Infection in man appears to be incidental to continued virus activity. Monath and Kemp (1973) have evidence of an enzootic cycle in nonhuman primates in Nigeria at restricted localized foci where *Stegomyia* mosquitoes are rare, and *Mansonia africana* may be the enzootic vector.

In Panama, YF virus has also been isolated from *Sabethes chlorop-*

terus (de Rodaniche *et al.*, 1957). It appears to be more than a second-level vector to assist *Haemagogus* in transmission to monkeys; like *A. africanus* it can survive through long, dry periods in the adult stage, which fits it well for carrying the virus over from one season to the next.

Why YF does not occur in eastern Asia, Indonesia, and India has not been fully explained. *Aedes aegypti* is abundantly present, and certainly, over the years the virus must have been introduced on many occasions. The most tenable theory for its absence is that the flavivirus antibody rate (dengue, JE, WN) in the indigenous human population is too high and that a dampening effect due to low-level cross-protection reduces the YF virus infection and transmission efficiency. This theory has been discounted by some workers because studies with experimental animals have failed to demonstrate that heterologous antibody will protect against disease. It should be borne in mind, however, that even a slight effect, such as a partial depression of YF viremia in the host or a slightly reduced susceptibility, could greatly lessen the likelihood of the serial transmission necessary for successful establishment of the virus in the area. Another theory can also be presented—that natural jungle cycles occurring in Africa and South America are lacking in the Orient and that without such a maintenance system the virus cannot survive.

Uganda S and Zika viruses have similar geographic distributions (Berge, 1975). Both have been found, either by virus isolation or by detection of antibody, in much of Africa and in India, Malaya, and Borneo. In addition, antibodies to Zika have been reported from the Philippines, North Vietnam, and Thailand. Uganda S virus was isolated from a pool of mixed *Aedes* (*A. longipalpis, A. ingrami, A. natronius*) in Uganda; from a bird (*Saxicola rubetra*) in the Central African Empire; and from sentinel mice in Senegal. Zika virus was isolated from a sentinel monkey and *Aedes africanus* in Uganda, *A. africanus* in the Central African Empire, *A. luteocephalus* in Nigeria, *A. aegypti* in Malaya, and the blood of man in Senegal. Neither virus appears to be a significant human disease agent, although high antibody rates in man (near 50%) have been detected in some areas. If, indeed, heterologous antibodies of YF-related viruses tend to buffer YF infections, their main significance may be in this capacity.

Wesselsbron virus is the cause of an important disease of sheep in South Africa, and in that country it has been isolated from sheep, man, *Aedes caballus, A. circumluteolus, A. minutus,* and *C. univittatus. Aedes caballus* and *A. circumluteolus* appear by far to be the most involved. In Rhodesia, the virus has been repeatedly found in *Aedes lineatopennis*; it has also been isolated from *A. lineatopennis* and *A. mediolineatus* in Thailand. In Cameroon, it was isolated from a member of the *Aedes tarsalis* group and from *Culex tellesilla* (Berge, 1975).

In the absence of any other apparent mechanism to ensure persistence of Wesselsbron virus, the capabilities of these primary vectors to transmit transovarially should be thoroughly investigated. The presence of both Wesselsbron virus and *A. lineatopennis* in such widely separated areas as Rhodesia and Thailand is probably more than coincidental.

Banzi virus (Berge, 1975) occurs in southern and eastern Africa and has been isolated on two occasions from man, once in South Africa and once in Tanganyika. *Culex rubinotus* is apparently the main vector; the virus has been isolated from this species in Transvaal and northern Natal, South Africa, on many occasions, and also in Rhodesia and Kenya. Other species taken infected are *C. nakurensis* and *M. africana* in Kenya. Banzi was also isolated from a sentinel hamster in Mozambique. Antibody rates of 5–15% have been found in humans in southern Africa; and antibodies have been detected in cattle and sheep, but rates are low. The natural cycle is still unknown.

Spondweni has been isolated in South Africa from man and a variety of mosquitoes: *Aedes circumluteolus, A. cumminsii, Mansonia uniformis, M. africana, Eretmapodites silvestris,* and *C. univittatus.* It has also been isolated from *Aedes fryeri* and *A. fowleri* in Mozambique and *Eretmapodites* spp. in the Cameroons. Antibody has been detected in small numbers of humans and domestic animals (cattle, sheep, goats) in the southern part of Africa (Berge, 1975). Spondweni is not of significance as a cause of human disease and is not of veterinary importance.

Bouboui virus (Berge, 1975) is known only in central Africa (Central African Empire, Senegal, Cameroon, and Zaire) where it has been isolated from *Aedes africanus, A. furcifer taylori,* and *Anopheles paludis.* It has also been found on one occasion in the blood of a baboon. Antibody was detected in less than 2% of over 1000 human blood samples tested, and not at all in nearly 400 various wild mammals, birds, dogs, and rodents.

Edge Hill virus was isolated in secondary forest in northern Australia on a number of occasions from *A. vigilax,* and also from *C. annulirostris* and *Anopheles meraukensis.* Antibodies believed to be specific have been detected in about one-third of wallabies tested in southeastern Queensland, in some domestic fowl and cattle in southern Queensland, and in bandicoots in northeastern Queensland. Even less is known of Sepik virus than of Edge Hill; it has been found only in mosquitoes in the Sepik District of New Guinea. It was isolated once from a pool of *Mansonia septempunctata*, twice from *Ficalbia* spp., and once from a mixed mosquito pool (Berge, 1975).

d. Ntaya Subgroup. Serological relationships of Israel turkey meningoencephalitis, Ntaya, and Tembusu viruses by neutralization test were shown by de Madrid and Porterfield (1974), and a cross-reaction of Bagaza virus with Ntaya was similarly shown by Robin (in Karabatsos, 1978). Little is known of Bagaza virus other than that it has been isolated in the Central African Empire one time each from a pool of mixed *Culex*, a pool of *Culex perfuscus*, and a pool of *C. guiarti* (Digoutte, 1966, 1968, 1969). It was also isolated from *Culex guiarti* plus *C. ingrami* in Cameroon and *C. thalassius* in Senegal. Ntaya virus, by isolation, is known only from Uganda, Cameroon, and the Central African Empire, where it was isolated from a pool of unidentified *Culex* spp. and from *C. guiarti*. By serological survey, neutralization test antibody reactions have been detected throughout the Far East (Philippines, Malaya, North Vietnam, Thailand, Singapore, Borneo, and India) and also in Egypt. Those in the Far East may be broad secondary reactions of individuals who have had multiple flavivirus infections (Berge, 1975).

Tembusu virus has been isolated in Malaya, Sarawak, and Thailand from a number of different mosquitoes: in Malaya, from *Culex tritaeniorhynchus*, *C. gelidus*, *C. vishnui*, *A. lineatopennis*, and *Anopheles philippinensis*; in Thailand, from *C. gelidus*, *C. tritaeniorhynchus*, *C. annulus*, and *C. sitiens*; and in Sarawak, from *C. gelidus*, *C. pseudovishnui*, and *C. tritaeniorhynchus* (Berge, 1975). Antibodies were found in over half of the humans and chickens sampled in Malaya, and two isolations of the virus were also made from sentinel chickens. The largely *Culex* vector pattern plus the isolations from chickens and the chicken antibody reactors suggest a bird cycle similar to those of members of the JE virus complex.

Israel turkey meningoencephalitis virus has been isolated on a number of occasions in Israel from turkeys in disease outbreaks in domestic flocks. It had been relegated to the "no-vector-demonstrated" grouping because it has never been isolated from an arthropod in the field. However, the evidence that it is a mosquito-transmitted virus of birds is now very convincing. Komarov and Kalmar (1960) experimentally infected *A. aegypti* by feeding them the virus. Nir (1972) infected *C. molestus* by feeding them a mixture of defibrinated rabbit blood, virus, and glucose, and 28 days later showed that some of them could transmit by bite to infant mice. Ianconescu *et al.* (1973) experimentally infected turkeys and found that they developed a viremia of 5–8 days' duration; these workers also demonstrated viremia in naturally infected turkeys. Furthermore, Japanese quail were experimentally infected by inoculation, and virus was recovered from brain and spleen for up to 21 days (Ianconescu *et al.*, 1974).

2. *The Tick-Borne Flaviviruses*

The RSSE (or tick-borne encephalitis) virus complex is by far the largest of the tick-borne flavivirus subgroups (Table III) and contains viruses of considerable medical significance. Central European encephalitis (CEE) and RSSE viruses are so closely related that they are judged by some workers to be the same agent or, at best, CEE is considered to be a subtype of RSSE. However, CEE virus is less virulent for man. RSSE ranges from Siberia to the western USSR; there CEE takes over, extends through central Europe, and ends up in Finland as the nearly identical Kumlinge virus. Louping ill virus, which is also antigenically close to CEE virus, occurs in Scotland, northern England, and northern Ireland (Berge, 1975).

Other members of the RSSE-related subgroup include Kyasanur Forest disease (KFD) virus, a highly pathogenic form known only in Mysore, India; Langat virus, an essentially nonpathogenic virus from Malaya and Thailand; and Omsk hemorrhagic fever (OHF) virus, a virulent agent in western Siberia.

Ixodes persulcatus is the main vector of typical RSSE, and various wild rodents and forest birds of the order Passeriformes serve as the primary vertebrate hosts. However, domestic animals of various types are also frequently infected. There is good evidence that transovarial transmission of the virus can take place in the tick; Chumakov (in Burgdorfer and Varma, 1967) has reported such passage for at least three generations in *I. persulcatus*. Stage-to-stage transmission (from larva to nymph to adult) apparently occurs with a high order of efficiency.

In western Russia and Europe, where the virus changes to the CEE subtype, *Ixodes ricinus* becomes the predominant tick species involved. It is possible that passage in this tick species has had a part in tempering the virulence of the ancestral RSSE. Or it might have been the other way around; if CEE were the ancestral type, virulence may have been increased through passage in *I. persulcatus*. The vertebrate hosts of CEE virus are similar to those of RSSE—chiefly various wild rodents and ground-feeding birds—and considerable infection of domestic goats also occurs. Besides transmission by the bite of ticks, the virus can be passed in goat's milk and can infect people who ingest it (Theiler and Downs, 1973, pp. 193–204). Transovarial passage of the virus in ticks has been demonstrated in about 6% of experimentally infected *I. ricinus,* and infected larvae have been collected in Austria, evidencing transovarial transmission in nature (Burgdorfer and Varma, 1967).

Kumlinge virus in Finland is also transmitted by *I. ricinus* and has been isolated from squirrels on a number of occasions and from birds. Cattle commonly have antibodies but are not known to be important in virus maintenance other than as a food source for the vector tick species. A number of human Kumlinge encephalitis cases are noted annually in southwest Finland.

Louping ill virus is quite close antigenically to RSSE and CEE viruses (Berge, 1975). The tick vector is again *I. ricinus,* and the wild hosts are primarily wild rodents. The virus has been isolated from the bank vole, the wood mouse, and the common shrew in Scotland (Smith *et al.*, 1964; Theiler and Downs, 1973, pp. 193–204), and also on one occasion from a red grouse in Ireland (Timoney, 1972). Evidence for transovarial transmission in the tick vector is conflicting (Burgdorfer and Varma, 1967). Louping ill is only a minor cause of disease in man but causes significant disease in sheep. It is believed that sheep can serve as amplifying hosts.

The vector–host relations of KFD virus are clearly and succinctly reviewed by Theiler and Downs (1973, pp. 193–204) and Boshell (1969) and are also listed by Berge (1975). Briefly, *Haemaphysalis spinigera* appears to be the principal vector, but the virus has been isolated in the field from several other tick species as well: on many occasions from *H. turturis*; less frequently from *H. papuana, H. cuspidata, H. kyasanurensis, H. minuta, H. bispinosa, H. wellingtoni,* and species of *Ixodes, Dermacentor, Rhipicephalus,* and *Ornithodoros.* Animals from which the virus has been isolated in the field include man and monkeys on many occasions, wild rats, two species of shrews, and a bat. Antibodies have been found in a wide variety of rodents and in ground-feeding birds. The latter, however, are believed to be of lesser importance as hosts than rodents. Man quite clearly plays no important role in the virus cycle. Monkeys, on the other hand, have been shown to develop very high viremias and to be proficient accumulators of larval ticks from the forest floor; therefore they may serve as effective virus amplifiers (Boshell, 1969). Kyasanur Forest disease virus has been detected in field-collected larvae of *H. spinigera* and *Ixodes* sp., which proves that transovarial transmission can occur in nature; however, it was detected on only two occasions in several years of searching, an indication that it occurs but rarely (Rao, 1963). Attempts to demonstrate transovarial passage experimentally by *H. spinigera* were unsuccessful, but Singh *et al.* (1968), working with *Ixodes petauristae,* were able to demonstrate virus in the larval progeny of 12 out of 16 females which had been infected by feeding on viremic squirrels. The larvae transmitted the virus by bite to mice. Transstadial transmission occurs readily in *Haemaphysalis spingera, H. turturis,* and *H. minuta,* but less readily in *H. papuana kinneari* (Singh *et al.*, 1964). Infected *Haemaphysalis* nymphs have been shown to survive through the monsoon from the preceding dry season and to overlap with the next generation of ticks. This appears to be an important mechanism for survival of KFD virus from year to year (Rajagopalan and Anderson, 1970).

Langat virus has been isolated on a number of occasions from *Ixodes granulatus* in Malaya (Berge, 1975) and once from *H. papuana* in Thailand (Bancroft *et al.*, 1976). Antibodies have rarely been found in man. The natural hosts are apparently species of forest rats; neutralizing antibodies have been detected in *Rattus sabanus, R. mülleri, R. rajah,* and *R. annandeli* in Malaya. Little else is known of its vector–host relations. Experimentally, larvae of *I. ricinus* were infected by feeding them on a viremic rat, and transstadial passage to nymphs and adults was demonstrated. Nymphs retained infection at high titer for at least 100 days after larval infection, and adults for at least 285 days (Varma and Smith, 1972).

Omsk hemorrhagic fever virus is an important disease agent for man in western Siberia. It has repeatedly been isolated from *Dermacentor pictus* ticks and less frequently from *I. persulcatus* and *D. silvarum.* It has also been found in the field mouse, *Microtus stenocranius gregalis,* and many times in muskrats (Berge, 1975). The natural cycle is not completely understood, but a rodent–tick cycle in the usual manner is suspected. Also, direct infection of man from muskrats is possible, and long-lasting infections in the tissues of muskrats are apparent. Federova and Sizemova (1964) reported an outbreak of OHF in seven of nine hunters who had come in contact with muskrats during November, December, or February when no ticks were active. Omsk hemorrhagic fever virus was isolated from brain tissue of some of the muskrats. Cold-blooded animals have also been suspected of being reservoirs; Rad'kova and Vorob'eva (1971) recovered five strains of OHF virus from the brain and inner organs of frogs. They also noted viremia in frogs which they had infected experimentally.

Powassan virus is related to the RSSE complex (Casals, 1960) but appears to be antigenically different enough to deserve a separate subgrouping. Its range extends from Ontario, Canada, to New York, South Dakota, Colorado, California, and Russia. It is not considered to be of major medical importance but can cause severe encephalitis in man. It has been isolated from woodchucks and a fox in upstate New York, a deer mouse in South Dakota, and a spotted skunk in California (Berge, 1975). It has also been isolated from brain tissue of humans in Canada (McLean and Donohue, 1959) and New York (Deibel *et al.*, 1975). Isolations from ticks have been much more numerous: from *Ixodes cookei* and *I. marxi* in Ontario, Canada; *I. cookei* in upstate New York; *I. spinipalpus and Dermacentor andersoni* in the Black Hills of South Dakota; *D. andersoni* in Colorado (Berge, 1975), and *Haemaphysalis neumanni* in the southern Primorye, USSR (L'vov *et al.*, 1974). High antibody rates in woodchucks and various other wild rodents in known areas of infection in Canada and the United States reinforce the isolation findings.

I have arbitrarily placed Royal Farm virus in the same subgroup as Powassan, although it actually may be more closely related antigenically to Langat virus (Williams *et al.*, 1972). Little is known of Royal Farm virus other than that it has been isolated on two occasions from *Argas hermanni* ticks collected in pigeon houses and pigeon bazaars in Kabul Province, Afghanistan.

Tyuleniy and Saumarez Reef viruses (Table III) are more closely related to each other than to the other tick-borne flaviviruses. Tyuleniy has been isolated from *Ixodes putus* (=*I. uriae*) collected in the USSR in seabird nesting grounds on Tyuleniy Island of the Sea of Okhotsk and Bering Island, and from *L. uriae* at Three Arch Rock in Oregon. Among the seabirds, the common murre, tufted puffin, and black-legged kittywake have been shown to have HI antibody. Antibody was also detected in 14 of 64 fur seals on Commodore Island. This virus is of interest as one of the few tick-borne viruses shown experimentally to be transmissible by mosquitoes. L'vov (in Berge, 1975) transmitted Tyuleniy virus to suckling mice by the bites of *A. aegypti* and *Culex pipiens molestus* which had been infected by feeding upon a Tyuleniy virus suspension.

Saumarez Reef virus has been isolated from *Ornithodoros capensis* collected in the nest of a sooty tern on Saumarez Reef, offshore Queensland, Australia, and from *Ixodes eudyptitis* in Tasmania. Antibodies were also found in one-fourth of the silver gulls sampled at the Tasmania site (Berge, 1975).

The two remaining catalogued tick-borne viruses listed in Table III, Kadam and Karshi, have been placed in an undefined "other" subgroup because they do not appear to fit antigenically with the rest of the tick-borne flaviviruses. Kadam crosses by neutralization test with vectorless Apoi and Entebbe bat viruses (de Madrid and Porterfield, 1974), and Karshi is most closely related to mosquito-borne WN virus (L'vov *et al.*, 1976).

Little is known of the ecology of either virus. Kadam was isolated from adult *Rhipicephalus pravus* ticks handpicked from cattle pastured on an open savannah in South Karamoja, Uganda. Attempts to infect *A. aegypti* experimentally by feeding them upon the virus were unsuccessful; however, *Dermacentor variabilis* ticks infected by injection were able to transmit by bite (Berge, 1975). Karshi virus was isolated from *Ornithodoros papillipes* ticks collected in the burrows of the great gerbil in the environs of Beshkent, Karshinsk steppe, Uzebek, USSR.

3. *Flaviviruses with no Apparent Vector*

The cataloged flaviviruses for which no vector is yet known are listed in Table IV. The factors which may have influenced their vector status have already been mentioned and will not be discussed further. However, two of the viruses deserve special mention. Saboya virus HI antibodies of a primary infection pattern have been found in a number of animals besides rodents in Senegal (Robin, in Berge, 1975); this broad host involvement is compatible with the existence of an arthropod vector. Neither a natural host nor a vector is known for Negishi virus, which originally was isolated from the spinal fluid of a patient with a neurological disease clinically diagnosed as JE. However, its close antigenic relationship to RSSE virus (Berge, 1975) suggests that it may be tick-borne.

REFERENCES

Aitken, T. H. G., Downs, W. G., Spence, L., and Jonkers, A. H. (1964). *Am. J. Trop. Med. Hyg.* **13**, 450–451.

Allen, R., Taylor, S. K., and Sulkin, S. E. (1970). *Am. J. Trop. Med. Hyg.* **19**, 851–859.

Anderson, J. R., Lee, V. H., Vadlamudi, S., Hanson, R. P., and DeFoliart, G. R. (1961). *Mosq. News* **21**, 244–248.

Anderson, S. G. (1954). *J. Hyg.* **52**, 447.

Bailey, C. L., Eldridge, B. F., Hayes, D. E., Watts, D. M., Tammariello, R. F., and Dalrymple, J. M. (1978). *Science* **199**, 1346–1349.

Bancroft, W. H., Scott, R. M., Snitbhan, R., Weaver, R. E., and Gould, D. J. (1976). *Am. J. Trop. Med. Hyg.* **25**, 500–504.

Bellamy, R. E., Reeves, W. C., and Scrivani, R. P. (1968). *Am. J. Epidemiol.* **87**, 484–495.

Belle, E. A., Grant, L. S., and Page, W. A. (1964). *Am. J. Trop. Med. Hyg.* **13**, 452–454.

Bennington, E. E., Sooter, C. A., and Baer, H. (1958). *Mosq. News* **18,** 299–304.
Berge, T. O., ed. (1975). "International Catalogue of Arboviruses," 2nd ed., DHEW Publ. No. (CDC) 75-8301. U.S. Dep. Health, Educ. Welfare, Public Health Serv., Washington, D.C.
Bond, J. O., Quick, D. T., Witte, J. J., and Oard, H. C. (1965). *Am. J. Epidemiol.* **81,** 392–404.
Boshell, M. (1969). *Am. J. Trop. Med. Hyg.* **18,** 67–80.
Bowen, E. T. W., Simpson, D. I. H., Platt, G. S., Way, H. J., Smith, C. E. G., Ching, C. Y., and Casals, J. (1970). *Ann. Trop. Med. Parasitol.* **64,** 263–268.
Bowen, G. S. (1977). *Am. J. Trop. Med. Hyg.* **26,** 171–175.
Buescher, E. L., and Scherer, W. F. (1959). *Am. J. Trop. Med. Hyg.* **8,** 719–722.
Burgdorfer, W., and Varma, M. G. R. (1967). *Annu. Rev. Entomol.* **12,** 347–376.
Burton, A. N., McLintock, J., and Rempel, J. G. (1966). *Science* **154,** 1029–1031.
Calisher, C. H., Maness, K. S. C., Lord, R. D., and Coleman, P. H. (1971). *Am. J. Epidemiol.* **94,** 172–178.
Casals, J. (1957). *Trans. N.Y. Acad. Sci.* **19,** 219–235.
Casals, J. (1960). *Can. Med. Assoc. J.* **82,** 355–358.
Casals, J. (1961). *Abstr. Symp. Pap., Pac. Sci. Congr., 10th, Honolulu* p. 458.
Casals, J. (1964). *J. Exp. Med.* **119,** 547–565.
Causey, O. R., Casals, J., Shope, R. E., and Udomsakdi, S. (1963). *Am. J. Trop. Med. Hyg.* **12,** 777–781.
Chamberlain, R. W. (1958). *Ann. N.Y. Acad. Sci.* **70,** 312–319.
Chamberlain, R. W. (1980). *In* "St. Louis Encephalitis" (T. P. Monath, ed.), pp. 3–64. Am. Public Health Assoc., New York.
Chamberlain, R. W., and Sikes, R. K. (1955). *Am. J. Trop. Med. Hyg.* **4,** 106–118.
Chamberlain, R. W., and Sudia, W. D. (1960). Unpublished observations.
Chamberlain, R. W., Nelson, D. B., and Sudia, W. D. (1954). *Am. J. Hyg.* **60,** 278–285.
Chamberlain, R. W., Sudia, W. D., and Reeves, W. C. (1959). Unpublished observations.
Chamberlain, R. W., Sudia, W. D., and Gogel, R. H. (1964). *Am. J. Hyg.* **80,** 254–265.
Chamberlain, R. W., Sudia, W. D., Newhouse, V. F., and Johnston, J. G., Jr. (1964–1965). Unpublished data from studies in Ware County (Waycross area), southeast Georgia.
Chamberlain, R. W., Sudia, W. D., Work, T. H., Coleman, P. H., Newhouse, V. F., and Johnston, J. G., Jr. (1969). *Am. J. Epidemiol.* **89,** 197–210.
Clarke, D. H., and Casals, J. (1965). *In* "Viral and Rickettsial Infections of Man" (F. M. Horsfall and I. Tamm, eds.), pp. 606–658. Lippincott, Philadelphia, Pennsylvania.
Coz, J., LeGonidec, G., Cornet, M., Valade, M., Lemoine, M. O., and Gueye, A. (1975). *CAH O.R.S.T.O.M. Ser. Entomol. Med. Parasitol.* **13,** 57–62.
Crans, W. J. (1977). *Mosq. News* **37,** 85–89.
Daniels, J. (1960). Personal communication.
Davies, J. B. (1972). *In* "Venezuelan Encephalitis," Sci. Publ. No. 243, pp. 258–260. Pan Am. Health Organ., Washington, D.C.
Deibel, R., Woodall, J. P., and Lyman, D. O. (1975). *Morbid. Mortal. Wkly. Rep.* **24,** 379.
de Madrid, A. T., and Porterfield, J. S. (1974). *J. Gen. Virol.* **23,** 91–96.
de Rodaniche, E., Galindo, R., and Johnson, C. M. (1957). *Am. J. Trop. Med. Hyg.* **6,** 681–685.
de Sousa Lopez, O., Coimbra, T. L. M., de Abrea Sacchetta, L., and Calisher, C. H. (1978). *Am. J. Epidemiol.* **107,** 444–449.
Digoutte, J. P. (1966). *Annu. Rep. Inst. Pasteur. Bangui* p. 40.
Digoutte, J. P. (1968). *Annu. Rep. Inst. Pasteur. Bangui* pp. 48–52.

Digoutte, J. P. (1969). *Annu. Rep. Inst. Pasteur. Bangui* p. 45.
Doherty, R. L. (1976). *Annu. Rep. Queensl. Inst. Med. Res., 31st* p. 5.
Doherty, R. L., Carley, J. G., Mackerras, M. J., and Marks, E. N. (1963a). *Aust. J. Exp. Biol. Med. Sci.* **41,** 17–40.
Doherty, R. L., Whitehead, R. H., Gorman, B. M., and O'Gower, A. K. (1963b). *Aust. J. Sci.* **26,** 183–184.
Doherty, R. L., Gorman, B. M., Whitehead, R. H., and Carley, J. G. (1964). *Aust. Ann. Med.* **13,** 322–327.
Doherty, R. L., Gorman, B. M., Whitehead, R. H., and Carley, J. G. (1966). *Aust. J. Exp. Biol. Med. Sci.* **44,** 365–378.
Doherty, R. L., Whitehead, R. H., Wetters, E. J., and Gorman, B. M. (1968). *Trans. R. Soc. Trop. Med. Hyg.* **62,** 430–438.
Doherty, R. L., Standfast, H. A., Domrow, R., Wetters, E. J., Whitehead, R. H., and Carley, J. G. (1971). *Trans. R. Soc. Trop. Med. Hyg.* **65,** 504–513.
East African Virus Research Institute (1959–1963). *Annu. Rep.*
East African Virus Research Institute (1962–1963). *Annu. Rep.*
Edridge, B. F. (1966). *Science* **151,** 226.
Federova, T. N., and Sizemova, G. A. (1964). *Zh. Mikrobiol., Epidemiol. Immunobiol.* **41,** 134–136.
Forattini, O. P. (1977). Cited by his permission from the *Arthropod-Borne Virus Inf. Exch.* No. 32, p. 154.
Francy, D. B. (1976). Personal communication.
Gaidamovich, S. Y., Melnikova, E. E., Agafonov, V. I., Lokhova, M. D., Rodina, V. Y., Goldin, R. B., and Klisenko, G. A. (1975). *Vopr. Virusol.* **3,** 317–320.
Galindo, P. (1963). *An. Microbiol.* **11,** 83–87.
Galindo, P. (1977). Personal communication.
Galindo, P., Srihongse, S., de Rodaniche, E., and Grayson, M. A. (1966). *Am. J. Trop. Med. Hyg.* **15,** 385–400.
Gebhardt, L. P., Stanton, G. J., Hill, D. W., and Collett, G. C. (1964). *N. Engl. J. Med.* **271,** 172–176.
Goldfield, M., and Sussman, O. (1964). *Annu. Epidemiol. Conf., Cent. Dis. Control, Atlanta, Ga.*
Grayson, M. A., and Galindo, P. (1968). *Am. J. Epidemiol.* **88,** 80–96.
Haddow, A. J., Davies, C. W., and Walker, A. J. (1960). *Trans. R. Soc. Trop. Med. Hyg.* **54,** 517–522.
Henderson, B. E., Cheshire, P. P., Kirya, G. B., and Lule, M. (1970). *Am. J. Trop. Med. Hyg.* **19,** 110–118.
Henderson, B. E., Chappell, W. A., Johnston, J. G., Jr., and Sudia, W. D. (1971). *Am. J. Epidemiol.* **93,** 194–205.
Henderson, J. R. (1964). *J. Immunol.* **93,** 452–461.
Henderson, J. R., Karabatsos, N., Bourke, A. T. C., Willis, R., and Taylor, R. M. (1962). *Am. J. Trop. Med. Hyg.* **11,** 800–810.
Howitt, B. F., Dodge, H. R., Bishop, L. K., and Gorrie, R. H. (1948). *Proc. Soc. Exp. Biol. Med.* **68,** 622–625.
Hurlbut, H. S. (1956). *Am. J. Trop. Med. Hyg.* **5,** 76–85.
Ianconescu, M., Aharonovici, A., Samberg, Y., Hornstein, K., and Merdinger, M. (1973). *Avian Pathol.* **2,** 251–262.
Ianconescu, M., Aharonovici, A., and Samberg, Y. (1974). *Refu. Vet.* **31,** 100–108.
Institute Pasteur, Dakar, Senegal (1972). *Annu. Rep.*
Karabatsos, N. (1975). *Am. J. Trop. Med. Hyg.* **24,** 527–532.

Karabatsos, N. (1978). *Am. J. Trop. Med. Hyg.* **27**, 372–440.
Karstad, L. H. (1961). *Trans. North Am. Wildl. Nat. Resour. Conf., 26th* pp. 186–202.
Karstad, L. H., Fletcher, O. K., Spalatin, J., Roberts, R., and Hanson, R. P. (1957). *Science* **125**, 395–396.
Kissling, R. E., Chamberlain, R. W., Eidson, M. E., Sikes, R. K., and Bucca, M. A. (1954). *Am. J. Hyg.* **60**, 237–250.
Kissling, R. E., Chamberlain, R. W., Nelson, D. B., and Stamm, D. D. (1956). *Am. J. Hyg.* **63**, 274–287.
Kitselman, C. H., and Grundman, A. W. (1940). *Kans. Agric. Exp. Stn., Tech. Bull.* No. 50, pp. 5–15.
Kokernot, R. H., Hayes, J., Will, R. L., Templis, C. H., Chan, D. H. M., and Radivojevic, B. (1969). *Am. J. Trop. Med. Hyg.* **18**, 750–761.
Komarov, A., and Kalmar, E. (1960). *Vet. Rec.* **72**, 257–261.
Le Duc, J. W., Suyemoto, W., Eldridge, B. F., Russell, P. K., and Barr, A. R. (1975). *Am. J. Trop. Med. Hyg.* **24**, 124–126.
Lord, R. D., and Calisher, C. H. (1970). *Am. J. Epidemiol.* **92**, 73–78.
Lord, R. D., Calisher, C. H., and Doughty, W. P. (1974). *Am. J. Epidemiol.* **99**, 364–367.
L'vov, D. K., Leonova, G. N., Gromashevskii, V. L., Belikova, N. P., Berezina, L. K., Safronov, A. V., Veselovskaya, O. V., Gofman, Y. P., and Klimenko, S. M. (1974). *Vopr. Virusol.* **5**, 538–541.
L'vov, D. K., Neronov, V. M., Gromashevsky, V. L., Skvortsova, T. M., Berezina, L. K., Siderova, G. A., Zhmaeva, Z. M., Gofman, Y. A., Klimenko, S. M., and Fomina, K. B. (1976). *Arch. Virol.* **50**, 29–36.
McLean, D. M., and Donohue, W. L. (1959). *Can. Med. Assoc. J.* **80**, 708–711.
Maguire, T., Miles, J. A. R., and Casals, J. (1967). *Am. J. Trop. Med. Hyg.* **16**, 371–373.
Miles, V. I., Howitt, B. F., Gorrie, R., and Cockburn, T. A. (1951). *Proc. Soc. Exp. Biol. Med.* **77**, 395–396.
Monath, T. P. (1977). Personal communication.
Monath, T. P. (1980). *In* "St. Louis Encephalitis" (T. P. Monath, ed.), pp. 239–312. Am. Public Health Assoc., New York.
Monath, T. P., and Kemp, G. E. (1973). *Trop. Geogr. Med.* **25**, 28–38.
Morbidity and Mortality Weekly Report (1977). **26**, No. 45, p. 370.
Nir, Y. (1972). *Arch. Gesamte Virusforsch.* **36**, 105–114.
Rad'kova, O. A., and Vorob'eva, N. N. (1971). *Izv. Sib. Otd. Akad. Nauk SSSR, Ser. Biol. Med. Nauk* **2**, 180–182.
Rajagopalan, P. K., and Anderson, C. R. (1970). *Indian J. Med. Res.* **58**, 1184–1187.
Rao, T. R. (1963). *An. Microbiol.* **11**, 95–98.
Reeves, W. C. (1974). *Prog. Med. Virol.* **17**, 193–220.
Reeves, W. C., Hammon, W. McD., Doetschman, W. H., McClure, H. E., and Sather, G. (1955). *Am. J. Trop. Med. Hyg.* **4**, 90–105.
Reeves, W. C., Hutson, G. A., Bellamy, R. E., and Scrivani, R. P. (1958). *Proc. Soc. Exp. Biol. Med.* **97**, 733–736.
Rosen, L., Tesh, R. B., Lien, J. C., and Cross, J. H. (1978). *Science* **199**, 909–911.
Ross, R. W., Miles, J. A. R., Austin, F. J., and Maguire, T. (1964). *Aust. J. Exp. Biol. Med. Sci.* **42**, 689–702.
Rudnick, A. (1976). Personal communication.
Scherer, W. F., Buescher, E. L., Flemings, M. B., Noguchi, A., and Scanlon, J. (1959a). *Am. J. Trop. Med. Hyg.* **8**, 665–667.
Scherer, W. F., Buescher, E. L., and McClure, H. E. (1959b). *Am. J. Trop. Med. Hyg.* **8**, 689–697.

Scherer, W. F., Funkenbusch, M., Buescher, E. L., and Izumi, T. (1962). *Am. J. Trop. Med. Hyg.* **11**, 255–268.
Scherer, W. F., Dickerman, R. W., and Ordonez, J. V. (1970). *Am. J. Trop. Med. Hyg.* **19**, 703–711.
Scherer, W. F., Anderson, K., Pancake, B. A., Dickerman, R. W., and Ordonez, J. V. (1976a). *Am. J. Epidemiol.* **103**, 576–588.
Scherer, W. F., Dickerman, R. W., Ordonez, J. V., Seymour, C., III, Kramer, L. D., Jahrling, P. B., and Powers, C. D. (1976b). *Am. J. Trop. Med. Hyg.* **25**, 151–162.
Shah, K. V., Johnson, H. N., Rao, T. R., Rajagopalan, P. K., and Lamba, B. S. (1960). *Indian J. Med. Res.* **48**, 300–308.
Shope, R. E. (1963). *An. Microbiol.* **11**, 167–171.
Shope, R. E., Causey, O. R., de Andrade, A. H. P., and Theiler, M. (1964). *Am. J. Trop. Med. Hyg.* **13**, 723–727.
Singh, K. R. P., Pavri, K. M., and Anderson, C. R. (1964). *Indian J. Med. Res.* **52**, 566–573.
Singh, K. R. P., McGoverdhan, M. K., and Bhat, H. R. (1968). *Indian J. Med. Res.* **56**, 628–631.
Smith, C. E. G., Varma, M. G. R., and McMahon, D. (1964). *Nature (London)* **203**, 992–993.
Smith, M. G., Blattner, R. J., and Heys, F. M. (1945). *Proc. Soc. Exp. Biol. Med.* **59**, 136–138.
Srihongse, S., and Johnson, C. M. (1971). *Trans. R. Soc. Trop. Med. Hyg.* **65**, 541–542.
Stamm, D. D., and Newman, R. J. (1963). *An. Microbiol.* **11**, 123–133.
Stamm, D. D., Chamberlain, R. W., and Sudia, W. D. (1962). *Am. J. Hyg.* **76**, 61–81.
Stone, A., and Hair, J. A. (1968). *Mosq. News* **28**, 39–41.
Strode, G. K., ed. (1951). "Yellow Fever." McGraw-Hill, New York.
Sudia, W. D. (1970). Personal communication.
Sudia, W. D., and Newhouse, V. F. (1975). *Am. J. Epidemiol.* **101**, 1–13.
Sudia, W. D., Stamm, D. D., Chamberlain, R. W., and Kissling, R. E. (1956). *Am. J. Trop. Med. Hyg.* **5**, 802–808.
Sudia, W. D., Coleman, P. H., and Grant, L. S. (1966). *Mosq. News* **26**, 39–42.
Sulkin, S. E., Wisseman, C. L., Izumi, E. M., and Zarafonetis, C. (1955). *Am. J. Trop. Med. Hyg.* **4**, 119–135.
Sulkin, S. E., Sims, R., and Allen, R. (1966). *Science* **152**, 223–225.
Taylor, R. M., Hurlbut, H. S., Work, T. H., Kingston, J. R., and Frothingham, T. E. (1955). *Am. J. Trop. Med. Hyg.* **4**, 844–862.
Taylor, R. M., Work, T. H., Hurlbut, H. S., and Rizk, F. (1956). *Am. J. Trop. Med. Hyg.* **5**, 579–620.
Theiler, M., and Downs, W. G. (1973). "The Arthropod-Borne Viruses of Vertebrates." Yale Univ. Press, New Haven, Connecticut.
Thomas, L. A., and Eklund, C. M. (1960). *Proc. Soc. Exp. Biol. Med.* **105**, 52–55.
Thomas, L. A., Eklund, C. M., and Rush, W. A. (1959). *Proc. Soc. Exp. Biol. Med.* **99**, 698–701.
Timoney, P. J. (1972). *Br. Vet. J.* **128**, 19–23.
Trapido, H. (1972). *In* "Venezuelan Encephalitis," Sci. Publ. No. 243, pp. 163–165. Pan Am. Health Organ., Washington, D.C.
Van Peenen, P. F. D., Irsiana, R., Saroso, J. S., Joseph, S. W., Shope, R. E., and Joseph, P. L. (1974). *Mil. Med.* **139**, 821–823.
Varma, M. G. R., and Smith, C. E. G. (1972). *Acta Virol. (Engl. Ed.)* **16**, 159–167.

Walton, T. E., Alvarez, O., Jr., Buckwalter, R. M., and Johnson, K. M. (1973). *J. Infect. Dis.* **128**, 271–282.

Watts, D. M., Thompson, W. H., Yuill, T. M., DeFoliart, G. R., and Hanson, R. P. (1974). *Am. J. Trop. Med. Hyg.* **23**, 694–700.

Whitehead, R. H., Doherty, R. L., Domrow, R., Standfast, H. A., and Wetters, E. J. (1968). *Trans. R. Soc. Trop. Med. Hyg.* **62**, 439–445.

Williams, M. C., Woodall, J. P., Corbet, P. S., and Gillett, J. D. (1965a). *Trans. R. Soc. Trop. Med. Hyg.* **59**, 300–306.

Williams, M. C., Woodall, J. P., and Gillett, J. D. (1965b). *Trans. R. Soc. Trop. Med. Hyg.* **59**, 186–197.

Williams, R. E., Casals, J., Moussa, M. I., and Hoogstraal, H. (1972). *Am. J. Trop. Med. Hyg.* **21**, 582–586.

Young, N. A., and Johnson, K. M. (1969). *Am. J. Epidemiol.* **89**, 286–307.

7

Arthropods as Hosts and Vectors of Alphaviruses and Flaviviruses—Experimental Infections

LEON ROSEN

I. INTRODUCTION

A review of all the data on experimental infection of arthropods with togaviruses obviously is beyond the scope of this or any other single review. Beginning with the demonstration of yellow fever transmission by *Aedes aegypti* at the turn of the century, the amount of information accumulated during the past 75 years has become so enormous that it can be digested only by concentrating on a single or a few related viruses, or on a single or a few related arthropods. The following discussion attempts only to outline some of the factors that should be considered in undertaking experimental infection of arthropods

THE TOGAVIRUSES

ISBN 0-12-625380-3

with togaviruses—or with any other viruses. Technical methods are not described but can be found in the literature cited.

Experimental infection of arthropods with viruses usually is undertaken with one of three objectives in mind. In the past, the most common objective has been to evaluate the ability of a particular species of arthropod to serve as a vector of a given virus to man or to economically important animals. The following questions can thus be answered: (1) Is the virus capable of replicating in the arthropod? (2) Can the arthropod be infected orally by the levels of viremia likely to be encountered in nature? (3) Is the arthropod capable of transmitting the virus to a susceptible vertebrate host? If the answer to any of these questions is "no," then that arthropod can be excluded from further consideration—saving investigative effort in the field. On the other hand, an answer of "yes" to all three questions does not incriminate the arthropod species as a vector in nature, since such important considerations as its abundance, feeding pattern, etc., can only be determined in the field.

Another reason that arthropods are infected experimentally is to study the interactions of a virus with an intact arthropod host. Since togaviruses may have originated in arthropods (Andrewes, 1957; Hurlbut and Thomas, 1960; Schlesinger, 1971), data on arthropod host range, on tropisms for various arthropod tissues, on dynamics of viral replication in arthropods, on effects of the virus on the arthropod, and of the latter on the virus, are important facets of their biology.

The third major reason that arthropods are infected experimentally is to detect, quantitate, and identify viruses. In instances where cell cultures or common vertebrate laboratory animals are relatively unsatisfactory for such purposes, arthropods sometimes can be utilized as convenient and inexpensive laboratory hosts for viruses that replicate in them. A disadvantage of such use of arthropods is that, unlike many vertebrate hosts, they ordinarily do not exhibit gross signs of infection. Moreover, and again unlike vertebrate hosts, infected arthropods do not produce specific antibodies which can be utilized as indirect evidence of infection. Consequently, other means of detecting the presence of viral replication must be employed. Balanced against these disadvantages, however, is the sensitivity to infection of an appropriate arthropod host, the large amount of virus produced, and the economy of raising arthropods in large numbers—as compared to vertebrates.

In undertaking experimental infection of arthropods a number of factors must be considered. These will be discussed below as they relate to the three objectives described.

II. CONSIDERATIONS IN UNDERTAKING EXPERIMENTAL INFECTION OF ARTHROPODS

A. Choice of Arthropod

1. *Oral Infection*

Although the choice of arthropod species is dictated in studies of vector competence, it should be kept in mind that not all populations of a species necessarily respond similarly to a given viral agent. Variation in susceptibility to oral infection among individual arthropods of the same species is well known (Chamberlain and Sudia, 1961). In addition, there may be differences in the percentage of relatively susceptible individuals among populations of the same species from different geographic areas (Hardy *et al.*, 1976) and, perhaps, among populations from the same geographic area at different times of the year. Differences in the percentage of susceptible individuals also have been demonstrated among laboratory colonies of the same arthropod species (Gubler and Rosen, 1976a; Hardy *et al.*, 1976; Tesh *et al.*, 1976). Also, there is no assurance that susceptibility remains unchanged in the process of colonizing a population for laboratory study. Thus, it cannot be assumed that the susceptibility of an arthropod colony originating from a certain geographic area is representative of the wild population in that area. As yet, relatively few data are available on the variation in viral susceptibility likely to be encountered among different field populations of the same arthropod species. One point that is clear, however, is that an arthropod population relatively susceptible to one viral agent is not necessarily similarly susceptible to another (Tesh *et al.*, 1976).

Little is known of the reasons for variation in susceptibility of an arthropod species to viral infection, either for individuals or for populations. Among the possibilities are genetic factors (Gubler and Rosen, 1976a), presence of interfering indigenous viruses in the arthropod (Clark *et al.*, 1965, 1969; Federici and Lowe, 1972; Lebedeva *et al.*, 1973; Wagner *et al.*, 1974; Richardson *et al.*, 1974), and "environmental" factors. At the present state of knowledge, the best procedure for evaluating the vector competence of a given arthropod species is the use of arthropods collected in the field—with an attempt, if possible, to sample populations collected in different geographic areas.

2. *Parenteral Infection*

The situation is entirely different with respect to the choice of an arthropod as a laboratory host for a virus. The differences in susceptibility mentioned above apply largely to infection by the oral route. When the gut is bypassed by parenteral inoculation, most if not all differences between individuals and populations of the same arthropod species disappear (Gubler and Rosen, 1976a; Hardy *et al.*, 1976). In the choice of a species for laboratory use such considerations as ease of propagation, size, hardiness, and safety considerations play a major role. Fortunately, in the case of mosquitoes at least, males appear as susceptible to parenteral infection as females (Rosen and Gubler, 1974) and, since they cannot pierce skin, can be employed without the safety precautions necessary for the latter. Moreover, even arthropods that have no relationship to a given virus in nature may prove to be valuable laboratory animals. Many togaviruses have surprisingly wide arthropod host ranges, such as Japanese encephalitis virus in beetles and moths (Aitken *et al.*, 1958; Hurlbut and Thomas, 1960, 1969) and Sindbis virus in *Drosophila* (Bras-Herreng, 1976). Mosquitoes of the genus *Toxorhynchites* are examples of arthropods that have proved useful in the laboratory despite their lack of relationship to a virus in nature (L. Rosen, unpublished observations). Neither sex of these very large mosquitoes is capable of piercing skin to obtain blood, and yet both are excellent hosts of dengue viruses which, surprisingly, have a restricted host range—even by parenteral inoculation—among genera of blood-feeding mosquitoes (Hurlbut and Thomas, 1960; L. Rosen and L. E. Rozeboom, unpublished observations). Another consideration in the choice of an arthropod as a laboratory host is the acceptability of the introduction of the species into an area where it does not naturally occur. There is no problem in this regard with respect to mosquitoes of the genus *Toxorhynchites*. Since their larvae are carnivorous, *Toxorhynchites* have been introduced deliberately in areas where they do not occur naturally in an attempt to control other mosquitoes. On the other hand, the introduction of *Aedes aegypti* into an area where it does not occur, and where it could become established if it escaped from the laboratory, obviously would not be desirable. Very little work has been done on the possible use of large, easily reared arthropods as laboratory hosts. It is likely that further explorations in this regard, along with the type of selection and husbandry employed in the development of vertebrate laboratory animals, will provide useful tools for those interested in viruses replicating in arthropods. Of course, it also is likely that some of the problems encountered with laboratory vertebrates, such as the presence of "indigenous" viruses, will be duplicated in the development of laboratory arthropods.

B. Choice of Virus

It is well known that sequential passage in vertebrates will change the characteristics of a viral population with respect to virulence and tissue tropisms for such animals. Similarly, sequential passage of a viral population in cell cultures will modify its characteristics with respect to such cultures, as well as for vertebrates. Comparatively little is known, however, about the effect of passage of a viral population in vertebrates or cell cultures on its subsequent infectivity for, replication in, or tissue tropisms in arthropods. What data are available suggest that extensive cell culture or vertebrate passage may modify the infectivity of a viral population for arthropods by the oral, but not by the parenteral, route (Whitman, 1939; Pattyn and De Vleesschauwer, 1968, 1969; Takahashi *et al.*, 1969; Gaidamovich *et al.*, 1971; Peleg, 1975). It also appears that, once an arthropod is infected, cell culture- or vertebrate-passaged viral populations multiply to as high titers as others but, for unknown reasons, are not as readily transmitted. The situation with respect to infectivity by the oral route is not completely clear since, in some cases at least, the effect may be the result of lower viremia levels in the vertebrate host rather than of a direct diminution in oral infectivity for the arthropod. The uncertainty cannot be resolved entirely by feeding arthropods on virus–sucrose–erythrocyte mixtures, since these differ qualitatively in oral infectivity for arthropods as compared with exposure to intact viremic hosts (see below).

The extensive laboratory manipulation of the Asibi strain of yellow fever virus which resulted in the 17D vaccine virus produced an agent that yielded lower levels of viremia in primates than the parent virus. Although mosquitoes fed on primates infected with 17D virus did not become infected (perhaps, at least in part, because of low viremias), they could be infected in the larval state by immersing them in a suspension of the virus. Adult females from larvae so infected were not capable of transmitting their infection to primates, although virus infectious for primates could be recovered from them by trituration (Whitman, 1939). When 17D virus is inoculated parenterally into adult mosquitoes, it replicates to high titer (L. Rosen, unpublished observations). Similarly, two strains of Sindbis virus replicate to equally high titer in *A. aegypti* but infected females can transmit one strain to mice and not the other (Peleg, 1975).

The effect of sequential passage of a viral population in arthropods on its subsequent attributes in vertebrates, cell cultures, or arthropods also is largely unknown. What information is available suggests that such viral characteristics as plaque size and virulence are not changed by serial passage of homogenized, infected arthropods—at least not within the relatively low number of passages studied and with virus for passage harvested at relatively short intervals after infection (Mussgay and Suárez, 1962; Schaffer and Scherer, 1971; Taylor and Marshall, 1975).

In studies of vector competence, it seems logical to utilize virus material which has been passaged as little as possible in the laboratory. If passage is necessary, it should be done, in so far as possible, in natural vertebrate or arthropod hosts.

Just as variations exist in the susceptibility of different natural arthropod populations, variations also exist among viral strains of the same type in their tissue tropisms and ability to replicate in arthropods (L'Héritier, 1958; Kramer and Scherer, 1976). The possibility of such viral variation should be kept in mind when evaluating vector competence of an arthropod—especially if laboratory findings are not what was expected on the basis of observations in nature.

C. Choice of Means of Infection

Studies of vector competence obviously require that the arthropod eventually be exposed by ingestion of virus. Under certain circumstances, however, parenteral injection of virus can be carried out beforehand to determine whether or not the agent will replicate in that particular species of arthropod. If a virus will not replicate after parenteral infection, there is no point in trying to infect by ingestion.

Exposure of an arthropod to infection by ingestion of virus can be carried out by allowing it to feed either on a viremic vertebrate host or on an artificial mixture of virus, sucrose, and erythrocytes. It is usual to employ a viremic host undergoing viral infection. However, in instances where such a host is not available, it sometimes is possible to inject sufficient virus into a nonsusceptible vertebrate to obtain a viremia of sufficient magnitude and duration to test arthropod susceptibility (McLean *et al.*, 1974).

In artificial mixtures, sucrose is included to induce the arthropod to feed, and erythrocytes are added so that the meal goes to the midgut rather than to the diverticula (Day, 1954; Hosoi, 1954). Many different procedures have been employed to feed arthropods (principally mosquitoes) on such mixtures (see references in Chamberlain and Sudia, 1967; Galun, 1967; Behin, 1967). These involve the use of various types of membranes which the arthropod must pierce in feeding and a variety of devices to keep the mixture warm. Some *Aedes* species, however, feed readily on simple drops of unwarmed mixture (Gubler and Rosen, 1976b). Obviously, it is easier to control the amount of virus and to duplicate feedings on artificial mixtures than on intact vertebrates. Usually, however, a larger amount of virus is required to infect mosquitoes with an artificial mixture than with an intact viremic host (Chamberlain *et al.*, 1954; Jupp, 1976). The reason for this is not known. The results of vector competence studies carried out with artificial mixtures must be interpreted with caution. If possible, such studies should be carried out or repeated with viremic vertebrates.

Parenteral inoculation is the only practical route of infection when arthropods are employed as laboratory hosts to detect, assay, and identify viruses. Individual arthropods which vary in susceptibility to infection by ingestion generally are uniformly susceptible to infection by parenteral inoculation (Gubler and Rosen, 1976a; Hardy *et al.*, 1976). However, individual arthropods of the same species infected parenterally often yield grossly different final amounts of virus (Tesh *et al.*, 1976). A variety of complicated and crude techniques have been employed in the past to inoculate arthropods, but the procedures now available are as simple and precise as those employed to inoculate mice (Carley *et al.*, 1973; Rosen and Gubler, 1974; Boorman, 1975).

The foregoing discussion refers to infection of the adult forms of arthropods, but it should be noted that arthropods also can be infected transovarially (Hoogstraal, 1973) and in various developmental stages between the egg and imago. Mosquito larvae, for example, can be infected with togaviruses by immersing them in a virus suspension (Whitman and Antunes, 1938; Hodes, 1946; Collins, 1962, 1963; Collins *et al.*, 1966; Peleg, 1965). A technique also is available for inoculating mosquito larvae (Hasan *et al.*, 1971). Ticks can be infected by rectal inoculation (Stelmaszyk, 1975), as well as parenterally, and mosquitoes can be infected by tarsal contact with a virus suspension (Jupp *et al.*, 1966).

D. Choice of Environmental Conditions

Temperature is the most obvious environmental consideration in experimental infection of arthropods with viruses. In general, the higher the temperature the more rapidly a given togavirus will replicate in a particular arthropod (Hurlbut, 1973; McLean *et al.*, 1975a). The limits of temperature which can be utilized will depend on those tolerated by the arthropod. As yet, no one has reported a decrease at high temperature in the virus content of arthropods infected with togaviruses such as has been observed in *Drosophila* infected with the rhabdovirus sigma (L'Héritier, 1958). Perhaps this is not surprising, since togaviruses also must replicate at the higher temperatures of vertebrates, whereas sigma virus has no need to do so. Temperature-sensitive togaviruses unable to replicate at temperatures tolerated by arthropod hosts have not been described. Relatively little is known of the effects of varying, as compared with constant, temperature on the replication of viruses in arthropods.

Exposure to light has been observed to affect the rate of replication of a togavirus in an arthropod (Cates and Huang, 1969)—perhaps because of its effect on arthropod behavior. Photoperiod also has important effects in inducing diapause in arthropods (Tauber and Tauber, 1976), and it would not be surprising to find that it affects viral replication in a variety of ways.

Environmental factors other than temperature and light, in so far as they have been studied, have not been found to affect the experimental infection of arthropods with viruses to a significant extent (Takahashi, 1976).

E. Techniques for Detection of Viruses in Arthropods

Generally arthropods infected with togaviruses show no gross signs of infection and consequently other means must be employed in detecting infected specimens. It should be noted, however, that relatively little effort has been devoted to a search for subtle effects of infection and at least one species of arthropod (*A. aegypti*) has been found to be adversely affected by a togavirus (Semliki Forest) (Mims *et al.*, 1966). For many years the only known effect of sigma virus in *Drosophila* was to render this insect lethally sensitive to carbon dioxide anesthesia (L'Héritier, 1958). Later, however, other effects were discovered (Seecof, 1964). It is unlikely that the sigma–*Drosophila* combination is the only instance in which a virus affects the biochemistry of its host in a grossly detectable manner. Possibly analogous techniques could be developed for detecting togavirus infection in arthropods. Some viruses produce inapparent infection in some species of arthropods and lethal infections in others (Bailey and Scott, 1973).

The classic means of detecting a virus in an arthropod is to triturate the entire specimen, or parts thereof, and inoculate an aliquot of the clarified suspension into a suitable intact vertebrate or cell culture system (LaMotte, 1960; Miles *et al.*, 1973). This relatively cumbersome procedure now is being replaced in some instances by more rapid and simple techniques. The latter involve the direct visualization of virus and viral products in squash preparations of the arthropod by immunofluorescence staining (Sinha and Black, 1962; Kuberski and Rosen, 1977a), as well as by electron microscopy (Richardson *et al.*, 1974). In the case of some togaviruses and mosquitoes, such a large amount of virus is present in the cerebral ganglia that infected insects can be detected simply by examining a head squash with a fluorescent antibody procedure (Kuberski and Rosen, 1977a). This procedure obviates the time-consuming procedure of dissecting out salivary glands and the difficulties of examining squash preparations of entire arthropods.

Infected arthropods also can be detected by such *in vitro* procedures as complement fixation (Kuberski and Rosen, 1977b), precipitin ring tests (Whitcomb and Black, 1961), and agar gel immunodiffusion techniques (Crozier *et al.*, 1970). Generally, the entire specimen is triturated and the clarified suspension used in the test. However, it also may be possible to test aliquots of hemolymph (Boorman, 1960) without killing the arthropod.

When it is desired to study the sites and progression of virus replication in an arthropod, organs can be dissected out and examined, *in toto* or after sectioning, by immunofluorescence and immunoperoxidase techniques and by electron microscopy (Doi *et al.*, 1967; Chernesky and McLean, 1969; Doi, 1970; Janzen *et al.*, 1970; Whitfield *et al.*, 1971, 1973; Larsen and Ashley, 1971; Janzen and Wright, 1971; Nosek *et al.*, 1972; Peers, 1972; Gaidamovich *et al.*, 1973; McLean *et al.*, 1975b; Maguire, 1975; Gubler and Rosen, 1976b).

There are several ways to demonstrate that an arthropod not only is infected but also can transmit its infection. One is to allow the arthropod to feed on an intact susceptible vertebrate host and to examine the latter for viremia, antibody, or visible signs of infection. A number of other techniques also are available, however, which involve the detection (and sometimes quantitation) of virus in arthropod saliva (Devine *et al.*, 1965; Pattyn and De Vleesschauwer, 1970; Hayles, 1976; Aitken, 1977; Gubler and Rosen, 1976b; see references in the latter publication to earlier literature). These provide the means for examining large numbers of individual arthropods economically and for testing for transmission of viruses for which suitable hosts are not readily available in the laboratory. Togaviruses also have been detected in the anal discharge of infected mosquitoes (Muangman *et al.*, 1969).

REFERENCES

Aitken, T. H. G. (1977). *Mosq. News* **37**, 130–133.
Aitken, T. H. G., Downs, W. G., and Anderson, C. R. (1958). *Proc. Soc. Exp. Biol. Med.* **99**, 635–637.
Andrewes, C. H. (1957). *Adv. Virus Res.* **4**, 1–24.
Bailey, L., and Scott, H. A. (1973). *Nature (London)* **241**, 545.
Behin, R. (1967). *Mosq. News* **27**, 87–90.
Boorman, J. (1960). *Trans. R. Soc. Trop. Med. Hyg.* **54**, 362–365.
Boorman, J. (1975). *Lab. Pract.* **24**, 90.
Bras-Herreng, F. (1976). *Ann. Microbiol. (Paris)* **127b**, 541–565.
Carley, J. G., Standfast, H. A., and Kay, B. H. (1973). *J. Med. Entomol.* **10**, 244–249.
Cates, M. D., and Huang, W. C. (1969). *Mosq. News* **29**, 620–623.
Chamberlain, R. W., and Sudia, W. D. (1961). *Annu. Rev. Entomol.* **6**, 371–390.
Chamberlain, R. W., and Sudia, W. D. (1967). *In* "Methods in Virology" (K. Maramorosch and H. Koprowski, eds.), Vol. 1, pp. 63–103. Academic Press, New York.
Chamberlain, R. W., Sikes, R. K., Nelson, D. B., and Sudia, W. D. (1954). *Am. J. Hyg.* **60**, 278–285.
Chernesky, M. A., and McLean, D. M. (1969). *Can. J. Microbiol.* **15**, 1399–1408.
Clark, T. B., Kellen, W. R., and Lum, P. T. M. (1965). *J. Invertebr. Pathol.* **7**, 519–520.
Clark, T. B., Chapman, H. C., and Fukuda, T. (1969). *J. Invertebr. Pathol.* **14**, 284–286.
Collins, W. E. (1962). *Am. J. Trop. Med. Hyg.* **11**, 535–538.
Collins, W. E. (1963). *Ann. Entomol. Soc. Am.* **56**, 237–239.
Collins, W. E., Harrison, A. J., and Jumper, J. R. (1966). *Mosq. News* **26**, 364–367.
Croizier, G., Plus, N., and Veyrunes, J. D. (1970). *C. R. Acad. Sci., Ser. D* **270**, 1185–1189.
Day, M. F. (1954). *Aust. J. Biol. Sci.* **7**, 515–524.
Devine, T. L., Venard, C. E., and Myser, W. C. (1965). *J. Insect Physiol.* **11**, 347–353.
Doi, R. (1970). *Jpn. J. Exp. Med.* **40**, 101–115.
Doi, R., Shirasaka, A., and Sasa, M. (1967). *Jpn. J. Exp. Med.* **37**, 227–238.
Federici, B. A., and Lowe, R. E. (1972). *J. Invertebr. Pathol.* **20**, 14–21.
Gaidamovich, S. Y., Tsilinsky, Y. Y., Lvova, A. I., and Khutoretskaya, N. V. (1971). *Acta Virol. (Engl. Ed.)* **15**, 301–308.
Gaidamovich, S. Y., Khutoretskaya, N. V., Lvova, A. I., and Sveshnikova, N. A. (1973). *Intervirology* **1**, 193–200.
Galun, R. (1967). *Bull. W.H.O.* **36**, 590–593.
Gubler, D. J., and Rosen, L. (1976a). *Am. J. Trop. Med. Hyg.* **25**, 318–325.
Gubler, D. J., and Rosen, L. (1976b). *Am. J. Trop. Med. Hyg.* **25**, 146–150.
Hardy, J. L., Reeves, W. C., and Sjogren, R. D. (1976). *Am. J. Epidemiol.* **103**, 498–505.
Hasan, S., Vago, C., and Kuhl, G. (1971). *Bull. W.H.O.* **45**, 268–269.
Hayles, L. B. (1976). *Res. Vet. Sci.* **21**, 358–359.
Hodes, H. L. (1946). *Bull. Johns Hopkins Hosp.* **79**, 358–359.
Hoogstraal, H. (1973). *In* "Viruses and Invertebrates" (A. J. Gibbs, ed.), pp. 349–390. North-Holland Publ., Amsterdam-London.
Hosoi, T. (1954). *Annot. Zool. Jpn.* **27**, 82–90.
Hurlbut, H. S. (1973). *J. Med. Entomol.* **10**, 1–12.
Hurlbut, H. S., and Thomas, J. I. (1960). *Virology* **12**, 391–407.
Hurlbut, H. S., and Thomas, J. I. (1969). *J. Med. Entomol.* **6**, 423–427.
Janzen, H. G., and Wright, K. A. (1971). *Can. J. Zool.* **49**, 1343–1345.
Janzen, H. G., Rhodes, A. J., and Doane, F. W. (1970). *Can. J. Microbiol.* **16**, 581–586.

Jupp, P. G. (1976). *Mosq. News* **36**, 166–173.
Jupp, P. G., Brown, R. G., and McIntosh, B. M. (1966). *S. Afr. J. Med. Sci.* **31**, 51–53.
Kramer, L. D., and Scherer, W. F. (1976). *Am. J. Trop. Med. Hyg.* **25**, 336–346.
Kuberski, T. T., and Rosen, L. (1977a). *Am. J. Trop. Med. Hyg.* **26**, 533–537.
Kuberski, T. T., and Rosen, L. (1977b). *Am. J. Trop. Med. Hyg.* **26**, 538–543.
LaMotte, L. C., Jr. (1960). *Am. J. Hyg.* **72**, 73–87.
Larsen, J. R., and Ashley, R. F. (1971). *Am. J. Trop. Med. Hyg.* **20**, 754–760.
Lebedeva, O. P., Kuznetsova, M. A., Zelenko, A. P., and Gudz-Gorban, A. P. (1973). *Acta Virol. (Engl. Ed.)* **17**, 253–256.
L'Héritier, P. (1958). *Adv. Virus Res.* **5**, 195–245.
McLean, D. M., Clarke, A. M., Coleman, J. C., Montalbetti, C. A., Skidmore, A. G., Walters, T. E., and Wise, R. (1974). *Can. J. Microbiol.* **20**, 255–262.
McLean, D. M., Grass, P. N., Miller, M. A., and Wong, K. S. K. (1975a). *Arch. Virol.* **49**, 49–57.
McLean, D. M., Gubash, S. M., Grass, P. N., Miller, M. A., Petric, M., and Walters, T. E. (1975b). *Can. J. Microbiol.* **21**, 453–462.
Maguire, T. (1975). *Proc. Univ. Otago Med. Sch.* **53**, 75–76.
Miles, J. A. R., Pillai, J. S., and Maguire, T. (1973). *J. Med. Entomol.* **10**, 176–185.
Mims, C. A., Day, M. F., and Marshall, I. D. (1966). *Am. J. Trop. Med. Hyg.* **15**, 775–784.
Muangman, D., Frothingham, T. E., and Spielman, A. (1969). *Am. J. Trop. Med. Hyg.* **18**, 401–410.
Mussgay, M., and Suárez, O. (1962). *Arch. Gesamte Virusforsch.* **12**, 387–392.
Nosek, J., Čiampor, F., Kožuch, O., and Rajčáni, J. (1972). *Acta Virol. (Engl. Ed.)* **16**, 493–497.
Pattyn, S. R., and De Vleesschauwer, L. (1968). *Acta Virol. (Eng. Ed.)*, **12**, 347–354.
Pattyn, S. R., and De Vleesschauwer, L. (1969). *Ann. Soc. Belge Med. Trop.* **49**, 63–68.
Pattyn, S. R., and De Vleesschauwer, L. (1970). *Acta Virol. (Engl. Ed.)* **14**, 510–512.
Peers, R. R. (1972). *Can. J. Microbiol.* **18**, 741–745.
Peleg, J. (1965). *Am. J. Trop. Med. Hyg.* **14**, 158–164.
Peleg, J. (1975). *Ann. N.Y. Acad. Sci.* **266**, 204–213.
Richardson, J., Sylvester, E. S., Reeves, W. C., and Hardy, J. L. (1974). *J. Invertebr. Pathol.* **23**, 213–224.
Rosen, L., and Gubler, D. (1974). *Am. J. Trop. Med. Hyg.* **23**, 1153–1160.
Schaffer, P. A., and Scherer, W. F. (1971). *Am. J. Epidemiol.* **93**, 68–74.
Schlesinger, R. W. (1971). *Curr. Top. Microbiol. Immunol.* **55**, 241–248.
Seecof, R. L. (1964). *Virology* **22**, 142–148.
Sinha, R. C., and Black, L. M. (1962). *Virology* **17**, 582–587.
Stelmaszyk, Z. J. (1975). *Wiad. Parazytol.* **21**, 29–36.
Takahashi, M. (1976). *J. Med. Entomol.* **13**, 275–284.
Takahashi, M., Yabe, S., and Okada, T. (1969). *Jpn. J. Med. Sci. Biol.* **22**, 163–174.
Tauber, M. J., and Tauber, C. A. (1976). *Annu. Rev. Entomol.* **21**, 81–107.
Taylor, W. P., and Marshall, I. D. (1975). *J. Gen. Virol.* **28**, 73–83.
Tesh, R. B., Gubler, D. J., and Rosen, L. (1976). *Am. J. Trop. Med. Hyg.* **25**, 326–335.
Wagner, G. W., Webb, S. R., Paschke, J. D., and Campbell, W. R. (1974). *J. Invertebr. Pathol.* **24**, 380–382.
Whitcomb, R. F., and Black, L. M. (1961). *Virology* **15**, 136–145.
Whitfield, S. G., Murphy, F. A., and Sudia, W. D. (1971). *Virology* **43**, 110–122.
Whitfield, S. G., Murphy, F. A., and Sudia, W. D. (1973). *Virology* **56**, 70–87.
Whitman, L. (1939). *Am. J. Trop. Med.* **19**, 19–26.
Whitman, L., and Antunes, P. C. A. (1938). *Proc. Soc. Exp. Biol. Med.* **37**, 664–666.

8

Togavirus Morphology and Morphogenesis

FREDERICK A. MURPHY

I. INTRODUCTION

The term "togavirus" emerged while efforts were being made to assimilate all the arthropod-borne viruses into a universal scheme of virus taxonomy—that is, the scheme based upon physicochemical properties of virus particles which is presently in use (Wildy, 1971). The term was advanced to emphasize the similarities between group-A and group-B arboviruses (now the *Alphavirus* and *Flavivirus* genera of the family Togaviridae), but at the time this was done few

THE TOGAVIRUSES

ISBN 0-12-625380-3

comparative physicochemical data were available on the large number of serologically and ecologically related viruses. Many electron microscopic studies were being carried out, however, and it was becoming clear that it was not possible to distinguish morphologically the member viruses of the *Alphavirus* genus or of the *Flavivirus* genus, but that members of the two genera were distinguishable from each other and from all other known viruses.

In retrospect, it seems that electron microscopic evidence, often with minimal resolution of detail, was used to prejudge relationships which should have been built only upon comprehensive physicochemical data. Nevertheless, it has turned out that definition of the major serogroupings of the arthropod-borne viruses has been confirmed by more detailed electron microscopic and physiochemical studies—that is, despite the placement of arthropod-borne viruses into several virus families, the serogroups themselves remain intact as distinct genera (Casals, 1971; Murphy, 1974, 1977; Berge, 1975; Fenner, 1976; Porterfield *et al.*, 1978). Even though there is no criticism of the precision of definitions at the genus level, it may be that the original morphological similarities between the *Alphavirus* and *Flavivirus* genera will stand as the strongest basis for tying the two genera together into a family. Much of the nonmorphological evidence (nucleic acid and protein constituency, mode of replication) which denies a close phylogenetic relationship between the *Alphavirus* and *Flavivirus* genera is presented elsewhere in this volume (see summary in Chapter 1 and details in Chapters 9–19). Just as these differences softened the definition of the Togaviridae family, it has been recognized that several non-vector-borne viruses have similar morphological or morphogenetic characteristics and, therefore, belong in this taxon. At present two further genera have been constructed, *Rubivirus* (rubella virus) and *Pestivirus* (hog cholera and bovine virus diarrhea viruses), and there are still several uncategorized viruses of similar morphology. Thus, the Togaviridae family has been expanded to contain all the small, enveloped, RNA-containing viruses which have proved or suspected cubic nucleocapside symmetry and a single-stranded genome of about 4×10^6 daltons which serves directly as a messenger species (i.e., without a virion-associated RNA transcriptase). The taxon is more encompassing, and as such it may be more useful. It is with an acceptance of the general similarities and precise differences among member genera that in this chapter morphological details of representative viruses are compared and extended, where possible, into concepts of structure. These comparisons suffer mostly from our technical inability to resolve structural details.

Experiences and successes with *Alphavirus* members have not yet led to parallel resolution of the member viruses of the other genera.

A comprehensive review of the structure of togaviruses has been published by Horzinek (1973a) and is a very valuable source of comparative ultrastructural data.

II. THE *ALPHAVIRUS* GENUS

Although the first member viruses of the *Alphavirus* genus were brought together strictly on the basis of serological interrelationships (arbovirus serogroup A) (Casals and Brown, 1954), and additional members have been added on the same basis only (Berge, 1975), physicochemical and biological characteristics have always suggested that the grouping would be narrowly constructed. Such precise uniformity has been fully substantiated by electron microscopic studies which, in more or less detail, have been carried out on most of the 20 presently recognized member viruses. Virion morphology and morphogenetic characteristics of alphaviruses are so uniform, and yet distinct from those of flaviviruses and other viruses, that a new virus could easily be recognized as an alphavirus by thin-section or even negative-contrast electron microscopic methods. Other physicochemical characteristics, as described elsewhere in this volume, confirm in several ways this precision in the construction and definition of the genus. Minor differences in ultrastructural characteristics of member alphaviruses have been emphasized in attempts to explain the unique biological characteristics of individual viruses (e.g., neurotropism versus nonneurotropism and high virulence versus low virulence). However, such morphological or morphogenetic differences are more likely to reflect varying technical conditions of virus growth and examination than fundamental differences. Critical analysis and precise comparisions should be demanded of any claimed ultrastructural variances among alphaviruses.

Studies carried out before electron microscopy was applied accurately predicted some characteristics which aided in the first identifications of alphavirus particles. Ether and detergent sensitivity predicted the presence of an envelope, and filtration data indicated a particle size between 40 and 55 nm. Consequently, in even the first electron microscopic studies there seemed to be an unambiguous identification of particles; this was not the case with flaviviruses and several other togaviruses. Shadowcasting techniques were used in the first electron microscopic studies; in a remarkable study for its time,

Sharp and colleagues (1943) used these techniques to show the general spherical shape of Eastern and Western equine encephalitis virus particles which had been partially purified by differential ultracentrifugation. Virus particle size varied from 40 to 53 nm, according to the suspension medium and preparation method. Bang and Gey (1949, 1952) also visualized similar particles in shadow-cast preparations from Eastern equine encephalitis virus-infected cell cultures. Mussgay and Weibel (1963) obtained such good resolution of partially purified Venezuelan equine encephalitis virus by the shadow-cast method that they were able to resolve the membrane and core of particles. Faulkner and McGee-Russell (1968) also observed these structures in shadow-cast preparations and even suggested that the core structure had cubic symmetry.

These observations were complemented by thin-section studies starting with the benchmark work of Morgan and colleagues (1961). In this study of Western equine encephalitis virus infection of cultured cells, viral precursor particles were identified and correctly related to the process of virus maturation by budding. Virus particle morphology was illustrated with detail that has hardly been surpassed in the ensuing years. At the same time Mussgay and Weibel (1962) found similar evidences of Venezuelan equine encephalitis virus replication in cultured cells. Negative-contrast methods were first applied to the study of alphaviruses—Sindbis virus by Mussgay and Rott (1964), Semliki Forest virus by Osterrieth and Calberg-Bacq (1966), chikungunya virus by Chain *et al.* (1966) and Igarashi *et al.* (1966), and Venezuelan equine encephalitis virus by Klimenko *et al.* (1965), Horzinek and Munz (1969), and Murphy and Harrison (1971).

A. Negative-Contrast Electron Microscopy

Negative-contrast studies of alphaviruses have been concentrated on Semliki Forest, Sindbis, and chikungunya viruses; fewer studies have dealt with Eastern equine encephalitis, Middelburg, and Venezuelan equine encephalitis viruses. In all cases, virus particles have been found to be spherical, to have a membranous envelope with surface projections, and to have a dense core.

In general, particle diameters have been reported as being between 45 and 80 nm. In most studies, particles have been found to have nucleocapsid diameters between 28 and 40 nm. These rather wide ranges of dimensions most likely resulted from a combination of four factors. First, it is likely that some variations stemmed from the flattening of particles upon their substrates during preparation for electron microscopy. Second, some variances likely stemmed from errors in the calibration of microscope and photographic magnification (Pfefferkorn and Shapiro, 1974). Third, some variations likely derived from a phenomenon studied in detail by von Bonsdorff (1973). He found that, of the factors involved in the preparation, treatment, and staining of virus samples, the pH and content of the negative-contrast medium had most effect upon particle size. For example, potassium phosphotungstate stain at pH 7.0–7.2 yielded Semliki Forest virus particles with a mean diameter of 65 nm but, when the pH of the same stain was lowered to 6.2, or when uranyl acetate (pH 4.5) or ammonium molybdate (pH 7.2) was used as the staining medium, mean diameters were significantly reduced (usually to 50–57 nm). In more recent studies Söderlund *et al.* (1975, 1979) found that it was the nucleocapsid size which was changed; they found that Semliki Forest virus nucleocapsids were reduced from 39 to 33 nm by exposure to a pH of 6.2 (the same treatment, however, does not affect Sindbis nucleocapsids). Similarly, treatment with RNase (5 μg/ml, 10 minutes) reduces the nucleocapsid size of Semliki Forest virus capsids (and Sindbis virus capsids) from 39 to 31–35 nm. The outer virus particle layers may become deformed as the nucleocapsid becomes condensed. It is not possible to try to reinterpret published size estimations of various alphaviruses in relation to this phenomenon, but this must be considered in all future studies. The fourth effect that may have contributed to the published variations in alphavirus particle size is that of genetic heterogeneity. Tsilinsky *et al.* (1971a,b) found that the variation in the size of Venezuelan encephalitis virus particles was greater in uncloned than in cloned virus stocks; each individual clone yielded particles with a distinct, yet reproducible, narrow size range. Average particle size could be correlated with plaque size. It remains to be seen whether other alphaviruses express similar genetic variance patterns.

In any case, when an alphavirus is examined under controlled conditions, a narrow and reproducible size range is found; when the same kinds of controlled treatment and staining conditions are used to compare size estimations between laboratories or between different alphaviruses, then the same narrow size range is obtained. If, in a most arbitrary fashion, particle dimension data are extracted from only the most carefully controlled published studies, the best estimation for the mean alphavirus particle diameter is 60–65 nm, and for the nucleocapsid diameter 35–39 nm. These figures closely match estimations made by x-ray diffraction (Harrison *et al.*, 1971).

A complete description of alphavirus morphology requires that separate attention be given to (1) the surface projection layer, (2) the envelope layer, and (3) the nucleocapsid.

The surface projection layer may vary in thickness (6–10 nm) and in appearance, according to the degree of virus purification (the degree of removal of nonviral proteinaceous background from the interstices of structural elements) and to varying detrimental effects of staining and drying conditions. In their most commonly illustrated form, alphavirus particles appear to have a finely textured, fuzzy surface layer; at the margins of particles this surface layer is seen as a rather ragged halo in which individual spikes are not resolved (Simpson and Hauser, 1968a; Igarashi *et al.*, 1970; Compans, 1971; Murphy and Harrison, 1971; von Bonsdorff, 1973) (Fig. 1). This form is seen in the diagnostic setting where minimal purification is justified; even in this form the presumptive identification of a virus as an alphavirus can be made with confidence.

The ultimate nature of the alphavirus surface projection layer is the subject of some controversy. In some instances, when a high degree of purification has been achieved, individual spikes 3–3.5 nm thick and 6–10 nm long have been found spaced at intervals of about 4–6 nm (Osterrieth, 1968; von Bonsdorff, 1973). These rodlike spikes appeared to be inserted radially, but usually were bent and often crossed one another in a complex pattern. In an alternate interpretation of observations, Horzinek (1973a) considered that the alphavirus surface projection layer was in the form of "loops" of continuously filamentous material. This surface model was supported by microdensitometer tracings across micrographs of Sindbis virus particles; influenza virus particles were used to exemplify a virus surface with true spikes.

Even though the configuration of alphavirus surface projections may be controversial, there is good evidence that their arrangement is symmetric. Simpson and Hauser (1968a) first found evidence of symmetrically placed subunits on the surface of Formalin-fixed alphavirus particles. Von Bonsdorff and Harrison (1975) studied the surface of Sindbis virus particles further and determined that the glycoprotein subunits were organized in trimer clusterings to form icosahedral particles with $T = 4$ (42-subunit) surface lattice symmetry (Fig. 2–4). The same structure was found on the surface of Semliki Forest virus particles and was also confirmed by use of the freeze-etch technique (Brown *et al.*, 1972; Horzinek, 1973a; von Bonsdorff, 1977) (Fig. 5). The convincing interpretation of alphavirus surface organization presented by von Bonsdorff and his colleagues has important bearing upon considerations of the structure of the nucleocapsid; it must also be considered in our attempts to unravel the fundamental nature of the budding process.

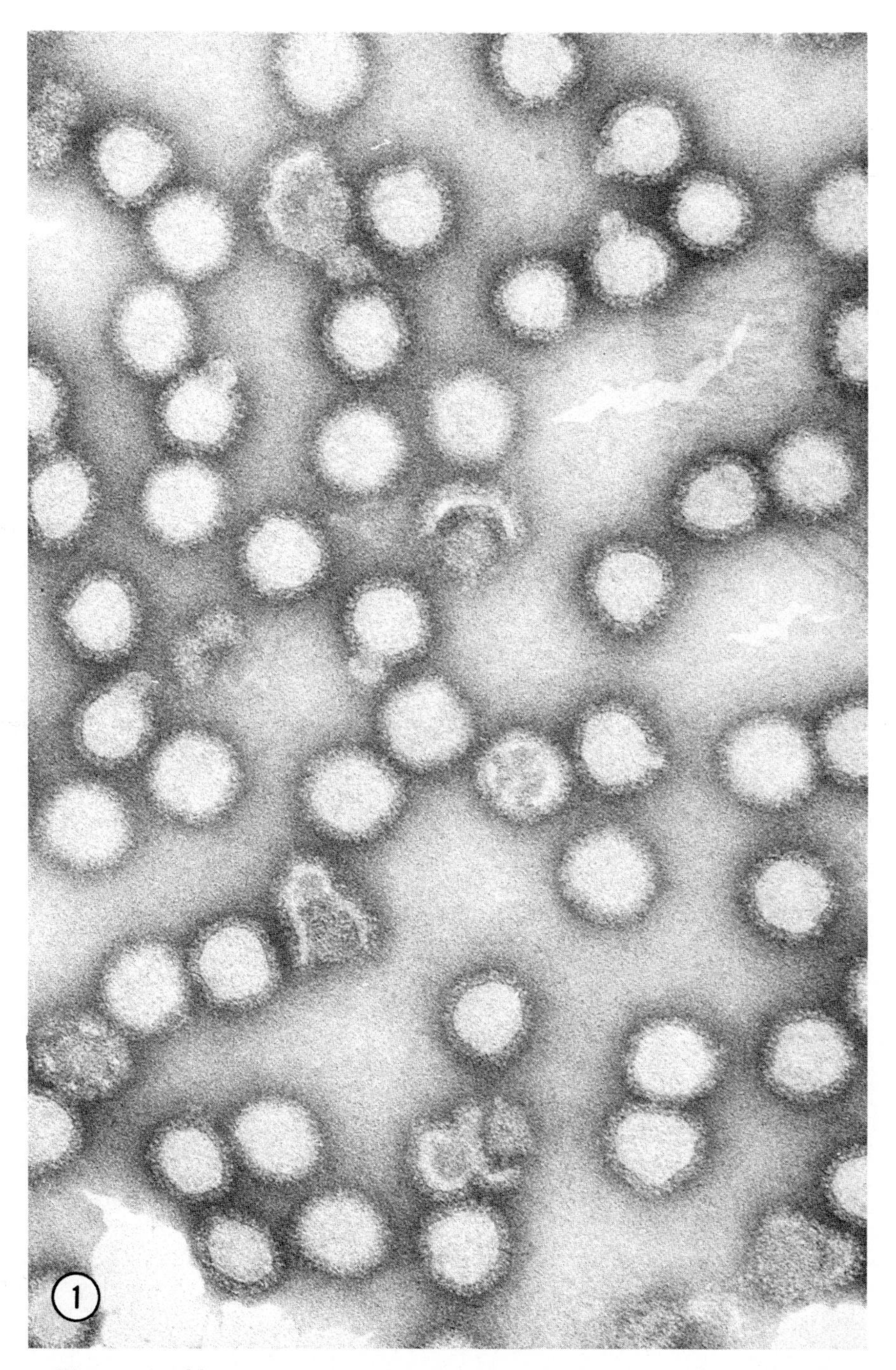

Fig. 1. Semliki Forest virus, negative-contrast with potassium phosphotungstate. This typical form of intact alphavirus particles is characterized by a fuzzy surface projection layer. ×200,000. Micrograph courtesy of Dr. C.-H. von Bonsdorff.

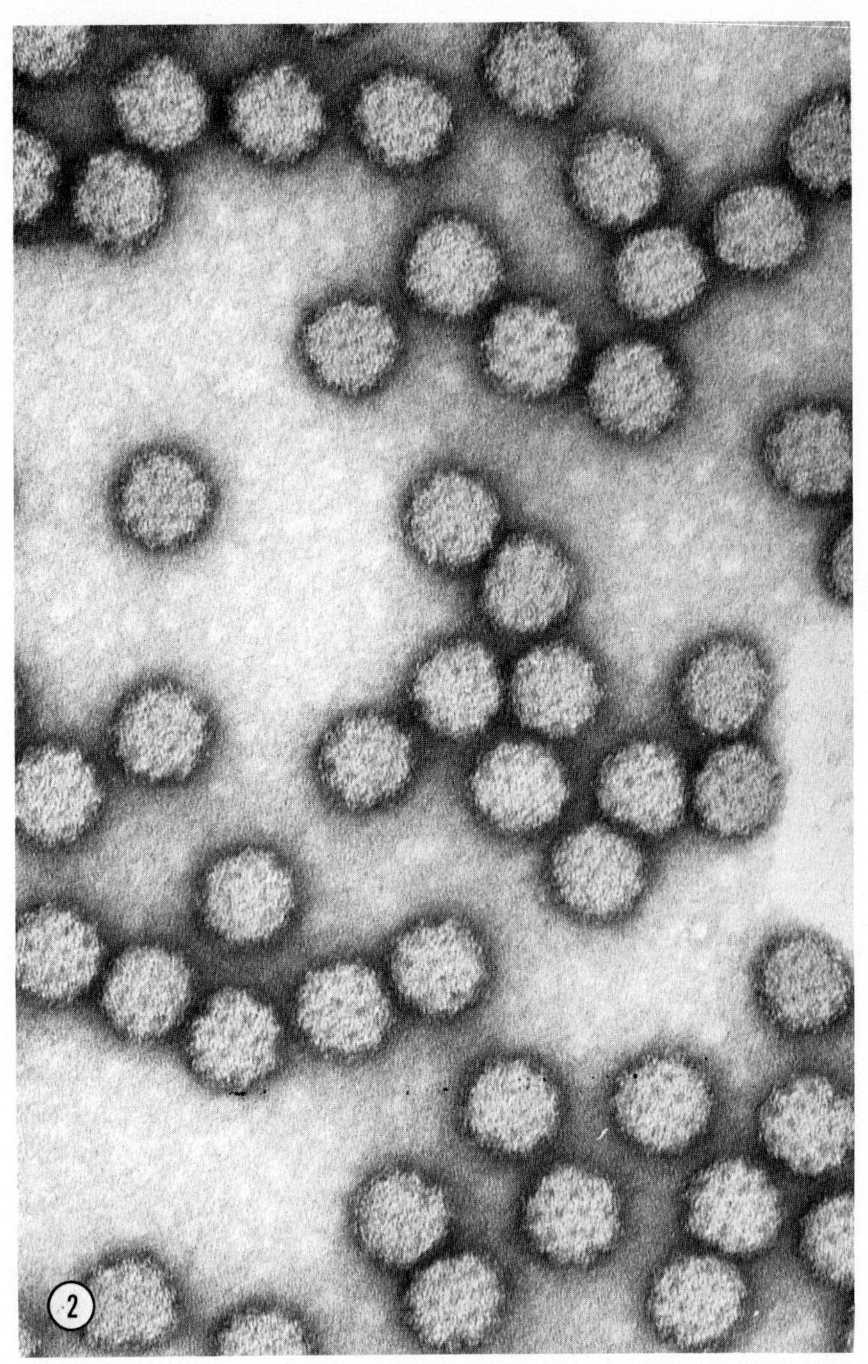

Fig. 2. Semliki Forest virus, negative-contrast with uranyl acetate (2-s contact with stain). This typical $T = 4$ icosahedral structure of alphavirus surface is seen when highly purified virus preparations are used. ×200,000. Micrograph courtesy of Dr. C.-H. von Bonsdorff.

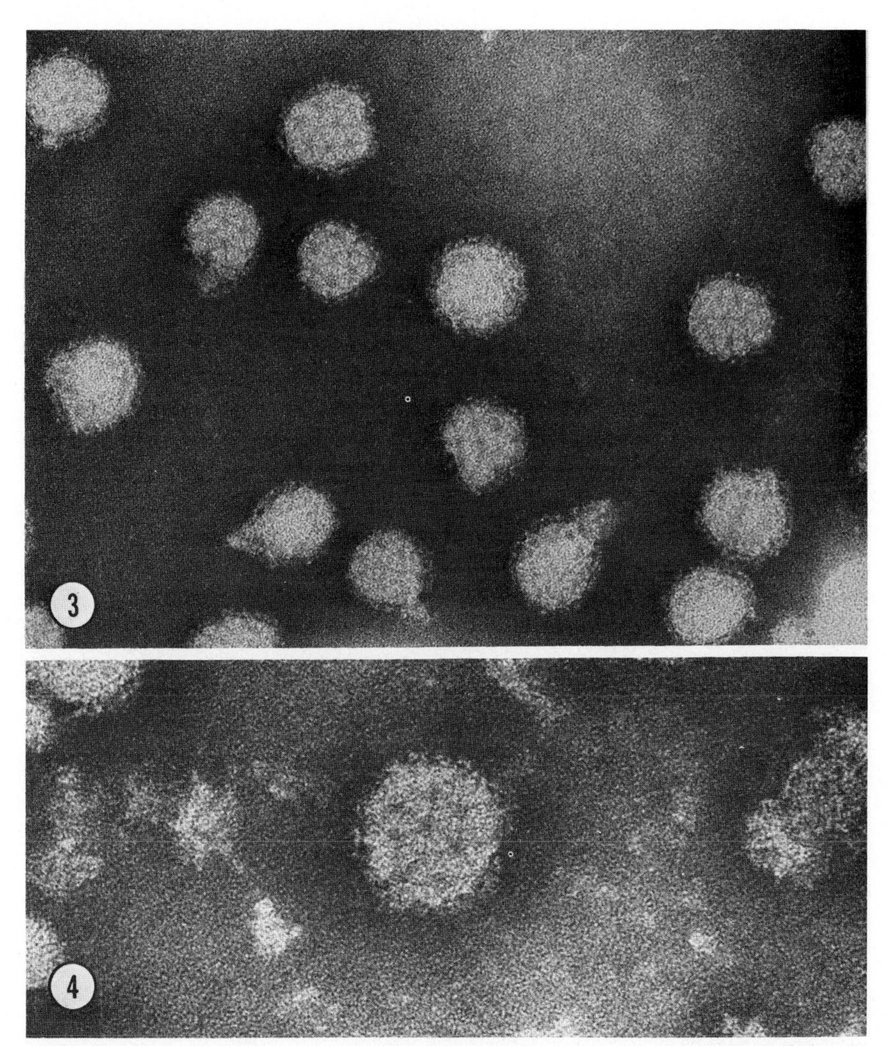

Figs. 3 and 4. Sindbis virus, negative-contrast with potassium phosphotungstate at pH 7.0 (2-s contact with stain). The triangular network demarcating surface subunits is evident on several particles; these images correspond to a $T = 4$ icosahedral surface lattice. Fig. 3: ×180,000. Fig. 4: ×300,000. Micrographs courtesy of Dr. C.-H. von Bonsdorff.

The envelope layer of alphaviruses is a modified unit membrane derived from the host cell membrane during the process of budding. The lipid constituency is similar to that of the host cell plasma membrane (see Chapter 10), but the complex pattern of plasma membrane proteins is replaced by virus-specific glycoproteins (see Chapters 9 and 17). Because the inner face of the lipid bilayer is in direct apposi-

Fig. 5. Sindbis virus, freeze-etch preparation. The same $T = 4$ surface symmetry as determined by negative-contrast microscopy has been found by analysis of images like this. ×200,000. Micrograph courtesy of Drs. C.-H. von Bonsdorff and S. C. Harrison.

tion with the nucleocapsid and the outer face is covered with glycoprotein projections, it is difficult to determine whether the membrane layer per se is morphologically modified (Fig. 6). The best evidence is that, although viral glycoproteins are present in the envelope in high concentration and extend completely through the lipid bilayer, there is little morphological difference between envelope and host cell membrane (von Bonsdorff and Harrison, 1975). It is rather characteristic of alphavirus envelopes to form blebs, that is, extrusions from the membrane layer which are free of surface projections. These have been considered to be osmotic artifacts (Murphy and Harrison, 1971), or to be the location of "excess" lipid of particles undergoing a nucleocapsid size decrease under the influence of low pH or enzymatic activity (von Bonsdorff, 1973). The native membranous envelope layer is approximately 5–7 nm thick, so that particles freed of their surface projections (e.g., with bromelin) have a diameter of about 50 nm.

The nucleocapsid of alphaviruses has been studied extensively: *in situ*, in association with disrupted virus particles, and in purified preparations derived from concentrated virus suspensions or from infected cells. In general, nucleocapsids have a spherical shape and an appearance of flexibility or deformability, unlike the rigid appearance of picornavirus and parvovirus nucleocapsids (Igarashi *et al.*, 1966; Acheson and Tamm, 1970; Horzinek and Mussgay, 1971). Within their centers, nucleocapsids have a core component which is 12–16 nm in diameter and is thought to be rich in RNA (Horzinek and Mussgay, 1971). The nucleocapsid surface is usually mottled and its edges are ragged, which suggests the presence of a subunit structure with cubic symmetry (Fig. 7). The only exceptional interpretation is that of Klimenko *et al.* (1965) and Bykovsky *et al.* (1977), who have considered that the nucleocapsid of Venezuelan equine encephalitis virus is formed from a helically wound ribonucleoprotein strand wound into a quasi-icosahedral form. With this exception, and despite great difficulty in resolving enough nucleocapsid surface detail, there has been an accumulation of evidence in recent years that alphavirus core symmetry is icosahedral.

Simpson and Hauser (1968a) found that formaldehyde-fixed Sindbis virus nucleocapsids exhibited sixfold radial symmetry, and Osterrieth (1968) demonstrated that Semliki Forest virus nucleocapsids, released by treatment of virus with a proteolytic enzyme, had polygonal outlines with edge lengths of 12 nm. Osterrieth concluded that surface symmetry was that of a $T = 3$ (32-capsomer) or a $T = 4$ (42-capsomer) icosahedron. Horzinek and Mussgay (1969), using sodium deoxycholate to release Sindbis nucleocapsids, determined that the particle

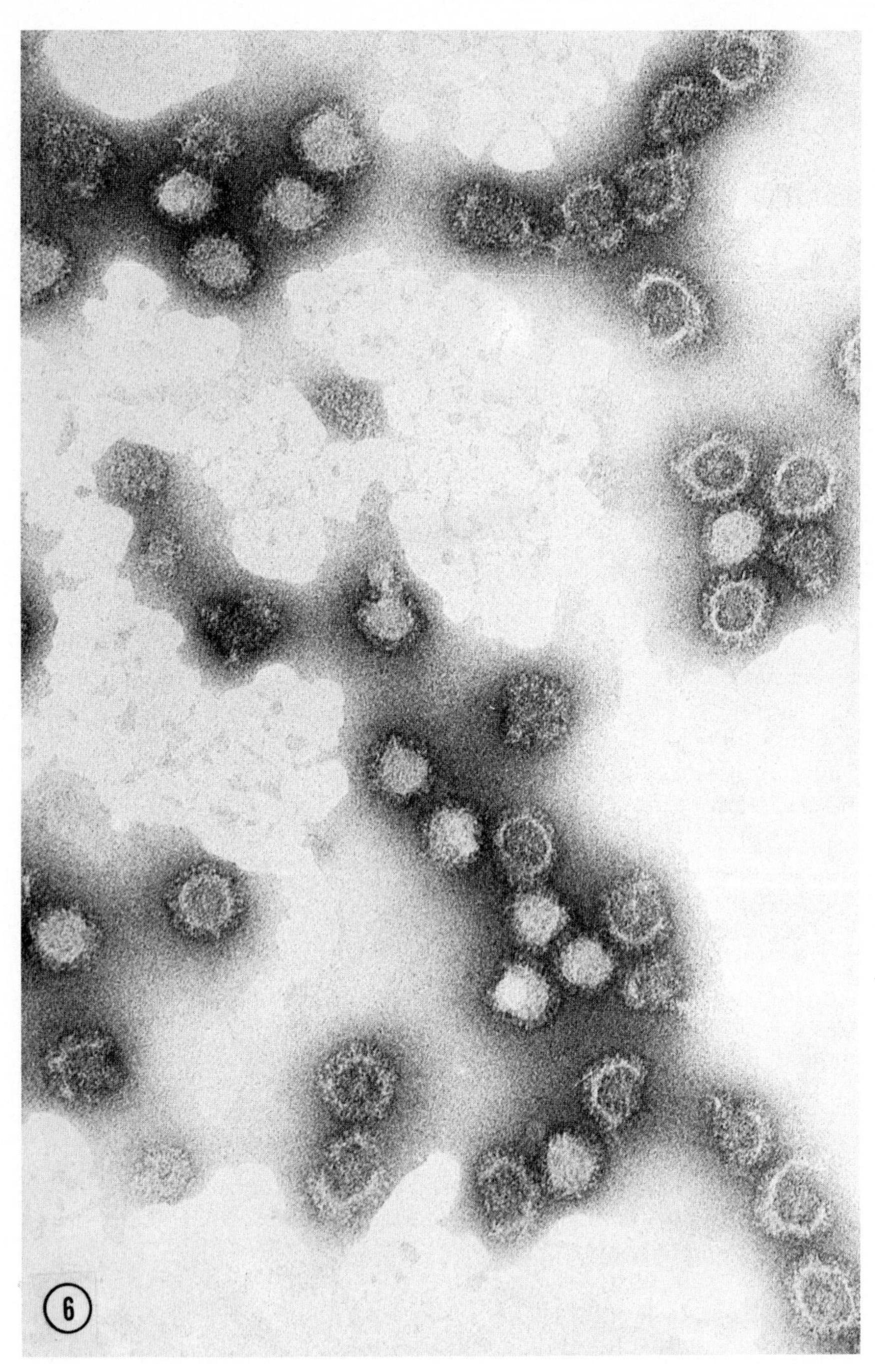

Fig. 6. Semliki Forest virus, negative-contrast with ammonium molybdate at pH 7.0. Stain penetration reveals the unit membrane envelope layer which is covered with surface projections and is in close apposition to the nucleocapsid. ×170,000.

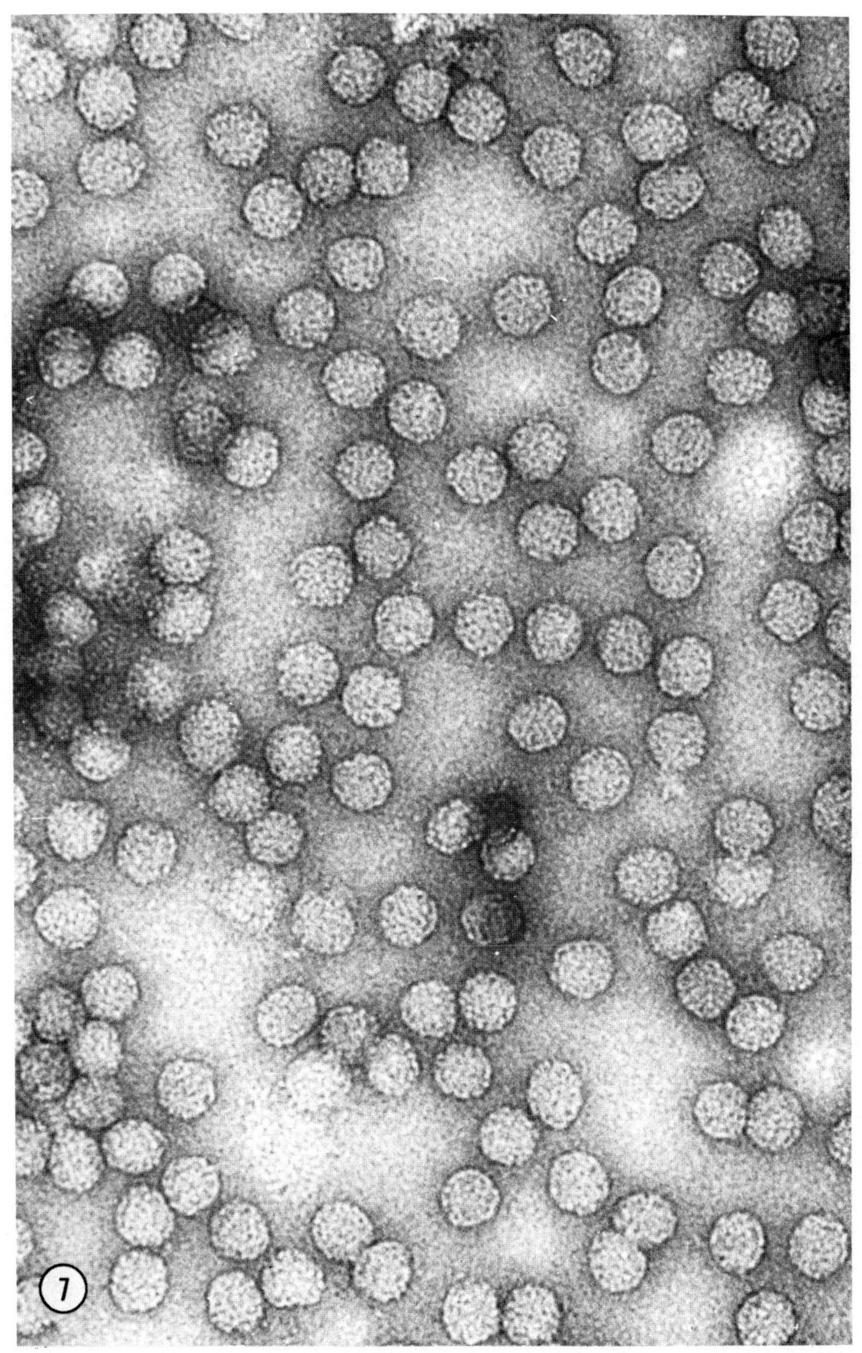

Fig. 7. Sindbis virus nucleocapsids, negative-contrast with uranyl acetate. This typical appearance of alphavirus nucleocapsids suggests the presence of subunits, but their organization is not clearly symmetric. ×200,000. Micrograph courtesy of Drs. C.-H. von Bonsdorff and S. C. Harrison.

surface had 12- to 14-nm ringlike morphological subunits arranged in a $T = 3$ (32-capsomer) icosahedral lattice. Von Bonsdorff (1973) also found symmetrically arranged subunits upon the surface of Semliki Forest virus nucleocapsids but, because he could not with certainty identify pentameric subunits, estimated capsomer numbers on the basis of subunit spacing. This approach favored a $T = 4$ (42-capsomer) icosahedral structure. As mentioned above, an intimacy in the relation between surface projection and nucleocapsid symmetry is intimated by the known penetration of the tails of surface glycoproteins through the lipid envelope. The convincing evidence that surface symmetry is that of a $T = 4$ icosahedral lattice (see above) furthers the position that nucleocapsid symmetry may be the same, that is, a $T = 4$ icosahedral design (von Bonsdorff and Harrison, 1975). Von Bonsdorff (1977) also studied the surface of Sindbis virus nucleocapsids by the freeze-etch method; subunits were visualized, but their organization was not resolved well enough to draw conclusions (Fig. 8). However, in a similar study of Sindbis virus nucleocapsids by the freeze-etch method, Brown *et al.* (1972) did find that surface subunits were 7 nm in diameter. From this observation they proposed that the alphavirus nucleocapsid was an icosahedron with $T = 9$ symmetry. It seems from Fig. 8 and the published micrographs of Brown *et al.* (1972) that interpretation of the results of freeze-etch microscopy will require further careful comparison with negative-contrast findings.

There is now evidence (see Chapter 17) that the alphavirus nucleocapsid is unique in the intimacy of its relationship to viral RNA and protein. It is not a "sealed" capsid shielding RNA from the effects of acid or RNase; several nonmorphological lines of evidence suggest that the RNA functions as an integral structural element of the nucleocapsid. One model presumes that loops of RNA extend near the surface of the nucleocapsid and lend the property of flexibility (Kääriäinen and Soderlund, 1971). This construction would explain the nucleocapsid flexibility observed by electron microscopy; it would also help explain observed variances in surface features of nucleocapsids. The larger, native nucleocapsid forms may have smooth surfaces, and structural subunits may only become visible as RNA is depleted and some reorganization takes place. Because of this possibility, it does not seem prudent, at this time, to attempt to decide between the published alternatives—between $T = 3$ or $T = 4$ construction. It seems clear that final resolution of symmetry questions will require that methods be found for achieving better subunit resolution while continuing to maintain nucleocapsids in their "native state" through the stages of preparation and examination.

Fig. 8. Sindbis virus nucleocapsids, freeze-etch preparation. Although images such as this have been confirmatory of negative-contrast findings in indicating that alphavirus nucleocapsids are constructed from subunits, they have not allowed the resolution of subunit architecture. ×200,000. Micrograph courtesy of Drs. C.-H. von Bonsdorff and S. C. Harrison.

In summary, negative-contrast microscopy has shown that alphavirus particles are composed of a surface projection layer, a modified unit membrane envelope, and an isometric nucleocapsid. An understanding of the fundamental construction of each of these components will require new ultrastructural approaches—rather than the repetitive superficial characterization that is tempting to investigators studying a genus with 20 member viruses.

B. Thin-Section Electron Microscopy

1. Virus Morphology in Thin Section

In the years since the first thin-section study of Western equine encephalitis virus infection by Morgan *et al.* (1961), many other similar studies of other alphaviruses have been reported; in most of them, virus morphology and morphogenesis are considered together. As with the negative-contrast studies, the viruses most frequently studied have been Semliki Forest (Acheson and Tamm, 1967; Erlandson *et al.*, 1967; Grimley and Friedman, 1970; Murphy *et al.*, 1970; Tan, 1970; von Bonsdorff, 1973), Sindbis (Simpson and Hauser, 1968b; Birdwell *et al.*, 1973; Gil-Fernandez *et al.*, 1973; Pedersen and Sagik, 1973), chikungunya (Chain *et al.*, 1966; Chatterjee and Sarkar, 1965; Higashi *et al.*, 1967), and Venezuelan equine encephalitis (Bykovsky *et al.*, 1969; Murphy and Harrison, 1971; Garcia-Tamayo, 1971). Fewer studies have been carried out on Aura virus (Lascano *et al.*, 1969), Mayaro virus (Saturno, 1963), Ross River virus (Raghow *et al.*, 1973; Murphy *et al.*, 1973), and Eastern equine encephalitis virus (Murphy and Whitfield, 1970).

The range of alphavirus particle diameters given in the various published descriptions has been rather wide—wider than can be explained by variations in shrinkage due to fixation and preparation differences. For example, in Horzinek's (1973a) tabulation of published alphavirus diameter values, the range is from 38 to 69 nm. In contrast, in the author's laboratory, where the same thin-section technique was used for 13 years (buffered glutaraldehyde fixation, ethanol dehydration, and Araldite–Epon embedment) (Murphy and Whitfield, 1970; Murphy and Harrison, 1971), no size variation was noted. In studies of 12 different alphaviruses we consistently obtained virus particle diameters of 55–58 nm and nucleocapsid diameters of 28–30 nm (Fig. 9). Similarly, von Bonsdorff (1973) measured the center-to-center spacing of Semliki Forest virus particles present in crystalline arrays in ultracentrifuge pellets and obtained a mean value of 52 nm and a range only between 48 and 55 nm (Fig. 10). Nucleocapsids had diameters of 30 nm.

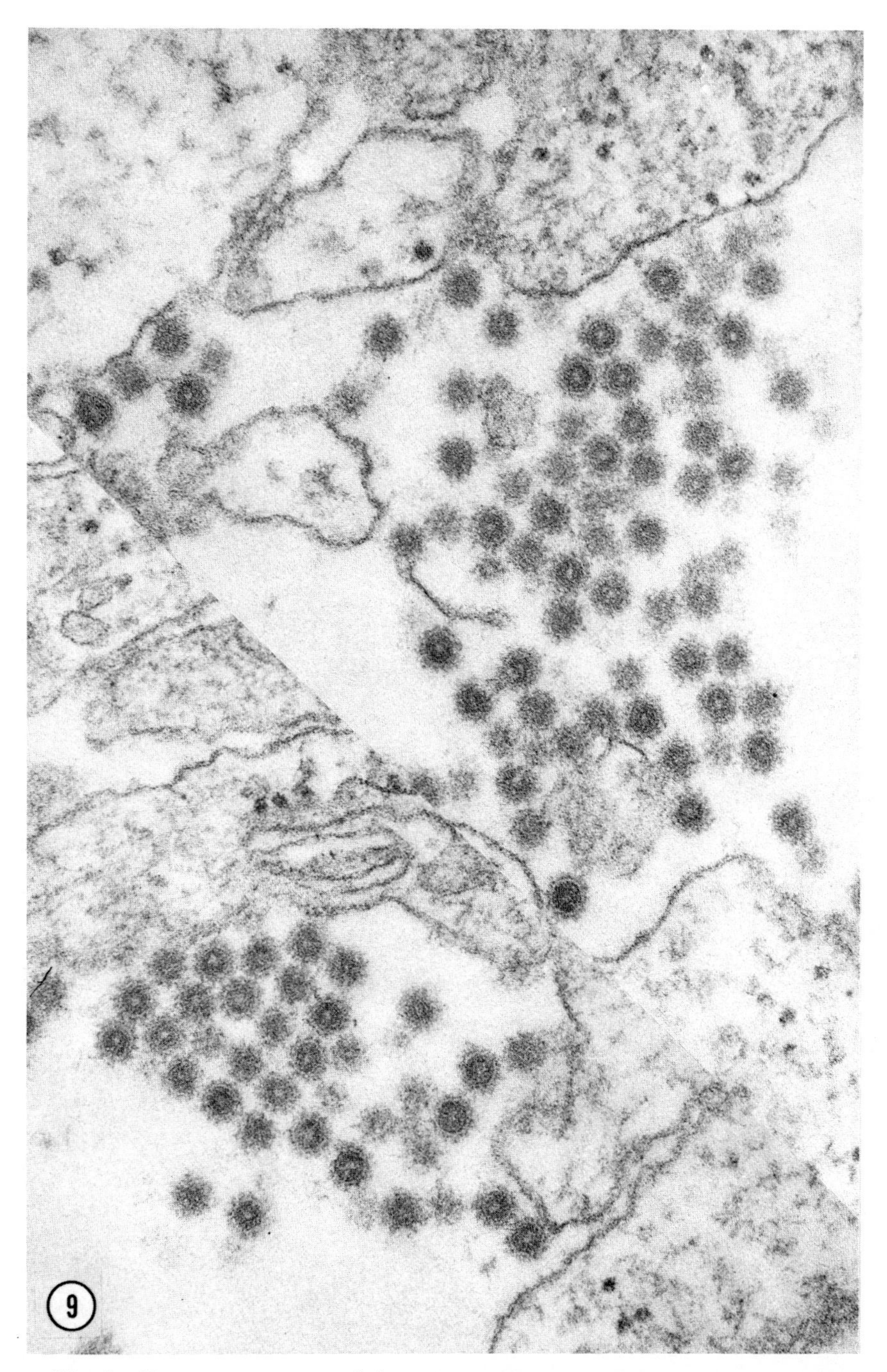

Fig. 9. Eastern equine encephalitis virus in the extracellular space of an infected mouse brain. Typical alphavirus particles in thin section exhibit a dense center and a ragged surface layer; in this case the surface layer is resolved into regular subunits. Composite. ×91,000.

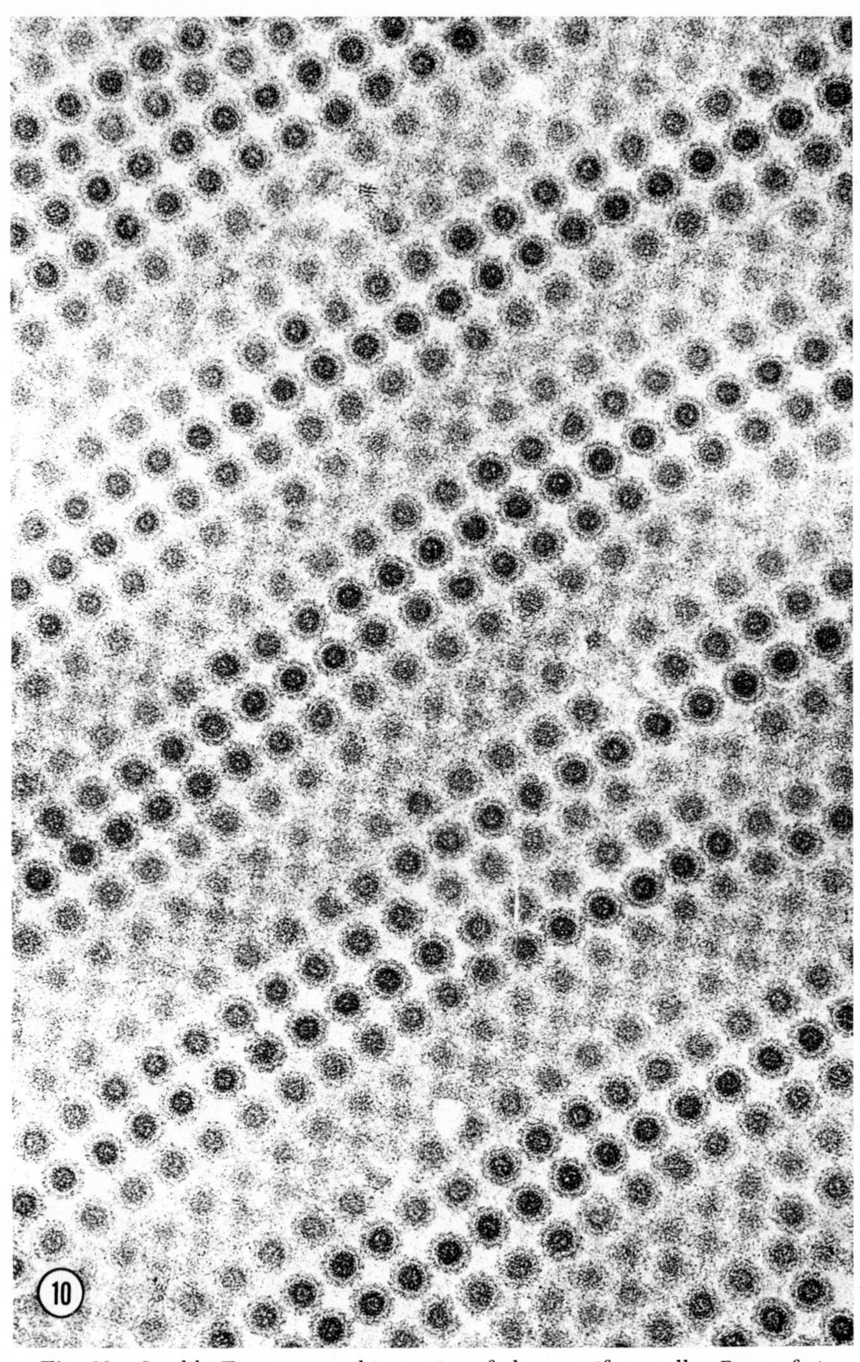

Fig. 10. Semliki Forest virus; thin section of ultracentrifuge pellet. Rows of virus particles in the plane of section have a center-to-center spacing of 52 nm (range, 48–55 nm) and show the precise uniformity characteristic of alphaviruses. ×120,000. Micrograph courtesy of Dr. C.-H. von Bonsdorff.

With the exception of size, the other aspects of alphavirus particle morphology have been described with remarkable consistency. Despite variations in the nature of infected cells *in vivo* and in culture, and differences in preparation and staining techniques, mature alphavirus particles have invariably been found to be spherical and to consist of a clearly defined electron-dense core structure and a surrounding rather lucent zone. In some cases more structural detail has been resolved; details of the surface projection layer, the envelope, and the nucleocapsid follow.

The surface projection layer of alphaviruses is never very well resolved in thin sections. Usually, particles have a fuzzy margin which blends into the envelope layer. Less commonly, this margin may be seen to be made up of distinct, radially set subunits; there are approximately 14 subunits at the margin of several of the Eastern equine encephalitis virus particles illustrated in Fig. 9.

The envelope layer is usually obscured in thin sections because the outer dense zone of the lipid bilayer membrane blends with the surface projection layer, as stated above, and the inner dense zone blends with the dense nucleocapsid. The observed single, thin, lucent zone, which is the central zone of the envelope membrane, does not represent a real junction but, because it is so clear, has often been taken as the margin of the nucleocapsid. Virus particles in specimens obtained early in the virus-shedding cycle (i.e., particles presumed to be newly formed) often exhibit a separation of the envelope layer from the nucleocapsid (Figs. 11 and 12). It seems most likely that continued

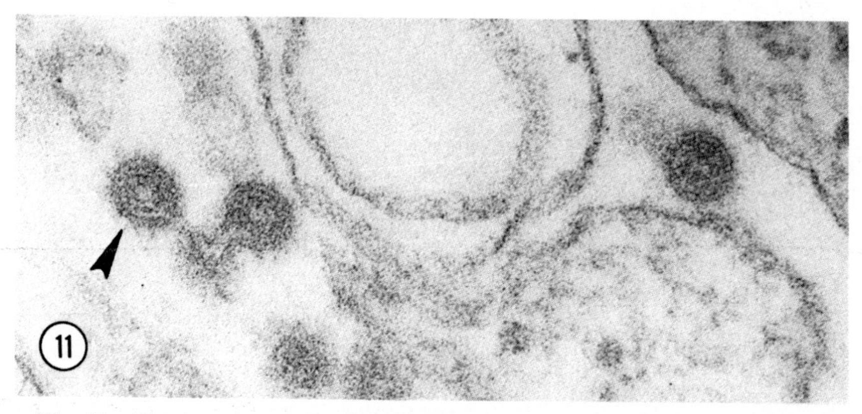

Fig. 11. Eastern equine encephalitis virus infection in mouse brain. In these presumably newly formed particles a delineation between the nucleocapsid and the envelope layer is evident (arrowhead). ×187,000.

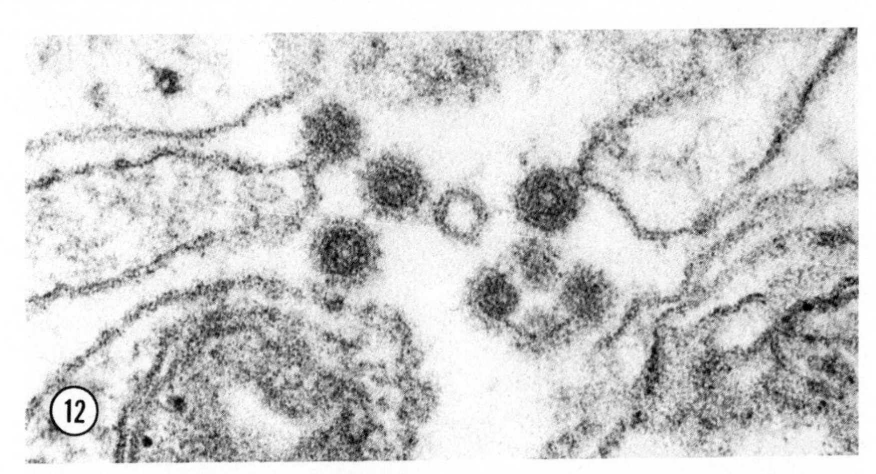

Fig. 12. Eastern equine encephalitis virus infection in mouse brain. In these virus particles the nucleocapsids are clearly seen to have a lucent central zone; this characteristic and the delineation between the nucleocapsid and the envelope are obscured in late cell culture and animal tissue harvests. ×187,000.

maturation after particle formation results in the usual close apposition of envelope upon nucleocapsid.

The nucleocapsid in thin sections, whether within a mature virus particle or unenveloped in cytoplasm, usually appears as a simple spherical form (Fig. 13). However, in early harvests which yield particles with loosely apposed envelopes, as described above, both free and enveloped nucleocapsids often have lucent centers (Figs. 11, 12, and 14). Again, continued maturation may result in the filling of this central area.

Overall, thin-section microscopy has increased our understanding of alphavirus morphology. It has been particularly useful in confirming negative-contrast observations, but further value will depend upon new approaches, perhaps centered upon cytochemical methods.

2. *Virus Morphogenesis in Thin Section*

The first morphologically discernible events in alphavirus morphogenesis take place in the cytoplasm before the beginning of cytopathic changes. Nucleocapsids first appear individually, either free in the cytosol or in close apposition to host cell membranes. Acheson and Tamm (1967) first realized that in cultured cells producing virus at their maximal rate there was very little to be seen by thin-sectioning other than the budding of virus upon plasma membranes and the accumulation of mature particles in extracellular space. Clearly, nucleocapsid envelopment can be very efficient during this

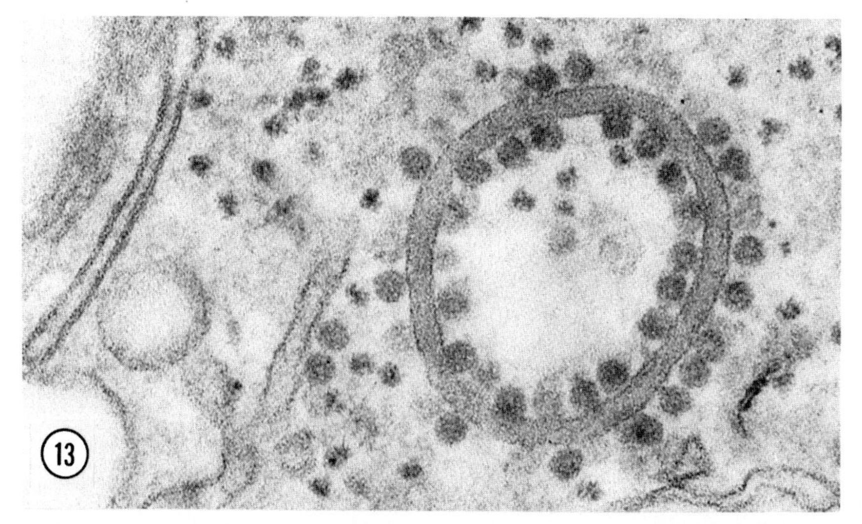

Fig. 13. Eastern equine encephalitis virus infection in mouse brain. Nucleocapsids, present in neuronal cytoplasm, are typical simple spherical forms associated with a membranous structure. ×147,000.

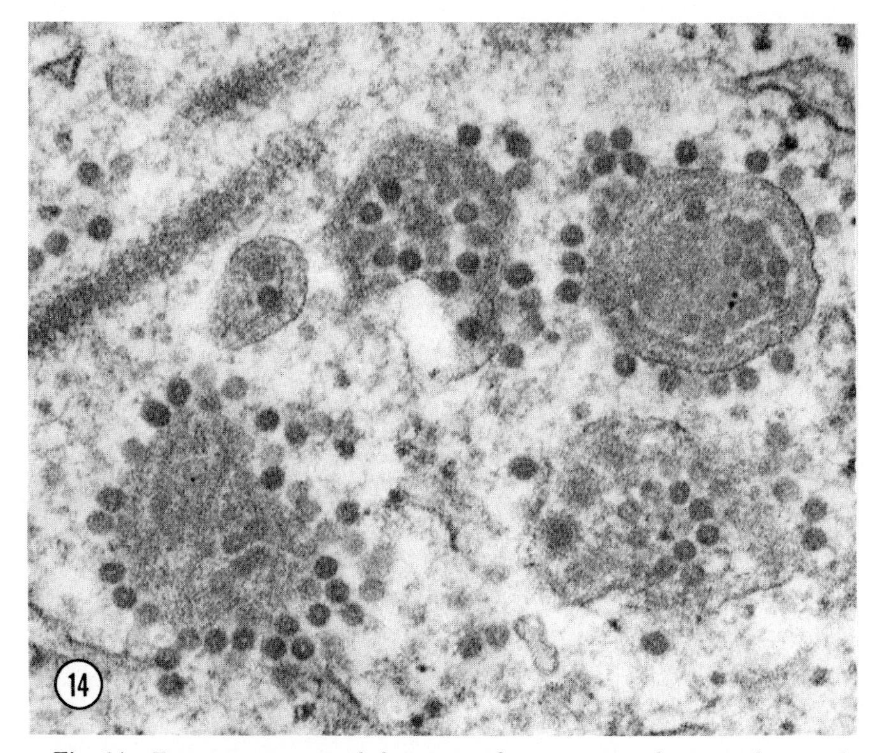

Fig. 14. Eastern equine encephalitis virus infection in mouse brain. Nucleocapsids in this early stage of infection have lucent centers but are already associated with intracytoplasmic membranous structures. ×114,000.

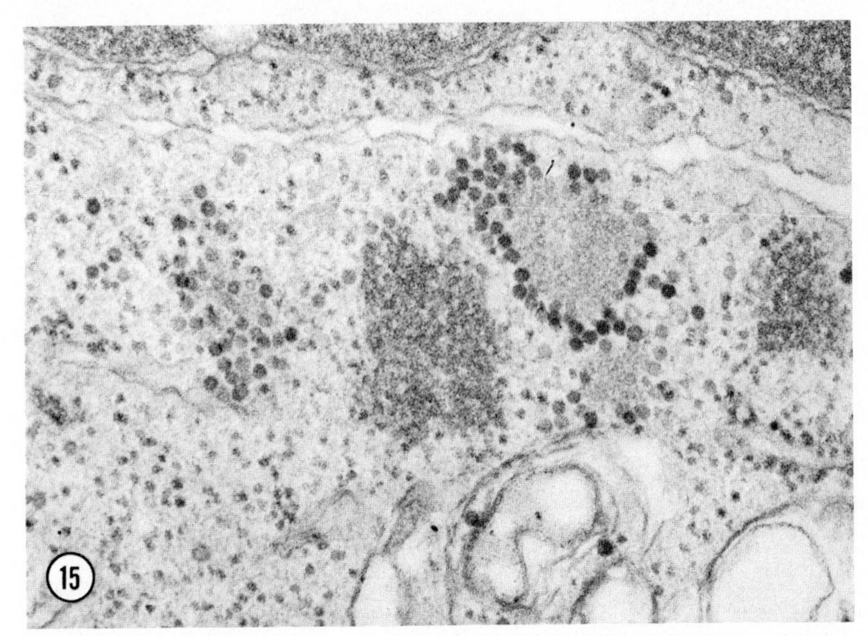

Fig. 15. Venezuelan equine encephalitis virus infection in mouse brain. In this case, intracytoplasmic nucleocapsid accumulation is concentrated at the margins of granular "viroplasmic" masses. ×58,000.

productive phase of infection, so that few leftover nucleocapsids accumulate. The same characteristic is observed in early stages of *in vivo* alphavirus infections (Murphy *et al.*, 1970; Murphy and Whitfield, 1970; Murphy and Harrison, 1971). Later in the course of infection of cultured cells or target organs of experimental animals, increasing numbers of nucleocapsids accumulate, usually in conjunction with the start of cytopathology. These nucleocapsids are (1) free and randomly distributed in groups, (2) upon cylindrical or distended membranous structures (Figs. 13 and 14), or (3) in association with granular or fibrillar "viroplasm" (Fig. 15). In the terminal stages of infection the accumulation of nucleocapsids and the cylindrical structures upon which they occur can become spectacular (Fig. 16). In some cases in cultured cells, nucleocapsids may also accumulate in late stages of infection within nuclei. Bykovsky *et al.* (1969) found Venezuelan equine encephalitis virus nucleocapsids in nuclei, and recently Monath *et al.* (1977) demonstrated very large numbers of Western equine encephalitis, Venezuelan equine encephalitis, Bijou Bridge, and Fort Morgan virus nucleocapsids at the margins of nucleoli in the nuclei of infected Vero cells (Fig. 17).

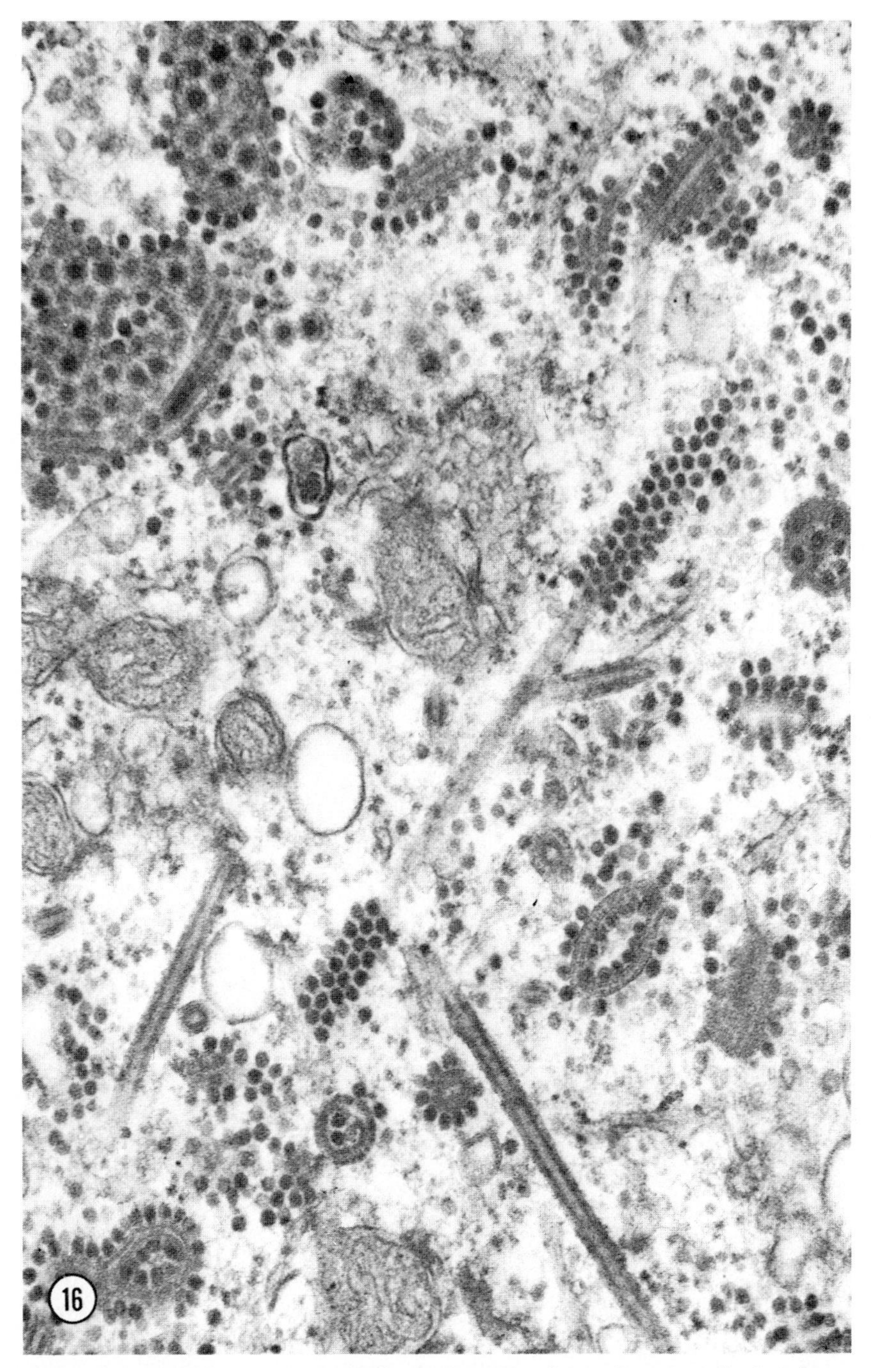

Fig. 16. Semliki Forest virus infection in mouse brain. Late in infection the accumulation of nucleocapsids and associated membranous structures within cytoplasm, as in this neuron, may be spectacular. ×72,000.

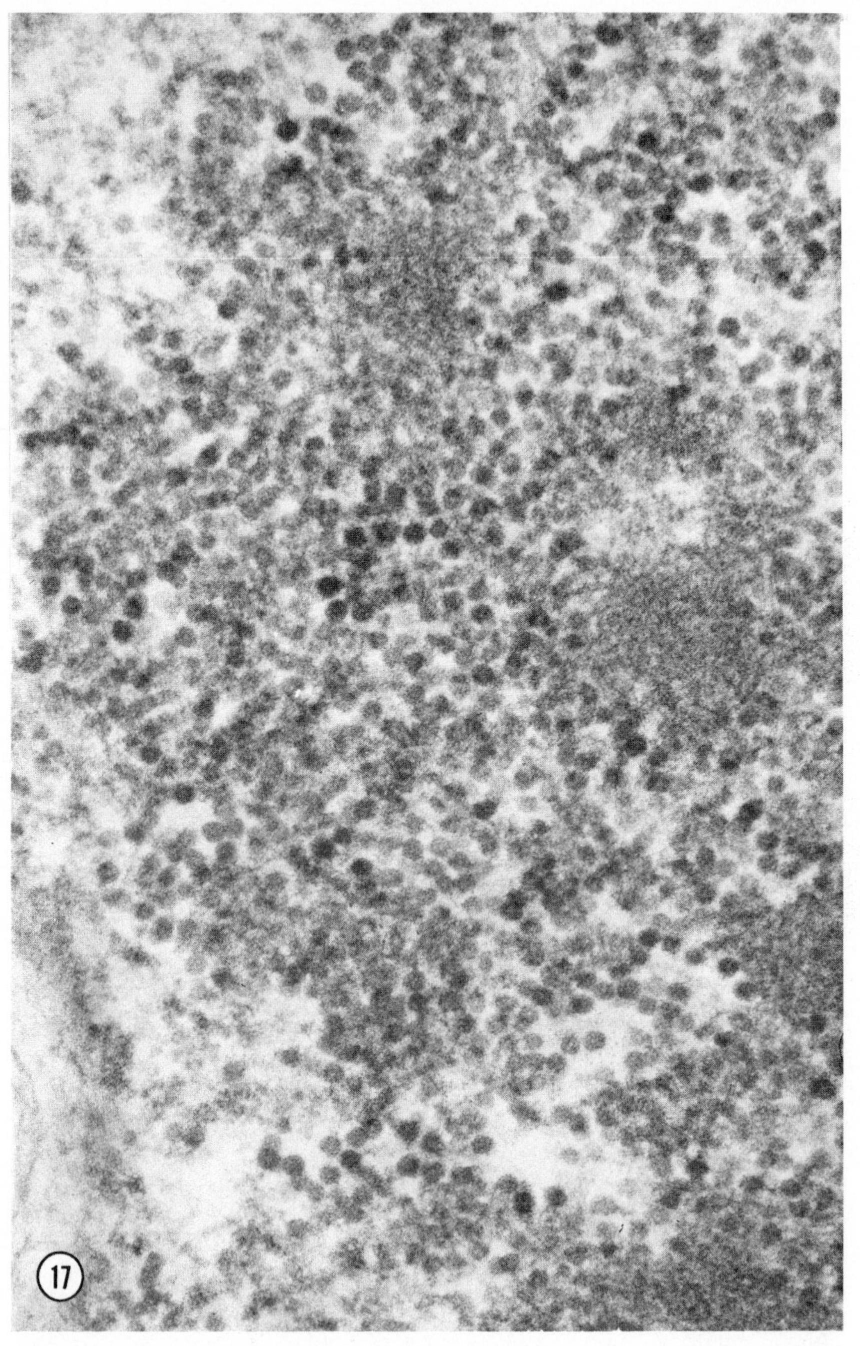

Fig. 17. Western equine encephalitis virus infection in a Vero cell culture. Nuclei with large numbers of nucleocapsids at the margins of nucleoli have been found late in the course of cell culture infection with several alphaviruses. ×90,000.

Alphavirus morphogenesis takes place via passage of the described nucleocapsids through host cell membranes. This budding process usually occurs upon plasma membranes in both cell culture and *in vivo* infection. Membrane deformation into the nascent bud seems to be coordinated with surface projection insertion, although the fuzzy nature of surface projections generally makes it impossible to visualize projections upon membranes except at bud sites. Immunoelectron microscopy has confirmed that discrete sites on plasma membrane are involved in the budding process via the insertion of glycoproteins (Higashi *et al.*, 1967; Pedersen and Sagik, 1973). Unlike the mysterious process of flavivirus maturation, every stage of the alphavirus budding process is easily seen (Figs. 18–20).

Intracytoplasmic membranous structures of a unique type, referred to as type-1 cytopathic vacuoles (CPV-1), have been associated with the replication of several alphaviruses. They were first described in infected cultured cells by Friedman and Berezesky (1967) and Acheson and Tamm (1967) and then were studied in detail by Grimley and colleagues (1968, 1972). They have also been found in infected cells *in vivo* (Murphy *et al.*, 1970; Murphy and Harrison, 1971). The structures are membrane-bound and are characteristically lined by regular membranous spherules, measuring 50–100 nm in diameter, which in their lucent interiors often contain delicate filaments. The same spherules may also occur upon the plasma membrane of cells which are budding virus (Fig. 21). Grimley *et al.* (1972) found that the detection of CPV-1 coincided with the stage of maximum virus yield; they concluded that these structures were associated with viral RNA synthesis (Grimley *et al.*, 1968) and with related modification of host cell membranes. The question of whether such structures are necessary for virus replication has not been answered but, because they have *not* been found at some sites of prolific alphavirus growth, there is skepticism about their true nature and role (Murphy and Whitfield, 1970; Murphy and Harrison, 1971).

Much interest has been taken in the question of whether there are differences in the growth of alphaviruses in arthropod and in vertebrate cells which might explain the differences in cytopathology noted *in vivo* and in cell cultures. This question has been addressed by ultrastructural study, both of infected arthropod cells in culture (Raghow *et al.*, 1973; Gliedman *et al.*, 1975) and of mosquito organs (Janzen *et al.*, 1970; Whitfield *et al.*, 1971; Raikova *et al.*, 1974). In culture, arthropod cells produce alphaviruses as well as most vertebrate cells do. Some nucleocapsids form in cytoplasm and bud from plasma membranes, but most activity is associated with complex

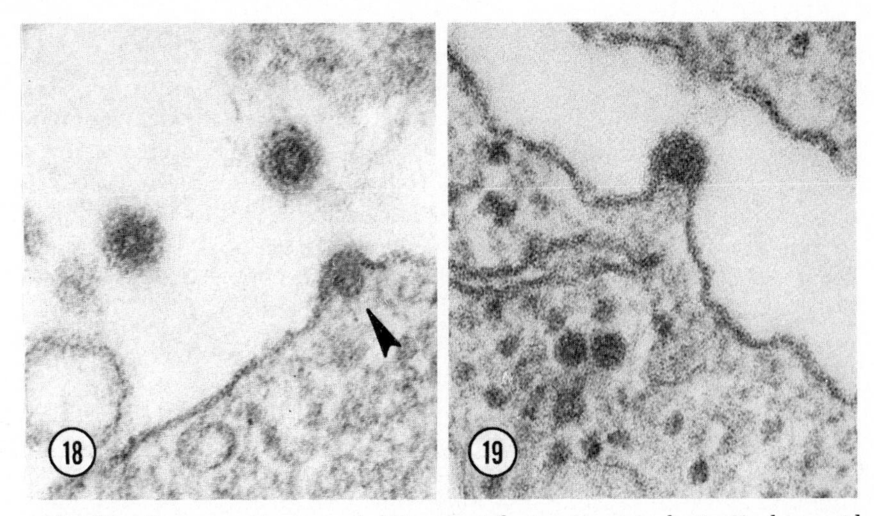

Fig. 18. Eastern equine encephalitis virus infection in mouse brain. Nucleocapsid (arrowhead) extruding through plasma membrane in the earliest stage of the budding process. ×187,000.

Fig. 19. Eastern equine encephalitis virus infection in mouse brain. Virus particle at a further stage of budding, showing the contiguity between the neuronal plasma membrane and the virus envelope. ×180,000.

intracytoplasmic membranous structures which have no counterpart in vertebrate cell infections. These structures apparently serve to contain the infection until it is released or digested via lysosomal fusion and endophagocytosis. In contrast, in the target organs of infected mosquitoes, alphaviruses bud from reticular and plasma membranes just as in vertebrate cells; there has been no ultrastructurally evident reason for the absence of cytopathology (Fig. 22). Further study of this subject is needed if we are to understand the basis for the lifelong transmitting capacity of arthropods infected with alphaviruses.

III. THE *FLAVIVIRUS* GENUS

Nonserological criteria for placing viruses in the *Flavivirus* genus are untested and arbitrary, since in reality candidates have only been included on the basis of antigenic interrelationships. Virion morphology and morphogenetic characteristics would be valuable adjuncts in placing serologically unrelated viruses, but we would first have to

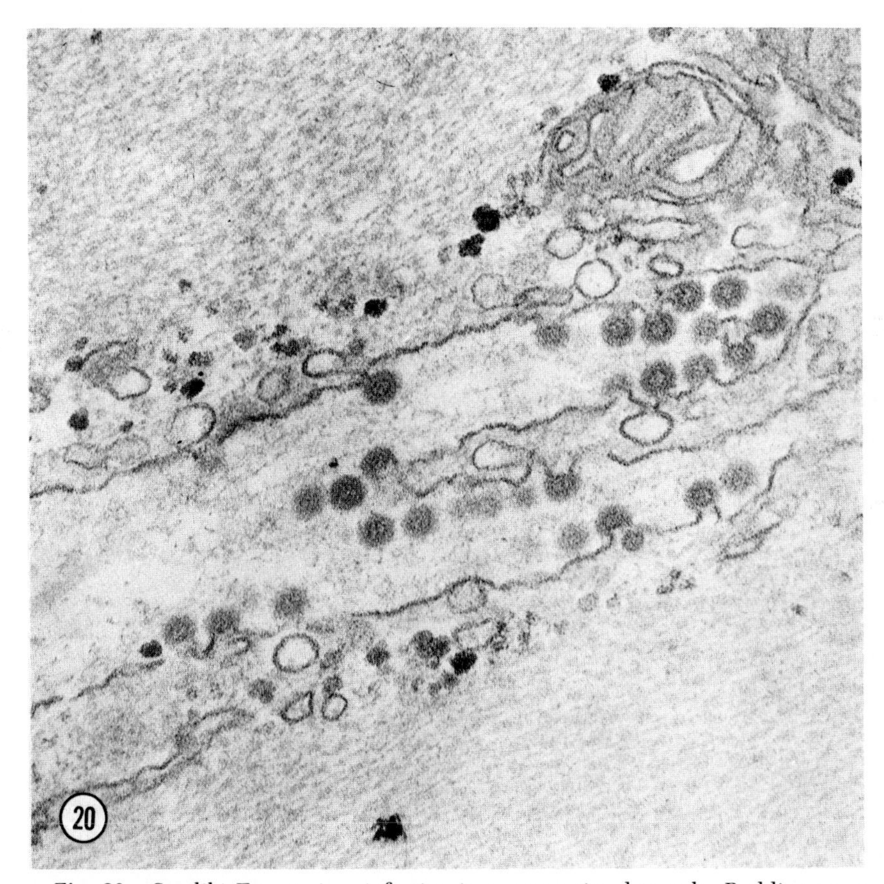

Fig. 20. Semliki Forest virus infection in mouse striated muscle. Budding upon plasma membrane delivers virus to extracellular spaces and ensures the high viremia necessary for transmission to a blood-feeding arthropod. ×85,000.

decide just how elastic the taxon's definition could be without detracting from the precision in describing the present member viruses. Such consideration of the precise morphological and morphogenetic attributes of flaviviruses is important in this regard because serologically unrelated viruses are known which are very similar to flaviviruses. In all studies to date it has been concluded that flaviviruses are so similar to each other morphologically (and in several physicochemical characteristics) that data obtained with a few viruses are predictive of the genus as a whole (Murphy, 1978).

Filtration studies which preceded all electron microscopic observations had suggested that flavivirus particles are extremely small—so

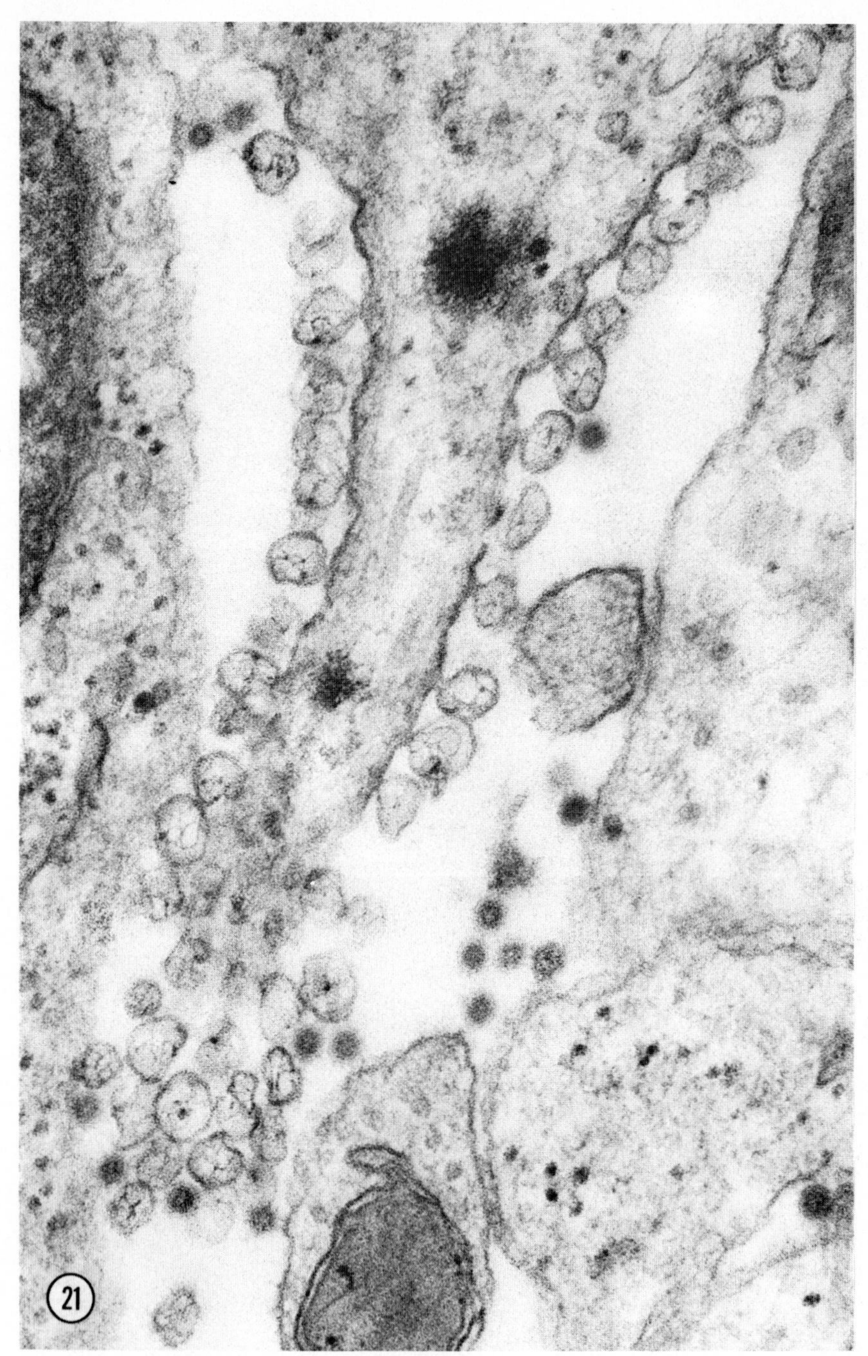

Fig. 21. Venezuelan equine encephalitis virus infection in mouse brain. Type-1 cytopathic vacuoles (CPV-1) are budding from the plasma membrane of an infected neuron. More commonly these spherules line intracytoplasmic membrane-bound structures. ×74,000.

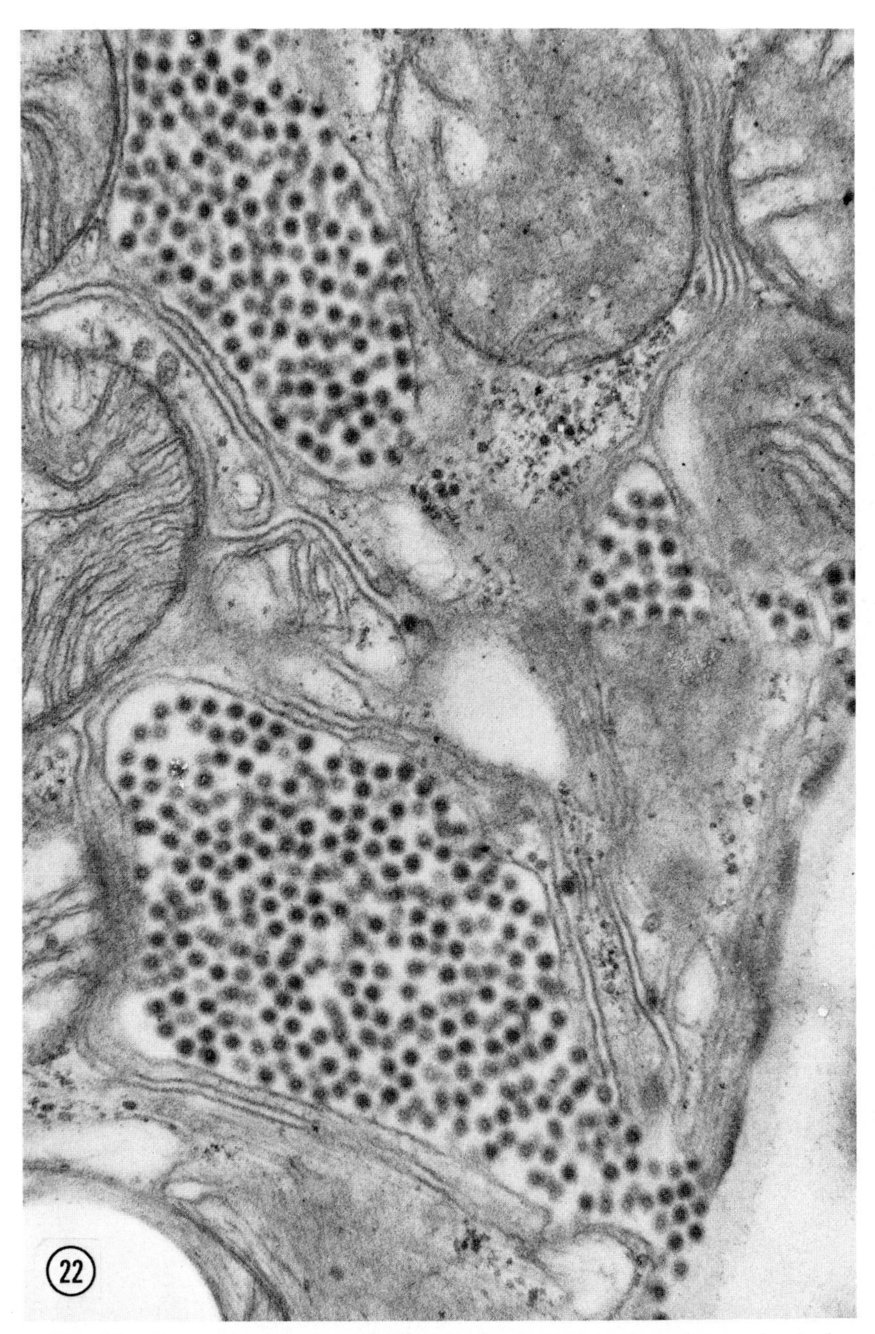

Fig. 22. Eastern equine encephalitis virus infection in the salivary gland of an *Aedes triseriatus* mosquito. Virus particles accumulate via budding in reticulum cisternae and in diverticula of the plasma membrane of the glandular epithelium; very large numbers of particles enter the saliva where they are free to be transmitted during blood feeding. ×55,000.

small that most of these size estimations cannot be explained in the light of present knowledge (Schlesinger, 1977). Perhaps biased by such filtration data, the earliest electron microscopic studies variously claimed identification of flaviviruses as very small unstructured particles. Particles consistent with our present knowledge were first described in shadow-cast preparations by Hotta (1953) working with dengue virus, by Lavillaureix *et al.* (1959; Lavillaureix, 1960) with West Nile virus, by Sokol (1961) with tick-borne encephalitis virus, and by Nozima *et al.* (1964) with Japanese encephalitis virus. The limits of resolution of the shadow-casting method were reached by Bergold and Weibel (1962); in their study, yellow fever virus was shown to be spherical, to be 38 nm in diameter, and to have a surrounding membrane. The resolution achieved by them virtually matched that of most published negative-contrast micrographs of flaviviruses.

A. Negative-Contrast Electron Microscopy

Negative-contrast studies of flaviviruses were at first concentrated on dengue (Brandt *et al.*, 1970; Smith *et al.*, 1970; Matsumura *et al.*, 1971; Yoshinaka and Hotta, 1971; Kitaoka *et al.*, 1971), Japanese encephalitis (Nishimura *et al.*, 1968; Yoshinaka and Hotta, 1971; Kitano *et al.*, 1974), and tick-borne encephalitis viruses (Weckstrom and Nyholm, 1965; Slavik *et al.*, 1967, 1970). Single studies appeared on Powassan (Abdelwahab *et al.*, 1964), Langat (Boulton *et al.*, 1971), West Nile (Chippaux-Hyppolite *et al.*, 1970), and Banzi viruses (Calberg-Bacq *et al.*, 1975). In these studies virus particles were found to be spherical and to have a membranous envelope and surface projections. In general, virus particles were stated to have diameters between 37 and 50 nm when intact, and between 45 and 60 nm when flattened on the substrate. Reported size estimations outside these ranges probably represent discrepancies in early methods of calibration of magnification (Pfefferkorn and Shapiro, 1974). The real uniformity of size of flaviviruses is evident in the very narrow range (±10%) of diameters reported by investigators using a single technique on a single virus. Few further structural details were described in these early negative-contrast studies: Dense-core structures were found in a few cases, and these were generally considered to be spherical and without symmetry. Two different groups of investigators claimed that core particles were constructed from symmetrically arrayed subunits (Slavik *et al.*, 1970; Abdelwahab *et al.*, 1964), one going so far as to describe a 92-subunit construction (Slavik *et al.*,

1970). Looking back at the published micrographs in these early papers, we find that the overall morphology was accurately described, but that the resolution obtained did not allow rigorous analysis of symmetry and substructure in accord with criteria established with other viruses (Caspar, 1965). In particular, evidence pertaining to the nature of core structures did not prove the presence of symmetric nucleocapsids. Since even the most recent studies have failed to be convincing on this point, the term "core particle" or "core structure" will be used in this chapter so as not to prejudice the matter further. Characterization of the interior structure of flaviviruses is one of the major unanswered questions in viral ultrastructural research.

If micrographs are selected from the negative-contrast studies which achieved the highest resolution, each virion constituent can be described separately: The virus envelope has been well resolved; its diameter is between 35 and 45 nm and its modified unit membrane nature has been shown consistently (Nishimura *et al.*, 1968; Smith *et al.*, 1970; Matsumura *et al.*, 1971; Kitano *et al.*, 1974; Hayashi *et al.*, 1978). Pronase and bromelin treatment were used to strip off surface projections, leaving the smooth-surfaced envelope intact (Kitano *et al.*, 1974). Osmotic shock was used to open envelopes and allow resolution of the interior of particles (Matsumura *et al.*, 1971). In some cases taillike processes extended from virus particles, indicating the presence of an osmotic barrier in the membrane layer.

The surface projection layer of flaviviruses has appeared to vary considerably in thickness and detail (Nishimura *et al.*, 1968; Smith *et al.*, 1970; Matsumura *et al.*, 1971; Kitano *et al.*, 1974). In some cases this layer has seemed very thin (<5 nm) and so unstructured that construction from individual spikes seems unlikely; in other cases the projection layer has appeared to be rather thick (10 nm) and to be constructed from radially arranged and uniformly spaced units. Detergents have been used to release projections, but aggregates or rosettes formed in this way have not revealed further structural detail (Kitano *et al.*, 1974). One important variation in surface detail was first reported by Smith *et al.* (1970) (Fig. 23). In their study of dengue virus, they showed that the virion surface was made up of rings (or doughnuts), 7 nm in diameter, with a 2- to 3-nm central hole. Large numbers of these units covered virus particles without any discernible symmetry. In addition, morphologically similar 7-nm units were found free in a nonvirion soluble complement-fixing (SCF) fraction when infected cell culture materials were subjected to gradient ultracentrifugation. This convincingly illustrated characterization of the surface of dengue virus correlates best with studies of other fla-

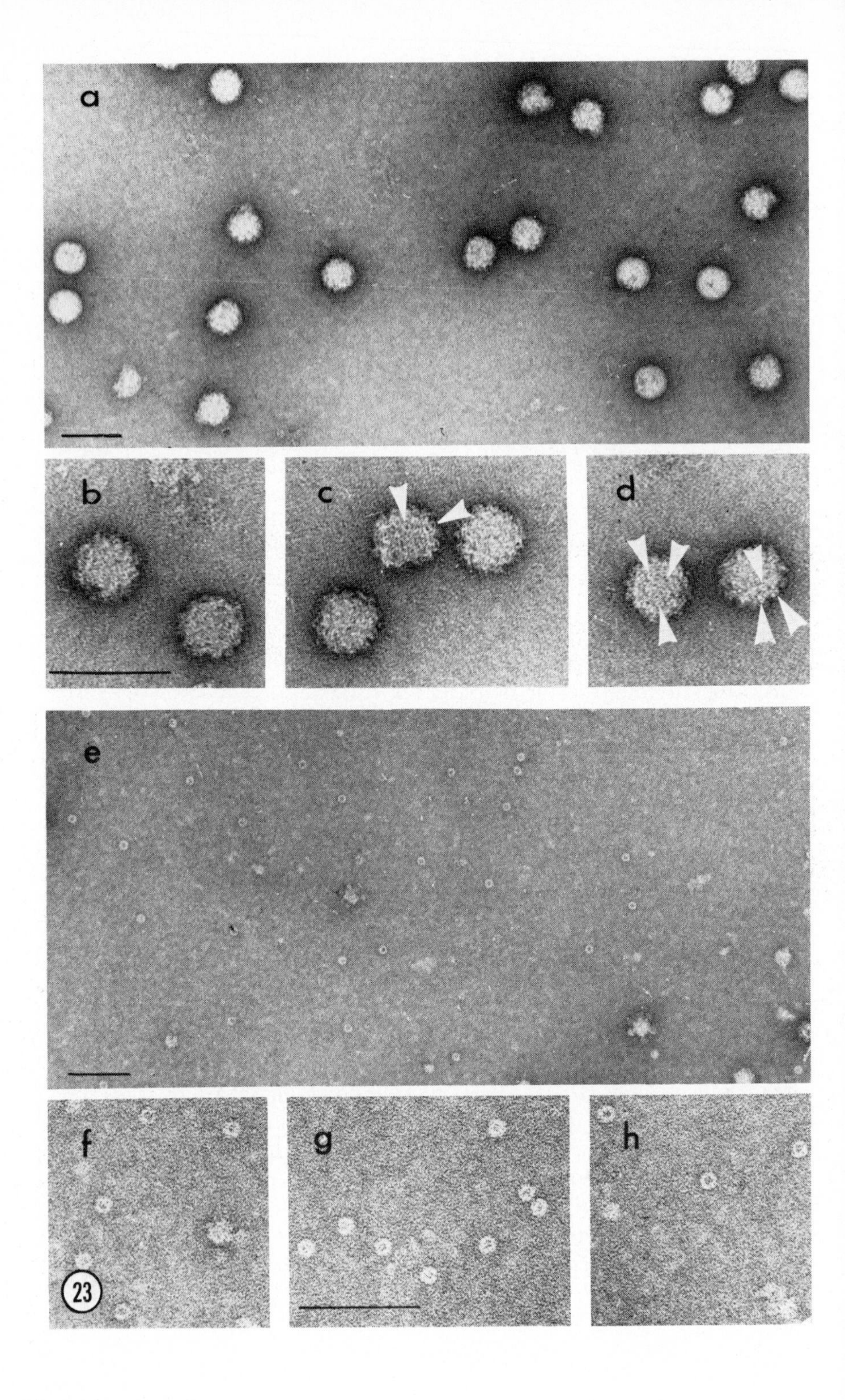
a
b
c
d
e
f
g
h
23

viviruses in which large, distinct surface units have been depicted (Slavik *et al.*, 1967; Nishimura *et al.*, 1968), but it is hard to relate this structure to studies in which a closely apposed, delicate surface layer has been found (Matsumura *et al.*, 1971; Kitano *et al.*, 1974). Varying effects of different negative staining materials may have been the source of some of the described differences in different laboratories, but it is difficult to say which, if any, of these variations represent the native state of the flavivirus surface.

The TBH-28 strain of St. Louis encephalitis virus was examined in the author's laboratory after it was stained with various negative-contrast materials (Bauer *et al.*, 1977; Murphy, 1978). The virus was propagated in the CER line of baby hamster kidney cells according to the recommendations of Trent (1977); it was partially purified from 72-h supernatant fluids by precipitation with polyethylene glycol (PEG 6000) and centrifugation to equilibrium in gradients of glycerol and potassium tartrate (Obijeski *et al.*, 1974). Uranyl acetate (0.5%, pH 4.5) was the harshest staining material used; as staining times were increased, virus particle structure deteriorated. This stain usually penetrated virus particles and delineated the membranous envelope layer (Fig. 24). Core particles (30 nm) were visible inside some envelopes, but no details could be resolved. Large numbers of smooth, spherical particles, also 30 nm in diameter, appeared free in these preparations (Fig. 24); these particles were identical to those found in dengue virus preparations by Matsumura *et al.* (1971) and, as in the latter case, were smoother than core structures seen within virus envelopes. Identification of these free particles is not settled, but this discrepancy between free particles and enveloped core particles is exemplary of our need for further study and for continued caution in identifying structural elements. When stained with uranyl acetate, the surface projections on St. Louis encephalitis virus particles formed a thin (5 nm) layer, or halo, in which individual projections were not resolved. Trent (1977) observed similar characteristics with uranyl acetate staining.

Staining of St. Louis encephalitis virus with sodium silicotungstate or phosphotungstate (2%, pH 7.0) for 30 s resulted in stain penetration of some particles, but core structures were not resolved any better

Fig. 23. Flavivirus structure as exemplified by dengue-2 virus in a study by Smith *et al.* (1970). (a) Rapidly sedimenting hemagglutinin is intact virus. ×70,000. (b–d) Intact virus particles with 7-nm subunits on their surfaces (arrowheads). ×140,000. (e) Slowly sedimenting hemagglutinin units are ring-shaped structures like those on the surface of intact virus particles. ×70,000. (f–h) Higher magnification of isolated 7-nm ring-shaped surface sununits. ×140,000. Micrographs courtesy of Dr. W. E. Brandt.

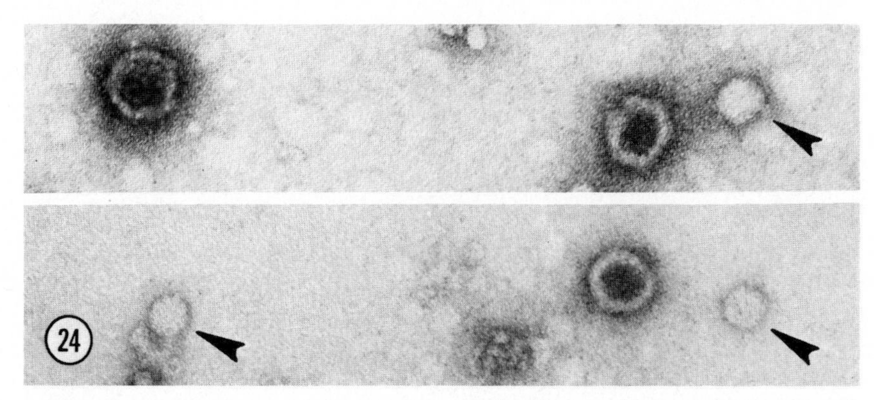

Fig. 24. St. Louis encephalitis virus stained with uranyl acetate (0.5%, pH 4.5). Virus particles are penetrated, revealing membranous envelope. Smooth 30-nm particles (arrowheads) always present with partially purified virions have not been identified with certainty. Composite. ×228,000.

with this stain than with uranyl acetate (Fig. 25). Some adverse effects of the tungstate stains were indicated by deterioration of the structural quality of particles found after the 2-s staining technique of von Bonsdorff and Harrison (1975). With the latter method, the surface projection layer was rather thick (8 nm) and unstructured, the unit membrane envelope was clear (6 nm), and core particles within virions were fuzzy and 30 nm in diameter. However, in no instance did this technique resolve symmetry in envelope or projection layers as it has done with alphaviruses.

The least damaging negative staining material used on St. Louis encephalitis virus was ammonium molybdate (3%, pH 5.5); virus particles were not noticeably affected by exposure times of up to 2 min. Core structures were visible within virus particles, and in most instances these appeared loosely organized like a ball of string (Fig. 26). Some particles contained two core structures (Fig. 27). Smoothly margined 30-nm particles similar to those found with uranyl acetate were seen with the molybdate stain, and in a similar way they appeared to differ from cores within virion envelopes (Fig. 28). The viral envelope layer was similar with molybdate staining as with the other stains, and osmotic blebbing of the envelope was common (Fig. 29). The surface structure was extremely delicate with the molybdate stain; the projection layer was very thin (<5 nm) and closely apposed to the envelope, and individual projections were not resolved (Fig. 30). The mean diameter of 100 intact virus particles stained with molybdate was 48.8 nm, and the size range was such that 99% of the

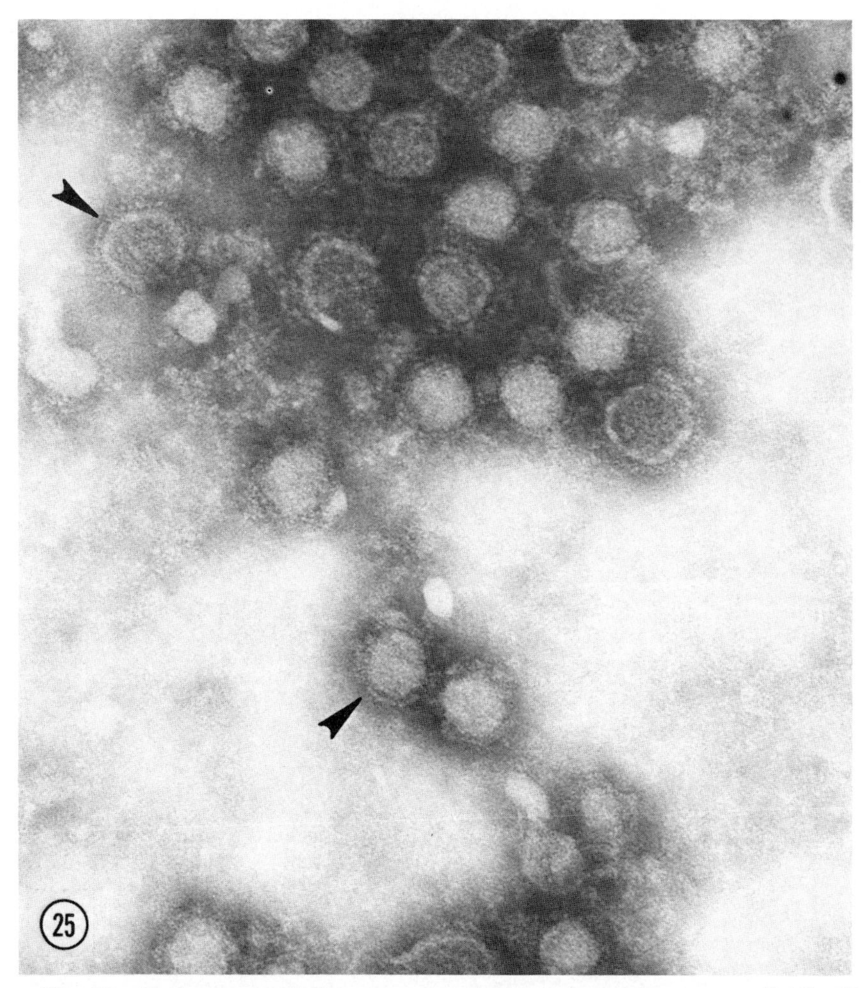

Fig. 25. St. Louis encephalitis virus stained with sodium silicoungstate (2.0%, pH 7.0). Membranous envelope and surface projections are delineated (arrowheads), but core structure is indistinct. ×234,000.

particles had diameters within 5% of the mean (measurements were made with catalase calibration of magnification).

These observations on St. Louis encephalitis virus compare generally with those of other flaviviruses, but when an attempt is made to consider the basic virion structure of flaviviruses, agreement extends only to the envelope layer. Core structures within flavivirus particles do not have a hexagonal outline and do not appear to consist of

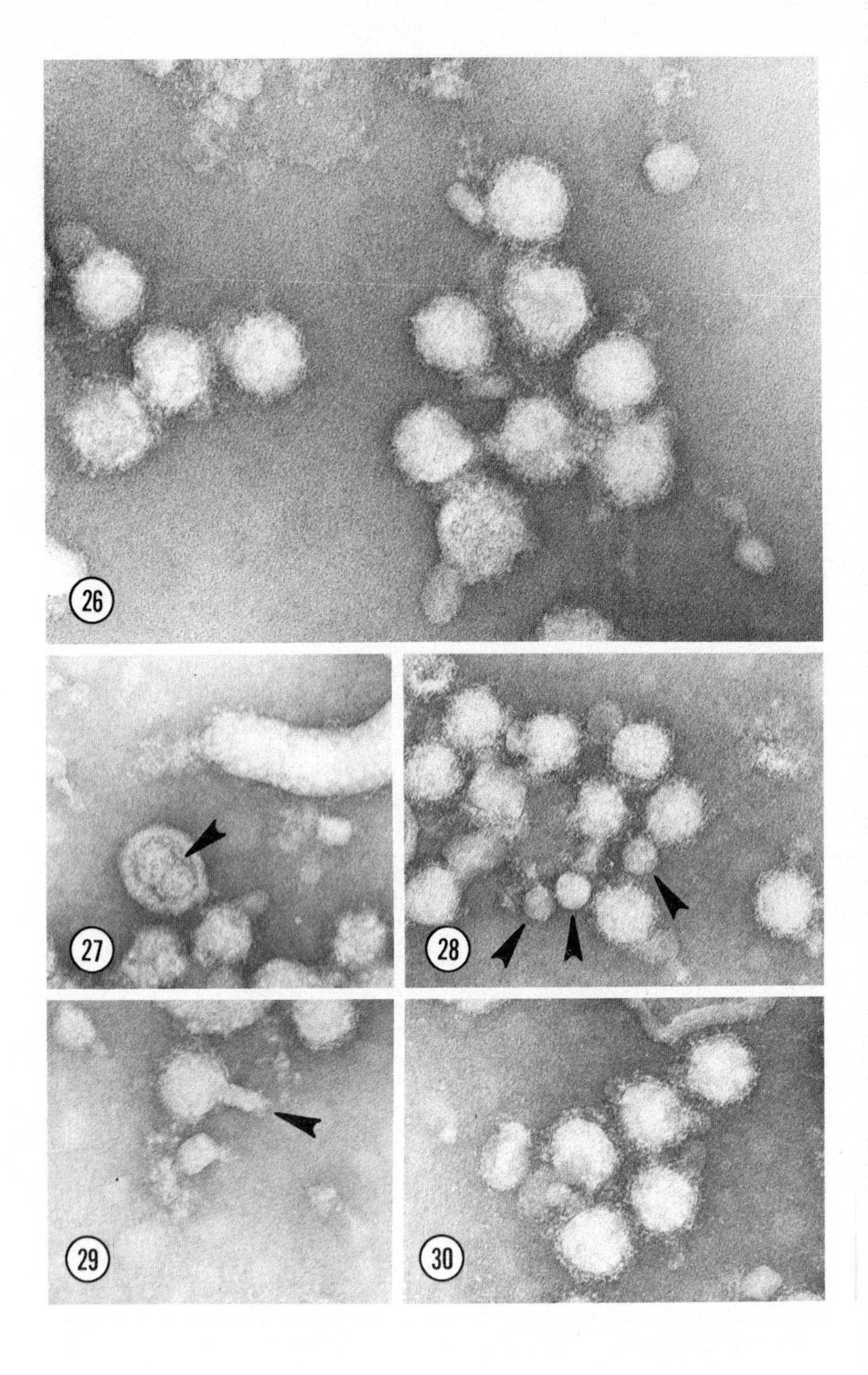
26
27
28
29
30

regularly arrayed subunits; the nature of the core organization remains unresolved. This matter should not be prejudged—a similar indeterminate structure was considered for alphavirus nucleocapsids for several years before icosahedral symmetry was proposed by Horzinek and Mussgay (1969). Since then, extension and confirmation of alphavirus nucleocapsid symmetry has been frustrating (see Section II), and we are left with a lesson of caution. Concentration, purification, stabilization, and biological identification of free flavivirus core particles must be accomplished before proper structural analysis can be anticipated. Similarly, the surface structure of flavivirus particles should not be prejudged; any model of the surface of flavivirus particles must deal with (1) the 7-nm ring-shaped units found on dengue virus particles by Smith *et al.* (1970), (2) the indistinct, delicate surface appearance seen in several instances, and (3) the progressively adverse effects of most negative-contrast materials on viral morphology. The question here is, What is the true native state? For high-resolution electron microscopy, it is necessary to remove contaminating nonviral materials from interstices between structural elements. However, if viral structural elements are also removed by harsh purification methods or changed by harsh staining methods, then any extra surface detail may be artifactual. Since this consideration has not been important with some nonenveloped viruses (e.g., papovaviruses and adenoviruses) and some enveloped viruses (e.g., orthomyxoviruses and rhabdoviruses), it is probable that methods yielding indistinct, delicate surface characteristics might never be recognized for their real worth. Is it possible that, in their native state, flaviviruses have 7-nm ring structures embedded in an indistinct matrix? Is it possible that structural elements which normally form a fuzzy surface can be "polymerized" into 7-nm rings by a common step of

---

Fig. 26. St. Louis encephalitis virus stained with ammonium molybdate (3.0%, pH 5.5). Surface projections appear as a very thin, indistinct layer. ×300,000.

Fig. 27. St. Louis encephalitis virus stained with ammonium molybdtate. Penetrated particle (arrowhead) has two cores, each appearing as a loosely organized sphere. ×234,000.

Fig. 28. St. Louis encephalitis virus stained with ammonium molybdate. The 30-nm particles (arrowheads) associated with virus particles are smoother than core particles seen within virions. ×210,000.

Fig. 29. St. Louis encephalitis virus stained with ammonium molybdate. The osmotic bleb or tail (arrowhead) of the virion envelope appears similar to that seen commonly with alphaviruses. ×234,000.

Fig. 30. St. Louis encephalitis virus stained with ammonium molybdate. Surface projections at the *edge* of these particles do not seem to be arrayed in any orderly fashion. ×234,000.

purification? This subject seems far from solution, but perhaps other techniques will contribute to our knowledge in the future. For example, the morphology of one flavivirus, dengue-2 virus, was studied by the freeze-fracture technique (Demsey *et al.*, 1974). A surface-grazing fracture plane revealed surface projections 7 nm in diameter. Deeper fracture planes showed the unit membrane envelope and a core structure without apparent symmetric subunits. Further work with freeze-fracturing may provide further structural insights.

B. Thin-Section Electron Microscopy

Structural analysis of many small enveloped viruses has been aided by thin-section electron microscopy. For example, there can be no doubt about the modified unit membrane nature of the envelope of togaviruses after visualizing the viral budding process in thin sections. Many studies on the morphology and morphogenesis of flaviviruses *in vivo* and in cell cultures have been reported. Studies on mosquito-borne flaviviruses have included St. Louis encephalitis virus (Murphy *et al.*, 1968b, 1975; Whitfield *et al.*, 1973; Murphy, 1975) and the other members of its serological subgroup, Japanese encephalitis virus (Kitaoka and Nishimura, 1963; Yasuzumi *et al.*, 1964; Yasuzumi and Tsubo, 1965a,b; Ota, 1965; Filshie and Rehacek, 1968; Oyanagi *et al.*, 1969), West Nile virus (Southam *et al.*, 1964; Shestopalova *et al.*, 1966a; Tamalet *et al.*, 1969), and Murray Valley encephalitis virus (Filshie and Rehacek, 1968). Other mosquito-borne viruses have also been studied: dengue (Nii *et al.*, 1970; Matsumura *et al.*, 1971; Cardiff *et al.*, 1973; Sriurairatna *et al.*, 1973, 1974; Stohlman *et al.*, 1975; Ko, 1976), yellow fever (Bergold and Weibel, 1962; McGavran and White, 1964; David-West *et al.*, 1972), Wesselsbron (Parker and Stannard, 1967; Lecatsas and Weiss, 1969; Parker *et al.*, 1969), Banzi (Calberg-Bacq *et al.*, 1975), Zika (Bell *et al.*, 1971), and Kunjin viruses (Boulton and Westaway, 1976). Tick-borne flavivirus studies have included the tick-borne encephalitis complex (Kovac and Stockinger, 1961; Blinzinger and Müller, 1970; Blinzinger, 1972; Blinzinger *et al.*, 1973) and Omsk hemorrhagic fever (Shestopalova *et al.*, 1965, 1966b, 1972), Kyasanur Forest disease (Jelinkova *et al.*, 1974), Powassan (Abdelwahab *et al.*, 1964), and Langat (Tikhomirova *et al.*, 1968; Boulton and Webb, 1971; Dalton, 1972) viruses. Nonarthropod-borne flaviviruses have also been studied: Entebbe bat salivary gland virus (Peat and Bell, 1970) and Cowbone Ridge virus (Calisher *et al.*, 1969). Even though the overall morphology of the flaviviruses as compiled from these thin-section studies has been remarkably consistent, important questions concerning morphogenesis and it relation to structure remain unanswered.

1. *Virus Morphology in Thin Section*

In all the thin-section studies listed above, virus particles have been found to have dense cores and lucent envelopes (Fig. 31). In most cases no further virion detail has been evident, but occasionally envelopes have been resolved as unit membranes despite their close apposition to the dense core. Surface projections have never been clearly visualized in thin sections. In most studies, particles have been described as spherical, but in three cases polygonal or hexagonal cores have been reported (Shestopalova *et al.*, 1966a; Jelinkova *et al.*, 1974; Calberg-Bacq *et al.*, 1975). In a compilation of particle dimensions in thin sections, as reported in 21 of the papers cited in this chapter dealing with 11 different flaviviruses, virion diameters ranged narrowly between 36 and 44 nm (mean, 40 nm), with few exceptional measurements as large as 50–55 nm. Core diameters ranged between 25 and 30 nm. Size estimations of St. Louis encephalitis virus made in the author's laboratory were the same whether the virus was in cell culture, mouse brain, or mosquito organs; the mean virion diameter was 38–39 nm, and the mean core diameter was 27 nm. In the lumen of the salivary gland of *Culex pipiens pipiens* mosquitoes, St. Louis encephalitis virus formed massive crystalline arrays in which center-

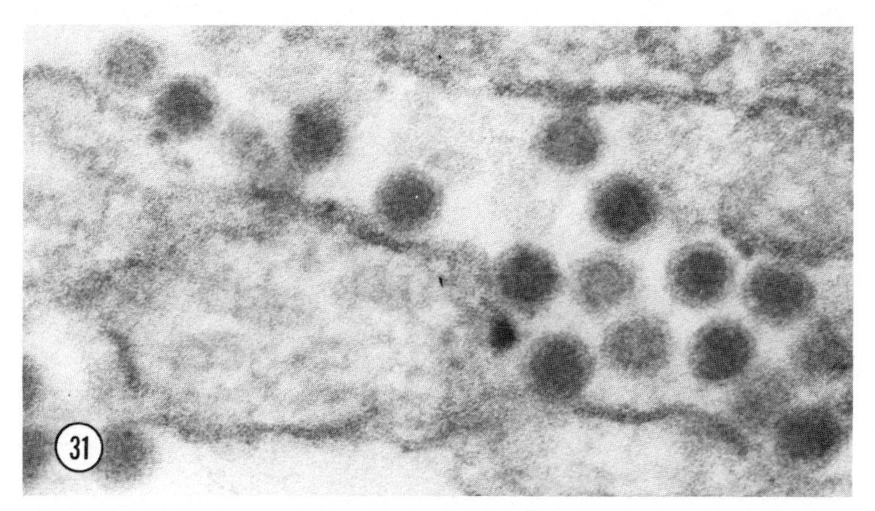

Fig. 31. St. Louis encephalitis virus infection of suckling mouse brain. Particles have a uniform core density and a surrounding lucent zone which is not resolved into envelope and projection layers. ×280,000.

to-center distances were 41 nm (Whitfield *et al.*, 1973) (Fig. 32). When these size estimations are considered together, it is clear that there is a great deal of uniformity in the shrinkage obtained with different fixation and embedment methods used in various laboratories. This uniformity may be of value in comparing observations made in remote laboratories.

2. *Virus Morphogenesis in Thin Section*

The first changes in flavivirus-infected cells occur in perikaryonic membranous organelles, mostly in endoplasmic reticulum. Hypertrophy of these membrane systems occurs in association with the

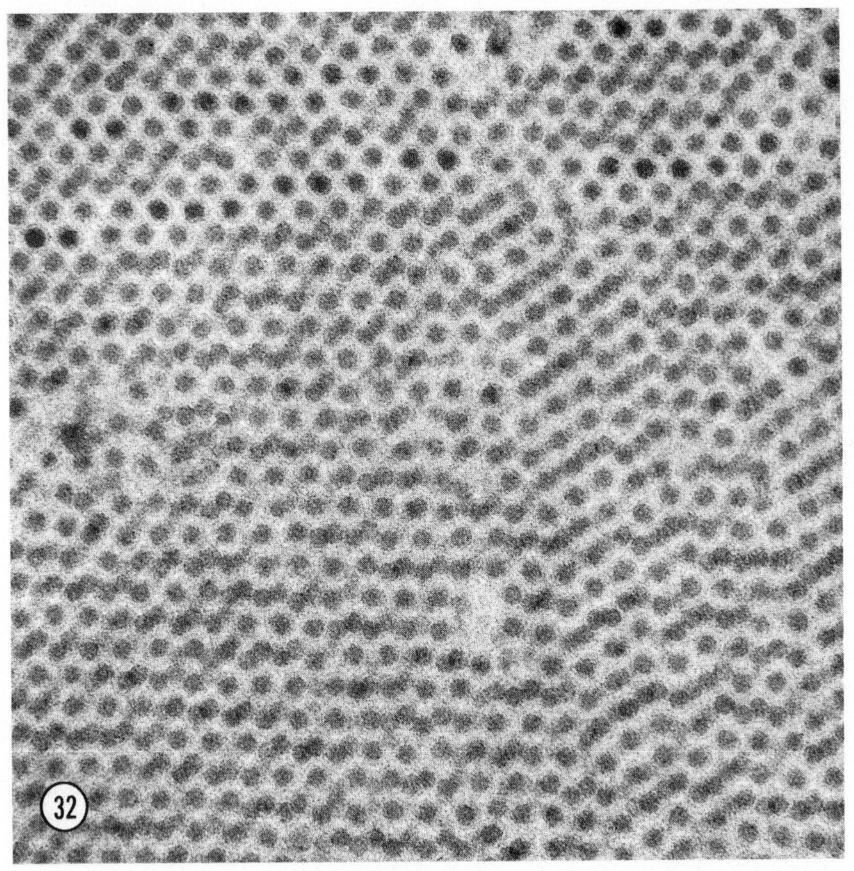

Fig. 32. St. Louis encephalitis virus in the lumen of a mosquito salivary gland. This virus particle crystalline array has a center-to-center interval of 41 nm, closely matching the 39-nm diameter measured from individual particles. ×97,000.

accumulation of virus particles within their cisternae (Murphy *et al.*, 1968b; Murphy, 1978) (Fig. 33). These events expand to involve peripheral cytoplasm and even the nuclear envelope system, but usually there are no large focal accumulations of virus particles in any one place until late in the infection (Murphy, 1978) (Figs. 34 and 35). The morphogenetic events leading to this accumulation have been the subject of continuing controversy, because sequential stages of virus particle maturation have not been clearly demonstrated. Some investigators have considered it adequate to build an analogy between events observed in flavivirus-infected cells and those so clearly documented in alphavirus infections. As shown in Section II, alphaviruses initially form a nucleocapsid, and this is followed by a clearly discernible budding process in which modified membrane becomes the virion envelope in precise coordination with surface projection insertion (Murphy and Whitfield, 1970). Building a valid analogy between flavivirus and alphavirus morphogenesis requires identification of preformed flavivirus core structures *in situ,* and of the various stages of a budding process. In some of the cited thin-section studies on flavivirus infections, these identifications were claimed, but critical examination does not allow clear distinction between structures called viral cores or nucleocapsids and host cell ribosomes. Similarly, the distinction between terminal budding stages and the incidental positioning of mature virus particles in contact with host cell membranes is unconvincing. As an example of this controversy, Ota in 1965 stated (and illustrated) that "Japanese encephalitis virus particles develop by budding from the membrane of cytoplasmic vacuoles" in porcine kidney cells. In 1968, Filshie and Rehacek repeated Ota's experiments and stated: "We have observed close contacts and pedicular attachments between vesicular membranes and virus particles similar to those found by Ota. However, we cannot confirm that any continuity exists between the capsid membrane of the virion and the unit membrane structure of the vesicle." The same two positions are voiced at present with little additional objective evidence available. In our studies on St. Louis encephalitis virus in cell culture and in mouse brain (Murphy *et al.*, 1968b; Murphy, 1978), the stages of a budding process were not seen either. Instead, the presumed sites of virus formation appeared as a rather indistinct membrane deformation with roughening (Fig. 36). In the cytoplasm at these sites the typical ribosome structure was replaced by larger, more irregular dense bodies (Fig. 37). Although these were not identified as preformed core structures, they did appear to be associated with some kind of nascent budding in which initial and

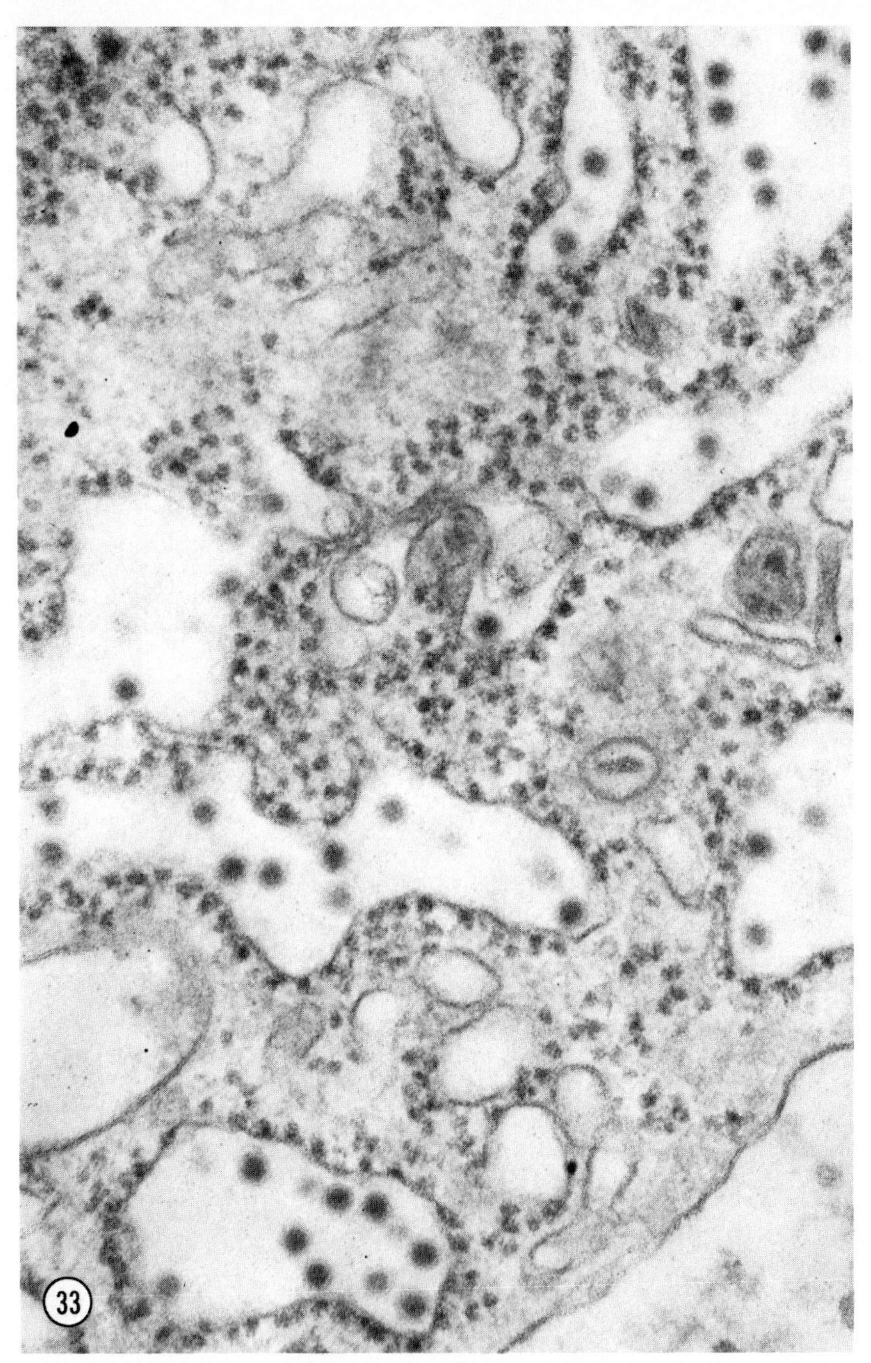

Fig. 33. St. Louis encephalitis virus infection of BHK-21 cell. Dispersed virus accumulation within distended endoplasmic reticulum at an early stage of infection. ×114,000.

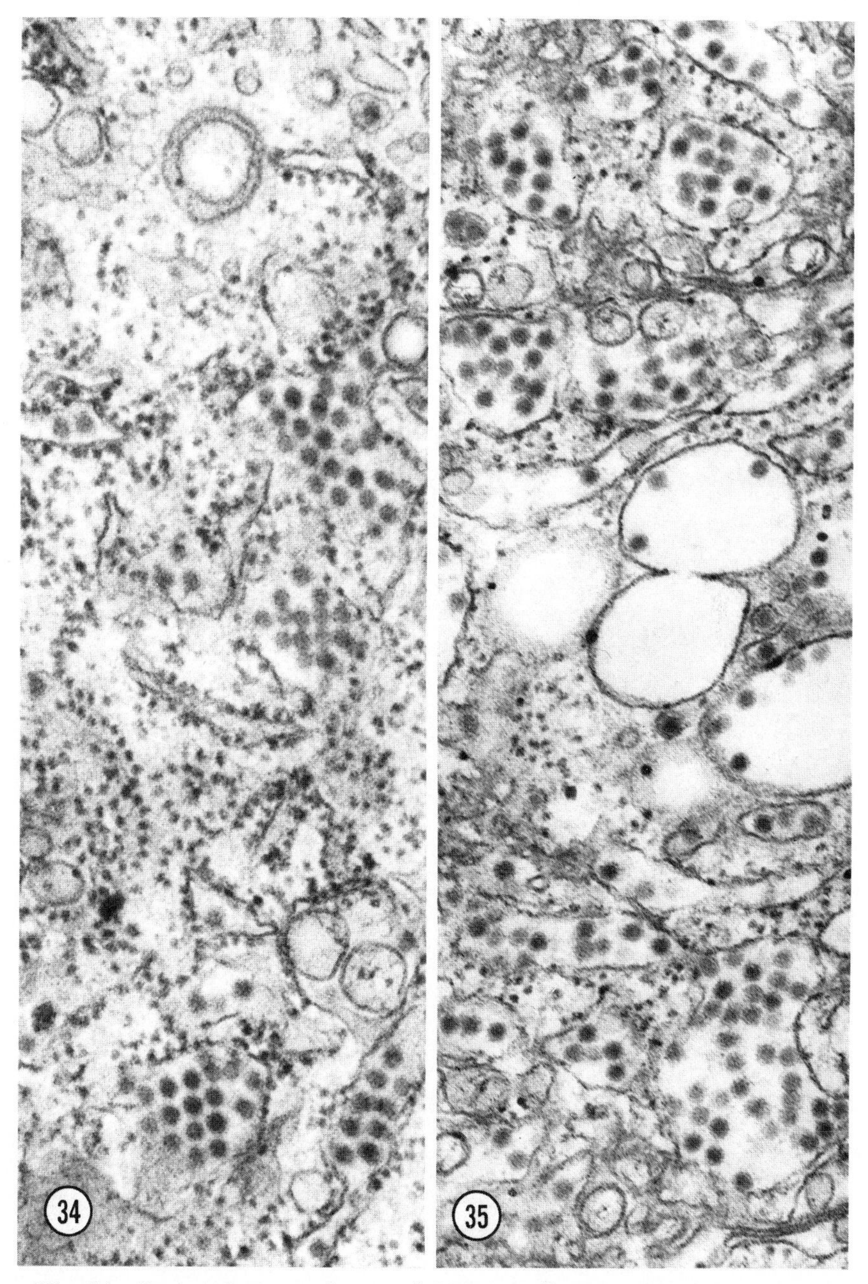

Fig. 34. Dengue-2 virus infection of BHK-21 cell. Virus particle accumulation within the endoplasmic reticulum is associated with a proliferation of host cell organelle membrane systems. ×70,000.

Fig. 35. Russian spring-summer encephalitis virus in the cytoplasm of a mouse neuron. The same intracisternal accumulation of virus particles found in cells in culture is typical of flavivirus infections of the central nervous system. ×70,000.

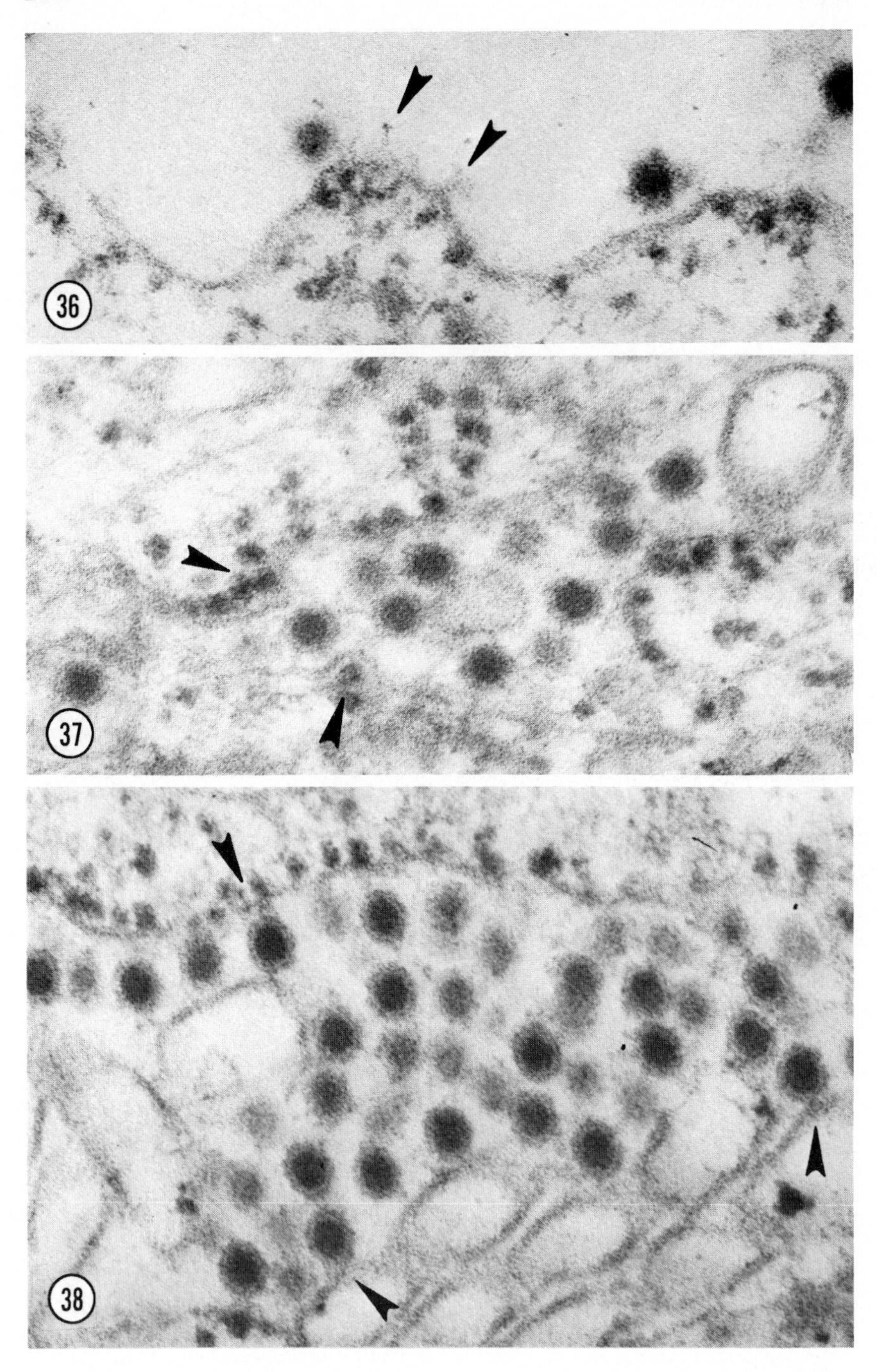
36
37
38

terminal stages, but not the crucial intermediate stages, were identified. Complete particles apparently connected to host membrane by stalks or indistinct shards were common, as described by Ota (1965) and Filshie and Rehacek (1968) (Fig. 38). We considered that visualization of continuity between host cell membrane and virus envelope—if budding did occur—would always be difficult because the convolutions of reticular membranes would rarely provide a long profile in an optimal plane of section. Similarly, we considered that identification of preformed core structures, if they did occur, would always be difficult because of the typically high ribosome concentration upon the reticulum in infected cells and because of the presence in infected cells of several other kinds of round forms with the expected size of virus cores. The most common of these round forms derive anomalously from proliferating membranes in infected cells and actually are membranous tubules cut transversely (Fig. 39); in several instances these have been misidentified as flavivirus cores.

Several hypotheses have been advanced to explain the usual failures to visualize sequential stages of a budding process. In one such case the term "modified form of budding" was used to qualify findings—but not to explain them (Nii *et al.*, 1970). Schlesinger (1977) explored possible alternatives in some detail: (1) He considered the possibility that the flavivirus envelope might not be derived from preformed host cell membrane, but might be laid down *de novo* in a way analogous to that of poxvirus envelope synthesis. In this case, budding would not be seen. (2) He reconsidered the possibility that the maturation of *infectious* virions might occur at plasma membranes. In this case, the lack of ultrastructural evidence of budding might be related to the usually late time of harvest for electron microscopic studies. (3) Finally, he considered that, if both premises were true, then formation of the characteristic virus-filled intracytoplasmic vacuoles may only represent a "dead-end" accumulation in cells in which normal maturation was inefficient. These hypothetical alternatives must be considered experimentally.

Fig. 36. St. Louis encephalitis virus infection of a BHK-21 cell. Even though a progressive budding process is not seen, membrane deformations and roughening (arrowheads) occur at sites where virion formation must be occurring. ×174,000.

Fig. 37. St. Louis encephalitis virus infection of BHK-21 cell. Identification of preformed cores has never been clear at sites of virion formation, but typical ribosomes are replaced upon reticulum membranes by larger, more irregular dense bodies (arrowheads). ×150,000.

Fig. 38. St. Louis encephalitis virus infection of BHK-21 cell. Connection of mature virus particles to reticulum membranes (arrowheads) has not been distinct enough to follow unit membrane leaflets across a "stalk." ×174,000.

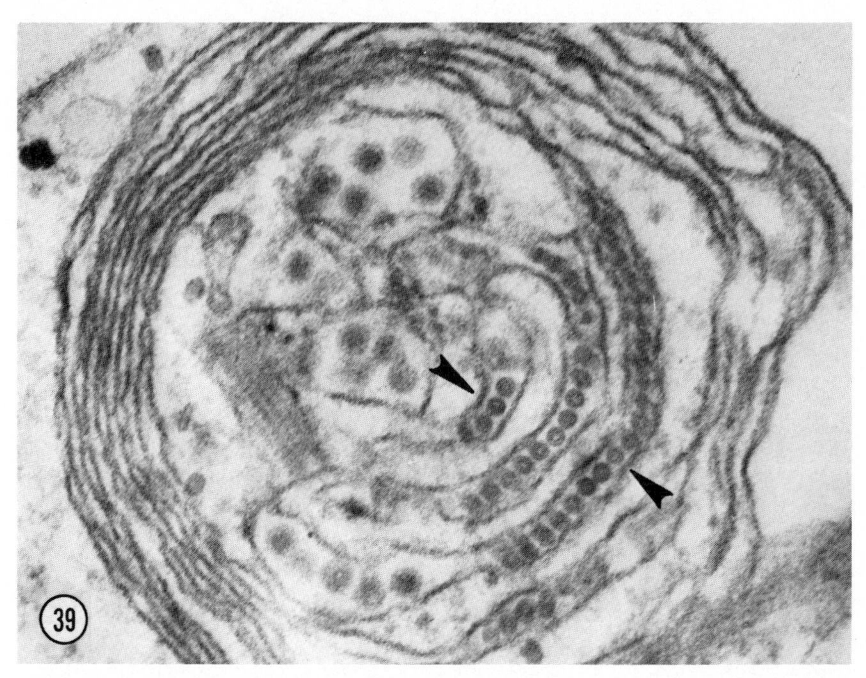

Fig. 39. Russian spring-summer encephalitis virus infection of mouse brain. The proliferation of membranes associated with flavivirus infection often appears as "myelinlike figures"; membranous tubules (arrowheads) in such figures have occasionally been misidentified as virus cores. ×88,000.

If flavivirus morphogenetic events are extremely rapid, then budding or any of the hypothetical alternates might occur without leaving evidence in the form of intermediate stages. Recently, the search for budding has been extended to host cell systems not previously used with togaviruses. Infections in mosquitoes and in mosquito cell cultures have been studied (Whitfield *et al.*, 1973; Igarashi *et al.*, 1973; reviewed in Schlesinger, 1977). Matsumura *et al.* (1977) reported finding intracytoplasmic viral nucleoids and budding upon plasma membranes in a human leukemic monocyte line (J-111) infected with dengue virus. Sung *et al.* (1975), working with dengue carrier cultures of human lymphoblastoid cells (Raji), found crystalline arrays of particles which they considered "incomplete viral nucleoids" together with intracisternal mature dengue virus particles. These identifications must be confirmed, but the use of novel host cells has been established as one hopeful approach. The use of lymphoid and reticuloendothelial cells is also important in relation to flavivirus disease pathogenesis (Halstead *et al.*, 1973; Theofilopoulos *et al.*, 1976).

Other approaches, such as variations in host cell temperature and the ionic strength of media, must also be correlated with ultrastructural observations. For example, a temperature lower than usual has been used with picornaviruses to slow virus growth and delay cytopathology, thereby greatly favoring electron microscopic observations. Similarly, Matsumura *et al.* (1972) used increased salt concentrations to increase rapidly dengue-2 virus release from Vero cells. Manipulation of this phenomenon to "synchronize virus maturation" for electron microscopic study would be of interest.

In 1968, in a study from the author's laboratory, dense intranuclear particles (25–38 nm in diameter) were described as occurring in mouse neurons infected with St. Louis encephalitis virus (Murphy *et al.*, 1968b). Similar particles were reported in the nuclei of brain cells infected with Japanese encephalitis virus (Yasuzumi and Tsubo, 1965a) and tick-borne encephalitis virus (Tikhomirova *et al.*, 1968). Our subsequent attempts to confirm identification of these particles as virus particles or virus core structures have failed. With improved resolution (via thinner sections) and more comprehensive searching, it now seems probable that elements of the nucleolar ribosome complex were misidentified.

Lucent vesicles are present within reticular cisternae of cells infected with many flaviviruses (Fig. 40). In keeping with the terminology used with alphaviruses, they have been called "cytopathic vacuoles." They are spherical or elongated, with a diameter of 250–500 nm and a maximum length of about 5000 nm. They often contain a very fine reticular web in their otherwise lucent interior. They are formed by evagination from reticular membranes; although their nature and function are presently unknown, it will be necessary to consider their relation to flavivirus morphogenesis. Their analog in alphavirus-infected cells have been shown to be active sites of viral RNA synthesis (Grimley *et al.*, 1972). Another characteristic of flavivirus-infected cells is the exuberant proliferation of smooth reticular membranes (Fig. 41). In some cases these membranes are randomly massed, but often they appear in large, ordered arrays, the most complex of which have sixfold symmetry constructed from precise foldings of membrane sheets (Fig. 42). The mechanism underlying this proliferation is unknown, but further study is needed to see if there is a relationship between membrane proliferation and virus formation (Murphy *et al.*, 1968b; Murphy 1978).

Release of flaviviruses from infected cells has been considered to involve at least two pathways. First, vesicles containing virus migrate to the host cell periphery and fuse with the plasma membrane,

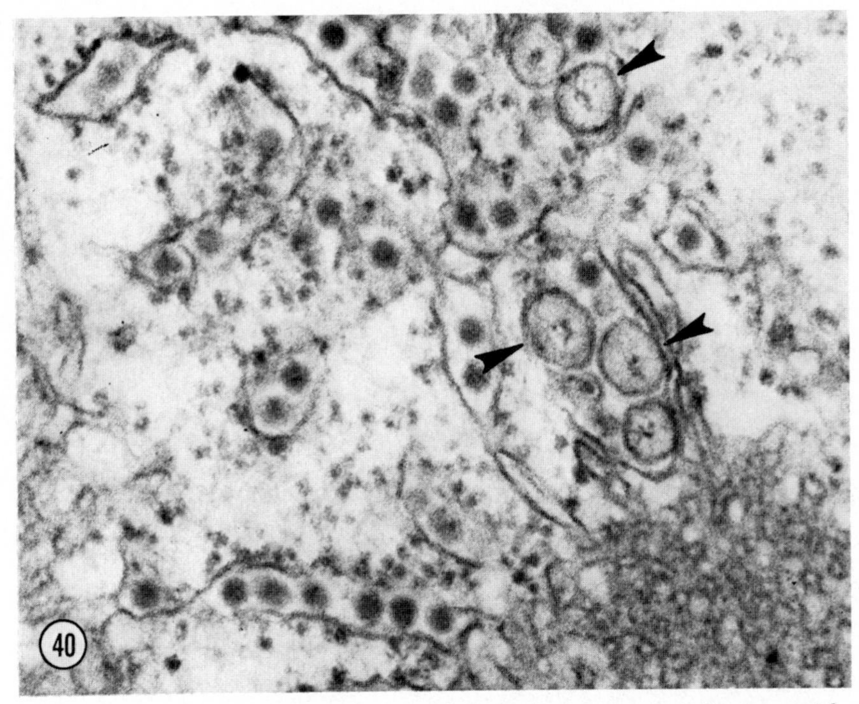

Fig. 40. St. Louis encephalitis virus infection of mouse neuron. Lucent vesicles (arrowheads) with internal reticular webs occur within cisternae of endoplasmic reticulum together with virus particles in all flavivirus infections. Their nature is unknown. ×99,000.

thereby releasing free virus particles before there is extensive cell damage. Second, cytopathology reaches the point of plasma membrane rupture with the release of free virus and virus still entrapped within vesicular membranes (Murphy *et al.*, 1968b). Both these mechanisms of virus release have been observed *in vivo*. If virus release also occurs directly from plasma membranes of intact infected cells, then the spread of virus in target organs would be further facilitated, especially in the crucial early phases of infection. This matter has been given new credence by the work of Stohlman *et al.* (1975), who

Fig. 41. St. Louis encephalitis virus infection of mouse brain. Hypertrophy of endoplasmic reticulum membranes of neurons is immediately associated with virus particle accumulation within lumina. ×32,000.

Fig. 42. St. Louis encephalitis virus infection of mouse brain. Membrane proliferation continues at such an intensity in infected neurons that convolutions form into ordered masses with sixfold symmetry. Such membranous structures are characteristic of flavivirus infections. ×45,000.

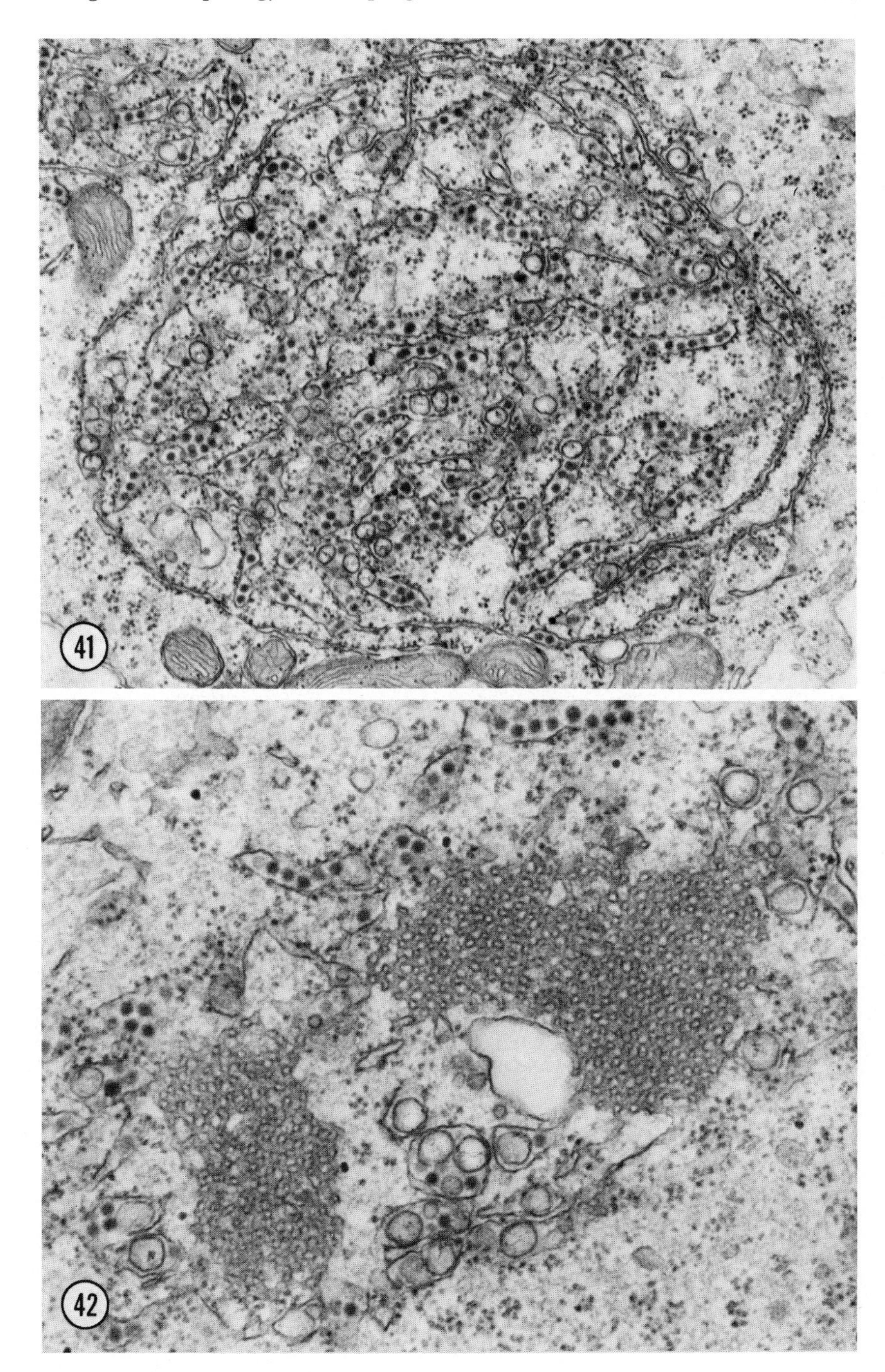
41
42

found dengue virus antigens in the "plasma membrane fraction" after fractionating infected cells, and by Boulton and Westaway (1976), who found the same with Kunjin virus-infected cells. Similarly, Catanzaro *et al.* (1974) found dengue antigens in plasma membranes with immunoperoxidase and immune cytolysis techniques. Brandt and Russell (1975) found the same localization of Japanese encephalitis virus antigens by an immune cytolysis method. Finally, Cardiff and Lund (1976) recently used immunoperoxidase reagents specific for dengue virion (RHA) and nonvirion (SCF) antigens to show ultrastructurally that both kinds of antigens occurred on the surface of infected cells. This kind of evidence of plasma membrane involvement in flavivirus infections must now be related directly to evidences of the biological maturation of virus particles, that is, maturation to *infectious* virus particles (Schlesinger, 1977). Development of such concepts must include ultrastructural studies, particularly more immunoelectron microscopic studies.

In conclusion, attempts to synthesize flavivirus structure from negative-contrast and thin-section studies of virion morphology and from thin-section studies of virion morphogenesis have proved frustrating. The one constituent of flavivirus particles that is best defined in negative contrast is the envelope, but in the absence of visualization of budding in a thin section, characterization of the envelope must be considered incomplete. Characterization of the surface projection layer is clearly incomplete from both negative-contrast and thin-section observations. Finally, given the poor resolution of core structure in negative contrasts, the absence of any preformed core structures within infected cells, and the poor resolution of dense cores within thin-sectioned particles, it is clear that structural analysis would be presumptuous if based upon present evidence.

IV. THE *RUBIVIRUS* GENUS

Rubella was the first non-arthropod-borne virus to be shown to have morphological and morphogenetic similarities to "arboviruses"; in fact, when these similarities were noted in 1968–1969, Holmes and his colleagues brought widespread attention to the matter with the rhetorical question, "Is rubella an arbovirus?" (Holmes and Warburton, 1967; Holmes *et al.*, 1969). Because arbovirus classification was still based only upon ecological and serological relationships at that time, the question was unresolvable. Since then, characterization studies have led to the placement of rubella in a separate genus,

Rubivirus in the Togaviridae family. So far, it is the only member of the genus.

The morphological characteristics of rubella virus were established, after several false starts, over a very short time period during which papers appeared from several laboratories. Although it now seems clear that Magnusson and Norrby (1965) and Norrby (1969) first illustrated rubella particles, the degradative effect of purification and negative staining procedures was appreciated and the native state of the virus was left unresolved. Best *et al.* (1967) published the first description of rubella particles that is consistent with our present understanding. In perhaps the first major use of immunoelectron microscopy for virus identification, they found that particles which were aggregated specifically with rubella antiserum "ranged in size between 50 and 75 nm and appeared to be ragged, with an amorphous pleomorphic envelope and an internal spherical component." Holmes and Warburton (1967) then found similar particles by negative staining of preparations purified by adsorption and elution from pigeon erythrocytes. Particles which were larger (85–100 nm) and obviously osmotically damaged were obtained by Palmer and Murphy (1968) from high-titered rubella hemagglutinin-purified by isopycnic ultracentrifugation in cesium chloride.

A. Negative-Contrast Electron Microscopy

Our present understanding of the morphology of rubella virus has been advanced by successes in propagating, purifying, and stabilizing virions, as described in Chapter 21. Unfixed particles obtained from multiple-step purification procedures have a mean diameter of 75 nm, with a range of 50–100 nm depending upon the degree of spreading upon the substrate. As an indication of the magnitude of this spreading effect, Vaheri *et al.* (1969) compared unfixed and glutaraldehyde-fixed particles; the latter had a mean diameter of 60 nm (range 50–75 nm), the same size range reported with thin-sectioning. The difference between 60-nm particles and those which appear larger must represent spreading due to loss of osmotic turgor caused by preparative methods. The common presence of osmotic blebs of the virion envelope and deformations of virion shape likely reflect fragility also (Vaheri *et al.*, 1969; Liebhaber *et al.*, 1969). Virus particles consist of a membranous envelope with a surface projection layer (Fig. 43). In some cases no internal structure has been evident within virus particles (Fig. 44), whether intact and penetrated by stain or disrupted either spontaneously or by deoxycholate (Palmer and Murphy, 1968;

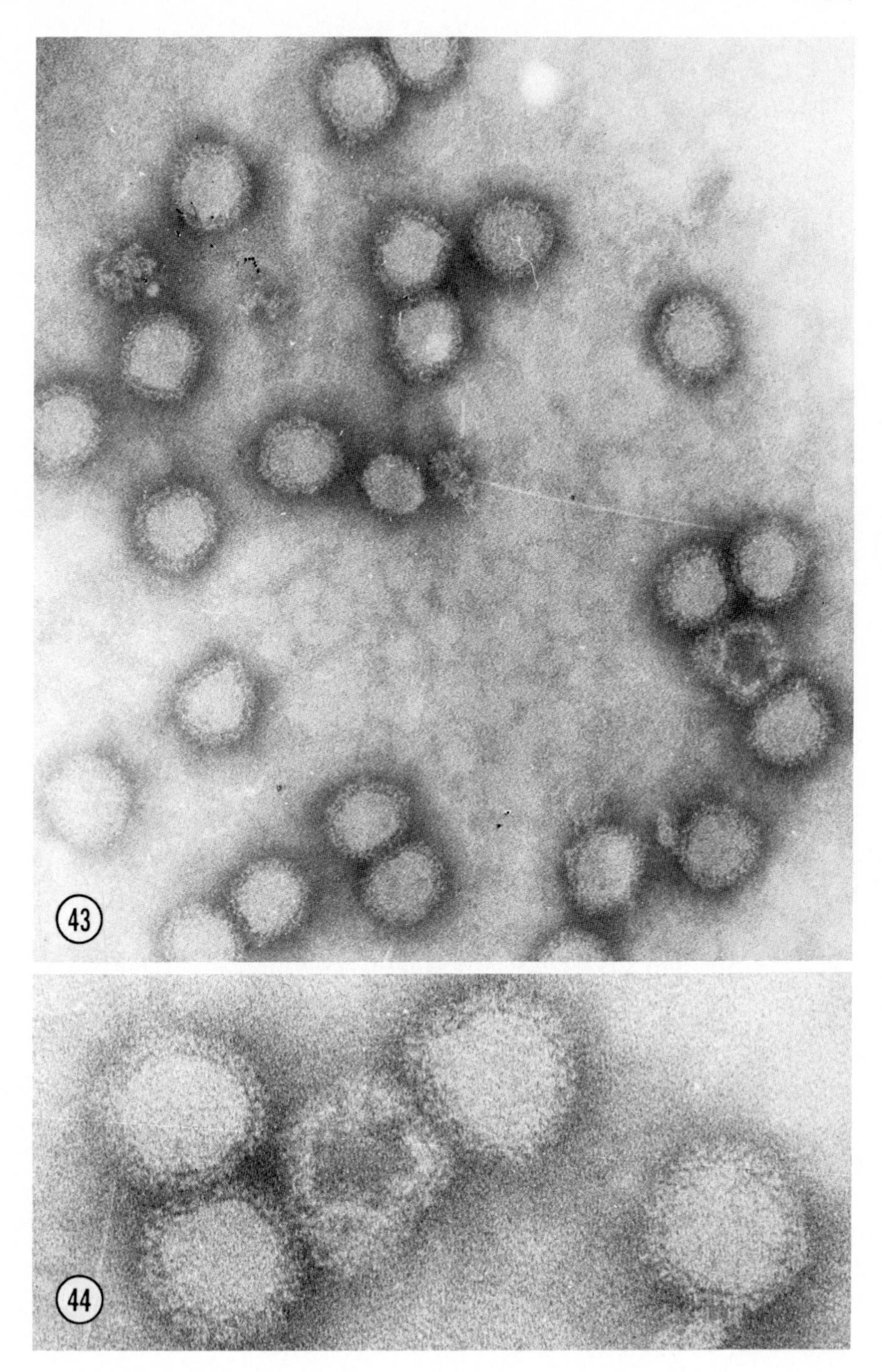
43
44

Vaheri *et al.*, 1969). However, in other cases core structures have been found, just as in the original study of Best *et al.* (1967). Horzinek *et al.* (1971) observed core particles within virions; these were released spontaneously or by treatment with 4 *M* urea. The cores were isometric, sometimes hexagonal in profile and sometimes "Hollow." They were 33 nm in diameter and had 11-nm ringlike subunits covering their surface. Similar observations were made by Payment *et al.* (1975a); degradation of virus by heat treatment (37°–56°C, 1–7 h) revealed 30-nm core structures made up of 10-nm circular subunits (Fig. 45). In both cases, even though precise symmetry could not be determined, it was considered that the core structures were true nucleocapsids. The illustrations support this contention, but at the same time they make it seem that ultimate analysis of symmetry will be most difficult.

Rubella envelope derivation from host cell membrane is clear, but there is little information on how the membrane is modified or how the surface projection layer is inserted. In most cases the virion surface has appeared fuzzy without resolution of individual spikes (Figs. 43 and 44), but Bardeletti *et al.* (1975) resolved 6- to 8-nm-long spikes which were considered to have enlarged distal ends. Payment *et al.* (1975a) found, again with heat denaturation, that the envelope layer contained 5- to 6-nm spherical sununits which formed hexagonal and pentagonal arrays. Just how these subunits, which were considered to be the soluble hemagglutinin moieties, relate to spikelike surface projections remains to be seen.

B. Thin-Section Electron Microscopy

In the years since the first depiction of rubella virus, a series of thin-section studies has appeared in which remarkably similar findings have been obtained in several host cell systems (Holmes and Warburton, 1967; Holmes *et al.*, 1968; Bonissol and Sisman, 1968; McCombs *et al.*, 1968; Murphy *et al.*, 1968a; Oshiro *et al.*, 1969; Liebhaber *et al.*, 1969; Banatvala *et al.*, 1969; Tuchinda *et al.*, 1969; Hamvas *et al.*, 1969; von Bonsdorff and Vaheri, 1969; Edwards *et al.*,

Fig. 43. Rubella virus purified according to the techniques of Vaheri *et al.* (1969) and stained with potassium phosphotungstate (pH 7.2). The surface projection layer is about 8-nm thick, and the total particle diameter is 75 nm. ×150,000. Micrograph courtesy of Dr. C.-H. von Bonsdorff.

Fig. 44. Rubella virus, as in Fig. 43, showing that stain penetration does not necessarily reveal a core structure. ×300,000. Micrograph courtesy of Dr. C.-H. von Bonsdorff.

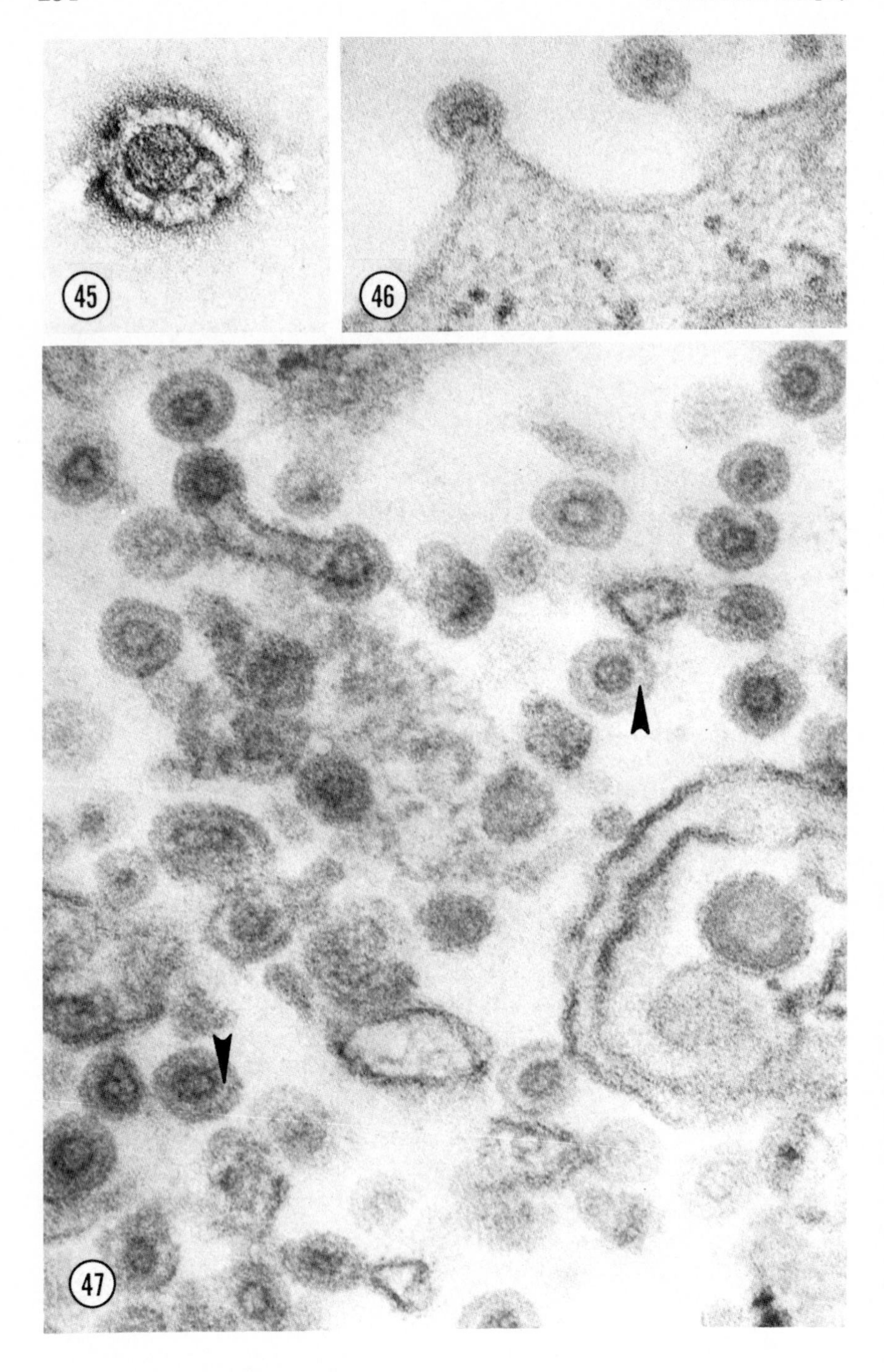
45
46
47

1969; Higashi *et al.*, 1969; Korolev *et al.*, 1973; Matsumoto and Higashi, 1974; Kouri *et al.*, 1974; Payment *et al.*, 1975b).

The findings from these thin-section studies may be compiled into an overview of rubella morphology and morphogenesis. Virus particles are round or oval and consist of a substantial electron-dense core structure surrounded by a rather lucent zone and a unit membrane envelope with a thin projection layer (Figs. 46 and 47). The mean particle diameter is 60 nm (the range is approximately 50–75 nm), and the mean core diameter is 30 nm. The smaller ratio of core diameter to total diameter and the space between the core and the envelope account for the characteristic difference between the appearance of rubella and alphaviruses or flaviviruses in thin section (Fig. 47). Although anomalously shaped alphavirus and flavivirus particles are rare, rubella particle shape variations are common. Particles may be elongated, occasionally reaching lengths of more than 200 nm while maintaining the usual 60-nm diameter. Particles may have multiple cores or imperfectly formed cores (crescent-shaped, open ring-shaped, etc.). The proportion of anomalous particles is affected by host cell variations (Tuchinda *et al.*, 1969).

One attempt was made to further analyze rubella virion structure in thin section (Matsumoto and Higashi, 1974). Because core structures within virus particles appeared to be formed with "a radial association of several morphological subunits," rotational enhancement was used to judge symmetry. Images with 12-fold symmetric enhancement were interpreted as representing a $T = 3$ icosahedral arrangement of 32 capsomers. The authors, however, state that confirmation from negative staining is needed before the merits of this approach can be judged.

Rubella virions mature via a clearly discernible budding process from host cell membranes; core formation occurs coincidentally with the process of membrane deformation in early budding (Fig. 46). This characteristic is similar to the coordinated nucleocapsid formation and budding of type-C oncoviruses and not like that of alphaviruses in which nucleocapsid formation is separate from virion maturation. The modification of host cell membrane at budding sites has been studied

Fig. 45. Rubella virus, degraded by heat treatment, reveals a 30-nm core structure. ×260,000. Micrograph courtesy of Dr. P. Payment.

Fig. 46. Rubella virus infection of a BHK-21 cell. Virus particle budding from plasma membrane is in synchrony with core formation. ×178,000.

Fig. 47. Rubella virus infection of a BHK-21 cell culture. The electron-dense core structure is separated by a lucent zone (arrowheads) from the unit membrane envelope; this lucent zone gives rubella virus a distinct appearance in thin section. ×192,000.

with ferritin-labeled antibodies (Oshiro *et al.*, 1969; Matsumoto and Higashi, 1974). In the large series of thin-section studies of rubella virus, the only significant inconsistency concerns host cell membrane systems involved in virus maturation. In several studies the Golgi, endoplasmic reticulum, and plasma membranes budded virus but, in other cases, despite the use of the same few host cell lines (BHK-21, Vero, and BS-C-1) virus maturation was limited to internal membrane systems only. This variance has not been explained, but Korolev (1977) has proposed that serial passage of virus in cultured cells results in a shift of most budding from plasma membranes to the endoplasmic reticulum.

As an overall impression of the descriptions and interpretations of rubella morphology and morphogenesis from many laboratories, there is a sense of more uniformity than is the case with any of the other non-arthropod-borne togaviruses. Perhaps it is this uniformity of findings which has had a dampening effect upon continuing study, but whatever the reason, there have been few recent reports. This is disappointing, since the descriptive morphological studies have all fallen short of resolution of fundamental virion structure. As with other togaviruses, new approaches are needed.

V. THE *PESTIVIRUS* GENUS

Primarily as a result of the work of Horzinek and his colleagues (Horzinek *et al.*, 1971; Horzinek, 1973a,b), several important pathogens of domestic animals are known to share physicochemical and morphological properties with arthropod-borne togaviruses. In recognition of this, the agent of bovine virus diarrhea (formerly called mucosal disease virus) and its serorelative, hog cholera virus (formerly also called European swine fever virus), have been placed in a separate genus, the *Pestivirus* genus, of the Togaviridae family (Fenner, 1976).

A. Negative-Contrast Electron Microscopy

Despite problems of virion fragility in concentration–purification procedures, the size and morphology of hog cholera virus in negative-contrast preparations were reported nearly simultaneously from several laboratories in 1967–1968 (Horzinek *et al.*, 1967; Mayr *et al.*, 1967, 1968; Ritchie and Fernelius, 1967, 1968; Cunliffe and Rebers, 1968a,b; Ushimi *et al.*, 1969). As presently characterized, hog cholera virus particles are essentially spherical (Fig. 48) (Laude, 1977; Enzmann and Weiland, 1978), but asymmetrically shaped particles are often seen as a result of envelope deformation. Particle diameter is between 40 and 60 nm and, as with other togaviruses, the larger particles likely represent spreading upon the substrate and the smaller particles may represent the virion native state. A surface projection layer has been described; it has usually appeared to consist of indeterminate projection units 6- to 8-nm long (Ritchie and Fernelius, 1967; Enzmann and Rehberg, 1977; Enzmann and Weiland, 1978) (Fig. 49). In some cases only smooth, membrane-bound particles have been seen, probably indicating the ease with which projections are lost. The envelope layer of hog cholera virus has been well resolved, usually appearing as a smooth unit membrane, but in one case it was beaded (a "rosary" envelope structure; Horzinek *et al.*, 1971). Core particles, 25–30 nm in diameter, are observed both within virions and free as a result of spontaneous or chemically induced disruption of virions. These particles are round, or occasionally polygonal, and in some cases have a central space 13 nm in diameter (Horzinek *et al.*, 1971).

Bovine virus diarrhea virus is similar in morphology to hog cholera virus (Hafez *et al.*, 1968; Ritchie and Fernelius, 1969; Maess and Reczko, 1970; Horzinek *et al.*, 1971) (Fig. 50). The diameter of the virus is 40–60 nm, although a wider range was reported in some earlier reports. Spontaneously or chemically (4 *M* urea or saponin) released core particles, 25–30 nm in diameter, appear isometric or hexagonal in profile and have a 13-nm central hole (Horzinek *et al.*, 1971). The surface of the virus has indistinct surface projections which are easily lost, as is the case with hog cholera virus. The virus envelope has in one case been resolved into beaded subunits (Horzinek *et al.*, 1971).

B. Thin-Section Electron Microscopy

In thin-sectioned ultracentrifuge pellets from bovine diarrhea virus preparations, Hermodsson and Dinter (1962) found 40-nm particles. Similar particles, 40–53 nm in diameter, have been found in thin sections of hog cholera virus-infected cells in culture and in infected swine tissues (Cheville and Mengeling, 1969; Scherrer *et al.*, 1970; Schultz, 1971). The excellent micrographs published by Scherrer *et al.* (1970) of pig kidney (PK-15) cells infected with hog cholera virus indicate that this virus is remarkably similar to flavivirus in morphology and morphogenetic characteristics (Fig. 51). Virus particles

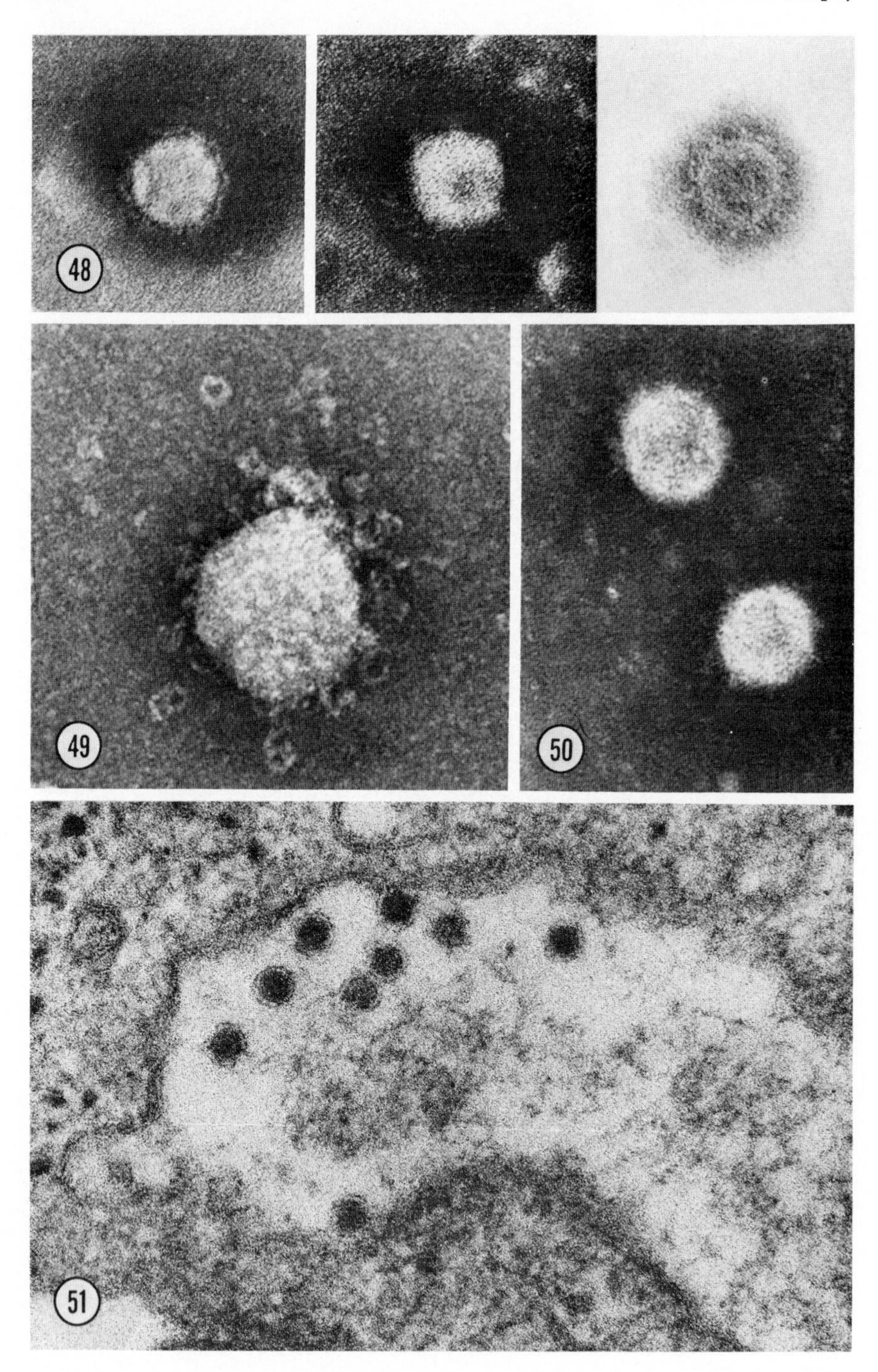
48
49
50
51

clearly consist of a dense nucleoid (29–33 nm), often of polygonal shape, and a narrow closely adherent envelope layer. Virus particles have been found extracellularly and within the cisternae of membranous organelles of infected cells; the evidence of budding is as suggestive but as elusive as with flaviviruses; that is, it is impossible to detect continuity between the lamellae of host cell membranes and virus envelopes (Fig. 52).

Taken together, these observations were considered by Horzinek *et al.* (1971) to be indicative of icosahedral symmetry of both member viruses of the *Pestivirus* genus. However, characterizing these viruses this far has been so difficult that it is unlikely that further structural analysis will be forthcoming soon. Horzinek (1973a) stated: "The difficulty in studying the symmetry of nucleocapsid surface patterns is due to the scarcity of one-sided images in negatively contrasted preparations of purified core particles and to their poor staining; since the viral lipid interacts directly with the capsid protein, remaining hydrophobic groups probably interfere with exhaustive embedding [of stain]." The question of how to offset this difficulty has not been resolved.

VI. UNCLASSIFIED TOGAVIRUSES

A. Equine Arteritis Virus

Equine arteritis virus has been compared directly with the other non-arthropod-borne togaviruses and, when taken together, its morphological and physicochemical properties have not indicated place-

---

Fig. 48. Hog cholera virus, fixed with formaldehyde and stained with sodium phosphotungstate, showing the envelope layer and surface projections. Composite. ×280,000. Micrograph courtesy of Dr. P.-J. Enzmann.

Fig. 49. Hog cholera virus particle with the surface projection layer resolved into circular structures such as those seen with flaviviruses and with rubella, equine arteritis, and lactic dehydrogenase viruses. ×350,000. Micrograph courtesy of Dr. A. E. Ritchie.

Fig. 50. Bovine virus diarrhea virus particles with a typical fuzzy surface which does not allow the resolution of individual projections. ×300,000. Micrograph courtesy of Dr. A. E. Ritchie.

Fig. 51. Hog cholera virus infection of a PK (15) (porcine kidney) cell culture. Virus particles, 44 nm in diameter, closely resemble flaviviruses in having a very dense core and a closely adherent thin envelope. ×160,000. Micrograph courtesy of Dr. R. Scherrer.

ment in one of the presently recognized genera (Horzinek *et al.*, 197-; Horzinek, 1973a,b). An apparent increased stability of equine arteritis virus, relative to that of hog cholera or bovine virus diarrhea virus, has led to better resolution and to more uniformity among descriptions (Magnusson *et al.*, 1970; Maess *et al.*, 1970a,b; Maess and Reczko, 1970; Horzinek *et al.*, 1971). Virus particles are spherical and 50–70 nm in diameter (Fig. 53). Core structures within virions penetrated by stain (spontaneously or after neuraminidase treatment) are 35 nm in diameter with a 15- to 17-nm central space or central substructure (Horzinek *et al.*, 1971). Isolated core particles (detergent treatment and rate zonal ultracentrifugation) clearly show a central space penetrated by negative-contrast media (Fig. 54). The virus envelope is clearly discerned to have a unit membrane structure, and ringlike structures 12–14 nm in diameter have been found situated upon the envelope layer (Hyllseth, 1973; Ellens, 1977).

In thin sections of infected cell cultures, equine arteritis virus has been found within cytoplasmic vacuoles; this virus is 40–45 nm in diameter with a core diameter of 35 nm (Breese and McCollum, 1970, 1971, 1973; Estes and Cheville, 1970) (Fig. 55). Particles accumulate in cytoplasmic vesicles as a result of budding (Crawford and Davis, 1970).

The information available emphasizes the predicament faced in fitting equine arteritis virus into a genus of the Togaviridae family. Although the virus is larger and has a more substantial core structure than pestiviruses, it does not quite resemble either alphaviruses or rubella virus. After comparing the morphological, morphogenetic, physiocochemical, and biological characteristics of non-arthropod-borne togaviruses, Horzinek (1977) recently noted that the similarities between equine arteritis virus, lactic dehydrogenase virus, and simian hemorrhagic fever virus might eventually lead to the construction of another genus in the family Togaviridae.

Fig. 52. Hog cholera virus infection of PK (15) cells. As with flaviviruses, virus maturation occurs upon host cell membranes without evidence of preformed cores or any intermediate stages of a budding process. Composite. ×200,000. Micrograph courtesy of Dr. R. Scherrer.

Fig. 53. Equine arteritis virus, stained with uranyl acetate, showing that the surface layer is composed of ring structures (arrowheads). ×300,000 (approximate). Micrograph courtesy of Dr. D. J. Ellens.

Fig. 54. Equine arteritis virus core structures, prepared from NP-40-treated virus and isolated on a sucrose gradient. The interiors of the 36-nm spherical structures are penetrated by the uranyl acetate stain. ×180,000. Micrograph courtesy of Dr. M. C. Horzinek.

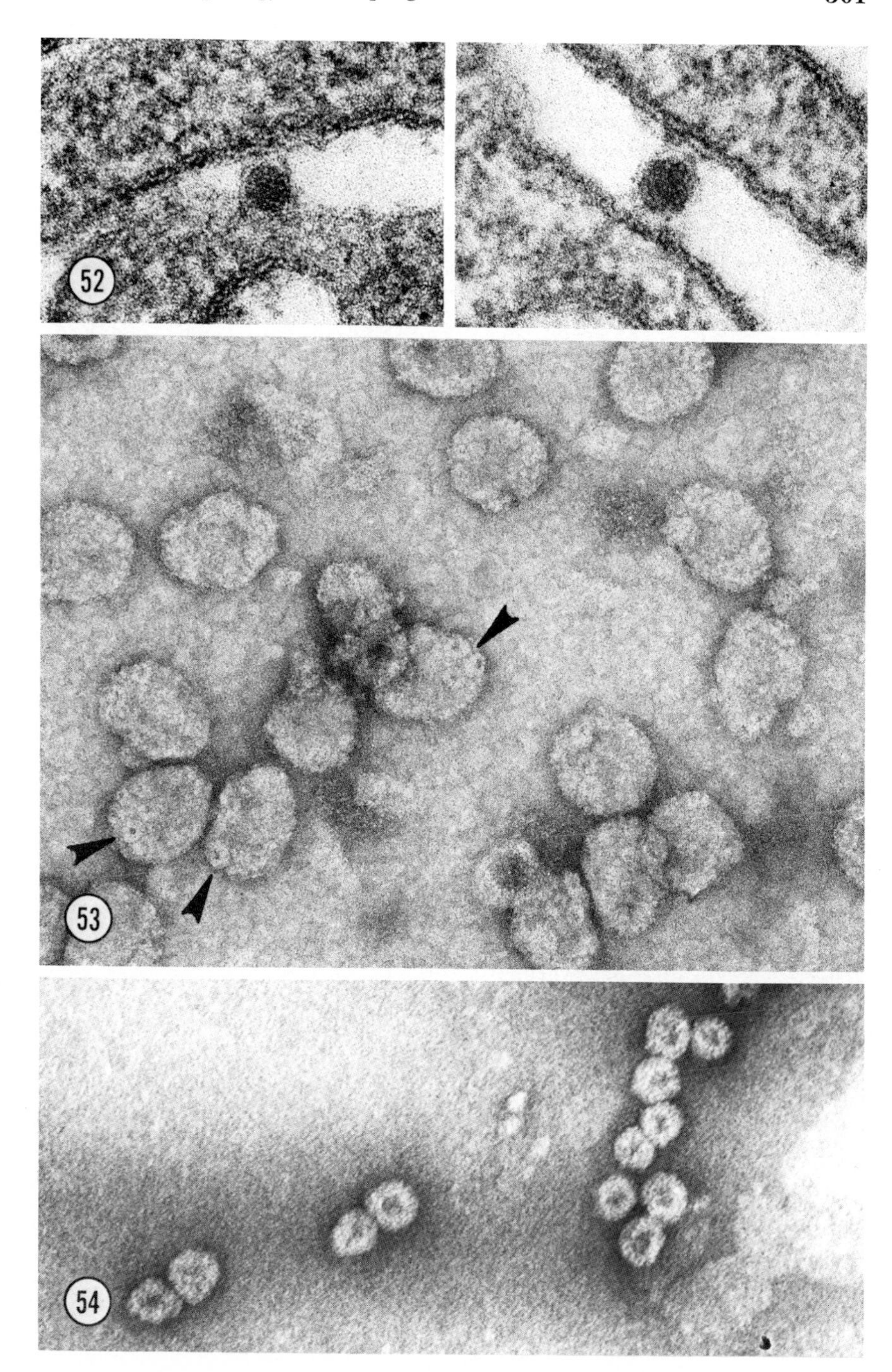
52
53
54

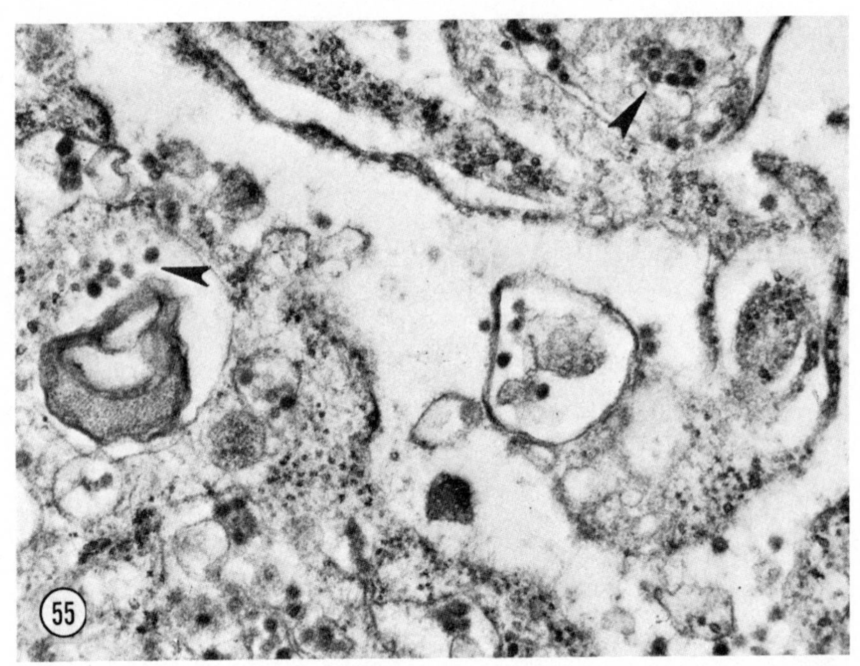

Fig. 55. Equine arteritis virus infection of an equine dermis (NBL-6) cell culture. Virus particles (arrowheads) are present within vacuoles of a cell undergoing severe cytopathic change. ×55,000. Micrograph courtesy of Dr. S. S. Breese, Jr.

B. Lactic Dehydrogenase Virus

Lactic dehydrogenase virus morphology and morphogenesis have been studied in detail because of interesting immunological and pathophysiological aspects of the murine infection (see Chapter 21). The extremely high titers reached in *in vitro* and *in vivo* infection (>10^{10} ID_{50}/ml) have made this virus somewhat easier to characterize than other non-arthropod-borne togaviruses (Bladen and Notkins, 1963; Notkins, 1965; DeThe and Notkins, 1965; Almeida and Mims, 1974; Horzinek *et al.*, 1975; Brinton-Darnell and Plagemann, 1975; Brinton-Darnell *et al.*, 1975). Nevertheless, virus morphology is still debated, since shape variations have occurred in some instances whether virus has originated from mouse plasma, mouse ascitic fluid, or primary cultures of mouse tissues; this shape variation does not seem to be simply related to either concentration and purification procedures or to negative staining conditions. When virions have been spherical, diameters have been between 70 and 85 nm. These particles have been shown to consist of a unit membrane envelope and a spherical core structure 35 nm in diameter (Fig. 56). In most cases virus particles have appeared to have a smooth surface, but Brinton-Darnell and Plagemann (1975) have found 8- to 14-nm rings at the surface of disintegrating particles. No symmetric subunits have been evident on core particles (Horzinek *et al.*, 1975). When virions appeared elongated or bottle-shaped, the particle surface was also smooth, but core structure was not resolved.

Thin-section electron microscopy has added to the controversy over virion shape; again, in some cases elongated particles have been found in infected cell cultures and in sectioned ultracentrifuge pellets, but in other cases only spherical particles have been seen (Fig. 57–59). Sectioned spherical particles are 45–55 nm in diameter. Elongated particles are of the same diameter and up to 85 nm in length. Evidence of the reality of elongated particles includes (1) the observa-

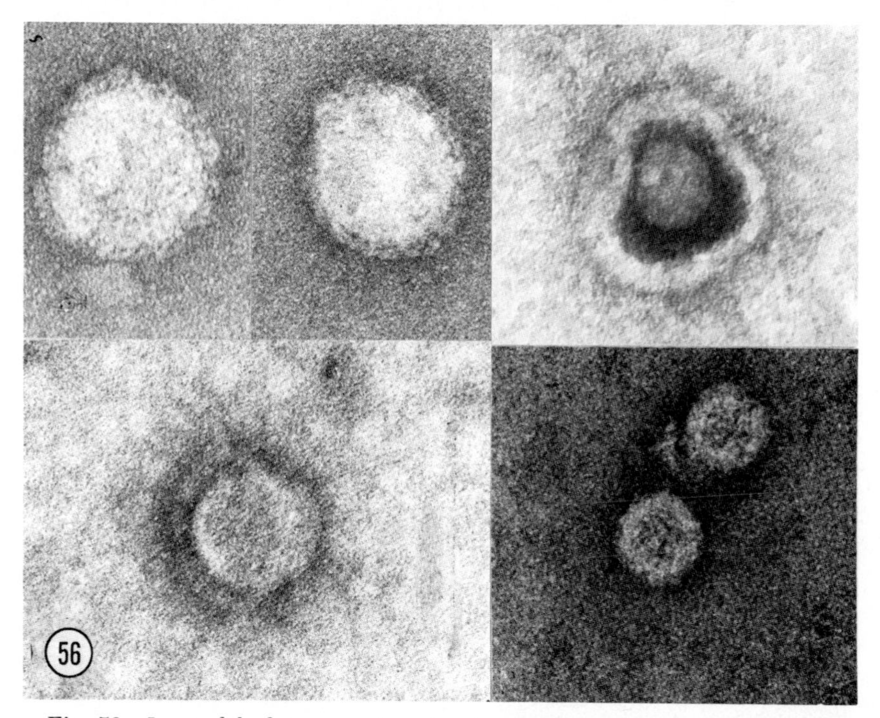

Fig. 56. Lactic dehydrogenase virus showing surface projections (left), core structure within envelope devoid of projections (right, top), and free core structures released by treatment with NP-40 and purified by centrifugation in a sucrose gradient (right, bottom). Composite. ×300,000 (approximate). Micrographs courtesy of Dr. M. C. Horzinek.

tion that the dense core within sectioned elongated particles is also elongated, and (2) the observation of elongated particles *in situ* upon host cell membranes of thin-sectioned cells in culture—a circumstance in which fixation artifacts are considered minimal. Evidence that elongated particles are artifactual includes (1) the prevention of shape variation by prefixation before ultracentrifugation and sectioning of virus pellets, and (2) the possibility that envelope deformations are easily caused by storage or by the action of antibody (Almeida and Mims, 1974).

Lactic dehydrogenase virus morphogenesis involves clearly discernible budding from the lining membranes of cytoplasmic vesicles but not from plasma membranes (Brinton-Darnell *et al.*, 1975). Core structure construction appears to occur at the bud site; the presence of cores within nascent budding virions, but not elsewhere, most closely resembles rubella morphogenesis. However, as stated above, overall physicochemical and biological characteristics favor the placement of lactic dehydrogenase virus together with equine arteritis and simian hemorrhagic fever viruses in a separate genus of the family Togaviridae; further ultrastructural study would be welcome to complement nonmorphological data.

C. Cell-Fusing Agent of *Aedes* Cell Cultures

An agent isolated from *Aedes aegypti* cell cultures by Stollar and Thomas (1975) has been shown to be antigenically distinct and yet like flaviviruses in several biological properties. Virus particles are 45–52 nm in diameter and contain a 23- to 28-nm core structure (Igarashi *et al.*, 1976) (Fig. 60). Particles have a unit membrane envelope and surface projections. In thin sections, virus particles have been found within cisternae of the endoplasmic reticulum of infected

---

Fig. 57. Lactic dehydrogenase virus. Thin section of viremic mouse plasma showing the elliptical or oblong shape of most particles. ×60,000. Micrograph courtesy of Dr. A. L. Notkins.

Fig. 58. Lactic dehydrogenase virus. Thin section of a virus pellet from a multistep purification scheme starting with viremic plasma. The uniformly spherical particles have a diameter of 55 nm. ×120,000. Micrograph courtesy of Dr. M. Brinton.

Fig. 59. Lactic dehydrogenase virus. Thin section of a virus pellet from a multistep purification scheme starting with mouse macrophage culture fluid. Particles exhibit an elliptical shape and contain a core structure. ×240,000. Micrograph courtesy of Dr. M. Brinton.

Fig. 60. Cell-fusing agent of *Aedes* cells, purified and stained with uranyl acetate. The 52-nm particles have surface projections, a unit membrane envelope, and a 28-nm core structure. ×140,000. Micrograph courtesy of Drs. A. Igarashi and V. Stollar.

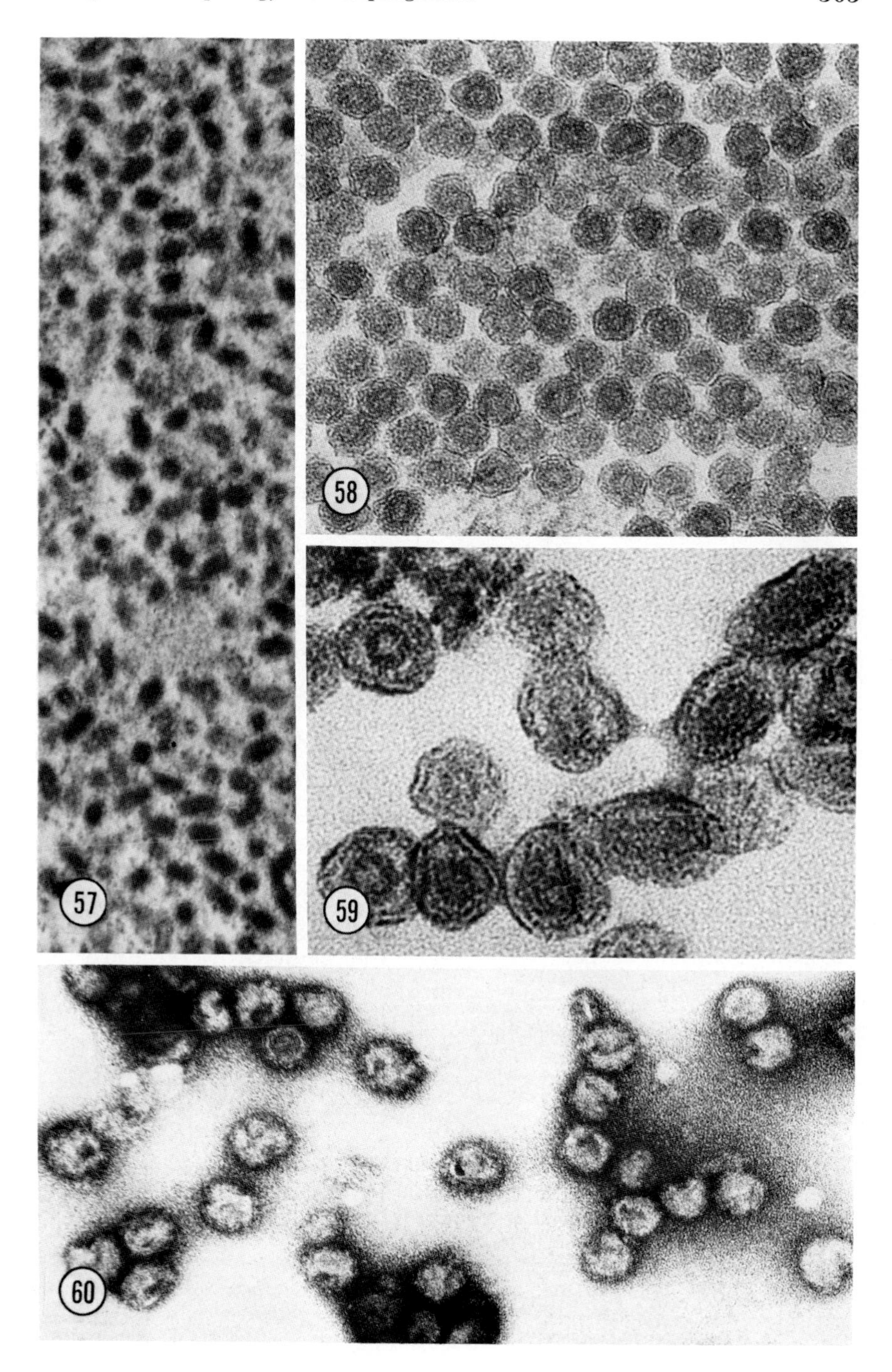
57
58
59
60

mosquito cells, but budding events have appeared as elusive as with flaviviruses (Fig. 61). Igarashi and his colleagues concluded that this virus was not a flavivirus, and its ultimate placement in the Togaviridae family has not been considered further.

D. Simian Hemorrhagic Fever Virus

Several large epizootics of hemorrhagic fever in *Macaca* monkey species colonized in several countries have been attributed to an antigenically independent togavirus (Shelokov *et al.*, 1971). Despite physicochemical studies which indicate closest relationship to lactic dehydrogenase virus, formal placement of the virus within the Togaviridae family has not yet been considered. Virus particles, propagated in monkey cell cultures and purified by isopycnic centrifugation, are spherical, enveloped, and 45–50 nm in diameter (Trousdale *et al.*, 1975) (Fig. 62). They have a core structure of 25-nm diameter which does not have apparent subunit symmetry. By thin-sectioning, virus particles occurring within cytoplasmic vacuoles of cultured monkey cells are seen to be 40–45 nm in diameter and have a 22- to 25-nm lucent center (Wood *et al.*, 1970; Tauraso *et al.*, 1971). Further thin-section electron microscopic study would be valuable to support the contention that simian hemorrhagic fever virus shares fundamental properties with lactic dehydrogenase and equine arteritis viruses.

VII. COMPARISONS AND CONCLUSIONS

When the Togaviridae family was formed, an overriding consideration was that the two groups of arthropod-borne viruses, the alphaviruses (group-A arboviruses) and the flaviviruses (group-B arboviruses) should be kept together in some way. The many biological, ecological, and morphological similarities between the viruses supported this contention, and there was little contradictory physicochemical evidence (Casals, 1971; Mussgay *et al.*, 1975). However, when the family was formalized (Fenner, 1976; Porterfield *et al.*,

Fig. 61. Cell-fusing agent infection of *Aedes* cells. Virus particles accumulating within cisternae of endoplasmic reticulum apparently as a result of a budding process. ×60,000. Micrograph courtesy of Drs. A. Igarashi and V. Stollar.

Fig. 62. Simian hemorrhagic fever virus purified from an infected rhesus monkey (MA-104) cell culture and stained with uranyl acetate. Spherical particles, 48 nm in diameter, consist of a rather smooth envelope surrounding a 25-nm hollow core structure (inset). ×150,000. Inset: 300,000. Micrograph courtesy of Dr. M. D. Trousdale.

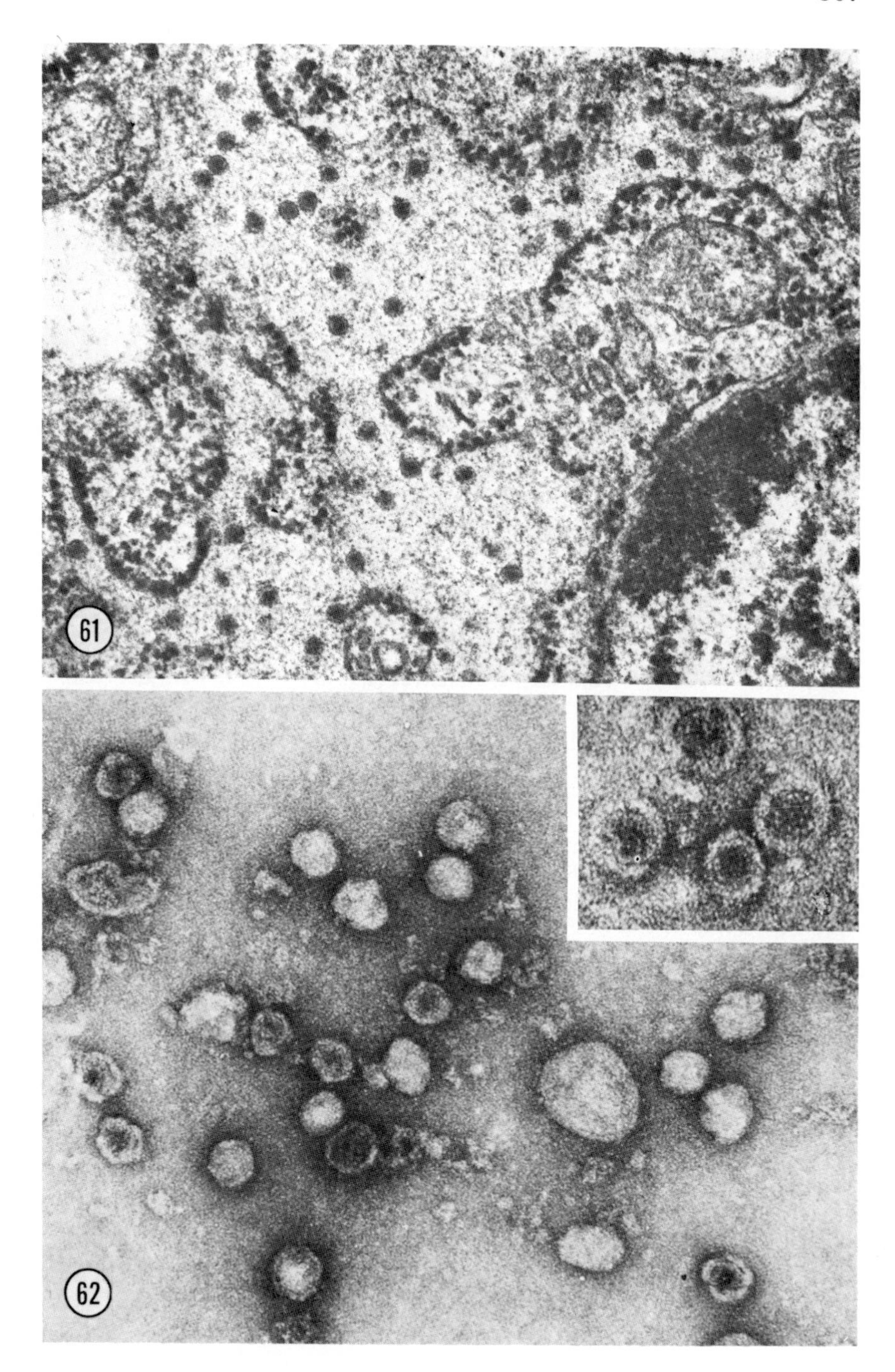
61
62

1978), characteristics were included in the definition which had not been proved to be entirely representative. Rather, the definition showed a strong sense of anticipated uniformity even as the family was being expanded to include all the small, enveloped RNA-containing viruses with proved or suspected cubic nucleocapsid symmetry and a single-stranded genome of about 4×10^6 daltons with messenger polarity (Fenner, 1976).

At the time the Togaviridae family was created the definition had been addressed in relation to only one virus, Sindbis virus (Horzinek and Mussgay, 1969). In the ensuing years, it has been shown repeatedly that other alphaviruses have a very distinct core structure, but resolution of a $T = 3$, 32-capsomer icosahedral nucleocapsid construction, as proposed by Horzinek and Mussgay, has been elusive (Söderlund *et al.*, 1975). This difficulty pales next to the lack of success in resolving the fundamental nature of the core structure of flaviviruses, rubella, or pestiviruses. A question must, therefore, be asked regarding the amount of structural detail that should be required before definitions for taxa are established. On the one hand, standards should be set in line with the reality of observations—all viruses will not appear as structured as papovaviruses or adenoviruses. On the other hand, overinterpretation of observations to imply structural detail that does not exist has no value. This question may relate more pointedly to the core structure of togaviruses than to any other consideration in virus anatomy at this time.

How much structural detail must be illustrated before it is prudent to use the term "icosahedral nucleocapsid"? The principles of icosahedral symmetry have been clearly expounded (Caspar, 1965); the morphological subunits of an isometric particle must be identified and the arrangement of these subunits around two adjoining 5-coordinated subunits (pentamers) must be resolved. In such an analysis, the presence of icosahedral symmetry is proved at the same time as is the detailed construction (triangulation number). There seems to be no halfway analysis possible; virus construction is either resolved precisely or the terms used to describe symmetric construction have no meaning. This becomes a real problem with togavirus core structures. In their native state most togaviruses, even most alphaviruses, have cores which appear as spherical or polyhedral forms with rather smooth surfaces. Even when outline shapes are hexagonal or pentagonal, the lack of morphological subunit distinction makes further analysis difficult. Horzinek (1973a) has commented upon these core characteristics: One-sided images of core particles in negatively stained preparations are scarce, core particles stain poorly, hydrophobic groups interfere with stain penetration, and core particles are rather flexible. These characteristics have made the definition of morphological subunits most difficult, and the failures of methods other than negative staining add to the sense of frustration.

Controversy regarding the core of togaviruses does not extend to the envelope. It has generally been agreed that togavirus envelopes are derived from host cell membranes and that the character of envelopes is similar throughout the family. This is the case even with viruses which have not been proved to bud via progressive stages of membrane deformation (flaviviruses and probably pestiviruses).

Surface projections are present upon the envelopes of all togaviruses which exhibit pH-dependent hemagglutination (alphaviruses, flaviviruses, and rubella) (Horzinek, 1975). Other togaviruses often appear to be smooth-surfaced, but at present there seems to be agreement that projection loss may be an artifact of storage, purification, or other adverse conditions (Horzinek, 1975). The character of surface projections has appeared to vary considerably, but at present it is not possible to distinguish real differences among viruses from artifactual effects caused by purification and staining. Comparative study of the surface of several representative togaviruses should be carried out under controlled conditions in a single laboratory.

The details of virion morphology relate directly to comprehensive physicochemical characterizations upon which new relationships among viruses are built. Viruses which are identical morphologically might be expected to have very similar physicochemical properties, but physicochemical similarities cannot be predicted when viruses are only generally similar in morphology. Thus, alphaviruses and flaviviruses were each found to form a precise group or taxon, whether comparisons were morphological or physicochemical. Serological cross-reactivity was found to be evidence of only the closest degree of relationship. Morphological similarities between hog cholera and bovine virus diarrhea virus, and differences between these viruses and other togaviruses (except perhaps flaviviruses), supported construction of the *Pestivirus* genus. The morphological (and morphogenetic) difference between rubella virus and the other togaviruses supported construction of a separate *Rubivirus* genus. This approach can be extended (see Chapter 21; Horzinek, 1975): Morphological similarities between lactic dehydrogenase virus, equine arteritis virus, and simian hemorrhagic fever virus favor consideration that these viruses may form another genus of the Togaviridae family (Horzinek, 1977). Finally, the close morphological similarity between the cell-fusing agent and flaviviruses must be recognized, but caution must be exercised before a first addition to the *Flavivirus* genus on nonserological grounds is considered (Igarashi *et al.*, 1976).

ACKNOWLEDGMENTS

The author wishes to express particular thanks to Dr. C. H. von Bonsdorff for his help with this chapter. He also wishes to thank Drs. June D. Almeida, Walter E. Brandt, S. S. Breese, Margo Brinton-Darnell, F. Bürki, P. J. Enzmann, Scott B. Halstead, Marian C. Horzinek, Susumu Hotta, Harvey Liebhaber, Bernd Liess, Erling Norrby, Abner L. Notkins, Lyndon S. Oshiro, Pierre Payment, A. E. Ritchie, Raoul Scherrer, Victor Stollar, Dennis W. Trent, and M. D. Trousdale for their help in gathering information and illustrations.

Finally, he wishes to thank Mrs. Alyne K. Harrison, Miss Patricia Echols, and Mrs. Barbara Gary for their help in preparing the manuscript.

REFERENCES

Abdelwahab, K. S. E., Almeida, J. D., Doane, F. W., and McLean, D. M. (1964). *Can. Med. Assoc. J.* **90**, 1068.
Acheson, N. H., and Tamm, I. (1967). *Virology* **32**, 128.
Acheson, N. H., and Tamm, I. (1970). *Virology* **41**, 306.
Almeida, J. D., and Mims, C. A. (1974). *Microbios* **10**, 175.
Banatvala, J. E., Best, J. M., and Kistler, G. S. (1969). *Symp. Ser. Immunobiol. Stand.* **11**, 161.
Bang, F. B., and Gey, G. O. (1949). *Proc. Soc. Exp. Biol. Med.* **71**, 78.
Bang, F. B., and Gey, G. O. (1952). *Bull. Johns Hopkins Hosp.* **91**, 427.
Bardeletti, G., Kessler, N., and Aymard-Henry, M. (1975). *Arch. Virol.* **49**, 175.
Bauer, S. P., Obijeski, J. F., and Murphy, F. A. (1977). Unpublished observations.
Bell, T. M., Field, E. J., and Narang, H. K. (1971). *Arch. Gesamte Virusforsch.* **35**, 183.
Berge, T. O., ed. (1975). "International Catalogue of Arboviruses," 2nd ed., DHEW Publ. No. (CDC) 75-8301. U.S. Dep. Health, Educ. Welfare, Public Health Serv., Washington, D.C.
Bergold, G. H., and Weibel, J. (1962). *Virology* **17**, 554.
Best, J. M., Banatvala, J. E., Almeida, J. D., and Waterson, A. P. (1967). *Lancet* **ii**, 237.
Birdwell, C. R., Strauss, E. G., and Strauss, J. H. (1973). *Virology* **56**, 429.
Bladen, H. A., Jr., and Notkins, A. L. (1963). *Virology* **21**, 269.
Blinzinger, K. (1972). *Ann. Inst. Pasteur, Paris* **123**, 497.
Blinzinger, K., and Müller, W. (1970). *Dtsch. Z. Nervenheilkd.* **197**, 18.
Blinzinger, K., Schaltenbrand, G., Müller, W., and Anzil, A. P. (1973). *Ann. Microbiol. (Paris)* **124A**, 123.
Bonissol, C., and Sisman, J. (1968). *C. R. Acad. Sci., Ser. D.* **267**, 1337.
Boulton, P. S., and Webb, H. E. (1971). *Brain* **94**, 411.
Boulton, P. S., Webb, H. E., Fairbairn, G. E., and Illavia, S. J. (1971). *Brain* **94**, 403.
Boulton, R. W., and Westaway, E. G. (1976). *Virology* **69**, 416.
Brandt, W. E., and Russell, P. K. (1975). *Infect. Immun.* **11**, 330.

Brandt, W. E., Cardiff, R. D., and Russell, P. K. (1970). *J. Virol.* **6,** 500.
Breese, S. S., and McCollum, W. H. (1970). *Proc. Int. Conf. Equine Infect. Dis., 2nd, Paris, 1969* pp. 133–139.
Breese, S. S., and McCollum, W. H. (1971). *Arch. Gesamte Virusforsch.* **35,** 290.
Breese, S. S., and McCollum, W. H. (1973). *Proc. Int. Conf. Equine Infect. Dis., 3rd, Paris, 1972* pp. 273–281.
Brinton-Darnell, M., and Plagemann, P. G. W. (1975). *J. Virol.* **16,** 420.
Brinton-Darnell, M., Collins, J. K., and Plagemann, P. G. W. (1975). *Virology* **65,** 187.
Brown, D. T., Waite, M. R. F., and Pfefferkorn, E. R. (1972). *J. Virol.* **10,** 524.
Bykovsky, A. F., Yershov, F. I., and Zhdanov, V. M. (1969). *J. Virol.* **4,** 496.
Bykovsky, A. F., Zhdanov, V. M., Vryvaev, L. V., and Ershov, F. I. (1971). *Vopr. Virusol.* **16,** 196.
Calberg-Bacq, C.-M., Rentier-Delrue, F., Osterrieth, P. M., and Duchesne, P. Y. (1975). *J. Ultrastruct. Res.* **53,** 193.
Calisher, C. H., Davie, J., Coleman, P. H., Lord, R. D., and Work, T. H. (1969). *Am. J. Epidemiol.* **89,** 211.
Cardiff, R. D., and Lund, J. K. (1976). *Infect. Immun.* **13,** 1699.
Cardiff, R. D., Russ, S. B., Brandt, W. E., and Russell, P. K. (1973). *Infect. Immun.* **7,** 809.
Casals, J. (1971). *In* "Comparative Virology" (K. Maramorosch and E. Kurstak, eds.), pp. 307–333. Academic Press, New York.
Casals, J., and Brown, L. V. (1954). *J. Exp. Med.* **99,** 429.
Caspar, D. L. D. (1965). *In* "Viral and Rickettsial Infections of Man" (F. L. Horsfall and I Tamm, eds.), 4th ed., pp. 51–93. Lippincott, Philadelphia, Pennsylvania.
Catanzaro, P. J., Brandt, W. E., Hogrefe, W. R., and Russell, P. V. (1974). *Infect. Immun.* **10,** 381.
Chain, M. M. T., Doane, F. W., and McLean, D. M. (1966). *Can. J. Microbiol.* **12,** 895.
Chatterjee, S. N., and Sarkar, J. K. (1965). *J. Exp. Biol.* **3,** 227.
Cheville, N. F., and Mengeling, W. L. (1969). *Lab. Invest.* **20,** 261.
Chippaux-Hyppolite, C., Choux, R., Olmer, H., and Tamalet, J. (1970). *C. R. Acad. Sci., Ser. D* **270,** 3162.
Compans, R. W. (1971). *Nature (London), New Biol.* **229,** 114.
Crawford, T. B., and Davis, W. C. (1970). *Fed. Proc., Fed. Am. Soc. Exp. Biol.* **29,** 286.
Cunliffe, H. R., and Rebers, P. A. (1968a). *Can. J. Comp. Med.* **32,** 409.
Cunliffe, H. R., and Rebers, P. A. (1968b). *Can. J. Comp. Med.* **32,** 486.
Dalton, S. (1972). *Ann. Inst. Pasteur, Paris* **123,** 489.
David-West, T. S., Labzoffsky, N. A., and Hamvas, J. J. (1972). *Arch. Gesamte Virusforsch.* **36,** 372.
Demsey, A., Steeve, R. L., Brandt, W. E., and Veltri, B. J. (1974). *J. Ultrastruct. Res.* **46,** 103.
DeThe, G., and Notkins, A. L. (1965). *Virology* **26,** 512.
Edwards, M. R., Cohen, S. M., Bruno, M., and Deibel, R. (1969). *J. Virol.* **3,** 439.
Ellens, D. J. (1977). Personal communication.
Enzmann, P.-J., and Rehberg, H. (1977). *Z. Naturforsch., Teil C* **32c,** 456–458.
Enzmann, P.-J., and Weiland, F. (1978). *Arch. Virol.* **57,** 339.
Erlandson, R. A., Babcock, V. I., Southam, C. M., Bailey, R. B., and Shipkey, F. H. (1967). *J. Virol.* **1,** 996.
Estes, P. C., and Cheville, N. F. (1970). *Am. J. Pathol.* **58,** 235.
Faulkner, P., and McGee-Russell, S. M. (1968). *Can. J. Microbiol.* **14,** 153.
Fenner, F. (1976). *Intervirology* **7,** 1.

Filshie, B. K., and Rehacek, J. (1968). *Virology* **34,** 435.
Friedman, R. M., and Berezesky, I. K. (1967). *J. Virol.* **1,** 374.
Garcia-Tamayo, J. (1971). *J. Virol.* **8,** 232.
Gil-Fernandez, C., Ronda-Lain, C., and Rubio-Huertos, M. (1973). *Arch. Gesamte Virusforsch.* **40,** 1.
Gliedman, J. B., Smith, J. F., and Brown, D. T. (1975). *J. Virol.* **16,** 913.
Grimley, P. M., and Friedman, R. M. (1970). *Exp. Mol. Pathol.* **12,** 1.
Grimley, P. M., Berezesky, I. K., and Friedman, R. M. (1968). *J. Virol.* **2,** 1326.
Grimley, P. M., Levin, J. G., Berezesky, I. K., and Friedman, R. M. (1972). *J. Virol.* **10,** 492.
Hafez, S. M., Petzoldt, K., and Reczko, E. (1968). *Acta Virol. (Engl. Ed.)* **12,** 471.
Halstead, S. B., Shotwell, H., and Casals, J. (1973). *J. Infect. Dis.* **128,** 15.
Hamvas, J. J., Vgovsek, S., Iwakuta, S., and Labzoffsky, N. A. (1969). *Arch. Gesamte Virusforsch.* **26,** 287.
Harrison, S. C., David, A., Jumblatt, J., and Darnell, J. E. (1971). *J. Mol. Biol.* **60,** 523.
Hayashi, K., Akashi, M., and Ueda, Y. (1978). *Trop. Med.* **20,** 1.
Hermodsson, S., and Dinter, Z. (1962). *Nature (London)* **194,** 893.
Higashi, N., Matsumoto, A., Tabata, K., and Nagatomo, Y. (1967). *Virology* **33,** 55.
Higashi, N., Arimura, H., and Fujiwara, E. (1969). *Annu. Rep. Inst. Virus Res. Kyoto Univ.* **12,** 100.
Holmes, I. H., and Warburton, M. F. (1967). *Lancet* **ii** 1233.
Holmes, I. H., Wark, M. C., Jack, I., and Grutzner, J. (1968). *J. Gen. Virol.* **2,** 37.
Holmes, I. H., Wark, M. C., and Warburton, M. F. (1969). *Virology* **37,** 15.
Horzinek, M. C. (1973a). *Prog. Med. Virol.* **16,** 109.
Horzinek, M. C. (1973b). *J. Gen. Virol.* **20,** 87.
Horzinek, M. C. (1975). *Med. Biol.* **53,** 406.
Horzinek, M. C. (1977). Personal communication.
Horzinek, M. C., and Munz, K. (1969). *Arch. Gesamte Virusforsch.* **27,** 94.
Horzinek, M. C., and Mussgay, M. (1969). *J. Virol.* **4,** 514.
Horzinek, M. C., and Mussgay, M. (1971). *Arch. Gesamte Virusforsch.* **33,** 296.
Horzinek, M. C., Reczko, E., and Petzoldt, K. (1967). *Arch. Gesamte Virusforsch.* **21,** 475.
Horzinek, M. C., Maess, J., and Laufs, R. (1971). *Arch. Gesamte Virusforsch.* **33,** 306.
Horzinek, M. C., vanWielink, P. S., and Ellens, D. J. (1975). *J. Gen. Virol.* **26,** 217.
Hotta, S. (1953). *Acta Sch. Med. Univ. Kioto* **31,** 7.
Hyllseth, B. (1973). *Arch. Gesamte Virusforsch.* **40,** 177.
Igarashi, A., Fukuoka, T., and Fukai, K. (1966). *Biken J.* **12,** 245.
Igarashi, A., Fukuoka, T., Nithiuthai, P., Hsu, L.-C., and Fukai, K. (1970). *Biken J.* **13,** 93.
Igarashi, A., Sasao, F., Wungkobkiat, S., and Fukai, K. (1973). *Biken J.* **16,** 17.
Igarashi, A., Harrap, K. A., Casals, J., and Stollar, V. (1976). *Virology* **74,** 174.
Janzen, H. G., Rhodes, A. J., and Doane, F. W. (1970). *Can. J. Microbiol.* **16,** 581.
Jelinkova, A., Danes, L., and Novak, M. (1974). *Acta Virol. (Engl. Ed.)* **18,** 254.
Kääriäinen, L., and Söderlund, H. (1971). *Virology* **43,** 291.
Kitano, T., Suzuki, K., and Yamaguchi, T. (1974). *J. Virol.* **14,** 631.
Kitaoka, M., and Nishimura, C. (1963). *Virology* **19,** 238.
Kitaoka, M., Shimizu, A., Tuchinda, P., and Kim-anaki, C. (1971). *Biken J.* **14,** 361.
Klimenko, S. M., Yershov, F. I., Gofman, Y. P., Nabatnikov, A. P., and Zhdanov, V. M. (1965). *Virology* **27,** 125.
Ko, K. K. (1976). *Biken J.* **19,** 43.
Korolev, M. B. (1977). Personal communication.

Korolev, M. B., Kuri, G., Aguilera, A., and Rodriguez, P. (1973). *Vopr. Virusol.* **2**, 194.
Kouri, G., Aguilera, A., Rodriguez, P., and Korolev, M. (1974). *J. Gen. Virol.* **22**, 73.
Kovac, W., and Stockinger, C. K. L. (1961). *Arch. Gesamte Virusforsch.* **11**, 544.
Lascano, E. F., Berria, M. I., and Barrera Oro, J. G. (1969). *J. Virol.* **4**, 271.
Laude, H. (1977). *Arch. Virol.* **54**, 41.
Lavillaureix, J. (1960). *Arch. Anat. Histol. Embryol.* **43**, 5.
Lavillaureix, J., Gruner, J. E., and Vendrely, R. (1959). *C. R. Soc. Biol.* **152**, 1241.
Lecatsas, G., and Weiss, K. E. (1969). *Arch. Gesamte Virusforsch.* **27**, 332.
Liebhaber, H., Pajot, T., O'Ryan, E. M., and Wood, O. L. (1969). *Symp. Ser. Immunobiol. Stand.* **11**, 149.
McCombs, R. M., Brunschwig, J. P., and Rawls, W. E. (1968). *Exp. Mol. Pathol.* **9**, 27.
McGavran, M. H., and White, J. D. (1964). *Am. J. Pathol.* **45**, 501.
Maess, J., and Reczko, E. (1970). *Arch. Gesamte Virusforsch.* **30**, 39.
Maess, J., Reczko, E., and Böhm, H. O. (1970a). *Proc. Int. Conf. Equine Infect. Dis., 2nd, Paris, 1969* pp. 130–132.
Maess, J., Reczko, E., and Böhm, H. O. (1970b). *Arch Gesamte Virusforsch.* **30**, 47.
Magnusson, P., and Norrby, E. (1965). *Arch. Gesamte Virusforsch.* **16**, 412.
Magnusson, P., Hyllseth, B., and Marusyk, H. (1970). *Arch. Gesamte Virusforsch.* **30**, 105.
Matsumoto, A., and Higashi, N. (1974). *Ann. Rep. Inst. Virus Res. Kyoto Univ.* **17**, 11.
Matsumura, T., Stollar, V., and Schlesinger, R. W. (1971). *Virology* **46**, 344.
Matsumura, T., Stollar, V., and Schlesinger, R. W. (1972). *J. Gen. Virol.* **17**, 343.
Matsumura, T., Shiraki, K., Sashikata, T., and Hotta, S. (1977). *Microbiol. Immunol.* **21**, 329.
Mayr, A., Bachmann, P. A., Sheffy, B. E., and Siegl, G. (1967). *Arch. Gesamte Virusforsch.* **21**, 113.
Mayr, A., Bachmann, P. A., Sheffy, B. E., and Siegl, G. (1968). *Vet. Rec.* **82**, 745.
Monath, T. P., Whitfield, S. G., and Murphy, F. A. (1977). Unpublished observations.
Morgan, C., Howe, C., and Rose, H. M. (1961). *J. Exp. Med.* **113**, 219.
Murphy, F. A. (1974). *Proc. Annu. Meet. U.S. Anim. Health Assoc., 78th* pp. 425–434.
Murphy, F. A. (1975). *Ann. N.Y. Acad. Sci.* **266**, 197.
Murphy, F. A. (1977). *In* "Handbook of Microbiology" (H. Lechevalier, ed.), 2nd ed., Vol. 1, pp. 623–637. CRC Press, Cleveland, Ohio.
Murphy, F. A. (1978). *In* "St. Louis Encephalitis Virus" (T. P. Monath, ed.), Thomas, Springfield, Illinois.
Murphy, F. A., and Harrison, A. K. (1971). *In* "Venezuelan Encephalitis," Sci. Publ. No. 243, pp. 28–29. Pan Am. Health Organ., Washington, D.C.
Murphy, F. A., and Whitfield, S. G. (1970). *Exp. Mol. Pathol.* **13**, 131.
Murphy, F. A., Halonen, P. E., and Harrison, A. K. (1968a). *J. Virol.* **2**, 1223.
Murphy, F. A., Harrison, A. K., Gary, G. W., Jr., Whitfield, S. G., and Forrester, F. T. (1968b). *Lab. Invest.* **19**, 652.
Murphy, F. A., Harrison, A. K., and Collin, W. K. (1970). *Lab. Invest.* **22**, 318.
Murphy, F. A., Taylor, W. P., Mims, C. A., and Marshall, I. D. (1973). *J. Infect. Dis.* **127**, 129.
Murphy, F. A., Whitfield, S. G., Sudia, W. D., and Chamberlain, R. W. (1975). *In* "Invertebrate Immunity" (K. Maramorosch and R. E. Shope, eds.), pp. 25–48. Academic Press, New York.
Mussgay, M., and Rott, R. (1964). *Virology* **23**, 573.
Mussgay, M., and Weibel, J. (1962). *Virology* **16**, 52.
Mussgay, M., and Weibel, J. (1963). *Virology* **19**, 109.

Mussgay, M., Enzmann, P.-J., Horzinek, M. C., and Weiland, E. (1975). *Prog. Med. Virol.* **19,** 257.
Nii, S., Naito, T., and Tuchinda, P. (1970). *Biken J.* **13,** 43.
Nishimura, C., Nomura, M., and Kitaoka, M. (1968). Jpn. J. Med. Sci. Biol. **21,** 1.
Norrby, E. (1969). *Virol. Monogr.* **7,** 115.
Notkins, A. L. (1965). *Bacteriol. Rev.* **29,** 143.
Nozima, T., Mori, H., Minobe, Y., and Yamamoto, S. (1964). *Acta Virol. (Engl. Ed.)* **8,** 97.
Obijeski, J. F., Marchenko, A. T., Bishop, D. H. L., Cann, B. W., and Murphy, F. A. (1974). *J. Gen. Virol.* **22,** 21.
Oshiro, L. S., Schmidt, N. J., and Lennette, E. H. (1969). *J. Gen. Virol.* **5,** 205.
Osterrieth, P. M. (1968). *Extrait Mem. Soc. R. Sci. Liege.* **16,** 1.
Osterrieth, P. M., and Calberg-Bacq, C. M. (1966). *J. Gen. Microbiol.* **43,** 19.
Ota, Z. (1965). *Virology* **25,** 372.
Oyanagi, S., Ikuta, F., and Ross, E. R. (1969). *Acta Neuropathol.* **13,** 169.
Palmer, E. L., and Murphy, F. A. (1968). *Appl. Microbiol.* **16,** 437.
Parker, J. R., and Stannard, L. M. (1967). *Arch. Gesamte Virusforsch.* **20,** 469.
Parker, J. R., Wouters, A. G., and Smith, M. S. (1969). *Arch. Gesamte Virusforsch.* **26,** 305.
Payment, P., Ajdukovic, D., and Pavilanis, V. (1975a). *Can. J. Microbiol.* **21,** 703.
Payment, P., Ajdukovic, D., and Pavilanis, V. (1975b). *Can. J. Microbiol.* **21,** 710.
Peat, A., and Bell, T. M. (1970). *Arch. Gesamte Virusforsch.* **31,** 230.
Pedersen, C. E., and Sagik, B. P. (1973). *J. Gen. Virol.* **18,** 375.
Pfefferkorn, E. R., and Shapiro, D. (1974). *In* "Comprehensive Virology" (H. Frankel-Conrat and R. W. Wagner, eds.), Vol. 2, pp. 171–230. Plenum, New York.
Porterfield, J. S., Casals, J., Chumakov, M. P., Gaidamovich, S. Y., Hannoun, C., Holmes, I. H., Horzinek, M. C., Mussgay, M., Oker-Blom, N., Russell, P. K., and Trent, D. W. (1978). *Intervirology* **9,** 129.
Raghow, R. S., Grace, T. D. C., Filshie, B. K., Bartley, W., and Dalgarno, L. (1973). *J. Gen. Virol.* **21,** 109.
Raikova, A. P., Khutoretskaya, N. V., Tsilinsky, Y. Y., Lvov, D. K., and Klimenko, S. M. (1974). *Vopr. Virusol.* **3,** 283.
Ritchie, A. E., and Fernelius, A. L. (1967). *Vet. Rec.* **81,** 417.
Ritchie, A. E., and Fernelius, A. L. (1968). *Arch. Gesamte Virusforsch.* **23,** 292.
Ritchie, A. E., and Fernelius, A. L. (1969). *Arch. Gesamte Virusforsch.* **28,** 369.
Saturno, A. (1963). *Virology* **21,** 131.
Scherrer, R., Aynaud, J.-M., Cohen, J., and Bic, E. (1970). *C. R. Acad. Sci., Ser. D* **271,** 620.
Schlesinger, R. W. (1977). *Virol. Monogr.* **16,** 1.
Schultz, P. (1971). *Arch. Exp. Vet. Med.* **25,** 413.
Sharp, D. G., Taylor, A. R., Beard, D., and Beard, J. W. (1943). *Arch. Pathol.* **36,** 167.
Shelokov, A., Tauraso, N. M., Allen, A. M., and Espana, C. D. (1971). *In* "Marburg Virus Disease" (G. A. Martini and R. Siegert, eds.), pp. 203–207. Springer-Verlag, Berlin and New York.
Shestopalova, N. M., Reingold, V. N., Gavrilovskaya, I. N., Belyaeva, A. P., and Chumako, M. P. (1965). *Vopr. Virusol.* **4,** 425.
Shestopalova, N. M., Reingold, V. N., Belyaeva, A. P., Butenko, A., and Chumakov, M. P. (1966a). *Proc. Int. Congr. Electron Microsc., 6th, Kyoto* p. 169.
Shestopalova, N. M., Reingold, V. N., Belyaeva, A. P., Gavrilovskaya, I. N., and Chumakov, M. P. (1966b). *Proc. Int. Congr. Electron Microsc., 6th Kyoto* p. 171.

Shestopalova, N. M., Reingold, V. N., Gagarina, A. V., Kornilova, E. A., Popov, G. V., and Chumakov, M. P. (1972). *J. Ultrastruct. Res.* **40**, 458.
Simpson, R. W., and Hauser, R. E. (1968a). *Virology* **34**, 358.
Simpson, R. W., and Hauser, R. E. (1968b). *Virology* **34**, 568.
Slavik, I., Mayer, V., and Mrena, E. (1967). *Acta Virol. (Engl. Ed.)* **11**, 66.
Slavik, I., Mrena, E., and Mayer, V. (1970). *Acta Virol. (Engl. Ed.)* **14**, 8.
Smith, T. J., Brandt, W. E., Swanson, J. L., McCown, J. M., and Buescher, E. L. (1970). *J. Virol.* **5**, 524.
Söderlund, H., Kääriäinen, L., and von Bonsdorff, C.-H. (1975). *Med. Biol.* **53**, 412.
Söderlund, H., Ulmanen, P., and von Bonsdorff, C.-H. (1979). Unpublished observations.
Sokol, F. (1961). *In* "Biology of Viruses of the Tick-Borne Encephalitis Complex" (H. Libiková, ed.), pp. 86–97, 113–117. Czech. Acad. Sci., Bratislava.
Southam, C. M., Shipkey, F. H., Babcock, V. I., Bailey, R., and Erlandson, R. A. (1964). *J. Bacteriol.* **88**, 187.
Sriurairatna, S., Bhamarapravati, N., and Phalavadhtana, O. (1973). *Infect. Immun.* **8**, 1017.
Sriurairatna, S., Bhamarapravati, N., and Onishi, S. (1974). *Biken J.* **17**, 183.
Stohlman, S. A., Wisseman, C. L., Jr., Eylar, O. R., and Silverman, D. J. (1975). *J. Virol.* **16**, 1017.
Stollar, V., and Thomas, V. L. (1975). *Virology* **64**, 367.
Sung, J. S., Diwan, A. R., Falkler, W. A., Yang, H.-Y., and Halstead, S. (1975). *Intervirology* **5**, 137.
Tamalet, J., Toga, M., Chippaux-Hyppolite, C., Choux, R., and Cesarini, J.-P. (1969). *C. R. Acad. Sci., Ser. D* **269**, 668.
Tan, K. B. (1970). *J. Virol.* **5**, 632.
Tauraso, N. M., Aulisio, C. G., Espana, C. D., Wood, O. L., and Liebhaber, H. (1971). *In* "Marburg Virus Disease" (G. A. Martini and R. Siegert, eds.), pp. 208–215. Springer-Verlag, Berlin and New York.
Theofilopoulos, A. N., Brandt, W. E., Russell, P. K., and Dixon, F. T. (1976). *J. Immunol.* **117**, 953.
Tikhomirova, T. I., Karpovich, L. G., Reingold, V. N., and Levkovich, E. N. (1968). *Acta Virol. (Engl. Ed.)* **12**, 529.
Trent, D. (1977). Unpublished observations.
Trousdale, M. D., Trent, D. W., and Shelokov, A. (1975). *Proc. Soc. Exp. Biol. Med.* **150**, 707.
Tsilinsky, Y. Y., Gushchin, B. V., Klimenko, S. M., and Lvov, D. K. (1971a). *Arch. Gesamte Virusforsch.* **34**, 301.
Tsilinsky, Y. Y., Gushchin, B. V., Klimenko, S. M., and Lvov, D. K. (1971b). *Acta Virol. (Engl. Ed.)* **15**, 177.
Tuchinda, P., Nii, S., Sasada, T., Naito, T., Ono, N., and Chatiyanon, K. (1969). *Biken J.* **12**, 201.
Ushimi, C., Tajima, S., Tanaka, S., Nakajima, H., Shimizu, J., and Furnuchi, S. (1969). *Natl. Inst. Anim. Health Q.* **9**, 28.
Vaheri, A., von Bonsdorff, C.-H., Vesikari, T., Hovi, T., and Väänänen, P. (1969). *J. Gen. Virol.* **5**, 39.
von Bonsdorff, C.-H. (1973). *Commentat. Biol., Soc. Sci. Fenn.* **74**, 1.
von Bonsdorff, C.-H. (1977). Personal communication.
von Bonsdorff, C.-H., and Harrison, S. C. (1975). *J. Virol.* **16**, 141.
von Bonsdorff, C.-H., and Vaheri, A. (1969). *J. Gen. Virol.* **5**, 47.

Weckstrom, P., and Nyholm, M. (1965). *Nature (London)* **205,** 211.
Whitfield, S. G., Murphy, F. A., and Sudia, W. D. (1971). *Virology* **43,** 110.
Whitfield, S. G., Murphy, F. A., and Sudia, W. D. (1973). *Virology* **56,** 70.
Wildy, P. (1971). *Monogr. Virol.* **5,** 1.
Wood, O., Tauraso, N., and Liebhaber, H. (1970). *J. Gen. Virol.* **7,** 129.
Yasuzumi, G., and Tsubo, I. (1965a). *J. Ultrastruct. Res.* **12,** 304.
Yasuzumi, G., and Tsubo, I. (1965b). *J. Ultrastruct. Res.* **12,** 317.
Yasuzumi, G., Tsubo, I., Sugihara, R., and Nakai, Y. (1964). *J. Ultrastruct. Res.* **11,** 213.
Yoshinaka, Y., and Hotta, S. (1971). *Proc. Soc. Exp. Biol. Med.* **137,** 1047.

Alphavirus Proteins

KAI SIMONS
HENRIK GAROFF
ARI HELENIUS

I. PROTEIN COMPOSITION OF ALPHAVIRUSES

Alphaviruses consist of an icosahedral nucleocapsid surrounded by a membrane. Sodium dodecyl sulfate (SDS) can be used to dissociate viruses into their constituent polypeptide chains, and their protein composition can subsequently be analyzed by polyacrylamide gel electrophoresis (Shapiro *et al.*, 1967). Table I shows the protein composition of different alphaviruses. One polypeptide occurs in the nucleocapsid, and two in the envelope. Semliki Forest virus (SFV) contains a third small membrane polypeptide not detected in Sindbis virus. Whether this polypeptide is present in other alphaviruses is not known.

The apparent molecular weights determined by SDS polyacrylamide gel electrophoresis are listed in Table I. The exact molecular

THE TOGAVIRUSES

ISBN 0-12-625380-3

TABLE I

Structural Proteins of Alphaviruses

Virus	Capsid protein (MW × 10^{-3})	Envelope glycoproteins (MW × 10^{-3})	References
Semliki Forest	30	49 (E1), 52 (E2), 10 (E3)	Hay *et al.* (1969); Kääriäinen *et al.* (1969); Garoff *et al.* (1974)
Sindbis	30	47–50 (E2), 56–59 (E1)	Strauss *et al.* (1968); Schlesinger and Schlesinger (1972)
Venezuelan equine encephalomyelitis	30	50–53, 56–59	Pedersen *et al.* (1974)
Western equine encephalitis	30	47–52, 55–59	Pedersen *et al.* (1974)
Eastern equine encephalitis	30	45–48, 52–56	Pedersen *et al.* (1974)
Chikungunya	30	52–55[a]	Pedersen *et al.* (1974)

[a] Only one envelope polypeptide was seen.

weights are not known for any of the proteins. The nomenclature for the envelope polypeptides has been established for Sindbis virus and SFV on the basis of biosynthetic and structural studies. The two large polypeptides with apparent molecular weights of about 50 × 10^3 are called E1 and E2, and the third small envelope polypeptide in SFV, E3 (Table I). It is important to note that Sindbis virus E1 migrates more slowly than E2, whereas for SFV the reverse is true. The separation of E1 and E2 in SFV by SDS polyacrylamide gel electrophoresis is, however, dependent on the reduction of disulfide bonds. If the polypeptides are fully reduced, E1 and E2 migrate together as one band in the discontinuous polyacrylamide electrophoresis systems in our laboratory. Trace amounts of the p62 protein (the precursor of E2 and E3 in the infected cell; see Chapter 13) can also be detected in SFV (our unpublished findings). Despite careful studies, no other proteins have been detected in pure preparations of SFV and Sindbis virus. Host proteins are, therefore, not present in detectable amounts.

II. ISOLATION OF ALPHAVIRUS PROTEINS

Each spike protein in SFV is a three-chain structure containing one chain each of E1, E2, and E3, whereas in Sindbis virus the most probable structure is a complex of E1 and E2 (see below). Various methods have been used to isolate the alphavirus proteins on a preparative scale. The choice of method depends on whether one wants to obtain the proteins in a native or a denatured state. With mild nonionic detergents the spike proteins can be detached in apparently nondenatured form. With the use of sucrose gradient centrifugation in Triton X-100 at neutral pH and physiological ionic strength, separation of the nucleocapsid from the spike protein is achieved (Kääriäinen *et al.*, 1969; Helenius and Söderlund, 1973). In a 24-h centrifugation at 195,000 *g*, spike proteins have been isolated from SFV as a lipid-free 4.5 S complex containing one spike monomer (E1, E2, and E3) bound to 75 molecules of Triton X-100 (Simons *et al.*, 1973). In Sindbis virus, the association between E1 and E2 is so weak that they can be separated from each other in Triton X-100 by isoelectric focusing (Dalrymple *et al.*, 1976) or by ion-exchange chromatography (Burke and Keegstra, 1976). When the same conditions are used for SFV, spike proteins stay intact with no dissociation of E1, E2, and E3 from each other (H. Garoff and K. Simons, unpublished observations). With deoxycholate, a somewhat harsher detergent, the spike protein in SFV can be dissociated into its constituent polypeptide chains (Helenius *et al.*, 1976). E1 and E3 are dissociated as monomeric peptide chains, whereas E2 aggregates after dissociation from the spike protein, presumably to an octameric species. E1, E2, and E3 can therefore be separated using sucrose gradient centrifugation in the presence of the bile salt.

The hemagglutinin activity of the virus spike protein is preserved after exposure to both deoxycholate and Triton X-100. Hemagglutination is a property of E1 (Dalrymple *et al.*, 1976; Helenius *et al.*, 1976). E1 presumably aggregates in the hemagglutination assay when the detergent is diluted out, since it could hardly be active as a monomer. When SDS, a strongly denaturing detergent, is used to solubilize the virus, the polypeptide chains are dissociated and denatured and can then be separated from each other either by polyacrylamide gel electrophoresis (for small quantities) or by hydroxylapatite chromatography (for larger quantities) (Garoff *et al.*, 1974).

A. The Nucleocapsid Protein

After synthesis in the infected cell, the capsid protein binds to the 42 S RNA to form the virus nucleocapsid (Friedman, 1968; Scheele and Pfefferkorn, 1969; Söderlund, 1973). The mechanism of nucleocapsid assembly is poorly understood. The capsid protein binds to the 60 S ribosomal subunit in infected cells (Söderlund and Ulmamen, 1977). Whether complexes between the capsid protein and the ribosomal subunit act as intermediates in the assembly of the nucleocapsid is not yet known. By analogy with other cytoplasmic proteins in the cell, the capsid protein contains no carbohydrate. All proteins located internal to the lipid bilayer in enveloped viruses are devoid of carbohydrate (Lenard and Compans, 1974). The amino acid compositions of the capsid proteins from SFV and Sindbis virus are shown in Table II. The SFV capsid protein is rich in lysine, whereas the Sindbis virus capsid protein has a fairly high content of both lysine and arginine. The possible subunit structure of the capsid protein in the nucleocapsid has not yet been studied. Nothing is known about the sites at which the protein binds to the RNA or about the postulated

TABLE II

Amino Acid Composition of the Nucleocapsid Protein[a]

Amino acid	SFV[b]	Sindbis virus[c]
Lysine	12.70	8.59
Histidine	2.62	2.58
Arginine	5.16	8.17
Aspartate	8.05	7.19
Threonine	5.88	6.64
Serine	4.14	5.12
Glutamate	10.75	10.33
Proline	8.29	9.73
Glycine	9.17	10.41
Alanine	8.79	8.59
Half cystine	1.64	N.D.[d]
Valine	6.46	5.54
Methionine	2.64	2.07
Isoleucine	3.95	3.34
Leucine	3.84	6.09
Tyrosine	2.50	1.86
Phenylalanine	2.41	3.64
Tryptophan	1.01	N.D.

[a] Values are moles per 100 moles.

[b] Virus grown in BHK cells (Simons and Kääriäinen, 1970).

[c] Virus grown in BHK cells (Burke and Keegstra, 1976).

[d] N.D., not determined.

binding sites on the capsid protein to which spike proteins bind during virus budding (see below).

B. Envelope Glycoproteins

Alphavirus envelope proteins have been studied in detail. They all contain carbohydrate. E1 and E2 are integral membrane proteins attached to the lipid bilayer by hydrophobic peptide segments which span the membrane.

1. *Amino Acid Compositions and N-Terminal Amino Acid Sequences*

The amino acid compositions of E1, E2, and E3 isolated from SFV and of E1 and E2 from Sindbis virus are shown in Table III. None of the proteins is especially hydrophobic, as the polarity indices of these proteins are in the same range as for water-soluble globular proteins.

TABLE III

Amino Acid Composition of Membrane Glycoproteins[a]

	SFV[b]			Sindbis virus[c,d]	
Amino Acid	E1	E2	E3	E1	E2
Lysine	6.36	5.20	1.99	6.60	5.97
Histidine	3.38	4.58	2.16	2.80	3.24
Arginine	3.62	4.87	6.02	3.26	4.39
Aspartic acid	8.80	8.48	11.49	9.05	10.42
Threonine	8.13	7.67	7.22	7.76	9.35
Serine	8.11	5.81	3.32	9.23	8.19
Glutamic acid	7.98	9.40	9.22	9.10	8.00
Proline	6.32	6.91	7.63	6.93	7.60
Glycine	7.63	7.51	5.73	7.59	6.87
Alanine	8.74	8.67	11.57	10.24	8.79
Half cystine	2.93	3.21	6.76	N.D.	N.D.
Valine	8.92	7.82	5.71	7.26	7.21
Methionine	1.71	2.00	2.16	0.66	0.31
Isoleucine	3.75	5.44	2.17	5.51	5.60
Leucine	5.80	5.99	9.58	6.83	6.79
Tyrosine	4.10	3.64	3.95	3.21	4.34
Phenylalanine	3.72	2.80	3.32	3.94	2.93

[a] Values are moles per 100 moles.
[b] Virus grown in BHK cells (Garoff *et al.*, 1974).
[c] Virus grown in BHK cells (Burke and Keegstra, 1976).
[d] N.D., not determined.

Figure 1 shows the N-terminal amino acid sequence of E1 and E2 from Sindbis virus (Bell *et al.*, 1978). No long stretches of hydrophobic amino acids are seen. From other studies it is known that the N-termini are located outside the membrane (see below).

2. *Carbohydrate Composition*

The oligosaccharide moieties of the envelope glycoproteins have been extensively studied both in Sindbis virus and in SFV. Two types of oligosaccharide units have been found, one complex consisting of sialic acid, fucose, galactose, mannose, and *N*-acetylglucosamine residues, and another that is simpler in composition and consists of *N*-acetylglucosamine and mannose (Burge and Strauss, 1970; Sefton and Keegstra, 1974; Mattila *et al.*, 1976). In Sindbis virus, both E1 and E2 probably contain one unit of each of these oligosaccharides (Burke and Keegstra, 1976), whereas in SFV E1 and E3 contain one to two oligosaccharide units of the complex type, and E2 an additional unit of the simple type (Mattila *et al.*, 1976, Pesonen and Renkonen, 1976). The oligosaccharides are attached to the polypeptide most probably through β-*N*-glycosidic linkages between asparagine and *N*-acetylglucosamine (Burke, 1976).

Preliminary sequence analysis by Burke (1976) for Sindbis virus and by Renkonen and co-workers (personal communication) for SFV suggest the structures shown in Fig. 2 for the oligosaccharide units. The terminal sialic acid residues may be completely missing, or present on only one of the two antennas. In SFV the complex oligosaccharide E3 is heterogeneous. Some molecules have the biantennary structure shown in Fig. 2, and others have a tri- or tetraantennary structure (K. Rasilo and O. Renkonen unpublished observations). Both in Sindbis virus and SFV the high-mannose units are heterogeneous and may contain structures with fewer or more mannose residues than the structure shown. The variations are observed in virus grown in the same host cell. Carbohydrate variability is accentuated when virus preparations in different cells are compared (Strauss *et al.*, 1970; Keegstra *et al.*, 1975). Semliki Forest and Sindbis virus grown in *Aedes albopictus* cells lack sialic acid (Renkonen *et al.*, 1974; Stollar *et al.*, 1976).

3. *Subunit Structure*

The E1 and E2 proteins in Sindbis virus and the E1, E2, and E3 proteins in SFV form the spikelike projections on the external surface of the virus particles (Calberg-Bacq and Osterrieth, 1966; Compans, 1971; Gahmberg *et al.*, 1972). In electron micrographs the surface

```
                    5             10             15             20             25             30
E1  Tyr-Glu-His-Ala-Thr-Thr-Val-Pro-Asn-Val-Pro-Gln-Ile-Pro-Tyr-Lys-Ala-Leu-Val-Glu-Arg-Ala-Gly-Tyr-Ala-Pro-Leu-Asn-Leu-Glu
E2  Ser-Val-Ile-Asp-Gly-Phe-Thr-Leu-Thr-Ser-Pro-Tyr-Leu-Gly-Thr-Cys-Ser-Tyr-Cys-His-His-Thr-Glu-Pro-Cys-Phe-Ser-Pro-Val-Lys

                   35             40                45                50
E1  Ile-Thr-Val-Met-Ser-Ser-Glu-Val-Leu-Pro-Ser-(Thr)-Asn-Gln-Glu-Tyr-Ile-(Ser)-(Trp)-Lys-Phe-(Ser)-(Thr)-
E2  Ile-Glu-Gln-Val-Trp-Asp-Glu-Ala-Asp-Asp-Asn- ? - Ile- ? -Ile-Gln- ? - ? -Ala-(Gln)-Phe-
```

Fig. 1. Amino acid sequence determined from the N-terminus of E1 and E2 from Sindbis virus. (From Bell *et al.*, 1978.)

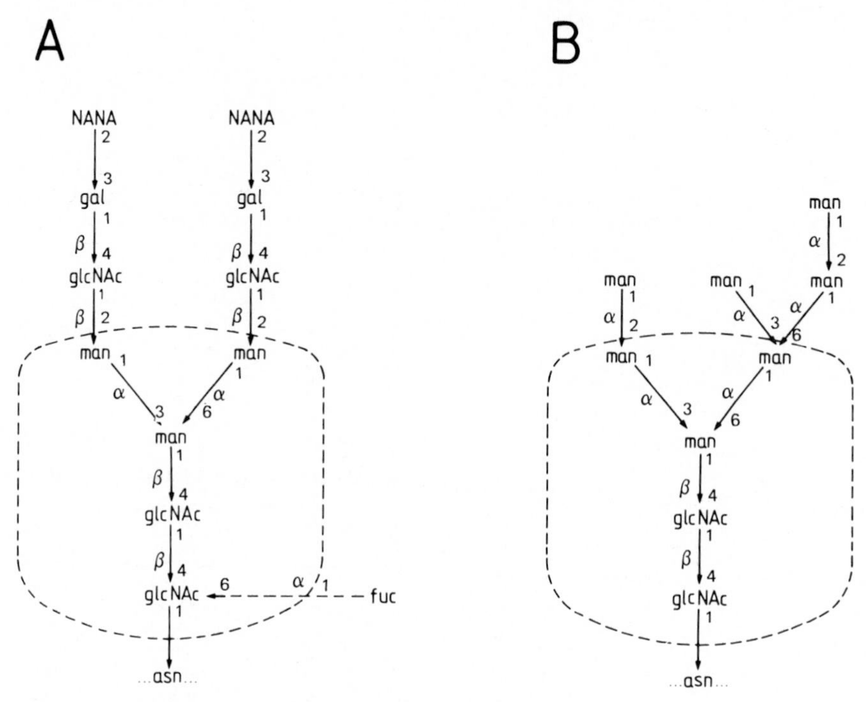

Fig. 2. Structures proposed for alphavirus glycopeptides. (A) Complex type. (B) High-mannose type. (From Burke, 1976.)

projections are not sufficiently resolved to show individual spikes. The quaternary structure of the membrane glycoproteins in SFV have been studied by chemical cross-linking using reversible cross-linking reagents (Ziemiecki and Garoff, 1978). The spikes are three-chain structures composed of one polypeptide each of E1, E2, and E3. The association of E1, E2, and E3 in the spike protein is not very strong. Triton X-100 at physiological salt conditions, at pH 7–8, solubilizes the spike with its subunit structure intact, but the binding of anti-E1 or -E2 to the complex suffices to cause dissociation (Ziemiecki and Garoff, 1978). As mentioned above, deoxycholate dissociates the spike into its constituent polypeptide chains.

In Sindbis virus it seems likely that E1 and E2 form a two-chain structure (Schlesinger and Schlesinger, 1972; Bracha and Schlesinger, 1976; Jones *et al.*, 1977). The association seems to be even weaker than in SFV. Triton X-100 alone can dissociate the two chains from each other under conditions where the SFV spikes remain intact (Dalrymple *et al.*, 1976; Burke and Keegstra, 1976).

III. TOPOGRAPHY OF THE SPIKE PROTEINS IN THE MEMBRANE

A. Penetration of Membrane

When intact SFV is treated with proteases, most of the three membrane polypeptides (E1, E2, and E3) are digested, leaving intact virus particles devoid of spikes (Utermann and Simons, 1974). All the carbohydrate is removed. Peptide segments representing about 10% of the membrane proteins are, however, left in the membrane. These peptides are hydrophobic. Their polarity index (hydrophilic amino acids/total amino acids) calculated from the amino acid composition is as low as 0.28, compared to 0.46, 0.46, and 0.41 for E1, E2, and E3, respectively. The peptides are soluble in chloroform–methanol. Peptide mapping shows that both E1 and E2 have hydrophobic fragments; E3 does not. Electrophoresis in a polyacrylamide gradient gel in the presence of SDS reveals three major bands with apparent molecular weights of 9K, 6K, and 5K (Garoff and Söderlund, 1978) of which the 9K and 5K fragments derive from E2 and the 6K fragment from E1. By labeling the polypeptide chains with a gradient of radioactivity increasing from one end of the molecule to the other, it has been possible to assign the location of the hydrophobic segments within the polypeptide chains of E1 and E2. The 6K peptide is derived from the C-terminal end of E1, whereas both the 9K and the 5K peptides are in the C-terminal region of E2 (Garoff and Söderlund, 1978). The two peptides obtained from E2 probably overlap, the 9K peptide presumably resulting from incomplete proteolytic cleavage. Thus both E1 and E2 are attached by their C-terminal regions to the bilayer and have their N-terminal portion on the outside. This is the orientation that has been found for most other surface glycoproteins studied so far (Segrest *et al.*, 1973; Bretscher, 1975; Skehel and Waterfield, 1975; Henning *et al.*, 1976).

The spike protein not only penetrates into the lipid bilayer but also spans the virus membrane. Most of the SFV spikes can be cross-linked to the nucleocapsid when intact virus particles are treated with the cross-linking reagent dimethyl suberimidate (Garoff and Simons, 1974). As dimethyl suberimidate maximally bridges a distance of 11 Å, which is about one-fourth of the lipid bilayer thickness in SFV, extensive cross-linking is possible only if the viral glycoproteins lie close to the nucleocapsid, i.e., span the lipid membrane. It is also possible to cross-link the spike glycoproteins to the nucleocapsid with formaldehyde (Brown *et al.*, 1974).

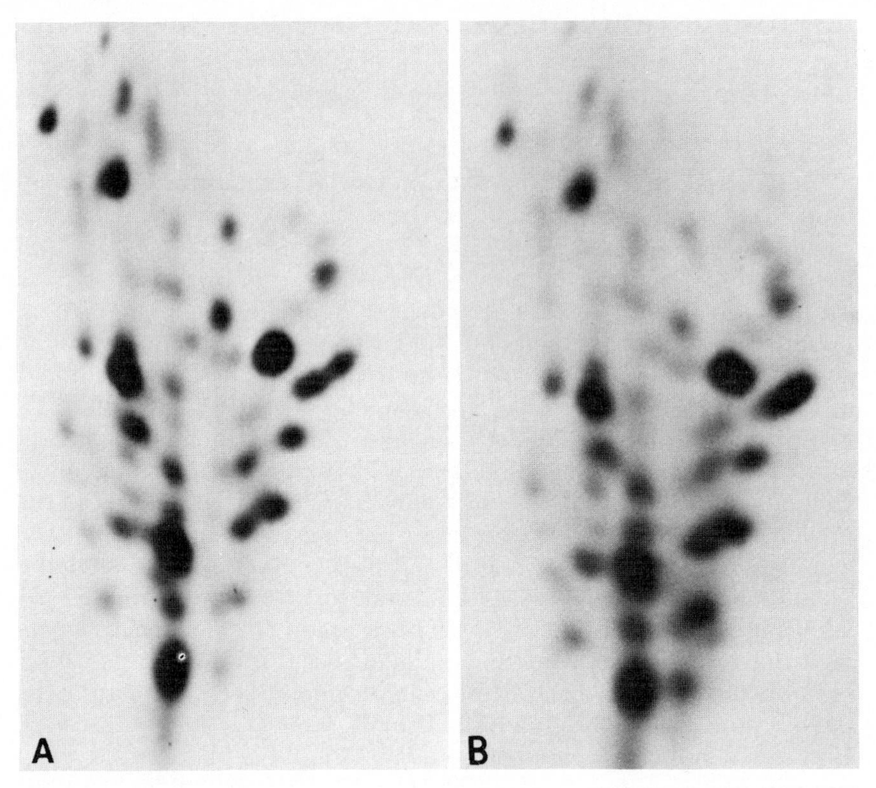

Fig. 3. ^{35}S-labeled formylmethionylsulfone methyl phosphate was reacted either with intact virus or with virus lysed with small amounts of Triton X-100 as described previously (Garoff and Simons, 1974). The E1 and E2 polypeptides were separated by SDS polyacrylamide gel electrophoresis and eluted from the gel into 0.1% SDS. The SDS was removed, and after performic acid oxidation the proteins were digested with a mixture of chymotrypsin and trypsin (Utermann and Simons, 1974). The peptides were separated on cellulose thin-layer plates (Polygram 300, 20 × 20 cm, from Macherey and Nagel, Düren, Germany). The first dimension was electrophoresis at pH 4.4 in pyridine–acetic acid–acetone–water (2:4:15:79) at 400 V for 100 min. The second dimension was chromatography in pyridine–acetic acid–acetone (in butanol)–water (10:3:15:12). The peptide maps were visualized by autoradiography. Peptide map of E1 from intact SFV (A), E1 from SFV lysed with Triton X-100 (B), E2 from intact SFV (C), and E2 from SFV lysed with Triton X-100 before labeling with [^{35}S]formylmethionylsulfone methyl phosphate (D). One significant difference is seen between (C) and (D) (see arrow)—a heavily labeled peptide is absent from the peptide map of E2 isolated from intact virus and present on the map of E2 labeled from both sites of the membrane.

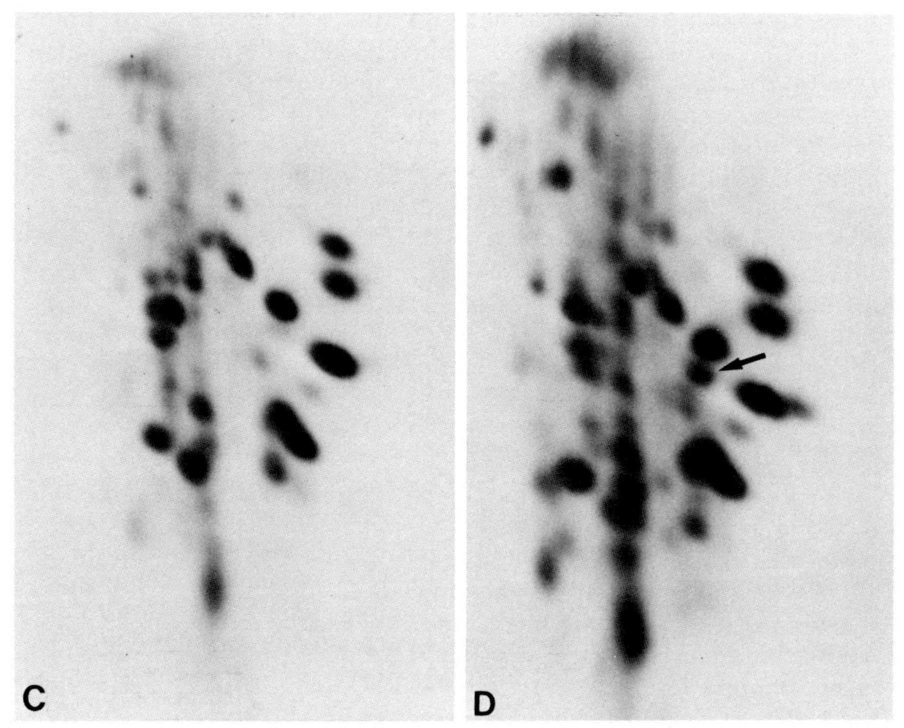

Fig. 3. *Continued*

Direct evidence for an interaction between the spike proteins and the nucleocapsid has been obtained from octylglucoside solubilization (Helenius and Kartenbeck, 1980). This nonionic detergent solubilizes the lipid from the virus but leaves most of the spikes attached to the nucleocapsid. If the salt concentration is increased above 50 mM, the spikes dissociate, but they attach again if the ionic strength is lowered. Further studies will be required to define the specificity of the binding.

To find out whether both E1 and E2 span the membrane, the glycoproteins in SFV were labeled with the radioactive surface label [^{35}S]formylmethionylsulfone methyl phosphate (Bretscher, 1971; Garoff and Simons, 1974) from the outside in intact virus. In order to allow labeling from both sides, membrane "ghosts" were prepared by adding enough Triton X-100 to the virus barely to lyse the membrane (Helenius and Söderlund, 1973). Labeling from both sides gave on the peptide map of the E2 glycoprotein one additional basic, clearly labeled peptide not seen on the peptide map derived from E2 glycoproteins labeled only from the outside (Fig. 3). The E1 peptide maps were identical (Fig. 3). This suggests that E2 penetrates the lipid bilayer. It cannot, however, be excluded that E1 spans; a negative result (no additional peptide labeled from the inside) only indicates the absence of a reactive amino group. In keeping with the labeling results it has been shown that about 3000 daltons of the C-terminal end of E2 in both Sindbis and SFV is accessible to proteases on the cytoplasmic side of the endoplasmic reticulum (Wirth *et al.*, 1977; Garoff and Söderlund, 1978). It is not clear whether the E2 polypeptide spans the membrane only once. The C-terminus of E2 is large enough (about 30 amino acid residues) to form a loop into the bilayer, bringing the C-terminal end to the external side of the membrane (Fig. 4).

B. Amphiphilic Properties

To find out whether a protein is amphiphilic or not, the properties of the solubilized protein can be studied in different ways. Amphiphilic proteins are known to bind Triton X-100. The SFV spike proteins bind 75 molecules of Triton X-100, which corresponds roughly to one micelle of the detergent (Simons *et al.*, 1973). The nucleocapsid does not bind Triton X-100 (Helenius and Söderlund, 1973). In charge-shift electrophoresis (Helenius and Simons, 1977), which is a rapid screening method for distinguishing between amphiphilic and hydrophilic proteins based on their Triton X-100 binding capacity, the E1 and E2 chains behave as amphiphilic proteins, while E3 is hydrophilic (Simons *et al.*, 1978a).

Another property of amphiphilic proteins is their ability to form

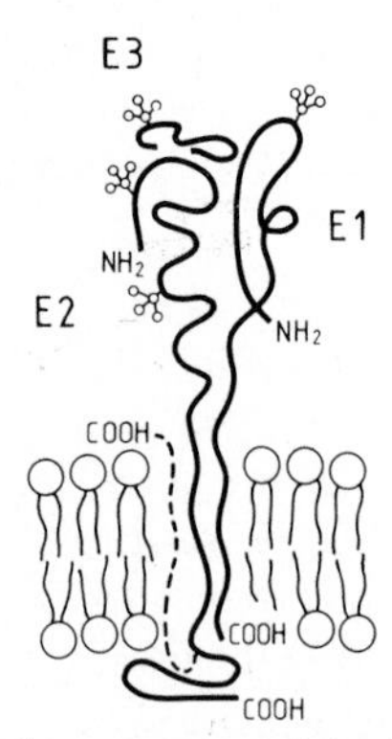

Fig. 4. Schematic model of the structure of the SFV spike protein. The possible loop of the C-terminus of E2 is indicated by a dashed line.

micelles. These are protein aggregates free of detergent and lipid in which the proteins are held together by their hydrophobic regions with the hydrophilic domains at the surface, making the whole complex sufficiently polar to remain soluble (Simons *et al.*, 1978b). The SFV spike protein can be made to form monodisperse complexes containing eight spike proteins virtually devoid of lipid and detergent (Helenius and von Bonsdorff, 1976). Micelle formation appears to be a property typical of amphiphilic membrane proteins having a large polar domain and a relatively small hydrophobic tail in the lipid bilayer.

A further characteristic of amphiphilic proteins is that they can be recombined with phospholipids to form lipid vesicles into which proteins are inserted. If the spike proteins of SFV are solubilized with Triton X-100 and subsequently subjected to sucrose gradient centrifugation in the presence of octylglucoside, the spike proteins are obtained lipid-free complexed to the detergent. By combining this preparation with egg lecithin, vesicles are obtained studded with virus spikes (Helenius *et al.*, 1977).

C. Lipid–Protein Interactions

The lipids of alphaviruses are derived from the host cell, in contrast to the virus structural proteins which are all virus-coded. The lipid composition of the virus depends on the host cell used, and it is similar to the composition of the host surface membrane from which the virus envelope is formed (Renkonen *et al.*, 1971, 1974; Quigley *et al.*, 1971; Hirschberg and Robbins, 1974) (see Chapter 10).

Whether virus glycoproteins exert any influence on the physical state of the lipids of the virus envelope has been studied in several ways (Sefton and Gaffney, 1974; Moore *et al.*, 1976) but, owing to limitations of the available methodology, interpretation of the results obtained is difficult (Lenard, 1978; see also Chapter 10). The significance of differences found between the virus envelope and the host cell plasma membrane is unknown.

IV. FUNCTIONS OF THE ALPHAVIRUS SPIKE GLYCOPROTEINS

Virus spike glycoproteins are necessary for virus entry into the cell to initiate infection (Compans, 1971; Utermann and Simons, 1974) and for virus exit from the host cell during budding.

The first stage in the entry process is the binding of virions to the cell surface (Birdwell and Strauss, 1974). Receptors for SFV have recently been identified (Helenius *et al.*, 1978). They are the major histocompatibility antigens, HLA-A and HLA-B antigens on human cells and H2-K and H-2D antigens on mouse cells. Alphaviruses agglutinate goose red blood cells at pH 5.8 (Clarke and Casals, 1958). The virus protein responsible for binding to red blood cells (the hemagglutinin) is E1 in both Sindbis virus (Dalrymple *et al.*, 1976) and SFV (Helenius *et al.*, 1976). Avian erythrocytes contain histocompatibility antigens, but whether they are the receptors to which E1 binds is not yet known.

Following adsorption of the virus to the cell, there are two ways the virus could enter the cell, either by fusion with the plasma membrane or by endocytosis. In the latter case the nucleocapsid or the RNA would have to be delivered through the endocytotic vesicle membrane in order to initiate virus replication in the cytoplasm. We have recently found that Semliki Forest virus uses the endocytotic route to infect the host cell (Helenius *et al.*, 1980).

After synthesis, the spike glycoproteins are transported from the endoplasmic reticulum to the cell surface (see Chapters 13 and 17). During the transport the carbohydrate chains of the virus glycopro-

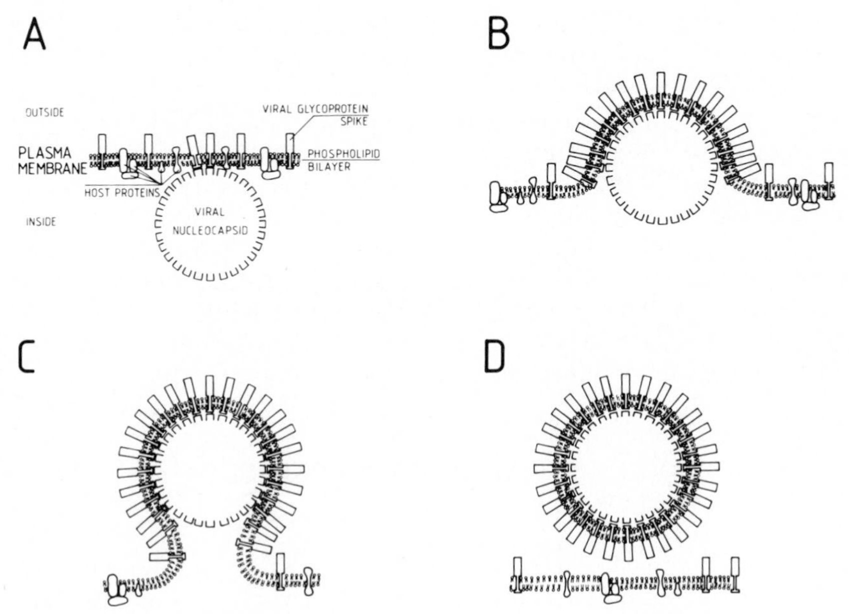

Fig. 5. Model for the budding of alphaviruses. (Taken from Simons *et al.*, 1978a.)

teins are processed. It is possible that both secretory proteins and surface membrane glycoproteins (including virus glycoproteins) are glycosylated by a common oligosaccharide unit transferred to asparagine residues during protein translation (Waechter and Lennarz, 1976; Rothman and Lodish, 1977; Sefton, 1977) and that this unit is subsequently processed to yield the high-mannose and complex oligosaccharides (Spiro *et al.*, 1976; Robbins *et al.*, 1977; Tabas *et al.*, 1978).

In the final stage of alphavirus maturation (see also Chapter 17) the nucleocapsid buds out from the host cell surface. We have postulated that this process is driven by spike protein–nucleocapsid interactions (Garoff and Simons, 1974) (Fig. 5). The nucleocapsid binds to the cytoplasmic face of the plasma membrane via the transmembrane segments of the spike proteins. More spikes move into the bud by diffusion and are trapped by binding to capsid protein sites. The release of the virus may result from the bend imposed on the bilayer by the nucleocapsid. Compositional studies have shown that the stoichiometric ratio of the spike protein to the capsid protein is 1:1 (Garoff *et al.*, 1974), in keeping with the postulated model.

REFERENCES

Bell, J. R., Hunkapiller, M. W., Hood, L. E., and Strauss, J. H. (1978). *Proc. Natl. Acad. Sci. U.S.A.* **75**, 2722–2726.
Birdwell, C. R., and Strauss, J. H. (1974). *J. Virol.* **14**, 366–374.
Bracha, M., and Schlesinger, M. J. (1976). *Virology* **74**, 441–449.
Bretscher, M. S. (1971). *J. Mol. Biol.* **58**, 775–781.
Bretscher, M. S. (1975). *J. Mol. Biol.* **98**, 831–833.
Brown, F., Smale, C. J., and Horzinek, M. (1974). *J. Gen. Virol.* **22**, 455.
Burge, B. W., and Strauss, J. H. (1970). *J. Mol. Biol.* **47**, 449–460.
Burke, D. J. (1976). Ph.D. Thesis, State Univ. of New York, Stony Brook.
Burke, D. J., and Keegstra, K. (1976). *J. Virol.* **20**, 676–686.
Calberg-Bacq, C. M., and Osterrieth, P. M. (1966). *Acta Virol. (Engl. Ed.)* **10**, 266.
Clarke, D. H., and Casals, J. (1958). *Am. J. Trop. Med. Hyg.* **7**, 561–573.
Compans, R. W. (1971). *Nature (London) New Biol.* **229**, 114–116.
Dalrymple, J. M., Schlesinger, S., and Russell, P. K. (1976). *Virology* **69**, 93–103.
Friedman, R. M. (1968). *J. Virol.* **2**, 26–32.
Gahmberg, C. G., Utermann, G., and Simons, K. (1972). *FEBS Lett.* **28**, 179–182.
Garoff, H., and Simons, K. (1974). *Proc. Natl. Acad. Sci. U.S.A.* **71**, 3988–3992.
Garoff, H., and Söderlund, H. (1978). *J. Mol. Biol.* **124**, 535–549.
Garoff, H., Simons, K., and Renkonen, O. (1974). *Virology* **61**, 493–504.
Hay, A. J., Skehel, J. J., and Burke, D. C. (1969). *J. Gen. Virol.* **3**, 175–184.
Helenius, A., and Kartenbeck, J. (1980). *Eur. J. Biochem.*, in press.
Helenius, A., and Simons, K. (1977). *Proc. Natl. Acad. Sci. U.S.A.* **74**, 529–532.

Helenius, A., and Söderlund, H. (1973). *Biochim. Biophys. Acta* **307**, 287–300.
Helenius, A., and von Bonsdorff, C. -H. (1976). *Biochim. Biophys. Acta* **436**, 895–899.
Helenius, A., Fries, E., Garoff, H., and Simons, K. (1976). *Biochim. Biophys. Acta* **436**, 319–334.
Helenius, A., Fries, E., and Kartenbeck, J. (1977). *J. Cell Biol.* **75**, 866–880.
Helenius, A., Kartenbeck, J., Simons, K., and Fries, E. (1980). *J. Cell Biol.* **84**, 404–420.
Helenius, A., Morein, B., Fries, E., Simons, K., Robinson, P., Schirrmacher, V., Terhorst, C., and Strominger, J. L. (1978). *Proc. Natl. Acad. Sci. U.S.A.* **75**, 3846–3850.
Henning, R., Milner, R. J., Reske, K., Cunningham, B. A., and Edelman, G. M. (1976). *Proc. Natl. Acad. Sci. U.S.A.* **73**, 118–122.
Hirschberg, C. B., and Robbins, P. W. (1974). *Virology* **61**, 602–608.
Jones, K. J., Scupham, R. K., Pfeil, J. A., Wan, K., Sagik, P., and Bose, H. R. (1977). *J. Virol.* **21**, 778–787.
Kääriäinen, L., Simons, K., and von Bonsdorff, C. -H. (1969). *Ann. Med. Exp. Biol. Fenn.* **47**, 235–248.
Keegstra, K., Sefton, B., and Burke, D. (1975). *J. Virol.* **16**, 613–620.
Lenard, J. (1978). *Annu. Rev. Biophys. Eng.* **7**, 139–165.
Lenard, J., and Compans, R. W. (1974). *Biochim. Biophys. Acta* **344**, 51–94.
Mattila, K., Luukkainen, A., and Renkonen, O. (1976). *Biochim. Biophys. Acta* **419**, 435–444.
Moore, N. F., Barenholz, Y., and Wagner, R. R. (1976). *J. Virol.* **19**, 126–135.
Pedersen, C. E., Jr., Marker, S. C., and Eddy, G. A. (1974). *Virology* **60**, 312–314.
Pesonen, M., and Renkonen, O. (1976). *Biochim. Biophys. Acta* **455**, 510–525.
Quigley, J. P., Rifkin, D. B., and Reich, E. (1971). *Virology* **46**, 106.
Renkonen, O., Kääriäinen, L., Simons, K., and Gahmberg, C. G. (1971). *Virology* **46**, 318–326.
Renkonen, O., Luukkonen, A., Brotherus, J., and Kääriäinen, L. (1974). *In* "Control of Proliferation in Animal Cells" (B. Clarkson and R. Baserga, eds.), pp. 495–504. Cold Spring Harbor Lab., Cold Spring Harbor, New York.
Robbins, P. W., Hubbard, S. C., Turco, S. J., and Wirth, D. F. (1977). *Cell* **12**, 893–900.
Rothman, E. J., and Lodish, H. F. (1977). *Nature (London)* **269**, 775–780.
Scheele, C. M., and Pfefferkorn, E. R. (1969). *J. Virol.* **3**, 369–375.
Schlesinger, S., and Schlesinger, M. J. (1972). *J. Virol.* **10**, 925–932.
Sefton, B. M. (1977). *Cell* **10**, 659–668.
Sefton, B. M., and Gaffney, B. J. (1974). *J. Mol. Biol.* **90**, 343–358.
Sefton, B. M., and Keegstra, K. (1974). *J. Virol.* **12**, 1366–1374.
Segrest, J. P., Kahane, I., Jackson, R. L., and Marchesi, V. T. (1973). *Arch. Biochem. Biophys.* **155**, 167–183.
Shapiro, A. L., Vinuela, E., and Maizel, J. V. (1967). *Biochem. Biophys. Res. Commun.* **28**, 815–520.
Simons, K., and Kääriäinen, L. (1970). *Biochem. Biophys. Res. Commun.* **38**, 981–988.
Simons, K., Helenius, A., and Garoff, H. (1973). *J. Mol. Biol.* **80**, 119–133.
Simons, K., Garoff, H., Helenius, A., and Ziemiecki, A. (1978a). *In* "Frontiers of Physicochemical Biology" (B. Pullman, ed.), pp. 387–407. Academic Press, New York.
Simons, K., Helenius, A., Leonard, K., Sarvas, M., and M. J. Gething (1978b). *Proc. Natl. Acad. Sci. U.S.A.* **75**, 5306–5310.
Skehel, J. J., and Waterfield, M. D. (1975). *Proc. Natl. Acad. Sci. U.S.A.* **72**, 93–97.
Söderlund, H. (1973). *Intervirology* **1**, 1–8.

Söderlund, H., and Ulmanen, I. (1977). *J. Virol.* **24,** 907–909.
Spiro, M. J., Spiro, R. G., and Bhoyroo, V. D. (1976). *J. Biol. Chem.* **251,** 6400–6408.
Stollar, V., Stollar, B. D., Koo, R., Harrap, K. A., and Schlesinger, R. W. (1976). *Virology* **69,** 104–115.
Strauss, J. H., Jr., Burge, B. W., Pfefferborn, E. R., and Darnell, J. E., Jr. (1968). *Proc. Natl. Acad. Sci. U.S.A.* **59,** 533–537.
Strauss, J. H., Jr., Burge, B. W., and Darnell, J. E., Jr. (1970). *J. Mol. Biol.* **47,** 437–448.
Tabas, I., Schlesinger, S., and Kornfeld, S. (1978). *J. Biol. Chem.* **253,** 716–722.
Utermann, G., and Simons, K. (1974). *J. Mol. Biol.* **85,** 569–587.
Waechter, C. J., and Lennarz, W. J. (1976). *Annu. Rev. Biochem.* **45,** 95–112.
Wirth, D. F., Katz, F., Small, B., and Lodish, H. F. (1977). *Cell* **10,** 253–263.
Ziemiecki, A., and Garoff, H. (1978). *J. Mol. Biol.* **122,** 259–269.

10

Lipids of Alphaviruses

JOHN LENARD

I. INTRODUCTION

Many of the chapters in this book provide ample documentation for the assertion that alphaviruses constitute a reasonably well-defined group of viruses, possessing properties that set it apart from the rest of the virus world. In considering the nature, structure, and function of the lipids that are an integral part of their structure, however, the distinction between alphaviruses and other enveloped viruses seems quite arbitrary. Neither the overall structural organization of the viral lipid nor the nature of its interaction with viral glycoproteins (see Chapter 9) serves to distinguish alphaviruses from any of the other well-characterized classes of enveloped viruses. Although this chapter will be limited to observations on alphaviruses only, a more general discussion of lipids in viral envelopes can be found in several recent reviews of virus envelope structure (Klenk, 1974; Lenard, 1978; Lenard and Compans, 1974).

II. STRUCTURE OF LIPIDS

Lipids constitute about 30% of the dry weight of purified Sindbis virus (SV) (Pfefferkorn and Hunter, 1963a; Strauss *et al.*, 1970) and

THE TOGAVIRUSES

ISBN 0-12-625380-3

Semliki forest virus (SFV) (Laine *et al.*, 1973; Renkonen *et al.*, 1971). The best evidence that these lipids exist as a bilayer comes from x-ray diffraction studies on SV (Harrison *et al.*, 1971). The radially averaged electron density profile of the virus particle shows a deep minimum centered at 232 Å from the center of the virion, at approximately the position at which a unit membrane structure is seen in electron micrographs. The electron density at the minimum is characteristic of the hydrocarbon region of a bilayer. The width of the trough (48 Å) is similar to that observed in other bilayers. It has been concluded that at least 90% of the structure at 232 Å is hydrocarbon in nature, so that the protein which apparently traverses it (Chapter 9) is a minor component of the bilayer structure.

Spin labeling has provided additional evidence for a bilayer structure in alphaviruses. Early spin-labeling studies demonstrated that artificial pure lipid bilayers were characterized by a relatively fluid interior, with a progressive increase in rigidity toward the glycerol moiety (Hubbell and McConnell, 1969; Jost *et al.*, 1971; reviewed in Lenard and Landsberger, 1976). Thus spin-labeled fatty acids or phospholipids with nitroxide groups attached to carbons high on the fatty acid chain, close to the glycerol or carboxyl group, appear to be in a highly constrained, rigid environment when incorporated into bilayers. Fatty acids or phospholipids containing nitroxide groups further down the chain probe closer to the center of the bilayer and are in a much more fluid environment. By demonstrating the presence of such a flexibility gradient, the presence of a bilayer has been shown in particles of Venezuelan equine encephalitis (VEE) virus (Hughes and Pedersen, 1975) and has been confirmed in SV (Sefton and Gaffney, 1974).

The lipid bilayer of SV, as of other enveloped viruses (Bittman *et al.*, 1976), is continuous and forms a permeability barrier that is osmotically responsive. The osmotic barrier in SV is not destroyed by repeated freezing and thawing of the viral particle (D. Madoff and J. Lenard, unpublished observations).

The question of how viral proteins modify the properties of the viral bilayer has been approached experimentally with SV by comparing different viral samples using physical measurements. The following samples have been compared: (1) intact virions, (2) particles treated with proteolytic enzymes to remove all accessible glycoprotein, (3) liposomes prepared from extracted SV lipids, and (4) plasma membranes from the host cell. Measurements have been made using spin labels (Sefton and Gaffney, 1974) and a hydrophobic fluorescent probe (Moore *et al.*, 1976). The results from the two techniques were in general agreement, showing that, of the samples tested, intact SV virions were the most rigid; i.e., the motions of the probe were the most severely restricted in the bilayer of these particles. Thus, removing the accessible portion of the glycoprotein from the particle, extracting the lipids, or substituting plasma membrane proteins for viral proteins all resulted in measurable increases in the motion of the probe, hence in the fluidity of the bilayer. It therefore appears that SV proteins may act to alter the physical properties of the bilayer by inhibiting molecular motions of the lipid molecules (Moore *et al.*, 1976; Sefton and Gaffney, 1974).

A general property of enveloped viruses, which is presumed to reflect a corresponding property of the host cell plasma membrane, is that the viral bilayer is asymmetric (reviewed in Lenard, 1978). Although data on alphaviruses are quite incomplete, it is clear that they share this property. Thus, all the glycolipids of SV are accessible to modification by neuraminidase added to the viral suspension and are therefore located on the external surface of the bilayer (Stoffel and Sorgo, 1976). In SFV grown in *Aedes albopictus* cells, an asymmetry of phospholipid composition is suggested by the finding that relatively more ceramide phosphoethanolamine and less glyceryl phosphoethanolamine is labeled by a nonpenetrating reagent than is present in the whole virion. The population labeled by the nonpenetrating reagent, although only a small part of the total, is presumed to reflect that present on the outer surface of the viral bilayer (Luukkonen *et al.*, 1976).

It has recently been shown that all the cholesterol in the SV envelope can be exchanged with cholesterol present in externally added lipid vesicles. The most likely explanation of this finding is that translocation of cholesterol occurs between the inner and outer surfaces of the bilayer and that this is rapid relative to exchange (Sefton and Gaffney, 1979).

III. ORIGIN AND DIVERSITY OF LIPIDS

In common with other enveloped viruses, the lipids of SV are not synthesized *de novo* after infection but are incorporated into mature virions from the pool of preexisting host cell lipids (Pfefferkorn and Hunter, 1963b). However, the cellular location utilized as the lipid source varies in different host cells. Both electron microscope (Acheson and Tamm, 1967) and biochemical (Richardson and Vance, 1976) evidence have shown that, when SFV is grown in baby hamster kidney (BHK) or chick embryo (CE) cells, budding occurs predominantly or exclusively from the plasma membrane of the infected cells. However, when cells from the mosquito *A. albopictus* are infected with SV, electron microscope evidence suggests that budding is predominantly from internal membranes into vacuoles which appear in the cytoplasm as a result of infection (Gliedman *et al.*, 1975) (see Chapter 17).

In cells in which alphaviruses bud from the plasma membrane, the lipid composition of the mature virions resembles quite closely the lipid composition of the plasma membrane of the host cell. The most thoroughly studied virus has been SFV grown in BHK cells. A comparison between the isolated plasma membranes of infected cells and purified SFV showed close similarity in overall lipid composition, in the distribution of phospholipid classes (Renkonen *et al.*, 1971), in the fatty acid chain composition of each phospholipid class, and in the nature of the glycerol–fatty acid attachment in each phospholipid class (Laine *et al.*, 1972; Renkonen *et al.*, 1972). Both the glycolipid and phospholipid compositions of SV grown in CE cells resemble those of plasma membranes from uninfected CE cells (Hirschberg and Robbins, 1974; Quigley *et al.*, 1971). Also, VEE virus possesses different phospholipids when grown in two different cell types (Heydrick *et al.*, 1971).

A significantly higher cholesterol/phospholipid ratio was found in BHK-grown SFV, as compared with BHK plasma membranes (Renkonen *et al.*, 1971, 1972). However, high cholesterol is not required for virion formation, since SFV formed from *A. albopictus* cells contains very little cholesterol. The virions are infective and possess the same phospholipid/protein ratio found in virions from BHK cells (Luukkonen *et al.*, 1977). In addition, the phospholipid composition of SFV is profoundly changed when it is grown in mosquito cells, with about two-thirds of the polar structures being different from those found in SFV grown in BHK cells (Luukkonen *et al.*, 1976). The comparative studies cannot rule out the selection of a population of phospholipids by the envelope proteins, since 100% substitution of each phospholipid was not achieved. However, they do set an upper limit on the size of the pool that can be thus selected and suggest that such a pool cannot be very large. Most of the viral lipids are passively incorporated into the budding virion from those present in the plasma membrane.

Compositional differences between genetically identical viruses grown in different cell types is reflected in physical changes as well. Fluorescence measurements using a lipid-soluble probe have shown that SV and SFV grown in BHK cells have bilayers with very similar measured properties. These properties differ greatly, however, from those of SV grown in *A. albopictus* cells (Moore *et al.*, 1976). Part or all of this difference may be attributed to the very low content of cholesterol in the mosquito-grown virions (Luukkonen *et al.*, 1977).

Sindbis virus grown in human fibroblasts from patients with I-cell disease (mucolipidosis II) becomes more sensitive to inactivation by freeze-thawing or by minute concentrations of nonionic detergents, in comparison with virions grown in normal human fibroblasts (Sly *et al.*, 1976). This property is shared by vesicular stomatitis virus grown in the same cells but not by influenza virus, suggesting that sialic acid may be responsible for these altered properties. I-cell-grown SV was found to possess excess sialic acid, although the glycopeptides from the viral glycoproteins were normal. Further, sensitivity to freeze-thawing was abolished by treatment of the I-cell grown SV with neuraminidase (Schlesinger *et al.*, 1978). These changes probably reflect an increase in the content of plasma membrane glycolipids (most notably the ganglioside G_{D_3}) in diseased as compared with normal fibroblasts (Dawson *et al.*, 1972).

Infection of CE cells by SV (Waite and Pfefferkorn, 1970) or of BHK cells by SV or SFV (Vance and Burke, 1974; Vance and Dahlke, 1975; Vance and Lam, 1975; cf. Waite and Pfefferkorn, 1970) results in a decrease in cellular phospholipid synthesis within a few hours after infection. Disruption of phospholipid metabolism by SFV infection has been shown to occur by quite different mechanisms in BHK cells and in CE cells. Infection of CE cells by different enveloped viruses also has quite different effects on phospholipid metabolism (Caric-Lazar *et al.*, 1978). However, the phospholipid composition of BHK plasma membranes is not altered by infection with SFV (Renkonen *et al.*, 1971).

IV. ROLE OF LIPIDS

In all enveloped viruses, lipids provide a matrix that mediates the interactions between the external glycoproteins and the internal core or capsid components of the intact virion. This role is established during viral biogenesis and budding from the host cell (Chapters 9 and 17). However, the interaction between the internal components and the lipid–glycoprotein envelope is distinctive in alphaviruses, as is evident from the effects of organic solvents and various detergents on the structure of alphavirions. Such treatments disrupt lipid structures generally, but in alphaviruses they have the specific effect of separating the virus cores cleanly from the lipid–glycoprotein envelopes (Appleyard *et al.*, 1970; Arif and Falkner, 1971; Goldblum *et al.*, 1972; Igarashi *et al.*, 1970; Kääriäinen *et al.*, 1969; Karabatsos, 1973; Pedersen *et al.*, 1973; Strauss *et al.*, 1968; Ventura and Scherer, 1970). The first effect of stepwise treatment of SFV with various detergents is thus detachment of the viral envelope from the core. This is followed by the gradual replacement of lipids associated with glycoproteins by detergent molecules, until finally lipid-free glycoproteins embedded in detergent micelles are obtained (Helenius and Soderlund, 1973; Helenius *et al.*, 1976; Simons *et al.*, 1973) (see Chapter 9).

The role of the lipid bilayer in maintaining the stability of SFV was investigated by Friedman and Pastan (1969). Semliki Forest virus was treated with phospholipase C, and it was found that nearly 60% of the total phospholipid was hydrolyzed. In light of more recent experiments, in which influenza virus was treated with the same enzyme (Rothman *et al.*, 1976; Tsai and Lenard, 1975), it seems certain that the hydrolyzed phospholipid was restricted to the outer half of the SFV bilayer. Infectivity of SFV was not affected by phospholipase digestion, but inactivation of the treated virus at 37°C was more rapid than that of an untreated control (Friedman and Pastan, 1969). Thus, the integrity of the outer leaflet of the SFV bilayer is not required for infectivity but apparently serves to stabilize the viral structure.

REFERENCES

Acheson, N. H., and Tamm, I. (1967). *Virology* **32,** 128–143.

Appleyard, G., Oram, J. D., and Stanley, J. L. (1970). *J. Gen. Virol.* **9**, 179–189.

Arif, B., and Faulkner, P. (1971). *Can. J. Microbiol.* **17,** 161–169.

Bittman, R., Majuk, S., Honig, D. S., Compans, R. W., and Lenard, J. (1976). *Biochim. Biophys. Acta* **433,** 63–74.

Caric-Lazar, M., Schwarz, R. T., and Scholtissek, C. (1978). *Eur. J. Biochem.* **91,** 351–361.

Dawson, G., Matalon, R., and Dorfman, A. (1972). *J. Biol. Chem.* **247,** 5951–5958.

Friedman, R. M., and Pastan, I. (1969). *J. Mol. Biol.* **40,** 107–115.

Gliedman, J. B., Smith, J. F., and Brown, D. T. (1975). *J. Virol.* **16,** 913–926.

Goldblum, N., Ravid, R., Hanoch, A., and Porath, Y. (1972). *Adv. Exp. Med. Biol.* **31,** 71–85.

Harrison, S. C., David, A., Jumblatt, J., and Darnell, J. E. (1971). *J. Mol. Biol.* **60**, 523–528.

Helenius, A., and Soderlund, H. (1973). *Biochim. Biophys. Acta* **307,** 287–300.

Helenius, A., Fries, E., Garoff, H., and Simons, K. (1976). *Biochim. Biophys. Acta* **436** 319–334.

Heydrick, F. P., Comer, J. F., and Wachter, R. F. (1971). *J. Virol.* **7**, 642–645.
Hirschberg, C. B., and Robbins, P. W. (1974). *Virology* **61**, 602–608.
Hubbell, W. L., and McConnell, H. M. (1969). *Proc. Natl. Acad. Sci. U.S.A.* **64**, 20–27.
Hughes, F., and Pedersen, C. E. (1975). *Biochim. Biophys. Acta* **394**, 102–110.
Igarashi, A., Fukuoka, T., Withiuthai, P., Hsu, L., and Fukai, K. (1970). *Biken J.* **13**, 93–110.
Jost, P., Libertini, L. J., Hebert, V. C., and Griffith, O. H. (1971). *J. Mol. Biol.* **59**, 77–98.
Kääriäinen, L., Simons, K., and von Bonsdorff, C. H. (1969). *Ann. Med. Exp. Fenn.* **47**, 235–248.
Karabatsos, N. (1973). *Arch. Gesamte Virusforsch.* **40**, 222–235.
Klenk, H. D. (1974). *Curr. Top. Microbiol. Immunol.* **68**, 29–58.
Laine, R., Kettunen, M., Gahmberg, C. G., Kääriäinen, L., and Renkonen, O. (1972). *J. Virol.* **10**, 433–438.
Laine, R., Soderland, H., and Renkonen, O. (1973). *Intervirology* **1**, 110–118.
Lenard, J. (1978). *Annu. Rev. Biophys. Bioeng.* **7**, 139–166.
Lenard, J., and Compans, R. W. (1974). *Biochim. Biophys. Acta* **344**, 51–94.
Lenard, J., and Lansberger, F. R. (1976). *In* "Mammalian Cell Membranes" (G. A. Jamieson and D. A. Robinson, eds.), Vol. 1, pp. 244–264. Butterworth, London.
Luukkonen, A., Kääriäinen, L., and Renkonen, O. (1976). *Biochim. Biophys. Acta* **450**, 109–120.
Luukkonen, A., von Bonsdorff, C., and Renkonen, O. (1977). *Virology* **78**, 331–335.
Moore, N. F., Barenholz, Y., and Wagner, R. R. (1976). *J. Virol.* **19**, 126–135.
Pedersen, C. E., Slocum, D. R., and Eddy, G. A. (1973). *Infect. Immun.* **8**, 901–906.
Pfefferkorn, E. R., and Hunter, H. S. (1963a). *Virology* **20**, 433–445.
Pfefferkorn, E. R., and Hunter, H. S. (1963b). *Virology* **20**, 446–456.
Quigley, J. P., Rifkin, D. B., and Reich, E. (1971). *Virology* **46**, 106–116.
Renkonen, O., Kääriäinen, L., Simons, K., and Gahmberg, C. G. (1971). *Virology* **46**, 318–326.
Renkonen, O., Kääriäinen, L., Gahmberg, C. G., and Simons, K. (1972). *Biochem. Soc. Symp.* **35**, 407–422.
Richardson, C. D., and Vance, D. E. (1976). *J. Biol. Chem.* **251**, 5544–5550.
Rothman, J. E., Tsai, D. K., Dawidowicz, E. A., and Lenard, J. (1976). *Biochemistry* **15**, 2361–2370.
Schlesinger, S., Sly, W. S., and Schulze, I. (1978). *Virology* **89**, 409–417.
Sefton, B. M., and Gaffney, B. J. (1974). *J. Mol. Biol.* **90**, 343–358.
Sefton, B. M., and Gaffney, B. J. (1979). *Biochemistry* **18**, 436–442.
Simons, K., Helenius, A., and Garoff, H. (1973). *J. Mol. Biol.* **80**, 119–133.
Sly, W. S., Lagwinska, E., and Schlesinger, S. (1976). *Proc. Natl. Acad. Sci. U.S.A.* **73**, 2443–2447.
Stoffel, W., and Sorgo, W. (1976). *Chem. Phys. Lipids* **17**, 324–335.
Strauss, J. H., Burge, B. W., Pfefferkorn, E. R., and Darnell, J. E. (1968). *Proc. Natl. Acad. Sci. U.S.A.* **58**, 533–537.
Strauss, J. H., Burge, B. W., and Darnell, J. E. (1970). *J. Mol. Biol.* **47**, 437–448.
Tsai, K. H., and Lenard, J. (1975). *Nature (London)* **253**, 554–555.
Vance, D. E., and Burke, D. C. (1974). *Eur. J. Biochem.* **43**, 327–336.
Vance, D. E., and Dahlke, R. M. (1975). *Can. J. Biochem.* **53**, 950–957.
Vance, D. E., and Lam, J. (1975). *J. Virol.* **16**, 1075–1076.
Ventura, A. K., and Scherer, W. F. (1970). *Proc. Soc. Exp. Biol. Med.* **133**, 711–717.
Waite, M. R. F., and Pfefferkorn, E. R. (1970). *J. Virol.* **6**, 637–643.

11

The Genome of Alphaviruses

S. IAN T. KENNEDY

Two members of the genus *Alphavirus*—Sindbis virus and Semliki Forest virus—have been intensively studied biochemically, particularly during the last decade. Other members of the genus, which contains about 20 members (Berge, 1975), have received much less attention. Despite this, and even though little nucleic acid sequence homology exists between the genetic material of different alphaviruses (Wengler *et al.*, 1977), it is likely that the concepts which have emerged from studies on Semliki Forest and Sindbis viruses, in particular those relating to the physical properties, genetic content, and organization of the genome, will apply to all alphaviruses.

The genetic material of alphaviruses is single-stranded RNA. Each virion contains a single molecule of this RNA complexed with about 300 molecules of a single protein species, the core protein, which together form the virus nucleocapsid (Horzinek and Mussgay, 1969; Laine *et al.*, 1973).

I. PHYSICAL PROPERTIES OF THE ALPHAVIRUS GENOME

Several techniques have been employed to determine the molecular weight of the alphavirus genome. Polyacrylamide gel electrophoresis gave estimates of 4.0–4.4 × 10^6 (Dobos and Faulkner,

THE TOGAVIRUSES

ISBN 0-12-625380-3

1970; Levin and Friedman, 1971; Simmons and Strauss, 1972; Martin and Burke, 1974), electron microscopy gave a value of 4.6×10^6 (Hsu *et al.*, 1973), sedimentation analysis of the duplex form of the genome gave a value for the component strands of 4.4×10^6 (Simmons and Strauss, 1972), and oligonucleotide fingerprinting gave an estimate of 4.4×10^6 (Lomniczi and Kennedy, 1977). Although each of these techniques suffers from one or more drawbacks, a molecular weight of 4.4×10^6 seems to be a reasonable working value.

The sedimentation coefficient of alphavirus RNA has been reported to be between 42 and 50 S. In view of the effect of several parameters, such as salt concentration, pH, and temperature, on the sedimentation behavior of single-stranded RNAs, together with the inherent inaccuracies of the determination of sedimentation coefficients by velocity sedimentation through density gradients in the preparative ultracentrifuge (particularly where such studies involve, in the main, extensive extrapolation from marker s values), it is not surprising that different estimates of the sedimentation coefficient of alphavirus RNA have appeared in the literature. Perhaps the most accurate sedimentation analysis was performed by J. H. Strauss (personal communication), who examined Sindbis virus RNA in the analytical ultracentrifuge and obtained an s_{20} of 49 S in 0.2 M salt. However, in the following discussion (see also Chapter 12), the alphavirus genome will be designated 42 S. This notation will be used because the author, in common with other workers in the field, uses 42 S in an identifying capacity.

Several early experiments using mild denaturing conditions suggested that the alphavirus genome might have subunit structure and be dissociable into several fragments (Dobos and Faulkner, 1970; Dobos *et al.*, 1971). More recently, however, carefully controlled extraction from either infected cells or virus particles has resulted in the preparation of RNA molecules which, by a variety of tests, including treatment with chaotropic agents such as urea, dimethyl sulfoxide, and methylmercuric hydroxide (Arif and Faulkner, 1972; Simmons and Strauss, 1972; Hsu *et al.*, 1973; S. I. T. Kennedy, unpublished observations), clearly consist of a single continuous polynucleotide chain which, taking a molecular weight of 4.4×10^6, contains about 13,000 nucleotides.

The alphavirus genome is infectious (Wecker, 1959; Friedman *et al.*, 1966); i.e., it is of positive polarity and can serve as mRNA both in infected cells and in cell-free protein-synthesizing systems (Mowshowitz, 1973; Söderlund *et al.*, 1973; Simmons and Strauss, 1974a,b; Wengler and Wengler, 1974; Smith *et al.*, 1974; Cancedda *et al.*, 1975; Glanville *et al.*, 1976). Expression of the genome therefore occurs directly upon infection, and the virus does not contain an associated polymerase.

In common with many eukaryotic mRNAs the alphavirus genome contains a tract of adenylic acid residues at its 3′-terminus. Estimates of the size of this poly(A) tract suggest a length of between 50 and 120 nucleotides (Eaton *et al.*, 1972; Johnston and Bose, 1972a,b; Clegg and Kennedy, 1974a). The biological significance of this poly(A) tract is unknown, but it seems likely that it plays a role either in translation or in replication of the RNA.

Again, in common with many eukaryotic mRNAs, the 5′-end of the alphavirus genome is capped and methylated. The structure of this cap in Sindbis virus RNA is 7mG(5′)ppp(5′)Ap, with uridylic acid as the next nucleotide (Dubin *et al.*, 1977). The base compositions of various alphavirus RNAs are similar, with about 29% adenine, 25% guanine, 25% cytosine, and 21% uracil (Pfefferkorn and Hunter, 1963; Kääriäinen and Gomatos, 1969).

When Sindbis virus RNA is examined in the electron microscope, circles are observed (Hsu *et al.*, 1973; J. Simons and S. I. T. Kennedy, unpublished observations). These circles contain panhandles which correspond to a double-stranded region of RNA formed by hydrogen bonding between the two ends of the genome. Thus there exists a 5′ → 3′ nucleotide sequence near the 5′-end of 42 S RNA, which is complementary to a 3′ → 5′ sequence near the 3′-end. Measurements in the electron microscope indicate that the length of these complementary ends is on the order of 150 nucleotides. Since the poly(A) tract at the extreme 3′-end of the genome appears to be outside the panhandle, this tract is probably not directly involved in hydrogen bonding of the ends. In a recent study Frey *et al.* (1979) presented evidence from melting studies on Sindbis virus RNA that the complementary sequences are relatively short, on the order of 10 to 20 nucleotides. More recently we have performed limited sequence analysis of fragments up to 150 nucleotides in length derived from the 5′- and 3′-ends of Semliki Forest virus RNA. These studies, although not yet complete, clearly confirm the existence of complementary sequences at the ends of Semliki Forest virus RNA. Although the biological significance of these complementary regions has yet to be determined, one possible role is in initiation or termination of RNA synthesis.

II. GENETIC CONTENT AND ORGANIZATION OF THE ALPHAVIRUS GENOME

Since the alphavirus genome is infectious, it must, ipso facto, encode all the virus-specific polypeptides required for multiplication.

These polypeptides include not only the structural proteins of the virion but also polypeptides with the enzymatic functions of replicating the viral RNA and of transcribing a subgenomic fragment—26 S RNA—which, as will be described in detail in Chapters 12 and 13, acts as mRNA for the structural proteins. For convenience therefore, the alphavirus genes have been divided into two groups—those encoding structural polypeptides (structural genes) and those encoding nonstructural polypeptides (NS polypeptides, nonstructural genes).

Three approaches have been employed in determining the number of genes present in the alphavirus genome and their organization therein. These are (1) analysis of virus-specific polypeptides in infected cells, (2) characterization of the virus-specific products formed in cell-free protein-synthesizing systems programmed by isolated Semliki Forest virus or Sindbis virus 42 S or 26 S RNA, and (3) complementation studies between temperature-sensitive (*ts*) mutants of Sindbis virus. Each of these topics will be covered in detail in other parts of this book, and therefore only a synthesis of the main observations will be presented here.

The observation that infection of vertebrate cells with alphaviruses results in the marked reduction of host cell macromolecular synthesis has greatly facilitated study of the synthesis of alphavirus-specific polypeptides in infected cells. Because much more structural than nonstructural protein is made during the multiplication cycle, elucidation of the mechanism of synthesis of structural proteins preceded studies on synthesis of NS polypeptides, and only recently has a fairly clear picture emerged of the formation of these polypeptides. First, concerning the structural polypeptides there is general agreement from biochemical and genetic experiments that the alphavirus genome contains four structural polypeptide genes (Strauss *et al.*, 1969; Acheson and Tamm, 1970; Simons and Kääriäinen, 1970; Garoff *et al.*, 1974; Burge and Pfefferkorn, 1966; Strauss *et al.*, 1975). These comprise one gene (*C*) coding for the nucleocapsid polypeptide, and three genes (*E1, E2,* and *E3*) coding for the peptide portions of glycopolypeptides. These glycopolypeptides form part of the virus envelope. Since, as mentioned previously, the mRNA for these polypeptides is a subgenomic fragment of 42 S RNA, namely, 26 S RNA (molecular weight, $1.8 \pm 0.1 \times 10^6$; Simmons and Strauss, 1972), the recent observation that the nucleotide sequence of 26 S RNA is located inward from the 3′-end of the genome (Wengler and Wengler, 1976; Kennedy, 1976) positions the structural protein genes in the 3′-terminal one-third of 42 S RNA. With the use of salt synchronization of translation initiation in Semliki Forest virus-infected cells (Clegg,

1975) and tryptic peptide mapping of aberrant polypeptides formed in Semliki Forest virus *ts* mutant-infected cells (Lachmi *et al.*, 1975), the 5′ → 3′ order of the structural genes has been determined. This order is *C-E3-E2-E1*. Studies on the translation of 26 S RNA both *in vivo* and *in vitro* show that initiation of synthesis of the structural proteins on 26 S RNA occurs at a single site at or near the 5′-end of this RNA and that the structural proteins are fashioned through a series of proteolytic cleavages (Clegg and Kennedy, 1975a,b; Cancedda and Schlesinger, 1974; Cancedda *et al.*, 1974; Simmons and Strauss, 1974b; Wengler *et al.*, 1974; Schlesinger and Schlesinger, 1973; Morser *et al.*, 1973; Simons *et al.*, 1973; Morser and Burke, 1974).

Let us now consider nonstructural genes. In contrast to the cell-free translation of isolated 26 S RNA, which produces discrete structural polypeptides of or close to the size of their respective *in vivo* counterparts, translation of 42 S RNA produces a spectrum of polypeptides ranging in size from 20,000 to 200,000 daltons. Few, if any, correspond to polypeptides found in the infected call (Simmons and Strauss, 1974b; Cancedda *et al.*, 1975; Glanville *et al.*, 1976). However, using carefully controlled conditions, it is clear from tryptic peptide mapping of the 42 S translation products that all these products contain amino acid sequences which correspond to NS polypeptide material. This observation indicates that the translation of 42 S RNA does not, per se, produce structural proteins. Thus, in common with other eukaryotic polycistronic mRNAs, the internal site for the initiation of synthesis of structural polypeptides is not functional in 42 S RNA. Cell-free studies also show that the translation of nonstructural genes is initiated at a single site (Cancedda *et al.*, 1975; Glanville *et al.*, 1976) and imply therefore that, like structural polypeptides, NS polypeptides are formed by posttranslational cleavage of a polyprotein precursor. Analysis of cells infected with Sindbis virus have recently shown that there exist three NS polypeptides designated NS p89, NS p82, and NS p60 (numbers refer to molecular weight in kilodaltons) and that these are components of the virus-specified RNA-dependent RNA polymerase (Clewley and Kennedy, 1976; Brzeski and Kennedy, 1977). In addition a fourth NS polypeptide (polypeptide X) of unknown function has been inferred (Brzeski and Kennedy, 1977). These observations, taken in conjunction with the finding that there are three, possibly four, RNA$^-$ *ts* mutant complementation groups (Strauss *et al.*, 1975), argue that there exist at least four nonstructural genes. With the use of salt synchronization the 5′ → 3′ order of the genes coding for these NS polypeptides has been established as X-NS p60-NS p89-NS p82. Since the structural genes occupy the 3′-terminal

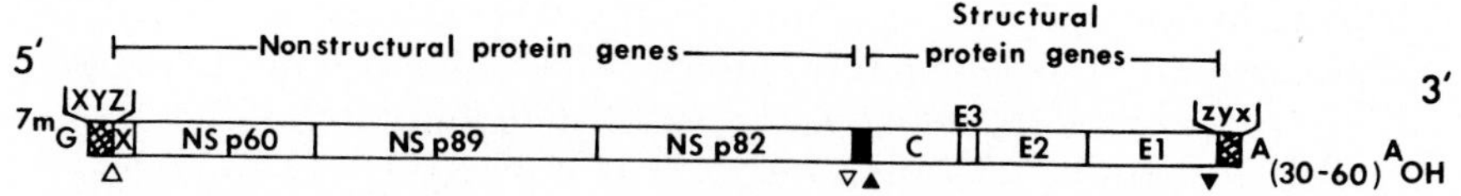

Fig. 1. Genetic content and organization of the alphavirus genome. ▭, Gene regions, approximately to scale; ■, noncoding region of unknown length; ▩, complementary nucleotide regions, presumed to be noncoding, where XYZ at the 5′-end denotes a nucleotide sequence complementary to xyz at the 3′ end; △, translational initiation site; ▽, translational termination site; ▲, cryptic translational initiation site; ▼, cryptic translational termination site.

one-third of the genome, these nonstructural genes must be located in the 5′ two-thirds of 42 S RNA. Moreover, since the 5′ two-thirds and the 3′ one-third are expressed by different mechanisms—the former directly and the latter only after transcription of 26 S RNA—there must exist a noncoding region between the nonstructural and structural genes. This region contains the nucleotide sequence which, in 26 S RNA, binds ribosomes and which, as discussed in the following chapter, also contains the complement of the binding site for the polymerase for 26 S RNA synthesis. The length of this noncoding region is unknown.

Based on a combination of the structural and nonstructural gene studies, the genetic content and organization of the alphavirus genome can be depicted as shown in Fig. 1. For details on both the expression and replication of 42 S RNA and synthesis and translation of 26 S RNA the reader is directed to the appropriate chapters in this volume.

REFERENCES

Acheson, N. H., and Tamm, I. (1970). *Virology* **41**, 321–331.

Arif, B. M., and Faulkner, P. (1972). *J. Virol.* **9**, 102–107.

Berge, T. O., ed. (1975). "International Catalogue of Arboviruses," 2nd ed., DHEW Publ. No. (CDC) 75–8301. U.S. Dep. Health, Educ. Welfare, Public Health Serv., Washington, D.C.

Brzeski, H., and Kennedy, S. I. T. (1977). *J. Virol.* **22**, 420–429.

Burge, B. W., and Pfefferkorn, E. R. (1966). *Virology* **30**, 214–223.

Cancedda, R., and Schlesinger, M. J. (1974). *Proc. Natl. Acad. Sci. U.S.A.* **71**, 1843–1847.

Cancedda, R., Swanson, R., and Schlesinger, M. J. (1974). *J. Virol.* **14**, 652–663.

Cancedda, R., Villa-Komaroff, L., Lodish, H., and Schlesinger, M. J. (1975). *Cell* **6**, 215–222.

Clegg, J. C. S. (1975). *Nature (London)* **254**, 454–455.

Clegg, J. C. S., and Kennedy, S. I. T. (1974a). *J. Gen. Virol.* **22**, 331–340.

Clegg, J. C. S., and Kennedy, S. I. T. (1974b). *FEBS Lett.* **42**, 327–330.

Clegg, J. C. S., and Kennedy, S. I. T. (1975a). *Eur. J. Biochem.* **53,** 175–183.
Clegg, J. C. S., and Kennedy, S. I. T. (1975b). *J. Mol. Biol.* **97,** 401–411.
Clewly, J. P., and Kennedy, S. I. T. (1976). *J. Gen. Virol.* **32,** 395–411.
Dobos, P., and Faulkner, P. (1970). *J. Virol.* **6,** 145–147.
Dobos, P., Arif, B. M., and Faulkner, P. (1971). *J. Gen. Virol.* **10,** 103–106.
Dubin, D. T., Stollar, J., Hsuchen, C., Timko, K., and Guild, G. M. (1977). *Virology* **77,** 457–470.
Eaton, B. T., Donaghue, P., and Faulkner, P. (1972). *Nature (London) New Biol.* **238,** 109–111.
Friedman, R. M., Levy, H. B., and Carter, W. B. (1966). *Proc. Natl. Acad. Sci. U.S.A.* **65,** 440–446.
Garoff, H., Simons, K., and Renkonen, O. (1974). *Virology* **61,** 493–502.
Glanville, N., Ranki, M., Morser, J., Kääriäinen, L., and Smith, A. E. (1976). *Proc. Natl. Acad. Sci. U.S.A.* **73,** 3059–3063.
Horzinek, M., and Mussgay, M. (1969). *J. Virol.* **4,** 514–521.
Hsu, M.-T., Kung, H.-J., and Davidson, N. (1973). *Cold Spring Harbor Symp. Quant. Biol.* **38,** 943–948.
Johnston, R. E., and Bose, H. R. (1972a). *Biochem. Biophys. Res. Commun.* **46,** 712–717.
Johnston, R. E., and Bose, H. R. (1972b). *Proc. Natl. Acad. Sci. U.S.A.* **69,** 1514–1519.
Kääriäinen, L., and Gomatos, P. J. (1969). *J. Gen. Virol.* **5,** 251–260.
Kennedy, S. I. T. (1976). *J. Mol. Biol.* **108,** 491–512.
Lachmi, B. E., Glanville, N., Karänen, S., and Kääriäinen, L. (1975). *J. Virol.* **16,** 1615–1629.
Laine, R., Söderlund, H., and Renkonen, O. (1973). *Intervirology* **1,** 110–114.
Levin, J. G., and Friedman, R. M. (1971). *J. Virol.* **7,** 504–510.
Lomniczi, B., and Kennedy, S. I. T. (1977). *J. Virol.* **24,** 99–107.
Martin, B. A. B., and Burke, D. C. (1974). *J. Gen. Virol.* **24,** 45–66.
Morser, M. J., and Burke, D. C. (1974). *J. Gen. Virol.* **22,** 395–409.
Morser, M. J., Kennedy, S. I. T., and Burke, D. C. (1973). *J. Gen. Virol.* **21,** 19–29.
Mowshowitz, D. (1973). *J. Virol.* **11,** 535–543.
Pfefferkorn, E. R., and Hunter, H. S. (1963). *Virology* **20,** 446–465.
Schlesinger, M. J., and Schlesinger, S. (1973). *J. Virol.* **11,** 1013–1016.
Simmons, D. T., and Strauss, J. H. (1972). *J. Mol. Biol.* **71,** 599–608.
Simmons, D. T., and Strauss, J. H. (1974a). *J. Virol.* **14,** 552–563.
Simmons, D. T., and Strauss, J. H. (1974b). *J. Mol. Biol.* **86,** 397–409.
Simons, K., and Kääriäinen, L. (1970). *Biochem. Biophys. Res. Commun.* **38,** 981–992.
Simons, K., Karänen, S., and Kääriäinen, L. (1973). *FEBS Lett.* **29,** 87–91.
Smith, A. E., Wheeler, T., Glanville, N., and Kääriäinen, L. (1974). *Eur. J. Biochem.* **49,** 101–112.
Söderlund, H., Glanville, N., and Kääriäinen, L. (1973). *Intervirology* **2,** 100–107.
Strauss, E. G., Lenches, E. M., and Strauss, J. H. (1975). *Virology* **68,** 191–223.
Strauss, J. H., Burge, B. W., and Darnell, J. E. (1969). *Virology* **37,** 367–373.
Wecker, E. (1959). *Virology* **7,** 241–249.
Wengler, G., and Wengler, G. (1974). *Virology* **59,** 21–30.
Wengler, G., and Wengler, G. (1976). *Virology* **73,** 190–199.
Wengler, G., Beato, M., and Hackemack, S. A. (1974). *Virology* **61,** 120–128.
Wengler, G., Wengler, G., and Filipe, A. R. (1977). *Virology* **78,** 124–134.

12

Synthesis of Alphavirus RNA

S. IAN T. KENNEDY

This chapter will review current knowledge of the mechanism and regulation of alphavirus-specified RNA synthesis and of the topographical arrangement of the virus RNA-synthesizing apparatus within the infected vertebrate cell.

I. INTRODUCTION

In common with other single-stranded positive-polarity RNA viruses (with the exception of the retroviruses), multiplication of alphaviruses involves the intermediate formation of RNA of negative polarity. The role of this species of RNA is to act as template for plus-strand synthesis, and it does not leave the infected cell in virus particles.

Several techniques, notably molecular hydridization, oligonucleotide fingerprinting, and analysis of the phenotypic expression of temperature-sensitive (*ts*) mutants, have contributed to our present

THE TOGAVIRUSES

ISBN 0-12-625380-3

knowledge of alphavirus RNA synthesis. However, despite intensive efforts and in common with many other animal viruses, there exists no technology for reconstructing viral RNA synthesis *in vitro*. This severe limitation relegates to the level of hypothesis many aspects of the mechanism of synthesis of alphavirus RNA.

Although suggested by many early studies, it is only the recent development of efficient enucleation techniques that has clearly shown that the alphavirus multiplication cycle in vertebrate cells is wholly independent of a functioning cell nucleus (Follett *et al.*, 1975). This observation, however, merely shows that neither the environment of the nucleus nor any host-coded polypeptide with an extremely short half-life and unstable mRNA is required during the 4- to 6-h alphavirus multiplication cycle.

II. EARLY EVENTS

As outlined in Chapter 11 and described in detail elsewhere in this volume, the incoming 42 S viral plus-strand RNA is translated to form nonstructural polypeptides some or all of which are components of the virus-specified RNA-dependent RNA polymerase (Lachmi *et al.*, 1975; Clewley and Kennedy, 1976; Brzeski and Kennedy, 1977). This polymerase then functions in mediating synthesis of 42 S minus-strand RNA, using the parental strand as template. Although these early events seem to be fairly clearly established, many aspects of them are obscure. For example, assuming that a single 42 S RNA molecule enters a given cell, what determines its partition of labor between translation on the one hand and template status on the other? Three possible explanations, not necessarily independent of one another, can be envisaged to solve this problem.

First, since synthesis of the polymerase polypeptides involves translation of only the 5′-terminal two-thirds of 42 S RNA (Clegg *et al.*, 1976; Simmons and Strauss, 1974a; Bracha *et al.*, 1976), polymerase, once made, is free to associate with the 3′-end of the RNA and commence minus-strand synthesis. Such could be the affinity of the enzyme for the template that, when the enzyme reaches the termination region for the nonstructural polypeptide genes, encumbent ribosomes are displaced and RNA synthesis continues, forming 42 S minus-strand RNA. Second, if input core protein remains associated with the genome, it may function in regulating the translational and template roles of 42 S RNA. The third possibility is that the circularization status of the RNA (see Chapter 11; see also Hsu *et al.*, 1973) may determine to what extent ribosomes and/or polymerase bind to the genome.

A second unanswered question relating to these early events involves precisely where they occur in the cytoplasm of the infected cell. Studies on the expression of 26 S RNA—the subgenomic species representing the 3′-terminal one-third of 42 S RNA and the mRNA for the virion structural proteins (Clegg and Kennedy, 1974a; Cancedda and Schlesinger, 1974; Cancedda *et al.*, 1974; Simmons and Strauss, 1974a; Wengler *et al.*, 1974)—have clearly shown that this RNA is translated on membrane-bound ribosomes (Kennedy, 1972; Mowshowitz, 1973; Martire *et al.*, 1977). A single report (Martire *et al.*, 1977) suggests that 42 S RNA is translated on free ribosomes, but confirmation of this is required.

It is clear from several studies that the polymerase is associated with the internal smooth membranes of the infected cell (Grimley *et al.*, 1972; Friedman *et al.*, 1972; Sreevalsan, 1972; Clewley and Kennedy, 1976), and indeed it has been suggested that there exist "RNA factories" consisting of complexes of polymerase and membrane which together form what have been described as cytopathic vacuoles (Grimley *et al.*, 1972; Friedman *et al.*, 1972). In view of this close association between the polymerase and internal membranes, and in light of the finding that the primary translation product of the nonstructural genes of 42 S RNA contains, at its N-terminus, a sequence of about 80 amino acid residues which have been likened to a signal sequence (Brzeski and Kennedy, 1977; Blobel and Dobberstein, 1975), it is tempting to speculate either that 42 S RNA is translated by membrane-bound ribosomes and that the nascent polypeptide chain is anchored in the membrane, or that a high affinity exists between the completed primary translation product of the nonstructural genes—synthesized by free ribosomes—and smooth membrane.

Another aspect of the virus-specified polymerase requiring investigation is the possible involvement of host cell-coded polypeptide components in its functioning. For present purposes these components will be considered distinct from the polypeptides which form part of the intracellular skeletal framework to which the polymerase is anchored. "Functional" host cell polypeptide involvement could take two forms. First, host cell-coded polypeptide(s) could play a direct role in the catalytic activity of the polymerase. We have already shown that highly purified, enzymatically active polymerase isolated from Semliki Forest virus (SFV)- or Sindbis virus (SV)-infected hamster cells contains two, or possibly three, virus-specified polypeptides (Clewley and Kennedy, 1976; Kennedy, unpublished observations).

Thus there is evidence that the polymerase (or polymerases if more than one exists; see later) may be a multisubunit enzyme, and it is conceivable that one or more of these subunits is host cell-specified. Indeed, purified preparations of polymerase do contain substantial amounts of a single host-coded polypeptide (Clewley and Kennedy, 1976), but whether this is a subunit of the polymerase is undetermined. On the other hand, host cell-coded polypeptide(s) may play an indirect rather than a direct role in the catalytic mechanism of the polymerase. They may determine the rate of RNA synthesis or, as will be discussed in more detail elsewhere in this volume, they may influence the fidelity of synthesis and play a role in the generation and/or replication of defective interfering (DI) RNA.

III. NATURE AND SYNTHESIS OF ALPHAVIRUS MINUS-STRAND RNA

As described above, incoming (parental) 42 S RNA is copied by the virus polymerase to form 42 S RNA of negative polarity. This species of RNA is not found free in the infected cell but rather is tightly associated with the RNA-synthesizing apparatus where, as will be described in Section IV, it functions as template for plus-strand synthesis. Thus 42 S negative-strand RNA, when isolated from infected cells by phenol extraction, is in the form of multistranded complexes. These complexes take two forms: Either they consist of a duplex of 42 S plus and minus strands hydrogen-bonded together (replicative form, RF) or they consist of a 42 S negative strand together with several nascent plus strands of varying length, each bound to the negative strand by regions of hydrogen bonding (replicative intermediate, RI). With the use of chaotropic agents it is possible to isolate the 42 S negative-strand RNA from these multistranded complexes (Bruton and Kennedy, 1975). Indeed, these studies show that the only size of negative-strand RNA formed in alphavirus-infected cells at any time during multiplication is 42 S and, although it has not been accurately sized, it is presumed that this minus strand is an essentially complete complement of 42 S plus-strand RNA. Certainly it comigrates with 42 S plus-strand RNA on polyacrylamide gels (Bruton and Kennedy, 1975). Moreover, the observation that 42 S minus-strand RNA contains a 5′-terminal tract of poly(U) complementary to the 3′-terminal poly(A) tract in 42 S plus-strand RNA (Sawicki and Gomatos, 1976) also supports the proposition that 42 S plus- and minus-strand RNAs are of equal size.

Possibly because of its exclusive template role, the amount of 42 S negative-strand RNA synthesized in alphavirus-infected cells is extremely small compared to the amount of plus-strand RNA. This observation, together with the difficulty and tedium of isolating 42 S minus-strand RNA free of plus-strand material, has severely restricted detailed characterization of this species of RNA. However, assuming that 42 S minus-strand RNA is the product of copying the entire nucleotide sequence (possibly excluding the 5′-terminal cap) of 42 S plus-strand RNA, and given that 42 S plus-strand RNA contains regions of sequence complementarity near its 3′- and 5′-ends (see Chapter 11), then it is possible to speculate on the mechanism of minus-strand synthesis. If it is assumed that copying is initiated at or near the 3′-end of 42 S plus-strand RNA, then the complementary nucleotide sequences present at the 3′- and 5′-ends (Hsu *et al.*, 1973) will also be present in the progeny minus strands. This situation is depicted in Fig. 1. If, as seems probable from studies on DI RNA (see later), polymerase binds to and initiates synthesis close to the region represented by XYZ (Fig. 1a) on the plus strand, then the identical sequence in the minus strand may act as a binding–initiation region for plus-strand synthesis using the negative strand as template. As a consequence, the enzyme which copies plus strand to minus strand could also copy minus strand to plus strand. It is, of course, possible that the 42 S plus-strand template RNA exists in the infected cell with its complementary ends hydrogen-bonded together (Fig. 1b). If this were the case, then the only modifications of the idea of one enzyme mediating both 42 S minus- and 42 S plus-strand synthesis would be that the enzyme binding–initiation sites would be double-stranded, namely $\begin{smallmatrix}Xx\\Yy\\Zz\end{smallmatrix}$, and that the minus strand would also exist with its ends juxtaposed. However, confirmation of this conservation model must await the *in vitro* reconstruction of systems capable of synthesizing 42 S plus- and minus-strand RNA.

IV. NATURE AND SYNTHESIS OF ALPHAVIRUS PLUS-STRAND RNA

Cells infected with alphaviruses contain two major species of single-stranded RNA of positive polarity. These are 42 S RNA, which is indistinguishable from the virion RNA, and RNA with a sedimentation coefficient of 26 S, which was originally designated "interjacent" RNA (Sonnabend *et al.*, 1967; Friedman *et al.*, 1966; Sreevalsan and Lockart, 1966). The molecular weight of 26 S RNA has been estimated

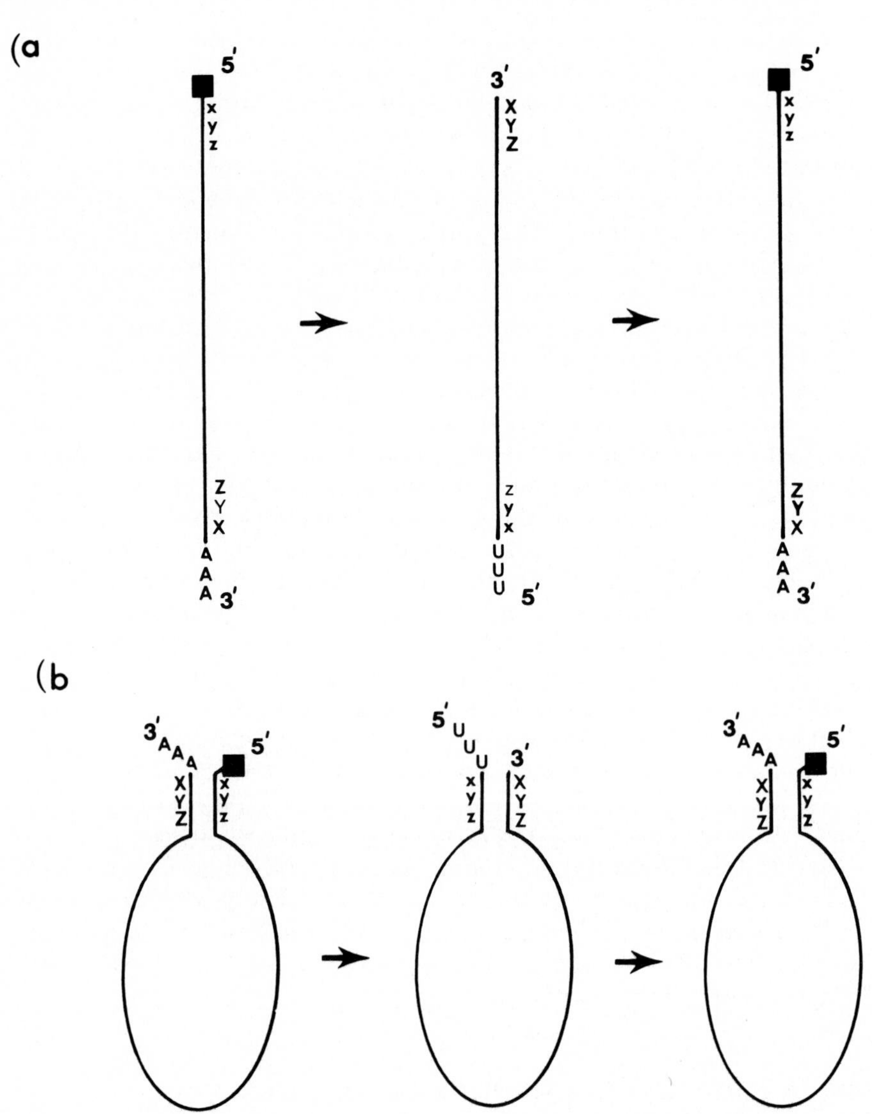

Fig. 1. Diagrammatic representation of the synthesis of alphavirus plus- and minus-strand 42 S RNA.

to be $1.8 \pm 0.2 \times 10^6$ by polyacrylamide gel electrophoresis (Kennedy, 1972; Martin and Burke, 1974; Levin and Friedman, 1971) and $1.75 \pm 0.1 \times 10^6$ by oligonucleotide fingerprinting (Kennedy, unpublished observations) and, like the genome, 26 S RNA is capped and methylated at its 5′-terminus (Dubin *et al.*, 1977) and contains a 3′-tract of poly(A) (Clegg and Kennedy, 1974b). From molecular hybridization studies (Simmons and Strauss, 1972a) and oligonucleotide fingerprinting (Wengler and Wengler, 1976; Kennedy, 1976), it is now clearly established that the nucleotide sequence of 26 S RNA represents about one-third of that of the 42 S RNA genome. Synthesis of 26 S RNA therefore represents a mechanism of selectively amplifying the genetic content of one-third of the virion RNA. Based on the alphavirus used and the host cell type employed, the 42 S/26 S RNA molar ratio of synthesis ranges from 0.2 to 0.8. For the AR339 wild-type strain of Sindbis virus in chicken fibroblasts the ratio varies by only 0.03 during the entire duration of the multiplication cycle, suggesting that the ratio is under close control.

In addition to 42 and 26 S RNA, two other species of virus-specified RNA are found in alphavirus-infected cells. These RNAs, which have been designated 38 and 33 S RNA, have apparent molecular weights of $3.1 \pm 0.2 \times 10^6$ and $2.4 \pm 0.2 \times 10^6$, respectively (Levin and Friedman, 1971; Martin and Burke, 1974). They are found in low and variable amounts during multiplication. Based on several studies taken together these minor species comprise no more than 10% of total plus-strand RNA synthesis. With the use of molecular hybridization (Simmons and Strauss, 1974b), denaturing gel electrophoresis, and oligonucleotide fingerprinting (Kennedy, 1976) it now seems clear that these minor RNA species are not unique RNAs but rather conformational variants: 38 S RNA appears to be a conformational variant of 42 S RNA, and 33 S RNA a conformational variant of 26 S RNA. Thus only two unique virus-specified single-stranded RNAs of positive polarity—42 and 26 S RNA—are synthesized during the multiplication cycle of standard infectious alphaviruses in vertebrate cells.

Over recent years the role of 42 and 26 S RNA in the multiplication cycle of alphaviruses has become clear. 42 S RNA has two roles. The first is to act as mRNA for the nonstructural polypeptides, some or all of which are components of the virus polymerase (Lachmi *et al.*, 1975; Lachmi and Kääriäinen, 1976; Clegg *et al.*, 1976; Clewley and Kennedy, 1976; Bracha *et al.*, 1976; Brzeski and Kennedy, 1977). The second role, as described in Section III, is to act as template for the synthesis of 42 S negative-strand RNA. 26 S RNA, however, plays only a single role—that of mRNA for the structural polypeptides of the virus particle. For a detailed consideration of the messenger functions of 42 and 26 S RNA the reader is directed to Chapter 13 of this volume.

The observation that 26 S RNA represents a discrete subgenomic fragment of 42 S RNA has prompted the question concerning which part of the nucleotide sequence of 42 S RNA it represents. The answer to this question was provided by oligonucleotide fingerprinting of 3′-terminal fragments of 42 and 26 S RNA generated by ultrasonic or mild alkali cleavage (Wengler and Wengler, 1976; Kennedy, 1976). Since these oligonucleotide fingerprints were identical, the nucleotide sequence of 26 S RNA must be located inward from the 3′-end of 42 S RNA. We shall now consider how 26 S RNA is synthesized in infected cells.

V. MECHANISM OF SYNTHESIS OF 26 S RNA

Despite careful searches by several groups, no evidence of a 26 S RNA of negative polarity has ever been found, and the current hypothesis for 26 S RNA synthesis is that it is transcribed directly from 42 S negative-strand RNA. Thus 42 S negative-strand RNA acts as template for both 42 and 26 S plus-strand RNA. At least two models can be envisaged for 26 S RNA formation. In the first, 26 S RNA synthesis is initiated internally at a point one-third of the way from the 5′-end of the 42 S minus-strand template, and the polymerase terminates synthesis at the 5′-end of the template strand, releasing 26 S RNA. In the second model, 26 S RNA is formed by degradative processing of a percentage of newly synthesized 42 S plus-strand RNA molecules. Two approaches have been used to distinguish between these two models. The first of these constitutes the elegant studies of Simmons and Strauss (1972b). Briefly, infected cells were pulse-labeled for varying times in the presence of actinomycin D (added to inhibit host cell RNA synthesis), and nucleic acid extracts prepared. These extracts were then treated with RNase to destroy single-stranded RNA and to trim off the single-stranded tails of the RIs, so converting them to structures akin to RFs. Three species of RF designated RFI, RFII, and RFIII were generated by this procedure. With the use of molecular hybridization the negative-strand component of RFI was shown to be complementary to the entire 42 S RNA genome, that of RFIII to be complementary to 26 S RNA, and that of RFII to be complementary to the portion of 42 S RNA not represented in 26 S RNA. By carefully analyzing the kinetics of labeling of each of these RFs and their molar ratios, Simmons and Strauss (1972b) tentatively concluded that RFI was derived from an RI-producing 42 S RNA and that RFII and RFIII were joined together in the infected cells, forming an RI-containing 42 S minus-strand RNA as template and devoted to 26 S RNA synthesis by transcription of only the "26 S RNA region of the template." Confirmation that 26 S RNA is formed by the transcription of one-third of the 42 S RNA negative-strand template has recently been presented by Brzeski and Kennedy (1978) who showed that the uv target size for the synthesis of 26 S RNA was the same as the physical size of 26 S RNA and not, as in the degradative processing model, the size of 42 S RNA. These uv inactivation studies also showed that in SV-infected chicken cells there existed a population of plus-strand RNA-synthesizing complexes whose members were each capable of forming both 42 and 26 S RNA and that, on a time-averaged basis, each complex contained one virus polymerase mediating 42 S RNA synthesis and three mediating 26 S RNA synthesis (Brzeski and Kennedy, 1979).

VI. REGULATION OF THE SYNTHESIS OF 42 AND 26 S PLUS-STRAND RNA

Let us now consider what is known about regulation of the synthesis of 42 and 26 S RNA. Since reconstruction experiments using purified polymerase components and 42 S negative-strand template RNA are not yet possible, the available information on the regulation of 42 and 26 S RNA synthesis comes from (1) studies using *ts* mutants, (2) fractionation studies on crude cell-free extracts capable of 42 and 26 S RNA synthesis, and (3) a characterization of the RFs and RIs isolated from infected cells. Scheele and Pfefferkorn (1969) and Waite (1973) reported that two of the *ts* mutants originally isolated by Burge and Pfefferkorn (1966) ceased to produce 26 S RNA but continued to form 42 S RNA after a shift from the permissive (30°C) to the restrictive 39.5°C) temperature. These findings were interpreted as evidence for two virus-specified RNA polymerases—one mediating the synthesis of 42 S RNA, and the other that of 26 S RNA. Additional support for two polymerases was provided by the finding that a crude *in vitro* system capable of 42 and 26 S RNA synthesis could be fractionated to give two classes of replication complex, one forming 42 S RNA and the other 26 S RNA (Michel and Gomatos, 1973). From a study of RFs and RIs, Segal and Sreevalsan (1974) also concluded that 42 and 26 S RNA were made by different polymerases.

Since these early *ts* mutant, cell fractionation, and RF–RI studies, the mechanism of synthesis of the three alphavirus-specified nonstructural polypeptides has been defined (Clegg *et al.*, 1976; Brzeski and Kennedy, 1977). These polypeptides (designated NS p89, NS p60, and NS p82 for SV are formed by a series of proteolytic cleavages from a large polyprotein and are all components of highly purified, enzymatically active polymerase (Clewley and Kennedy,

1976; Brzeski and Kennedy, unpublished observations; see also Chapter 13).

Against this background of knowledge of the mechanism of synthesis of the polymerase polypeptides, we have recently analyzed the RNA and polypeptide phenotype of several *ts* mutants of SV strain AR339 which, when continuously incubated at the restrictive temperature, make less than 10% of the wild-type level of RNA (RNA^- mutants). Infected cells were incubated at the permissive temperature for several hours, to allow virus-specified RNA synthesis to become established, and then shifted to the restrictive temperature. Cultures were pulsed and pulsed-chased before and/or after the temperature shift, and labeled virus-specified RNA and protein were analyzed by polyacrylamide gel electrophoresis. Several points emerged from these experiments.

First, the RNA^- mutants could be divided into two classes on the basis of their RNA phenotype before and after the shift-up. The first class behaved essentially like the wild-type and continued to synthesize both 42 and 26 S RNA. The second class eventually ceased the production of 26 S RNA after the shift-up but continued to form 42 S RNA. Moreover, the fall in 26 S RNA synthesis was paralleled by an increase in 42 S RNA formation. Interestingly, however, no mutant was found (35 were examined) which ceased 42 S RNA synthesis but continued 26 S RNA production after the shift-up. Indeed, no mutant capable of 26 S formation but incapable of 42 S formation has ever been described for any alphavirus.

Second, analysis of the polypeptide phenotype of mutants which showed an altered RNA synthesis in favor of 42 S RNA formation after the shift-up were shown to have a defect in the gene coding for NS p89. This defect was manifest either in failure to process the polyprotein precursor to the nonstructural polypeptide or, if formed, NS p89 was unstable and was rapidly degraded at the restrictive temperature.

Taking these RNA and polypeptide studies together, we have proposed that two of the nonstructural polypeptides, NS p60 and NS p82, are components of a "core" polymerase and that this core polymerase functions in the synthesis of *both* 42 and 26 S RNA. The third virus-specified polymerase component, NS p89, is associated with the core enzyme in either an active or an inactive form. If NS p89 is inactive, the polymerase mediates synthesis of only 42 S RNA. Activation of NS p89 alters the specificity of the polymerase to mediate 26 S RNA formation (Brzeski and Kennedy, 1978). These ideas are summarized in Fig. 2. According to this model, NS p89, possibly playing a role akin to that of a sigma factor, alters the initiation specificity of the core

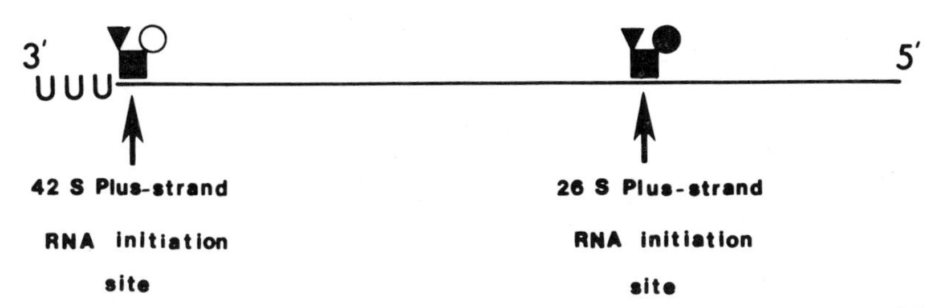

Fig. 2. Diagrammatic representation of the state of the subunit structure of the alphavirus polymerase and the effect of this state on 42 and 26 S plus-strand RNA synthesis. Squares, NS p60; triangles, NS p82; solid circles, active NS p89; open circles, inactive NS p89.

enzyme from one involving initiation at the 3′-end of the 42 S RNA negative-strand template to one involving initiation at a point one-third of the way from the 5′-end of the template strand, i.e., to the initiation site for 26 S RNA synthesis. This model, although tentative, offers an explanation for all that is presently known about 42 and 26 S RNA synthesis. For example, it explains the increase in 42 S RNA synthesis with the concomitant fall in 26 S RNA formation observed with several RNA^- mutants after a shift-up to the restrictive temperature under conditions when no further polymerase is made. Second, it predicates that any mutation in NS p82 or NS p60—the virus-specified components of the core enzyme—will affect the synthesis of *both* 42 and 26 S RNA, and therefore that no mutant exists which synthesizes 26 S RNA in the total absence of 42 S RNA production. To date no such mutant has been described.

Clearly the next question relating to modulation of the core enzyme by NS p89 is what controls this modulation. One possible answer to this question has come from studies on *ts* mutants of SV which form between 10 and 150% of the wild-type level of RNA during continuous incubation at the restrictive temperature (RNA^+ mutants). Several of these mutants fall into a class first recognized by Strauss *et al.* (1969), which form the polyprotein precursor of the structural polypeptides of the virus particle at the restrictive temperature. These mutants also show an altered RNA phenotype in favor of 26 S RNA synthesis at the expense of 42 S RNA formation (Brzeski *et al.*, 1978). Although the change in the RNA ratio is relatively small compared to that observed for the RNA^- mutants, it suggests that either production of the structural protein precursor or failure to form one of the structural proteins of the virus particle accounts for this altered RNA ratio in favor of 26 S RNA formation. When Zn^{2+} inhibition was used to prevent formation of the three envelope glycoproteins but to allow synthesis of the nucleocapsid polypeptide, the RNA ratio was indistinguishable from that of the wild type. This, in turn, has suggested that failure to form the nucleocapsid polypeptide results in augmented synthesis of 26 S RNA and therefore that the nucleocapsid polypeptide plays a role analogous to that of a feedback inhibitor. Since no free 42 S RNA or nucleocapsid polypeptide is found in infected cells and since the association of 42 S RNA and nucleocapsid polypeptide is rapid (Söderlund, 1973), it is possible that nucleocapsid polypeptide synthesis (and therefore all structural protein synthesis) and 42 S RNA synthesis are coordinated and that coordination is achieved by the nucleocapsid polypeptide interacting rapidly but transiently with NS p89.

This situation is depicted diagrammatically in Fig. 3. In the presence of excess nucleocapsid polypeptide the equilibrium between "active" and "inactive" NS p89 alters in favor of the inactive conformation, and RNA synthesis alters in favor of 42 S RNA formation (Brzeski, unpublished observations). Conversely, under conditions of depleted nucleocapsid polypeptide synthesis the active–inactive equilibrium of NS p89 alters in favor of the active conformation, and RNA synthesis swings toward 26 S RNA formation.

Although many aspects of this simple model for the regulation of alphavirus RNA synthesis await further experimentation, the model offers a mechanism which is not only consistent with all the available data but which also indicates that interactions between virus-specified polypeptides for modulating and even controlling viral-specific events are not restricted to prokaryotic viruses.

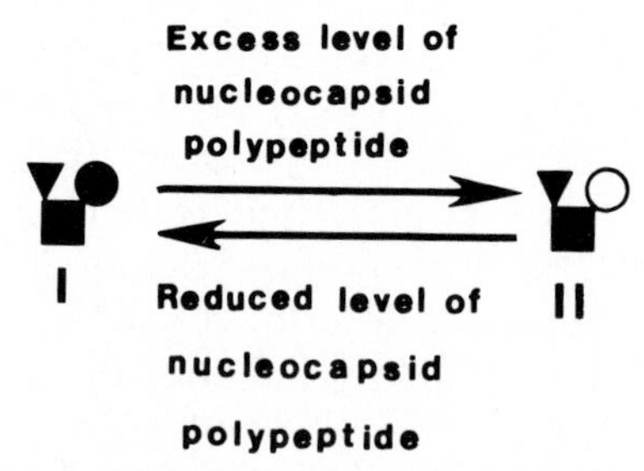

Fig. 3. Diagrammatic representation of the effect of the nucleocapsid polypeptide on the state of the alphavirus polymerase. State I catalyzes the synthesis of 26 S RNA and state II the synthesis of 42 S RNA. Squares, NS p60; triangles, NS p82; solid circles, active NS p89; open circles, inactive NS p89.

VII. KINETICS OF ALPHAVIRUS PLUS- AND MINUS-STRAND SYNTHESIS AND OF POLYMERASE FORMATION

Figure 4 shows the kinetics of alphavirus plus- and minus-strand synthesis during multiplication. Synthesis of RNA of both polarities is detectable between 1 and 1.5 h postinfection and rapidly rises during the following hour. At this time the rate of minus-strand synthesis falls, and shortly thereafter the rate of plus-strand synthesis becomes constant (Bruton and Kennedy, 1975). If minus-strand synthesis is equated with the formation of 42 S negative-strand RNA, and plus-strand synthesis with the formation of 42 and 26 S positive-strand RNA, an explanation for the cessation of synthesis of 42 S negative-strand RNA, shortly followed by attainment of a constant rate of synthesis of 42 and 26 S RNA, must be sought. One, but by no means the only, explanation centers on encapsidation by newly formed nucleocapsid polypeptide. Since this process is known to be rapid and complete (Söderlund, 1973), newly formed 42 S RNA in the presence of a continuing supply of nucleocapsid polypeptide may not be available to act as template for further 42 S RNA negative-strand synthesis, and formation of this species of RNA would therefore decline, perhaps reaching zero. This in turn would reduce the rate of synthesis of 42 S positive-strand RNA, since the number of 42 S negative-strand template RNAs would remain constant, and, if no more 42 S negative-strand RNA was made, the rate of synthesis of plus-strand RNA would become constant. Put another way, the fall in the rate of plus-strand synthesis—from rapidly increasing to constant—may be due to the cessation of negative-strand formation due in turn to rapid encapsidation of newly formed 42 S RNA.

Another aspect of the kinetics of alphavirus RNA synthesis which is worthy of note is the effect of cycloheximide on the synthesis of virus RNA. Friedman and Grimley (1969), Scheele and Pfefferkorn (1969), and Wengler and Wengler (1975) reported that virus RNA synthesis was sensitive to cycloheximide early in the multiplication cycle, but that this compound was without effect when added at later times. In the light of present knowledge, we have reexamined this observation, and Fig. 5 shows the effect of cycloheximide added at various times to SFV-infected hamster cells. What seems clear from this experiment is that cycloheximide does not inhibit RNA synthesis per se but rather that the level of synthesis remains essentially identical to that attained immediately before addition of the drug. The simplest explanation of

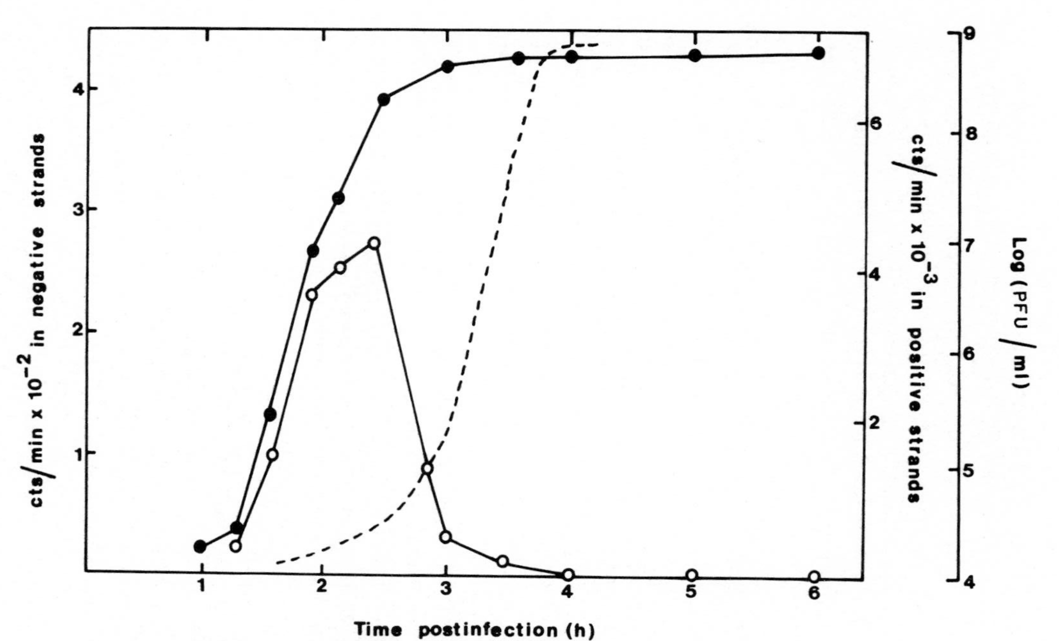

Fig. 4. Kinetics of alphavirus positive- and negative-strand synthesis. Portions of a suspension culture of SFV-infected BHK cells were pulsed for 5 min with [^{32}P]orthophosphate at the times indicated, and total nucleic acids extracts prepared. These extracts were fractionated into RI plus RF and single-stranded (positive polarity) RNA. Radioactivity in negative strands (open circles) was determined after displacement hybridization of RF and RI using an excess of unlabeled virus particle RNa. Radioactivity in total positive strands (solid circles) was the sum of that in purified free strands together with that released after denaturing RI plus RF. Samples were also taken for determination of the titer of extracellular virus (dashed line). (Modified from Bruton and Kennedy, 1975; reproduced by courtesy of Cambridge University Press.)

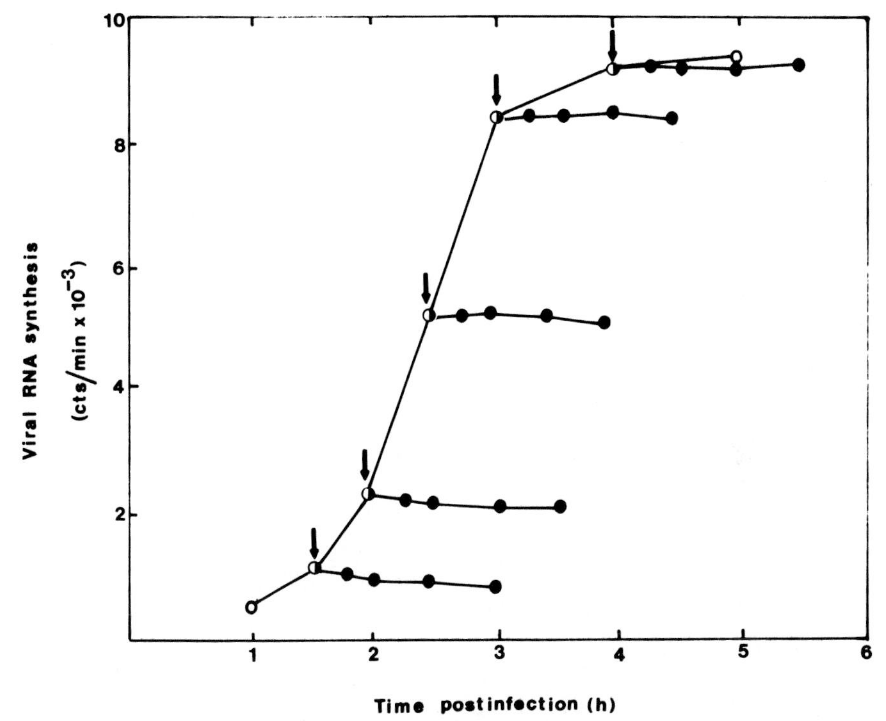

Fig. 5. Effect of cycloheximide on the development of alphavirus RNA synthesis. Portions of an actinomycin D-treated suspension culture of SFV-infected BHK cells, incubated from 30 min postinfection with [^{3}H]uridine, were removed at the times indicated by arrows and incubated in the presence of 100 μg/ml cycloheximide. At the times specified triplicate samples were trichloroacetic acid (TCA)-precipitated and counted (solid circles). In addition control samples (open circles), incubated in the absence of drug were also TCA-precipitated and counted. ◒—◒ indicates that both the control sample and the treated sample gave the same value. (Unpublished data of H. Brzeski and S. I. T. Kennedy.)

this finding is that virus polymerase is synthesized up until about 3 h postinfection and that little or no additional polymerase is made thereafter. Moreover, polymerase activity does not depend on continued protein synthesis; i.e., it does not depend on a rapidly turning over polypeptide. Direct support for the idea that polymerase is made early and hardly, if at all, at later times, comes from an analysis of the kinetics of virus-specified polypeptide formation, which clearly shows that at early times nonstructural polypeptide synthesis predominates over structural protein formation and that this situation is markedly reversed later in infection (Clegg *et al.*, 1976). To account for this change in virus-specified polypeptide synthesis, Wengler and Wengler (1975) have suggested that, as 26 S RNA formation increases, this species of RNA competes not only with cellular mRNAs but also with 42 S RNA for the translational apparatus of the cell. In addition, as we have already seen, given a continuing supply of nucleocapsid polypeptide, there is little available free 42 S RNA late in infection to act as mRNA for continued polymerase formation.

When the kinetics of plus- and minus-strand synthesis are combined with the effect of cycloheximide and the kinetics of virus-specified polypeptide formation, the following temporal phases in the alphavirus multiplication cycle can be envisaged.

Phase 1. Translation of 42 S RNA to form the polymerase polypeptides. Onset of 42 S minus-strand formation.

Phase 2. Onset of 42 and 26 S strand synthesis. Continued synthesis of polymerase and therefore a rapid rise in the rate of 42 S negative-stranded RNA formation.

Phase 3. Increasing synthesis of 42 and 26 S RNA and therefore increasing formation of nucleocapsid polypeptide and a reduction in the availability of unencapsidated 42 S RNA to act (a) as mRNA for further polymerase and (b) template for 42 S negative-strand synthesis.

Phase 4. Cessation of 42 S negative-strand formation and therefore continuation of RNA synthesis at a constant rate.

The essential regulating element in this scheme is the nucleocapsid polypeptide. As its synthesis increases, so the formation of additional polymerase and 42 S negative-strand RNA decreases. The extent to which this "mass action" effect of the nucleocapsid polypeptide accounts for the observed alterations in the kinetics of synthesis of virus-specified RNA and polymerase and whether or not other factors are responsible will only become established after further experimentation.

VIII. MECHANISTIC ASPECTS OF THE SYNTHESIS OF ALPHAVIRUS RNA

In this final section we shall consider what little is known about the mechanism of the catalysis of formation of alphavirus RNA. If the direction of strand copying for the alphavirus polymerase is like that for other nucleic acid polymerases, namely, $3' \rightarrow 5'$, the initial event in this process is the association or binding of the polymerase (1) at or near the $3'$-end of 42 S plus-strand RNA, (2) at or near the $3'$-end of 42

S negative-strand RNA, and (3) at a position about one-third of the way from the 5′-end of 42 S negative-strand RNA. Little is known about the nucleotide sequences or conformations of the portions of the templates involved in these binding events. However, recent studies on the nucleotide sequence relationships between DI RNAs and 42 S RNA have cast some light on this question (Kennedy, 1976). As the mechanism of generation and replication of DI particles is considered in Chapter 15, only those aspects which provide information on the mechanism of alphavirus RNA synthesis will be considered here. Briefly, it is now clear that each species of DI RNA represents an internally deleted form of 42 S RNA. Each DI RNA species therefore contains the same nucleotide sequences at its 5′- and 3′-extremities as 42 S RNA, but only part of the internal sequence (Kennedy, 1976; Guild and Stollar, 1977). Moreover, DI RNA is replicated in cells coinfected with DI particles and standard (helper) virus using polymerase encoded by the genome of the helper virus. In an attempt to define the minimum sequence–conformation requirements for an alphavirus RNA to be replicated we have surveyed the generation and replication of DI RNA in several cell lines. To date the smallest DI RNA of SFV we have identified has a molecular weight of 0.27×10^6. Oligonucleotide fingerprinting and limited sequencing have established that this DI RNA contains about 750 nucleotides; about 500 of these are derived from the 5′-end of 42 S RNA, and about 200 from the 3′-end (S. I. T. Kennedy, unpublished observations). These studies indicate three things. First, in order to replicate alphavirus RNA both extremities of 42 S RNA are required. Second, over 95% of the nucleotide sequence of 42 S RNA can be deleted, and yet the resultant RNA can still be replicated. This indicates that neither the primary nor secondary structure of the portion of 42 S RNA deleted in the DI RNA plays an essential role in polymerase binding, elongation, or termination. Third, in the synthesis of 42 S minus-strand RNA, binding of the polymerase must occur within 200 nucleotides of the 42 S plus strand. This last conclusion is not, of course, unexpected since, if synthesis did not commence at the 3′-end of the template, progeny strands would be shorter than the template. Perhaps the significance of this DI RNA lies in its future use. With the development of rapid methods for RNA sequencing (Simoncsits *et al.*, 1977) and continued efforts to solubilize and reconstruct polymerase activity *in vitro* it should soon be possible to define the essential sequence requirements for the replication of alphavirus RNA and, from the medical standpoint, of interfering with this virus-unique process. As already discussed in Section III and in Chapter 11, there is evidence that there exists a 3′ → 5′ nucleotide sequence at the 3′-end of 42 S RNA which is complementary to a 5′ → 3′ sequence at the 5′-end of the molecule. The observation that these complementary sequences are present in *all* DI RNA species (Kennedy, unpublished observations) again suggests that they play an essential role in RNA synthesis (see also Chapter 15).

Finally let us consider some of the complete unknowns of alphavirus RNA synthesis. First, what is the nature and where is the exact location of the polymerase binding site for 26 S RNA synthesis on 42 S negative-strand RNA? Second, as with many nucleic acid polymerases, is a primer involved in alphavirus RNA replication and is this primer an oligonucleotide or an oligonucleotide–protein? Third, what is the mechanism of 5′-capping and methylation of 42 and 26 S RNA and are these processes carried out by host cell- or virus-specified enzymes? Fourth, what is the significance of the poly(A) tract at the 3′-end of 42 and 26 S (and DI) RNA? Fifth, as we have discussed previously, do host cell specified polypeptides form part of the polymerase and, if so, what is the exact nature of their role? Sixth, what is the molecular mechanism of generation of DI RNA?

Let us hope that when revision of this book becomes necessary, a contributory cause will be that some of these questions will have been answered.

REFERENCES

Blobel, G., and Dobberstein, B. (1975). *J. Cell Biol.* **67**, 835–851.
Bracha, M., Leone, A., and Schlesinger, M. S. (1976). *J. Virol.* **20**, 612–620.
Bruton, C. J., and Kennedy, S. I. T. (1975). *J. Gen. Virol.* **28**, 111–127.
Brzeski, H., and Kennedy, S. I. T. (1977). *J. Virol.* **22**, 420–429.
Brzeski, H., and Kennedy, S. I. T. (1978a). *J. Virol.* **25**, 630–640.
Brzeski, H., Clegg, J. C. S., Atkins, G. A., and Kennedy, S. I. T. (1978a). *J. Gen. Virol.* **38**, 461–470.
Burge, B. W., and Pfefferkorn, E. R. (1966). *Virology* **30**, 214–231.
Cancedda, R., and Schlesinger, M. J. (1974). *Proc. Natl. Acad. Sci. U.S.A.* **71**, 1843–1847.
Cancedda, R., Swanson, R. W., and Schlesinger, M. S. (1974). *J. Virol.* **14**, 664–671.
Clegg, J. C. S., and Kennedy, S. I. T. (1974a). *FEBS Lett.* **42**, 327–330.
Clegg, J. C. S., and Kennedy, S. I. T. (1974b). *J. Gen. Virol.* **22**, 331–345.
Clegg, J. C. S., Brzeski, H., and Kennedy, S. I. T. (1976). *J. Gen. Virol.* **32**, 413–430.
Clewley, J. P., and Kennedy, S. I. T. (1976). *J. Gen. Virol.* **32**, 395–411.
Dubin, D. T., Stollar, V., Hsuchen, C., Timko, K., and Guild, G. M. (1977). *Virology* **77**, 457–470.
Follett, E. A. C., Pringle, C. R., and Pennington, T. H. (1975). *J. Gen. Virol.* **26**, 183–196.
Friedman, R. M., and Grimley, P. M. (1969). *J. Virol.* **4**, 292–299.

Friedman, R. M., Levy, H., and Carter, W. B. (1966). *Proc. Natl. Acad. Sci. U.S.A.* **56,** 440–447.

Friedman, R. M., Levin, J. G., Grimley, P. M., and Berezesky, I. K. (1972). *J. Virol.* **10,** 492–503.

Grimley, P. M., Levin, J. G., Berezesky, I. K., and Friedman, R. M. (1972). *J. Virol.* **10,** 504–515.

Guild, G. M., and Stollar, V. (1977). *Virology* **77,** 175–188.

Hsu, M.-T., Kung, H.-J., and Davidson, N. (1973). *Cold Spring Harbor Symp. Quant. Biol.* **38,** 943–948.

Kennedy, S. I. T. (1972). *Biochem. Biophys. Res. Commun.* **48,** 1254–1258.

Kennedy, S. I. T. (1976). *J. Mol. Biol.* **108,** 491–512.

Lachmi, B. E., and Kääriäinen, L. (1976), *Proc. Natl. Acad. Sci. U.S.A.* **73,** 1936–1940.

Lachmi, B. E., Glanville, N., Keränen, S. and Kääriäinen, L. (1975). *J. Virol.* **16,** 1615–1629.

Levin, J. G., and Friedman, R. M. (1971). *J. Virol.* **7,** 504–513.

Martin, B. A. B., and Burke, D. C. (1974). *J. Gen. Virol.* **24,** 45–66.

Martire, G., Bonatti, S., Aliperti, G., De Giuli, C., and Cancedda, R. (1977). *J. Virol.* **21,** 610–618.

Michel, M. R., and Gomatos, P. J. (1973). *J. Virol.* **11,** 900–914.

Mowshowitz, D. (1973). *J. Virol.* **11,** 535–543.

Sawicki, D. L., and Gomatos, P. J. (1976). *J. Virol.* **20,** 446–464.

Scheele, C. M., and Pfefferkorn, E. R. (1969). *J. Virol.* **4,** 117–122.

Segal, S., and Sreevalsan, T. (1974). *Virology* **59,** 428–442.

Simmons, D. T., and Strauss, J. H. (1972a). *J. Mol. Biol.* **71,** 599–613.

Simmons, D. T., and Strauss, J. H. (1972b). *J. Mol. Biol.* **71,** 615–631.

Simmons, D. T., and Strauss, J. H. (1974a). *J. Mol. Biol.* **86,** 397–409.

Simmons, D. T., and Strauss, J. H. (1974b). *J. Virol.* **14,** 552–559.

Simoncsits, A., Brownlee, G. G., Brown, R. S., Rubin, J. R., and Guilley, H. (1977). *Nature (London)* **269,** 833–836.

Söderlund, H. (1973). *Intervirology* **1,** 1–8.

Sonnabend, J. A., Martin, E. M., and Mécs, E. (1967). *Nature (London)* **213,** 365–372.

Sreevalsan, T. (1972). *J. Virol.* **6,** 438–442.

Sreevalsan, T., and Lockart, R. Z. (1966). *Proc. Natl. Acad. Sci. U.S.A.* **55,** 974–981.

Strauss, J. H., Burge, S. W., and Darnell, J. E., Jr. (1969). *Virology* **37,** 367–375.

Waite, M. R. F. (1973). *J. Virol.* **11,** 198–211.

Wengler, G., and Wengler, G. (1975). *Virology* **66,** 322–326.

Wengler, G., and Wengler, G. (1976). *Virology* **73,** 190–199.

Wengler, G., Beato, M., and Hackemack, S. A. (1974). *Virology* **61,** 120–128.

13

Translation and Processing of Alphavirus Proteins

MILTON J. SCHLESINGER
LEEVI KÄÄRIÄINEN

I. INTRODUCTION

The genome of alphaviruses has a molecular weight of about 4.3×10^6, a size that can code for 8–10 polypeptides averaging 40,000 in molecular weight. Genetic complementation studies of Sindbis virus temperature-sensitive mutants predict that four genes are required for virus-specific RNA synthesis and three for viral structural proteins (Burge and Pfefferkorn, 1966b; Strauss *et al.*, 1976). Analyses based on sodium dodecyl sulfate (SDS) polyacrylamide gel electrophoresis of virus-specific proteins formed during the replication of Semliki Forest virus (SFV) or Sindbis virus in tissue culture cells have shown that there are 3 or 4 distinct viral nonstructural proteins (Lachmi and Kääriäinen, 1976; Clegg *et al.*, 1976; Brzeski and Kennedy, 1977), 2 of

THE TOGAVIRUSES

ISBN 0-12-625380-3

which can be detected in a purified preparation of SFV replicase (Clewley and Kennedy, 1976), and three to four distinct proteins in the virion structure (Strauss *et al.*, 1968; Simons and Kääriäinen, 1970; Schlesinger *et al.*, 1972; Garoff *et al.*, 1974). A variety of experimental procedures utilizing various host cells infected with wild-type virus or temperature-sensitive mutants have clearly shown that both nonstructural and structural virus coded proteins are derived from higher-molecular-weight polypeptides by a sequence of post- or co-translational proteolytic cleavage reactions. Results of these investigations have established that alphaviruses utilize two species of mRNA for translation of these proteins: the virion-sized 42 S RNA for nonstructural proteins, and the subgenomic 26 S RNA for structural polypeptides. For both polycistronic mRNAs, translation initiates at a single site near the 5′-end of the nucleic acid, and subsequent protease activities generate the individual proteins. This mechanism for gene expression appears to be a general one for viral replication in eukaryotic cells.

In this chapter we describe the specific pathways elucidated thus far in alphavirus protein translation and processing. The latter includes not only a series of proteolytic cleavages but also a sequence of glycosylations and fatty acid acylations, activities that appear to be essential for the final maturation, assembly, and budding of newly formed virions. We also discuss possible mechanisms that might function in regulating and controlling expression of the alphavirus genes and indicate some of the unusual features in the translation of virus-specific proteins.

II. TRANSLATION OF VIRAL PROTEINS

A. Translation of Nonstructural Proteins

Alphaviruses were among the first viruses shown to have infectious RNA (Cheng, 1958; Wecker and Schafer, 1957; Wecker, 1959). This finding suggested that parental viral proteins were not needed for the initiation of infection and that the viral genome directed the synthesis of necessary components required for the replication of viral RNA. If this "early protein" synthesis was prevented by a protein synthesis inhibitor such as parafluorophenylalanine (Wecker *et al.*, 1962), puromycin (Wecker, 1963; Sreevalsan and Lockart, 1964; Scheele and Pfefferkorn, 1969) or cycloheximide (Friedman and Grimley, 1969; Friedman and Sreevalsan, 1970), or by pretreatment with interferon

(Taylor, 1965; Friedman *et al.*, 1967), no viral RNA synthesis could be observed. When the puromycin was removed, viral replication started, but there was a delay respective to the length of treatment (Sreevalsan and Lockart, 1964). Virus-specific proteins must be responsible for the early events of replication, as evidenced by the isolation of temperature-sensitive RNA negative mutants which are unable to start viral RNA replication at the restrictive temperature (Burge and Pfefferkorn, 1966a,b; Tan *et al.*, 1969; Atkins *et al.*, 1974; Keränen and Kääriäinen, 1974; Strauss *et al.*, 1976). All proteins needed for normal yields of RNA are synthesized during the first 3 h after infection, since the addition of protein synthesis inhibitors at 3 h postinfection does not affect viral RNA synthesis (Friedman and Grimley, 1969; Scheele and Pfefferkorn, 1969; Ranki and Kääriäinen, 1970; Wengler and Wengler, 1975b). Identification of the virus-specific early proteins has been a difficult task because of the ongoing host protein synthesis during the first hours of infection (Strauss *et al.*, 1969; Wengler and Wengler, 1976b; Keränen and Kääriäinen, 1975; Atkins, 1976; Lachmi and Kääriäinen, 1977; see also reviews in Pfefferkorn and Shapiro, 1974; Strauss and Strauss, 1976). Specific inhibition of host protein synthesis has been attempted by simultaneous infection in the presence of guanidine (Friedman, 1968) or by preceding infection with fowl plague virus (Kaluza, 1976; Kaluza *et al.*, 1976). These techniques have failed to give unambiguous identification of early proteins. Several apparently virus-specific proteins have been described by a number of investigators (Strauss *et al.*, 1969; Hay *et al.*, 1968; Scheele and Pfefferkorn, 1970; Pfefferkorn and Boyle, 1972; Snyder and Sreevalsan, 1974; Ranki *et al.*, 1972; Morser *et al.*, 1973; Morser and Burke, 1974; Burrell *et al.*, 1969); among them are probably nonstructural proteins and aberrant cleavage products of the structural polyprotein.

Two nonstructural proteins with apparent molecular weights of 86,000 and 72,000 were first identified in cells infected with a temperature-sensitive mutant, *ts*-1, of SFV (Keränen and Kääriäinen, 1975). These proteins were not related to the virion structural proteins, since they had different amino acid sequences as evidenced by tryptic peptide mapping of [^{35}S]methionine-labeled proteins (Lachmi *et al.*, 1975). When *ts*1-infected cells were released from a hypertonic block affecting the initiation of translation (Saborio *et al.*, 1974), sequential labeling of four *stable* nonstructural proteins with apparent molecular weights of 70,000 (ns 70), 86,000 (ns 86), 72,000 (ns 72), and 60,000 (ns 60) took place (Lachmi and Kääriäinen, 1976). Two short-lived proteins with molecular weights of about 155,000 (ns 155) and

135,000 (ns 135) were labeled sequentially; ns 155 after 30 s and ns 135 after 10 min labeling. A precursor–product relationship between the large short-lived and the smaller stable nonstructural proteins was suggested from results of pulse-chase experiments (Lachmi and Kääriäinen, 1976). Evidence for existence of the ns 60 nonstructural protein was indirect and based on the labeling of a protein of this size in pactamycin-treated cells in which the synthesis of structural proteins had been inhibited by the drug. Comparison of [^{35}S]methionine-labeled tryptic peptides of the different nonstructural proteins has to a large extent confirmed the proposed precursor–product relationship: (1) The primary structures of ns 155 and ns 135 are different and nonoverlapping, indicating that the total molecular weight of the nonstructural proteins must be close to 300,000 (Glanville and Lachmi, 1977). (2) The tryptic peptides of ns 70 and ns 86 are found in ns 155, and those of ns 72 in ns 135. In ns 135 there are peptides which may be detected in ns 60 (Glanville *et al.*, 1978). The viral specificity of these *ts*1-induced nonstructural proteins has been substantiated by changes in their synthesis during temperature-shift experiments (Kääriäinen *et al.*, 1976). More direct evidence that these are virus-specific nonstructural proteins has been obtained from analyses of the tryptic peptides derived from proteins made *in vitro* with RNA from wild-type SFV. All the peptides of ns 155 (i.e., ns 70 and ns 86) and some of the peptides of ns 135 (ns 72) were generated in a wheat germ cell-free extract programmed with SFV 42 S RNA (Glanville and Lachmi, 1976).

A third nonstructural putative precursor protein with a molecular weight of about 220,000 has been detected in cells infected with a temperature-sensitive RNA negative mutant (*ts*4) of SFV. This protein appeared only when the infection with *ts*4 was started at 28°C and radioactive amino acid added after a shift to 39°C (Kääriäinen *et al.*, 1978). Short pulses of isotype given after release of the cells from a hypertonic block of initiation indicate that the 220,000-dalton protein originates from the N-terminal part of the hypothetical nonstructural polyprotein. It presumably contains the ns 70, ns 86, and ns 72 protein sequences, but this has not been verified by tryptic peptide mapping. The presence of ns 155, ns 135, ns 86, ns 72, and ns 70 has also been observed in SFV wild type-infected cells (Lachmi and Kääriäinen, 1977). Clegg *et al.* (1976) have demonstrated virus-specific proteins with molecular weights of 200,000, 184,000, 150,000, 90,000, and 63,000 in SFV wild type-infected cells. Only the latter two were stable proteins, and they had tryptic peptides different from those of structural proteins. Pulse-chase experiments suggested that they were derived from p200, p150, and possibly p184. Short pulses with [^{35}S] methionine given after release from a hypertonic block showed that ns 63 was at the N-terminal end of the precursor.

A 200,000-dalton protein is probably also a precursor protein of Sindbis virus nonstructural proteins. Bracha *et al.* (1976) detected a protein of this size in cells infected with Sindbis virus temperature-sensitive RNA negative mutants *ts*21 and *ts*24, it yielded tryptic peptides different from those of viral structural proteins. The large protein accumulated only at the restrictive temperature after a shift-up of the mutant-infected cultures. Brzeski and Kennedy (1977) found two virus specific proteins with molecular weights above 200,000 in cells infected with wild-type Sindbis virus. In addition they detected nonstructural proteins with molecular weights of 150,000, 89,000, 82,000, 76,000, and 60,000; only the 89,000, 82,000, and 60,000 proteins were stable. The large proteins accumulated when zinc ions were used to inhibit cellular proteolytic enzymes (Bracha and Schlesinger, 1976a). The stable nonstructural proteins were labeled sequentially after synchronous initiation of protein synthesis in the following order: ns 60, ns 89, ns 82. Tryptic peptide analysis indicated that all the stable products were derived from p230 and possibly also from p215, and pulse-chase experiments suggested that ns 82 was derived from p76.

Several different pathways for the processing of alphavirus nonstructural proteins have emerged from these studies. Figure 1 represents the possible models currently proposed. In all of them, a protein of 60,000–70,000 molecular weight is derived from the 5′-end of the 42 s mRNA, followed by a protein of 86,000–90,000 molecular

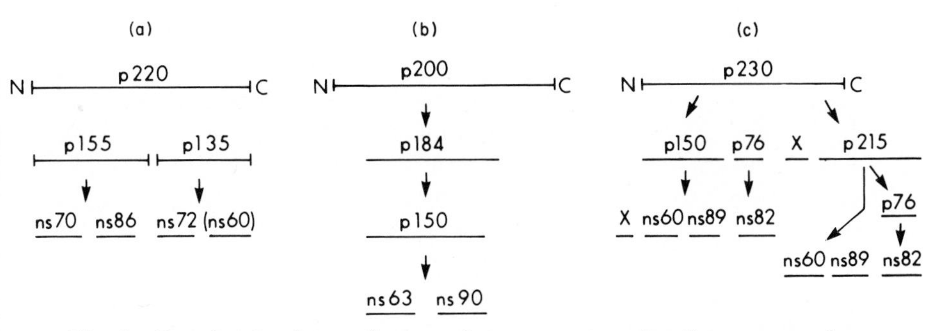

Fig. 1. Postulated pathways for proteolytic processing of viral nonstructural proteins. (a) Semliki Forest virus—p220 was found in *ts*4-infected cells (Kääriäinen *et al.*, 1978). (b) Semliki Forest virus according to Clegg *et al.*, (1976). (c) Sindbis virus. (From Brzeski and Kennedy, 1977.)

weight. The other products (two in the case of SFV, and one for Sindbis virus) are not yet clearly resolved.

B. 42 S RNA as a Messenger

The messenger for nonstructural proteins is the viral 42 S RNA genome, as evidenced by results from both *in vivo* and *in vitro* experiments. The 42 S RNA has been found on polysomes by several groups of investigators (Mowshowitz, 1973; Söderlund *et al.*, 1973–1974; Simmons and Strauss, 1974a; Martire *et al.*, 1977; Wengler and Wengler, 1974, 1975a) and has been translated in different cell-free systems derived from reticulocytes (Simmons and Strauss, 1974b; Cancedda *et al.*, 1974a, 1975), Krebs ascites cells (Smith *et al.*, 1974), and wheat germ (Glanville *et al.*, 1976a,b). The most complete product has been reported by Simmons and Strauss (1974b), consisting of six polypeptides with molecular weights between 60,000 and 180,000. Tryptic peptide analysis of the wheat germ *in vitro* product of SFV 42 S RNA has yielded peptides with mobilities identical to those derived from nonstructural proteins isolated from *ts*1-infected cells (cf. above).

Evidence for the translation of structural proteins from 42 S RNA has also been reported by Smith *et al.* (1974) and Glanville *et al.* (1976a) and can be accounted for by translation of those sequences which code for the structural proteins in 42 S RNA (equivalent to the 26 S RNA; Kennedy, 1976; Wengler and Wengler, 1976a).

The formation of structural gene products by *in vitro* translation of 42 S RNA could arise by (1) internal initiation at the capsid gene, (2) read-through starting from the 5′-end of 42 S RNA, followed by proteolytic processing, (3) contamination of the 42 S RNA with 26 S RNA, or (4) nicking of the 42 S RNA in regions allowing for new 5′-end initiation sites. Different preparations of 42 S RNA yield different amounts of capsid protein *in vitro* (Cancedda *et al.*, 1975), and internal initiation could not be shown with the wheat germ extract used by Glanville *et al.* (1976b). Preparations of 42 S RNA from virions harvested early after infection fail to make capsid protein *in vitro* but can form polypeptides up to 200,000 in molecular weight (D. Sagher and M. J. Schlesinger, unpublished results).

In the wheat germ cell-free system of Glanville *et al.* (1976b) only one initiation peptide was detected when 42 S RNA was translated in the presence of [^{35}S]fMet-tRNA$_f$. Analysis of the dipeptide gave fMet-Ala which is different from the initiation dipeptide fMet-Asn of the 26 S RNA *in vitro* product (Glanville *et al.*, 1976b; Clegg and Kennedy, 1975b). One major and one minor initiation peptide was reported for Sindbis virus genome 42 S RNA, employing reticulocyte lysate in the presence of [^{35}S]fMet-tRNA$_f$ (Cancedda *et al.*, 1975). The minor peptide had a mobility similar to that detected for 26 S RNA. Studies with a Sindbis virus *ts* mutant which shuts off 26 S mRNA synthesis but allows 42 S RNA to be translated after a shift to the nonpermissive temperature have suggested that internal initiation of 42 S RNA does not occur *in vivo* to any significant extent (Bracha *et al.*, 1976).

C. Translation of Structural Proteins

The polypeptide chains found in the alphavirus structure are translated from a subgenomic species of viral RNA with a molecular weight of 1.6×10^6 and a sedimentation coefficient of 26 S. This 26 S mRNA, which represents the one-third portion of the genome RNA located at the 3′-end of the latter molecule (see Chapters 11 and 12), is the major virus-specific RNA on polysomes of the infected cell at the time of maximum virus formation (Simmons and Strauss, 1974a; Mowshowitz, 1973; Söderlund *et al.*, 1973–1974; Wengler and Wengler, 1974, 1975a; Kennedy, 1972; Rosemond and Sreevalsan, 1973). The 26 S mRNA has a coding capacity of about 150,000 daltons of protein—a value that closely equals the sum of the molecular weights of the viral structural proteins. These proteins consist of the capsid (C) with a molecular weight of 33,000, glycoproteins E1 and E2 with molecular weights each of about 50,000 and E3, with a molecular weight of 14,000 for the polypeptide chain. Proteins E1 and E2 contain about 8% carbohydrate (Sefton and Keegstra, 1974), and E3 has 45% carbohydrate (Garoff *et al.*, 1974).

The pathway leading to formation of these individual proteins has been inferred from three kinds of data: (1) analysis of proteins accumulating in cells infected with temperature-sensitive mutants or treated with inhibitors designed to block the proteolytic and glycosylating activities, (2) changes in patterns of radioactive label in proteins during pulse-chase conditions and in cells treated so that initiation of protein synthesis can be manipulated, and (3) translation of viral proteins *in vitro* with cell-free extracts supplemented with purified preparations of virus-specific 26 S RNA. From these studies, we can formulate the series of reactions described in Fig. 2. Initiation near the 5′-end of 26 S RNA first leads to capsid protein which is released from the nascent polysome. Some modification may occur at the N-terminal part of the capsid polypeptides (Clegg and Kennedy, 1975a). Continued translation produces a protein of 62,000 molecular

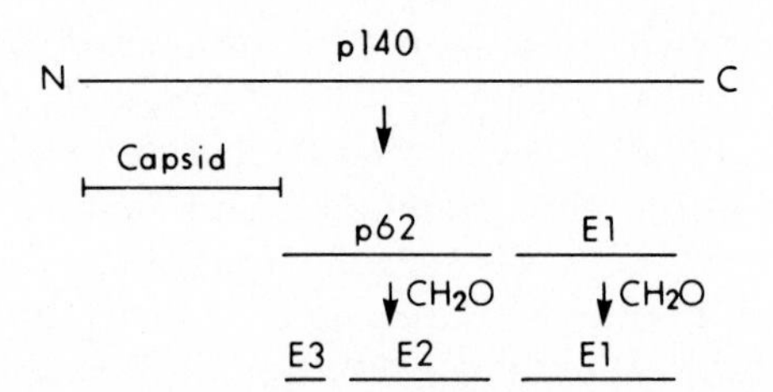

Fig. 2. Pathway for the processing of viral structural proteins.

weight (p62) destined to become the E2 and E3 glycoproteins of the virion envelope. The p62 protein also is cleaved from the nascent polysome which by this point in the translation process has become associated with the membranes of the rough endothelium reticulum (see Fig. 3; see also Wirth *et al.*, 1977). The last protein translated from 26 S RNA is E1 which also inserts into the cell membrane. Modifications at the N-terminal portions of p62 and E1 may occur, since some secreted proteins have been shown to lose a hydrophobic "leader" amino acid sequence after initial insertion into a cellular membrane (Bloebel and Dobberstein, 1975). A p110 (p97 in SFV) protein has been detected in quite large amounts in Sindbis virus-infected baby hamster kidney (BHK) cells (Strauss *et al.*, 1969) but not in great yields in chick embryo fibroblasts (CEFs). A protein of this molecular weight also accumulates in *ts*-mutant-infected cells (Scheele and Pfef-

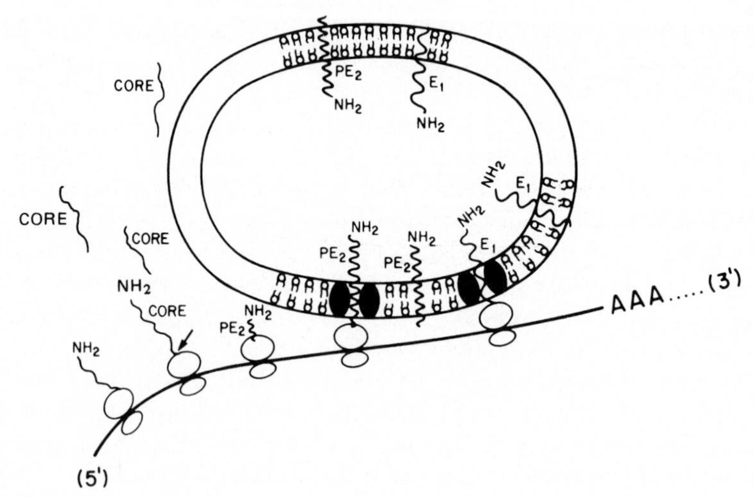

Fig. 3. Model for the synthesis of Sindbis virion proteins, subsequent nascent cleavages, and sequestering of the envelope proteins. (From Wirth *et al.*, 1977.)

ferkorn, 1970; Bracha and Schlesinger, 1976b; Waite, 1973; Keränen and Kääriäinen, 1975). Addition of the protein inhibitor, tosylphenylalanylchloromethyl ketone (TPCK) or amino acid analogs of virus-infected cells interfered with the total synthesis and changed the distribution of label in virus-specific protein such that p110 protein accumulated (Morser and Burke, 1974; Pfefferkorn and Boyle, 1972; Ranki *et al.*, 1972). This protein contains the amino acid sequences of the viral envelope proteins (Schlesinger and Schlesinger, 1973; Lachmi *et al.*, 1975) and is metabolically unstable in pulse-chase experiments (Schlesinger and Schlesinger, 1973). However, it is not glycosylated to any significant extent (Sefton and Burge, 1973; Duda and Schlesinger, 1975), and conditions have been described in which the smaller glycoproteins could be formed in the absence of the larger species (Duda, 1975). This latter result strongly suggested that p110 was a by-product and not a true precursor of the envelope proteins. Thus the nascent proteolytic cleavage reaction that forms p62 and E1 probably occurs prior to completion of translation of the 26 S RNA. Söderlund (1976) has published data from kinetic experiments on puromycin-treated HeLa cells infected with SFV that shows E1 translation is almost completed before its cleavage is detected. A p86 protein accumulates in BHK cells infected with the SFV *ts*3 mutant at nonpermissive temperatures and contains tryptic peptides of the capsid and p62 proteins (Lachmi *et al.*, 1975). Its presence has been explained by a defect in the cleavage reaction between capsid and p62 while retaining the p62 E1 protease activity (see Fig. 2).

The formation of these aberrant cleavage products, p110 and p86, have helped to establish the map for the three structural protein genes on 26 S RNA. The gene sequence of 5′-capsid-p62-E1-3′ has been firmly established from results of experiments in which the initiation of viral structural protein synthesis was altered. In one series of experiments, the antibiotic pactamycin was added to cells, followed by a pulse of radioactive amino acids. Under conditions where the drug preferentially inhibited initiation of translation, viral capsid protein was the protein that showed the greatest decrease in label (Schlesinger and Schlesinger, 1972). Capsid was also the only protein detected in very short pulses of radioactive labeling of cells released from a hypertonic condition in which initiation of protein synthesis was inhibited (Clegg and Kennedy, 1975b). With longer labeling, p62 appeared and was followed by E1. When a chase period was incorporated in these experiments, the E2 and E3 proteins were detected, and their labeling pattern indicated that the E3 was located at the N-terminal end of p62 (Clegg, 1975). Labeling of cells immediately after a shift to a hypertonic medium confirmed that capsid protein was the first one initiated by translation of 26 S RNA (Clegg and Kennedy, 1975b). Other kinetic experiments utilizing puromycin to block translation led to a labeling pattern which showed that capsid release occurred before the translation of 26 S RNA was terminated (Söderlund, 1976).

These kinetic experiments not only delineated the gene order of the structural protein but showed that cotranslational proteolysis operated for the first two proteolytic processing events. They further established that a single site existed on the 26 S mRNA for translation *in vivo*. Other *in vivo* pulse-chase experiments provided data that showed p62 was a relatively long-lived precursor of E2 and E3 (Schlesinger and Schlesinger, 1972; Simons *et al.*, 1973). In addition, tryptic peptide maps confirmed that p62 was the precursor of E2 (Schlesinger and Schlesinger, 1972; Lachmi *et al.*, 1975; Jones *et al.*, 1974).

Formation of E2 and E3 occurs after glycosylation of p62 (Duda and Schlesinger, 1975) and probably after the glycoprotein has moved through the internal cell membranes to the plasma membranes. This cleavage of p62 is a relatively slow one and appears to be closely coupled to budding of virions from the host cells. Inhibitors of protease block the formation of E2 and stop viral particle secretion (Bracha and Schlesinger, 1976a). The defect in several *ts* mutants of Sindbis virus and SFV has been localized to an inability to cleave p62 at the nonpermissive temperature (Jones *et al.*, 1974; Bracha and Schlesinger, 1976b, 1978; Simons *et al.*, 1973). It is of interest that Sindbis virus *ts* mutants in two distinct complementation groups (D and E of the Burge–Pfefferkorn set) show the same defect. To explain how a mutation in E1 could affect E2 formation and allow for complementation, a model has been proposed in which p62 is complexed with E1 prior to the cleavage reaction (Bracha and Schlesinger, 1976b). Consistent with this model is the observation that antibodies directed against E1 protein inhibit the cleavage of p62 when antiserum is added to intact cell monolayers (Bracha and Schlesinger, 1976b). Conversion of p62 to E2 and E3 appears to involve glycosylation reactions (Sefton and Burge, 1973) and probably changes the conformation of the E2 glycoprotein. The p62 precursor is inaccessible to surface iodination reaction, whereas the E2 protein can be iodinated (Sefton *et al.*, 1973). Shortly before the proteolytic conversion of p62 to E2, p62 becomes acylated with fatty acid (Fig. 4; see also Schmidt and Schlesinger, 1980).

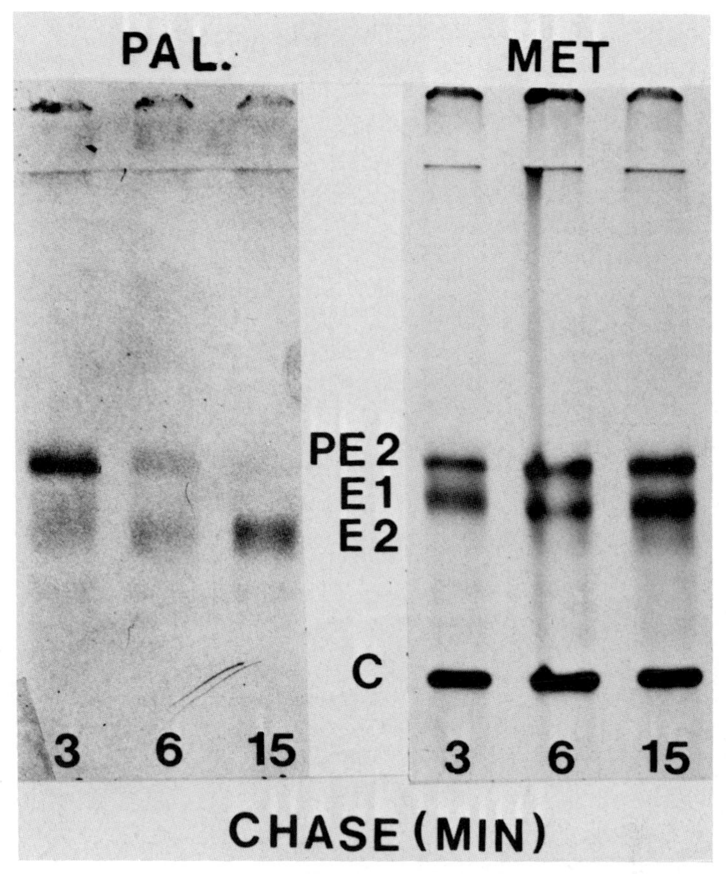

Fig. 4. Kinetics of the conversion of fatty acid-labeled and amino acid-labeled Sindbis virus PE2 to E2. Virus-infected CEFs were pulse-labeled for 3 min with either [^{3}H]palmitic acid or [^{35}S]methionine at 5 h postinfection. Cell extracts were analyzed on 10% SDS polyacrylamide slab gels, and autoradiograms prepared by fluorography.

D. *In Vitro* Translation of 26 S mRNA

The definitive role of the subgenomic 26 S RNA as mRNA for alphavirus structural proteins has been demonstrated in experiments with cell-free protein-synthesizing systems. When purified preparations of 26 S RNA extracted from cells infected with either Sindbis virus or SFV were added to extracts of rabbit reticulocytes, Krebs ascites cells, mouse L cells, wheat germ, or HeLa cells, virus-specific polypeptides were identified among the reaction products (Cancedda *et al.*, 1974a; Simmons and Strauss, 1974b; Clegg and Kennedy, 1974,

1975a; Wengler *et al.*, 1974; Glanville *et al.*, 1976b; Garoff *et al.*, 1978). In all these systems, the addition of [^{35}S]methionine to extracts supplemented with reagents essential for protein synthesis and viral 26 S RNA produced relatively large amounts of a polypeptide that was equivalent in molecular weight and contained tryptic peptide sequences identical to those of the capsid protein derived from virions. Some discrete lower-molecular-weight proteins that contained viral capsid tryptic peptides were also generated, but relatively small amounts of higher-molecular-weight proteins were detected when the *in vitro* reaction products were separated by SDS polyacrylamide gel electrophoresis. Among the larger proteins was one of molecular weight 100,000 that was equivalent in size to the p110 detected in infected BHK cells (Simmons and Strauss, 1974b). In the L-cell *in vitro* system primed with 26 S RNA from SFV-infected BHK cells, proteins antigenically related to viral envelope proteins and containing the tryptic peptides of these polypeptides could be isolated. A virus-specific immunoglobulin adsorbed to Sephadex was utilized in these experiments (Clegg and Kennedy, 1974). With HeLa cell extracts supplemented with membranes from dog pancreas, both proteolytic processing and glycosylation of SFV membrane proteins were detected (Garoff *et al.*, 1978). In this system, the glycoproteins formed *in vitro* were bound to the microsomal fraction. Kinetic studies showed that the insertion of nascent protein into membranes was coupled to glycosylation and cleavage of the p62–E1 site, since these latter activities failed to occur if membranes were added to the *in vitro* system after the synthesis of about 100 amino acids of the p62 protein had proceeded.

Analysis from the *in vivo* studies on the formation of viral structural proteins predicted that the polycistronic 26 S mRNA was translated from a single initiation site followed by posttranslation proteolytic cleavages. These *in vivo* studies assigned the viral capsid cistron to the 5′-end of the mRNA, and this protein should contain the only initiator peptide for the entire messenger. Three laboratories have examined the initiator peptides formed *in vitro* from 26 S RNA-primed systems by adding to the extracts a formylated derivative of [^{35}S]methionine bound to the appropriate fMet-tRNA species. Under these conditions, the initiator methionine residue is not removed, and the peptide and protein containing this residue can be identified. When Sindbis virus 26 S RNA was translated in extracts of rabbit reticulocytes or ascites cells, only the capsid contained the formylmethionine label and only a single virus-specific species of [^{35}S]Met peptide was detected (Cancedda *et al.*, 1975). In an L-cell system primed with 26 S RNA of SFV, a single tryptic peptide was detected and only capsid was labeled with the [^{35}S]formylmethionine. In this latter system, a single Met-Asp dipeptide was isolated when the *in vitro* system was given diphtheria toxin and a NAD reagent which blocks elongation of polypeptide synthesis yet allows the initial depeptide to form (Clegg and Kennedy, 1975b). A fMet-Asp dipeptide was also isolated from an *in vitro* wheat germ system containing SFV 26 S RNA (Glanville *et al.*, 1976b). Thus, the *in vitro* translation of 26 S RNA also initiates from a single site on the viral capsid gene.

All these cell-free extracts have shown a remarkably efficient capacity for cleaving the capsid polypeptide from the growing polypeptide chain. Because the widely different extracts might be considered to contain variable amounts of a host-specified protease activity, many investigators studying 26 S RNA translation have speculated that viral capsid protein itself could provide the cleavage activity. One set of experimental results that provides support for the self-cleavage model was obtained with 26 S RNA isolated from cells infected with a Sindbis virus temperature-sensitive mutant defective at the nonpermissive temperature in posttranslational cleavage (Scheele and Pfefferkorn, 1970). When this 26 S RNA was used as messenger in rabbit reticulocyte extracts, normal capsid was formed at 30°C, but very low amounts of this protein were produced at 40°C (Simmons and Strauss, 1974b; Cancedda *et al.*, 1974b). At the higher temperatures, higher-molecular-weight polypeptides were made *in vitro*, and some of them contained tryptic peptides of the capsid (Cancedda *et al.*, 1974b). In one system, one of the larger proteins had a mobility in SDS polyacrylamide gel electrophoresis identical to that of the p140 formed by the *ts* mutant *in vivo* (Simmons and Strauss, 1974b). Once made in this larger form, there was no subsequent cleavage *in vitro* at permissive temperatures. This *in vitro* reaction mimics in a remarkable fashion the *ts* defect observed in the infected cell. One possible explanation for these results is that the amino acid substitution resulting from a missense mutation distorts the growing polypeptide chain so that it is unable to function as a protease. Another possibility is that the missense mutation leads to a misfolding which affects the site of proteolytic cleavage.

Strong support for a capsid autoprotease activity has come from results of *in vitro* experiments in which amino acid analogs were used together with [^{35}S]methionine during the initial stages of translation but were then replaced with normal amino acids during a chase period with excess unlabeled methionine (Aliperti and Schlesinger, 1978). In the presence of analogs, normal capsid cleavage was inhibited by

50%, and higher-molecular-weight proteins that contained capsid polypeptides were found. The latter were converted to capsid during a chase period that contained normal amino acids, but not if analogs alone were present during the chase. These results suggested that the addition of normal amino acids during the chase period allowed for *de novo* formation of a protease that acted on the preformed, analog-containing polypeptides, and the authors postulated that this protease was the capsid itself.

Although all the *in vitro* systems are extremely efficient in producing the capsid protein, they form relatively low amounts of the envelope gene product—p62, E1, or p100. The difficulty in producing viral envelope proteins *in vitro* may be due to the fact that these polypeptides are formed *in vivo* on polysomes associated with host cell membranes (Wirth *et al.*, 1977; Matire *et al.*, 1977) and are glycosylated during synthesis. In extracts not supplemented with membranes or detergents the viral envelope proteins might easily aggregate, denature, or be degraded. However, there was even a poor yield of envelope polypeptides when membrane-associated polysomes from SFV-infected cells were incubated *in vitro* (Wirth *et al.*, 1977).

Nevertheless, this latter system has provided important information about synthesis of these envelope polypeptides. After incubation *in vitro* and generation of the 26 S RNA individual gene products (C and the two envelope proteins), the extract was treated with chymotrypsin in order to determine if any of the proteins formed *de novo* were protected from the protease action by virtue of being transported into membrane vesicles. Newly synthesized capsid was totally degraded, but with the exception of a small portion of the p62 the envelope proteins were resistant to protease action, indicating they had been transported into the membrane vesicles. An *in vitro* system containing extracts of HeLa cells, microsomes from dog pancreas, and SFV 26 S mRNA produced capsid as well as glycosylated forms of p62 and E1 and nonglycosylated p100 in good yields (Garoff *et al.*, 1978). By varying the time of addition of the microsome fraction to synchronized mRNA translation, these workers showed that insertion of the N-terminal portion of p62 into membranes was essential for glycosylation and cleavage of the p62–E1 site. These latter reactions did not occur if membranes were added to the reaction after about 100 amino acids of p62 had been incorporated into polypeptide. In this system, too, the newly formed p62 and E1 were localized in a microsome fraction and protected from chymotryptic digestion, while capsid and p100 were degraded. A small part of the carboxy "tail" of p62 is initially synthesized as a transmembranal protein. Wirth *et al.* (1977) reported a similar result for p62, but neither system has produced evidence that the putative signal sequence at the N-terminal portion of p62 or E1 is processed. Partial N-terminal sequences of p62 and p100 produced *in vitro* and p62 made *in vivo* showed that this signal sequence was not cleaved during translation (Bonatti and Blobel), 1979).

In these *in vitro* systems as well as that described by Clegg and Kennedy (1975a), the proteolytic cleavage that initially generates the two glycoproteins is also apparently functioning. The model suggests that the cleavage that produces p62 and E1 occurs on the polysome side of the membrane. This situation would allow both E1 and E2 to remain as transmembranal polypeptides. The detection of an aberrant protein with capsid and p62 sequences in cells infected with *ts* mutants is consistent with the model. Other data indicate that both E1 and E2 can span the membrane (Utermann and Simons, 1974) and can be cross-linked to capsid protein *in situ* (Garoff and Simons, 1974). The protease that cleaves E1 from nascent membrane-associated polypeptide has not been identified, but Aliperti and Schlesinger (1978) have proposed that the autoprotease activity of capsid can also act at the p62–E1 site.

III. GLYCOSYLATION OF VIRAL ENVELOPE PROTEINS

The two larger alphavirus glycoproteins E1 and E2 contain from 8 to 14% carbohydrate, and the smaller E3 protein is 45% carbohydrate by weight (Strauss *et al.*, 1970; Laine *et al.*, 1973; Garoff *et al.*, 1974; Keegstra *et al.*, 1975). Two types of oligosaccharide chains have been identified, presumably bound via an asparagine residue to the polypeptide; they are referred to as a high-mannose chain and a complex A-type structure (Kornfeld and Kornfeld, 1976; Sefton and Keegstra, 1974; Burge and Strauss, 1970; Mathita *et al.*, 1976). The glycosylation activities responsible for affixing oligosaccharides to the viral polypeptide are host cell enzymes (Burge and Huang, 1970; Grimes and Burge, 1971; Keegstra *et al.*, 1975; Schlesinger *et al.*, 1976). Initial glycosylation occurs early in envelope polypeptide synthesis, possibly on nascent chains (Sefton, 1977), and consists of *en bloc* transfer of a mannose and *N*-acetylglucosamine oligosaccharide from a dolichol pyrophosphate intermediate (Waechter and Lennarz, 1976). Some of the sugar residues are removed or "trimmed" from the initial glycosylated chains, and additional, different glycosides added

(Tabas *et al.*, 1977; Robbins *et al.*, 1977). In this process, these glycosylated proteins move from interior membrane sites of synthesis to peripheral plasma membrane regions where viral assembly occurs.

The initial proteolytic cleavage of the alphavirus envelope proteins that produces p62 and E1 can occur in the absence of glycosylation (Duda and Schlesinger, 1975; Leavitt *et al.*, 1977; Kaluza, 1975; Garoff and Schwarz, 1978) but the conversion of p62 to E2 and E3 appears to require glycosylation. Data from Sefton and Keegstra (1974) showed that fucose and sialic acid were added to E2 during or subsequent to formation from p62. Failure to glycosylate p62 and E1 of Sindbis virus adequately interfered with their transport to the cell's surface (Leavitt *et al.*, 1977), and budding of virus was blocked (Duda and Schlesinger, 1975; Schwarz *et al.*, 1976).

Most of the current data show that the presence of some oligosaccharide is essential for alphavirus envelope protein processing and localization, but that outer sugars can be altered without influencing viral infectivity (Schlesinger *et al.*, 1976). Whether alphavirus development can proceed in the total absence of glycosylation remains an interesting question. Apparently, vesicular stomatis virus can assemble, bud from the cell, and form infectious particles under conditions where an inhibitor has shut off all glycosylation of the viral envelope protein (Gibson *et al.*, 1978).

IV. FATTY ACID ACYLATION OF ENVELOPE PROTEINS

In addition to a proteolytic cleavage and the processing of oligosaccharide, there is another kind of posttranslational modification that occurs during the maturation of alphavirus glycoproteins. Schmidt and Schlesinger (1979) found that E1 and E2 of Sindbis virions contained small amounts of fatty acid that appeared to be covalently attached directly to the polypeptide chain by an ester bond. By following the labeling pattern of [^{3}H]- or [^{14}C]palmitic acid added to virus-infected cells, these investigators showed that fatty acid became attached to p62 shortly before the p62 cleavage to E2 (Schmidt and Schlesinger, 1980). Results from a double-labeled [^{35}S]methionine, [^{3}H]palmitic acid pulse-chase experiment showed that the fatty acid ^{3}H-label appeared in virus particle glycoproteins about 6 min after addition of the isotope, while the amino acid ^{35}S-label did not appear in the particle glycoproteins until 20–25 min later. The latter data are consistent with earlier results (Schlesinger *et al.*, 1972), but the earlier appearance of fatty acid in particles clearly shows that the addition of fatty acids is a late step in the maturation process and probably occurs as the glycoproteins move through the cell's Golgi apparatus.

V. STOICHIOMETRY OF VIRAL STRUCTURAL GENE PRODUCTS

If translation of a polycistronic mRNA is initiated at a single site on the mRNA and individual proteins form by posttranslational cleavages, there should be equal numbers of the individual gene products. Analysis of the virion structure shows that capsid and envelope proteins occur in a 1:1:1 ratio (Garoff *et al.*, 1974; Schlesinger *et al.*, 1972). However, in almost all cases examined, there is a greater amount of capsid protein than envelope protein in the infected cell. In BHK cells infected with Sindbis virus, a substantial portion of the envelope proteins appears as an uncleaved polypeptide of about 100,000 molecular weight (Strauss *et al.*, 1968). This protein has been detected in SFV-infected cells as well and accumulates in cells infected with several *ts* mutants incubated at the nonpermissive temperature (Bracha and Schlesinger, 1976b). This protein is metabolically unstable in pulse-chase experiments, but it is unlikely that it serves as a normal precursor of the PE2 and E1 proteins.

Effective maturation of the envelope proteins requires that polyribosomes carrying 26 S mRNA associate properly with internal cell membranes. Inability to establish this interaction could alter the fate of these polypeptides destined for transport to the plasma membrane. They could either accumulate as an unstable p100 protein or as fragments that could be highly susceptible to proteolytic degradation. The capsid protein, however, would have been processed from the nascent polysome and might fold into a protease-resistant form which would continue to accumulate. Nonglycosylated proteins have been shown to be degraded (Schwarz *et al.*, 1976).

Studies on the amounts of viral gene products from cells infected with rhinoviruses have shown clearly how aberrant cleavages and polypeptide instabilities can account for varying ratios of mRNA products despite the presence of only a single initiation site of translation (McLean *et al.*, 1976). It is likely that similar types of aberrant cleavages occur in alphavirus-infected cells and can account for the frequently observed nonstoichiometric distribution of structural gene products.

VI. CONCLUSION

Translation of alphavirus mRNAs conforms to the general scheme established for protein formation in eukaryotic cells in which mRNA has a single site for the initiation of translation. Furthermore, the individual gene products encoded in the alphavirus polycistronic messengers arise from polyprotein precursors after post- or cotranslational proteolytic processing and glycosylation, a mechanism now well established for a wide variety of eukaryotic viral mRNAs (reviewed in Korant, 1975). There is also with alphaviruses evidence for temporal control of genome expression, a phenomenon which occurs during the replication cycle of most viruses.

Data from experiments with cells infected with certain alphavirus *ts* mutants strongly suggest that a protein produced from the nonstructural segment of the genome is required for formation of the subgenomic 26 S RNA species (Bracha *et al.*, 1976; Saraste *et al.*, 1977; Kääriäinen *et al.*, 1978; Sawicki *et al.*, 1978). *In vivo,* the maximum rate of synthesis of nonstructural protein is achieved between 3 and 4 h after infection, at a time when about 50% of the 26 S RNA has been made (Lachmi and Kääriäinen, 1977). At this time about equal amounts of structural proteins are synthesized. Later the rate of synthesis of structural proteins increases in parallel to the amount of their messenger, the 26 S RNA, suggesting that their synthesis is controlled at the transcriptional level. The synthetic rate of nonstructural proteins declines in spite of an increase in the amount of 42 S RNA, which is their potential messenger. The mechanism of this "shut-off" of nonstructural protein synthesis is not understood (Kääriäinen *et al.*, 1976; Kääriäinen and Söderlund, 1978), although an increased rate of encapsidation of 42 S RNA is the most plausible explanation. In any case there seems to be an early phase of protein synthesis consisting of prevalent translation of nonstructural proteins and a late phase during which mostly structural proteins are made. This shift, as well as its dependence on an early viral gene protein product for later gene expression, has been observed with many viruses replicating in both eukaryotic and prokaryotic cells.

There are, however, some unusual features in the translation and processing of alphavirus mRNA that have emerged from some of the studies described in this chapter. For example, there is the apparent requirement for formation of a subgenomic mRNA species (here, 26 S RNA) in order to translate the viral structural protein genes, even though they are present in the genomic 42 S RNA. Although these genes may be expressed by translation of the latter mRNA, all data thus far show that 42 S RNA translation terminates in the region of RNA prior to the structural protein RNA sequences. A similar kind of internal termination and the formation of separate subgenomic species seem to occur during expression of the RNA tumor virus genome (Weiss *et al.*, 1977) and also during replication of RNA plant viruses (Knowland *et al.*, 1975; Shih and Kaesberg, 1973). In fact this mechanism may prevail for some eukaryotic host cell RNAs.

Alphavirus 26 S RNA contains information for both membrane and nonmembrane proteins. The former requires close interaction between polysome and internal cell membrane. In many viral systems a distinct mRNA exists for membrane protein-encoded genes, and it is not clear what, if any, functional qualities result from the translation of alphavirus capsid polypeptide on membrane-associated polysomes.

Finally, there is in the maturation of alphavirus structural proteins a late processing event closely coupled to the budding of virions. This proteolytic cleavage of glycosylated precursor to one of the envelope proteins appears to be analogous to similar alterations in other viral envelope proteins. Such conformational changes are important, if not essential, for viral infectivity (Scheid and Choppin, 1974; Nagai and Klenk, 1977; Lazarowitz *et al.*, 1973; Witte *et al.*, 1977). With alphaviruses, however, the proteolytic event is also required for particle secretion. Thus far, data from studies on other enveloped viruses have not indicated a similar effect on the budding process.

The proteolytic activities required for the processing and maturation of alphavirus proteins have not been rigorously identified, but existing data suggest that at least one of them is virus-coded. This activity appears to be similar to those described for the processing of other RNA viruses (Von der Helm, 1977; Yoshinaka and Luftig, 1977; Pelham, 1978, 1979; Vogt *et al.*, 1979), and it will be of considerable interest to determine if the alphavirus activities are encoded by a single virus-specified enzyme or by a combination of different host and virus activities.

REFERENCES

Aliperti, G., and Schlesinger, M. J. (1978). *Virology* **90**, 366–369.
Atkins, G. J. (1976). *Virology* **71**, 593–597.
Atkins, G. J., Samuels, J., and Kennedy, S. I. T. (1974). *J. Gen. Virol.* **25**, 371–380.
Blobel, G., and Dobberstein, B. (1975). *J. Cell Biol.* **67**, 835–851.
Bonatti, S., and Blobel, G. (1979). *J. Biol. Chem.* **254**, 12,261–12,264.
Bracha, M., and Schlesinger, M. J. (1976a). *Virology* **72**, 272–277.
Bracha, M., and Schlesinger, M. J. (1976b). *Virology* **74**, 441–449.

Bracha, M., and Schlesinger, M. J. (1978). *J. Virol.* **26**, 126–135.
Bracha, M., Leone, A., and Schlesinger, M. J. (1976). *J. Virol.* **20**, 612–620.
Brzeski, H., and Kennedy, S. I. T. (1977). *J. Virol.* **22**, 420–429.
Burge, B. W., and Huang, A. S. (1970). *J. Virol.* **6**, 176–182.
Burge, B. W., and Pfefferkorn, E. R. (1966a). *Virology* **30**, 204–213.
Burge, B. W., and Pfefferkorn, E. R. (1966b). *Virology* **30**, 214–223.
Burge, B. W., and Strauss, J. H. (1970). *J. Mol. Biol.* **47**, 449–466.
Burrell, C. J., Martin, E. M., and Cooper, P. D. (1969). *J. Gen. Virol.* **6**, 319–323.
Cancedda, R., Swanson, R., and Schlesinger, M. J. (1974a). *J. Virol.* **14**, 652–663.
Cancedda, R., Swanson, R., and Schlesinger, M. J. (1974b). *J. Virol.* **14**, 664–671.
Cancedda, R., Villa-Komaroff, L., Lodish, H., and Schlesinger, M. (1975). *Cell* **6**, 215–222.
Cheng, P.-Y. (1958). *Nature (London)* **181**, 1800.
Clegg, J. C. S. (1975). *Nature (London)* **254**, 454.
Clegg, J. C. S., and Kennedy, S. I. T. (1974). *FEBS Lett.* **42**, 327–330.
Clegg, C., and Kennedy, S. I. T. (1975a). *Eur. J. Biochem.* **53**, 175–183.
Clegg, J. C. S., and Kennedy, S. I. T. (1975b). *J. Mol. Biol.* **97**, 401–411.
Clegg, J. C. S., Brzeski, H., and Kennedy, S. I. T. (1976). *J. Gen. Virol.* **32**, 413–430.
Clewley, J. P., and Kennedy, S. I. T. (1976). *J. Gen. Virol.* **32**, 395–411.
Duda, E. (1975). *Med. Biol.* **53**, 368–371.
Duda, E., and Schlesinger, M. J. (1975). *J. Virol.* **15**, 416–419.
Friedman, R. M. (1968). *J. Virol.* **2**, 1076–1080.
Friedman, R. M., and Grimley, P. M. (1969). *J. Virol.* **4**, 292–299.
Friedman, R. M., and Sreevalsan, T. (1970). *J. Virol.* **6**, 169–175.
Friedman, R. M., Fantes, K. H., Levy, H. B., and Carter, W. B. (1967). *J. Virol.* **1**, 1168–1173.
Garoff, H., and Schwarz, R. T. (1978). *Nature (London)* **274**, 487–489.
Garoff, H., and Simons, K. (1974). *Proc. Natl. Acad. Sci. U.S.A.* **71**, 3988–3992.
Garoff, H., Simons, K., and Renkonen, O. (1974). *Virology* **61**, 493–504.
Garoff, H., Simons, K., and Dobberstein, B. (1978). *J. Mol. Biol.* **124**, 587–600.
Gibson, R., Leavitt, R., Kornfeld, S., and Schlesinger, S. (1978). *Cell* **13**, 671–679.
Glanville, N., and Lachmi, B. (1977). *FEBS Lett.* **81**, 399–402.
Glanville, N., Morser, J., Uomala, P., and Kääriäinen, L. (1976a). *Eur. J. Biochem.* **64**, 167–175.
Glanville, N., Ranki, M., Morser, J., Kääriäinen, L., and Smith, A. E. (1976b). *Proc. Natl. Acad. Sci. U.S.A.* **73**, 3059–3063.
Glanville, N., Lachmi, B., Smith, A. E., and Kääriäinen, L. (1978). *Biochim. Biophys. Acta* **518**, 497–506.
Grimes, W. J., and Burge, B. W. (1971). *J. Virol.* **7**, 309–313.
Hay, A. J., Skehel, J. J., and Burke, D. C. (1968). *J. Gen. Virol.* **3**, 175–184.
Jones, K. J., Waite, M. R. F., and Bose, H. R. (1974). *J. Virol.* **13**, 809–817.
Kääriäinen, L., and Söderlund, H. (1978). *Curr. Top. Microbiol. Immunol.* **82**, 15–69.
Kääriäinen, L., Lachmi, B., and Glanville, N. (1976). *Ann. Microbiol. (Paris)* **127A**, 197–203.
Kääriäinen, L., Sawicki, D., and Gomatos, P. J. (1978). *J. Gen. Virol.* **39**, 463–473.
Kaluza, G. (1975). *J. Virol.* **16**, 602–612.
Kaluza, G. (1976). *J. Virol.* **19**, 1–12.
Kaluza, G., Kraus, A. A., and Rott, R. (1976). *J. Virol.* **17**, 1–9.
Keegstra, K., Sefton, B., and Burge, B. W. (1975). *J. Virol.* **16**, 613–620.
Kennedy, S. I. T. (1972). *Biochem. Biophys. Res. Commun.* **48**, 1254–1258.

Kennedy, S. I. T. (1976). *J. Mol. Biol.* **108**, 491–511.
Keränen, S., and Kääriäinen, L. (1974). *Acta Pathol. Microbiol. Scand., Sect. B* **82**, 810–820.
Keränen, S., and Kääriäinen, L. (1975). *J. Virol.* **16**, 388–396.
Knowland, J. Hunter, T., Hunt, T., and Zimmern, D. (1975). *Inserm* **47**, 211–216.
Korant, B. D. (1975). "Proteases and Biological Control," pp. 621–644. Cold Spring Harbor Lab., Cold Spring Harbor, New York.
Kornfeld, R., and Kornfeld, S. (1976). *Annu. Rev. Biochem.* **41**, 217–237.
Lachmi, B., and Kääriäinen, L. (1976). *Proc. Natl. Acad. Sci. U.S.A.* **73**, 1936–1940.
Lachmi, B., and Kääriäinen, L. (1977). *J. Virol.* **22**, 142–149.
Lachmi, B., Glanville, N., Keränen, S., and Kääriäinen, L. (1975). *J. Virol.* **16**, 1615–1629.
Laine, R., Söderlund, H., and Renkonen, O. (1973). *Intervirology* **1**, 110–118.
Lazarowitz, S., Compans, R., and Choppin, P. (1973). *Virology* **52**, 199–212.
Leavitt, R., Schlesinger, S., and Kornfeld, S. (1977). *J. Virol.* **21**, 375–385.
McLean, C., Matthews, T. J., and Rueckert, R. R. (1976). *J. Virol.* **19**, 903–914.
Martire, G., Bonatti, S., Aliperti, G., DeGiuli, C., and Cancedda, R. (1977). *J. Virol.* **21**, 610–618.
Mathita, A., Luukkonen, A., and Renkonen, O. (1976). *Biochim. Biophys. Acta* **419**, 435–444.
Morser, M. J., and Burke, D. C. (1974). *J. Gen. Virol.* **23**, 395–409.
Morser, M. J., Kennedy, S. I. T., and Burke, D. C. (1973). *J. Gen. Virol.* **21**, 19–29.
Mowshowitz, D. (1973). *J. Virol.* **11**, 535–543.
Nagai, Y., and Klenk, H. (1977). *Virology* **77**, 125–134.
Pelham, H. R. B. (1978). *Eur. J. Biochem.* **85**, 457–462.
Pelham, H. R. B. (1979). *Virology* **97**, 256–265.
Pfefferkorn, E. R., and Boyle, M. K. (1972). *J. Virol.* **9**, 187–188.
Pfefferkorn, E. R., and Shapiro, D. (1974). In "Comprehensive Virology" (H. Fraenkel-Conrat, and R. R. Wagner, eds.), Vol. 2, pp. 171–230. Plenum, New York.
Ranki, M., and Kääriäinen, L. (1970). *Ann. Med. Exp. Biol. Fenn.* **48**, 238–245.
Ranki, M., Kääriäinen, L., and Renkonen, O. (1972). *Acta Pathol. Microbiol. Scand., Sect. B* **80**, 760–768.
Robbins, P. W., Krag, S., and Lieu, T. (1977). *J. Biol. Chem.* **252**, 1780–1785.
Rosemond, H., and Sreevalsan, T. (1973). *J. Virol.* **11**, 399–415.
Saborio, J. L., Pong, S. S., and Koch, G. (1974). *J. Mol. Biol.* **85**, 195–211.
Saraste, J., Kääriäinen, L., Söderlund, H., and Keränen, S. (1977). *J. Gen. Virol.* **37**, 399–406.
Sawicki, D., Kääriäinen, L., Lambeck, C., and Gomatos, P. J. (1978). *J. Virol.* **25**, 19–27.
Scheele, C. M., and Pfefferkorn, E. R. (1969). *J. Virol.* **4**, 117–122.
Scheele, C. M., and Pfefferkorn, E. R. (1970). *J. Virol.* **5**, 329–337.
Scheid, A., and Choppin, P. (1974). *Virology* **57**, 475–490.
Schlesinger, M. J., and Schlesinger, S. (1973). *J. Virol.* **11**, 1013–1016.
Schlesinger, M. J., Schlesinger, S., and Burge, B. W. (1972). *Virology* **47**, 539–541.
Schlesinger, S., and Schlesinger, M. J. (1972). *J. Virol.* **10**, 925–932.
Schlesinger, S., Gottlieb, C., Feil, P., Gelb, N., and Kornfeld, S. (1976). *J. Virol.* **17**, 239–246.
Schmidt, M. F. G., and Schlesinger, M. J. (1979). *Cell* **17**, 813–819.
Schmidt, M. F. G., and Schlesinger, M. J. (1980). *J. Biol. Chem.*, **255**, 3334–3339.
Schmidt, M. F. G., Bracha, M., and Schlesinger, M. J. (1979). *Proc. Natl. Acad. Sci. U.S.A.* **76**, 1687–1691.

Schwarz, R. T., Rohrschneider, J. M., and Schmidt, M. F. G. (1976). *J. Virol.* **19**, 782–791.
Sefton, B. M. (1977). *Cell* **10**, 659–668.
Sefton, B. M., and Burge, B. W. (1973). *J. Virol.* **12**, 1366–1374.
Sefton, B. M., and Keegstra, K. (1974). *J. Virol.* **14**, 522–530.
Sefton, B. M., Wikus, G. G., and Burge, B. W. (1973). *J. Virol.* **11**, 730–735.
Shih, D. S., and Kaesberg, P. (1973). *Proc. Natl. Acad. Sci. U.S.A.* **70**, 1799–1805.
Simmons, D. T., and Strauss, J. H. (1974a). *J. Virol.* **14**, 552–559.
Simmons, D. T., and Strauss, J. H. (1974b). *J. Mol. Biol.* **86**, 397–409.
Simons, K., and Kääriäinen, L. (1970). *Biochem. Biophys. Res. Commun.* **38**, 981–988.
Simons, K., Keränen, S., and Kääriäinen, L. (1973). *FEBS Lett.* **29**, 87–91.
Smith, A. E., Wheeler, T., Glanville, N., and Kääriäinen, L. (1974). *Eur. J. Biochem.* **49**, 101–110.
Snyder, H. W., and Sreevalsan, T. (1974). *J. Virol.* **13**, 541–544.
Söderlund, H. (1976). *FEBS Lett.* **63**, 56.
Söderlund, H., Glanville, N., and Kääriäinen, L. (1973–1974). *Intervirology* **2**, 100–113.
Sreevalsan, T., and Lockart, R. Z. (1964). *Virology* **24**, 91–96.
Strauss, E. G., Lenches, E. M., and Strauss, J. H. (1976). *Virology* **74**, 154–168.
Strauss, J. H., and Strauss, E. G. (1977). In "The Molecular Biology of Animal Viruses" (D. P. Nayak, ed.), pp. 111–166. Dekker, New York.
Strauss, J. H., Burge, B. W., Pfefferkorn, E. R., and Darnell, J. E. (1968). *Proc. Natl. Acad. Sci. U.S.A.* **59**, 533–537.
Strauss, J. H., Burge, B. W., and Darnell, J. E. (1969). *Virology* **37**, 367–376.
Strauss, J. H., Burge, B. W., and Darnell, J. E. (1970). *J. Mol. Biol.* **47**, 437–448.
Tabas, I., Schlesinger, S., and Kornfeld, S. (1978). *J. Biol. Chem.* **253**, 716–722.
Tan, K. B., Sambrook, J. F., and Bellet, A. J. P. (1969). *Virology* **38**, 427–439.
Taylor, J. (1965). *Virology* **25**, 340–349.
Utermann, G., and Simons, K. (1974). *J. Mol. Biol.* **85**, 569–587.
Vogt, V. M., Wight, A., and Eisenman, R. (1979). *Virology* **98**, 154–167.
Von der Helm, K. (1977). *Proc. Natl. Acad. Sci. U.S.A.* **74**, 911–915.
Waechter, C. J., and Lennarz, W. J. (1976). *Annu. Rev. Biochem.* **41**, 95–112.
Waite, M. R. F. (1973). *J. Virol.* **11**, 198–206.
Wecker, E. (1959). *Virology* **7**, 241–243.
Wecker, E. (1963). *Nature (London)* **197**, 1277–1279.
Wecker, E., and Schafer, S. (1957). *Z. Naturforsch., Teil B* **12**, 415–417.
Wecker, E., Hummeler, K., and Goetz, O. (1962). *Virology* **17**, 110–117.
Weiss, S. R., Varmus, H. E., and Bishop, J. M. (1977). *Cell* **12**, 983–992.
Wengler, G., and Wengler, G. (1974). *Virology* **59**, 21–35.
Wengler, G., and Wengler, G. (1975a). *Virology* **65**, 601–605.
Wengler, G., and Wengler, G. (1975b). *Virology* **66**, 322–326.
Wengler, G., and Wengler, G. (1976a). *Virology* **73**, 190–199.
Wengler, G., and Wengler, G. (1976b). *J. Virol.* **17**, 10–19.
Wengler, G., Beato, M., and Hackemach, B. A. (1974). *Virology* **61**, 120.
Wirth, D. F., Katz, F., Small, B., and Lodish, H. F. (1977). *Cell* **10**, 253–263.
Witte, O. N., Tsukamuto-Adey, A., and Weissman, I. L. (1977). *J. Virol.* **76**, 539–553.
Yoshinaka, Y., and Luftig, R. B. (1977). *Cell* **12**, 709–719.

14

Mutants of Alphaviruses: Genetics and Physiology

ELLEN G. STRAUSS
JAMES H. STRAUSS

I. INTRODUCTION

Genetic analysis of alphaviruses has been a major tool for studying viral replication on a molecular level and has been the basis for many of the discoveries about alphavirus replication reported in other chap-

THE TOGAVIRUSES

ISBN 0-12-625380-3

ters of this volume. By analysis of various conditional lethal mutants, it has been possible to delineate the number of necessary virus functions, to separate some of the events essential for replication of the genomic RNA, to study the primary translation products of the two viral messages, to determine the sequence of processing steps leading to the final virus-specific proteins, and to study the events involved in maturation of the virus. Furthermore, comparison of mutant behavior in parallel mammalian or avian and arthropod cells, corresponding to the natural homeothermic and poikilothermic hosts, is just beginning and may ultimately reveal a great deal about the host cell functions involved in virus replication.

Alphaviruses appear to have a relatively simple genetic system. The RNA genome has a molecular weight of 4.5×10^6, enough to code for about nine polypeptides with an average size of 40,000 daltons. Three of the polypeptides are the structural proteins of the virion and are translated from the 26 S mRNA which contains the sequences of the 3′-terminal one-third of the genome. The remaining polypeptides are nonstructural, containing the replicase activity and possibly other enzymatic functions, and are translated from 49 S RNA.[1] Most of the coding capacity of the genome appears to be translated into protein, and little RNA appears to be present in genetic control elements.

The genetic diversity among members of the alphavirus group has not been studied in any detail, other than the serological grouping and immunological cross-reactions between the viruses discussed in Chapter 2. One recent study to determine the relationships among these viruses on a molecular level was that of Wengler *et al.* (1977) who examined the RNAs of Semliki Forest, Sindbis, chikungunya, and o'nyong-nyong viruses. By comparing the T1 digests of the various RNAs and by RNA–RNA hybridization, they found that chikungunya and o'nyong-nyong viruses contained approximately 13% base sequence homology, but that there was less than 1% homology between any other pair tested. Admittedly, as measures of relatedness between viruses, the methods used were very stringent. The redundancy in the genetic code permits significant evolutionary divergence in the RNA

[1] The genomic RNA of the alphaviruses has been referred to variously as 49, 45, and 42 S RNA. The discrepancies in the reported sedimentation coefficients appear to be due in part to the differences in sedimentation coefficient between the circular and linear forms of the RNA (Frey *et al.*, 1979) and in part to the use of different ionic strength solutions and different markers for sedimentation analysis. In our hands the circular form of the RNA sediments at 49 S in solutions containing 0.06–0.2 *M* NaCl, as determined both by centrifugation in an analytical ultracentrifuge and in sucrose gradients.

sequences, enough so that no homology would be detected by nucleic acid hybridization, before the encoded polypeptides are materially altered.

Another approach toward studying the relations among alphaviruses has been to study the structural proteins of the viruses. The amino acid compositions of the core and envelope proteins of Sindbis virus (Burke and Keegstra, 1974; Bell *et al.*, 1979) and Semliki Forest virus (SFV) (Simons and Kääriäinen, 1970; Kennedy and Burke, 1972; Garoff *et al.*, 1974) have shown that the proteins are similar but by no means identical. For example, one difference between the capsid proteins of these two viruses is the apparent substitution of eight arginines in Sindbis virus by lysine in SFV (Bell *et al.*, 1979). However, determination of primary amino acid sequences, such as those for the N-terminal regions of E1 and E2 of Sindbis virus (Bell *et al.*, 1978), may reveal more similarities. It is anticipated that functionally important domains, such as points of interaction between proteins or between protein and RNA, will be highly conserved.

Finally, relationships among alphaviruses can be explored by comparing the mutants of different alphaviruses for functional similarities. To date, only the genetics of Sindbis virus and SFV have been studied in any detail (see reviews in Pfefferkorn, 1969, 1977; Pfefferkorn and Burge, 1967). Recently, however, more data are becoming available on mutants of Western equine encephalitis (WEE) virus, which will be valuable for evolutionary comparisons.

II. TYPES OF MUTANTS

A. Plaque Morphology Mutants

Ever since the earliest articles characterizing alphaviruses there have been persistent reports of virus stocks which were heterogeneous with respect to plaque morphology. By repeated single plaque isolations, stocks of small-plaque and large-plaque variants of Middelburg (Pattyn and de Vleesschauwer, 1966), Sindbis (Brés, 1965; Hannoun *et al.*, 1964; Pedersen *et al.*, 1973), Getah (Kimura and Ueba, 1978), and WEE (Quersin-Thiry, 1961; Marshall *et al.*, 1962) viruses were produced which were genetically relatively stable, although small-plaque variants are thought to arise continuously from large-plaque stocks, and vice versa. Curiously enough, there are no reports of spontaneously arising large- and small-plaque variants of SFV except for virus passed in mosquito cells. The heat-resistant (HR) strain of Sindbis virus, which was isolated by Burge and Pfefferkorn (1966a) by repeated cycles of thermal inactivation at 56°C and is a multiple mutant of wild-type Sindbis virus, also produces large-plaque and small-plaque variants.

Several studies have attempted to correlate virulence of alphaviruses with plaque morphology. For example, Rasmussen *et al.* (1973) isolated an avirulent large-plaque strain of Sindbis virus from a virulent strain originally containing both large- and small-plaque variants. On the other hand, some authors report that the small-plaque variant of Sindbis virus is more capable of producing a chronic (asymptomatic) infection than the large-plaque variant (Inglot *et al.*, 1973; Schwöbel and Ahl, 1972). With WEE virus, Jahrling (1976) found that a population which was heterogeneous with respect to virulence could be separated into an avirulent small-plaque strain and a virulent large-plaque variant. With Venezuelan equine encephalitis (VEE) virus, naturally occurring epizootic strains (which are virulent for horses and numerous other mammals including guinea pigs) generally form small plaques on Vero monolayers, while benign (enzootic) strains form large plaques (Johnson and Martin, 1974). However, small-plaque variants cloned from large-plaque, avirulent VEE were still avirulent for guinea pigs (Jahrling and Eddy, 1977). Brown (1963) also reported that an attenuated variant of VEE virus formed small plaques. Kimura and Ueba (1978) found that the large-plaque strain of Getah virus was more virulent for suckling mice than the small-plaque variant. Furthermore, large-plaque Getah virus possessed hemagglutinating activity, while small-plaque Getah virus did not. Thus it appears that virulence and plaque morphology are not necessarily correlated in alphaviruses (see Chapter 4).

There seems to be no disagreement that passage in arthropod cells strongly selects for small-plaque variants. This has been shown for SFV in *Aedes albopictus* cell cultures by Davey and Dalgarno (1974), for Sindbis virus passaged in whole fruit flies (*Drosophila melanogaster*) by Bras-Herreng (1973), for Middelburg virus in *Aedes aegypti* mosquitoes by Pattyn and de Vleesschauwer (1968), and for VEE virus in mosquitoes by Gaidamovich *et al.* (1971). In each case the virus recovered from insect cells, regardless of the plaque morphology of the input inoculum, was 90–100% of the small-plaque phenotype. It has also been reported that virus from chronically infected mosquito cells is temperature-sensitive (discussed in more detail in Section II,C) (Igarashi *et al.*, 1977; see Chapter 20). The small-plaque phenotype is independent of the temperature-sensitive mutation in most cases, however. For example, Shenk *et al.* (1974) isolated a small-plaque *ts* mutant of Sindbis virus from persistently infected *A. albopictus* cells and found that ts^+ revertant clones of the stock retained the small-plaque phenotype. Because such small-plaque variants from alphaviruses grown in arthropod cells are usually avirulent and/or temperature-sensitive, such stocks have been proposed as a source of attenuated virus for vaccine strains (Peleg, 1971).

We have shown in our laboratory that there is a definite selection for the small-plaque Sindbis HR variant when the virus is passaged in BHK-21 cells. The small-plaque variant grows to higher titers in BHK than in monolayers of chick embryo fibroblasts, while the reverse is true of the large-plaque variant, which is preferentially replicated in chick cells (Strauss *et al.*, 1980). In contrast, Jahrling (1976) found that the large-plaque strain of WEE virus gave rise to predominantly small-plaque virus after five successive passages in duck embryo cells.

The molecular basis for the difference in plaque size appears to be related in some cases to the surface charge on the particle. Variants with a higher net negative charge interact electrostatically with sulfated polysaccharides present in the agar overlay and produce small plaques; the addition of polycations to the assay results in large plaques (Liebhaber and Takemoto, 1963). Furthermore, more negatively charged virions tend to bind more tightly to hydroxylapatite (calcium phosphate); in many cases large-plaque formers elute more readily from these columns than small-plaque formers. For example, Bose *et al.* (1970) found that large-plaque and small-plaque variants of Sindbis AR339 virus could be separated by chromatography on calcium phosphate. It is interesting, however, that attempts in the same laboratory to separate these strains by isoelectric focusing were unsuccessful (Pedersen *et al.*, 1973). Jahrling and Beall (1977) and Jahrling and Eddy (1977) found that large-plaque strains of both WEE and VEE viruses eluted from hydroxylapatite earlier than the corresponding small-plaque strains. Furthermore, small-plaque variants isolated from naturally occurring large-plaque (avirulent) strains of VEE virus acquired the chromatographic profile of the naturally occurring small-plaque strain while retaining their antigenic specificities and remaining avirulent (Jahrling and Eddy, 1977). On the other hand, Symington and Schlesinger (1978) found that a host range mutant of Sindbis virus which adsorbed more readily to mouse plasmacytoma cells had glycoproteins which were more negatively charged than those of standard virus while having no effect on plaque size.

In some cases the plaque size probably reflects in part the relative growth rates of the large-plaque and small-plaque strains. It is clear that the minute plaques formed by morphological mutants of Sindbis

HR virus (Strauss *et al.*, 1977) and of WEE virus (Simizu *et al.*, 1973) are due to the slow growth rate of these mutants (see Section II,B).

We have found that the relative sizes of plaques produced by naturally occurring large-plaque and small-plaque strains of Sindbis HR virus depend upon both the host cell used for the assay and the solidifying material, agar or agarose, in the overlay (Table I). Agarose is a purified neutral galactan hydrocolloid which should have minimal electrostatic interaction with the virions. When plaqued on chick cell monolayers, the two strains give large plaques or small plaques, respectively, under agar; the size of the plaques is larger under agarose than under agar. When plaqued on BHK monolayers, the two strains give the same (small) size plaques under agar; under agarose the plaques from the small-plaque strain are again small, whereas those of the large-plaque strain are large. From these results it appears that both the surface charge on the particle and the relative growth rate of a particular strain in cells of a particular host influence the size of the plaques produced.

B. Host Range Mutants

Although alphaviruses grow in a wide variety of mammalian, avian, and arthropod hosts, both in the whole organism and in tissue culture cells derived from them (Pfefferkorn and Shapiro, 1974; Strauss and Strauss, 1977), no useful collection of host range mutants has ever been isolated or exploited. As discussed in Section II,A, small-plaque and large-plaque variants have different selective advantages in various cell lines, but the advantage is relatively slight. Of particular interest in elucidating the mechanisms of replication of alphaviruses in arthropod cells would be mutants capable of growing in insect cells but not vertebrate cells, and vice versa. Unfortunately, none of the more commonly used alphaviruses except chikungunya forms plaques

TABLE I

Large-Plaque and Small-Plaque Strains of Sindbis Virus

Sindbis strain	Monolayer for plaque assay	Size of plaques	
		Agar	Agarose
Small plaque	BHK	Small	Small
Large plaque	BHK	Small	Large
Small plaque	Chick	Small	Large
Large plaque	Chick	Large	Very large

on mosquito cell monolayers (Yunker and Cory, 1975), making mutant isolation very difficult. A mutant clone of mosquito cells in which Sindbis virus causes a cytopathic infection has been described (Sarver and Stollar, 1977), which may obviate this difficulty (V. Stollar, personal communication). In the reverse direction, the temperature-sensitive alphavirus mutants examined for replication in mosquito cells (Renz and Brown, 1976) appear to retain the same phenotype by which they were selected on vertebrate cells. Additional studies along these lines are needed to determine whether alphaviruses encode proteins whose function is exclusively required in either the vertebrate or invertebrate host (see Chapter 20).

C. Particle Morphology Mutants

Two alphavirus mutants which produce predominantly virions containing more than one nucleocapsid within a single envelope (multinucleoid or multiploid virions) have been studied in some detail. Both were originally isolated as small-plaque mutants; the small-plaque morphology is probably due to the low yield of mature virus produced by these mutants and the delayed maturation of these virions compared to wild-type infection. One was a mutant of WEE virus isolated after γ-ray mutagenesis (Simizu *et al.*, 1973; Hashimoto *et al.*, 1975), and the other a nitrous acid-induced mutant of Sindbis virus, *ts*103 (Strauss *et al.*, 1977). (The *ts*103 mutant is not truly temperature-sensitive, but it forms minute plaques under any circumstances and at 40°C under less than optimal plaque-forming conditions the plaques vanish, hence the original *ts* designation.) Although these two mutants share several characteristics, they also exhibit fundamental differences. Both mutants produce a heterogeneous collection of virions; and the majority of the particles are multiploid, although some virions appear to have only one nucleocapsid and are not separable from the parental virus by physical means. In both cases RNA synthesis is completely normal. The nucleocapsid protein and envelope glycoproteins are indistinguishable from those of the parental strain by polyacrylamide gel electrophoresis, and the membrane proteins are inserted at the usual time into the cell surface membrane. Infected cells hemadsorb, and nearly normal concentrations of intracellular nucleocapsids are formed. However, the defect appears to lie in the final juxtaposition of the nucleocapsid with the plasmalemma and the subsequent budding events. In *ts*103 the budding of the virus is extremely aberrant. No virus particles are seen budding directly from the cell surface; instead they appear in long, filamentous projections extending from the cell surface, which are reminiscent of those seen in the late stages of Sindbis HR infection (Birdwell *et al.*, 1973). Some regions of these processes are bloated with nucleocapsids, while other regions appear almost empty (Strauss *et al.*, 1977). All the nucleocapsids found in *ts*103 virions, except for those in large multiploids of low isopycnic density (particles with many cores and a "loose" envelope), are indistinguishable from wild-type nucleocapsids and, in fact, all the various-sized virus particles released are highly infectious. In the cytoplasm, however, it is possible to find abnormal capsids which are evidently selected against in the final maturation. In the WEE mutant, on the other hand, intracellular cores appear normal (although nucleocapsids accumulate in pseudocrystalline arrays within the cytoplasm), while nucleocapsids isolated from multiploid virions are pleomorphic. These multiploid particles are also infectious, however (Simizu *et al.*, 1973). In both these mutants it appears that a rather subtle alteration in the nucleocapsid protein produces a broad spectrum of pleiotropic effects, presumably as a result of improper interactions between the altered capsids and the viral glycoproteins in the cell plasmalemma during budding.

D. Temperature-Sensitive Mutants

1. Temperature-Sensitive Mutants Produced by Chemical Mutagenesis

Temperature-sensitive mutants have been extremely useful in the study of alphaviruses. Since these viruses replicate over a wide range of temperatures both in nature and in the laboratory, it is easy to select widely separated permissive (28°C–30°C) and nonpermissive (38°–41.5°C) conditions. In all cases stocks can be grown at the permissive temperature, and radiolabeled mutant virions can be prepared in good yield. On the other hand, the nature of the *ts* defect can be studied at the nonpermissive temperature. Temperature-shift experiments permit temporal location of the defect (whether the gene product is required early or late or continuously for normal infection), and complementation between mutants determines the number of necessary functions in the alphavirus genome. The various series of *ts* mutants of alphaviruses isolated after chemical mutagenesis are summarized in Table II. Within a series they have been listed by mutagen and subdivided by their RNA phenotype.

One of the earliest series of animal virus mutants consisted of the 23 *ts* mutants of Sindbis virus (HR strain) isolated by Burge and Pfefferkorn (1966a). These mutants, which were classified according to their ability to synthesize RNA at the nonpermissive temperature as RNA^+ or RNA^- and grouped into five complementation groups (Section III,A), have become the "type mutants" of alphavirus research. A second series of mutants consisted of the 38 *ts* isolates of SFV characterized by Tan *et al.* (1969). Subsequently, an independent series of 16 SFV mutants was isolated by Keränen and Kääriäinen (1974), as well as two additional large catalogs of Sindbis mutants: a series of 89 *ts* mutants of the HR strain (Strauss *et al.*, 1976) and 104 mutants of Sindbis AR339 virus (Atkins *et al.*, 1974a). Recently 13 *ts* mutants of WEE virus have been described (Hashimoto and Simizu, 1978).

As shown in Table II, a variety of chemical mutagens have been used to produce mutants of alphaviruses. The mutagens fall into two classes: (1) those which produce a direct chemical modification of the viral nucleic acid, and (2) base analogs which are incorporated into the RNA and cause mispairing. The former group includes nitrous acid, hydroxylamine, *N*-methyl-*N'*-nitro-*N*-nitrosoguanidine, and ethyl methane sulfonate. Nitrosoguanidine has been a very useful mutagen because it produces a relatively high ratio of mutagenic events to lethal events. Nitrous acid has also been used with good effect, despite the fact that acidic conditions are lethal to alphaviruses (Burge and Pfefferkorn, 1966a). Hydroxylamine is more difficult to use effectively because it is extremely toxic to cell monolayers, but success with this reagent has been reported by several groups. Ethyl methane sulfonate was used successfully by Atkins *et al.* (1974a); but other authors have failed to obtain mutants using this reagent (Pfefferkorn and Burge, 1967; Strauss *et al.*, 1976).

Base analogs used include azacytidine and 5-fluorouracil. Azacytidine has proved to be a powerful mutagen for alphaviruses (Halle, 1968). 5-Fluorouracil must be used at very high concentrations for mutagenesis (up to 100 times that of other mutagens), which may reflect either the inefficiency of conversion of 5-fluorouracil to 5-fluorouridine by the cell or the fact that an impurity in the 5-fluorouracil is the actual mutagen. In this regard 5-fluorouridine has been shown to be mutagenic at low concentrations (Strauss *et al.*, 1976), but the mutants induced have been unstable.

Various authors have found an apparent bias toward obtaining mutants with a particular RNA phenotype upon using different mutagens. This appears to be due more to the procedures and virus strains used by different laboratories than to any intrinsic property of the mutagen, however. For example, using nitrosoguanidine as a mutagen, Tan *et al.* (1969) with SFV and Strauss *et al.* (1976) with Sindbis HR virus

TABLE II

Temperature-Sensitive Mutants of Alphaviruses

Virus	Selection procedure	Mutagen[a]	Number of mutants	Phenotype[b]			References
				RNA⁺	RNA⁻	RNA±	
Sindbis virus, HR strain	Plaque efficiency at 30° versus 40°C	HNO_2	5	2	3		Burge and Pfefferkorn (1966a)
		NNG	17	4	13		
		EMS	1	1	—		
			23	7	16	—	
Sindbis virus, HR strain	Plaque efficiency at 30° versus 40°C	HNO_2	17	7	5	5	Strauss *et al.* (1976)
		NNG	24	15	2	7	
		AzaC	35	12	16	7	
		5-FU	2		2		
			78	34	25	19	
Sindbis virus, HR strain	Plaque efficiency at 30° versus 37°C	AzaC	32	7	23	2	Strauss (unpublished observations)
Sindbis virus, AR339 strain	Plaque enlargement technique	HNO_2	20		15	5	Atkins *et al.* (1974a)
		NNG	19		17	2	
		5-FU	15	4	9	2	
		EMS	23	3	20		
		HA	27	5	14	8	
			104	12	75	17	

Semliki Forest virus, strain 25639	Plaque efficiency at 23° versus 38.5°C	5-FU	11	1	8	2	Tan *et al.* (1969)
		HA	13	4	3	6	
		NNG	14	11	2	1	
			38	16	13	9	
Semliki Forest virus	Cytopathic effect at 27° versus 39°C	NNG	16	6	8	2	Keränen and Kääriäinen (1974)
Western equine encephalitis virus	Cytopathic effect at 37° versus 41.5°C	NNG	10	1	8	1	Hashimoto and Simizu (1978)
		HNO_2	3		2	1	
			13	1	10	2	
Eastern equine encephalitis virus	Plaque efficiency at 30° versus 42°C	HNO_2	1	1			Zebovitz and Brown (1970)

[a] HNO_2, Nitrous acid; NNG, *N*-methyl-*N'*-nitro-*N*-nitrosoguanidine; EMS, ethylmethane sulfonic acid; AzaC, 5-azacytidine; 5-FU, 5-fluorouracil; HA, hydroxylamine.

[b] RNA^+ mutants make at least 60% as much RNA at the nonpermissive temperature as the parental strain, $RNA^{\pm}$ mutants make 10–60% as much RNA, and RNA^- mutants make less than 10% as much RNA.

obtained predominantly RNA^+ mutants, whereas Keränen and Kääriäinen (1974) with SFV and Burge and Pfefferkorn (1966a) with Sindbis HR virus obtained predominantly RNA^- mutants. It appears, however, that the selection technique used for isolation of the mutants affects the relative numbers of RNA^+ and RNA^- mutants obtained. Blind selection for isolates which make plaques at 30°C but not at 40°C appears to give a reasonable balance between RNA^+ and RNA^- mutants (Strauss *et al.*, 1976), whereas the plaque enlargement technique used by Atkins *et al.* (1974a) or the selection by cytopathic effects used by Hashimoto and Simizu (1978) gives rise predominantly to RNA^- isolates.

2. *Temperature-Sensitive Mutants from Persistently Infected Mosquito Cells*

Temperature-sensitive mutants have also been reported to arise from persistently infected *A. albopictus* cell lines, presumably because there is no selection against such mutants in mosquito cells, which are maintained at about 28°C. Shenk *et al.* (1974) found that virus produced by mosquito cells chronically infected with Sindbis virus for 2–3 months consisted of *ts* virions of small-plaque phenotype. From this population they isolated a series of 20 clones and found all of them to be temperature-sensitive for replication in chick or BHK cells: Neither infectious virus nor hemagglutinin was released into the culture fluid during infection at the nonpermissive temperature (39.5°C). Nineteen out of 20 of these clones were RNA^+ and produced heat-labile virions, indicating a mutation in a structural protein. Complementation assays with various pairs of these mutants, including RNA^-–RNA^+ pairs, were all negative. Since these isolates were all initially members of the same population, it is possible that all the RNA^+ clones possessed the same defect (i.e., were really one mutant) and that the RNA^- mutant, which also produced heat-labile particles, was a double mutant.

Both single and multiple mutants arising from mosquito cell cultures chronically infected with WEE virus have been recently described by Maeda *et al.* (1979). They found that singly mutated *ts* mutants (both RNA^+ and RNA^-) could be isolated from cultures persistently infected for up to 30 days, but that thereafter (up to 170 days postinfection) only multiple mutants which were temperature-sensitive for RNA synthesis as well as defective in the structural proteins were isolated. In addition these isolates all gave small plaques on chick monolayers, although the input WEE virus was a large-plaque strain. These mutants and their physiological characteristics are summarized in Table III (for further discussion see Chapter 20).

III. GENETIC INTERACTIONS BETWEEN ALPHAVIRUSES

A. Complementation

Complementation tests between mutants of RNA viruses have been frequently employed to sort isolates into groups and to enumerate the number of essential functions encoded in the viral genome. For complementation between two *ts* mutants, a cell culture is mixedly infected with a high enough multiplicity of both mutants to ensure that all cells receive at least one infectious virus of each type. The infection is carried out at the nonpermissive temperature, and the resultant virus yield is assayed under permissive conditions. For simple systems where recombination is undetectable complementation is defined as positive if the yield of such a mixed infection is greater than the sum of the yields of parallel infections with each mutant separately. If two mutants complement, it is almost certain that they possess defects in different essential functions. If two mutants fail to complement, they are normally assigned to the same complementation group. However, failure to complement does not necessarily mean that two mutants possess defects in the same polypeptide but may result from an inability of the products of one mutant to interact productively with the products of the other. Also, some mutants are known which interfere with the growth of other mutants or of the parental strain of the virus. Thus *ts*9 of Burge and Pfefferkorn (1966b) and *ts*125 of Strauss *et al.* (1976) not only fail to complement but also suppress the final yield in a mixed infection with another mutant or with the parental virus (Sindbis HR). Another interfering mutant, *ts*11 of SFV, has been studied by Keränen (1977), who found that interference by *ts*11 did not require expression of the *ts*11 genome, was a direct function of the multiplicity of infection, and was not due to interferon. Remarkably, superinfection with *ts*11 up to 4 h after infection by wild-type SFV resulted in depression of the wild-type yield. Keränen (1977) concluded that a structural component of the *ts*11 virion, either the RNA itself or one of the structural proteins, was responsible for the interference. It is unclear at present whether this is a general phenomenon or a peculiarity of this one mutant.

A different type of interference has also been described (Zebovitz

TABLE III

Temperature-Sensitive Mutants of Western Equine Encephalitis Virus

Mutant	Derivation or mutagen[a]	Synthesis at 41.5°C[b,c]		Stability at 50°[d]		Complementation group(s)[e]	Defect[f]
		RNA	Nucleocapsids	PFU	HA		
Single mutants							
*ts*2	HNO_2	–	ND	+	+	II	Unknown
*ts*3	HNO_2	–	ND	+	+	III	Unknown
A106	H (5)	–	ND	+	+	III	Unknown
*ts*39	NNG	+	+	–	–	I	Defective E1 Group D
A108	B (20)	+	+	–	–	IV	Defective E2
A114	G (30)	+	+	–	–	IV	Group E
A117	A (30)	+	+	+	+	IV	
Double mutants							
A125–A134[g]	P (80)	–	ND	–	–	III, IV	Multiple lesions
A135–A143[g]	P (170)	–	ND	–	–	III, IV	Multiple lesions
Triple mutant							
A124	P (80)	–	ND	–	–	I, III, IV	Multiple lesions

[a] Mutants were isolated from five separate persistently infected cultures of *A. albopictus* cells (called A, B, G, H, and P) at the number of days after infection indicated by the number in parentheses (Maeda *et al.*, 1979). Other mutants were isolated following mutagenesis with *N*-methyl-*N'*-nitro-*N*-nitrosoguanidine (NNG) or nitrous acid (HNO_2) (Hashimoto and Simizu, 1978).

[b] (+) indicates synthesis at 41.5°C.

[c] ND, not determined.

[d] Virus stocks were inactivated at 50°C for 30 min. Residual infectivity (in plaque forming units, PFU) was determined by plaque assay, and hemagglutination (HA) by standard assays. (+) indicates the activity was stable.

[e] Complementation groups as indicated in Hashimoto and Simizu (1978) and Maeda *et al.* (1979).

[f] Physiological defect and corresponding complementation group of Burge and Pfefferkorn (1966b).

[g] Nineteen clones from culture P; probably all the same mutant.

and Brown, 1968; Johnston *et al.*, 1974) in which infection of cells with one strain of virus followed by superinfection with a second strain leads to a depressed yield of the superinfecting strain. This interference is independent of the multiplicity in that infection of a cell with one infectious particle is sufficient to establish maximum interference, *does* require expression of the viral genome, and is established in less than 1 h (Johnson *et al.*, 1974). Existence of this phenomenon makes it important that mixed infections be synchronously initiated if complementation is to result.

In Table IV the entire catalog of Sindbis HR virus mutants which have been unambiguously assigned by complementation is listed. Note that among RNA^+ mutants the majority of the mutants belong to group C, whereas among RNA^- mutants most belong to group A. Not all mutants complement equally well. For example, *ts*6 (RNA^-, group F) complements well with mutants of almost all other groups, whereas *ts*20 (RNA^+, group E) complements poorly with members of all other groups. In general RNA^+–RNA^- pairs of mutants complement most efficiently. This is presumably due to the fact that RNA^+ mutants are defective in the structural proteins, which are translated from 26 S RNA, while RNA^- mutants are defective in nonstructural proteins, which are translated from 49 S RNA (see Section IV), and thus complementation is uncomplicated by protein-processing abnormalities. With the exception of *ts*6, RNA^- mutants are in general less efficient at complementation and require very high multiplicities of infection (Pfefferkorn and Burge, 1967).

Complementation has also been demonstrated among several mutants of WEE virus (Hashimoto and Simizu, 1978; Maeda *et al.*, 1979), and a total of four complementation groups has been defined (Table III). Also shown in Table III are the corresponding complementation groups of Sindbis HR virus where this is known from physiological studies (see Section IV,C). It is of interest that the *ts* mutants obtained from chronically infected mosquito cells can be grouped (in fact, one of these mutants defines the fourth complementation group) and that double mutants can be characterized by complementation; this is one of the few instances in which multiple mutants can be grouped by complementation.

Mutants of the Sindbis virus AR339 strain isolated by Atkins *et al.* (1974a) failed to complement, and all series of SFV mutants obtained to date fail to complement (Tan *et al.*, 1969; Keränen and Kääriäinen, 1974). An attempt has been made in Table IV to correlate many of these AR339 mutants and SFV mutants with the corresponding group of Sindbis HR mutants upon the basis of the physiological defects shown by the mutants (see Section IV).

TABLE IV

Classification of Sindbis Virus and Semliki Forest Virus Temperature-Sensitive Mutants

Parental virus	Mutants assigned by complementation		Mutants assigned by physiological defects[b]		
	Sindbis HR (Burge and Pfefferkorn, 1966a,b)[a]	Sindbis HR (Strauss *et al.*, 1976)	Sindbis AR339 (Atkins *et al.*, 1974a)	Semliki Forest (Tan *et al.*, 1969)	Semliki Forest (Keränen and Kääriäinen, 1974)
RNA⁺ groups					
C	*ts*2, *ts*5, *ts*13	*ts*106, *ts*112, *ts*114, *ts*122, *ts*127[d], *ts*128, *ts*131, *ts*152, *ts*153, *ts*113[e], [*ts*158, *ts*170, *ts*174, *ts*191, *ts*226][f]	A5, H133, N4, A120, A93, F36	*ts*18[b], *ts*21[b], *ts*23[c], *ts*31, *ts*32, *ts*33, *ts*35	*ts*3
D	*ts*10, *ts*23	*ts*104, *ts*126, *ts*129, *ts*132, *ts*156	ID[g]	*ts*4, *ts*23[c], *ts*26, (*ts*36)	*ts*1, *ts*7
E	*ts*20		ID	*ts*12, *ts*15, *ts*25, *ts*28	*ts*2, *ts*5, *ts*15, *ts*16
RNA⁻ groups[h]					
A	*ts*4, *ts*14, *ts*15, *ts*16, *ts*17, *ts*19, *ts*21, *ts*24	*ts*133, *ts*138	N2, N7	ID	*ts*4
B	*ts*11		F294, A82[i]	ID	ID
F	*ts*6	*ts*110, *ts*118	ID	ID	ID
G	*ts*7, *ts*18	*ts*134	ID	ID	ID

[a] Reference given is for the original isolation of the mutants; in some cases preliminary characterization of the mutants is found in the same paper. [b] See text for discussion of phenotype. [c] *ts*23 is a double mutant of groups C and D. [d] The typographical error in the assignment of *ts*127 in the original reference has been corrected here. [e] *ts*113 is a double mutant of group C and an RNA⁻ group. [f] These additional mutants in brackets have been put in group C because they accumulate the p130 protein at the non-permissive temperature. (Strauss, unpublished observations). [g] ID, Insufficient data to assign any isolates to this group. [h] Complementation groups of Strauss *et al.* (1976). Burge and Pfefferkorn (1966b) originally used *ts*11 to define group B, as in the case here, but later defined a group A′ with this mutant and used *ts*6 to define group B (Pfefferkorn, 1977). [i] A82 is a double mutant of groups B and C (Atkins, 1979).

The majority of complementation analyses have been carried out in chick embryo fibroblasts. Renz and Brown (1976) have also examined complementation among the Sindbis HR mutants of Burge and Pfefferkorn (1966a,b) in both BHK cells and in mosquito cells. Six mutants representing six complementation groups were found to give the same complementation results in BHK cells as had been previously obtained in chick cells. In *A. albopictus* cells, nine mutants representing six complementation groups were tested, and only two pairs, *ts*6 with *ts*13 and *ts*11 with *ts*13, gave detectable complementation. It is possible that the noncomplementing mutants contain, in addition to the defect which defines their complementation group in chick cells, additional mutations in functions only necessary in mosquito cells. On the other hand, failure to complement could reflect differences in the mode of replication in vertebrate and arthropod hosts. Raghow *et al.* (1973) and Gliedman *et al.* (1975) have reported that virus is produced in membrane-bound factories or vesicles within the cytoplasm of mosquito cells. There is disagreement as to whether these vacuoles are the major sites of replication of the infectious virus ultimately released into the culture fluid; but the fact that alphavirus replication is asymptomatic in arthropods suggests that viral synthesis is sequestered from host cytoplasmic events. Such compartmentalization could make complementation, which depends upon the interchange of virus-encoded products, too inefficient to be detectable.

In summary, complementation analysis has defined at least seven functions (proteins) of alphaviruses essential for replication in mammalian or avian cells. Three groups (all RNA^+) have been assigned to the three major structural proteins of the virion. Four groups are found among the RNA^- mutants, and the assignment of functions to these is more difficult; although all fail to synthesize RNA at the nonpermissive temperature, they may not all necessarily code for integral components of the viral replicase. The molecular biology of alphavirus *ts* mutants will be discussed in more detail in Section IV.

B. Recombination

Recombination has not been demonstrated for alphaviruses. Several investigators have attempted to demonstrate recombination between *ts* mutants of Sindbis HR virus (Burge and Pfefferkorn, 1966b), Sindbis AR339 virus (Atkins *et al.*, 1974a), and SFV (Tan *et al.*, 1969). There is one preliminary report of marker rescue between a uv-inactivated wild-type Sindbis virus and a nonirradiated *ts* mutant (Brawner and Sagik, 1971). In recent years there was little interest in further pursuing this question since many workers in the field believed that genetic recombination in alphaviruses was extremely unlikely on theoretical grounds. Most of the genetic recombination reported earlier in RNA-containing viruses has been shown, on further scrutiny, to be due either to reassortment of genes located on independent RNA molecules of a segmented genome, such as in reovirus (Cross and Fields, 1977) or orthomyxoviruses (Bratt and Hightower, 1977), or to the production of unstable heteropolyploids such as those from Newcastle disease virus infection (Simon, 1972). Yet another type of recombination, that seen with RNA tumor viruses (Vogt, 1977), appears to take place while the virus genome is present as a DNA copy. True recombination does occur in picornaviruses (Cooper, 1977), however, and it is essential to keep an open mind concerning the possibility of recombination between alphaviruses.

C. Phenotypic Mixing and Formation of Pseudotypes

Phenotypic mixing has been shown to occur between Sindbis virus and other enveloped viruses under the proper conditions. In a mixed infection with another related (or unrelated) enveloped virus, particles are produced with the genotype of one virus and an envelope whose glycoproteins are derived either wholly or partially from the second virus.

The first demonstration of phenotypic mixing in alphaviruses was a mixed infection at 30°C between a *ts* mutant of Sindbis HR virus, which was genotypically unable to grow at 40°C but whose virion was stable to thermal inactivation at 60°C (like the HR parental type), with wild-type Eastern equine encephalitis (EEE) virus, WEE virus, or wild-type Sindbis virus, all of which can replicate (Burge and Pfefferkorn, 1966c) as well at 40°C as at 30°C but are rapidly inactivated by incubation at 56°C.

Progeny of the mixed infection, which were capable of forming plaques at 40°C (and therefore contained the WEE, EEE, or Sindbis wild-type genome) were tested for thermal inactivation. Up to 50% of these progeny virions were only slowly inactivated at 60°C, indicating that they contained sufficient structural polypeptides encoded by the HR mutant to confer heat stability. After another cycle of replication, these mixed progeny regained the thermolabile phenotype.

In an investigation of mixed infection with Sindbis and lactic dehydrogenase viruses, it was found that Sindbis virus formed both pure pseudotypes (containing the RNA of Sindbis virus and the envelope of lactic dehydrogenase virus) and particles containing the envelope proteins of both viruses (Lagwinska *et al.*, 1975). Lactic dehydrogenase virus is structurally similar to alphaviruses in protein composition and in RNA size but is not serologically related to them.

The most intriguing example of phenotypic mixing was the asymmetric mixing recently observed between either vesicular stomatitis virus (VSV) or defective Rous sarcoma virus and Sindbis virus (Zavadova *et al.*, 1977). Many particles were found which contained VSV genomes with Sindbis antigens in their envelopes, but none of the reverse (Sindbis capsids with VSV envelopes). The same type of asymmetric virus assembly (Rous nucleoids with Sindbis antigens in the envelope but not the reverse) prevailed when cells were coinfected with Sindbis virus and Rous sarcoma virus. In the light of these results it is unclear why Burge and Pfefferkorn (1966c) failed to find phenotypic mixing between Sindbis virus and VSV, since their test would have detected primarily VSV genomes with Sindbis envelope elements.

This asymmetric phenotypic mixing is probably a reflection of differences in the budding and maturation processes of VSV and Sindbis virus. Vesicular stomatitis virus is known to form pseudotypes and/or phenotypically mixed particles with a wide variety of viruses from such diverse taxonomic groups as paramyxoviruses, orthomyxovirus, murine and avian retroviruses, and even herpes simplex virus (Pringle, 1977). Thus it appears that the interaction between the nucleocapsid and the envelope (probably mediated through the matrix protein) is not very specific in the case of VSV and that any configuration roughly equivalent to that of the VSV envelope will suffice. On the other hand, alphaviruses appear to be more stringent as to which glycoproteins can be used for virion assembly (Strauss, 1978), which may reflect the need for direct interaction of the capsid protein with the cytoplasmic regions of one or both glycoproteins during budding (Strauss and Strauss, 1977).

IV. FUNCTIONAL DEFECTS OF THE ALPHAVIRUS *ts*MUTANTS

A. Introduction

Much effort has been devoted to elucidation of the physiological defects of alphavirus *ts* mutants. As noted earlier (Section III and Table IV), there are three complementation groups of RNA$^+$ mutants corresponding to the three major structural proteins of the virion, and

TABLE V

Physiological Defects of Alphavirus Temperature-Sensitive Mutants

RNA^+ groups (RNA synthesis at 41°C greater than 60% of wild type; all defects are in late functions)	
C	Defect in nucleocapsid protein; no nucleocapsids formed
	*ts*2 protein (p130) accumulates in cytoplasm in many mutants
	Particles produced at 30°C are thermolabile
	Infected cells hemadsorb weakly
	Cellular DNA synthesis not inhibited
D	Defect in protein E1 (hemagglutinin); infected cells do not hemadsorb; some mutants have *ts* hemagglutinin
	Nucleocapsids present free in cytoplasm
	No cleavage of PE2 to E2
	Neither S1 nor E2 integrated into membrane
	Particles produced at 30°C may be either thermostable or thermolabile
E	Defect in PE2; no cleavage of PE2 to E2
	E1 integrated into membrane; infected cells hemadsorb
	Nucleocapsids formed and migrate to plasmalemma
	Particles produced at 30°C may be either thermostable or thermolabile
RNA^- groups (RNA synthesis at 41°C is <10% of wild type; all defects are in early functions; particles produced at 30°C are thermostable)	
A	Polyprotein p200 accumulates after shift-up
	Ratio of 26 to 49 S RNA altered
	Input RNA does not enter double-stranded form
	Defect in protein controlling initiation of 26 S transcription
	Interferon not induced on shift-up
	Unable to establish superinfection exclusion at the nonpermissive temperature?
B	Polyprotein intermediate p133 (or p150) accumulates on shift-up
	Interferon induced on shift-up
F	Input RNA does not enter double-stranded form
	Interferon not induced on shift-up
	Able to establish superinfection exclusion at the nonpermissive temperature
G	Ratio of 26S to 49S RNA reduced
	Interferon induced on shift-up

at least four complementation groups of RNA^- mutants (Strauss *et al.*, 1976). Table V summarizes the known phenotypic properties of each group and includes data not only from Sindbis HR mutants but also from mutants from other (noncomplementing) catalogs whose physiology is very similar. It is of interest that even in relatively small mutant catalogs the same types of mutants occur, pointing to the extensive similarities in the mode of replication of these serologically related viruses.

Mutant catalogs exist for Sindbis virus, SFV, and WEE virus; of these Sindbis and WEE viruses form one antigenic subgroup, while SFV is in another subgroup with Mayaro, chikungunya, and o'nyong-nyong viruses (Casals and Clarke, 1965). Mutants of Sindbis and WEE viruses do not appear more closely related than those of Sindbis virus and SFV in the sense that the same physiological defects occur in all three mutant catalogs. However, since complementation takes place between pairs of WEE mutants like it does for Sindbis HR mutants, it would be extremely interesting to determine whether interspecific complementation also occurs, i.e., whether the two viruses are closely enough related so that defects in Sindbis HR mutants could be complemented by WEE virus functions, and vice versa.

However, in preliminary experiments it has not been possible to demonstrate complementation between selected WEE and Sindbis mutants. Some mutant pairs gave complementation indices of unity, while others gave interference; i.e., the yield of the mixed infection was only a few percent of the sum of the yields of the mutants separately (B. Simizu and E. G. Strauss, unpublished observations). One interpretation of these experiments is that the gene products of WEE and Sindbis are not interchangeable, but are closely enough related to interfere at least in some instances. It is also possible that problems in the experimental protocol are responsible for the negative results.

B. Determination of RNA Phenotype

One of the first characterizations of a new series of alphavirus mutants in many cases has involved measurement of the incorporation of uridine into virus-specific RNA in cells infected at the permissive and nonpermissive temperatures. This simple test divides a catalog of mutants into (ideally) two classes, those which incorporate uridine at the nonpermissive temperature in amounts approximating that of the parental strain (RNA^+) and those which incorporate little or no uridine under these conditions (RNA^-). This test also divides mutants in functions translated from the 26 S RNA message (which have all proved to be RNA^+) from mutants in functions translated from the 49 S virion RNA message (RNA^-). Moreover, with a few exceptions which may be double mutants, RNA^+ mutants have been shown to be in late functions (i.e., they will replicate normally if cells are infected at the nonpermissive temperature and then shifted to the permissive temperature at 1–3 h postinfection), while RNA^- mutants are in early functions (i.e., they will replicate if incubated at the permissive temperature at the onset of infection and then subsequently shifted to the nonpermissive temperature for the remainder of the growth cycle).

Although the original series of mutants of Burge and Pfefferkorn (1966a) appeared to be clearly RNA^+ or RNA^-, all later catalogs have included an intermediate class ($RNA^\pm$) or indeed a smooth continuum from undetectable RNA synthesis at one end to 100% or more of wild-type incorporation at the other. As noted previously (Section II,C, Table II) we have somewhat arbitrarily defined RNA^- as 0–10%, $RNA^\pm$ as 10–60%, and RNA^+ as 60–100% of the wild-type RNA synthesis at the nonpermissive temperature. The nature of the defect in the $RNA^\pm$ class is unclear. These mutants may simply be leaky for RNA synthesis; they may be mutants in the structural proteins (normally RNA^+) which carry additional nonlethal but debilitating defects in the RNA-synthesizing capacity or in the ability to translate 49 S RNA; and/or they may be mutants in nonreplicase functions translated from 49 S RNA, which depress RNA synthesis by affecting the processing of the polyprotein precursor. Complementation analysis of these mutants as well as biochemical characterization would be useful in determining their defects.

C. Defects of RNA^+ Mutants

Mutants of complementation group C of Sindbis virus have a defect in the nucleocapsid protein. The large precursor of the three structural proteins (called the *ts*2 protein, p140, or p130 in various references) is not processed at the nonpermissive temperature. This precursor, which cannot be subsequently chased into virion proteins, accumulates in the cytoplasm (Strauss *et al.*, 1969; Scheele and Pfefferkorn, 1970) and no detectable nucleocapsids are formed (Burge and Pfefferkorn, 1968). Furthermore, cleavage of this precursor is inefficient even at the permissive temperature. The virus particles formed at 30°C have all been found to be very thermolabile (Strauss *et al.*, 1976).

Some leakage, i.e., a small amount of cleavage of the polyprotein to produce capsid protein (probably defective) plus functional E1 (and E2), must occur at the nonpermissive temperature, since cells infected with group C mutants will hemadsorb weakly, indicating that some functional E1 has been integrated into the membrane (Burge and Pfefferkorn, 1967). Although the p140 (p130) precursor is considered a dead-end product, it has been shown that in cells doubly infected at the nonpermissive temperature with *ts*2 (group C) and *ts*20 (group D) or *ts*2 and *ts*23 (group E) the precursor is found in smaller amounts, indicating that the mutant polyprotein, while nascent, can be cleaved by a diffusible factor (Scupham *et al.*, 1977). Either this cleavage in doubly infected cells and/or the small amount of cleavage that occurs naturally could explain the complementation observed (Burge and Pfefferkorn, 1966b; Strauss *et al.*, 1976) between group C mutants and members of the other two RNA$^+$ groups.

Further evidence that group C mutants are in the capsid protein is the isolation of a group C mutant, *ts* 106, which has a capsid protein of altered electrophoretic mobility (Strauss *et al.*, 1976). There is also a report of a group C mutant (*ts*2*) which has altered E2 protein, but this was demonstrated to be a second lesion, unrelated to the temperature-sensitive group C defect (Bracha and Schlesinger, 1978). This *ts*2* mutant was used in complementation studies to ascertain whether the virions produced by complementation at 41°C between group C (RNA$^+$) and group F (RNA$^-$) contained structural proteins encoded by both mutants. The complementation was found to be nonreciprocal, with all virions produced carrying E2 glycoprotein encoded by the RNA$^-$ parent.

Mutants analogous to the group C Sindbis mutants have been isolated from SFV (Tan *et al.*, 1969; Keränen and Kääriäinen, 1975). After infection by these mutants at the nonpermissive temperature no nucleocapsids are formed and the virus particles produced at the permissive temperature are thermolabile, indicating a defect in a structural protein. Mutant *ts*3 of Keränen and Kääriäinen (1975) also accumulates the precursor polyprotein of approximately 140,000 daltons. Semliki Forest virus mutants with a somewhat different phenotype but which also appear to be capsid protein mutants have been studied by Lomniczi and Burke (1970), who found that mutants *ts*21 and *ts*18 of Tan *et al.* (1969) produced thermolabile virions at the permissive temperature, that cells infected with these mutants hemadsorbed at 41°C, albeit weakly, and that virus-modified membrane was detected at the nonpermissive temperature, similar to the results with group C mutants. These SFV mutants do *not* produce the large precursor, however, but rather produce primarily capsid protein which is apparently incapable of successful interaction with the RNA, for no nucleocapsids are formed at the nonpermissive temperature. Thus while most capsid protein mutations block processing of the polyprotein precursor of the structural proteins, these two SFV mutants apparently allow normal processing to occur.

Group C mutants are easy to isolate from Sindbis HR or AR339 virus (Table IV), are somewhat less common for SFV (Table IV), and so far have not been detected for WEE virus (Table III; Maeda, *et al.* 1979). The reasons for this disparity are not known.

The mutants in Sindbis group D (Table IV) possess a defect in the viral hemagglutinin shown by Dalrymple *et al.* (1976) to be glycoprotein E1; *ts*10- and *ts*23-infected cells do not hemadsorb at 39°C (Burge and Pfefferkorn, 1967; Yin and Lockart, 1968) nor do they contain virus-specific polypeptides in the plasmalemma which are detectable by antibody binding (Bell and Waite, 1977) or lactoperoxidase iodination (Smith and Brown, 1977). However, a polypeptide migrating on polyacrylamide gels, like E1, can be found in whole-cell extracts (Jones *et al.*, 1977). Furthermore the hemagglutinin produced by *ts*23 at the permissive temperature is temperature-sensitive: Virus particles formed at 30°C will hemagglutinate at 30°C but not at 40°C (Yin, 1969). Another indication that functional E1 is not integrated into the cell surface at 39°C is the observation that at the nonpermissive temperature nucleocapsids are found distributed randomly throughout the cytoplasm, whereas with the group E mutant, *ts*20, the nucleocapsids line the inner surface of the plasmalemma, even under conditions when budding cannot take place (Brown and Smith, 1975). In addition to the mutation in the E1 glycoprotein, mutants of group D also fail to process the precursor of E2 (variously called PE2, p65, or p62). This cleavage is thought to require the presence of functional E1 in proper juxtaposition (Bracha and Schlesinger, 1976). The lesion in *ts*23 is reversible, at least in part: Upon a shift to the permissive temperature in the presence of cycloheximide to prevent further protein synthesis, PE2 (made at 39°C) is cleaved to E2, and infectious virus is released from the cells. The lesion in *ts*10 is apparently not reversible: Upon a shift to the permissive temperature in the presence of cycloheximide, PE2 is not cleaved and no infectious virus is released by the cells (Jones *et al.*, 1977).

Mutants of both SFV and of WEE virus have been isolated, which correspond phenotypically to group D of Sindbis virus. The SFV mutants *ts*4 and *ts*26 of Tan *et al.* (1969) (Table IV) and the WEE mutant of Hashimoto and Simizu (1978), *ts*39, belonging to complementation group I (Table III), are RNA$^+$ mutants which make nucleocapsids at the nonpermissive temperature but fail to hemadsorb. In addition *ts*39 of WEE virus makes a temperature-sensitive hemagglutinin at the permissive temperature, analogous to the results with *ts*23 of Sindbis virus, and nucleocapsids do not bind to the plasmalemma at 41°C, indicating that functional membrane protein is not integrated into the membrane at the nonpermissive temperature (Hashimoto *et al.*, 1977). There is insufficient biochemical data to assign any of the AR339 *ts* mutants to this group.

Keränen and Kääriäinen (1974) found that none of their SFV RNA$^+$ mutants produced a temperature-sensitive hemagglutinin. Recently, it has been shown by a combination of electron microscopy and immunofluorescence that *ts*1 and *ts*7 of SFV probably belong to group

D. At the nonpermissive temperature, both *ts*1 and *ts*7-infected cells had nucleocapsids distributed evenly throughout the cytoplasm and little or no envelope protein incorporated into the host plasmalemma. The defect in *ts*7 (but not in *ts*1), like that of Sindbis *ts*23, appears to be reversible after a shift to the permissive temperature (Saraste *et al.*, 1980).

Unlike the members of group C, almost all of which produce virions at 30°C that are quite sensitive to thermal inactivation at 50°–56°C, members of group D have mixed phenotypes. Sindbis *ts*10 and *ts*23 mutant virions are thermosensitive at 56°C, but *ts*104 and *ts*126 are thermostable. Among the SFV mutants, *ts*4 of Tan *et al.* (1969) is thermolabile, while *ts*26 and *ts*36 are not. Mutant *ts*39 of WEE virus, mentioned above, is very thermolabile and has only low virulence for mice (Simizu and Takayama, 1972). No generalities can be drawn because of the small number of mutants assigned to this group, but it is clear that group D mutants produce both thermostable and thermolabile particles.

The single member of Sindbis group E is *ts*20 which has a defect in E2 (or PE2). Cells infected with *ts*20 at the nonpermissive temperature will hemadsorb, indicating the presence of E1 in the surface, and produce nucleocapsids (Burge and Pfefferkorn, 1968). In addition, Smith and Brown (1977) found significant E1 present in preparations of the cell plasmalemma. However, *ts*20 fails to cleave the precursor PE2 to E2 (Bracha and Schlesinger, 1976; Jones *et al.*, 1974), and antiserum to E2 fails to react with *ts*20-infected cells at the nonpermissive temperature, indicating an absence of functional PE2 in the cell under these conditions (Bell and Waite, 1977).

Mutants analogous to *ts*20 have been isolated for SFV (Table IV) and for WEE virus (complementation group IV, Table III). These mutants are RNA$^+$, and produce nucleocapsids at the nonpermissive temperature, and infected cells hemadsorb at the nonpermissive temperature. It is of interest that, of the SFV mutants of this phenotype isolated by Tan *et al.* (1969), *ts*25 virions (produced at 30°C) are thermostable at 56°C (as is *ts*20 of Sindbis HR virus), whereas those of *ts*12, *ts*15, and *ts*28 are thermolabile at this temperature. Thus it appears that mutants of this group may produce either thermolabile or thermostable virions.

D. Defects of RNA$^-$ Mutants

In comparison to the RNA$^+$ mutants, relatively little is known about the phenotypes of the RNA$^-$ mutants. There are four known complementation groups in Sindbis virus (A, B, F, and G) as shown in

Table IV. Lack of biochemical characterization makes it impossible to assign noncomplementing mutants to RNA$^-$ groups with two exceptions, *ts*4 of SFV, and four AR339 mutants. Primarily Sindbis HR mutants are represented in the table. The RNA$^-$ mutants code for nonstructural polypeptides translated from 49 S mRNA as a single large polyprotein precursor of approximately 200,000 daltons. This is processed to give final products variously known as ns 70, ns 86, ns 72, and ns 60 for SFV or X, ns 60, ns 89, ns 82 for Sindbis virus (see Chapter 13).

Of 12 original Burge and Pfefferkorn (1966a,b) RNA$^-$ mutants, 8 belong to group A. Under proper conditions it is possible to find a 200,000-dalton polypeptide in cells infected at the permissive temperature with either of the group A mutants *ts*21 or *ts*24 and subsequently shifted to the nonpermissive temperature (Bracha *et al.*, 1976). Thus group A appears to involve a faulty first-cleavage step, analogous to group C, and it is of interest that group A appears to be the most common group among RNA$^-$ mutants while group C appears to be the most common group among RNA$^+$ mutants. Scheele and Pfefferkorn (1969) also reported that the synthesis of 26 S RNA in cells infected by *ts*24 was preferentially inhibited if the cells were shifted from the permissive to the nonpermissive temperature, or if the cultures were treated with puromycin early in infection.

Similar results have seen found for the SFV mutant *ts*4 (Saraste *et al.*, 1977; Kääriäinen *et al.*, 1978; Sawicki *et al.*, 1978); *ts*4 also accumulates the large polyprotein precursor (reported to be 220,000 daltons in this case) at the nonpermissive temperature. Pulse-labeling experiments using a hypertonic block to synchronize the initiation of translation have indicated that *ts*4 is located near the 5′-end of the 49 S RNA, and thus the "*ts*4 protein" is the N-terminus of the precursor. After a shift to the nonpermissive temperature *ts*4-infected cells continue to make 49 S RNA but cease the synthesis of 26 S RNA. The current theory is that *ts*4 has a defect in a virus-specific protein which is necessary to promote initiation of transcription of 26 S RNA at an internal site on the 49 S minus-strand template. Thus the *ts*4 protein affects RNA synthesis without necessarily being a component of the RNA polymerase per se.

Other *ts* mutants have also been examined for the ratio of intracellular 49 S RNA to 26 S RNA. Atkins *et al.* (1974a) examined a large number of isolates and found a continuum of ratios from greater amounts of 26 S than normal (in group C RNA$^+$ mutants) to much less than normal. Thus, although group A mutants affect the ratio of 49 S to 26 S RNA, this ratio cannot be used for grouping purposes in the absence of additional data.

Interference with the replication of superinfecting Sindbis virus was found by uv inactivation experiments to require expression of about 20% of the viral genome for establishment (Johnston *et al.*, 1974). Four *ts* mutants were examined for the ability to establish such interference at the nonpermissive temperature. Two RNA$^+$ mutants, *ts*2 (group C) and *ts*20 (group E), and one RNA$^-$ mutant, *ts*6 (group F), did establish interference. RNA$^-$ mutant *ts*24 (group A) did not, implying that gene-A function is required for exclusion. It will be necessary, however, to examine other RNA$^-$ mutants for this effect, especially those from the two remaining complementation groups, before firm conclusions can be drawn.

Following a slightly different experimental protocol, Atkins (1979) examined mutants of Sindbis AR339 for interference by simultaneous infection with the parental strain at the nonpermissive temperature. Five of the RNA$^-$ mutants, one which accumulates the p150 (or p133) protein (physiologically group B), two which accumulate p200 protein (physiologically group A) and two with unknown lesions failed to show interference. On the other hand, six RNA$^+$ mutants interfered markedly with the replication of the parental virus at 40°C. A double mutant in groups B and C showed only weak interference.

Sindbis HR mutants *ts*4 (group A) and *ts*6 (group F) were found not to form RNase-resistant RNA at 39°C, i.e., the (radiolabeled) input RNA from the infecting virus was not transformed to a double-stranded form by synthesis of a complementary RNA molecule (Pfefferkorn *et al.*, 1967; Pfefferkorn and Burge, 1968). Unfortunately, no other RNA$^-$ mutants of Sindbis virus have been looked at. In a related study, Atkins and Lancashire (1976) were unable to detect synthesis of double-stranded RNA at the nonpermissive temperature in any of numerous Sindbis AR339 RNA$^-$ mutants examined. However, in this case the input RNA strand was not labeled, and the authors estimated that synthesis of less than 5% of the wild-type amount of double-stranded RNA would have gone undetected.

The sole member of Sindbis group B, *ts*11, also appears to have an altered pattern of protein synthesis, producing a 133,000-dalton precursor after infection at 28°C followed by a shift to 41.5°C (Waite, 1973). This precursor is very similar in molecular weight to the *ts*2 protein, and at the time of its discovery *ts*11 was thought to be either a double mutant or a mutant in an enzyme responsible for protein processing. It now appears more likely that *ts*11 accumulates one of the polypeptide intermediates in the processing of nonstructural proteins, possibly the p150 described by Brzeski and Kennedy (1977).

Early studies on SFV RNA$^-$ mutants showed that in most cases RNA synthetase was not made at the nonpermissive temperature but, once made, the enzyme was active at 41°C. The optimum for synthetase activity from *ts*15-, *ts*5-, and *ts*6-infected cells, however, was a few degrees lower than that for the wild type. In only one instance (*ts*5) was the synthetase activity unstable to incubation at elevated temperatures (Martin, 1969).

E. Induction of Interferon by Alphavirus Mutants

Alphaviruses have been shown to be efficient inducers of interferon during the course of a normal infection (Lockart *et al.*, 1968). Several studies have attempted to determine which viral functions are required for interferon induction by examining cells infected with various *ts* mutants at the nonpermissive temperature for interferon production. The first such study gave negative results, and all Sindbis HR mutants tried, whether RNA^+ or RNA^- in phenotype, failed to induce interferon at the nonpermissive temperature (Lockart *et al.*, 1968). The mutants used were *ts*15 (group A), *ts*2 (group C), *ts*23 (group D) and *ts*20 (group E) isolated by Burge and Pfefferkorn (1966b).

A second study by Lomniczi and Burke (1970) with the SFV *ts* mutants of Tan *et al.* (1969) showed that the multiplicity of infection and the exact temperature chosen for the infection were important variables. Using high input multiplicities and 39°C as the nonpermissive temperature, they found that all RNA^+ mutants induced interferon, and the RNA^- mutants were divided into two groups: those which induced no measurable interferon and those which induced small amounts of interferon.

Reexamining the Sindbis HR mutants and comparing them to their own catalog of Sindbis AR339 mutants, Atkins *et al.* (1974b) found that only by careful control of temperature, multiplicity of infection, and length of the infection cycle could accurate comparisons be made. Of the HR mutants, they showed that *ts*2 and *ts*5 (group C), *ts*10 (group D), and *ts*20 (group E) were active inducers at the nonpermissive temperature, indeed in some cases producing more interferon at 39°C than at 30°C. Another member of group D, *ts*23, is noticeably deficient in interferon induction under nonpermissive conditions, which partially explains the initial results of Lockart *et al.* (1968). Of their AR339 RNA^+ isolates, all induced interferon, although in variable quantities from mutant to mutant, even at the permissive temperature.

In agreement with the results with the SFV mutants, the RNA^- mutants presented a more complicated story. Atkins *et al.* (1974b) tested 24 of their RNA^- mutants and found 7 which induced interferon at the nonpermissive temperature and 17 which did not. Of the Burge and Pfefferkorn mutants which they tested, *ts*6 (group F) and *ts*11 (group B) induced interferon at the nonpermissive temperature, even in the absence of measurable RNA synthesis. The group G mutant *ts*7 and two out of three group A mutants (*ts*15 and *ts*21) failed to induce interferon. This latter result is in agreement with the recent report of Marcus and Fuller (1978) who determined by uv inactivation studies that only one-fourth of the Sindbis genome was required for interferon induction. These authors also found that mutants *ts*11, *ts*6, and *ts*110 induced interferon (groups B and F), while representatives of groups A and G did not. Thus with the exceptions of *ts*1, a group A mutant which induces interferon (Atkins *et al.*, 1974b), and *ts*23, a group D mutant which fails to induce interferon, it appears that all RNA^+ mutants and the RNA^- groups B and F are capable of interferon induction at the nonpermissive temperature, while the RNA^- groups A and G fail to induce interferon under these conditions.

These results indicate that it is not simply the total amount of RNA synthesis which determines the induction of interferon. By analogy with investigations of interferon induction by other viruses it is probably the first double-stranded RNA complex which is important. However, we note that *ts*6 is reported not to make even the parental double-stranded RNA at the nonpermissive temperature (Pfefferkorn and Burge, 1968), although it induces interferon. Furthermore, Atkins and Lancashire (1976) were unable to detect synthesis of double-stranded RNA at the nonpermissive temperature in any RNA^- mutant, although some of these mutants induced interferon and some did not. The sensitivity of the method used to detect synthesis of double-stranded RNA in the latter study was not very high, however, and more studies of the various mutants will be required in order to understand why some RNA^- mutants can induce interferon while others cannot.

F. Suppression of Host Cell Macromolecular Synthesis by Alphavirus Mutants

Infection of vertebrate cells by alphaviruses leads to the suppression of host cell protein synthesis (Friedman, 1968; Strauss *et al.*, 1969), DNA synthesis (Atkins, 1976), and, to some extent, RNA synthesis (Taylor, 1965), as described in more detail in Chapter 16. Inhibition of cellular protein synthesis depends upon viral RNA synthesis: RNA^- mutants did not inhibit cellular protein synthesis at the nonpermissive temperature (Atkins, 1976). This is consistent with the hypothesis that the inhibition of cellular protein synthesis after alphavirus infection results from competition between viral mRNA and cellular messages for limited translation capabilities.

Inhibition of DNA synthesis apparently requires synthesis of one or more of the virion structural proteins (Atkins, 1976). RNA⁻ mutants (which make no virion polypeptides at the nonpermissive temperature) and a series of RNA⁺ mutants phenotypically similar to *ts*2 of Sindbis HR virus (which make a large precursor polyprotein that is not further processed to the virion polypeptides) fail to inhibit DNA synthesis at the nonpermissive temperature. Thus it appears that one or more of the final virion polypeptides, E1, E2, or C, or intermediate cleavage products such as p110 (the B protein) or p65 (PE2) is responsible for the inhibition of cellular DNA synthesis.

The effect of alphavirus infection on host cell RNA synthesis appears to be much less dramatic than the effects on either DNA or protein synthesis. Levy *et al.* (1966) found no effect on host RNA synthesis after Sindbis infection of chick monolayers, while Mussgay *et al.* (1970) and Taylor (1965) have reported a 15–20% inhibition of host RNA synthesis in SFV-infected monolayers. On the other hand, Tan *et al.* (1969) reported that four of their RNA⁻ mutants inhibited host RNA synthesis up to 80% at the nonpermissive temperature after a 1-h incubation at the permissive temperature. Interpretation of these latter results is difficult, and correlation of inhibition of cellular RNA synthesis with mutant phenotype is not possible at the present time.

V. CONCLUDING REMARKS

Genetic analysis of the alphaviruses has been an important tool for studying viral replication, and many of the details of the virus replication cycle have been obtained through the use of mutants. This is particularly true of the late events in the virus life cycle, production of the structural proteins and virus budding, which have used RNA⁺ groups. It is to be expected that a more complete characterization of the RNA⁻ mutants will lead to a better understanding of the early events in the viral replication cycle and of the RNA transcription events. Yet because of the inherent simplicity of the system, these studies have not revealed any new insights into genetics in general; genetic control elements, such as repressors, operators, and promoters, are conspicuously absent. There is also no evidence for the presence of suppressors of *ts* mutations, as has been shown by Ramig *et al.* (1977) for reovirus mutants. There is no evidence at the current time for spliced genes, i.e., genes containing elements not present contiguously on the viral genome, or for overlapping genes which are read in different "frames." It is possible that the complete sequencing of the RNA of an alphavirus will reveal complexities of genetic organization which have been hitherto undetectable.

Unlike many other groups of viruses, alphaviruses are remarkably stable in their physical and chemical structure, and the outlines of the replication process are quite similar from virus to virus and from host to host. The requirement that the viruses grow and replicate in alternate vertebrate and invertebrate hosts and the constraints of a small genome have obviously shaped the evolution of these viruses.

ACKNOWLEDGMENTS

We are grateful to Dr. M. J. Schlesinger and to Dr. S. Maeda for furnishing us with preprints prior to their publication. The work of the authors is supported by grants AI 10793 and GM 06965 from the U.S. Public Health Service and by grant PCM 77-26728 from the National Science Foundation.

REFERENCES

Atkins, G. J. (1976). *Virology* **71,** 593–597.
Atkins, G. J. (1979). *J. Gen. Virol.* **45,** 201–207.
Atkins, G. J., and Lancashire, C. L. (1976). *J. Gen. Virol.* **30,** 157–165.
Atkins, G. J., Samuels, J., and Kennedy, S. I. T. (1974a). *J. Gen. Virol.* **25,** 371–380.
Atkins, G. J., Johnston, M. D., Westmacott, L. M., and Burke, D. C. (1974b). *J. Gen. Virol.* **25,** 381–390.
Bell, J. R., Hunkapiller, M. W., Hood, L. E., and Strauss, J. H. (1978). *Proc. Natl. Acad. Sci. U.S.A.* **75,** 2722–2726.
Bell, J. R., Strauss, E. G., and Strauss, J. H. (1979). *Virology,* **97,** 287–294.
Bell, J. W., Jr., and Waite, M. R. F. (1977). *J. Virol.* **21,** 788–791.
Birdwell, C. R., Strauss, E. G., and Strauss, J. H. (1973). *Virology* **56,** 429–438.
Bose, H. R., Brundige, M. A., Carl, G. Z., and Sagik, B. P. (1970). *Arch. Gesamte Virusforsch.* **31,** 207–214.
Bracha, M., and Schlesinger, M. J. (1976). *Virology* **74,** 441–449.
Bracha, M., and Schlesinger, M. J. (1978). *J. Virol.* **26,** 126–135.
Bracha, M., Leone, A., and Schlesinger, M. J. (1976). *J. Virol.* **20,** 616–620.
Bras-Herreng, F. (1973). *Ann. Microbiol. (Paris)* **124a,** 507–533.
Bratt, M. A., and Hightower, L. E. (1977). *In* "Comprehensive Virology" (H. Fraenkel-Conrat and R. R. Wagner, eds.), Vol. 9, pp. 457–533. Plenum, New York.
Brawner, T. A., and Sagik, B. P. (1971). *Abstr., Am. Soc. Microbiol., 71st Annu. Meet.,* p. 218.
Brés, P. (1965). *Ann. Inst. Pasteur, Paris* **109,** 866–873.
Brown, A. (1963). *Virology* **21,** 362–372.

Brown, D. T., and Smith, J. F. (1975). *J. Virol.* **15**, 1262–1266.

Brzeski, H., and Kennedy, S. I. T. (1977). *J. Virol.* **22**, 420–429.

Burge, B. W., and Pfefferkorn, E. R. (1966a). *Virology* **30**, 204–213.

Burge, B. W., and Pfefferkorn, E. R. (1966b). *Virology* **30**, 214–223.

Burge, B. W., and Pfefferkorn, E. R. (1966c). *Nature (London)* **210**, 1397–1399.

Burge, B. W., and Pfefferkorn, E. R. (1967). *J. Virol.* **1**, 956–962.

Burge, B. W., and Pfefferkorn, E. R. (1968). *J. Mol. Biol.* **35**, 193–205.

Burke, D. C., and Keegstra, K. (1976). *J. Virol.* **20**, 676–686.

Casals, J., and Clarke, D. H. (1965). *In* "Viral and Rickettsial Infections of Man" (F. L. Horsfall and I. Tamm, eds.), pp. 583–605. Lippincott, Philadelphia, Pennsylvania.

Cooper, P. D. (1977). *In* "Comprehensive Virology" (H. Fraenkel-Conrat and R. R. Wagner, eds.), Vol. 9, pp. 133–207. Plenum, New York.

Cross, R. K., and Fields, B. N. (1977). *In* "Comprehensive Virology" (H. Fraenkel-Conrat and R. R. Wagner, eds.), Vol. 9, pp. 291–340. Plenum, New York.

Dalrymple, J. M., Schlesinger, S., and Russell, P. K. (1976). *Virology* **69**, 93–103.

Davey, M. W., and Dalgarno, L. (1974). *J. Gen. Virol.* **24**, 453–463.

Frey, T., Gard, D. L., and Strauss, J. H. (1979). *J. Mol. Biol.* **132**, 1–18.

Friedman, R. M. (1968). *J. Virol.* **2**, 26–32.

Gaidamovich, S. Y., Tslinsky, Y. Y., Lvova, A. I., and Khutoretskaya, N. V. (1971). *Acta Virol. (Engl. Ed.)* **15**, 301–308.

Garoff, H., Simons, K., and Renkonen, O. (1974). *Virology* **61**, 493–504.

Gliedman, J. B., Smith, J. F., and Brown, D. T. (1975). *J. Virol.* **16**, 913–926.

Halle, S. (1968). *J. Virol.* **2**, 1228–1229.

Hannoun, C., Asso, J., and Ardoin, P. (1964). *Ann. Inst. Pasteur, Paris* **107**, 598–603.

Hashimoto, K., and Simizu, B. (1978). *Virology* **84**, 540–543.

Hashimoto, K., Suzuki, K., and Simizu, B. (1975). *J. Virol.* **15**, 1454–1466.

Hashimoto, K., Suzuki, K., and Simizu, B. (1977). *Arch. Virol.* **53**, 209–219.

Igarashi, A., Koo, R., and Stollar, V. (1977). *Virology* **82**, 69–83.

Inglot, A., Albin, M., and Chudzio, T. (1973). *J. Gen. Virol.* **20**, 105–110.

Jahrling, P. B. (1976). *J. Gen. Virol.* **32**, 121–128.

Jahrling, P. B., and Beall, J. L. (1977). *J. Clin. Microbiol.* **6**, 238–243.

Jahrling, P. B., and Eddy, G. A. (1977). *Am. J. Epidemiol.* **106**, 408–417.

Johnson, K. M., and Martin, D. H. (1974). *Adv. Vet. Sci.* **18**, 79–116.

Johnston, R. E., Wan, K., and Bose, H. R. (1974). *J. Virol.* **14**, 1076–1082.

Jones, K. J., Waite, M. R. F., and Bose, H. R. (1974). *J. Virol.* **13**, 809–817.

Jones, K. J., Scupham, R. K., Pfeil, J. A., Wan, K., Sagik, B. P., and Bose, H. R. (1977). *J. Virol.* **21**, 778–787.

Kääriäinen, L., Sawicki, D., and Gomatos, P. J. (1978). *J. Gen. Virol.* **39**, 463–473.

Kennedy, S. I. T., and Burke, D. C. (1972). *J. Gen. Virol.* **14**, 87–98.

Keränen, S. (1977). *Virology* **80**, 1–11.

Keränen, S., and Kääriäinen, L. (1974). *Acta Pathol. Microbiol. Scand., Sect. B* **82**, 810–882.

Keränen, S., and Kääriäinen, L. (1975). *J. Virol.* **16**, 388–396.

Kimura, T., and Ueba, N. (1978). *Arch. Virol.* **57**, 221–229.

Lagwinska, E., Stewart, C. C., Adles, C., and Schlesinger, S. (1975). *Virology* **65**, 204–214.

Levy, H. B., Snellbaker, L. R., and Baron, S. (1966). *Proc. Soc. Exp. Biol. Med.* **121**, 630–632.

Liebhaber, H., and Takemoto, K. K. (1963). *Virology* **20**, 559–566.

Lockart, R. A., Jr., Bayliss, N. L., Toy, S. T., and Yin, F. H. (1968). *J. Virol.* **2**, 962–965.

Lomniczi, B., and Burke, D. C. (1970). *J. Gen. Virol.* **8,** 55–68.
Maeda, S., Hashimoto, K., and Simizu, B. (1979). *Virology* **92,** 532–541.
Marcus, P. I., and Fuller, F. J. (1978). *Am. Soc. Microbiol., Annu. Meet., 78th,* Las Vegas, Nevada. *Abstr.* No. S200.
Marshall, I. D., Scriviani, R. P., and Reeves, W. C. (1962). *Am. J. Hyg.* **76,** 216–224.
Martin, E. M. (1969). *Virology* **39,** 107–117.
Mussgay, M., Enzmann, P. J., and Horst, J. (1970). *Arch. Gesamte Virusforsch.* **31,** 81–92.
Pattyn, S. R., and de Vleesschauwer, L. (1966). *Arch. Gesamte Virusforsch.* **19,** 176–189.
Pattyn, S. R., and de Vleesschauwer, L. (1968). *Acta Virol. (Engl. Ed.)* **12,** 347–354.
Pedersen, C. E., Jr., Barrera, C. R., and Sagik, B. P. (1973). *Arch. Gesamte Virusforsch.* **41,** 28–39.
Peleg, J. (1971). *Curr. Top. Microbiol. Immunol.* **55,** 155–161.
Pfefferkorn, E. R. (1969). *In* "Fundamental Techniques in Virology" (K. Habel and N. Salzman, eds.), pp. 87–93. Academic Press, New York.
Pfefferkorn, E. R. (1977). *In* "Comprehensive Virology" (H. Fraenkel-Conrat and R. R. Wagner, eds.), Vol. 9, pp. 209–238. Plenum, New York.
Pfefferkorn, E. R., and Burge, B. W. (1967). *In* "The Molecular Biology of Viruses" (J. S. Colter and W. Parenchych, eds.), pp. 403–426. Academic Press, New York.
Pfefferkorn, E. R., and Burge, B. W. (1968). *Perspect. Virol.* **6,** 1–14.
Pfefferkorn, E. R., and Shapiro, D. (1974). *In* "Comprehensive Virology" (H. Fraenkel-Conrat and R. R. Wagner, eds.), Vol. 2, pp. 171–230. Plenum, New York.
Pfefferkorn, E. R., Burge, B. W., and Coady, H. M. (1967). *Virology* **33,** 239–249.
Pringle, C. R. (1977). *In* "Comprehensive Virology" (H. Fraenkel-Conrat and R. R. Wagner, eds.), Vol. 9, pp. 239–289. Plenum, New York.
Quersin-Thiry, L. (1961). *Br. J. Exp. Pathol.* **42,** 511–522.
Raghow, R. S., Grace, T. D. C., Filshie, B. K., Bartley, W., and Dalgarno, L. (1973). *J. Gen. Virol.* **21,** 109–122.
Ramig, R. F., White, R. M., and Fields, B. N. (1977). *Science* **195,** 406–407.
Rasmussen, L. E., Armstrong, J. A., and Ho, M. (1973). *J. Infect. Dis.* **128,** 156–162.
Renz, D., and Brown, D. T. (1976). *J. Virol.* **19,** 775–781.
Saraste, J., Kääriäinen, L., Söderlund, H., and Keränen, S. (1977). *J. Gen. Virol.* **37,** 399–406.
Saraste, J., von Bonsdorff, C.-H., Hashimoto, K., Kääriäinen, L., and Keränen, S. (1980). *Virology* **100,** 229–245.
Sarver, N., and Stollar, V. (1977). *Virology* **80,** 390–400.
Sawicki, D. L., Kääriäinen, L., Lambek, C., and Gomatos, P. J. (1978). *J. Virol.* **25,** 19–27.
Scheele, C. M., and Pfefferkorn, E. R. (1969). *J. Virol.* **4,** 117–122.
Scheele, C. M., and Pfefferkorn, E. R. (1970). *J. Virol.* **5,** 329–337.
Schwöbel, W., and Ahl, R. (1972). *Arch. Gesamte Virusforsch.* **38,** 1–10.
Scupham, R. K., Jones, K. J., Sagik, B. P., and Bose, H. R., Jr. (1977). *J. Virol.* **22,** 568–571.
Shenk, T. E., Koshelnyk, K. E., and Stollar, V. (1974). *J. Virol.* **13,** 439–447.
Simizu, B., and Takayama, N. (1972). *Arch. Gesamte Virusforsch.* **38,** 328–337.
Simizu, B., Yamazaki, S., Suzuki, K., and Terasima, T. (1973). *J. Virol.* **12,** 1568–1578.
Simon, E. H. (1972). *Prog. Med. Virol.* **14,** 36–67.
Simons, K., and Kääriäinen, L. (1970). *Biochem. Biophys. Res. Commun.* **38,** 981–988.
Smith, J. F., and Brown, D. T. (1977). *J. Virol.* **22,** 662–678.
Strauss, E. G. (1978). *J. Virol.* **28,** 466–474.
Strauss, E. G., Lenches, E. M., and Strauss, J. H. (1976). *Virology* **74,** 154–168.

Strauss, E. G., Birdwell, C. R., Lenches, E. M., Staples, S. E., and Strauss, J. H. (1977). *Virology* **82,** 122–149.
Strauss, E. G., Lenches, E. M., and Martin, M. S. (1980). *J. Gen. Virol.* (in press).
Strauss, J. H., and Strauss, E. G. (1977). *In* "The Molecular Biology of Animal Viruses" (D. P. Nayak, ed.), Vol. 1, pp. 111–166. Dekker, New York.
Strauss, J. H., Burge, B. W., and Darnell, J. E. (1969). *Virology* **37,** 367–376.
Symington, J., and Schlesinger, M. J. (1978). *Arch. Virol.* **58,** 127–136.
Tan, K. B., Sambrook, J. F., and Bellett, A. J. D. (1969). *Virology* **38,** 427–439.
Taylor, J. (1965). *Virology* **25,** 340–349.
Vogt, P. K. (1977). *In* "Comprehensive Virology" (H. Fraenkel-Conrat and R. R. Wagner, eds.), Vol. 9, pp. 341–455. Plenum, New York.
Waite, M. R. F. (1973). *J. Virol.* **11,** 198–206.
Wengler, G., Wengler, G., and Filipe, A. R. (1977). *Virology* **78,** 124–134.
Yin, F. H. (1969). *J. Virol* **4,** 547–548.
Yin, F. H., and Lockart, R. Z., Jr. (1968). *J. Virol.* **2,** 728–737.
Yunker, C. E., and Cory, J. (1975). *Appl. Microbiol.* **29,** 81–89.
Zavadova, Z., Zavada, J., and Weiss, R. (1977). *J. Gen. Virol.* **37,** 557-567.
Zebovitz, E., and Brown, A. (1968). *J. Virol.* **2,** 1283–1289.
Zebovitz, E., and Brown, A. (1970). *J. Mol. Biol.* **50,** 185–198.

15

Defective Interfering Alphaviruses

VICTOR STOLLAR

I. INTRODUCTION

Defective interfering (DI) particles have been found in every group of animal viruses which has been carefully examined for their presence. Although specific features of DI particles and their genomes vary from one virus group to another, there are a number of properties which appear to be common to all (Huang, 1973): (1) They lack, to a greater or lesser extent, genomic material present in standard virus and are therefore deletion mutants; (2) they contain the same structural proteins as standard virus; (3) they are unable to replicate alone

THE TOGAVIRUSES

ISBN 0-12-625380-3

but can replicate in the presence of standard or "helper" virus; (4) they interfere with the replication of homologous standard virus; (5) they are commonly accumulated by the experimental artifact of serial undiluted (high-multiplicity) passage. All these statements are valid for DI particles of alphaviruses.

Most studies on DI particles of different viruses have been done *in vitro,* i.e., in cultured cells, and important information has been obtained with respect to DI genomes, their transcription and, in some cases, translation. The relationship between DI particles and persistent viral infections has also been examined *in vitro.*

The generation of DI particles *in vivo,* in the intact organism, is less well documented. However, it has been suggested (Huang and Baltimore, 1970) that, if they are produced *in vivo* along with standard virus, they may, by virtue of their interference with standard virus, substantially limit the progression of a viral infection.

Nearly all the work concerning DI particles of togaviruses has been done with either Sindbis virus (SV) or Semliki Forest virus (SFV), both alphaviruses. Very little has been reported about DI particles of flaviviruses.

II. FIRST REPORTS OF DEFECTIVE INTERFERING PARTICLES OF ALPHAVIRUSES

The ease with which DI particles can be demonstrated varies considerably from virus to virus. In the case of vesicular stomatitis virus (VSV) (Stampfer *et al.*, 1971) and influenza virus (Von Magnus, 1954), DI particles were rapidly produced, leading to a substantial fall in virus titer after only a few high-multiplicity passages. At the other extreme, DI particles of poliovirus, possibly because of a relatively weak interfering capacity, were detectable in HeLa cells only after about 16–18 passages (Cole *et al.*, 1971). Alphavirus DI particles are intermediate between these two extremes. They are produced with relative ease by serial undiluted passage of virus in chick embryo fibroblasts (CEFs) or baby hamster kidney (BHK) cells; they interfere well with the replication of standard virus, but during a passage series the variations in titer are often not as great as those seen, for example, when VSV is passaged without dilution.

The ability to separate DI particles from standard virus particles greatly facilitates many experimental approaches. Although this is possible for VSV, with most other animal viruses the separation is difficult because of minimal differences in size and density between standard and DI particles. For alphaviruses there have been reports of both success and failure.

One of the first reports which suggested the existence of DI particles of alphaviruses was that of Inglot and Chudzio (1972; see also Inglot *et al.*, 1973). After serial undiluted passage of SV in CEFs, they were able to separate by means of calcium phosphate column chromatography two populations of viral particles, one complete and the other noninfectious but capable of interfering with the replication of infectious virus.

The first detailed documentation for the production of alphavirus DI particles was provided by Schlesinger *et al.* (1972). Plaque-purified SV was serially passaged without dilution in BHK-21 cells, and after nine passages the levels of viral infectivity and hemagglutinin (HA) and the specific infectivity of extracellular virus all fell. The specific infectivity was expressed in terms of plaque-forming units (PFU)/cpm for virus labeled with uridine or amino acids. The decreased specific infectivity was consistent with the production of an increased proportion of noninfective or defective viral particles. Unlike what had been observed with influenza virus (von Magnus, 1954), the ratio of PFUs to HA did not show any marked decrease. When passage-9 SV was used to infect cells together with an early-passage stock, the yield of infective virus was reduced 10-fold compared to infection with the early-passage virus alone.

Subsequently it was reported by Shenk and Stollar (1972) that, when BHK-21 cells were infected with SV which had been serially passaged in BHK cells without dilution (SV_{BP}), the cells contained two species of viral double-stranded (ds) RNA with sedimentation constants of 22 and 12 S. This contrasted with the finding of a single species of dsRNA, 22 S, found in cells infected with standard SV (SV_{STD}) alone. Thus, infection with viral stocks presumably containing DI particles led to an altered pattern of viral RNA in the infected cells.

It had actually been observed even earlier (Stollar and Stollar, 1970; Stollar *et al.*, 1972) that, after infection with certain stocks of SV, CEFs, and BHK cells contained prominent species of viral dsRNA smaller than 22 S. It is clear now that the stocks used in those experiments must have contained DI particles.

These first reports demonstrated that SV DI particles were easily produced in BHK cells, and that their presence could be demonstrated either by their interference with SV_{STD} or by the production of an altered viral RNA pattern in infected cells.

III. PROPERTIES OF ALPHAVIRUS DEFECTIVE INTERFERING PARTICLES

A. Separation of Defective Interfering Particles from Standard Virus

Attempts to separate DI particles from standard virions have met with mixed success. At the present time it is fair to say, however, that it has not been possible to achieve separation easily and reproducibly.

As already noted, calcium phosphate gel chromatography was used successfully in one laboratory to resolve SV DI particles from standard virions (Inglot and Chudzio, 1972), but this has so far not been followed up by other workers.

Attempts using velocity gradients (Guild and Stollar, 1975; Bruton and Kennedy, 1976) or composite velocity and equilibrium gradients (Weiss and Schlesinger, 1973) have not been successful. Similarly, there has been no suggestion that more than one class of nucleocapsids could be resolved by velocity gradient centrifugation (Kennedy *et al.*, 1976; Guild *et al.*, 1977).

Two reports have described the successful separation of standard and defective particles by isopycnic centrifugation. In one case, SV DI particles produced in BHK cells were resolved by centrifugation in sucrose–D_2O (Shenk and Stollar, 1973a,b) and in the other case SFV DI particles, also produced in BHK cells, were resolved by centrifugation in CsCl adjusted to pH 9.0 (Bruton and Kennedy, 1976). In both instances the DI particles were more dense than the standard virus (1.22 versus 1.20 g/cm^3 in the case of SV, and 1.23 versus 1.20 g/cm^3 in the case of SFV). Since the particles in the more dense peak were not infectious and interfered well with homologous virus but not with a heterologous alphavirus, the assumption that they were DI particles appeared a valid one. For SFV a similar difference in densities was found when the DI nucleocapsids were compared with their standard counterparts (1.35 versus 1.34 g/cm^3) (Bruton and Kennedy, 1976).

Although the SV DI particles were resolvable in several consecutive stocks in three different passage series (Shenk and Stollar, 1973b), in other stocks prepared in BHK cells and known to contain DI particles, they could not be separated from standard virus. Weiss and Schlesinger (1973) were unable to separate SV DI particles from standard virus by any gradient technique; and in experiments with SV DI particles produced in CEFs, Guild and Stollar (1975) were not able to resolve the two types of particles. Although SFV DI particles could be resolved by equilibrium centrifugation in CsCl, they were not resolved by centrifugation in other media such as glycerol, sucrose, and potassium tartrate (Bruton and Kennedy, 1976).

The ability in some cases, but not in others, to separate DI particles by isopycnic centrifugation remains puzzling. The most likely explanation may be that alphavirus DI particles, depending on the host cell and the passage level, vary considerably in their precise composition (see below with reference to RNA content) and structure. There may be certain classes of DI particles which, for reasons still not understood, are assembled in such a way that they are of greater density than other DI particles or standard virions.

It should be emphasized that in experiments in which a 1.22 g/cm^3 peak of labeled SV DI particles was found, assays of gradient fractions revealed interfering activity not only at a density of 1.22 g/cm^3 but also in all fractions from 1.22 g/cm^3 to 1.20 g/cm^3 (the density of standard virus) (Shenk and Stollar, 1973b). Thus, even in instances in which a clear peak of DI particles could be resolved on the basis of labeling techniques, interference assays showed the DI particles to be quite heterogeneous with respect to density.

B. Morphology of Defective Interfering Particles

Johnston *et al.* (1975) have reported that, beginning with passage 6 of an undiluted passage series in BHK cells, smaller than normal-sized SV particles could be seen by electron microscopy. By passage 11 the small particles made up 40% of the total particles seen in concentrates prepared from the medium.

The appearance of the small particles coincided approximately with the appearance of interfering activity and small or truncated RNAs in the extracellular particles (see below). The smaller variant particles were round, enveloped, and had peripheral spikes but, in contrast to a diameter of 50 nm for standard virions, measured only 37 nm in diameter; these measurements did not include the peripheral spikes. Since in certain passages the variant particles represented up to 40% of all particles and, since they were significantly smaller than standard virus, it should have been possible by velocity gradient centrifugation to separate them from standard virions. This did not prove possible. It was suggested but not concluded that the small variant particles were DI particles.

Both in size and appearance, the variant particles were very similar to the small virus particles observed in the medium of mosquito cell cultures 3–5 days after infection with SV (Brown and Gliedman, 1973). It should be noted that this interval corresponds to the time when such cell cultures are in transition to a persistently infected state, with greatly reduced yields of infectious virus per unit time (see Chapter 20).

C. The RNAs of Defective Interfering Particles

The first reports of RNA from SV DI particles described either RNA of genome length (Shenk and Stollar, 1973a) or a heterogeneous population of small RNA molecules (Weiss and Schlesinger, 1973). It is evident now, however, that discrete subgenomic lengths of RNA are present in virus stocks which contain DI particles (Johnston *et al.*, 1975; Bruton and Kennedy, 1976; Guild *et al.*, 1977) and that some of the earlier results were probably caused by artifacts related to the RNA extraction (see Kennedy *et al.*, 1976).

It should be emphasized at this point that, because of the general inability to separate DI from standard particles (see above), "DI stocks" in fact contain a mixture of standard and DI particles. Therefore the RNA extracted from such stocks contains, in addition to the deleted or DI RNA, a varying amount of standard-genome-length 42 S RNA.

The small DI RNA species which can be extracted from extracellular particles correspond in size to the single stranded (ss) RNAs present in the cells during the same infection (see below). This has been shown for SFV DI particles produced in BHK cells (Bruton and Kennedy, 1976; Bruton *et al.*, 1976) as well as for SV DI particles produced in CEFs (Guild *et al.*, 1977). The latter system was especially notable because with increasing number of passages there was a progressive and stepwise decrease (33 S to 24 S to 22 S) in the size of the ssRNAs extracted from extracellular particles (Fig. 1) and a corresponding decrease in the size of the ssRNA species extracted from the cells (hereafter these RNA species will be referred to as ss DI RNAs).

According to two reports (Kennedy *et al.*, 1976; Bruton *et al.*, 1976), BHK cells infected with virus stocks containing DI particles of SV or SFV yielded progeny particles containing, in the first case, ss DI RNAs of sedimentation coefficient 20 S and, in the second case, RNAs with molecular weights of 0.81×10^6 and 0.75×10^6. However, in another report with an SV–BHK cell system (Johnston *et al.*, 1975) there was, as in the SV–CEF system (Guild *et al.*, 1977), a progressive stepwise decrease in the size of the ss DI RNAs extracted from extracellular particles. The molecular weights of these DI RNAs were estimated to be 3.3×10^6, 2.7×10^6, and 2.2×10^6. The intracellular DI RNAs were not examined in these experiments.

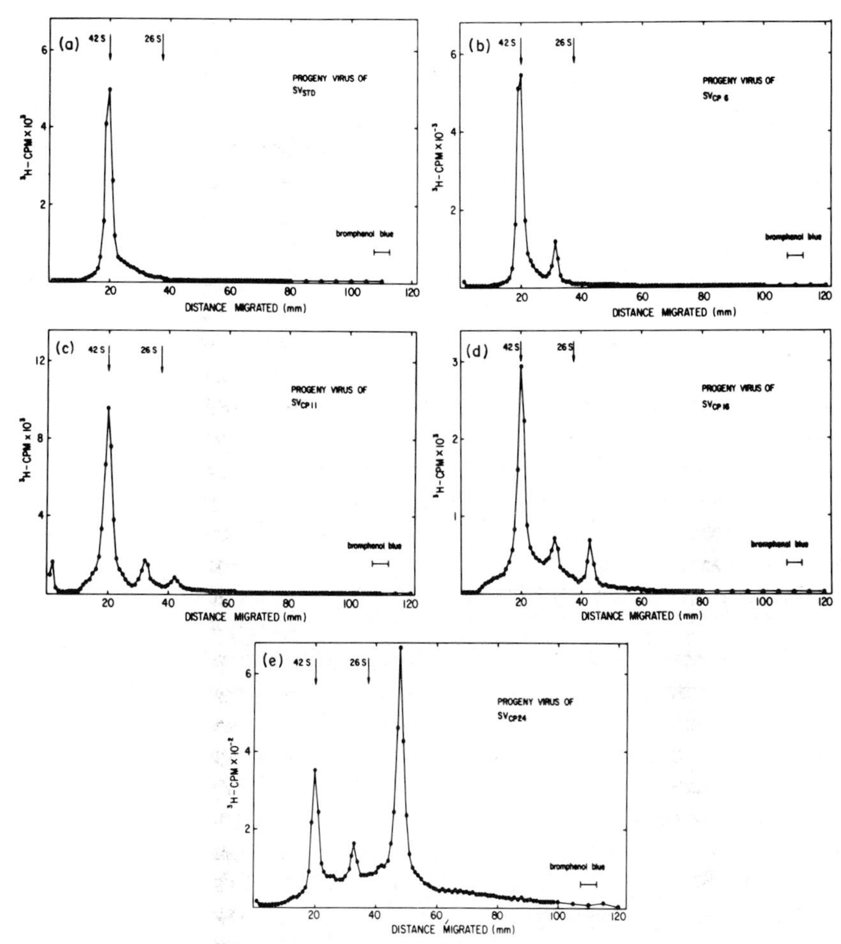

Fig. 1. RNA species present in virions released from cells infected with SV_{STD} (a), SV_{CP6} (b), SV_{CP11} (c), SV_{CP16} (d), or SV_{CP24} (e). Progeny virus was labeled with [^{3}H]uridine from 2 to 16 h after infection and then gradient-purified. RNA was extracted with SDS and mercaptoethanol and analyzed by gel electrophoresis. SV_{CP6}, SV_{CP11}, SV_{CP16}, and SV_{CP24} are stocks produced by serial undiluted passage of SV_{STD} in CEFs. The numbers indicate how many passages were used to prepare each stock. (From Guild *et al.*, 1977.)

Although there is no direct evidence that the DI RNAs extracted from extracellular particles contain poly(A), it has been shown both for SV and SFV that intracellular ss DI RNA is polyadenylated (Weiss *et al.*, 1974; Bruton *et al.*, 1976). No information is available as to whether the 5′-termini of the ss DI RNAs are capped.

The question of polarity of the defective ssRNAs has been addressed in the case of the SV–CEF system (Guild *et al.*, 1977). The ssRNAs from virus stocks obtained at various passage levels were labeled and incubated under annealing conditions with excess virion RNA (positive polarity). As measured by protection of the labeled RNA from RNase digestion, little significant annealing could be detected. Although because of experimental limitations it could not be concluded that no negative-strand sequences were present in the ss DI RNAs, it was calculated that less than 5% of the sequences (in one case less than 1%) were of negative polarity. Consistent with these results, oligonucleotide fingerprints of 20 S ss DI RNA extracted from extracellular particles of SV showed that all the characteristic oligonucleotides in the defective RNA were also present in the genome RNA (Kennedy *et al.*, 1976).

IV. INTRACELLULAR VIRAL DEFECTIVE INTERFERING RNAs

Cells infected with alphaviruses contain one major species of viral dsRNA, referred to here as 22 S, and two major species of viral ssRNA, usually referred to as 42 and 26 S RNA (see Chapter 12). Since these RNAs are found in cells infected with SV_{STD} alone, they will be referred to as standard RNAs.

As first described by Simmons and Strauss (1972) a certain proportion of the 22 S dsRNA molecules (23.5 S according to Simmons and Strauss) have an RNase-susceptible site such that they can be cleaved into 20 and 16 S fragments. Those molecules, which remain uncleaved or 22 S, are called RFI, and the 20 and 16 S species are referred to as RFII and RFIII, respectively. The 42 S ssRNA is equivalent to the genome RNA, and the 26 S ssRNA sequences are a subset of the 42 S RNA sequences and are located at the 3′-end of the 42 S RNA (Kennedy, 1976; Wengler and Wengler, 1976). RFIII is the ds equivalent of the 26 S RNA. The structure, function, and transcription of these RNAs is the subject of Chapter 12. Experimentally, as with certain other RNA viruses, viral RNAs can be conveniently labeled in cells pretreated with actinomycin D in order to eliminate host RNA synthesis.

When cells are infected with alphavirus stocks which contain DI particles, there are drastic changes in the patterns of viral RNA synthesized in the infected cells. In general, but depending on the ratio of DI to standard virus particles in the inoculum, the total amount of viral RNA synthesis is decreased (Fig. 2) when cells are coinfected with DI

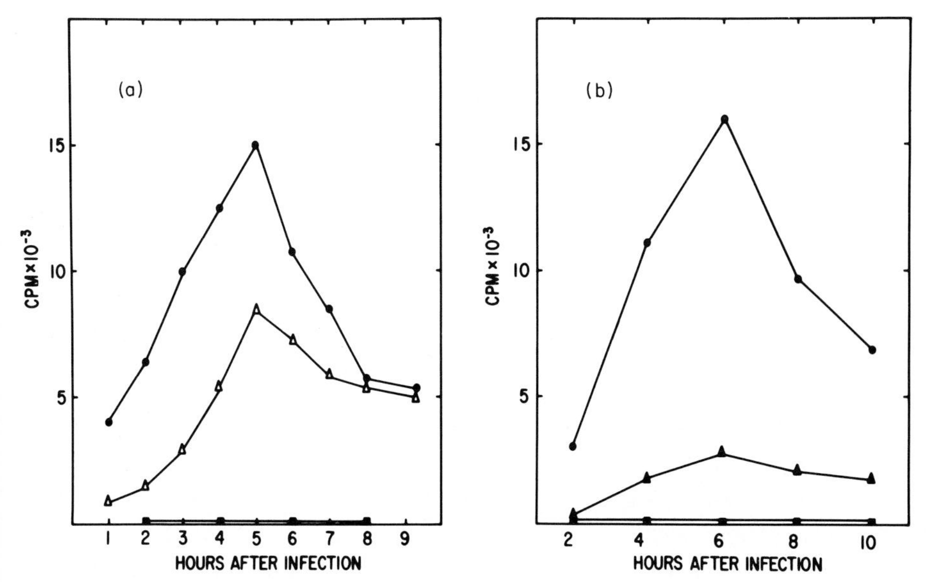

Fig. 2. RNA synthesis directed by SV_{STD} (circles) or SV_{CP6} (triangles) in CEFs (a) and by SV_{STD} (circles) or SV_{BP18} (triangles) in BHK-21 cells (b). At various times monolayer cultures of infected or mock-infected (squares) cells were pulse-labeled (50 min) with [^{3}H]uridine in the presence of actinomycin D. To terminate the pulse label cells were dissolved in 1% SDS, and the acid-precipitable radioactivity was measured. (From Guild and Stollar, 1975.)

particles and standard virus as compared to infection with standard virus alone (Guild and Stollar, 1975). Furthermore, "new" species of viral RNA (DI RNAs) are found which are smaller than standard RNAs. These changes, which affect both ss and ds viral RNA species are readily seen when the extracted RNAs are analyzed either by polyacrylamide gel electrophoresis or by sucrose velocity gradient centrifugation.

The first descriptions of intracellular DI RNAs of alphaviruses were obtained from experiments with BHK cells infected with SV or SFV. In general the results were similar with the two viruses. Infection of BHK cells with SV stocks containing DI particles led to the synthesis of ss and ds DI RNAs with sedimentation constants of 18–20 S (molecular weight about 0.8×10^6) and 12–15 S, respectively (Shenk and Stollar, 1972; Eaton and Faulkner, 1973; Weiss and Schlesinger, 1973; Weiss *et al.*, 1974). The 1973 report by Weiss and Schlesinger also described a larger ss DI RNA with a molecular weight of 2.2×10^6.

By working with BHK cells infected with SFV DI particles, it was possible by gel electrophoresis to resolve two species of ss DI RNA (Bruton *et al.*, 1976). They were, however, quite similar both in size (molecular weights of 0.81×10^6 and 0.75×10^6) and sequence content.

Another system, SV in CEF, has also been examined in detail and has given somewhat different results (Guild and Stollar, 1975; Guild *et al.*, 1977). In this case the first intracellular DI RNAs seen (clearly evident by passage 6) were much larger than the predominant species described above in BHK cells. Of greater interest, they showed a progressive stepwise decrease in size as the high-multiplicity passage series was continued (Fig. 3). Thus, the first DI RNAs seen were a 33 S ssRNA and an 18 S dsRNA. These gave way after a number of passages to a 24 S ssRNA and a 15 S dsRNA, and finally to a 22 S ssRNA and a 12 S dsRNA. This process of evolution took place during the course of about 20–24 passages.

The molecular weights of these DI RNAs were estimated in the case of the ds species by electron microscopy and in the case of the ss species by acrylamide gel electrophoresis (Guild *et al.*, 1977). These values are presented in Table I. The various DI RNAs are approximately one-half, one-third, and one-fourth to one-fifth the size of the full-length genome.

It should be noted that each ss DI RNA is always associated with a specific ds DI RNA, suggesting that, at any given passage level, there is in the infected cells a ss DI RNA and its replicative form. This interpretation is supported by the fact that each ss DI RNA has a molecular weight precisely one-half that of its associated ds DI RNA (Guild *et al.*, 1977). The smallest DI RNAs seen in the SV–CEF system approximated in size those described in BHK cells. Single-stranded alphavirus DI RNAs smaller than $0.7–0.8 \times 10^6$ in molecular weight have not been described in any system.

Experiments have already been described which showed that ss DI RNAs from extracellular viral particles contained very little if any negative-strand sequences. These experiments did not, however, exclude the incorporation of nonviral sequences, perhaps host-derived, into the defective RNAs. Evidence that the SV DI RNAs in CEFs contain only viral RNA sequences was obtained from experiments in which labeled ds DI 18, 15, or 12 S RNA was denatured and then annealed with an excess of unlabeled virion 42 S (Guild *et al.*, 1977). As was the case with labeled standard 22 S dsRNA, the labeled DI strands became increasingly resistant to RNase as the amount of virion RNA was increased until a plateau was reached at 50% resistance (Fig. 4). This is the expected theoretical value if the dsRNAs

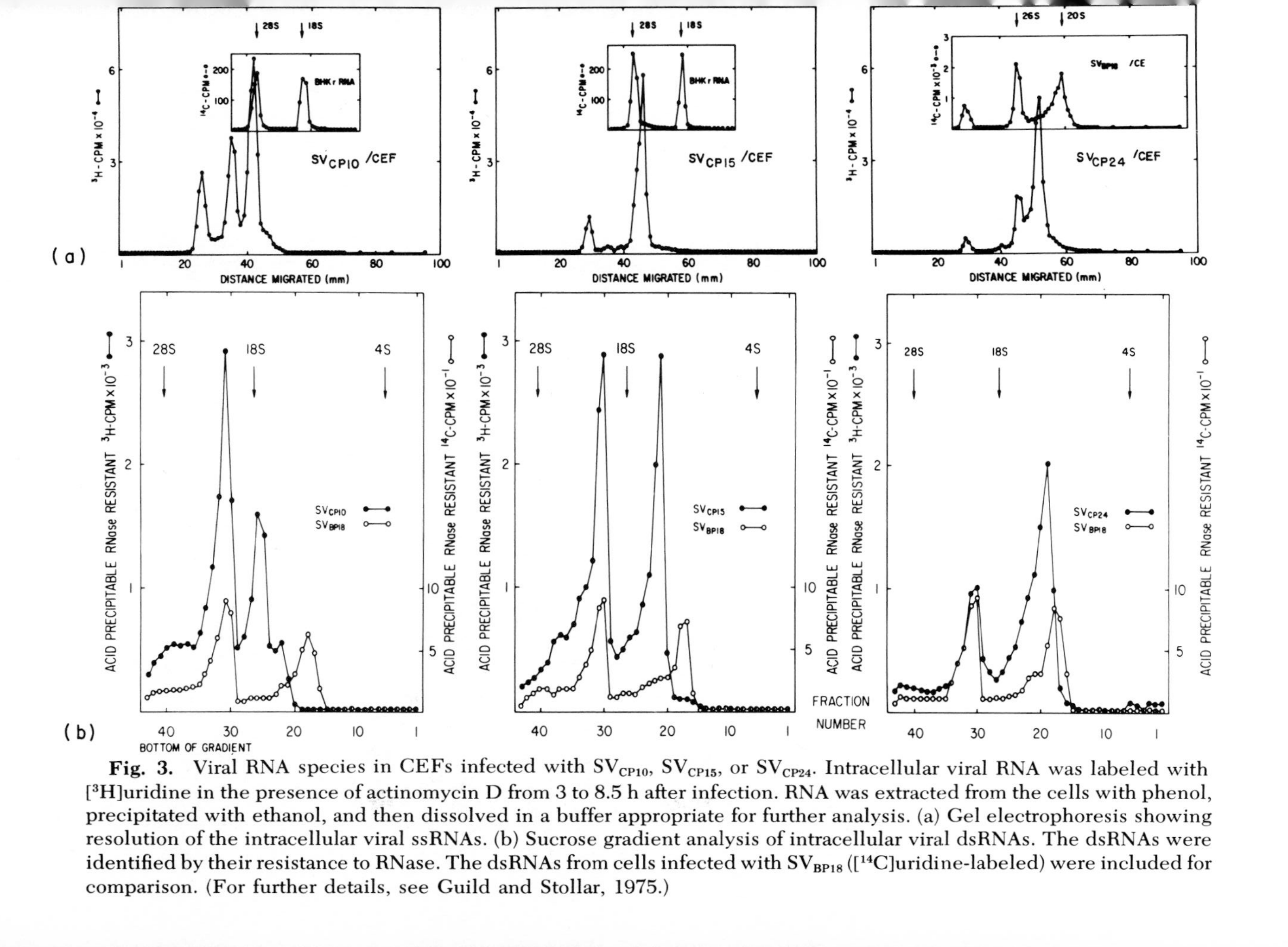

Fig. 3. Viral RNA species in CEFs infected with SV_{CP10}, SV_{CP15}, or SV_{CP24}. Intracellular viral RNA was labeled with [^{3}H]uridine in the presence of actinomycin D from 3 to 8.5 h after infection. RNA was extracted from the cells with phenol, precipitated with ethanol, and then dissolved in a buffer appropriate for further analysis. (a) Gel electrophoresis showing resolution of the intracellular viral ssRNAs. (b) Sucrose gradient analysis of intracellular viral dsRNAs. The dsRNAs were identified by their resistance to RNase. The dsRNAs from cells infected with SV_{BP18} ([^{14}C]uridine-labeled) were included for comparison. (For further details, see Guild and Stollar, 1975.)

TABLE I

Molecular Weights of ss and ds DI RNAs in CEF Infected with SV at Various Passage Levels[a]

Passage level	dsRNA	Molecular weight	ssRNA	Molecular weight	Approximate no. of nucleotides in ssRNA	MW ss DI RNA/ MW 42 S SSRNA
Standard RNAs						
SV_{STD}	22 S	8.7×10^6	42 S	4.3×10^6	12,600	
			26 S	1.6×10^6	4,600	
Defective RNAs						
SV_{CP6}	18 S	4.4×10^6	33 S	2.2×10^6	6,360	0.51
SV_{CP15}	15 S	2.8×10^6	24 S	1.4×10^6	4,040	0.33
SV_{CP24}	12 S	2.0×10^6	22 S	1.0×10^6	2,880	0.23

[a] These values are taken from Guild and Stollar (1975) and from Guild *et al.* (1977). The molecular weights of the ds DI RNAs were determined by electron microscopy and those of ss DI RNAs by acrylamide gel electrophoresis.

contain only viral RNA sequences. If the ds DI RNAs contain only viral RNA sequences, it is likely that the same is true for the ss DI RNAs.

Stark and Kennedy (1978) examined the patterns of DI RNAs synthesized in a variety of cell types [chick (CEF), mouse (3T3), rat (NRK), hamster (BHK), porcine (PK15), and human (HeLa strains H and D] during a serial passage of SFV. In these experiments the multiplicity of infection during the passage series was maintained constant at 50 PFU/cell until the limit passage (the first passage at which the titer dropped below 5×10^6 PFU/ml) was reached. No standard virus was added to the various passaged stocks in the primary passage series. In order to see which types of viral RNA were being synthesized, BHK cells were infected with the appropriate passaged stock and then labeled with inorganic [^{32}P]phosphate.

Two major points emerge from this work. As was the case in the SV–CEF system just described, in most of the cell types the size of the DI RNAs decreased as the passage series was continued. Second, it was clearly shown that the host cell strongly influenced the precise size classes of DI RNA produced. For example, after 4 passages of SFV in BHK cells a ss DI RNA with a molecular weight of 2.23×10^6 was observed; after 5 passages the most prominent DI had a molecular weight of 1.05×10^6, and after 6 or more passages the two species (0.81

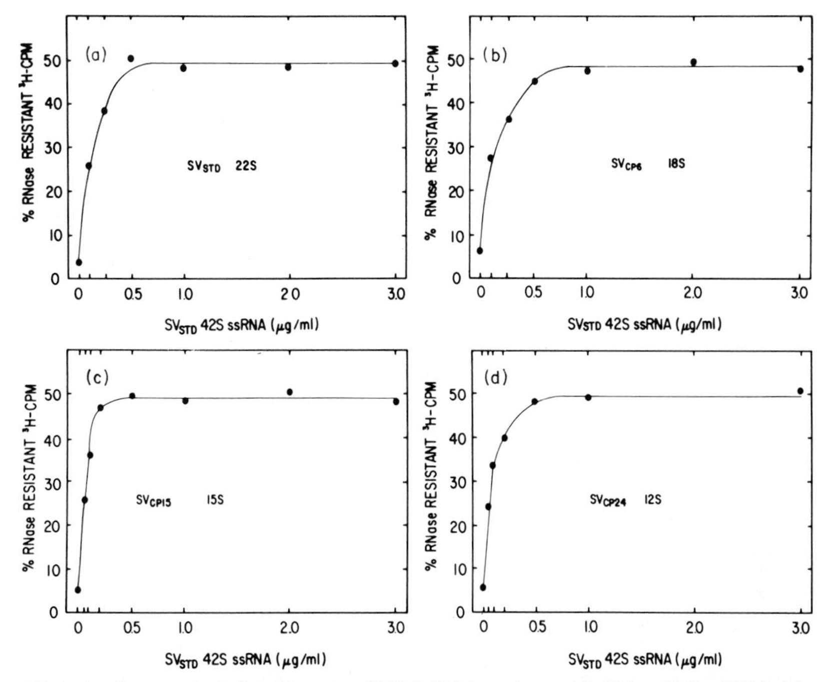

Fig. 4. Saturation hybridization of SV dsRNA probes with SV_{STD} 42 S ssRNA driver. The [^{3}H]uridine-labeled dsRNA probes were purified, denatured, and annealed (2.5 h, 70°C) to increasing amounts of unlabeled phenol-extracted SV_{STD} virion 42 S ssRNA and then assayed for acid-precipitable RNase radioactivity. (From Guild *et al.*, 1977.)

and 0.75 × 10^6) described earlier (Bruton *et al.*, 1976) came to predominate. In contrast, in porcine cells after 17 passages only the 2.23 × 10^6 and the 1.05 × 10^6 species were made, and in CEFs only the 2.23 × 10^6 was seen up to the limit passage (passage 5). If passage was subsequently continued in CEFs, but with the addition of standard virus, then the two smallest species (0.81 and 0.75 × 10^6) appeared. In addition to the DI species already mentioned, another of molecular weight 1.41 × 10^6 was seen during several passages in HeLa D cells. It is striking that the three largest ss DI RNAs (2.23, 1.41, and 1.05 × 10^6) observed by Stark and Kennedy during passage of SFV correspond precisely in size to those described in the SV–CEF system (see Table I). This suggests quite strongly that there exist important selective pressures or size constraints which determine whether or not alphavirus DI RNAs can be packaged and thus amplified (see below).

With reference to the question of the functional capacity of DI

RNAs, experiments with the 20 S ss DI RNA obtained from BHK cells infected with SV DI particles showed that, although it was polyadenylated, it did not function in an *in vitro* protein-synthesizing system (Weiss *et al.*, 1974).

There have been no published reports describing whether or not ss DI RNA is capped, however, in the same report Weiss *et al.* (1974) compared 26 S ss standard RNA (which is capped; see Dubin *et al.*, 1977) and 20 S ss DI RNA with respect to their association with polysomes. The interpretation of such experiments is complicated (1) by the overlapping of the nucleocapsid and polysome regions when cytoplasmic extracts are analyzed on sucrose velocity gradients and (2) because in cells producing DI particles ss DI RNAs but not 26 S standard RNA can be present in nucleocapsids. With these qualifications in mind, it was observed (Weiss *et al.*, 1974) that, whereas 57% of the 26 S RNA was found to be polysome-associated, only 13% of the 20 S DI RNA was. Their data also indicate, however, that when a lysate of cells infected with a specific high-passage virus stock was fractionated on a sucrose gradient, the "polysome region" contained twice as many 20 S ss DI as 26 S ss standard RNA molecules. As already noted, it is possible that a certain fraction of these DI RNA molecules were actually in nucleocapsids rather than polysome-associated. In one other experiment, Weiss *et al.* (1974) showed that DI RNAs failed to associate with reticulocyte ribosomes as efficiently as standard viral RNAs. Overall then, it appears that this specific ss DI RNA (20 S) and probably other DI RNAs of this size are not translated *in vivo*. The possibility still exists, however, that larger DI RNAs can be translated into polypeptides. Such polypeptides would probably be nonfunctional and would be rapidly degraded, as is the case with the polypeptides translated from defective poliovirus RNAs (Cole and Baltimore, 1973).

It is apparent from the preceding discussion that, depending on the virus–cell system, alphavirus DI RNAs can range in molecular weight from as small as 0.75×10^6 to as large as 2.2×10^6 and perhaps 3.3×10^6. It is also clear from certain of the reports, especially the earlier ones, that difficulties may be encountered with the extraction of RNA from DI particles, leading to inaccurate estimates of size. Although the precise reasons for this are unclear, it has been suggested (Kennedy *et al.*, 1976) that DI RNAs may tend to aggregate either because of their primary and secondary structure, or because of special interactions related to their packaging into nucleocapsids. Extraction of the DI RNAs from infected cells, on the other hand, has generally provided reproducible, easily interpretable results, whether the ss- or dsRNAs were examined. Because of their inherent resistance to RNase, there are often advantages to working with ds DI RNAs.

In any given infection the same ss DI RNA(s) found within the cells is (are) also found in released particles. Any discrepancy between the size of extracellular and intracellular ss DI RNAs probably indicates some type of experimental artifact.

It is notable that not only genome RNA with a molecular weight of 4.3×10^6 but, alternatively, defective RNAs of several different sizes can be packaged into viral particles. These include, for example, in the SV–CEF system RNAs with molecular weights of 2.2×10^6, 1.4×10^6, and approximately 1×10^6. Thus, all ss viral RNAs found in infected cells are packaged with the exception of the standard 26 S mRNA. This suggests that there are certain sequence or structural properties of an RNA required for packaging and that these requirements are met by the genome RNA and by all the ss DI RNAs but not by the standard 26 S ss RNA. Certain information obtained about the sequence content of DI RNAs (see below) is consistent with these ideas.

With respect to the functional capacity of DI RNAs, there is little if any evidence so far to suggest that any of the DI RNAs can be translated into polypeptides. On the contrary, as already noted, it has been demonstrated at least for certain of the smallest DI RNAs that, even though they are polyadenylated, they are not polysome-associated and do not direct *in vitro* protein synthesis.

It may be asked why, during the passage series in the SV–CEF system, the DI RNAs are found only in certain discrete size classes rather than in all sizes ranging from close to the size of the viral genome down to RNA species with molecular weights of only 0.7–0.8 $\times 10^6$. The fact that, in this particular system, it has not been possible to separate DI particles from standard virions either by means of velocity gradient centrifugation or by centrifugation to equilibrium suggests that, with respect to both size and density, the DI particles resemble the standard virions. It may be, therefore, that the total amount of RNA contained in each DI particle is very similar to that contained in the standard virion.

In describing the intracellular DI RNAs found in CEFs, it has been noted that the ss or ds DI RNAs represent 51, 32, or 23% of the length of the whole genome (see Table I). These values are very close to the integral fractions one-half, one-third, and one-fourth. Thus, if each DI particle contains the standard amount of RNA and the RNAs in question are one-half, one-third, or one-fourth of the genome, it is likely the DI particles contain two one-half-length molecules, three one-third-length molecules, or four one-fourth-length molecules. It therefore can be postulated that the DI RNAs we can detect are largely determined by constraints at the level of packaging the DI RNA into a nucleocapsid. Perhaps DI RNAs of random sizes are indeed generated all the time, but only those of a suitable size can be neatly packaged first into nucleocapsids and then into virionlike structures which are released by infected cells, enabling them to coinfect other cells along with standard virus. Thus, if DI RNAs are generated which happen to be of an inappropriate size, they would not be efficiently packaged, would not infect cells in a subsequent passage, and therefore would never become amplified.

V. SEQUENCES FOUND IN DEFECTIVE INTERFERING RNAs

Evidence has been presented that the ss DI RNAs produced in the SV–CEF system contain only viral RNA sequences and are of positive polarity, i.e., the same polarity as the standard ssRNAs found within the virions and within the infected cells. The question of which viral RNA sequences are contained in the alphavirus DI RNAs has been approached in several different laboratories. To begin, it is helpful to remember that, whenever cells are coinfected with SV or SFV DI particles plus helper virus, the cells produce *not only* ss DI RNAs of certain specific sizes *but also* the corresponding ds DI RNAs. The same ss DI RNA(s) is also found packaged in extracellular particles. Since, as already discussed, the ds DI RNA is the replicative form (RF) of the ss DI RNA, a given ds DI RNA should contain in its positive strand the same sequences as those in the corresponding ss DI RNA. Thus, sequence content can be ascertained by examination of ss DI RNAs or of ds DI RNAs. Two methods have been used to this end: (1) RNA–RNA hybridization and (2) oligonucleotide fingerprinting.

Weiss *et al.* (1974) studied a 20 S ss DI RNA produced in BHK cells infected with SV DI particles. A competition type of hybridization was used. ^{32}P-labeled standard 26 S RNA was annealed with excess denatured standard 22 S dsRNA; 26 S ss standard and 20 S ss DI RNAs were both tested for their abilities to compete with the ^{32}P-labeled 26 S RNA species. From these experiments it was concluded (1) that the 20 S ss DI RNA did not contain sequences from the whole genome but was a unique species of viral RNA, and (2) that it contained 50% of the sequences of the standard 26 S ssRNA.

Assuming molecular weights of 0.86×10^6 for the 20 S ss DI RNA

(Weiss and Schlesinger, 1973) and 1.6×10^6 for the 26 S ss standard RNA (Simmons and Strauss, 1972), then $[(0.5 \times 1.6 \times 10^6)/(0.86 \times 10^6)] \times 100$ or 93% of the 20 S sequences must be represented in the 26 S ss standard RNA. Thus it appeared that not more than about 5–10% of the 20 S ss DI RNA sequences could be derived from the non-26 S region of the genome. Oligonucleotide fingerprint analysis (Kennedy *et al.*, 1976) was consistent with these results (Fig. 5b). Of 12 large characteristic oligonucleotides present in the 20 S ss DI RNA,

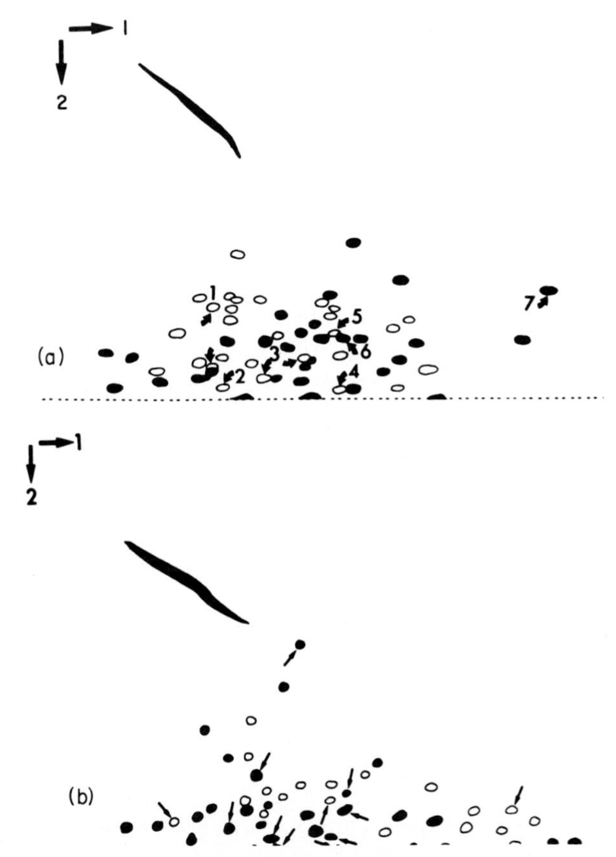

Fig. 5. Diagrammatic representations of the fingerprints of the characteristic oligonucleotides present in alphavirus DI RNAs. (a) RNAs from SFV. Oligonucleotides common to both 42 and 26 S ssRNAs are represented as solid spots and those unique to 42 S RNA as open spots. Arrows indicate spots representing characteristic oligonucleotides from the ss DI RNA (molecular weight 0.81×10^6) obtained from BHK cells infected with SFV DI particles (from Kennedy, 1976). (b) RNAs from SV. The 42 and 26 S ssRNAs were obtained from BHK cells infected with SV_{STD}, and the 20 S ss DI RNA from BHK cells infected with SV_{BP} which contained DI particles. The solid spots, open spots, and arrows have the same significance as in (a). (From Kennedy *et al.*, 1976.)

9 were also found in the 26 S ss standard RNA. The other 3 must have been from the non-26 S region of the viral genome.

Bruton *et al.* (1976) examined the DI RNAs produced in BHK cells infected with SFV. In this case, two kinds of ss DI RNA were found in extracellular particles as well as within the cells. Their molecular weights were 0.81 and 0.75 × 10^6 (Bruton and Kennedy, 1976). When competition hybridization was used to determine which sequences were present in these DI RNAs, the results indicated that none of their nucleotide sequences were present in the 26 S region of the genome (Bruton *et al.*, 1976). Thus, even though the SFV DI RNAs were similar in size to the SV DI RNAs just described, the former apparently contained no 26 S RNA sequences, whereas 93% of the latter was composed of 26 S RNA sequences. Another type of hybridization experiment, however, using labeled ds DI RNA as probe and annealing with an excess of various viral ssRNAs (see Table I, Bruton *et al.*, 1976) suggested that there actually was a small amount of annealing (4–6%) between standard 26 S ssRNA and the labeled ds DI RNA probe.

When the oligonucleotide fingerprints of these SFV ss DI RNAs were examined (Kennedy, 1976), the patterns for the two ss DI RNAs were very similar. Of nine characteristic oligonucleotide spots, seven were from the non-26 S region of the genome and two were from the 26 S region. Therefore, it seems that the results of the hybridization experiments just described (Bruton *et al.*, 1976), suggesting that the DI RNA contained some 26 S RNA sequences, were indeed significant. The oligonucleotide analysis also revealed that these DI RNAs contained oligonucleotides from the 5′-end of the 42 S RNA as well as oligonucleotides from the 26 S region (Fig. 5a). Given the small size of the DI RNAs, these observations could only be explained if the DI RNAs contained a large internal deletion.

The defective RNAs produced in the SV–CEF system differed from those generated in BHK cells by either SV or SFV. Not only were much larger DI RNAs seen but, as already noted, they progressively decreased in size as the passage series was continued. The sequence content of these DI RNAs was determined by RNA–RNA hybridization (Guild and Stollar, 1977). Constant amounts of highly ^{32}P-labeled ssRNA probes were annealed with varying amounts of lightly ^{3}H-labeled driver dsRNAs. The probes used were 42 and 26 S ss standard RNAs and the smallest of the ss DI RNAs, the 22 S DI species (Table II). The drivers included 22 S ds standard RNAs and RFIII (the ds equivalent of 26 S ss standard RNA), both derived from cells infected with SV_{STD}, and the 18, 15, and 12 S ds DI RNAs obtained from cells

TABLE II

Sequence Content of DI RNAs Produced in an SV–CEF System as Shown by Hybridization Experiments[a]

Experiment 1—^{32}P-labeled 42 S ss Standard RNA as probe
STD 22 S dsRNA (RFI): contains >95% of sequences of 42 S RNA
STD RFIII dsRNA: contains 36% of sequences of 42 S RNA
DI 18 S dsRNA: contains 50% of sequences of 42 S RNA
DI 15 S dsRNA: contains 35% of sequences of 42 S RNA
DI 12 S dsRNA: contains 21% of sequences of 42 S RNA
Experiment 2—^{32}P-labeled 26 S ss Standard RNA as probe
STD 22 S dsRNA (RFI): contains 100% of sequences of 26 S RNA
STD RFIII dsRNA: contains 100% of sequences of 26 S RNA
DI 18 S dsRNA: contains 20–25% of sequences of 26 S RNA
DI 15 S dsRNA: contains 20–25% of sequences of 26 S RNA
DI 12 S dsRNA: contains 20–25% of sequences of 26 S RNA
Experiment 3—Hybridization experiments with ^{32}P-labeled 22 S ss DI RNA as probe
STD 22 S dsRNA (RFI): contains 100% of sequences of 22 S RNA
STD RFIII dsRNA: contains 31% of sequences of 22 S RNA
DI 18 S dsRNA: contains 100% of sequences of 22 S RNA
DI 15 S dsRNA: contains 100% of sequences of 22 S RNA
DI 12 S dsRNA: contains 100% of sequences of 22 S RNA

[a] See Fig. 6 (experiment 1) and Guild and Stollar (1977).

infected with the appropriate virus stocks (Table I). Optimal conditions for annealing were determined (50% formamide, 6× SSC and 60°C), and reactions were carried out so as to reach a state of equilibrium. In order to minimize problems arising from cross-contamination of one species of dsRNA with another, experiments were done with low driver/probe ratios (between 1:1 and 10:1), and data were analyzed using a double reciprocal plot (Darby and Minson, 1973). Experimental data are presented here only for one experiment (Fig. 6), but the conclusions from the other experiments are summarized in Table II. For further details, see Guild and Stollar (1977).

In the first experiment (Fig. 6 and Table II) the five dsRNA drivers were each annealed with ^{32}P-labeled 42 S RNA. As expected, the 22 S dsRNA contained all the sequences of the 42 S RNA, whereas RFIII contained only about 36%. This last observation was consistent with previously published results concerning the size of 26 S ss standard RNA and RFIII (Simmons and Strauss, 1972). The 18, 15, and 12 S dsRNAs contained 50, 35, and 21%, respectively, of the sequences of the 42 S ss standard RNA. These values correspond very closely to their sizes relative to the genome-length dsRNA, the 22 S ds standard RNA. Thus, it was concluded that each of the various DI RNAs repre-

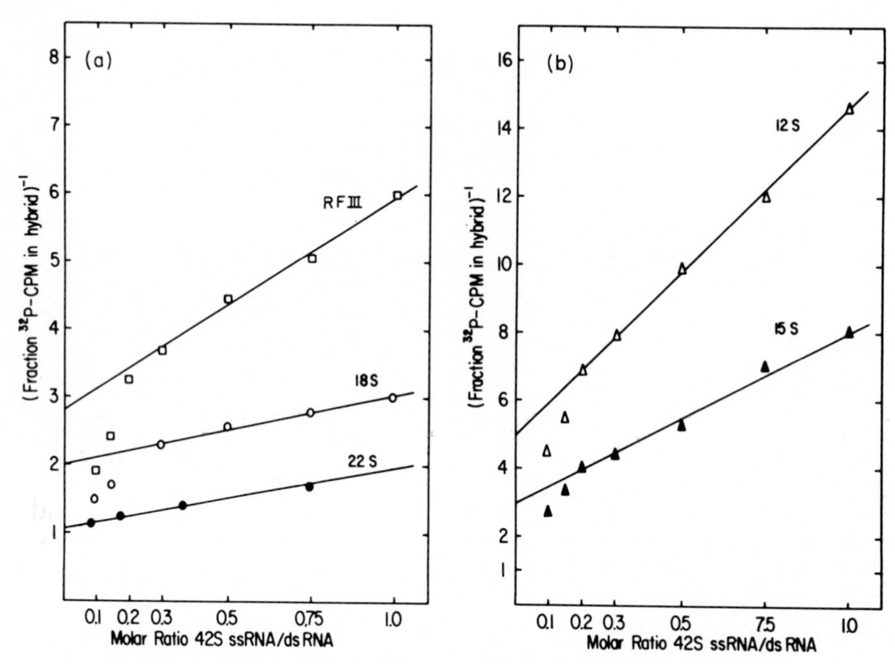

Fig. 6. Double reciprocal plot of hybridization of SV_{STD} 42 S ssRNA probe with SV_{STD} 22 S and RFIII dsRNAs and 18, 15, and 12 S ds DI RNAs. The 42 S ssRNA was obtained from CEFs which had been labeled with $^{32}PO_4^{3-}$ in the presence of actinomycin D. Each set of reactions was incubated for a time sufficient to allow the reaction containing the least amount of driver dsRNA (driver/probe molar ratio = 1:1) to reach a *Crt* value of 2×10^{-2} mol·s/liter. Hybridization was measured by determining the extent to which the 42 S viral RNA was converted to an RNase-resistant form. When the data are plotted in this fashion (Darby and Minson, 1973), the reciprocal of the intercept on the y axis is equal to the fraction of the probe (42 S RNA) which is homologous to a specific dsRNA driver. For example, with the 15 S dsRNA driver the y intercept is 3. Therefore one-third of the 42 S RNA sequences are represented in the 15 S dsRNA. (For further details, see Guild and Stollar, 1977.)

sented a unique sequence of the genome rather than a population of small viral RNAs containing sequences from the entire genome.

In the second experiment, the same dsRNA drivers were annealed with 26 S ss standard RNA probe (Table II). As expected, both the 22 S ds standard RNA and RFIII contained all the sequences present in the probe. Somewhat surprisingly, each of the ds DI RNAs contained about the same amount of the 26 S RNA sequences, between 20 and 25%. Thus, of the 4600 nucleotides in 26 S ss standard RNA, it was estimated that each of the DI RNAs contained about 1000. Each of the ss DI RNAs, however, contains many more than 1000 nucleotides

(Table I). These additional sequences must therefore be derived from the non-26 S region of the genome. Because the amount of the 26 S sequences remains constant, it is from the non-26 S part of the genome that sequences must be lost as the DI RNAs progressively diminish in size.

In the third experiment, the smallest of the ss DI RNAs, the 22 S species, was used as a probe (Table II). The 22 ds standard contained all the sequences of this probe, confirming that the probe was composed exclusively of viral sequences. Its ds homolog, the 12 S ds DI RNA also contained all the probe sequences, as did the two larger ds DI RNAs, the 15 and 18 S species. These results indicated that, at least in this system, the sequences in the smallest DI RNA are a subset of those in the larger DI RNAs. It appears, therefore, that as the size of the DI RNA decreases, sequences are lost but no new different sequences are acquired. RFIII only contained 31% of the sequences present in the 22 S ss DI RNA, confirming the conclusion drawn from the second experiment that these DI RNAs contain sequences from the non-26 S region of the genome.

By oligonucleotide fingerprint analysis Kennedy (1976) showed that sequences near both the 3′- and 5′-termini of the 42 S RNA were retained in the SFV DI RNAs. These results indicate that DI RNAs probably arise by means of a large internal deletion of sequences from the genome-length (42 S) ssRNA. This idea has further appeal because inverted complementary base sequences are found at the 3′- and 5′-termini of the viral RNA molecule (Hsu *et al.*, 1973; Kennedy, Chapter 11 in this volume). Such sequences are undoubtedly of important functional significance, perhaps with respect to replication and/or packaging and are therefore likely to be retained in the DI RNAs.

The hybridization experiments with the DI RNAs generated in the SV–CEF system could only indicate whether DI RNAs retained sequences from the 26 or non-26 S region of the genome but could not place the sequences within these regions. If it is assumed that the SV DI RNAs in this system are also characterized by an internal deletion, and that the sequences at the termini are retained, then the various DI RNAs can be represented in relation to the 42 and 26 S RNA as shown in Fig. 7. In the largest DI RNA, most of the sequences were from the non-26 S region of the genome. As the DI RNAs decreased in size, the internal deletion would be progressively extended, but only into the non-26 S region.

In comparing the sequence content of the smallest ss DI RNA in the SV–CEF system with that of similar-sized DI RNAs produced in BHK

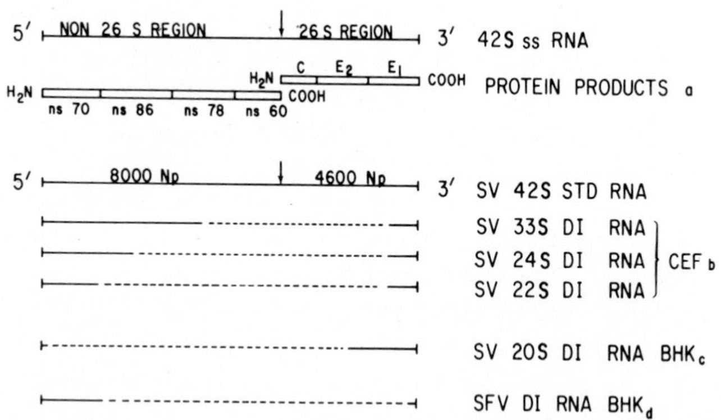

Fig. 7. A model for the sequence of organization of alphavirus DI RNAs (modified from Guild and Stollar, 1977). (a) The coding capacity of the 42 and 26 S RNAs and the probable order of protein gene products for the 26 S RNA have been established. Much less is known concerning the proteins coded for by the non-26 S region of the genome, and the findings from different laboratories are sometimes difficult to reconcile. These molecular weight estimates have been selected from the report of Lachmi *et al.* (1975) with SFV. The order of the polypeptides from the non-26 S region is somewhat arbitrary at this time. (See also Chapter 8 of this volume.) (b) From Guild and Stollar (1977). (c) Derived from Weiss *et al.* (1974) and from Kennedy *et al.* (1976). (d) Derived from Bruton *et al.* (1976) and from Kennedy (1976).

cells, either by SV or SFV (Fig. 7), it is seen that in one case largely non-26 S sequences were retained (Kennedy, 1976), in the other case, largely 26 S sequences (Kennedy *et al.*, 1976), and in the third, roughly equal proportions of both (Guild and Stollar, 1977). It is likely, however, that in each instance similar mechanisms account for the generation of DI RNAs and that in each case DI RNAs originate because of large internal deletions from the genome RNA. Although the exact location of these deletions may not be completely random, there appears to be considerable variation over where they may occur. The factors which influence the location of the sequence deletion are probably related both to the virus strain and to the host cell, but at present are largely unknown.

The model presented in Fig. 7 represents each DI RNA as having a single continuous internal deletion. Two recent papers, however, indicate that at least certain DI RNAs may have more than one deletion and that the deletions need not be contiguous (Stark and Kennedy, 1978; Dohner *et al.*, 1979). In the report of Stark and Kennedy the noncontiguous deletions were observed early in the passage series; but with continued passage this particular DI RNA was replaced by one with what was likely a single continuous deletion.

It is of interest to compare the deletions in alphavirus DI RNAs with those described for DI RNAs of another positive-strand RNA virus. In the case of poliovirus, the deletion was a relatively small one, involving only 10–13% of the genome, was internal, and was localized to the 5′-half of the genome (Cole *et al.*, 1971; Huang and Baltimore, 1977).

VI. ROLE AND INFLUENCE OF THE HOST CELL IN THE GENERATION AND REPLICATION OF ALPHAVIRUS DEFECTIVE INTERFERING PARTICLES

There is considerable variation not only in the ability of different cells to generate DI particles of a given virus but also in how different cells respond to infection with DI particles of the same virus. For example, DI particles of the WSN strain of influenza virus are produced in MDBK (bovine kidney) cells with a much lower efficiency than in other cell types (Choppin, 1969); and relatively little autointerference is seen with VSV grown in PK-15 or MDCK cells relative to what is observed in HeLa or BHK cells (Perrault and Holland, 1972).

Similar phenomena have been observed with alphavirus DI particles. Thus, when serial undiluted passage of SFV was carried out in CEFs, there was no significant drop in the titer of infective virus (Levin *et al.*, 1973). However, when the stock obtained after nine passages was tested in murine cells, it produced much lower yields of virus than it did in CEFs. It is likely, therefore, that DI particles of SFV were produced during passage in chick cells, but that their interfering ability in CEFs was much poorer than in murine cells.

A more striking demonstration that the host cell influences the production of alphavirus DI particles is contained in the report of Stark and Kennedy (1978) which shows that, whereas the D strain of HeLa cells produced DI particles (as indicated both by a fall in titer during high-multiplicity serial passage and by the synthesis of DI RNAs in infected cells), there was no such indication for their production by the H strain of HeLa cells even after 200 passages of virus. A strain of HeLa cells has also been described which is unable to produce DI particles of VSV (Holland *et al.*, 1976). The host factors related to the production of DI particles are not known.

The influence of the host cell was also demonstrated dramatically when mosquito cells were infected with SV or SFV stocks which were produced in BHK cells and contained DI particles (Stollar *et al.*, 1975; Eaton, 1975; Igarashi and Stollar, 1976). The viral RNA patterns in these cells were indistinguishable from those produced in cells infected with standard virus alone (Fig. 8). When CEF or BHK cells were infected with the same stocks containing DI particles, typical ss and ds DI RNAs were generated. These results strongly suggested that DI RNAs were not being replicated in mosquito cells. The lack of interference by such stocks (Table III) (Igarashi and Stollar, 1976) is consistent with the inability to observe replication of DI RNA in these cells.

In another experiment, SV was serially passaged in mosquito cells without dilution, in a manner similar to that used to produce DI particles in CEFs or BHK cells. The resultant virus stocks (SV_{AP}) contained no interfering activity (Igarashi and Stollar, 1976), and infection of mosquito cells or vertebrate cells with these stocks failed to show evidence of DI RNAs within infected cells. It was concluded from these experiments that, under conditions suitable for the production of large amounts of alphavirus DI particles in CEFs or BHK cells, *Aedes albopictus* cells either produced no DI particles or very much lower levels.

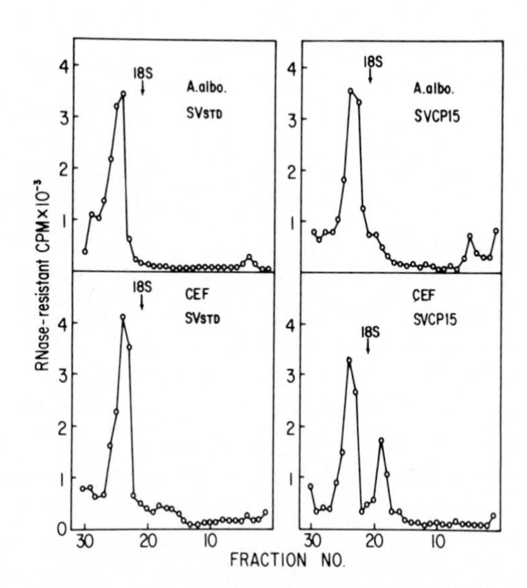

Fig. 8. Double-stranded RNA synthesized in *A. albopictus* and CEFs infected with SV_{STD} or SV_{CP15}. Cells were infected with SV_{STD} or SV_{CP15} (multiplicity of infection = 30 PFU/cell) and labeled with [^{3}H]uridine in the presence of actinomycin D. RNA was extracted and fractionated by sucrose gradient centrifugation. Double-stranded RNA was localized by counting RNase-resistant acid-insoluble radioactivity. (From Igarashi and Stollar, 1976.)

When, however, the serial undiluted passage in *A. albopictus* cells was carried out using 48-h harvests instead of 16-h harvests, good evidence for the production of DI particles was obtained (M. Waite, personal communication). This has since been confirmed in our laboratory (K. Kowal and V. Stollar, unpublished results); and Logan (1979), using 20-h harvests, has recently reported the generation of SFV DI particles in mosquito cells. These findings are also consistent with the production of SV DI particles in persistently infected (PI) cultures of *A. albopictus* cells (see Section VII).

The basis for the nonrecognition by mosquito cells of DI particles produced in BHK cells is unexplained. It may reflect a critical step involved in the recognition of viral RNA, standard or defective, by the viral RNA polymerase. If this were so, it would suggest differences in the viral polymerase, in the mosquito and vertebrate cells, possibly due to a specific type of host modification.

VII. DEFECTIVE INTERFERING PARTICLES IN PERSISTENTLY INFECTED CULTURES

Alphaviruses are in general strongly cytocidal for vertebrate cells. Persistently infected cultures are therefore not readily initiated, and little effort has so far been made to establish such cultures (see, however, Chambers, 1957; Schwöbel and Ahl, 1972; Eaton and Hapel, 1976) by infecting cells for example with temperature-sensitive viral mutants or mixtures of standard virus and DI particles.

On the other hand, since alphaviruses are in general not cytocidal for mosquito cells (but see also Chapter 20), cultures of mosquito cells persistently infected with alphaviruses are established without difficulty and can be maintained for many months, if not indefinitely. Examination of the viral RNAs in such cultures has shown that for about 8–9 weeks after *A. albopictus* cells are infected with S Virus, they contain only one well-defined species of viral dsRNA, the 22 S species (Igarashi *et al.*, 1977). Beginning after about 10 weeks and persisting thereafter, they contain in addition a smaller viral dsRNA (about 12–15 S) as well as a small viral ssRNA (about 20 S) (Stollar and Shenk, 1973; Eaton, 1977; Igarashi *et al.*, 1977). These intracellular viral RNA species are reminiscent of the small DI RNAs described earlier. Whether or not DI particles were present in the medium of PI cultures was at first difficult to ascertain. This was because (1) by 10 weeks after infection the virus produced by PI cultures usually becomes predominantly temperature-sensitive (Igarashi *et al.*, 1977) and

TABLE III

Test for Interference by SV_{CP15} in CEFs and *Aedes albopictus* cells[a]

Inoculum (PFU/cell)		24-h yield from infected cultures (PFU/ml)	
SV_{STD}	SV_{CP15}	*A. albopictus*	CEFs
28		3.3×10^9	1.6×10^9
	32	1.8×10^9	1.5×10^8
0.3		2.0×10^8	2.2×10^9
0.3	32	2.8×10^9	1.5×10^8

[a] Virus was inoculated at the input multiplicities indicated for the various experiments (from Igarashi and Stollar, 1976).

(2) this *ts* virus interferes as efficiently with the replication of SV_{STD} as do DI particles (Igarashi *et al.*, 1977).

The demonstration, however, that infection of BHK cells with medium from PI cultures leads to the synthesis of DI RNAs which correspond in size to those seen in the original PI cultures (Eaton, 1977) makes it very likely that DI particles are not only produced in PI cultures, but that they are released into the extracellular medium. Thus mosquito cells which initially could not be shown to produce DI particles of SV under conditions of serial undiluted passage (see above) are capable of doing so when persistently infected with SV. It is of special interest that these DI particles, in contrast to those produced in vertebrate cells, are perfectly capable of being replicated in vertebrate cells as well as in mosquito cells (Eaton, 1978).

A common property of PI cultures is that they are resistant to superinfection with the homologous virus but not to heterologous viruses. Thus mosquito cells persistently infected with SV are resistant to superinfection with SV_{STD} (Stollar and Shenk, 1973) but not with Eastern equine encephalitis virus. It is likely that replicating DI RNAs and extracellular DI particles play some role in the inhibition of superinfecting virus.

VIII. DEFECTIVE INTERFERING PARTICLES IN THE WHOLE ANIMAL

As with other virus groups, most of the work related to togavirus DI particles has been carried out in cultured cells. However, systems have also been described which point to an important role for togavirus DI particles in the whole animal.

Woodward and Smith (1975) worked with two strains of SFV which differed markedly in their virulence for mice. Both strains multiplied equally in most tissues except for brain. In the brain, the titer of virulent virus was approximately 100-fold higher than that of the avirulent strain, but there was very little, if any, significant difference in the amount of HA or complement-fixing antigen. These observations suggested to these workers that the avirulent virus strain was producing large amounts of defective particles (or at least incomplete virus) in the brain which, because of interference, reduced the final yield of infectious virus. When mouse brain-grown virulent virus was inoculated into organ cultures of mouse brain, the final yields of infective virus were higher with a high input of inoculum virus than with a low input. In contrast, the brain-grown avirulent strain produced moderate yields following infection with low-titered inocula, but diminishing yields and then no yield as the amount of inoculum was increased. When the same experiment was done with virus stocks prepared in chick cells, neither in the case of the virulent or the avirulent strain was the final yield affected by the amount of virus inoculated. These results raise the possibility that DI particles are produced in the brains of infected mice by the avirulent strain of SFV, and that their presence strongly influences the outcome of a virus infection. Why two strains of the same virus should differ with respect to the production of infectious virus in brain and not in other organs is unexplained. (See also Chapter 4 for discussion of virulence of togaviruses.)

In a more recent report from the same laboratory, however (Woodward *et al.*, 1978), it was concluded that, because brain homogenates from animals infected with avirulent virus did not induce the synthesis of DI RNAs in BHK cells, it was unlikely that DI particles were made in these brains. If so, the lack of virulence of this particular strain of SFV would be unrelated to the production of DI particles. The possibility still remains open that DI particles were present but could not be replicated in BHK cells (see Section VI).

In another system, Darnell and Koprowski (1974) studied a flavivirus, West Nile virus (WNV). Two inbred strains of mice were described, one of which was 100% susceptible to fatal infection with flaviviruses, whereas the other was completely resistant. With respect to yields of infectious virus, the resistant mice produced significantly less than the susceptible mice. In contrast, after infection with an alphavirus, both strains of mice produced similar amounts of virus. Embryos of the two strains were used to prepare cell cultures, some of which were transformed with SV40. Upon infection with WNV, cultures from the susceptible strain produced from three- to eightfold more virus than cultures produced from the resistant strain. This was true whether comparisons were made in untransformed or in transformed cells. Thus, these results *in vitro* reflect what was observed in the whole animal.

In a further extension of the *in vitro* experiments, serial undiluted passage of WNV was carried out in transformed cells derived from the two strains of mice. When the passage series was initiated at a high multiplicity of infection, the virus titer fell more sharply and rapidly when the passage was in cells from the resistant strain than when the passage was in cells derived originally from the susceptible strain of mice. Furthermore, when cells were infected with third-passage WNV from resistant cells, mixed with WNV prepared in hamster brain, there was a significant reduction in the yield of virus as compared to infection with hamster brain virus alone. No such effect was seen when the cells were infected with hamster brain virus mixed with third-passage virus from the susceptible cells.

The results of the *in vitro* experiments are suggestive of the production of DI particles by cells from the resistant strains of mice and illustrate a possible correlation between the ability to produce DI particles and resistance to flavivirus infection. In extrapolating from the *in vitro* experiments, it is suggested that following infection with WNV, the resistant mice produce large amounts of DI particles which, because of interference, lead to lower levels of infectious virus than are seen in the susceptible mice. The differences in the levels of infectious virus produced would then account for the differing outcomes of infection (see Chapter 4).

A more clear-cut demonstration of the protective role alphavirus DI particles can play in the whole animal is presented in a paper by Dimmock and Kennedy (1978). They showed that, when mice were inoculated intranasally with standard SFV plus homologous DI particles, there was (1) a marked reduction in mortality, compared to a group infected with standard virus alone—in one instance the mortality rate fell from 100 to 20%; (2) an approximately 10^5-fold reduction in the titer of infectious virus in the brain compared to that in a group infected with standard virus alone and (3) replication of DI particles in mouse brain as demonstrated by the ability of brain homogenates to induce the synthesis of DI RNAs in BHK cells. Earlier work (Doyle and Holland, 1973; Holland and Villarreal, 1975) had shown the protective effect of VSV DI particles in adult mice, but there was no conclusive evidence for the actual replication of DI particles.

IX. CONCLUDING REMARKS

The study of togavirus DI particles has so far focused primarily on alphaviruses. Aside from the suggestive experiments with WNV just described, there have been no other reports of the production of flavivirus DI particles in cultured cells. Considerable understanding has been attained about the properties of alphavirus DI particles, especially in relation to the DI RNAs found in extracellular particles and within infected cells.

A number of intriguing questions have been raised by the results so far obtained and remain to be answered. If, as seems likely, DI particles of alphaviruses arise because of a large internal deletion in the genome RNA, precisely how, in molecular terms, does this occur? What is the molecular basis for the differential ability of various cells not only to generate but also to respond to and replicate alphavirus DI particles?

Especially important in the future will be to extend the study of togavirus DI particles to model systems with whole animals. Are DI particles a general feature of togavirus infection in whole animals and, if so, what precise factors can influence their production? To what degree can DI particles modify the course of an infection, and what relationships do they have to the initiation and maintenance of persistent infections?

Certain generalizations apply to DI particles of all viruses. On the other hand, because the strategy of the viral genome varies so markedly among different groups of viruses, it is not surprising that the DI particles generated by each group of viruses also vary widely with respect to their properties and their functional capabilities.

REFERENCES

Brown, D. T., and Gliedman, J. B. (1973). *J. Virol.* **12,** 1534–1539.
Bruton, C. J., and Kennedy, S. I. T. (1976). *J. Gen. Virol.* **31,** 383–395.
Bruton, C. J., Porter, A., and Kennedy, S. I. T. (1976). *J. Gen. Virol.* **31,** 397–416.
Chambers, V. C. (1957). *Virology* **3,** 62–75.
Choppin, P. W. (1969). *Virology* **39,** 130–134.
Cole, C. N., and Baltimore, D. (1973). *J. Mol. Biol.* **76,** 325–343.
Cole, C. N., Smoler, D., Wimmer, E., and Baltimore, D. (1971). *J. Virol.* **7,** 478–485.

Darby, G., and Minson, A. C. (1973). *J. Gen. Virol.* **21**, 285–295.
Darnell, M. B., and Koprowski, H. (1974). *J. Infect. Dis.* **129**, 248–256.
Dimmock, N. J., and Kennedy, S. I. T. (1978). *J. Gen. Virol.* **39**, 231–242.
Dohner, D., Monroe, S., Weiss, B., and Schlesinger, S. (1979). *J. Virol.* **29**, 794–798.
Doyle, M., and Holland, J. J. (1973). *Proc. Natl. Acad. Sci. U.S.A.* **70**, 2105–2108.
Dubin, D. T., Stollar, V., Hsuchen, C., Timko, K., and Guild, G. M. (1977). *Virology* **77**, 457–470.
Eaton, B. T. (1975). *Virology* **68**, 534–537.
Eaton, B. T. (1977). *Virology* **77**, 843–848.
Eaton, B. T. (1978). *In* "Viruses and Environment" (E. Kurstak and K. Maramorosch, eds.), pp. 181–201. Academic Press, New York.
Eaton, B. T., and Faulkner, P. (1973). *Virology* **51**, 85–93.
Eaton, B. T., and Hapel, A. J. (1976). *Virology* **72**, 266–271.
Guild, G. M., and Stollar, V. (1975). *Virology* **67**, 25–41.
Guild, G. M., and Stollar, V. (1977). *Virology* **77**, 175–188.
Guild, G. M., Flores, L., and Stollar, V. (1977). *Virology* **77**, 158–174.
Holland, J. J., and Villarreal, L. P. (1975). *Virology* **67**, 438–449.
Holland, J. J., Villarreal, L. P., and Breindl, M. (1976). *J. Virol.* **17**, 805–815.
Hsu, M-T., Kung, H-J., and Davidson, N. (1973). *Cold Spring Harbor Symp. Quant. Biol.* **37**, 943–950.
Huang, A. S. (1973). *Annu. Rev. Microbiol.* **27**, 101–117.
Huang, A. S., and Baltimore, D. (1970). *Nature (London)* **226**, 325–327.
Huang, A. S., and Baltimore, D. (1977). *In* "Comprehensive Virology" (H. Fraenkel-Conrat and R. R. Wagner, eds.), Vol. 10, pp. 73–116. Plenum, New York.
Igarashi, A., and Stollar, V. (1976). *J. Virol.* **19**, 398–408.
Igarashi, A., Koo, R., and Stollar, V. (1977). *Virology* **82**, 69–83.
Inglot, A. D., and Chudzio, T. (1972). *Proc. Int. Congr. Virol., 2nd, Budapest*, p. 166.
Inglot, A. D., Albin, M., and Chudzio, T. (1973). *J. Gen. Virol.* **20**, 105–110.
Johnston, R. E., Tovell, D. R., Brown, D. T., and Fulkner, P. (1975). *J. Virol.* **16**, 951–958.
Kennedy, S. I. T. (1976). *J. Mol. Biol.* **108**, 491–511.
Kennedy, S. I. T., Bruton, C. J., Weiss, B., and Schlesinger, S. (1976). *J. Virol.* **19**, 1034–1043.
Lachmi, B., Glanville, N., Keränen, S., and Kääriäinen, L. (1975). *J. Virol.* **16**, 1615–1629.
Levin, J. G., Ramseur, J. M., and Grimley, P. M. (1973). *J. Virol.* **12**, 1401–1406.
Logan, K. B. (1979). *J. Virol.* **30**, 38–44.
Perrault, J., and Holland, J. (1972). *Virology* **50**, 148–158.
Schlesinger, S., Schlesinger, M., and Burge, V. W. (1972). *Virology* **48**, 615–617.
Schwöbel, W., and Ahl, R. (1972). *Arch. Gesamte Virusforsch.* **38**, 1–10.
Shenk, T. E., and Stollar, V. (1972). *Biochem. Biophys. Res. Commun.* **49**, 60–67.
Shenk, T. E., and Stollar, V. (1973a). *Virology* **53**, 162–173.
Shenk, T. E., and Stollar, V. (1973b). *Virology* **55**, 530–534.
Simmons, D. T., and Strauss, J. H. (1972). *J. Mol. Biol.* **71**, 615–631.
Stampfer, M., Baltimore, D., and Huang, A. S. (1971). *J. Virol.* **7**, 409–411.
Stark, C., and Kennedy, S. I. T. (1978). *Virology* **89**, 285–299.
Stollar, B. D., and Stollar, V. (1970). *Virology* **42**, 276–280.
Stollar, V., and Shenk, T. E. (1973). *J. Virol.* **11**, 592–595.
Stollar, V., Shenk, T. E., and Stollar, B. D. (1972). *Virology* **47**, 122–132.
Stollar, V., Shenk, T. E., Koo, R., Igarashi, A., and Schlesinger, R. W. (1975). *Ann. N.Y. Acad. Sci.* **266**, 214–231.

Stollar, V., Igarashi, A., and Koo, R. (1977). *In* "Microbiology, 1977" (D. Schlessinger, ed.), pp. 456–461. Am. Soc. Microbiol., Washington, D.C.
Von Magnus, P. (1954). *Adv. Virus Res.* **2,** 59–79.
Weiss, B., and Schlesinger, S. (1973). *J. Virol.* **12,** 862–871.
Weiss, B., Goran, D., Cancedda, R., and Schlesinger, S. (1974). *J. Virol.* **14,** 1189–1198.
Wengler, G., and Wengler, G. (1976). *Virology* **73,** 190–199.
Woodward, C. G., and Smith, H. (1975). *Br. J. Exp. Pathol.* **56,** 363–372.
Woodward, C. G., Marshall, I. D., and Smith, H. (1978). *Br. J. Exp. Pathol.* **58,** 616–624.

16

Effects of Alphaviruses on Host Cell Macromolecular Synthesis

GERD WENGLER

I. INTRODUCTION

This chapter will deal exclusively with alterations induced by primary alphavirus infection in tissue culture cells growing *in vitro*.

The replication of alphaviruses is supported by vertebrate and arthropod cell cultures. The outcome of infection of permissive vertebrate cells with alphaviruses typically is the production of infectious virus followed by cell death, whereas in permissive arthropod cells infectious progeny virus is synthesized without detectable inhibition of host cell multiplication (see Chapter 20). This dichotomy between the cytolytic infection of vertebrate cells and the noncytolytic infection of arthropod cells is not absolute, since it is possible to establish chronically infected vertebrate cell cultures from cells surviving primary infection (Schwöbel and Ahl, 1972; Zhdanov, 1975); conversely, replication of alphaviruses in certain clones of *Aedes albopictus* cells leads to cell death (Sarver and Stollar, 1977). In view of the very

THE TOGAVIRUSES

ISBN 0-12-625380-3

limited data available, no attempt will be made to describe the effect on host cell macromolecular synthesis of cytolytic infection of arthropod cells or of an established chronic infection of either arthropod or vertebrate cells. The two typical virus–cell interactions, cytolytic infection of permissive vertebrate cells and noncytolytic infection of permissive arthropod cells, will be dealt with separately in this chapter.

II. THE EFFECTS OF LYTIC INFECTION OF VERTEBRATE CELL CULTURES ON THE MACROMOLECULAR SYNTHESIS OF THE HOST CELL

Experimental analyses have been done almost exclusively using Sindbis virus, Semliki Forest (SF) virus, or mutants derived from these viruses, and the continuous hamster tissue culture cell line BHK-21 or primary chick embryo fibroblasts (CEFs). The data indicate many similarities among these four virus–cell systems; therefore an attempt is made in the following discussion to develop a unifying concept of the processes involved in the virus-induced interference with host cell macromolecular synthesis. On the other hand, it will become clear that the type and strain of virus and some cell species-specific quantitative variables of host cell metabolism may well have an important influence on the extent and time course of the expression of virus-induced alterations.

A. Effects on Cellular Protein Synthesis

Two phenomena have been observed in almost all studies on vertebrate cells productively infected with alphaviruses (Fig. 1a): (1) the development of an overall inhibition of protein synthesis as measured by a reduction in incorporation of radioactive amino acids into protein, in some studies complemented by analysis of the specific radioactivity of the intracellular amino acids and by taking into account the amino acid composition of the virus-specific and cellular proteins (Pfefferkorn and Clifford, 1964; Lust, 1966; Strauss *et al.*, 1969; Mussgay *et al.*, 1970; Wengler and Wengler, 1976); (2) an increase in the relative amounts of virus-specific versus cellular polypeptides synthesized; at the end of the exponential phase of virus growth the majority of the newly synthesized proteins are virus-specific (Friedman, 1968; Hay *et al.*, 1968; Strauss *et al.*, 1969; Schlesinger and Schlesinger, 1973).

Experimental data concerning the number of ribosomes actively involved in synthesizing protein in infected cells are much less abundant than those concerning points (1) and (2). Available data (Kennedy, 1972; Söderlund *et al.*, 1973–1974; Wengler and Wengler, 1976) suggest that the overall inhibition of protein synthesis is not reflected by a corresponding decrease in the number of ribosomes actively engaged in protein synthesis, as measured by analysis of sucrose density gradient profiles of the polyribosomes isolated from infected cells (Fig. 1b). That the polysomal ribosomes are actively translocating along the mRNA can be determined by analyzing the effect of inhibition of initiation of protein synthesis on the integrity of the polyribosomes: Under these conditions polyribosomes engaged in protein synthesis are converted to monomeric ribosomes and messenger ribonucleoprotein particles, whereas static, inactive polyribosomes remain unaltered. A transient decrease in polyribosomes early after infection also has not been detected. Pulse-labeling experiments have shown that the half-lives of most newly synthesized cellular proteins are not drastically altered in infected cells.

The data cited constitute indirect evidence for the conclusion that the majority of cellular mRNA in the polyribosomes is replaced by viral mRNA during virus multiplication. Since the viral structural polypeptides and their precursors are the major species of virus-specific proteins synthesized (except for the earliest times after infection), the 26 S virus-specific RNA which functions as mRNA for all viral structural proteins (see Chapter 13) should be the predominant species associated with polyribosomes late in infection. Two independent types of experiments have furnished direct evidence in support of this conclusion: (1) Polyribosomes have been isolated from infected cells at a time when the majority of newly synthesized proteins are virus-specific, and the relative amounts of 28 S rRNA, 18 S rRNA, and 26 S virus-specific RNA present in these polyribosomes have been determined (Wengler and Wengler, 1976). These analyses have lead to the conclusion that about 60% of the polyribosomes contain 26 S RNA as mRNA. (2) Using a ^{32}P equilibrium labeling technique, Tuomi *et al.* (1975) determined the number of ribosomes present in HeLa cells and BHK-21 cells and the amount of virus-specific RNA accumulating in these cells after infection with SF virus. They concluded that in a BHK-21 cell (which contains about 3×10^6 ribosomes) about 2×10^5 molecules of virus-specific 26 S RNA accumulated during virus multiplication, and that this number was comparable to the number of cellular mRNA molecules present in an uninfected cell. Since the 26 S RNA is almost quantitatively bound to

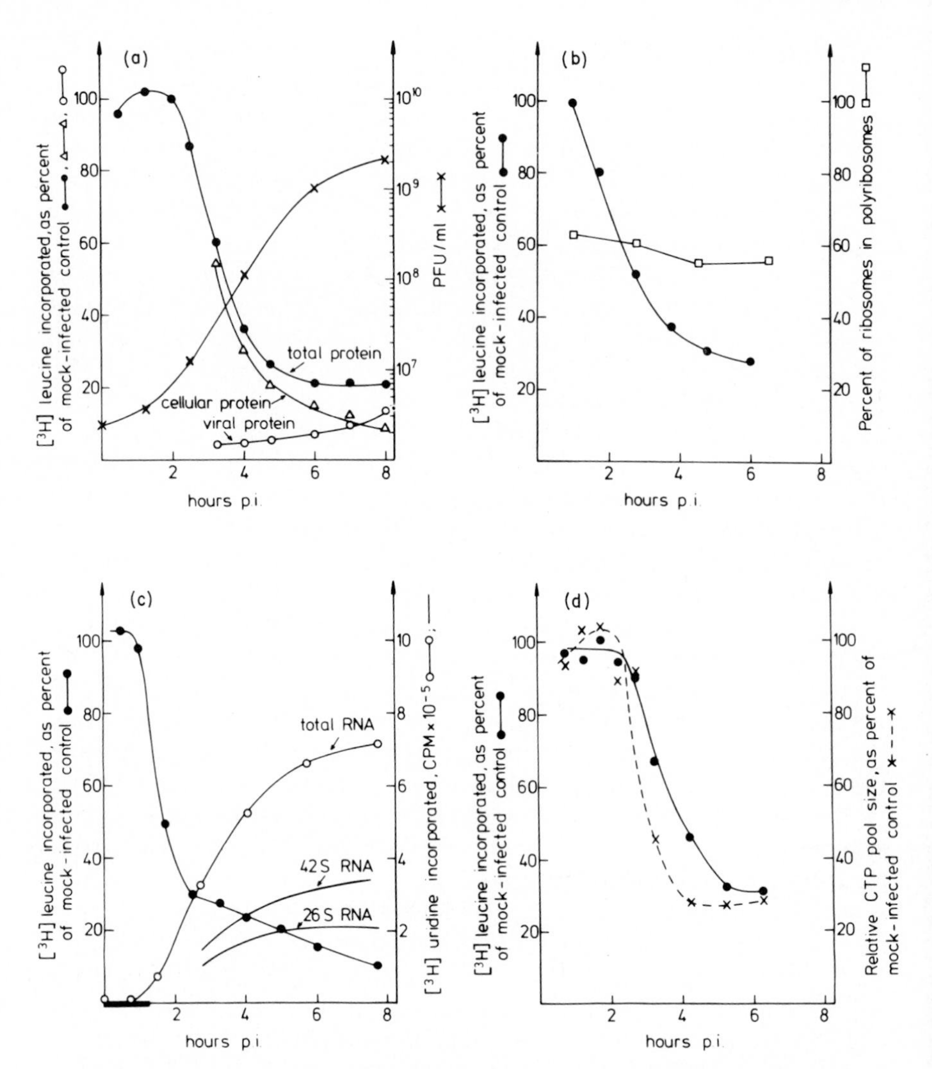

Fig. 1. Alterations induced in BHK cells by infection with SFV. Cells were infected at a multiplicity of infection of about 20 PFU per cell. Zero hours p.i. corresponds to the end of the 30 min virus adsorption period. a, b, and c from Wengler and Wengler (1976). The data of d are unpublished results obtained in our laboratory. In all experiments the overall inhibition of cellular protein synthesis, as determined by measuring the incorporation of [^{3}H]leucine into protein during 30 min labeling intervals is expressed as percentage of the radioactivity incorporated in infected relative to mock-infected cells. (a) Protein synthesis in infected cells. Cells were pulse labeled for 30 min periods with radioactive leucine at various times after infection. The relative amounts of virus-specific and cellular proteins were determined by difference analyses of PAGE profiles. (b) Polyribosome content of infected cells. Data were obtained by

polyribosomes and does not exist to any significant extent in a non-polyribosome-associated state (Mowshowitz, 1973), these data show that most cellular mRNA is replaced by viral mRNA in the polyribosomes. Furthermore, these data are compatible with the finding that no drastic reduction in the number of polyribosomes occurs during infection.

The molecular basis of the substitution of virus-specific 26 S RNA for cellular mRNA in the polyribosome is not known. The most straightforward mechanism by which this phenomenon could be brought about would be an especially high affinity for initiation of translation of the 26 S RNA, as compared to the majority of cellular mRNA. Increased osmolarity of the growth medium inhibits the initiation of protein synthesis *in vivo* (Wengler and Wengler, 1972). Nuss *et al.* (1975) have observed that different species of mRNA differ in their sensitivity to this type of inhibition of initiation, and they have suggested that these differences reflect different affinities of these mRNA species for translation *in vivo*. Experiments on the influence of osmolarity on the translation of viral and cellular mRNA in alphavirus-infected cells and comparative studies on the translation of viral 26 S RNA and cellular mRNA *in vitro* under conditions of mutual competition could help in analyzing the relative affinities of viral and cellular mRNA for initiation of translation. Nucleotide sequence analysis of the 5′-terminal region of the 26 S RNA will allow the determination of the sequence complementarity of this RNA segment to the 3′-terminal region of 18 S rRNA. A possible role of such complementarity in initiation of translation of eucaryotic mRNA has been

---

sucrose density gradient fractionation of cell lysates. (c) Synthesis of RNA in infected cells. [^{3}H]Uridine was added to the cells at 0 hr p.i. (in the presence of 1 μg/ml actinomycin D) and the radioactivity incorporated into acid precipitable material determined at various times after infection. Negligible amounts of radioactivity were incorporated in mock-infected cells. The amount of [^{3}H]uridine incorporated into 42 S and 26 S RNA, respectively, was determined after sucrose density gradient fractionation. The time interval during which inhibition of protein synthesis interferes with viral RNA synthesis is indicated by a bar on the abscissa. (d) Size of the cytidinetriphosphate pool in infected cells. Cells were grown in the presence of [^{32}P]orthophosphate (10 μCi/ml) for at least five cell doubling times prior to infection. Mock-infected and infected cells were then further incubated in growth medium containing [^{32}P]-orthophosphate. Nucleoside triphosphates were extracted at various times from cells with perchloric acid and fractionated as described by Cashel *et al.* (1969). The pool size of each triphosphate is determined as the radioactivity present in the triphosphate recovered from infected cells, expressed as percentage of the radioactivity recovered in the same triphosphate from mock-infected cells. The data concerning the CTP pool are given in this figure. Similar results were obtained for the other three ribonucleoside triphosphates (data not shown).

suggested for a number of different mRNA molecules (see Kozak, 1978, for a review).

The molecular basis for overall inhibition of protein synthesis in infected cells is also not known. In this context studies on the multiplication of temperature-sensitive (*ts*) mutants of alphaviruses are of great importance (Burge and Pfefferkorn, 1966; Tan *et al.*, 1969; Keränen and Kääriäinen, 1974; Strauss *et al.*, 1976; see also Chapter 14 in this volume). Three types of mutants have been described: RNA^- mutants, which are unable to synthesize detectable amounts of virus-specific single-stranded RNA (ssRNA) in cells infected and incubated at the nonpermissive temperature, and $RNA^{\pm}$ and RNA^+ mutants which, under nonpermissive conditions, produce about 10–20% and 20–100%, respectively, of the amount of virus-specific ssRNA synthesized in cells infected with wild-type virus. RNA^+ mutants are defective in the synthesis of viral structural proteins. The viral RNA polymerase synthesized at the permissive temperature in cells infected with the RNA^- mutants described so far continues to function after a shift-up to the nonpermissive temperature (Burge and Pfefferkorn, 1966; Tan *et al.*, 1969; Keränen and Kääriäinen, 1974).

In cells infected with RNA^- mutants no inhibition of overall protein synthesis is observed at the nonpermissive temperature (Keränen and Kääriäinen, 1975). This finding allows the important conclusion that the molecular constituents of the infecting virus particles are not directly responsible for overall inhibition of protein synthesis. Alternative mechanisms might be (1) inhibition of initiation of mRNA translation by double-stranded RNA (dsRNA), (2) lowering of the rate or mass of protein synthesis resulting from the substitution of viral 26 S RNA for cellular mRNA, and (3) a reduction in intracellular ribonucleoside triphosphate pools resulting from viral RNA synthesis. Discussion of these three alternatives follows.

1. *Inhibition by dsRNA*

Ehrenfeld and Hunt (1971) have shown that poliovirus dsRNA inhibits the initiation of protein synthesis in rabbit reticulocyte lysates. If a similar inhibition were to develop in cells in the course of alphavirus infection, it should be fully expressed when the total amount of virus-specific double-stranded and complex RNA has been synthesized. In SF virus-infected BHK-21 cells RNA of negative polarity is synthesized only early after infection, before the start of the phase of exponential viral multiplication (Bruton and Kennedy, 1975).

Normal accumulation of virus-specific ssRNA occurs in alphavirus-infected cells in which protein synthesis is inhibited after the beginning of the phase of linear accumulation of virus-specific RNA (Scheele and Pfefferkorn, 1969; Wengler and Wengler, 1975). It has been concluded from these results that all viral polymerase–template complexes and dsRNA molecules accumulating in infected cells should be present at this time. The time of accumulation of virus-specific dsRNA, as estimated from these experiments, is indicated by a horizontal bar on the abscissa in Fig. 1c. However, the figure also shows that the inhibition of overall protein synthetic activity develops after the synthesis of virus-specific dsRNA, i.e., during the phase of linear accumulation of virus-specific ssRNA and exponential virus multiplication (Strauss *et al.*, 1969; Wengler and Wengler, 1976). These results indicate that the depression of overall protein synthetic activity in alphavirus-infected cells is not due to the inhibition of initiation of translation by the viral dsRNA accumulating in infected cells. Evidence supporting this conclusion also comes from studies by Shenk and Stollar (1972) who have shown that initiation of protein synthesis in cell-free lysates of CEFs is not inhibited by virus-specific dsRNA isolated from Sindbis virus-infected cells.

2. *Substitution of Viral for Cellular mRNA*

A 26 S viral mRNA with a high affinity for initiation but a slow rate of translation could possibly explain both the mRNA substitution and a reduction in the rate of overall protein synthesis. The primary translation product of the 26 S RNA is proteolytically cleaved before the 26 S RNA molecule is fully translated (see Chapter 13), and the viral envelope proteins are glycosylated before they are released from the polyribosomes (Sefton, 1977). It seems reasonable to assume that these complex steps may retard the translation of 26 S viral mRNA compared to that of the average cellular mRNA. Unfortunately, no measurements of the time necessary to translate a 26 S RNA molecule *in vivo* have been published. The data presented in Fig. 1a, however, show that in SF virus-infected BHK-21 cells the inhibition of overall protein synthetic activity develops much more rapidly than the substitution of virus-specific for cellular protein synthesis, being almost fully expressed when the majority of newly synthesized proteins are still of cellular origin.

3. *Reduction of Ribonucleoside Triphosphate Pools*

The kinetic relationships between the accumulation of viral ds- and ssRNA and the development of inhibition of overall protein synthesis (Fig. 1c) can be formulated as follows: The protein synthetic activity rapidly declines when all virus-specific RNA molecules of negative polarity have been synthesized and all viral polymerase–template complexes are actively synthesizing virus-specific 42 and 26 S RNA. In SF virus-infected BHK-21 cells this relationship has been observed during virus multiplication at 37°, 30°, and 27°C (Wengler and Wengler, 1976). About 2×10^5 molecules of each of both species of virus-specific ssRNA (42 and 26 S) accumulate in a single SF virus-infected BHK-21 cell which contains about 3×10^6 ribosomes (Tuomi *et al.*, 1975). These virus-specific RNA molecules are synthesized during the 5 h of linear accumulation of virus-specific RNA in the infected cell, whereas 3×10^6 molecules of 28 and 18 S rRNA are synthesized during the 20-h cell generation time in the uninfected or mock-infected cell. Thus, in the infected cell a cytoplasmic viral RNA-synthesizing apparatus is built up whose activity is comparable to that of the nucleolar rRNA-synthesizing capacity of the uninfected cell. These considerations have led to the suggestion that the consumption of ribonucleoside triphosphates by viral RNA polymerase might reduce the concentration of these triphosphates in the cytoplasm and thereby inhibit the overall protein synthetic capacity of the infected cell (Wengler and Wengler, 1976). Hammer *et al.* (1976) have in fact shown that, at 5 h after infection, the intracellular concentration of UTP, CTP, and GTP in SFV-infected BHK-21 cells and CEFs is reduced to approximately one-third of the value found in mock-infected cells. Our analyses of the time course of this process have shown that this pool size reduction and the overall inhibition of cellular protein synthesis occur in parallel in SFV-infected BHK-21 cells (Fig. 1d). It will be of great interest to see whether it is possible, under suitable conditions, to increase the ribonucleoside triphosphate pools of infected cells and thereby stimulate their protein synthetic capacity.

Recently Koizumi *et al.* (1979) have detected a unique nucleoside triphosphate phosphohydrolase which could not be found in uninfected cells, in BHK cells infected with the alphavirus Western equine encephalitis virus. The enzyme did release inorganic phosphate from all eight major ribonucleoside—and deoxyribonucleoside triphosphates *in vitro*. Obviously this enzyme might be responsible for the reduction of the ribonucleoside triphosphate pool sizes observed in the experiments reported above. The role of the enzyme during virus replication is not known.

In none of the experimental analyses of the molecular biology of the replication of alphaviruses has evidence been obtained for the existence of a viral function specifically aimed at altering the cellular protein synthetic apparatus. In the above evaluation of the processes possibly involved in the substitution of viral for cellular mRNA and in virus-induced inhibition of overall protein synthesis the assumption has been made that no single such function exists in alphavirus-infected cells. More likely, multiple functions act cooperatively to bring about these changes.

Moreover, the profundity of protein inhibition may well depend on the type and strain of virus and on the host cell. For example, the relative affinities of the viral RNA polymerase and the cellular protein-synthesizing system for ribonucleoside triphosphates or the ability of the host cell to synthesize the triphosphates are parameters which may well be determined by such variables.

B. Effects on Cellular RNA Synthesis

The amount of radioactivity incorporated into RNA from nucleosides during 1-h labeling intervals is not significantly different in mock-infected CEFs and in CEFs infected with SF virus up to the end of the exponential phase of viral growth (Taylor, 1965). At this time, between 70 and 90% of the newly synthesized RNA is virus-specific, since its synthesis has been found to be actinomycin D-resistant. It has been concluded from these results that virus replication leads to a significant reduction in host cell RNA synthesis and that this reduction occurs concomitantly with the increase in the synthesis of virus-specific RNA in the infected cell (Taylor, 1965).

Pretreatment of CEFs with interferon drastically reduces the amount of SF virus produced after infection, and no inhibition of cellular RNA synthesis is detectable under these conditions (Taylor, 1965). Infection of cells with RNA$^-$ *ts* mutants under nonpermissive conditions also does not lead to an inhibition of cellular RNA synthesis (Tan *et al.*, 1969). These data show that the molecular constituents of the infecting virus particles are not directly responsible for the inhibition of cellular RNA synthesis.

No direct experimental data are currently available which would allow one to specify exactly the effects of infection on the synthesis of the different types of cellular RNA; i.e., rRNA and its precursors (pre-rRNA), mRNA, heterogeneous nuclear RNA, tRNA and its precursor, low-molecular-weight nuclear RNA, mitochondrial RNA, and other RNA species. However, since in alphavirus-infected vertebrate cells a viral RNA-synthesizing apparatus is built up whose activity is comparable to that of the rRNA-synthesizing capacity of the uninfected cell (see above), the concomitant synthesis of virus-specific RNA and rRNA in alphavirus-infected cells would lead to almost a doubling of the amount of long-lived RNA synthesized in infected cells as compared to uninfected cells. The experiments of Taylor discussed above, therefore, strongly suggest that the accumulation of rRNA is inhibited in alphavirus-infected cells.

Inhibition of protein synthesis leads to a cessation of the appearance of newly synthesized ribosomes in the cytoplasm of eukaryotic cells (for review, see Hadjiolov and Nikolaev, 1976). The molecular basis of this phenomenon has not been characterized in detail, but it has been shown that in some cells it is the synthesis of the ribosomal precursor RNA which is preferentially inhibited, whereas in others the processing of this RNA into mature rRNA is blocked. It has been indicated above that in alphavirus-infected cells the inhibition of cellular RNA synthesis occurs in parallel with the synthesis of virus-specific RNA. Since the virus-specific 26 and 42 S RNA molecules both accumulate in infected cells with a similar time course and in roughly equimolar amounts, and since the 26 S RNA is almost exclusively present in polyribosomes (see above), it is concluded that the inhibition of cellular RNA synthesis coincides with the substitution of viral 26 S for cellular mRNA in the polyribosomes. This leads to the hypothesis that the suppression of host cell protein synthesis caused by this mRNA substitution might lead to an inhibition of rRNA synthesis measured as inhibition of overall host cell RNA synthesis in the experiments described above. Two expectations follow from this hypothesis: (1) The virus-induced inhibition of cellular protein synthesis should precede the inhibition of RNA synthesis. (2) The type of abnormality introduced by alphavirus infection into the synthesis and/or processing of ribosomal precursor RNA in a certain cell line should also be obtained in uninfected cells of this cell line by treatment with an inhibitor of protein synthesis. Although both expectations are susceptible to experimental analyses, none have been published. However, the data on the time course of inhibition of cellular RNA synthesis and its dependence on the expression of viral functions described above are compatible with the first expectation. The experimental test of the second is complicated by the fact that the nature of the drug used to inhibit protein synthesis would influence the resulting alteration of cellular rRNA synthesis, as exemplified by the differences observed in the inhibition of rRNA synthesis induced in HeLa cells by puromycin or cycloheximide (Soeiro *et al.*, 1968; Warner *et al.*, 1966). The inhibition of cellular protein synthesis by puromycin (which permits continued synthesis of peptide bonds) seems to resemble the virus-induced inhibition of cellular protein synthesis more than the inhibition caused by cycloheximide. No comparative analysis of the effect of protein synthesis inhibitors and alphavirus infection on the synthesis and processing of pre-rRNA has been published. Our unpublished data indicate that 45 S pre-rRNA continues to be synthesized in SF virus-infected BHK-21 cells during the phase of linear accumulation of virus-specific RNA; on the other hand, mature rRNA does not accumulate under these conditions, and a similar alteration of pre-rRNA processing is found in mock-infected BHK-21 cells treated with puromycin. The available experimental data are therefore compatible with the hypothesis that the inhibition of cellular RNA synthesis is secondary to virus-induced inhibition of cellular protein synthesis.

A fundamentally different mechanism by which alphavirus replication could inhibit cellular RNA synthesis involves the synthesis of a virus-specific protein or polypeptide able to interfere specifically with the synthesis of one or more species of cellular RNA. No direct evidence has been obtained for the existence of such a molecule in the studies using wild-type or mutant alphaviruses. On the other hand, the effect of a defect in such a putative function on virus multiplication is unknown, and no procedure exists for the selection of viruses defective in such a function.

C. Effects on DNA Synthesis

Only a rather limited number of experimental analyses concerning this aspect of the virus-induced alterations of host cell macromolecular synthesis are available, and therefore this point cannot be discussed in great detail. Obviously the questions that could be asked and the experiments that could be done are analogous to those discussed above for the virus-induced alterations of host cell RNA synthesis. In a study of the effect of infection by Sindbis virus *ts* mutants on host cell protein and DNA synthesis, Atkins (1976) has shown that the drastic inhibition of cellular DNA synthesis observed in BHK cells infected by wild-type Sindbis virus can also be detected in mutant virus-infected cells in the absence of overall inhibition of protein synthesis. Recently, Koizumi *et al.* (1979) have isolated from BHK cells infected with Western equine encephalitis (WEE) virus a factor that inhibits DNA synthesis *in vitro.* This factor could not be detected in uninfected cells. Characterization of the factor has indicated that it is a nucleoside triphosphate phosphohydrolase which releases inorganic phosphate from all eight major ribonucleoside, and deoxyribonucleoside triphosphates *in vitro* (Koizumi *et al.*, 1979). The possible roles of this enzyme in the virus-induced inhibition of cellular DNA synthesis *in vivo* and in the replication of alphaviruses in general remain to be determined.

III. EFFECTS OF NONCYTOLYTIC INFECTION OF ARTHROPOD CELL CULTURES ON MACROMOLECULAR SYNTHESIS OF HOST CELL

Alphaviruses are able to replicate in a variety of arthropod tissue culture cell lines (see Weiss, 1971, for a collection of pertinent papers). Since the initial studies of these virus–cell systems, it has been observed that generally no overt cytopathic effect is manifest and that cell multiplication continues at similar rates in uninfected and infected cultures (Rehacek, 1968; Peleg, 1969; Buckley, 1969; Stevens, 1970; Yunker, 1971; Singh, 1972; Davey and Dalgarno, 1974; Shenk *et al.*, 1974; Yunker and Cory, 1975). Variable results have been obtained in analyses of the percentage of infected cells in such systems, but it has been concluded that at least in some of these systems more than 50% of the cells are infected (Davey and Dalgarno, 1974). No studies on the effect of virus infection on host cell macromolecular synthesis similar to those described above for vertebrate cells have been done in arthropod cell cultures, nor have the quantitative aspects of the accumulation of virus-specific macromolecules discussed above been analyzed in such cells. Therefore a detailed comparison of the effects of alphavirus infection on the macromolecular synthesis of vertebrate and arthropod cells is not possible.

Studies on the development of alphaviruses in cultured *A. albopictus* mosquito cells (Singh, 1967) by electron microscopy (Raghow *et al.*, 1973a,b; Gliedman *et al.*, 1975) have shown the presence in infected cells of complex vesicular structures containing viral capsids and mature virions. The role of these structures in the production of mature extracellular virus is not fully understood. Free nucleocapsids were detected only rarely in the cytoplasm of infected arthropod cells. Gliedman *et al.* (1975) therefore have suggested that virus assembly is sequestered in infected mosquito cells and that possibly also a sequestering of the synthesis of virus-specific macromolecules occurs in alphavirus-infected arthropod cells, which enables these cells to survive infection.

If, however, such a sequestering does not generally occur in alphavirus-infected arthropod cells, one may infer that a stable substitution of virus-specific mRNA for the majority of cellular mRNA in the polyribosomes, as described above for alphavirus-infected vertebrate cells, in all likelihood will not occur in infected arthropod cells, since it would presumably be incompatible with undiminished cell survival and division. Whether the absence of such mRNA substitution is a consequence of the synthesis in infected arthropod cells of rather limited amounts of virus-specific mRNA, of an instability of mRNA in these cells, or of a low affinity of the mRNA for the initiation of translation remains to be elucidated.

Quantitative differences between the virus cell systems might play an important role in determination of the fate of the infected host cell. Differences in the concentrations of intracellular ions in different host cell systems might, for example, influence the interaction of the viral core protein with the viral 42 S RNA and thereby the assembly of the viral ribonucleoprotein. Since the 42 S RNA presumably functions as mRNA for the viral RNA polymerase as long as it is not incorporated into the viral core ribonucleoprotein (see Chapter 13), such differences could have profound influences on the synthesis of virus-specific RNA polymerase and on the outcome of virus infection.

In view of the extensive involvement of host cell functions in the synthesis of viral structural and nonstructural proteins (see Chapter 13) and the modification and possibly also the synthesis of the virus-specific nucleic acids (see Chapter 12), it has to be expected that qualitative differences will be present between the virus-specific macromolecules synthesized in vertebrate and arthropod cells. The absence of sialic acid, which is present in vertebrate cell-grown Sindbis virus, in the glycoproteins of Sindbis virus grown in *A. albopictus* cells is an example of this type of difference (Stollar *et al.*, 1976; Luukkonen *et al.*, 1977). The experimental data currently available do not allow evaluation of the role of such differences in the ability of arthropod cells to survive infection.

More studies on the qualitative and quantitative aspects of alphavirus replication in arthropod tissue culture cells will be necessary before the outline of the molecular basis of the observation that replication of these viruses in a variety of arthropod tissue culture cell lines does not interfere with host cell multiplication is revealed.

Note Added in Proof. Recently, two important contributions by Garry *et al.* [Garry, R. F., Bishop, J. M., Parker, S., Westbrook, K., Lewis, G., and Waite, M. R. F. (1979) *Virology* **96,** 108; Garry, R. F., Westbrook, K., and Waite, M. R. F. (1979) *Virology* **99,** 179] have appeared in which it is shown that Sindbis virus infection of chick cells leads to changes in the intracellular concentrations of Na^+ and K^+. The data are interpreted by the authors to indicate that the virus-induced selective inhibition of host cell protein synthesis results from an ability of viral mRNA to initiate translation under these altered ionic conditions whereas initiation of translation of the great majority of cellular mRNA is drastically inhibited by these alterations.

REFERENCES

Atkins, G. J. (1976). *Virology* **71,** 593.

Bruton, C. J., and Kennedy, S. I. T. (1975). *J. Gen. Virol.* **28,** 111.

Buckley, S. M. (1969). *Proc. Soc. Exp. Biol. Med.* **131,** 625.

Burge, B. W., and Pfefferkorn, E. R. (1966). *Virology* **30,** 204.

Cashel, M., Lazzarini, R. A., and Kalbacher, B. (1969). *J. Chromatogr.* **40**, 103.
Davey, M. W., and Dalgarno, L. (1974). *J. Gen. Virol.* **24**, 453.
Ehrenfeld, E., and Hunt, T. (1971). *Proc. Natl. Acad. Sci. U.S.A.* **68**, 1075.
Friedman, R. M. (1968). *J. Virol.* **2**, 1076.
Gliedman, J. B., Smith, J. F., and Brown, D. T. (1975). *J. Virol.* **16**, 913.
Hadjiolov, A. A., and Nikolaev, N. (1976). *Prog. Biophys. Mol. Biol.* **31**, 95.
Hammer, G., Schwarz, R. T., and Scholtissek, C. (1976). *Virology*, **70**, 238.
Hay, A. J., Skehel, J. J., and Burke, D. C. (1968). *J. Gen. Virol.* **3**, 175.
Kennedy, S. I. T. (1972). *Biochem. Biophys. Res. Commun.* **48**, 1254.
Keränen, S., and Kääriäinen, L. (1974). *Acta Pathol. Microbiol. Scand., Sect. B* **82**, 810.
Keränen, S., and Kääriäinen, L. (1975). *J. Virol.* **16**, 388.
Koizumi, S., Simizu, B., Ishida, I., Oya, A., and Yamada, M. (1979). *Virology* **98**, 439.
Kozak, M. (1978). *Cell* **15**, 1109.
Lust, G. (1966). *J. Bacteriol.* **91**, 1612.
Luukkonen, A., von Bonsdorff, C.-H., and Renkonen, O. (1977). *Virology* **78**, 331.
Mowshowitz, D. (1973). *J. Virol.* **11**, 535.
Mussgay, M., Enzmann, P.-J., and Horst, J. (1970). *Arch. Gesamte Virusforsch.* **31**, 81.
Nuss, D. L., Oppermann, H., and Koch, G. (1975). *Proc. Natl. Acad. Sci. U.S.A.* **72**, 1258.
Peleg, J. (1969). *J. Gen. Virol.* **5**, 463.
Pfefferkorn, E. R., and Clifford, R. L. (1964). *Virology* **23**, 217.
Raghow, R. S., Davey, M. W., and Dalgarno, L. (1973a). *Arch. Gesamte Virusforsch.* **43**, 165.
Raghow, R. S., Grace, T. D. C., Filshie, B. K., Bartley, W., and Dalgarno, L. (1973b). *J. Gen. Virol.* **21**, 109.
Rehacek, J. (1968). *Acta Virol. (Engl. Ed.)* **12**, 241.
Sarver, N., and Stollar, V. (1977). *Virology* **80**, 390.
Scheele, C. M., and Pfefferkorn, E. R. (1969). *J. Virol.* **4**, 117.
Schlesinger, M. J., and Schlesinger, S. (1973). *J. Virol.* **11**, 1013.
Schwöbel, W., and Ahl, R. (1972). *Arch. Gesamte Virusforsch.* **38**, 1.
Sefton, B. M. (1977). *Cell* **10**, 659.
Shenk, T. E., and Stollar, V. (1972). *Biochim. Biophys. Acta* **287**, 501.
Shenk, T. E., Koshelnyk, K. A., and Stollar, V. (1974). *J. Virol.* **13**, 439.
Singh, K. R. P. (1967). *Curr. Sci.* **36**, 506.
Singh, K. R. P. (1972). *Adv. Virus Res.* **17**, 187.
Söderlund, H., Glanville, N., and Kääriäinen, L. (1973–1974). *Intervirology* **2**, 100.
Soeiro, R., Vaughan, M. H., and Darnell, J. E. (1968). *J. Cell Biol.* **36**, 91.
Stevens, T. M. (1970). *Proc. Soc. Exp. Biol. Med.* **134**, 356.
Stollar, V., Stollar, B. D., Koo, R., Harrap, K. H., and Schlesinger, R. W. (1976). *Virology* **69**, 104.
Strauss, J. H., Jr., Burge, B. W., and Darnell, J. E., Jr. (1969). *Virology* **37**, 367.
Strauss, S. G., Lenches, E. M., and Strauss, J. H. (1976). *Virology* **74**, 154.
Tan, K. B., Sambrook, J. F., and Bellett, A. J. D. (1969). *Virology* **38**, 427.
Taylor, J. (1965). *Virology* **25**, 340.
Tuomi, K., Kääriäinen, L., and Söderlund, H. (1975). *Nucleic Acids Res.* **2**, 555.
Warner, J. R., Girard, M., Latham, H., and Darnell, J. E. (1966). *J. Mol. Biol.* **19**, 373.
Weiss, E. (1971). *Curr. Top. Microbiol. Immunol.* **55**, 1.
Wengler, G., and Wengler, G. (1972). *Eur. J. Biochem.* **27**, 162.
Wengler, G., and Wengler, G. (1975). *Virology* **66**, 322.
Wengler, G., and Wengler, G. (1976). *J. Virol.* **17**, 10.
Yunker, C. E. (1971). *Curr. Top. Microbiol. Immunol.* **55**, 113.
Yunker, C. E., and Cory, J. (1975). *Appl. Microbiol.* **29**, 81.
Zhdanov, V. M. (1975). *Nature (London)* **256**, 471.

17

The Assembly of Alphaviruses

DENNIS T. BROWN

I. INTRODUCTION

Alphaviruses are composed of two morphologically distinct entities, an electron-dense nucleocapsid or core and a membranous envelope

THE TOGAVIRUSES

ISBN 0-12-625380-3

which bears spikelike projections on its outer surface (see Chapter 8). The assembly of these two viral structures takes place by separate pathways in infected cells, and the events occurring during the formation of these components are the subject of this chapter. It has been demonstrated in a variety of virus systems (animal, plant, and bacterial) that the process of virus assembly is dependent upon the interaction of viral structural components in a specified and predictable fashion (Casjens and King, 1975). A number of homologous and nonhomologous molecular species may interact with one another in the formation of a virus particle. Protein–protein, protein–nucleic acid, protein–lipid, and nucleic acid–lipid interactions may be essential in the formation of a given virus structure. The interactions of these components result, in most cases, in the consistent production of structurally identical virus particles, and this implies that virus components contain information essential to determination of the form of the completely assembled virion.

The alphavirion represents an ideal system for study of the processes involved in the assembly of a membrane-containing virus. Alphaviruses grow to very high titers in a variety of host cell systems, and their structural components can be readily distinguished from those of the host cell by virtue of their preferential labeling with precursors of protein and RNA synthesis, which can be achieved in the absence of host macromolecular synthesis (see Chapters 12, 13, and 16).

Both the envelope and the nucleocapsid of the alphavirion are structurally uncomplicated (see Chapter 8). The nucleocapsid contains components entirely of viral origin, the viral RNA and multiple copies of a single polypeptide, the capsid protein (C). The viral membrane, in contrast, is a hybrid structure derived from preexisting host cell lipids and structures which are in part the products of viral genetic information (the envelope glycoproteins). The glycoproteins form the spike structures seen on the surface of the virion and are themselves the products of both virus and host genetic information. The primary amino acid sequence of the envelope proteins is determined by the viral RNA. The proteins are glycosylated, however, by existing host cell glycosyltransferases and reflect in their carbohydrates the expression of host cell genetic information. Given these fundamental differences in the genetic origin of the components of viral nucleocapsids and envelopes, it is not surprising that they are, in part, assembled independently of one another.

Based on observations obtained by both morphological and

biochemical investigations, the following general summation regarding the process of alphavirus assembly can be made (Figs. 1 and 2). Viral RNA and capsid protein are aggregated together and form the nucleocapsid in the cytoplasm of the infected cell. The morphologically complete (but not necessarily mature) capsid then migrates to and associates with the cytoplasmic side of host cell membranes which have been modified by the insertion of virus-specified proteins destined to become the glycoproteins of the viral envelope. Envelopment of the nucleocapsid in the modified membrane follows and involves a sequential wrapping of the nucleocapsid in the membrane. Envelopment of the alphavirus nucleocapsid is accompanied by changes in the morphology of the membrane and in the molecular weight and topological arrangement of at least one of the two envelope proteins.

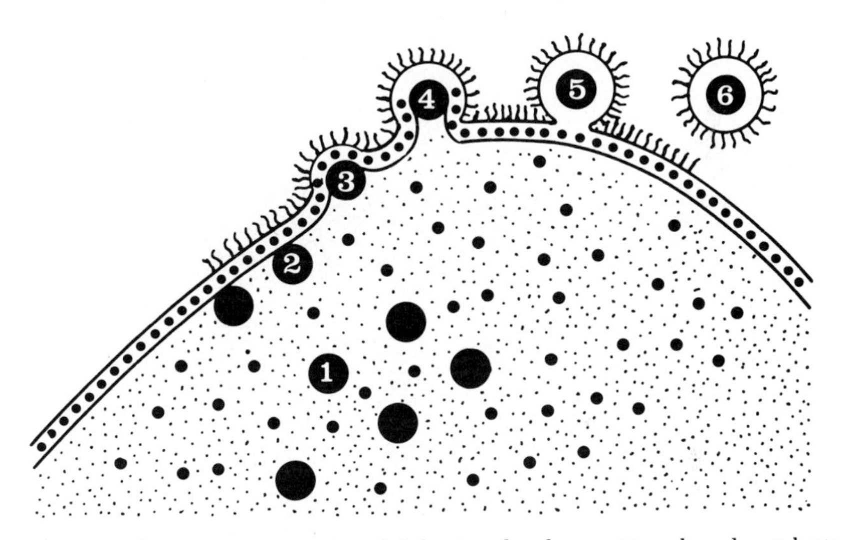

Fig. 1. Schematic representation of alphavirus development in cultured vertebrate cells. (1) Nucleocapsids are assembled in the cytoplasm of the infected cell. (2) Completed nucleocapsids attach to the inner surface of the host plasma membrane which has been altered by the addition of virus glycoproteins. (3 and 4) Envelopment is initiated and progresses as the nucleocapsid becomes wrapped in the modified plasma membrane; as envelopment takes place, the intramembranal particles (resolved by freeze-fracturing of cell membranes; see Figs. 4 and 5) characteristic of cell membrane interiors disappear. (5) Release of the virion is initiated by fusion of one of the two members of the developing envelope bilayer. (6) The fusion of the second (outer) member of the envelope bilayer with itself releases the completed virion into the surrounding environment. The numbers appearing in subsequent electron micrographs point out aspects of alphavirus assembly corresponding to those schematically represented here.

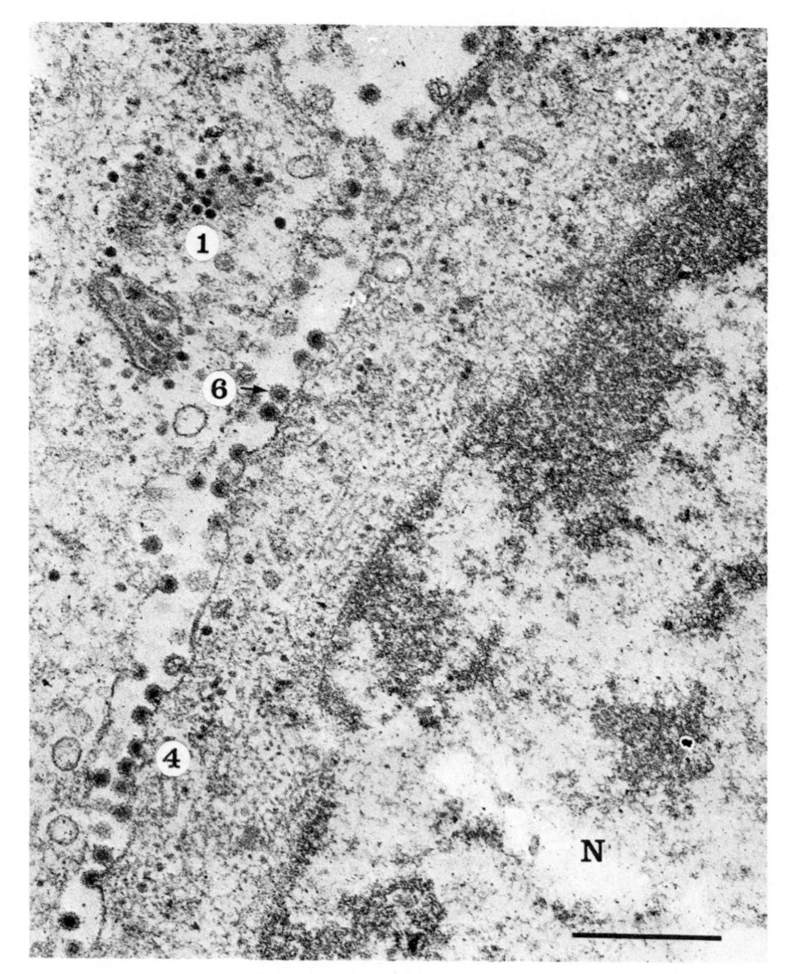

Fig. 2. Electron micrograph of an ultrathin section of a Sindbis virus-infected baby hamster kidney cell. Nucleocapsids are seen as completed structures in the cytoplasm of the cell to the left (1), and numerous virions appear in various stages of envelopment at the plasma membrane of both cells. The membrane of the virion is continuous with the host plasma membrane (4), and spike structures can be seen on the surface of some virions (6). N, Nucleus. Bar is 0.5 μm.

II. ASSEMBLY OF ALPHAVIRUS NUCLEOCAPSIDS

Although the alphavirus nucleocapsid is simple in composition, little is known regarding the process of its formation. The major reason for this lack of information is simply that little experimentation has been specifically directed toward answering questions related to this process. A number of experiments, however, bear indirectly upon this subject and must be taken into consideration in discussing nucleocapsid assembly.

A. Morphology and Form-Determining Functions

Direct observations of alphavirus nucleocapsids by electron microscopy have suggested that this structure has icosahedral symmetry (Brown *et al.*, 1972; Horzinek and Mussgay, 1969; Simpson and Hauser, 1968). Although such conclusions are particularly appealing when implicating a self-assembly mechanism in formation of the nucleocapsid, icosahedral symmetry has not been rigorously proved and much confusion exists concerning the reported organization of viral components in the nucleocapsid (see Chapter 8).

In the absence of a clearly demonstrated symmetric organization of the nucleocapsid, certain observations imply that this structure is not a simple condensation of the protein and RNA. The formation of a nucleocapsid-like condensate in the cytoplasm of the infected cell seems to require the presence of the capsid polypeptide (C) and 42 S plus-strand RNA. Empty capsids are unknown in the alphavirus system, suggesting that capsid protein is unable to form a structural aggregate in the absence of RNA. It thus appears that protein–protein interactions are not the primary structure-forming interactions in nucleocapsid development. Conversely, experiments in which 42 S RNA synthesis is allowed to occur in the absence of capsid protein synthesis do not result in the formation of RNA aggregations with a shape similar to that of capsids (Burge and Pfefferkorn, 1968; Friedman and Grimley, 1969). Thus, both 42 S RNA and capsid protein possess form-determining functions in the assembly of the nucleocapsid. The interaction of the capsid protein with RNA seems to be specific for 42 S plus-strand RNA. In a number of experiments characterizing the composition of cytoplasmic nucleocapsids, ribonucleoprotein complexes containing capsid protein and host RNA were not reported. Nucleocapsid-like structures containing capsid protein and 26 S RNA are also not known, even though this viral RNA species has the same polarity as, and a nucleotide sequence identical to, part of the 42 S viral RNA (see Chapters 11 and 12). It appears that the binding of capsid protein to RNA and the condensation of this complex into a structure which may ultimately be enveloped requires regions of the RNA sequence unique to the 42 S RNA species. The formation of defective interfering virions containing truncated 42 S RNA species suggests that, although certain RNA regions or sequences may be required for the formation of a nucleocapsid, RNA of a specific length is not essential (see Chapter 15; see also S. Schlesinger *et al.*, 1972; Shenk and Stollar, 1973; Weiss and Schlesinger, 1973; Johnston *et al.*, 1975; Guild and Stollar, 1975; Kennedy *et al.*, 1976).

B. Kinetics of Assembly

Interaction of the viral RNA with the capsid protein occurs very rapidly in the cytoplasm of infected cells. Pulse-chase experiments indicate that capsid protein is incorporated into ribonucleoprotein complexes with the sedimentation characteristics of completed nucleocapsids within 5 min of labeling (Söderlund, 1973). No pool of free capsid protein can be detected in the cytoplasm of infected cells, suggesting that the protein is the rate-limiting component in formation of the capsid (Friedman and Grimley, 1969). In this context it is interesting that studies directed toward determining the kinetics of nucleocapsid formation did not detect intermediates in the formation of this structure (Söderlund, 1973); indeed, no partially completed aggregates of capsid protein and RNA of any kind have been reported. The apparent absence of partially completed capsid structures is somewhat difficult to explain, given the observation that capsid protein is present in rate-limiting amounts in the infected cell, and suggests that the limited amount of capsid protein is not uniformly distributed among the large number of 42 S progeny RNA molecules present in the infected cell. It is possible that newly synthesized capsid protein is preferentially incorporated into the capsid complexes which have begun assembly. Formation of a nucleocapsid may require the initial interaction of a few molecules of capsid protein with a particular RNA sequence unique to 42 S plus-strand RNA. This initial interaction might be the most difficult to achieve in terms of thermodynamic considerations. Once established, however, this protein–RNA complex may be stabilized in a configuration favoring rapid addition of the remaining required capsid proteins at a rate which prevents the detection of partially completed structures in either pulse-labeling experiments or in electron microscopy of ultrathin-sectioned infected cells.

C. RNA–Protein Interaction and Implications for Nucleocapsid Symmetry

No information is available regarding the nature of the RNA–protein interactions which take place in formation of the nucleocapsid, nor is it known which portions of the RNA are associated with protein. Chemical analysis suggests that the nucleocapsid contains an average of 300 copies of C protein (Strauss *et al.*, 1969). This is certainly not enough protein to cover the linear 4.2–4.6 μm of viral RNA in the 42 S genome. Furthermore, the completed nucleocapsid has been demonstrated to be partially sensitive to degradation by RNase, suggesting that some portion of the RNA is exposed at the surface of the nucleocapsid (Acheson and Tamm, 1970; Kääriäinen and Söderlund, 1971). In addition, if the proposed models for the budding of the completed nucleocapsids through the plasma membrane (to be described below) are correct, each copy of capsid protein incorporated into the completed nucleocapsid must also be located on the surface of this structure to enable it to interact on a one-to-one basis with the precursors of the viral envelope proteins located in the modified cell membranes. If the nucleocapsid is indeed organized such that all the proteins and much of the RNA are exposed on the surface of the structure, RNA–protein complexes must make up the morphological units forming the icosahedron. Therefore, each of the morphological subunits observed in electron microscopic studies of purified nucleocapsids (Horzinek and Mussgay, 1969; Brown *et al.*, 1972) must contain a discrete number of molecules of capsid protein and a fixed number of folds of RNA.

III. ASSEMBLY OF ALPHAVIRUS MEMBRANES

A. Appearance of Viral Membrane Proteins in Host Cell Membranes

The alphavirus membrane is formed by modification of host cell membranes and incorporation of the altered membrane structures into mature virions. The entire process of virus maturation occurs very rapidly, radioactively labeled amino acids being detected in mature virions within 20 min after their addition to a culture of virus-producing cells (Scheele and Pfefferkorn, 1969). The process of adapting the membranes of the host cell for incorporation into virions is a complex one and seems to proceed more slowly than formation of the nucleocapsid.

Limited information is available regarding the pathway followed by alphavirus envelope proteins from the time they are synthesized until the moment they appear on the surface of the infected cell. This process seems to be initiated as the newly synthesized envelope proteins are transferred from membrane-bound polyribosomes into the membranes of the rough endoplasmic reticulum (ER) (Wirth *et al.*, 1977). Insertion of the viral membrane proteins into the host cell membranes seems to accompany their synthesis. Transfer of the membrane-associated viral proteins to the cell surface requires 20–30 min in a "wild-type" virus infection at 30°C (Brown and Renz, unpublished observations), and pulse-chase labeling experiments in combination with membrane fractionation techniques suggest that the polypeptides pass from the high-density membranes of the rough ER through the medium-density membranes to the plasma membrane (Brown and Renz, unpublished observations). It is not known whether the membranes of the Golgi apparatus are involved in the processing of viral membrane proteins or whether this organelle plays a role in the transport of these proteins in the infected cell.

B. Mutants Defective in Modification of Host Cell Membranes

One group of Sindbis virus temperature-sensitive mutants (complementation group D) appears to be defective in the events required for transportation of the viral proteins to the plasma membrane (Brown and Renz, unpublished observations). Mutant *ts*23 produces the viral membrane protein precursors (PE2 and E1) at nonpermissive temperatures, but neither viral antigens nor viral hemadsorbing activity can be detected on the surface of the mutant virus-infected cells (Burge and Pfefferkorn, 1967, 1968; Bell and Waite, 1977; Smith and Brown, 1977). Membrane fractionation studies have further revealed that the viral membrane proteins E1 and PE2 are trapped in the smooth ER of cells infected with this mutant at the nonpermissive temperature (Erwin and Brown, unpublished observations). Although these smooth ER-associated viral proteins were demonstrated to be glycosylated, it has not been determined whether glycosylation is complete or whether the carbohydrate residues are associated with the same protein sequences as in a wild-type infection. Interestingly, although the defect in this mutant is a single defect in one of the two viral membrane proteins (presumably E1) (Dalrymple *et al.*, 1976; Yin, 1969; see also Chapter 14 in this volume), both E1 and PE2 are trapped in the rough ER and neither migrates to the cell surface (Brown and Renz, unpublished observations). It has also been demonstrated that neither of the viral membrane proteins synthesized by this mutant at the nonpermissive temperature can be transferred efficiently to mature virions upon a shift to the permissive temperature (Brache and Schlesinger, 1976; Smith and Brown, 1977), nor do they appear in the plasma membranes of cells after such a shift (Erwin and Brown, unpublished observations). Thus it appears that (1) the defect in one of the two viral membrane proteins produced at the nonpermissive temperature in the *ts*23 mutant-infected cell system is not reversible upon a shift to the permissive temperature, and (2) the proteins PE2 and E1 are so complexed with one another shortly after insertion into the membranes of the rough ER that a temperature-sensitive defect, in preventing the normal processing and transport of one of the proteins, also blocks processing and transport of the other. The low levels of complementation obtained between mutant *ts*23 and mutants of other complementation groups is further evidence for the early establishment of a relatively stable complex between these two polypeptides (Burge and Pfefferkorn, 1966b; Renz and Brown, 1976).

C. Organization of Viral Membrane Proteins Prior to Envelopment

The exact arrangement of the alphavirus envelope proteins PE2 and E1 in the various cell membranes during processing and transport to the cell plasma membrane is not known. It appears that in wild-type infections as well as in infections with maturation-defective mutants such as *ts*23, the viral polypeptides are glycosylated as soon as they are integrated into the membrane of the rough ER. The presence of carbohydrate on these proteins may imply that they are exposed on the outer surface of the ER membranes (Blobel and Dobberstein, 1975a,b; Kriebich and Sabatini, 1973). Glycosylation plays no role in insertion of the viral membrane proteins (Wirth *et al.*, 1979); however, viral proteins prevented from completing the normal glycosylation process cannot be utilized in the formation of mature virions (Kaluza, 1976; Leavitt *et al.*, 1977).

When inverted vesicles of rough ER containing Sindbis virus glycoproteins are treated with proteolytic enzymes, a 3000-molecular-weight fragment is removed from the PE2 polypeptide (Wirth *et al.*, 1977), suggesting that this protein is exposed on the cytoplasmic as well as on the external side of the membrane and that it must be transmembranal. Polypeptide E1 was not degraded in this study and thus might not possess such a transmembranal configuration.

In another study on the organization of Sindbis virus polypeptide E1 and PE2 in plasma membranes of wild-type and mutant-infected cells, it was further demonstrated that E1 and PE2 were both exposed on the surface such that they could bind antibody prepared against purified E1 or E2 (Brache and Schlesinger, 1976; Jones *et al.*, 1977; Bell and Waite, 1977). Polypeptide E1 possesses, in addition, the same hemagglutinating capability in the plasma membrane of the infected cell as it does in the membrane of the mature virion (Burge and Pfefferkorn, 1968). E1 is also as susceptible to radioiodination by the lactoperoxidase procedure in the plasma membrane as it is in the membrane of the mature virion (Sefton *et al.*, 1973; Smith and Brown, 1977).

In contrast, the configuration of PE2 within the cell membrane seems to differ from that of its cleavage product E2 in the membrane of the mature virion. Polypeptide PE2 cannot be iodinated by the lactoperoxidase procedure, even though it can be demonstrated to interact with antibody on the cell surface (Brache and Schlesinger, 1976; Jones *et al.*, 1977; Bell and Waite, 1977). The presence of PE2 in the plasma membranes of Sindbis virus-infected cells has recently been clearly demonstrated utilizing improved methods of plasma membrane purification. Scheefers and Brown (unpublished observation) have used a technique of plasma membrane purification involving the reisolation from cells of latex spheres which have been ingested by pinocytosis. The membranes associated with the repurified spheres have been shown to be free of contaminating internal membranes, and polyacrylamide gel electrophoresis revealed PE2 to be present in the surface of both wild type and *ts*20 infected cells. After its conversion to E2, the tyrosine residues of the PE2 polypeptide are readily iodinated by this procedure (Sefton *et al.*, 1973; Smith and Brown, 1977). Treatment of virus-infected cells with low concentrations of nonionic detergents further indicate that E2 is readily, in fact quantitatively, released from the cell surface by this noncytolytic procedure (Smith and Brown, 1977). On the other hand, PE2 is not released by this treatment and is presumably more tightly associated or more deeply buried in the plasma membrane than its cleavage product. The E1 polypeptide, also released from the surface of the cell by this treatment, though in much lower quantities than E2, is presumed to be more loosely associated with the membrane than PE2. This conclusion agrees well with those derived from the protease digestions of isolated membrane vesicles described above (Wirth *et al.*, 1977).

D. Distribution of Viral Antigens in Host Cell Plasma Membranes Prior to Envelopment

The appearance of virus membrane proteins on the cell surface occurs as early as 2 h after infection and precedes the appearance of extracellular virus by several hours (Birdwell and Strauss, 1974; see also Fig. 3).

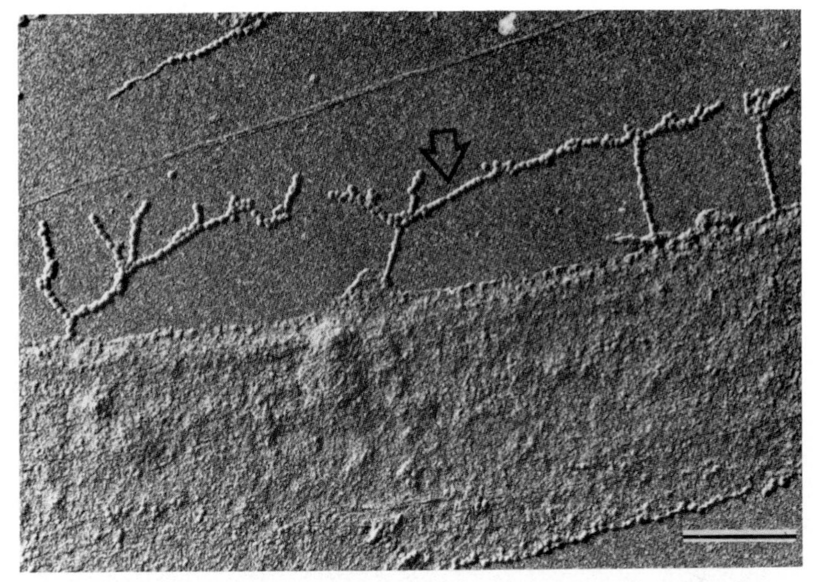

Fig. 3. A surface replica of a chick embryo fibroblast infected with Sindbis virus. The arrow points to a virus-induced extrusion in the cell surface, which contains a large number of partially enveloped nucleocapsids. At the edge of the cell a number of other nucleocapsids can also be seen as bumps under the cell membrane. These nucleocapsids may bud directly through the plasma membrane. Bar is 1 μm. (From Strauss *et al.*, 1977; with permission of Academic Press.)

The observations summarized above suggest that the proteins appearing on the surface of the plasma membranes in preparation for envelopment of the nucleocapsids consist of tightly associated pairs of viral membrane proteins E1 and PE2. The polypeptide PE2 is a transmembranal structure exposed on both the exterior and the cytoplasmic sides of the plasma membrane. Viral membrane protein E1 is also exposed on the surface of the cell plasma membrane in a configuration possibly identical to that in the mature virion.

Studies on the distribution of viral membrane proteins on the surface of cells early after infection suggest that the E1–PE2 complexes are randomly distributed on the cell surface prior to virus production and that their concentration in the surface membranes increases with time (Birdwell *et al.*, 1973; Birdwell and Strauss, 1973, 1974).

IV. ENVELOPMENT OF THE ALPHAVIRUS NUCLEOCAPSID

The assembly of the viral nucleocapsid in the cytoplasm and the appearance of the appropriately processed viral membrane proteins in the plasma membrane of the infected cell provide the prerequisites for envelopment of the nucleocapsid and release of the mature virion. This process is the terminal event in the assembly of the alphavirion; because of electron microscopic descriptions this envelopment process is also referred to as "budding" (see also Chapter 8).

A. Cellular Location of Envelopment

The process of envelopment is initiated as the completed nucleocapsid associates itself with the inner surface of the modified plasma membrane (Acheson and Tamm, 1967; Brown *et al.*, 1972). It is not clear how nucleocapsids are transported from their position in the cell cytoplasm to the plasma membrane. Surprisingly, the electron microscope has revealed that, although the surface of the infected cell is uniformly modified by the presence of viral membrane proteins (Birdwell and Strauss, 1974), the process of budding itself seems to occur in a nonrandom fashion. Virus maturation has been described in one study as occurring in localized areas or patches on the surface of infected cells (Brown *et al.*, 1972; see also Figs. 3–5) and in another study as occurring from virus-induced extrusions in the plasma membrane (Birdwell *et al.*, 1973). A third study has shown that virus maturation can also take place in internally situated vesicles (Grimley *et al.*, 1972). In many instances virions maturing in a particular area of the cell surface were found to be in the same stage of envelopment, suggesting that in such clusters the process of budding might occur with some degree of synchrony (Brown *et al.*, 1972; see also Figs. 3–5). The preferential clustering of virions in equivalent stages of envelopment has also been demonstrated in electron microscopic studies of ultrathin sections employing ferritin-conjugated antiviral antibody (Smith and Brown, 1977). Because of the static nature of the electron microscope in examinations of this type, it is not implied that such patches or processes are the only points on the cell at which virus production will take place; it is very likely that ultimately the entire cell surface will serve as a source of virus production. These studies do, however, imply that, although the surface of the cell contains randomly distributed viral proteins, nucleocapsids are found preferentially associated with restricted areas of the cell surface at any particular time the cells are examined. The morphological localization of virus envelopment, combined with the observation that capsid protein is incorporated into mature virions at the same rate as envelope protein in pulse-chase experiments (Scheele and Pfefferkorn,

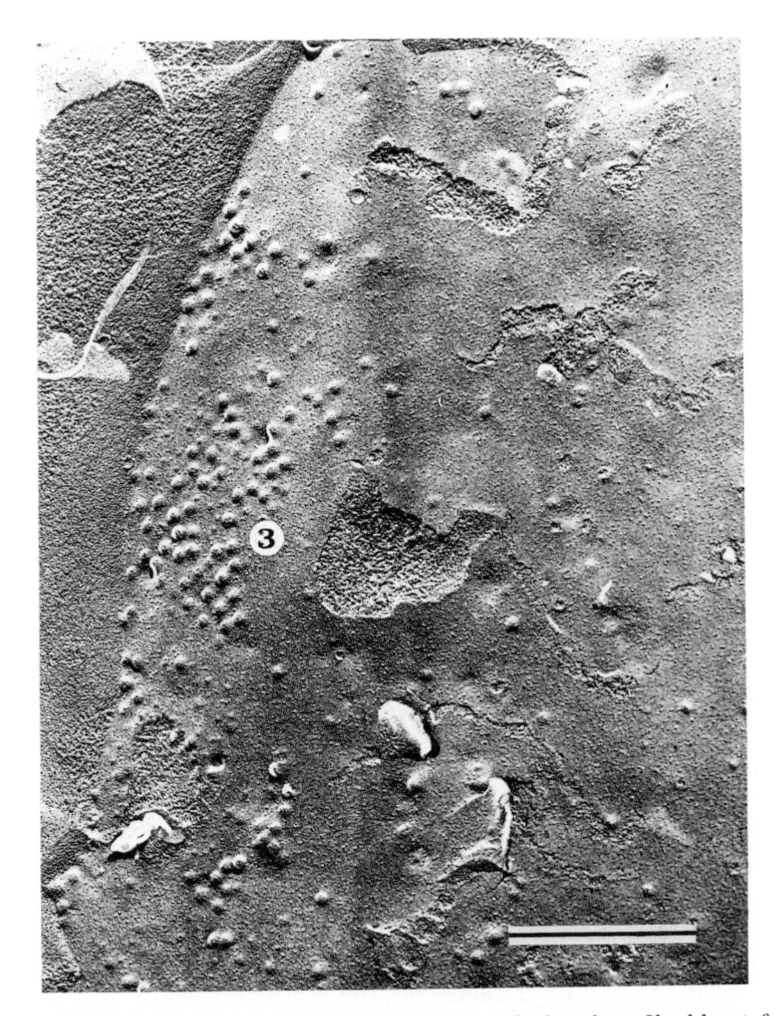

Fig. 4. Electron micrograph of a freeze-fractured chick embryo fibroblast infected with Sindbis virus. The virions are shown at an early stage in envelopment [see (3) in Fig. 1]. The outermost leaflet of the host plasma membrane bilayer has been removed by the fracture, and the view is of the surface of the inner leaflet of the bilayer which faces the cell exterior (protoplasmic fracture face). The cytoplasm of the cell lies underneath this portion of the bilayer. The exposed surface is covered with interior membrane particles (see Fig. 1), and these particles are also seen on the developing virus membranes (3). The developing virions are clustered in the area of the cell to the left, and the virions in this cluster are advanced to the same extent in the envelopment process. Bar is 0.5 μm. (From Brown *et al.*, 1977; with permission of the American Society of Microbiology.)

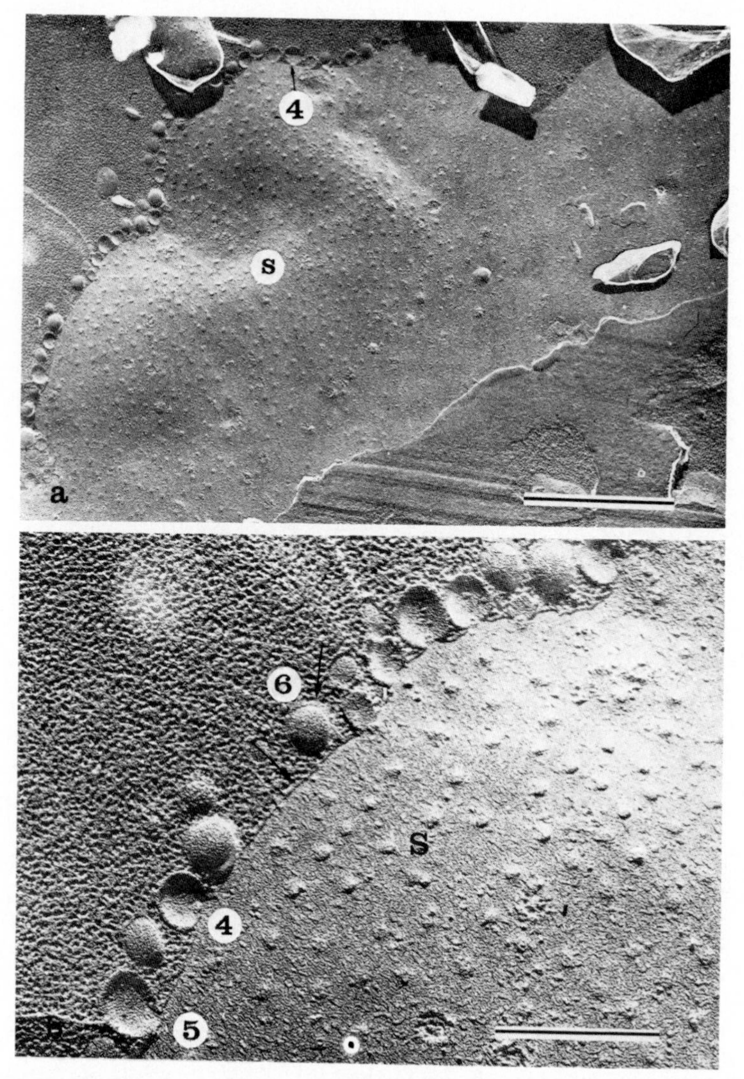

Fig. 5. Electron micrograph of a freeze-fractured chick embryo fibroblast infected with Sindbis virus. The exposed fracture face is similar to that in Fig. 2 (protoplasmic fracture face). (a and b) Low and high magnifications of the same cell surface, respectively. The virions are revealed in (a) to be developing from a localized area of the cell surface as in Fig. 2. These virions are in a more advanced stage of envelopment, and those seen at the edge of the infected cell are attached by only a small part of their envelope to the cell surface [see (4) and (5) in Fig. 1]. The exposed surface of the membrane has numerous large bumps or "stalks" located in one region of the membrane (S). These structures are larger than the interior membrane particles and are frequently separated from each other by a distance roughly equivalent to the center-to-

1969), suggests that movement of the nucleocapsids to the modified plasma membrane may not be a process of simple diffusion.

B. Changes in Membrane Morphology Accompanying Envelopment

Electron microscopy of freeze-fractured purified Sindbis virions and Sindbis virus-infected cells has demonstrated that a change in the morphology of the modified cell plasma membrane occurs as the plasma membrane is converted to the viral envelope during envelopment (Brown *et al.*, 1972; see also Figs. 1, 4, and 5). The membrane of the mature virion differs from the modified plasma membrane from which it is derived in that it lacks the intramembranal particles typically present in the hydrophobic interior of cell membranes. Evidence suggests that these intramembranal particles represent membrane protein elements located within or penetrating into the hydrophobic center of the bilayer (for an explanation of the freeze-etching technique, see Brown *et al.*, 1972). Thus the observation that intramembranal particles are excluded from developing viral membranes might be interpreted as resulting from the exclusion of host proteins from the developing viral envelope, a process which must occur, as chemical analysis of purified virions has shown them to be free of host proteins (Pfefferkorn and Clifford, 1964). The absence of intramembranal structures in the interior of the viral membrane further implies that, although the viral membrane proteins E1 and E2 have hydrophobic

---

center spacing of the virions seen at the edge of the cell. These large particles are possibly the result of removing the outer leaflet of the membrane of the cell with many virions in the final stages of envelopment. These virions are still attached to the cell membrane by the continuity of their envelope with the host plasma membrane [see (4) in Fig. 1]. This configuration produces a distortion in the inner leaflet of the plasma membrane, which is resolved after the outer leaflet is removed by the fracturing process. Some of the virions at the edge of the cell have been fractured such that the nucleocapsid and the inner leaflet of the envelope have been removed, providing a view of the inwardly facing surface of the outer portion of the bilayer [(4) and (5)]. Some of these particles have the stalk at the point of association of the envelope with the membrane (4), suggesting that they are still attached to the cell membrane by both portions of the bilayer. Other virions (5) do not have this structure and are attached by the continuity of only the outer leaflet of the envelope and plasma membrane [see (5) in Fig. 1]. The interiors of the viral membranes are with few exceptions free of the interior membrane particles seen in the host membrane and in the viral envelope during the early phase of envelopment (Fig. 2). In (b) it is possible to see, at the edge of the cell, the cross-fractured outer leaflet of both the host and viral membranes [arrows near (6)]. Bar is 1.0 μm for (a) and 0.2 μm for (b). (From Brown *et al.*, 1972; with permission of the American Society of Microbiology.)

regions which anchor them in the envelope (Garoff and Simons, 1974), this hydrophobic tail does not penetrate into or cross the center of the envelope bilayer. In this regard, the electron microscopic data obtained by freeze-fracture techniques are in agreement with conclusions produced from studies employing x-ray diffraction (Harrison *et al.*, 1971) and electron spin resonance (Sefton and Gaffney, 1974; see also Chapter 10 in this volume).

The interiors of plasma membranes of infected cells were never found to possess areas free of intramembranal particles in the absence of budding (Brown, unpublished observations). This observation probably results from random distribution of the virus membrane proteins among the proteins of the host prior to envelopment, and from the transmembranal nature of PE2. The latter might be expected to create a morphological disturbance in the interior of the modified membrane, producing an intramembranal particle indistinguishable from those of the host membrane proteins. It was also found that the change from the typical beaded appearance of the host membrane interior to the smooth nonstructured appearance found in the virus membrane occurred only after envelopment of the nucleocapsid was initiated (Brown *et al.*, 1972; see also Figs. 4 and 5). The transition from particle-bearing to particle-free morphology did not occur immediately upon initiation of budding. Rather a few intramembranal particles could be seen on budding structures which were in an early stage of envelopment, and these intramembranal particles disappeared as budding was completed. This situation was thus found to differ from that occurring during the envelopment of influenza (myxovirus) (Bächi *et al.*, 1969) and vesicular stomatitis (rhabdovirus) viruses (Riedel and Brown, 1977), where the transition in membrane morphology was found to occur at the stage when budding structures could first be detected. The possible meaning of this structural alteration in the interior of the membrane will be discussed further below.

C. Release of Virus from Infected Cells

Morphologically the final release of the virion was demonstrated by freeze-fracture studies to occur as a two-step process which began after the developing virion was found to be completely wrapped in the forming envelope membrane with both leaflets of the respective host membrane and viral envelope continuous with one another (Brown *et al.*, 1972; see also Figs. 1 and 5). Contact of the developing membrane with itself at the base of the developing virion was followed by fusion of the inner leaflets of the viral and host membranes with themselves, leaving the virion attached to the surface of the infected cell by only the continuity of the outer leaflet of its envelope with the outer leaflet of the plasma membrane. The subsequent fusion of the outer leaflet of the virus membrane with itself results in separation of the virus membrane from the host membrane and release of the mature virion into the medium.

D. Arrangement and Rearrangement of Viral Components during Envelopment

Although morphological aspects of the process of alphavirus envelopment are well described, the molecular biology and biochemistry of this process are less clear. A number of experiments examining the physical, chemical, and morphological properties of purified virions have provided important, though sometimes conflicting, information on the structure of the product of this process which has been used to generate models suggesting how, in molecular terms, envelopment may occur.

The envelope of the mature alphavirion contains only viral proteins (Pfefferkorn and Clifford, 1964; M. J. Schlesinger *et al.*, 1972). These proteins are present in the viral membrane in equimolar ratios to the capsid proteins in the nucleocapsid (see Chapter 9). The envelope proteins are anchored in the membrane by hydrophobic tails (Utermann and Simons, 1974), although there is some controversy regarding the extent to which these structures penetrate the envelope bilayer. Experiments employing chemical cross-linking agents have suggested that the envelope proteins E1 and E2 traverse the membrane completely and are intimately associated with the capsid proteins on the inner surface of the envelope (Garoff and Simons, 1974; see also Chapter 9 in this volume). Electron spin resonance studies of the membrane of the purified virion suggest that the hydrophobic anchor of the envelope proteins penetrates less than halfway across the membrane bilayer (Sefton and Gaffney, 1974; see Chapters 9 and 10 in this volume, agreeing with earlier conclusions made by x-ray diffraction and freeze-fracture studies of the virus membrane (Harrison *et al.*, 1971; Brown *et al.*, 1972). Studies on the properties of the membrane envelope lipid bilayer itself demonstrate that, although the viral membrane is similar in composition to the membrane of the host cell, it possesses a higher relative viscosity (Sefton and Gaffney, 1974; Moore *et al.*, 1976).

E. Models for the Process of Sindbis Envelopment

Information obtained from studies on purified virions has been extrapolated to the infected cell system in attempts to answer such important questions as: (1) How do viral nucleocapsids identify and attach to the inner surface of the modified plasma membrane? (2) How are host membrane proteins excluded from the developing envelope and how are viral membrane proteins incorporated into this structure in numbers equivalent to that of the capsid polypeptide? (3) What mechanism wraps the modified cell membrane about the nucleocapsid as envelopment progresses?

One can propose two different models which provide solutions to these questions. If one accepts the conclusion that the envelope proteins are both transmembranal structures and assumes that the same configuration of these polypeptides exists in the modified plasma membrane, the problems related to the budding process described above are readily solved (Fig. 6). The completed nucleocapsid would simply attach to the inner surface of the plasma membrane by binding of the capsid protein to one copy each of the two envelope proteins. A sequential interaction of the capsid protein with the tails of the envelope protein exposed on the inner surface of the plasma membrane would account for the stoichiometric relationship of the envelope glycoproteins to the core and would provide the energy to drive the budding process to completion. Thus the final event in the assembly of the alphavirion would be brought about by specific protein–protein interactions.

If, on the other hand, the viral membrane proteins are not transmembranal structures in the mature virion and are likewise not exposed on the inner surface of the plasma membrane of the infected cell, the problem becomes more difficult. A model can be suggested, however, by drawing on the observation that the viral membrane has a higher relative viscosity than the host plasma membrane (Fig. 7; see also Sefton and Gaffney, 1974; Moore *et al.*, 1975). The higher viscosity of the viral membrane may result from a disturbance in the fluidity of the lipids brought about by the presence of viral membrane proteins (see Chapter 10). In the modified membrane of the infected cell, the alteration in lipid viscosity might cause the aggregation of viral membrane proteins into patches on the cell surface, host membrane proteins being excluded from these domains because they are not adaptable to the reduced fluidity exhibited by lipids adjacent to viral proteins. The viral nucleocapsids would, in this model, identify and bind to the regions of reduced fluidity in the plasma membrane, and

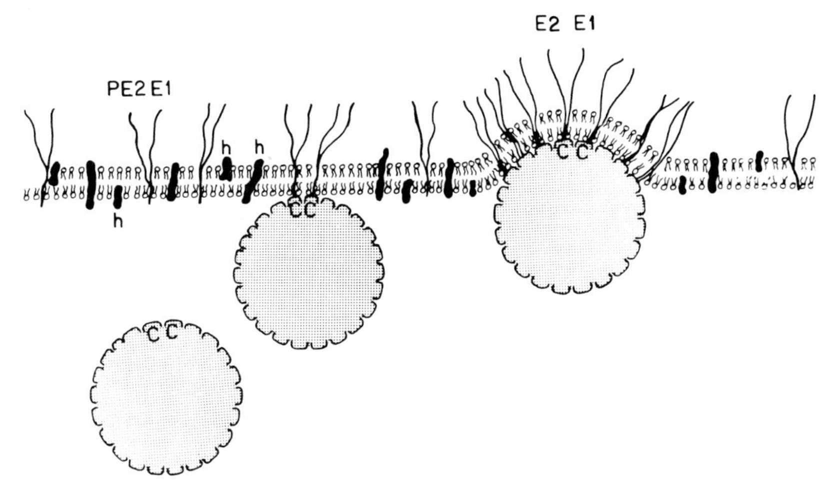

Fig. 6. A model for the envelopment of alphaviruses based on the assumption that the viral membrane proteins are transmembranal structures in both the mature virion and in the modified host plasma membrane. Envelopment proceeds from left to right. The completed nucleocapsid containing viral capsid protein (C), moves from a position in the cell cytoplasm to become attached to the inner surface of the plasma membrane, which contains host membrane proteins (h) intermingled with viral glycoproteins PE2 and E1. Attachment occurs through the interaction of a single copy of C with one copy each of PE2 and E1 exposed on the inner surface of the membrane. As this association is repeated, the plasma membrane is wrapped around the nucleocapsid. Host proteins are excluded from the developing viral envelope for steric reasons, and a mature particle is produced in which E2 and E1 maintain their transmembranal configuration and their close proximity to C. At some point in the envelopment PE2 is cleaved to E2 and E2. In the Sindbis virus system (represented here) E3 is lost, and in Semliki Forest virus it remains as a structural component of the virion.

this tight association would drive the budding process to completion. The number of copies of the envelope glycoproteins carried into mature virions would be a function of the closeness of packing of these components in the plasma membrane, their equivalence to the capsid protein in number being somewhat of a coincidence.

Although arguments can be put forth supporting either one of these models, the assumption that viral glycoproteins have the same configuration in the cell plasma membrane prior to budding as they have in the envelope of the mature virion ignores the fact that fundamental differences exist in the composition of the modified cell membrane relative to that of the membrane of the mature virion. Comparative studies of cells infected with wild-type virus or with mutant virions defective in the terminal stage of assembly have produced a

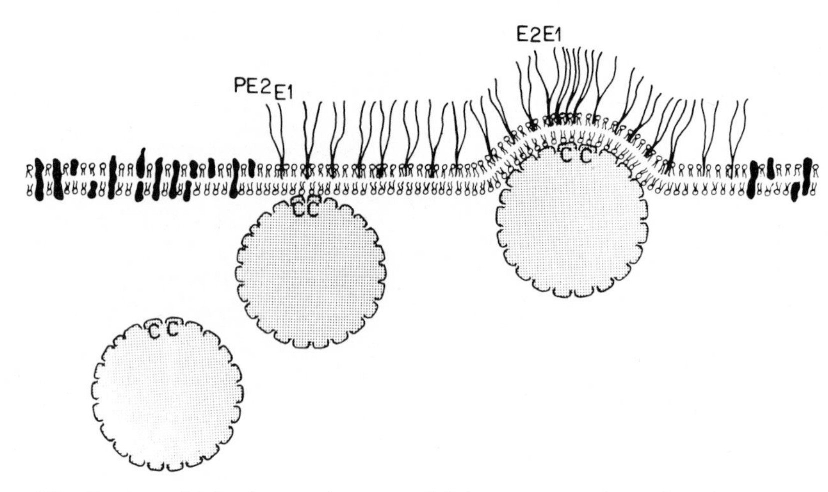

Fig. 7. A model for the envelopment of alphaviruses predicated on the assumption that the viral proteins are peripherally located in the membrane of both the host and the mature virion. The virus glycoproteins form patches on the cell surface by virtue of their constructing areas in the membrane of higher than normal viscosity. Host proteins are excluded from this area, favoring regions of the membrane which are less viscous. The virus capsids identify and bind to the regions of the membrane with the higher viscosity, and this tight binding results in envelopment. The cleavage of PE2 to E2 and E2 occurs during this process as in Fig. 6.

great deal of information regarding the organization of the envelope proteins in the membrane of the infected cell prior to and during envelopment of the nucleocapsid (S. Schlesinger and Schlesinger, 1972; M. J. Schlesinger and Schlesinger, 1973; Jones *et al.*, 1974; Smith and Brown, 1977). Although these studies indicate that the viral glycoprotein E1 is arranged in the cell membrane in the same configuration it will ultimately have in the viral envelope, they also demonstrate that, prior to initiation of the budding process, the modified cell membranes do not contain envelope protein E2 but rather its precursor, PE2. PE2 is converted to E2 by proteolytic cleavage, yielding polypeptide E2 and a smaller (ca. 10,000–13,000-molecular-weight) fragment, E3. In cells infected with Semliki Forest virus, E3 remains associated with the envelope of the mature virion, while in the Sindbis virus system it is released into the surrounding medium (Welch and Sefton, 1979).

The cleavage of PE2 is intimately associated with the process of virus envelopment, and evidence implies that envelopment cannot occur if this cleavage is blocked. Combining data obtained from a number of studies, one finds that PE2 undergoes a number of important changes as envelopment takes place. The Sindbis virus temperature-sensitive mutant *ts*20 appears to be defective in the cleavage of PE2 to E2 and has provided, together with studies on wild type-infected cells, information on the role of this precursor in the envelopment process. The membrane of cells infected with *ts*20 at the nonpermissive temperature contains E1 which has the same hemagglutinating activity (Burge and Pfefferkorn, 1967, 1968) and exposure on the surface of the infected cell (as determined by enzymatic iodination and specific antibody binding) (Smith and Brown, 1977) as it has in wild type-infected cells or in the membrane of the mature virion (Sefton *et al.*, 1973). PE2 is also located in the plasma membrane of *ts*20- and wild type-infected cells and is exposed on the surface of these cells such that it can be detected by anti-E2 antiserum (Bell and Waite, 1977). PE2 is, however, organized in the membrane of mutant- or wild type-infected cells in such a way that its tyrosine residues are completely protected from radioiodination by the lactoperoxidase procedure (Sefton *et al.*, 1973; Smith and Brown, 1977). E2 produced by the cleavage of PE2 is, on the other hand, very readily iodinated by the lactoperoxidase technique in either the membrane of wild type-infected cells (where envelopment is taking place) or in purified mature virions (Sefton *et al.*, 1973; Smith and Brown, 1977).

Cells infected with *ts*20 are defective in the cleavage of PE2 but are still capable of attaching nucleocapsids to the inner surface of the plasma membrane (Brown and Smith, 1975; Smith and Brown, 1977; see also Fig. 8). The binding of the nucleocapsids to the modified membrane is so strong that a membrane–nucleocapsid complex can be purified from cells infected with the *ts*20 mutant as well as from wild type-infected cells even after treatment of the infected cells with low concentrations of nonionic detergents (Smith and Brown, 1977). The detergent treatment quantitatively removes E2 from the membranes of wild type-infected cells. The removal of E2 from the membrane does not, however, result in a corresponding release of capsid protein from the membrane. E1 is also released from the cell membrane more readily than PE2 in either ts-20-infected or wild type-infected cells, but only small amounts are released compared to the quantitative release of E2 from wild type-infected cells. A detergent-stable nucleocapsid–membrane complex cannot be demonstrated in mature virions, which contain E2, using similar conditions of treatment (Simons *et al.*, 1973; see also Chapter 9 in this volume).

The finding that the association of PE2 and capsid proteins with the infected cell membrane is stable to treatment which removes some E1 and all of E2, and the lack of a similarly detergent-stable binding of

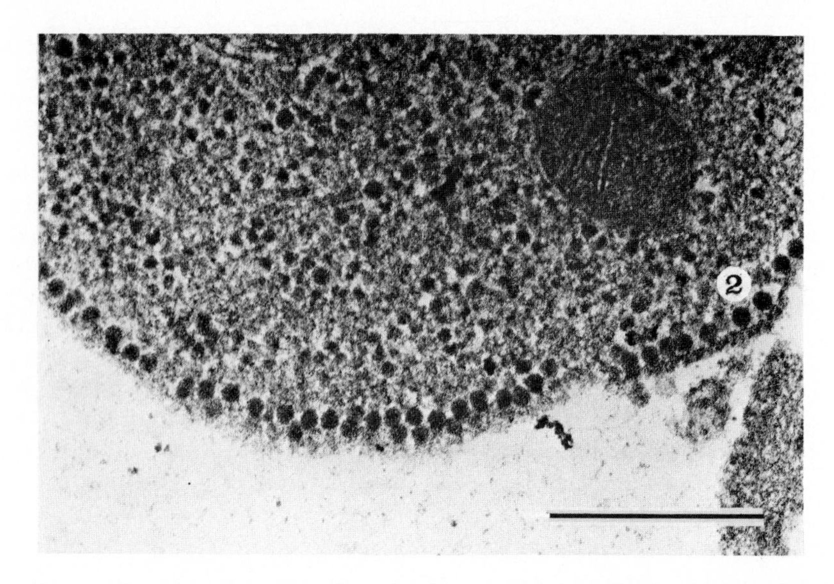

Fig. 8. Electron micrograph of an ultrathin section of a cell infected for 8 h at the nonpermissive temperature with the Sindbis mutant *ts*20. Numerous nucleocapsids are seen associated with the inner surface of the plasma membrane. In some areas the attachment of the capsid to the membrane can be seen (2). Although association of the nucleocapsid with the membrane occurs, and the binding can be demonstrated to be tight (see text), envelopment does not progress beyond this point and polypeptide PE2 is not cleaved to E2. Bar is 0.5 μm. (From Brown and Smith, 1975; with permission of the American Society of Microbiology.)

the capsid to the membrane of purified virions, combined with the observation (described above) (Wirth *et al.*, 1977) that PE2 is exposed on both the inner and outer surfaces of the modified membrane, suggests that the initial binding of the nucleocapsid to the modified membrane of the host cell is mediated by attachment of the capsid protein to the PE2 polypeptide.

The suggestion that envelopment is driven exclusively by the attachment of nucleocapsid protein to the tails of transmembranal envelope proteins or to the inner surface of a membrane of reduced fluidity seems to be incorrect, as such an interaction takes place in *ts*20-infected cells, even though no infectious or noninfectious particles are produced. If this binding were the only requirement for envelopment, one might expect virions to be produced which contain uncleaved PE2. This has never been demonstrated, and cleavage of the PE2 polypeptide seems to be essential for envelopment to progress beyond the initial attachment of the capsid to the host cell membrane.

The transient interaction of the capsid protein with the PE2 polypeptide suggests a third model for the envelopment of alphaviruses (Fig. 9). This model proposes that, prior to budding, the cell plasma membrane contains PE2 and E1 paired together in a tight complex. These pairs are randomly distributed in the plasma membrane. Within the complex E1 has the same configuration and red cell binding characteristics it will have in the membrane of the mature virion. The exact arrangement of E1 in the membrane of the infected cell is not clear. If, however, the mature virion contains no proteins which penetrate into the hydrophobic region of the viral membrane, and if E1 has the same configuration in both the cell and the viral membrane, then it may also be peripherally situated in the host cell membrane bilayer.

In this model, PE2 has a very different arrangement in the cell membrane than its product E2. It is exposed on both the inner and

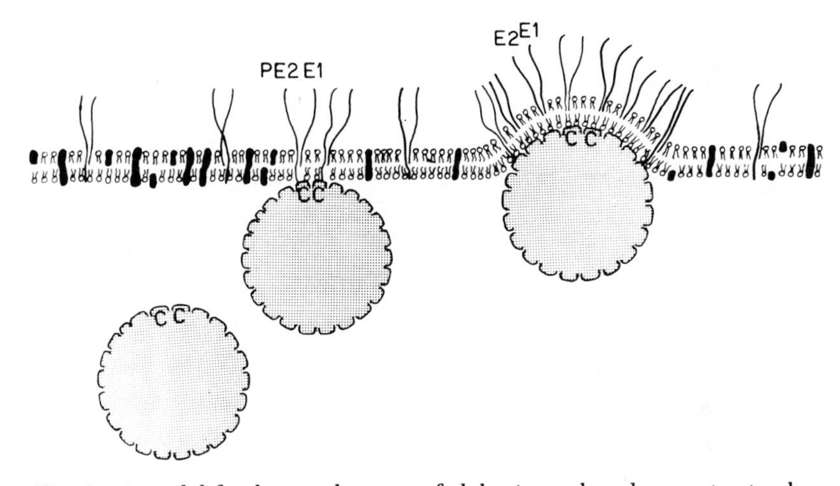

Fig. 9. A model for the envelopment of alphaviruses based upon structural reorganization of one of the viral glycoproteins during envelopment. PE2 and E1 are paired together and are intermixed with host proteins in the modified membrane. E1 is peripherally located (as in Fig. 6) and has the same configuration it will have in the viral envelope. PE2, however, is transmembranal and is exposed on the inner surface of the modified host membrane. The nucleocapsid attaches to the membrane through the interaction of C with the tail of the PE2 polypeptide (see Fig. 8). As this initial association of C with PE2 takes place, PE2 is cleaved to E2 and E2. The E2 molecule assumes a more peripheral location in the membrane bilayer, losing its transmembranal configuration and its association with C. The reorganization of E2 permits the membrane to contour itself around the capsid, allowing the interaction of more copies of C with PE2, and the sequence is repeated until envelopment is complete. The mature virion will contain no transmembranal protein elements.

outer surfaces of the plasma membrane and is thus a transmembranal structure. The mature nucleocapsids become attached to the modified cell plasma membrane through the interaction of a few of the capsid polypeptides with the tails of the PE2 polypeptide exposed on the inner surface of the plasma membrane. Before the membrane can be wrapped around the nucleocapsid and more copies of capsid protein bound to PE2, the PE2 proteins already complexed with the capsid must be cleaved. It is at this point that the *ts*20 mutant is likely defective. In a wild-type infection, this cleavage would result in breaking of the tight association of C protein with the PE2 polypeptide as it becomes E2. The E2 protein would then change its orientation in the membrane, losing its transmembranal configuration, and become exposed to iodination by lactoperoxidase and susceptible to release by detergents.

The implication that the cleavage of PE2 and the subsequent change in the configuration of E2 in the membrane are required before the plasma membrane can be contoured, bringing it into contact with other copies of C in the nucleocapsid, suggests that this rearrangement is essential for distortion of the plasma membrane as it is wrapped about the virus nucleocapsid. The change in the organization of the E2 protein in the membrane may facilitate slippage of the inner leaflet of the cell membrane relative to the outer leaflet. Such a sliding of one member of the membrane bilayer relative to the other must occur during envelopment, as the radius of the alphavirus is so small that during the process of maturation it incorporates about 45% more of the outer leaflet of the cell membrane than of the inner leaflet. This unequal incorporation of the two sides of the membrane bilayer of the host cell into the viral membrane would result in the preferential incorporation of lipid components existing in the outer leaflet relative to those comprising the inner leaflet. Such a preferential incorporation of one side of the host membrane bilayer would be expected to amplify any asymmetry in the lipid bilayer and might account, in part, for the increased viscosity of viral membrane lipids relative to the host cell membrane (Sefton and Gaffney, 1974; Moore *et al.*, 1976; see also Chapter 10 in this volume). Evidence for such an asymmetric distribution of the lipid components in the Sindbis virus membrane has been published (Stoffel and Sorgo, 1976).

The repeated interaction of capsid protein with PE2, subsequent cleavage and relocation of E2 in the membrane, and contouring of the modified membrane would result in the wrapping of the nucleocapsid in the membrane and ensure that envelope proteins would be incorporated in one-to-one molar ratios to the protein of the nucleocapsid.

The morphological change witnessed by electron microscopy in the interior of the membrane during envelopment (the loss of intramembranal particles) might therefore reflect the change in the organization of the E2 polypeptide in the developing envelope after its cleavage from PE2, resulting in the production of a mature virion with no protein elements in the center of the membrane bilayer.

von Bonsdorff and Harrison (1975) have demonstrated that the surface of the envelope of Sindbis virus is organized in triangles. These authors have suggested that the glycoprotein pairs E1 and E2 are organized as trimers, one trimer composing each of 80 triangles existing on the surface of this alphavirus membrane. More recently, these authors have demonstrated that the triangular arrangement of the glycoproteins can be maintained even when the viral envelopes are separated from the virus nucleocapsids by treatment with detergents (von Bonsdorff and Harrison, 1978). The observation that the triangular or trimeric arrangement of the glycoprotein pairs can be maintained even when the alphavirus membrane is removed from the nucleocapsid suggests that some interaction exists among the three pairs of E1 and E2 forming each of the triangles, which is not broken by a low concentration of detergent. It is possible that the interaction among these pairs is maintained in each of the triangles by an interaction between the E1 glycoproteins. Because the E1 glycoproteins have the same configuration in the membrane of the infected host cell as they have in the membrane of the mature virion, these trimeric arrangements may also exist in the membrane of the host cell prior to envelopment. Some mechanism must organize these triangles into the icosahedral structure which exists on the surface of the virion itself. This implies that some type of interaction between the trimers of E1 and E2 must be developed as morphogenesis occurs. The conversion of PE2, which results in the more peripherally located and externally exposed E2 polypeptide, may also expose a site on this polypeptide which allows it to interact with other E2 polypeptides in neighboring triangles to form the intertriangular association. Thus one might envision a situation in which the membrane of the Sindbis virus-infected cell contains numerous triangles consisting of trimers of paired PE2–E1 complexes. These triangles are dispersed uniformly on the surface of the infected cell and are brought into contact with one another when they are organized by the association of the transmembranal PE2 glycoprotein with the capsid protein located in the virus nucleocapsid. The cleavage of polypeptide PE2 to E2 would result in exposure of the E2 polypeptide on the surface of the infected cell and allow it to form a noncovalent interaction with E2 molecules located in neighboring triangles. This process would be repeated, resulting in formation of the icosahedral surface of the mature virus membrane.

It has been suggested that the brief interaction of the capsid protein with the PE2 polypeptide and the subsequent reorganization of E2 in the viral membrane during envelopment may be essential for the eventual penetration and uncoating of the mature virion during subsequent infection of another host cell (Smith and Brown, 1977). Relocation of the E2 protein to a more peripheral location in the cell membrane may expose a function associated with this glycoprotein essential for fusion of the viral envelope with a host plasma membrane during penetration (such a function might be disadvantageous on the surface of the infected cell, resulting in fusion of the infected cells with each other). The transient interaction of the capsid protein with the PE2 polypeptide may also result in reorganization of the nucleocapsid itself, allowing its disassembly during uncoating in the cytoplasm of a cell, after penetration, which is identical to the cytoplasm in which it was structurally stable after assembly. It has been demonstrated that nucleocapsids purified from virions are more sensitive to degradation by RNase than nucleocapsids purified from infected cell cytoplasm (B. Bode and D. T. Brown, unpublished observation).

F. Source of Proteolytic Cleavage of PE2

Particularly prominent in this discussion of the assembly of alphaviruses is the role of the proteolytic cleavage of the PE2 polypeptide in envelopment. The origin of the proteolytic activity responsible for this cleavage is unknown. A number of investigations have suggested that the capsid protein possesses such a proteolytic activity (Cancedda and Schlesinger, 1974; Simmons and Strauss, 1974; Scupham *et al.*, 1977) and that this protein may in fact be responsible for releasing itself from the high-molecular-weight precursor of all the structural proteins (see Chapter 13). The proximity of the capsid protein to the PE2 polypeptide proposed in the above model makes it an attractive candidate for the source of the enzymatic activity cleaving PE2. Attributing the cleavage of PE2 to the capsid protein raises certain topological difficulties in that it implies that cleavage takes place at some point on the inner surface of the cell plasma membrane. A very complex orientation of the PE2 polypeptide within the membrane would have to be suggested to allow it to be cleaved on the inner surface of the plasma membrane and (as is the case in the Sindbis virus system) released into the surrounding medium after this cleavage. Further experimentation is required to clarify this most important aspect of alphavirus morphogenesis.

G. Formation of Phenotypically Mixed Viruses

The morphological and biochemical events described above suggest a strong and specific interaction between the alphavirus nucleocapsid and its membrane proteins during assembly. That this specificity is not absolute is demonstrated by the fact that phenotypic mixing (pseudotype formation) occurs between Sindbis virus and Semliki Forest virus (Burge and Pfefferkorn, 1966a), as well as between Sindbis virus and lactic dehydrogenase (LDH) virus (Lagwinska *et al.*, 1975). The mechanism of the formation of such phenotypically mixed particles during virus assembly is not known. It is important to know if in the formation of these pseudotypes only one of the two viral membrane proteins is substituted for by the other virion during morphogenesis. Formation of the phenotypically mixed particles might be similar to complementation. It would be important to know if, for example, Sindbis capsids bound only to Sindbis PE2 but in phenotypically mixed particles contained a LDH or Semliki Forest equivalent of E1.

V. FUTURE RESEARCH IN ALPHAVIRUS ASSEMBLY

Particularly promising for further elucidation of the process of alphavirus envelopment is the discovery of *ts* mutants which, at a nonpermissive temperature, disturb the usually exact packaging of a single nucleocapsid per virion. These mutants allow envelopment to occur, but several nucleocapsids are enclosed in a single envelope, resulting in multinucleoid virions (Hashimoto *et al.*, 1975; Strauss *et al.*, 1977). With one mutant (*ts*103) (Strauss *et al.*, 1977) it has been demonstrated that all the normal processing of viral proteins takes place during the formation of these aberrant virions, including terminal processing of PE2 to E2, so that the multinucleoid virions contain no uncleaved precursor. It has been suggested that the defect in *ts*103 is in the capsid protein and that this defect does not allow the tight interaction of the nucleocapsid with the developing envelope demonstrated in infections with wild-type virus. It is interesting to note that multinucleoid particles are also produced when cultured mosquito cells are infected with wild-type virus (Gliedman *et al.*, 1975). These observations suggest that multinucleoid particles may result either from an alteration in one of the viral structural proteins or from a major change in the chemical composition of the host membrane employed in formation of the viral envelope. The examination of nucleocapsid–membrane interactions with *ts*103 in both vertebrate and mosquito cells, as well as wild-type virus in mosquito cells, may clarify the role of the nucleocapsid and viral protein–host lipid interaction in the envelopment of alphaviruses.

REFERENCES

Acheson, N. H., and Tamm, I. (1967). *Virology* **32**, 128.
Acheson, N. H., and Tamm, I. (1970). *J. Virol.* **5**, 714.
Bächi, T., Gehard, W., Lindenmann, J., and Muhlethaler, K. (1969). *J. Virol.* **4**, 739–776.
Bell, J. W., and Waite, M. R. F. (1977). *J. Virol.* **21**, 788–791.
Birdwell, C. R., and Strauss, J. H. (1973). *J. Virol.* **11**, 502.
Birdwell, C. R., and Strauss, J. H. (1974). *J. Virol.* **14**, 366.
Birdwell, C. R., Strauss, E. G., and Strauss, J. H. (1973). *Virology* **56**, 429.
Blobel, G., and Dobberstein, B. (1975a). *J. Cell Biol.* **67**, 835–851.
Blobel, G., and Dobberstein, B. (1975b). *J. Cell Biol.* **67**, 852–862.
Brache, M., and Schlesinger, M. J. (1976). *Virology* **74**, 441–449.
Brown, D. T., and Smith, J. F. (1975). *J. Virol.* **15**, 1262.
Brown, D. T., Waite, M. R., and Pfefferkorn, E. R. (1972). *J. Virol.* **10**, 524.
Burge, B. W., and Pfefferkorn, E. R. (1966a). *Nature (London)* **210**, 1397.
Burge, B. W., and Pfefferkorn, E. R. (1966b). *Virology* **30**, 214.
Burge, B. W., and Pfefferkorn, E. R. (1967). *J. Virol.* **1**, 956.
Burge, B. W., and Pfefferkorn, E. R. (1968). *J. Mol. Biol.* **35**, 193.
Cancedda, R., and Schlesinger, M. J. (1974). *Proc. Natl. Acad. Sci. U.S.A.* **71**, 1843.
Casjens, S., and King, J. (1975). *Annu. Rev. Biochem.* **40**, 555–611.
Dalrymple, J. M., Schlesinger, S., and Russell, P. K. (1976). *Virology* **69**, 93–103.
Friedman, R. M., and Grimley, P. M. (1969). *J. Virol.* **4**, 292–299.
Garoff, H., and Simons, K. (1974). *Proc. Natl. Acad. Sci. U.S.A.* **71**, 3988.
Gliedman, J. B., Smith, J. F., and Brown, D. T. (1975). *J. Virol.* **16**, 913–926.
Grimley, P. M., Levin, J. G., Berezesky, I. K., and Friedman, R. M. (1972). *J. Virol.* **10**, 492.
Guild, G. M., and Stollar, V. (1975). *Virology* **67**, 24–41.
Harrison, S. C., David, A., Jumblatt, J., and Darnell, J. E., Jr. (1971). *J. Mol. Biol.* **60**, 523.
Hashimoto, K., Suzuki, K., and Simizu, B. (1975). *J. Virol.* **15**, 1454.
Horzinek, M., and Mussgay, M. (1969). *J. Virol.* **4**, 514.
Johnston, R. E., Tovell, D. R., Brown, D. T., and Faulkner, P. (1975). *J. Virol.* **16**, 951–958.
Jones, K. J., Waite, M. R. F., and Bose, H. R. (1974). *J. Virol.* **13**, 809.
Jones, K. J., Scupham, R. K., Pfeil, J. A., Wan, K., Sagik, B. P., and Bose, H. R. (1977). *J. Virol.* **21**, 778–787.
Kääriäinen, L., and Söderlund, H. (1971). *Virology* **43**, 291.
Kaluza, G. (1976). *J. Virol.* **19**, 1–12.

Kennedy, S. I. T., Bruton, G. V., Weiss, B., and Schlesinger, S. (1976). *J. Virol.* **19,** 1034–1043.
Kriebich, G., and Sabatini, D. (1973). *Fed. Proc., Fed. Am. Soc. Exp. Biol.* **32,** 2133–2138.
Lagwinska, E., Stewart, C. C., Adles, C., and Schlesinger, S. (1975). *Virology* **5,** 204–214.
Leavitt, R., Schlesinger, S., and Kornfeld, S. (1977). *J. Virol.* **21,** 375–385.
Moore, N. F., Barenholz, Y., and Wagner, R. R. (1976). *J. Virol.* **19,** 126–135.
Pfefferkorn, E. R., and Clifford, R. L. (1964). *Virology* **23,** 217.
Renz, D., and Brown, D. T. (1976). *J. Virol.* **19,** 775–781.
Riedel, B., and Brown, D. T. (1977). *J. Virol.* **21,** 601–609.
Scheele, C. M., and Pfefferkorn, E. R. (1969). *J. Virol.* **3,** 369.
Schlesinger, M. J., and Schlesinger, S. (1973). *J. Virol.* **11,** 1013.
Schlesinger, M. J., Schlesinger, S., and Burge, B. W. (1972). *Virology* **47,** 539.
Schlesinger, S., and Schlesinger, M. J. (1972). *J. Virol.* **10,** 925.
Schlesinger, S., Schlesinger, M. J., and Burge, B. W. (1972). *Virology* **48,** 615.
Scupham, R. K., Jones, K. J., Sagik, B. P., and Bose, H. R. (1977). *J. Virol.* **22,** 568–571.
Sefton, B. M., and Gaffney, B. J. (1974). *J. Mol. Biol.* **90,** 343.
Sefton, B. M., Wickus, G. G., and Burge, B. W. (1973). *J. Virol.* **11,** 730.
Shenk, T. E., and Stollar, V. (1973). *Virology* **53,** 162.
Simmons, D. T., and Strauss, J. H. (1974). *J. Mol. Biol.* **86,** 397.
Simons, K., Helenius, A., and Garoff, H. (1973). *J. Mol. Biol.* **90,** 119–133.
Simpson, R. W., and Hauser, R. E. (1968). *Virology* **34,** 358.
Smith, J. F., and Brown, D. T. (1977). *J. Virol.* **22,** 662–678.
Söderlund, H. (1973). *Intervirology* **1,** 354–361.
Stoffel, W., and Sorgo, W. (1976). *Chem. Phys. Lipids* **17,** 324–335.
Strauss, J. H., Burge, B. W., and Darnell, J. E. (1969). *Virology* **37,** 367–376.
Strauss, E. G., Birdwell, C. R., Lenches, E. M., Staples, S. E., and Strauss, J. H. (1977). *Virology* **82,** 122–149.
Utermann, G., and Simons, K. (1974). *J. Mol. Biol.* **85,** 560.
von Bonsdorff, C.-H., and Harrison, S. C. (1975). *J. Virol.* **16,** 141–145.
von Bonsdorff, C.-H., and Harrison, S. C. (1978). *J. Virol.* **28,** 578–583.
Weiss, B., and Schlesinger, S. (1973). *J. Virol.* **12,** 862.
Welch, W. J., and Sefton, B. M. (1979). *J. Virol.* **29,** 1186–1195.
Wirth, D. F., Katz, F., Small, B., and Lodish, H. (1977). *Cell* **10,** 253–263.
Wirth, D. F., Lodish, H. F., and Robbins, P. N. (1979). *J. Cell Biol.* **81,** 154–162.
Yin, F. H. (1969). *J. Virol.* **4,** 547.

18

Chemical and Antigenic Structure of Flaviviruses

PHILIP K. RUSSELL
WALTER E. BRANDT
JOEL M. DALRYMPLE

I. INTRODUCTION

The detailed chemical structure of the majority of the more than 50 viruses in the genus *Flavivirus* is not specifically known from direct examination of each individual virus. Several viruses, thought to be

THE TOGAVIRUSES

ISBN 0-12-625380-3

representative of the genus, have been examined in some detail, e.g., Japanese encephalitis (JE), St. Louis encephalitis (SLE), Kunjin (KUN), dengue-1 (DEN-1), and dengue-2 (DEN-2) (abbreviations as listed in the International Catalogue of Arboviruses, 1975). From the results and from less detailed studies of yellow fever (YF), Langat (LAN), and tick-borne encephalitis (TBE) viruses, generalizations can be drawn which are thought to be valid for the entire genus. The International Committee on Taxonomy of Viruses (ICTV) has defined flaviviruses as containing a nonsegmented single-stranded infectious 42 S RNA and three virion polypeptides, one of which is a glycoprotein (Porterfield *et al.*, 1978). The published observations on various members of the genus are consistent with this definition, and flaviviruses with divergent antigenic and biological characteristics (i.e., mosquito-borne, tick-borne, neurotropic, and viscerotropic) have been shown to have these basic properties. In this chapter we make the underlying assumption that the genus is relatively uniform in the basic biochemical structure of the virions.

Measurements such as molecular-weight estimates of proteins vary somewhat among viruses. In part this reflects minor but real differences among viruses, and in part the lack of precision and the variation inherent in existing techniques for estimating such parameters as sedimentation, density, and molecular size. Most biochemical studies to date have been performed using established laboratory strains with varying passage history, undefined genetic composition, and an undefined relationship to the wild viruses from which they were derived. The similarity or difference between strains or clones has largely been unexplored; thus we are, for the most part, unable to define the extent of biochemical variation which may exist in genetically heterogeneous wild virus populations. Evidence exists that wild populations of flaviviruses are composed of subpopulations with a considerable range of biological properties; this may be reflected in significant variation in chemical structure.

The substrates in which flaviviruses are propagated affect biochemical and biophysical properties of the virions because the lipid and carbohydrate components are cell-derived and, although they may be virus-modified, are specified by the host cell genome. Thus the host cell must be considered when evaluating differences in composition among viruses. The most significant differences are seen among viruses grown in arthropod tissue. Most published work refers to viruses propagated in mammalian systems. In the following sections, data derived from studies on viruses propagated in arthropod tissue will be specifically designated. Previously extremely valuable and detailed reviews were written by Schlesinger (1977) and Pfefferkorn and Shapiro (1974).

II. PHYSICAL AND CHEMICAL CHARACTERISTICS

A. Physical Properties of Virions and Associated Particles

Flavivirus infection of cell culture or suckling mouse brain results in the production and release of progeny virus, noninfectious hemagglutinin, and soluble antigens. Physical separation methods, principally rate zonal and equilibrium centrifugation and molecular sieve chromatography, have been used to isolate and characterize these components. Initial studies relied on the hemagglutinating (HA) property of virions (designated RHA for rapidly sedimenting hemagglutinin) and smaller hemagglutinins (designated SHA for slowly sedimenting hemagglutinin) or on antigenic activity for assay of separated components.

1. Virions

Infectious virus and the RHA of DEN-1, DEN-2, JE, KUN, and SLE viruses cosediment in rate zonal sucrose gradients, forming a single peak (Kitoaka and Nishimura, 1965; Smith *et al.*, 1970; Stollar *et al.*, 1966; Trent and Qureshi, 1971; Westaway and Reedman, 1969). Dengue virions partially purified in this manner appear on electron microscopy as spherical particles approximately 50 nm in diameter (Matsumura *et al.*, 1971; Matsumura and Hotta, 1971; Smith *et al.*, 1970). Sedimentation coefficients reported for flaviviruses range from 175 to 218 S (Boulton and Westaway, 1972; Brandt *et al.*, 1970b; Rosato *et al.*, 1974; Russell and Brandt, 1973; Strohmaier *et al.*, 1965). These studies indicate that the host cell origin of a virus affects the sedimentation coefficient and density. Dengue-2 virions propagated in cell culture have a sedimentation coefficient of approximately 200 S, whereas virions propagated in mouse brain have a sedimentation coefficient of 175 S. Densities of DEN-2 virions propagated in suckling mouse brain ranged from 1.22 to 1.23 g/cm^3 by equilibrium centrifugation in cesium chloride, whereas those propagated in cell culture ranged from 1.23 to 1.24 g/cm^3 (Smith *et al.*, 1970; Stevens and Schlesinger, 1965). Differences in density have also been attributed to the particular method used; Heinz and Kunz (1977) reported that TBE virus propagated in cell culture had a density of 1.23 g/cm^3 in cesium chloride but a density of 1.195 g/cm^3 in sucrose gradients. However, Shapiro *et al.* (1971a) obtained a value of 1.23 g/cm^3 for cell culture-propagated JE virus in deuterium oxide–sucrose gradients. Finally, the surface structure of virions obtained from cesium chloride gradients appears smooth, in contrast to the projections easily discernible in virions obtained from sucrose gradients (Smith *et al.*, 1970; see also Chapter 8, Fig. 23 in this volume). This may be due to partial disruption, impregnation, or fixation by the cesium salts.

2. *Associated Particles*

In addition to virions, a slowly sedimenting peak of noninfectious hemagglutinin (SHA) is readily resolved on sucrose gradients (Igarashi *et al.*, 1963; Kitoaka and Nishimura, 1963, 1965; Smith *et al.*, 1970; Stollar *et al.*, 1966). After purification by repeated centrifugation on sucrose gradients or equilibrium centrifugation in cesium chloride, electron micrographs revealed ring or doughnut-shaped structures about 14 nm in diameter (Smith *et al.*, 1970). The optimal pH for hemagglutination by both RHA and SHA was 6.2. In contrast to the virions which had different sedimentation coefficients and densities depending on whether the virus was propagated in mouse brain or cell cultures, the SHA particle sedimented at 70 S regardless of the host system. Virus preparations obtained from mouse brain contained a high proportion of SHA compared to cell culture preparations. Equilibrium centrifugation in cesium chloride or deuterium oxide–sucrose gradients does not separate RHA and SHA. Direct density determinations of virus concentrates will result in a single peak ranging from 1.22 to 1.24 g/cm^3 containing both RHA and SHA. In some experiments, recentrifugation of this peak resolved two components (Stevens and Schlesinger, 1965). Prior rate zonal separation of RHA and SHA is required to analyze these two structures by equilibrium centrifugation, resulting in a value for SHA of 1.23 g/cm^3 regardless of the host cell. However, we obtained separation of RHA and SHA with extended periods of centrifugation in 50% potassium tartrate–30% glycerol gradients as described by Obijeski *et al.* (1974). Equilibrium established after an initial rate zonal separation resulted in an RHA that was denser than SHA, perhaps because of the higher viscosity at the top of the tube in the decreasing viscosity–increasing density type of gradient.

3. *Virion Components*

Virions can be converted to a 70 S peak of hemagglutinin by Tween 80 and ether treatment; this material was designated "derived" SHA since its physical form (fragments) and density (1.18 g/cm^3) was distinctly different from those of "native" SHA which are the ring-shaped structures described above (Brandt *et al.*, 1970a). Nonionic detergents fragment the virion envelope and release an intact core or nucleocapsid (Stollar, 1969; Westaway and Reedman, 1969) having a density of 1.30–1.31 g/cm^3 (Shapiro *et al.*, 1971a; Trent and Qureshi, 1971) and a sedimentation coefficient of 120–140 S (Heinz and Kunz, 1977; Stollar, 1969; Westaway and Reedman, 1969). Disruption of virions during equilibrium centrifugation in cesium chloride results occasionally in a dense 1.31 g/cm^3 peak detected by hemagglutination (Smith *et al.*, 1970). The small amount of envelope glycoprotein (see Section IV) that remains attached to the nucleocapsid may account for the hemagglutination. Virions that remain intact during centrifugation in cesium chloride form a small peak of hemagglutinin at 1.22–1.24 g/cm^3, and the fragmented envelopes are found at 1.19 g/cm^3 (Stevens and Schlesinger, 1965; Smith *et al.*, 1970). Envelope fragments removed by detergents have densities as low as 1.15 g/cm^3 (Shapiro *et al.*, 1971a).

4. *Nonvirion Antigen*

A soluble complement-fixing (SCF) antigen which is antigenically distinct from virions and SHA was found in dengue-infected mouse brain and serum of infected mice (Brandt *et al.*, 1970a; Cardiff *et al.*, 1971; Russell *et al.*, 1970), human lymphoblastoid cell lines (Sung *et al.*, 1975), and LLC-MK2 and *Aedes albopictus* cells (Cardiff *et al.*, 1973; Sinarachatanant and Olson, 1973). Japanese encephalitis SCF antigen was found in Vero and *A. albopictus* cells (Rai and Ghosh, 1976), and an SCF antigen of West Nile (WN) virus in infected mice was described by Lavrova (1977). Dengue-2 SCF has a sedimentation coefficient of approximately 4 S, and both DEN-2 and JE SCF antigens are resistant to inactivation by sodium dodecyl sulfate (SDS) and 2-mercaptoethanol (2ME), treatment which inactivates the complement fixation (CF) activity of the virions (Brandt *et al.*, 1970a; Eckels *et al.*, 1975; Rai and Ghosh, 1976). Dengue-2 and JE SCF antigens are separable from virion, SHA, and nonvirion antigens of higher molecular weight by molecular sieve chromatography (see Section IV).

B. Basic Chemical Composition

Very little work has been done in this area as compared to the alphaviruses. Ada *et al.* (1962) reported lipid/protein/RNA ratios of 11:81:8 for mouse brain-derived Murray Valley encephalitis (MVE)

TABLE I

Chemical Composition of St. Louis Encephalitis Virus[a]

Component	Milligrams equivalent[b,c]	Percentage of total dry weight
RNA	0.48	6
Protein	5.23	66
Carbohydrate	0.71	9
Neutral carbohydrate	(0.24)	(3)
Hexosamine	(0.48)	(6)
N-Acetylneuraminic acid	(0.24)	(3)
Lipids[d]	1.35	17
Phospholipid	(0.96)	(12)
Neutral lipid	(0.39)	(5)

[a] From Trent (1980).
[b] Based on known standards.
[c] The values given are the calculated mean from four separate experiments using SLE virus.
[d] Molar ratio of cholesterol to phospholipid: 0.29.

virus, and Trent (1980) reported values for cell culture-propagated SLE virus of 17:66:6 as well as a figure of 9% for carbohydrate. Nozima *et al.* (1964) reported an unusually large lipid value for mouse brain-derived JE virus (58:34:7), probably due to a difference in purification methods (see Section V). The best agreement among investigators was obtained with the RNA component—about 7%. Table I contains the most recent data obtained for SLE virus (Trent, 1980).

C. Physical and Chemical Stability

1. *Chemical Stability*

The sensitivity of enveloped viruses to chemical degradation provided a useful laboratory tool for early arbovirologists attempting to classify and identify viruses recovered in nature. Theiler (1957) reported that all arboviruses in his possession at the time were sensitive to deoxycholate. The infectivity of all togaviruses is destroyed by lipid solvents. Table II lists early studies on the sensitivity of infectious virus to lipid solvents and detergent. Tween 80 plus ether destroys dengue virions but leaves HA activity intact in the form of slower sedimenting particles (Brandt *et al.*, 1970a). The Tween 80–ether combination has been used to extract HA antigen of JE virus from tis-

TABLE II

Inactivation of Flaviviruses by Lipid Solvents and Detergents

Chemical	Virus[a]	Reference
Ether	DEN-1, DEN-2	Hotta and Evans (1956)
Chloroform	MVE	Anderson and Ada (1961)
Deoxycholate	Togaviruses	Theiler (1957)
	MVE	Anderson and Ada (1959b)
	JE, WN, ILH	Sunaga *et al.* (1960)
Deoxy-, glyco-, taurocholate	KFD, WN	Banerjee and Paul (1965)
Deoxycholate plus trypsin	JBE	Hardy and Scherer (1965)
Saponin, dodecyl sulfate	LI, RSSE	Bögel and Mayr (1962)

[a] ILH, Ilheus; KFD, Kyanasur Forest Disease; LI, Louping ill.

sue (Saturno, 1969). Acetone effectively degrades virions, and the standard sucrose–acetone procedure used for producing HA antigens for serological tests (Clarke and Casals, 1958) converts virions into smaller slowly sedimenting HA particles (Brandt *et al.*, 1970a). β-Propiolactone destroys infectivity of SLE, LAN, and YF viruses while preserving HA and CF activity of antigen preparations (Finkelstein and Sulkin, 1957; French and McKinney, 1964).

Urea, as well as SDS and 2ME, destroys the serological activity and the structural integrity of virions, but these chemicals do not affect dengue nonstructural SCF antigen (Brandt *et al.*, 1970a; Nicoll and Maydat, 1965). However, SDS-treated-JE virion preparations are still immunogenic in rabbits (K. H. Eckels, personal communication). The destructive effect of urea on the hemagglutinin in infected mouse brain suspensions was accompanied by an increase in dengue serospecificity by the CF test (Corneski *et al.*, 1972); the urea-treated type-specific antigen was excluded in the void volume of Sephadex G-200 columns, well separated from the SCF antigen. This particulate antigen probably consisted of host cell membrane fragments which contained all the virus-specified polypeptides (Kos *et al.*, 1975; Shapiro *et al.*, 1972b) of which the SCF antigen is the major antigen at late stages of infection (Stohlman *et al.*, 1975, 1978). Urea had a varied effect on clones of JE virus, some being more sensitive than others (Karpova *et al.*, 1972). Nishimura *et al.* (1967) showed that urea in less than 2 *M* concentration reversibly inactivated the JE virion, whereas in excess of 3 *M* virus was irreversibly destroyed. Partial reassociation was shown by gradient analysis.

Flaviviruses are inactivated by trypsin, chymotrypsin, papain (Cheng, 1958), and pancreatic lipase (Takehara and Hotta, 1961).

Nicoll and Maydat (1965) and Hannoun (1968) found flaviviruses to be significantly more sensitive to trypsin than alphaviruses.

2. *Physical Stability*

a. pH. The effect of pH depends on the suspending medium for the virus particles. Purified JE virus loses its structural integrity at the pH at which hemagglutination is carried out (6.2–6.6); its hemagglutinin is found at the top of a gradient, no longer sedimenting as a 200 S particle. Tick-borne flaviviruses, however, probably survive low pH in nature. Gresíková (1959) reported that TBE virus survived pH 3 in milk for 24 h. The virus can survive pH 1.5 for 4 h at 4°C and 24 h at pH 3 (Pogodina, 1958). Gastric juice of normal acidity inactivates TBE virus but only after 2 h at 37°C, and inactivation is inhibited by the addition of milk. Maximum stability of the Graz 1 strain of TBE virus is attained at pH 7.5, and the virus is more stable in Tris than in phosphate buffer (Bloedhorn, 1963). We found excellent preservation of the infectivity of JE virus during purification procedures using Tris–saline–EDTA at pH 8.2. Manning and Collins (1979) report that the stability of dengue virus is optimum at pH 8.0.

b. Photoinactivation. Anan'ev and Blinova (1960) reported that TBE could be reliably inactivated by treatment with 1:25,000 methylene blue for 10 min in saline, and 3 h in serum, following exposure to a 96-W bulb at 30 cm. Photoinactivation with visible light in the presence of proflavine inactivates JE and the extracted infectious RNA by single-hit kinetics. The integrity of the virions is not affected by the treatment, as judged by sedimentation kinetics, and they maintain their HA activity (Fukunaga *et al.*, 1967). Tomita and Prince (1963) found that JE, WN, SLE, MVE, YF, and DEN viruses were inactivated with 10^{-6} neutral red and visible light with a 10^4-fold reduction in titer after 20 min. We found that exposure to 2000 ergs/cm^2 of uv for 3 min totally inactivated 10^6 PFU of purified DEN-2 virus in a protein-free medium. Ultraviolet light also inactivated WN virus in frozen sections in a matter of minutes (Kundin and Liu, 1963).

c. Temperature. Stability studies with mouse brain suspensions are not applicable to purified virus preparations, therefore it is difficult to make meaningful comparisons of data from different laboratories. Taken together, however, the following references suggest that tick-borne viruses are somewhat more heat-stable than the mosquito-borne viruses (Andonov, 1957; Bloedhorn and Ackerman, 1961; Eylar and Wisseman, 1975; Manning and Collins, 1979; Marchette, 1973; Nir and Goldwasser, 1961; Schlesinger, 1977).

d. Centrifugation. Dengue virions disintegrate readily during equilibrium centrifugation in cesium chloride (Stevens and Schlesinger, 1965) but generally remain intact during rate zonal centrifugation through sucrose (Smith *et al.*, 1970), as well as during equilibrium centrifugation in deuterium oxide–sucrose (Shapiro *et al.*, 1971a). Tick-borne encephalitis virus withstands cesium chloride solutions better than WEE virus, an alphavirus (Gresíková and Vachlkova, 1971) and, from our observations, apparently better than DEN-2 virus. Thus, by both shearing forces and temperature, tick-borne viruses appear more resistant to disintegration than mosquito-borne viruses. The differences in stability may be due to differences in the V1 protein which may be a membrane protein (see Section IV).

III. VIRION RNA

Detailed virion RNA analyses have been performed with only a few representatives of the flaviviruses. Infectious phenol-extracted nucleic acid that is RNase-sensitive has been demonstrated for JE (Liu *et al.*, 1962; Igarashi *et al.*, 1963; Nakamura and Ueno, 1963), MVE (Anderson and Ada, 1959a), SLE (Trent *et al.*, 1969), WN (Colter *et al.*, 1957; Naude, 1963), and DEN (Stollar *et al.*, 1966, 1967) viruses. The RNase sensitivity of purified virion RNA indicates that it is a single polynucleotide chain. This is supported by base composition data from several investigators (Table III), which are inconsistent with known base-paired structure. The sedimentation constants of flavivirus virion RNAs have been variously reported from 38 to 45 S with a corresponding molecular weight estimate of $3–4 \times 10^6$ (Stollar *et al.*, 1966; Boulton and Westaway, 1972; Nishimura and Tsukeda, 1971).

TABLE III

Nucleotide Base Analysis of Flavivirus Genome RNA

	Nucleotides per 100 nucleotides				
Virus	Adenosine	Guanosine	Cytosine	Uridine	Reference
Dengue-2	30.6	26.4	21.3	21.6	Stollar *et al.* (1966)
Intracellular DEN-2	31.3	22.7	23.0	22.9	Stollar *et al.* (1967)
Japanese encephalitis	23.3	31.0	24.9	20.7	Blair (personal communication)
Murray Valley encephalitis	25.5	27.5	21.5	25.5	Ada *et al.* (1962)
St. Louis encephalitis	30.7	26.2	21.7	21.4	Trent *et al.* (1969)

Wengler *et al.* (1978) were unable to distinguish among the RNA sedimentation of the flaviviruses WN and Uganda S (UGS) and of an alphavirus, Semliki Forest (SF). It therefore appears logical to assume that the sedimentation constants (42 S) and molecular weight estimates (4.2×10^6) of alphaviruses approximate the virion RNAs of most flaviviruses.

The flavivirus RNA genome has been shown to be infectious, indicating a "plus" strand or positive polarity according to the convention of Baltimore (1971). Direct evidence of a messenger function for flavivirus RNA is lacking, although virus-specific RNA of genome size was detected as polysome-associated RNA present in vertebrate cells infected with DEN-2 virus (Cleaves and Schlesinger, 1977) or SLE virus (Naeve and Trent, 1978).

Flavivirus virion RNAs appear to differ from their alphavirus counterparts at the 3′-terminus. Brawner *et al.* (1977) found that SLE virus RNA contained only 2.4% enzyme-resistant ^{32}P-label, while Sindbis (SIN) virus RNA, which has messenger function and is poly(A) rich (Johnston and Bose, 1972), contained 9.2% enzyme-resistant label. Other indirect evidence suggests the absence of, or very little, poly(A) in flavivirus RNA. Two-dimensional chromatographic separation of SLE, WN, and UGS virus RNA digests failed to detect the "poly(A) tract" characteristic of similar digests of poly(A)-rich mRNAs (D. W. Trent, personal communication; Wengler *et al.*, 1978). Any evidence of poly(A) is similarly lacking in WN virus RNA (Wengler *et al.*, 1978). The 42 S RNA from either WN virions or infected BHK cells failed to bind to oligo(dT)-cellulose columns (Wengler *et al.*, 1978), suggesting either the absence of a 3′-poly(A) sequence or a short sequence of less than 10 bases. Kos (1977) was similarly unable to obtain binding of JE virion RNA to poly(U) columns under conditions in which SIN virion RNA was bound. These observations have been confirmed with JE virus RNA on both poly(U) and oligo(dT) columns (C. D. Blair, personal communication). Although it is perhaps premature to assume that the absence of poly(A) at the 3′-terminus is a characteristic of all flavivirus RNAs, the evidence to date is certainly convincing.

The initiation sequences of flaviviruses mRNA species are of particular interest primarily because of some rather unique aspects of flavivirus translation; i.e., synthesis of the structural glycosylated proteins is independent of the nonstructural, nonglycosylated proteins (Shapiro *et al.*, 1973b; see also Chapter 19 in this volume). Cleaves and Dubin (1979) examined the methylation status of intracellular DEN-2 RNA and found that it contained a single cap per molecule of the form $m^7G(5')ppp(5')Amp$; they also found penultimate ribose methylation.

This structure represents a "type-1 cap," using the nomenclature of Shatkin (1976). Less than one molecule in five contained adenine base methylation either in the cap or internally. Genome-sized RNA molecules isolated from both WN-infected cells and from WN virus particles were found to contain the structure $m^7GpppAmpN_1$ at their 5′-termini; furthermore, some of the molecules contained an additional methyl group, giving rise to a "type-2 cap" with the structure $m^7GpppAmpN_1mpN_2$ (Wengler *et al.*, 1978). It was calculated that the great majority and possibly all molecules of both species of WN virus-specific RNA contained at least one of the cap structures identified.

Fingerprinting or oligonucleotide mapping has only recently been applied to the study of flavivirus RNA. No similarity was detected between the large oligonucleotides generated by RNase T1 digestion of WN and UGS virus-specific 42 S plus-strand RNAs (Wengler *et al.*, 1978). Similar analysis of different isolates or strains of SLE virus RNA exhibit changes in approximately 10% of the large oligonucleotides, while no differences have been observed among preparations of the same strain passaged in different cell cultures (D. W. Trent, personal communication). Further analysis of virion RNA by oligonucleotide mapping and hybridization studies should provide real insight into the diversity of this large, antigenically complex group of flaviviruses.

IV. PROTEINS AND GLYCOPROTEINS

Virion structural polypeptides first identified for DEN-2 virus (Stollar, 1969) and KUN virus (Westaway and Reedman, 1969) have been shown to be similar for 11 members of the genus *Flavivirus* (Shapiro *et al.*, 1972c, 1973a; Westaway, 1973). In all instances, mature virions have a nucleocapsid or core protein V2, a large envelope glycoprotein V3, and a small nonglycosylated envelope protein V1. Electropherograms for several viruses are shown in Fig. 1. Coelectrophoresis studies indicate that this configuration is similar for both mosquito- and tick-borne members of the genus. Japanese encephalitis virions from mammalian cells (BHK-21) and mosquito cells (*Aedes albopictus*) have identical protein compositions (Igarashi *et al.*, 1973).

A. Envelope Proteins

Disruption of virions with nonionic detergents solubilizes the V3 and V1 proteins, leaving V2 associated with the nucleic acid (Stollar,

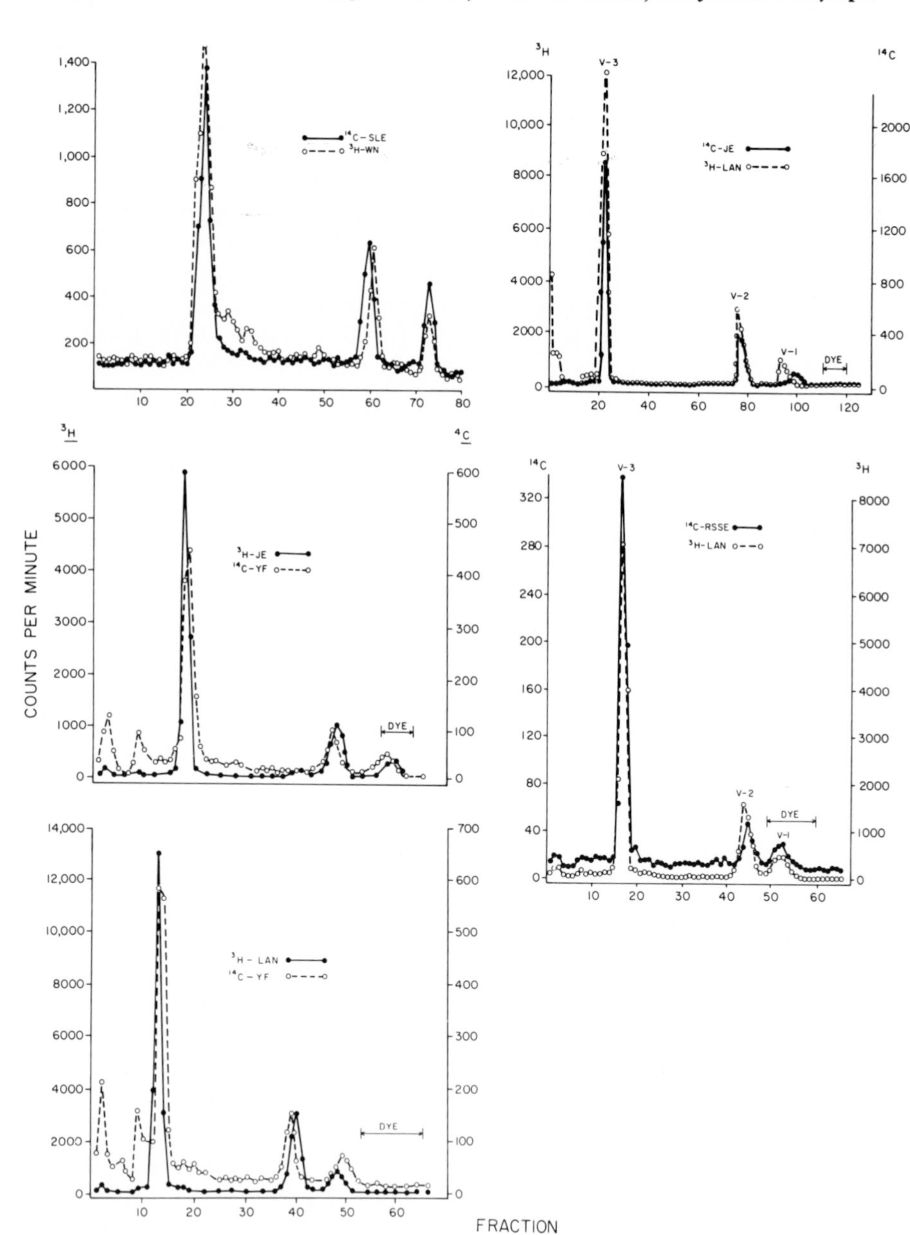

Fig. 1. Polyacrylamide gel coelectrophoresis of pairs of flaviviruses labeled with radioactive amino acids and degraded with 2ME, SDS, and heat. (Adapted from Shapiro *et al.*, 1972c.)

1969; Shapiro *et al.*, 1971a). The V3 protein is readily labeled with radioactive amino acids and glucosamine; V3 can also be extrinsically labeled with ^{125}I while on the intact virion (D. W. Trent, personal communication).

Significant differences in the molecular weight of V3 have been observed among flaviviruses; estimations vary from 51,000 for KUN to 59,000 for DEN-2 (Stollar, 1969; Westaway, 1973). Variations within antigenic subgroups are also significant, with a range of 51,000 for DEN-4 to 59,000 for DEN-2 and an intermediate value of 53,000–56,000 for DEN-3 (Westaway *et al.*, 1977). There is some evidence for size variation of V3 in different strains of SLE virus (Shapiro *et al.*, 1972c), although rigorous proof of this is lacking. Further evidence that V3 is a single glycoprotein is provided by the observation that isoelectric focusing of SLE V3 results in a single peak at pI 7.8 (Trent, 1977). The isoelectric points of JE and DEN-2 V3 glycoproteins are similar, 7.6 and 7.8, respectively.

The V1 protein is not glycosylated but is removed from virions along with V3 by treatment with NP-40 or Brij-58 (Shapiro *et al.*, 1971a; Stollar, 1969; Trent and Qureshi, 1971). However, treatment with deoxycholate leaves V1 associated with the sedimented particles (Trent and Qureshi, 1971; Westaway and Reedman, 1969). Thus, V1 appears to be associated with the envelope but is more intimately associated with the nucleocapsid than V3 and may not be exposed on the surface of the virion. The size range of V1 protein is narrow, approximately 7000–8000 daltons for most agents studied. However, the V1 proteins of three tick-borne viruses [Russian spring-summer encephalitis (RSSE), Powassan (POW), and LAN] have a significantly slower migration when coelectrophoresed by polyacrylamide gel electrophoresis (PAGE) with mosquito-borne agents (Shapiro *et al.*, 1972c). At present this is the only observation on chemical structure which correlates with biological differences. The possibility that V1 is a membrane protein analogous to M protein of influenza viruses is consistent with the above information on its location and with observed differences in chemical and thermal stability between tick- and mosquito-borne flaviviruses.

A third protein which appears similar or identical to the intracellular glycoprotein NV2 has been found on intracellular JE virions and on released virions under special culture or labeling conditions (Fig. 2) (Shapiro *et al.*, 1972a; Westaway and Shew, 1977). It has been postulated that this glycoprotein is a precursor of V1, but this has not been proved.

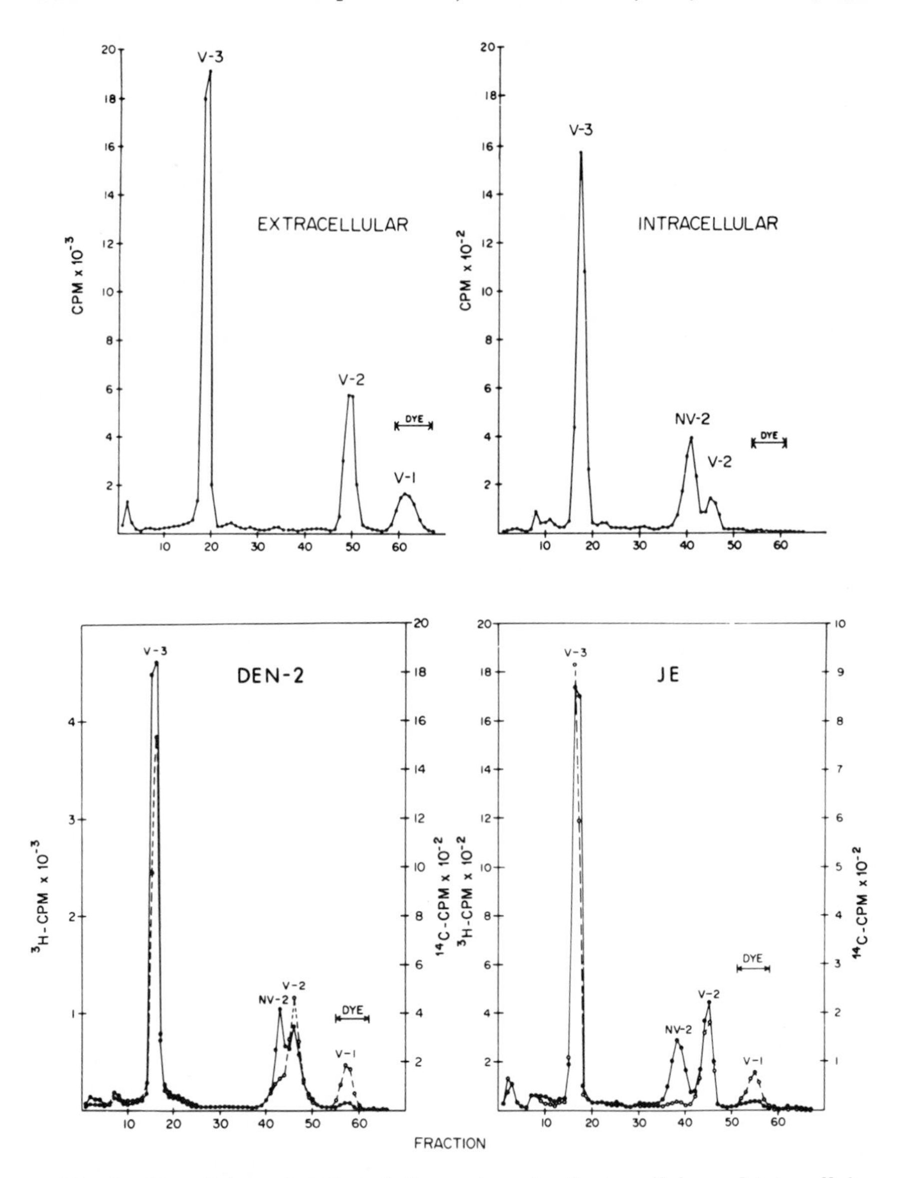

Fig. 2. Top: Polyacrylamide gel electrophoresis of extracellular and intracellular [^{3}H]amino acid-labeled JE virus treated as in Fig. 1. Bottom: Polyacrylamide gel coelectrophoresis of extracellular virus from cells cultured in "normal" ^{14}C-labeled amino acid medium (open circles) and from cells cultured in medium containing 6 m*M* Tris and ^{3}H-amino acids (solid circles). (Adapted from Shapiro *et al.*, 1972a.)

B. Nucleocapsid Proteins

The protein V2 is nonglycosylated, has a molecular weight of approximately 13,000, and remains associated with the virion RNA after detergent treatment of virions. Molecular-weight estimations and coelectrophoresis studies indicate minimal variation within the genus. The V2 protein is lysine-rich compared to V3 (Stollar, 1969). All observations to date are consistent with the interpretation that V2 is the capsid protein.

C. Nonvirion Proteins

Precise identification of nonvirion proteins produced by translation of the flavivirus genome has been difficult because of minimal suppression of protein synthesis in the host cell by the viral infection. Thus, analytic studies of viral gene products of necessity required suppression of host protein synthesis by the combined action of actinomycin D and cycloheximide (Shapiro *et al.*, 1971a; Trent and Qureshi, 1971) or manipulation of results of analyses to subtract host cell components using actinomycin D alone (Westaway, 1973). Both approaches are less than ideal. There is agreement among published reports on the virus-specific nature of the large intracellular proteins NV5, NV4, and NV3. These have been clearly identified for KUN (Westaway and Reedman, 1969), JE (Shapiro *et al.*, 1971a), SLE (Qureshi and Trent, 1973b), MVE (Westaway, 1975), and DEN-2 (Westaway, 1973) viruses. The precise number and molecular size of the smaller proteins remains controversial; a total of nine nonvirion proteins has been described by Westaway (see Chapter 19).

The SCF antigen of DEN-2 virus has been shown to be a nonvirion protein (Brandt *et al.*, 1970a; Cardiff *et al.*, 1971). The SCF antigens of the four dengue serotypes have identical molecular weights but different isoelectric points. Analysis by the Ferguson plot technique in polyacrylamide gel electrophoresis indicates that they are "charge isomers" (Cardiff *et al.*, 1970). Molecular sieve chromatography suggested a molecular weight of 39,000 for DEN-2 SCF antigen (Brandt *et al.*, 1970b). With discontinuous gel electrophoresis, a molecular weight of 39,000 was found for SCF antigens of all four dengue serotypes (Cardiff *et al.*, 1970). However, coelectrophoresis of mouse brain-derived SCF antigen with radioactive DEN-2 virion proteins in SDS-containing gels resulted in a molecular-weight estimation of 45,000 which coincided with that of the nonvirion protein NV3 (Fig. 3) (Cardiff *et al.*, 1971). Proof of the identity of SCF antigen and

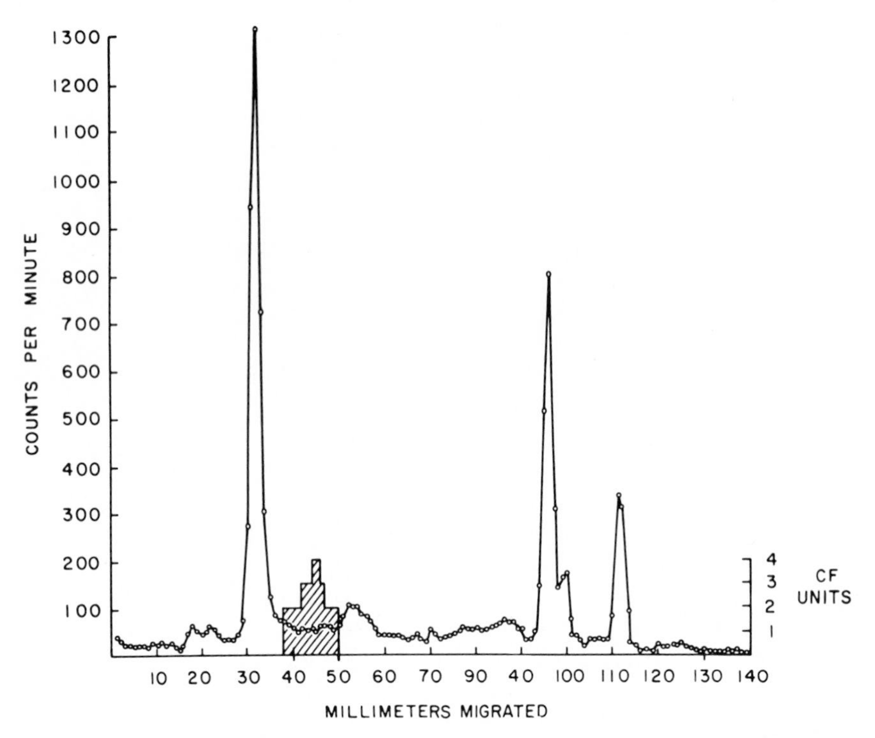

Fig. 3. Polyacrylamide gel coelectrophoresis of [^{3}H]amino acid-labeled DEN-2 virions harvested from LLC-MK$_2$ cells, and the SCF antigen harvested from infected mouse brain. (From Cardiff *et al.*, 1971.)

NV3 remains, however, to be developed. Small differences in migration in polyacrylamide gels have been observed with the SCF antigens of different strains of DEN-1 virus, allowing biochemical differentiation of very closely related strains (McCloud *et al.*, 1971). A JE virus SCF antigen has been described with biochemical characteristics similar to those of dengue SCF, i.e., stability to treatment with 2ME and SDS, but with an apparently higher molecular weight (approximately 53,000, estimated by molecular sieve chromatography) (Eckels *et al.*, 1975; Rai and Ghosh, 1976). No functions have been ascribed to SCF antigen proteins.

V. LIPIDS

Very limited data on the lipids of flaviviruses are available, as in the case of alphaviruses (see Chapter 10). The difference in morphogenesis of each group, particularly in the maturation steps, indicates that comparisons should not be made between alphavirus and flavivirus lipids; the lipid composition of alphaviruses is derived from plasma membranes and that of flaviviruses from internal membranes.

A. Viral Envelope

The lipid component of flaviviruses can only be estimated within 10–20% because of the different purification procedures utilized by various investigators. Nozima *et al.* (1964), using streptomycin to clarify JE virus-infected mouse brain suspensions, reported that the virus consisted of about 58% lipid. However, Ada *et al.* (1962), using protamine sulfate to clarify MVE virus-infected mouse brain suspensions, reported only 11% lipid in the virus. Trent (1979) examined SLE virus obtained from cell culture media which did not require a chemical clarification step prior to concentration and purification and reported 17% total lipid, 12% phospholipid, and 5% neutral lipid (Table I). Of the 11% total lipid of MVE virus, Ada *et al.* (1962) reported 0.8% phospholipid and 1% cholesterol. The phospholipid composition of SLE virus determined by Trent (1980) is listed in Table IV.

B. Relationship of Lipid Composition to the Host Cell

St. Louis encephalitis virions tend to reflect the phospholipid composition of the host cell (Table V). The phospholipid composition of flaviviruses more closely reflects the composition of the internal host

TABLE IV

Phospholipid Composition of St. Louis Encephalitis Virus and PS Cell Membranes[a]

	Percentage of total phospholipid			
Phospholipid	SLE virus	Whole cell	Endoplasmic reticulum	Plasma membrane
Sphingomyelin	7	8	10	15
Phosphatidylcholine	56	52	63	43
Phosphatidylserine	7	6	9	11
Phosphatidylinositol	3	1	1	2
Phosphatidylethanolamine	25	27	16	23
Other		6	1	6

[a] From Trent (1980).

TABLE V

Phospholipid Patterns of MA-104, Vero, PS, and BHK-21 Cells and St. Louis Encephalitis Virus Grown in These Cells[a]

Source	Percentage of total phospholipid				
	Sphingo-myelin	Phosphatidyl-choline	Phosphatidyl-inositol	Phosphatidyl-serine	Phosphatidyl-ethanolamine
Vero cells	0.6	52.0	0.7	14.9	27.7
SLE virus from Vero cells	2.7	46.2	2.7	9.6	23.9
MA-104 cells	5.2	45.5	2.2	0.5	26.8
SLE virus from MA-104 cells	11.3	57.1	4.6	4.3	20.9
PS cells	5.2	51.6	1.0	13.6	27.2
SLE virus from PS cells	7.2	55.6	3.2	7.1	25.1
BHK-21 cells	7.1	58.4	1.4	12.6	21.2
SLE from BHK-21 cells	9.6	57.6	4.3	6.1	24.6

[a] From Trent (1980).

cell membranes than of the plasma membrane (Table IV), which is consistent with the intracellular maturation of these viruses. Alphaviruses, on the other hand, bud from the cell surface and have a lipid composition similar to the composition of the plasma membrane (Renkonen *et al.*, 1971). While the lipid percentages of flaviviruses and endoplasmic reticulum tend to correspond, there are some changes in the percentages of phosphatidylethanolamine, phosphatidylcholine, and/or sphingomyelin (Table V).

C. Structural Considerations

Inactivation of flaviviruses by ether, desoxycholate, chloroform and butanol, and lipase indicates that lipids are essential for the structural stability of the virions. Complete inactivation of up to 10^8 LD_{50} of MVE virus with phospholipase A, ether, chloroform, or butanol affected only the viral envelope, since infectious RNA could be extracted from the inactivated preparations of MVE virus (Anderson and Ada, 1961). Thus, either attachment to or selective uncoating within the susceptible cell is prevented by disruption of the lipid envelope. Most probably, the lipid serves a structural role in maintenance of the V3 envelope glycoprotein in the outer structure of the virus similar to the role of structural lipid in alphaviruses (Chapters 9, 10, and 17).

VI. ANTIGENIC CHARACTERISTICS

A variety of test systems have been used to assay antibody response to flavivirus infection and immunization. These include complement fixation (CF), hemagglutination inhibition (HI), radioimmunoprecipitation (RIP), radioimmunoassay (RIA), immunoprecipitation, immunofluorescence, and neutralization. The several methods vary widely in specificity and sensitivity, and interpretation of results is often difficult because of cross-reactions reflecting antigenic relationships among the flaviviruses (Chapter 2) and because of varying complexity of the antigen preparations used in the various tests. The following sections attempt to describe the present knowledge of the antigenicity of individual virion components and, where possible, correlate these data with serological reactivity in various test systems.

A. Antibody Response to Flaviviruses

In man and laboratory animals, flavivirus infection stimulates both IgM and IgG antibodies (Bellanti *et al.*, 1965; Gresíková *et al.*, 1965;

Russell *et al.*, 1967; Otsuka *et al.*, 1967). The IgM response in man varies in duration, depending on the infecting virus and previous antigenic experience. In dengue infections of man the IgM response is transitory, disappearing in a few weeks; 17D yellow fever vaccine, however, induces a persistent IgM antibody detectable for more than a year after immunization (Monath, 1971). In man JE virus infection produces IgM antibody which persists from a few months to 2 years (Edelman *et al.*, 1976). IgM antibody is highly specific in HI and neutralization tests and has been used for virus-specific diagnosis; in contrast, IgG antibody is highly cross-reactive (Scott *et al.*, 1972; Edelman and Pariyanonda, 1973; Westaway, 1968a,b). IgG antibodies resulting from infection persist for many years, probably for life, in most persons.

Several of the viral proteins of flaviviruses contain antigenic determinants which are either identical or sufficiently similar to produce immunological cross-reactions between viruses. These similar antigenic determinants are responsible for anamnestic antibody responses when sequential infections with different virus species occur; broadly cross-reactive high-titered IgG antibodies are rapidly produced (Russell, 1970). The IgM response when it occurs in second infections is specific for the second virus; IgM antibody is often absent or difficult to detect when second infections occur with dengue viruses (Russell *et al.*, 1969; Russell, 1971; Westaway *et al.*, 1974).

B. Immunological Reactivity

1. *Virions*

The antibodies formed after either infection or experimental immunization of animals with suspensions of infected tissue react only with the protein components specified by the virus. There is no convincing evidence for immunological reactivity of the lipid or polysaccharide components of the virions which are host-determined, or of the viral RNA. The major antigens of the virion are the envelope glycoprotein V3 and the nucleocapsid protein V2. The small envelope (or membrane) protein V1 is very probably antigenic, but there is as yet no direct evidence for its immunogenicity.

Since the major envelope glycoprotein V3 may be the only exposed protein antigen on the virion and is responsible for hemagglutination, it is the antigen involved in HI reactions (Della-Porta and Westaway, 1977). Purified V3 from JE virions reacts by the CF test with antisera to heterologous flaviviruses as well as with homologous antisera (Eckels *et al.*, 1975). The V3 glycoprotein of JE, WN, SLE, and DEN-2 viruses derived from infected cell extracts is cross-reactive with heterologous flavivirus antisera by CF, immunoprecipitation, and RIA tests (Qureshi and Trent, 1973c; Trent *et al.*, 1976). The V3 glycoprotein is also the principal reactive antigen in the more specific virus neutralization reaction. Purified SLE V3 absorbs neutralization antibody from homologous antisera (Qureshi and Trent, 1973c), and antibody raised to purified JE V3 derived from virions neutralizes virus infectivity (K. H. Eckels, personal communication; Kitano *et al.*, 1974). Thus, there is evidence for at least two antigenic determinants on the envelope glycoprotein—one cross-reactive, the other virus-specific. Trent (1977) has proposed the hypothesis that an invariant segment exists in the primary sequence of V3 of all members of the genus and is the cross-reactive determinant. A variant portion which is the major portion of the molecule is responsible for the virus-specific antigenic determinant. This concept is probably an oversimplification and does not adequately explain the various levels of antigenic relationships among flaviviruses as revealed by HI tests.

The monosaccharides associated with the envelope glycoprotein of flaviviruses most likely vary as a function of the host cell as is the case with alphaviruses (Luukkonen *et al.*, 1977). Since the antigenic specificity of JE virus and dengue viruses by neutralization tests does not appear to change when viruses are propagated in different host cells, including arthropod cells which lack sialic acid, it follows that the carbohydrate constituents of the virion do not contribute significantly to the major antigenic determinants.

The nucleocapsid protein V2 has not been extensively studied regarding antigenic specificity or immunological reactivity. Qureshi and Trent (1973a) found the nucleocapsid protein to be immunologically distinct from other virion proteins and from nonstructural antigens by immunodiffusion, and Trent (1977) showed immunological identity of V2 proteins of JE, DEN-2, and SLE viruses by competitive RIA.

It is tempting to speculate on the basis of minimal data and analogies with alphaviruses and myxoviruses that the nucleocapsid protein V2 is antigenically similar for the entire genus, but definitive studies have yet to be done. The V2 protein is immunogenic in animals injected with suspensions of virus-infected tissues, but there is no information on anti-V2 antibody resulting from natural infection.

2. *Nonvirion Hemagglutinins*

The SHA particles formed during infection contain V3, V1, and NV2 (Cardiff *et al.*, 1971; Stollar, 1969; Westaway, 1975). Relative amounts of NV2 and V1 in SHA vary, and to some extent at least the apparent variation in chemical composition is due to differences in time of addition of radioactive label (Shapiro *et al.*, 1971b). It would be expected that virions and SHA particles would be antigenically similar, since V3 is a major surface component common to both. Analysis by HI tests indicated identical specificity between RHA and SHA for KUN, MVE, and DEN-2 (Westaway *et al.*, 1975; Della-Porta and Westaway, 1977) viruses. However, Cardiff *et al.* (1971) showed antigenic differences between DEN-2 RHA and SHA by RIP inhibition and by neutralization antibody absorption. There is as yet no explanation for antigenic differences between virions and SHA, except for a tentative hypothesis that there may be a difference in exposure of the antigenic determinants principally responsible for virus neutralization.

3. *Nonvirion Proteins*

Nonstructural intracellular virus-specified proteins are formed in significant quantities during infection, and some are exposed on the surface of infected cells (Cardiff and Lund, 1976) and may be significant in immunological and immunopathological responses of infected mammalian hosts. Antibody to only two nonvirion antigens, NV5 and SCF, has been demonstrated in convalescent serum of man, although only these two nonstructural antigens have been obtained in sufficiently purified form for immunological testing.

The largest nonvirion protein, NV5 of SLE, JE, WN, and DEN-2 viruses has been shown to be antigenically distinct from virion antigens and type-specific by CF and immunoprecipitation tests (Qureshi and Trent, 1973a). Whether cross-reactions exist among very closely related viruses, e.g., within the dengue and TBE groups, is as yet untested; however, even high-titered hyperimmune mouse ascitic fluids show no cross-reactions with NV5 antigens of the closely related JE, WN, and SLE viruses, indicating that this protein may be antigenically unique for each virus.

The SCF antigens of dengue viruses are antigenically as well as biochemically distinct from virion antigens (Cardiff *et al.*, 1971). The SCF protein contains two antigenic determinants: One determinant is similar or identical for all dengue serotypes and one is type-specific (Russell *et al.*, 1970). Minimal antigenic differences have been shown between SCF antigens of two strains of DEN-1 virus (McCloud *et al.*, 1971).

Antibody rises to SCF antigens have been demonstrated in human dengue infections by CF (Falkner *et al.*, 1973) and RIA (Brandt *et al.*, 1975). The SCF antigen has been shown to be present on the surface of dengue-infected cells (Cardiff and Lund, 1976). As previously indicated, dengue SCF antigen is thought to be the NV3 protein on the basis of molecular size obtained by PAGE (Cardiff *et al.*, 1971).

An antigen termed SCF has been isolated from JE-infected cells by Eckels *et al.* (1975) and Rai and Ghosh (1976). This JE SCF antigen is stable to treatment with SLS and 2ME, as are the dengue SCF antigens, although the molecular weight (53,000) differs from that of dengue SCF antigen (39,000) by molecular sieve chromatography. The JE SCF antigen cross-reacts strongly with MVE and WN antisera and is clearly distinct from virion antigens and from NV5. Whether it is analogous to the dengue SCF antigens is as yet unclear. Other nonvirion proteins, NV4, NV2, and NV1, would be expected to be antigenic and immunogenic because of their size and protein nature, but there is no direct evidence to support this.

VII. PRESENT AND FUTURE INVESTIGATIONS

An area that needs to be pursued is the analysis of flavivirus genomes to determine quantitatively the chemical basis for both antigenic and biological variation among strains; such data may reduce the requirement for the cumbersome detailed antigenic analysis and biological evaluation of new isolates and provide a better basis for comparison and classification. The nature of the complex antigenic relationships among the flaviviruses will become clear only when the precise chemical definition of specific and cross-reactive determinants of both virion and nonstructural antigens are known. A conclusive definition of these components as well as information on their function should provide some insight into the morphogenesis of these viruses and determinants of viral virulence.

Recent evidence of multiple internal initiation sites on the flavivirus genome and future work on the biochemical basis for these observations should also provide insight into the mechanism of replication of these viruses, which is poorly understood at the present time. The current state of knowledge encompassing the replication of flaviviruses will be covered in Chapter 19.

REFERENCES

Ada, G. L., Abbot, A., and Anderson, S. G. (1962). *J. Gen. Microbiol.* **29**, 165–170.

Anan'ev, V. A., and Blinova, M. I. (1960). *Zh. Mikrobiol., Epidemiol. Immunobiol.* **31**, 1219–1224.

Anderson, S. G., and Ada, G. L. (1959a). *Aust. J. Exp. Biol.* **37**, 353–364.
Anderson, S. G., and Ada, G. L. (1959b). *Virology* **8**, 270–271.
Anderson, S. G., and Ada, G. L. (1961). *J. Gen. Microbiol.* **25**, 451–458.
Andonov, P. S. (1957). *Vopr. Virusol.* **2**, 237–242.
Baltimore, D. (1971). *Bacteriol. Rev.* **35**, 235–241.
Banerjee, K., and Paul, S. D. (1965). *Indian J. Med. Res.* **53**, 304–308.
Bellanti, J. A., Russ, S. B., Holmes, G. E., and Buescher, E. L. (1965). *J. Immunol.* **94**, 1–11.
Bloedhorn, H. (1963). *Zentralbl. Bakteriol. Parasitenkd., Infektionskr. Hyg., Abt. 1: Orig.* **190**, 149–163.
Bloedhorn, H., and Ackermann, R. (1961). *Arch. Gesamte Virusforsch.* **10**, 522–528.
Bögel, K., and Mayr, A. (1962). *Zentralbl. Bakteriol. Parasitenkd. Infektionskr. Hyg., Abt. 1: Orig.* **186**, 134–138.
Boulton, R. W., and Westaway, E. G. (1972). *Virology* **49**, 283–289.
Brandt, W. E., Cardiff, R. D., and Russell, P. K. (1970a). *J. Virol.* **6**, 500–506.
Brandt, W. E., Chiewsilp, D., Harris, D. L., and Russell, P. K. (1970b). *J. Immunol.* **105**, 1565–1568.
Brandt, W. E., Dalrymple, J. M., Top, F. H., Jr., and Russell, P. K. (1975). *Int. Virol.* **3**, 187.
Brawner, T. A., Lee, J. C., and Trent, D. W. (1977). *Arch. Virol.* **54**, 147–151.
Cardiff, R. D., and Lund, J. K. (1976). *Infect. Immun.* **13**, 1699–1709.
Cardiff, R. D., McCloud, T. G., Brandt, W. E., and Russell, P. K. (1970). *Virology* **41**, 569–572.
Cardiff, R. D., Brandt, W. E., McCloud, T. G., Shapiro, D., and Russell, P. K. (1971). *J. Virol.* **7**, 15–23.
Cardiff, R. D., Russ, S. B., Brandt, W. E., and Russell, P. K. (1973). *Infect. Immun.* **7**, 809–816.
Cheng, P. (1958). *Virology* **6**, 129–136.
Clarke, D. H., and Casals, J. (1958). *Am. J. Trop. Med. Hyg.* **7**, 561–573.
Cleaves, G. R., and Dubin, D. T. (1979). *Virology* **96**, 159–165.
Cleaves, G. R., and Schlesinger, R. W. (1977). *Abstr., Am. Soc. Microbiol., Annu. Meet., 77th, New Orleans* p. 287.
Colter, J. S., Bird, H. H., Moyer, A. W., and Brown, R. A. (1957). *Virology* **4**, 522–532.
Corneski, R. A., Hammon, W. M., Atchison, W., and Sather, G. E. (1972). *Infect. Immun.* **6**, 952–957.
Della-Porta, A. J., and Westaway, E. G. (1977). *Infect. Immun.* **15**, 874–882.
Eckels, K. H., Hetrick, F. M., and Russell, P. K. (1975). *Infect. Immun.* **11**, 1053–1060.
Edelman, R., and Pariyanonda, A. (1973). *Am. J. Epidemiol.* **98**, 29–38.
Edelman, R., Schneider, R. J., Vejjajiva, A., Pornpibul, R., and Voodhikul, P. (1976). *Am. J. Trop. Med. Hyg.* **25**, 733–738.
Eylar, O. R., and Wisseman, C. L. (1975). *Acta Virol. (Engl. Ed.)* **19**, 167–168.
Falkler, W. A., Jr., Diwan, A. R., and Halstead, S. B. (1973). *J. Immunol.* **111**, 1804–1809.
Finkelstein, R. A., and Sulkin, S. E. (1957). *Proc. Soc. Exp. Biol. Med.* **95**, 112–115.
French, G. R., and McKinney, R. W. (1964). *J. Immunol.* **92**, 772–778.
Fukunaga, T., Ishikawa, N., Igarashi, A., and Fukae, K. (1967). *Biken J.* **10**, 1–9.
Gresíková, M. (1959). *Acta Virol. (Engl. Ed.)* **3**, 159–167.
Gresíková, M., and Vachlkova, A. (1971). *Acta Virol. (Engl. Ed.)* **15**, 143–147.
Gresíková, M., Hana, L., Sekeyova, M., and Styk, B. (1965). *Acta Virol. (Engl. Ed.)* **9**, 416–422.

Hannoun, C. (1968). *Nature (London)* **219**, 753–755.
Hardy, J. L., and Scherer, W. F. (1965). *Am. J. Epidemiol.* **82**, 73–84.
Heinz, F., and Kunz, C. (1977). *Acta Virol. (Engl. Ed.)* **21**, 308–316.
Hotta, S., and Evans, C. A. (1956). *Virology* **2**, 704–706.
Igarashi, A., Kitano, H., Fukunaga, T., and Fukai, K. (1963). *Biken J.* **6**, 165–179.
Igarashi, A., Fukuoka, T., Sasao, F., Surimarut, S., and Fukai, K. (1973). *Biken J.* **16**, 67–73.
"International Catalog of Arboviruses" (1975). (T. O. Berge, ed.), DHEW Publ. No. (CDC) 75-8301, pp. 34–38.
Johnston, R. E., and Bose, H. R. (1972). *Proc. Natl. Acad. Sci. U.S.A.* **69**, 1514–1516.
Karpova, E. F., Loginova, N. V., and L'vov, D. K. (1972). *Vopr. Virusol.* **17**, 207–210.
Kitano, T., Suzuki, K., and Yamaguchi, T. (1974). *J. Virol.* **14**, 631–639.
Kitoaka, M., and Nishimura, C. (1963). *Virology* **19**, 238–239.
Kitoaka, M., and Nishimura, C. (1965). *Jpn. J. Med. Sci. Biol.* **18**, 177–187.
Kos, K. A. (1977). Ph.D. Thesis, Univ. of Maryland, College Park.
Kos, K. A., Shapiro, D., Vaituzis, Z., and Russell, P. K. (1975). *Arch. Virol.* **47**, 217–224.
Kundin, W. D., and Liu, C. (1963). *Proc. Soc. Exp. Biol. Med.* **114**, 359–360.
Lavrova, N. A. (1977). *Vopr. Virusol.* **22**, 193–196.
Liu, Y.-Y., Chan, M. Y., Liu, H. C., Shu, C. S., and Kuan, H. C. (1962). *Acta Microbiol. Sin.* **8**, 231–235.
Luukkonen, A., Von Bonsdorff, C., and Renkonen, O. (1977). *Virology* **78**, 331–335.
McCloud, T. G., Cardiff, R. D., Brandt, W. E., Chiewsilp, D., and Russell, P. K. (1971). *Am. J. Trop. Med. Hyg.* **20**, 964–968.
Manning, J. S., and Collins, J. K. (1979). *J. Clin. Microbiol.* **10**, 235–239.
Marchette, N. J. (1973). *Am. J. Trop. Med. Hyg.* **22**, 242–243.
Matsumura, T., and Hotta, S. (1971). *Kobe J. Med. Sci.* **17**, 85–95.
Matsumura, T., Stollar, V., and Schlesinger, R. W. (1971). *Virology* **46**, 344–355.
Monath, T. P. C. (1971). *Am. J. Epidemiol.* **93**, 122–129.
Naeve, C. W., and Trent, D. W. (1978). *J. Virol.* **25**, 535–545.
Nakamura, M., and Ueno, Y. (1963). *Proc. Soc. Exp. Biol. Med.* **113**, 1037–1039.
Naude, W. D. T. (1963). *S. Afr. J. Lab. Clin. Med.* **9**, 94.
Nicoll, J., and Maydat, L. (1965). *Ann. Inst. Pasteur, Paris* **109**, 855–865.
Nir, Y. D., and Goldwasser, R. (1961). *Am. J. Hyg.* **73**, 294–296.
Nishimura, C., and Tsukeda, H. (1971). *Jpn. J. Microbiol.* **15**, 309–316.
Nishimura, C., Nomura, M., and Kitaoka, M. (1967). *Jpn. J. Med. Sci. Biol.* **20**, 377–386.
Nozima, T., Mori, H., Minobe, Y., and Yamamota, S. (1964). *Acta Virol. (Engl. Ed.)* **8**, 97–103.
Obijeski, J. F., Marchenko, A. T., Bishop, D. H. L., Cann, B. W., and Murphy, F. A. (1974). *J. Gen. Virol.* **22**, 21–33.
Otsuka, S., Manako, K., and Mori, R. (1967). *Saikingaku Zasshi Nippon* **22**, 321–325.
Pfefferkorn, E. R., and Shapiro, D. (1974). *In* "Comprehensive Virology" (H. Fraenkel-Conrat and R. R. Wagner, eds.), Vol. 2, pp. 171–230. Plenum, New York.
Pogodina, V. V. (1958). *Vopr. Virusol.* **3**, 295–299.
Porterfield, J. S., Casals, J., Chumakov, M. P., Gaidamovich, S. Y., Hannoun, C., Holmes, I. H., Horzinek, M. C., Mussgay, M., Oker-blom, N., Russell, P. K., and Trent, D. W. (1978). *Intervirology* **9**, 129–148.
Qureshi, A. A., and Trent, D. W. (1973a). *Infect. Immun.* **7**, 242–248.
Qureshi, A. A., and Trent, D. W. (1973b). *Infect. Immun.* **8**, 985–992.
Qureshi, A. A., and Trent, D. W. (1973c). *Infect. Immun.* **8**, 993–999.
Rai, J., and Ghosh, S. N. (1976). *Indian J. Med. Res.* **64**, 981–991.

Renkonen, O., Kääriäinen, L., Simons, K., and Gahmberg, G. G. (1971). *Virology* **46,** 318–326.
Rosato, R. R., Dalrymple, J. M., Brandt, W. E., Cardiff, R. D., and Russell, P. K. (1974). *Acta Virol. (Engl. Ed.)* **8,** 25–30.
Russell, P. K. (1970). *In* "Immunopathology" (P. A. Miescher, ed.), pp. 426–435. Grune & Stratton, New York.
Russell, P. K. (1971). *In* "Progress in Immunology" (B. Amos, ed.), pp. 831–838. Academic Press, New York.
Russell, P. K., and Brandt, W. E. (1973). *Virol.* **8,** 263–277.
Russell, P. K., Udomsakdi, S., and Halstead, S. B. (1967). *Jpn. J. Med. Sci. Biol.* **20,** 103–108.
Russell, P. K., Intavivat, A., and Kanchanapilant, S. (1969). *J. Immunol.* **102,** 412–420.
Russell, P. K., Chiewsilp, D., and Brandt, W. E. (1970). *J. Immunol.* **105,** 838–845.
Saturno, A. (1967). *Bull. W. H. O.* **36,** 347–349.
Schlesinger, R. W. (1977). "Dengue Viruses," Virology Monographs, Vol. 16. Springer-Verlag, Berlin and New York.
Scott, R. M., McCown, J. M., and Russell, P. K. (1972). *Infect. Immun.* **6,** 277–281.
Shapiro, D., Brandt, W. E., Cardiff, R. D., and Russell, P. K. (1971a). *Virology* **44,** 108–124.
Shapiro, D., Cardiff, R. D., Brandt, W. E., and Russell, P. K. (1971b). *Bacteriol. Proc.* p. 182.
Shapiro, D., Brandt, W. E., and Russell, P. K. (1972a). *Virology* **50,** 906–911.
Shapiro, D., Kos, K. A., Brandt, W. E., and Russell, P. K. (1972b). *Virology* **48,** 360–372.
Shapiro, D., Trent, D. W., Brandt, W. E., and Russell, P. K. (1972c). *Infect. Immun.* **6,** 206–209.
Shapiro, D., Kos, K. A., and Russell, P. K. (1973a). *Virology* **56,** 88–94.
Shapiro, D., Kos, K. A., and Russell, P. K. (1973b). *Virology* **56,** 95–109.
Shatkin, A. J. (1976). *Cell* **9,** 645–653.
Sinarachatanant, P., and Olson, L. C. (1973). *J. Virol.* **12,** 275–283.
Smith, T. J., Brandt, W. E., Swanson, J. L., McCown, J. M., and Buescher, E. L. (1970). *J. Virol.* **5,** 524–532.
Stevens, T. M., and Schlesinger, R. W. (1965). *Virology* **27,** 103–112.
Stohlman, S. A., Wisseman, C. L., Jr., Eylar, O. R., and Silverman, D. J. (1975). *J. Virol.* **16,** 1017–1026.
Stohlman, S. A., Wisseman, C. L., Jr., and Eylar, O. R. (1978). *Acta Virol. (Engl. Ed.)* **22,** 31–36.
Stollar, V. (1969). *Virology* **39,** 426–438.
Stollar, V., Stevens, T. M., and Schlesinger, R. W. (1966). *Virology* **30,** 303–312.
Stollar, V., Schlesinger, R. W., and Stevens, T. M. (1967). *Virology* **33,** 650–658.
Strohmaier, K., Streissle, G., and De Noronha, S. C. (1965). *Arch. Gesamte Virusforsch.* **17,** 300–303.
Sunaga, H., Taylor, R. M., and Henderson, J. R. (1960). *Am. J. Trop. Med. Hyg.* **9,** 419–424.
Sung, J. S., Diwan, A. R., Falkler, W. A., Jr., Yang, H.-Y., and Halstead, S. B. (1975). *Intervirology* **5,** 137–149.
Takehara, M., and Hotta, S. (1961). *Science* **134,** 1878–1880.
Theiler, M. (1957). *Proc. Soc. Exp. Biol. Med.* **96,** 380–382.
Tomita, Y., and Prince, A. M. (1963). *Proc. Soc. Exp. Biol. Med.* **112,** 887–890.
Trent, D. W. (1977). *J. Virol.* **22,** 608–618.

Trent, D. W. (1980). *In* "St. Louis Encephalitis" (T. Monath, ed.), pp. 159–199. Am. Public Health Assoc., Washington, D.C.
Trent, D. W., and Qureshi, A. A. (1971). *J. Virol.* **7**, 379–388.
Trent, D. W., Swenson, C. C., and Qureshi, A. A. (1969). *J. Virol.* **3**, 385–394.
Trent, D. W., Harvey, C. L., Qureshi, A. A., and LeStourgeon, D. (1976). *Infect. Immun.* **13**, 1325–1333.
Wengler, G., Wengler, G., and Gross, H. J. (1978). *Virology* **89**, 423–437.
Westaway, E. G. (1968a). *Nature (London)* **219**, 78–79.
Westaway, E. G. (1968b). *J. Immunol.* **100**, 569–580.
Westaway, E. G. (1973). *Virology* **51**, 454–465.
Westaway, E. G. (1975). *J. Gen. Virol.* **27**, 283–292.
Westaway, E. G., and Reedman, B. M. (1969). *J. Virol.* **4**, 688–693.
Westaway, E. G., and Shew, M. (1977). *Virology* **80**, 309–319.
Westaway, E. G., Della-Porta, A. J., and Reedman, B. M. (1974). *J. Immunol.* **112**, 656–663.
Westaway, E. G., Shew, M., and Della-Porta, A. J. (1975). *Infect. Immun.* **11**, 630–634.
Westaway, E. G., McKimm, J. L., and McLeod, L. G. (1977). *Arch. Virol.* **53**, 305–312.

19

Replication of Flaviviruses

E. G. WESTAWAY

I. INTRODUCTION

Flaviviruses are the most numerous of the togaviruses (60 of the total of 88; see Chapter 1) and are prominent in disease in man (e.g., yellow fever, dengue, encephalitis) and in domestic animals (e.g.,

THE TOGAVIRUSES

ISBN 0-12-625380-3

louping ill, Wesselsbron disease). Although the susceptibility of individual hosts may vary from one flavivirus to another, a wide range of species and corresponding cell cultures from mammals, birds, and vector mosquitoes and ticks support their replication (Catalogue of Arthropod-Borne Viruses of the World, 1967; Theiler and Downs, 1973). Nevertheless, the events of their replication are poorly understood. Efforts to improve this situation are beset by inherent difficulties not commonly encountered with most other RNA viruses because in cell cultures the yields of flaviviruses are usually lower, the latent period is much greater, and macromolecular syntheses of the host cell are not switched off. The last-mentioned is the most serious difficulty, resulting in problems of resolution and identification of intracellular virus-specified proteins against a high background of continuing host protein synthesis.

The initial reports of RNA synthesis (Stollar *et al.*, 1967) and protein synthesis (Westaway and Reedman, 1969) in flavivirus-infected cells indicated differences and/or greater complexity of replicative events compared to results with the better characterized alphaviruses. Further progress has continued to be hampered by the difficulties enumerated above (Pfefferkorn and Shapiro, 1974). Recent evidence indicates that the replication of flaviviruses differs profoundly, especially in regard to strategy of translation, not only from that of alphaviruses but also from that of all other RNA viruses in eukaryotic cells (Westaway, 1977). Hence the time is appropriate to review these and other recent developments in the broad context of all known replication events of flaviviruses.

The biochemical evidence which forms much of the basis for this chapter is confined mainly to studies with dengue type 2 (DEN-2), Japanese encephalitis (JE), Kunjin, West Nile, and St. Louis encephalitis (SLE) viruses, and comparisons of these data indicate such a close similarity that it seems reasonable to assume that all flaviviruses share a common strategy of replication.

II. TRANSCRIPTION

A. The Initial Template

Because of the infectivity of naked viral RNA, the genome must provide the template from which all transcription processes stem. The sedimentation coefficient of the genomic RNA is reported to be in the range 38–45 S (Igarashi *et al.*, 1964; Stollar *et al.*, 1966; Trent *et al.*,

1969; Boulton and Westaway, 1972). This range probably reflects a variation in the sedimentation markers or in the conformation of the RNA rather than any real difference in size. Measurement of the molecular weight of virion RNA by gel electrophoresis has been reported only for Kunjin virus. The size is identical to that of the alphavirus Sindbis, 4.2×10^6, but the Sindbis genome sediments slightly faster; if 42 S (or 49 S as measured by Strauss and Strauss, 1977) is assumed for the Sindbis genome, Kunjin RNA sediments at 38 S (or 45 S) (Boulton and Westaway, 1972). The genome of flaviviruses differs significantly from that of other positive-strand viruses in that the size of any poly(A) tract is either much reduced (Brawner *et al.*, 1977) or is virtually absent. In a definitive study Wengler *et al.* (1978) showed that West Nile virion RNA did not bind to oligo(dT)-cellulose, and that the T1 RNase fingerprints contained no oligo(A) sequences at the limit of detection (10 bases in length); the 5′-termini did contain the cap structure $m^7GpppAmpN_1$.

For the remainder of the chapter, the sedimentation coefficient of the flavivirus genome, and of the corresponding single-stranded RNA found in infected cells, will be assumed in all cases to be 44 S, which represents the mean of reported values.

B. Virus-Specified RNA

Inhibition by actinomycin D of host DNA-dependent RNA synthesis provides an extremely useful tool in analysis of flavivirus-specified RNA synthesis. A pulse of 2 h with an adequate but nontoxic dose is sufficient to reduce [^{3}H]uridine incorporation to 3–5% in uninfected cells compared to about 17% in flavivirus-infected cells at 24 h (Stollar *et al.*, 1967; Boulton and Westaway, 1977, and unpublished results). Attempts to enhance this effect are fraught with the hazard of induction of drug-mediated cytotoxicity. Thus reports of actinomycin D inhibition of replication of JE virus (Yamazaki, 1968; Zebovitz *et al.*, 1972) appear to be based on excessive use of the drug (dosage or period of exposure).

In DEN-2 virus-infected cells treated with actinomycin D, Stollar *et al.* (1967) found that [^{3}H]uridine was incorporated predominantly into two species of RNA with sedimentation coefficients of 44 and 20 S (Fig. 1). The 44 S form was RNase-sensitive; its identity as genomic RNA was supported by the close similarity of its base composition to that of virion RNA and by the high yield of infectious RNA extractable from infected cells. The 20 S RNA was RNase-resistant, hence double-stranded; after denaturation with dimethyl sulfoxide, it

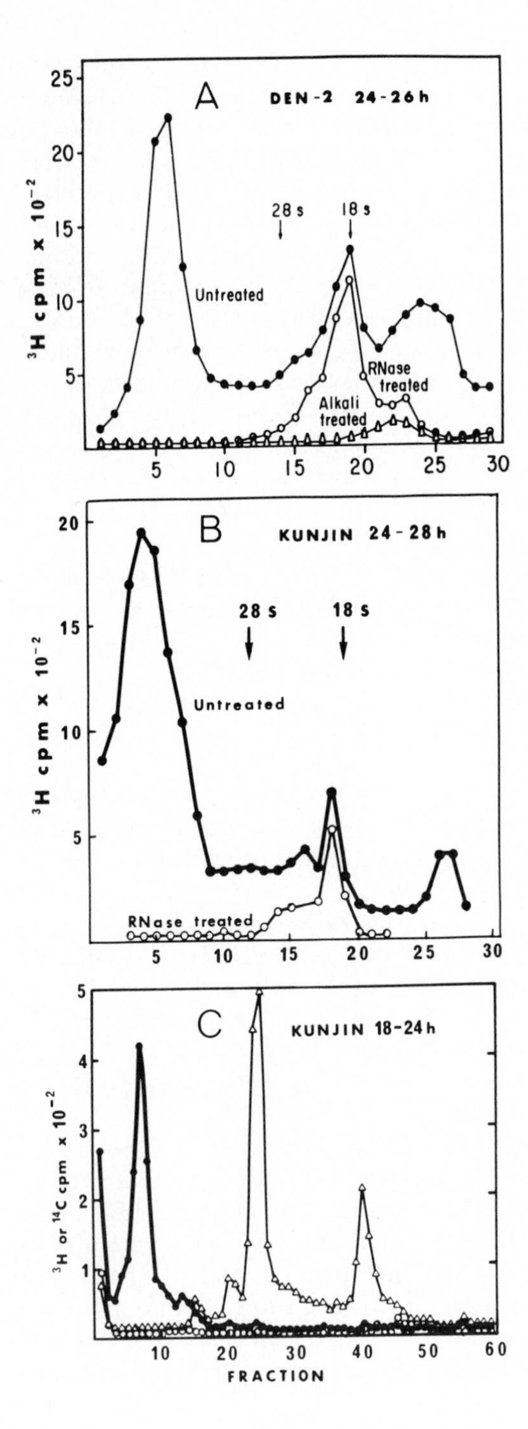
A
DEN-2 24-26h
25
20
15
10
5
^{3}H cpm x 10^{-2}
28 s
18 s
Untreated
RNase treated
Alkali treated
5 10 15 20 25 30
B
KUNJIN 24-28h
20
15
10
5
^{3}H cpm x 10^{-2}
28 s
18 s
Untreated
RNase treated
5 10 15 20 25 30
C
KUNJIN 18-24h
5
4
3
2
1
^{3}H or ^{14}C cpm x 10^{-2}
10 20 30 40 50 60
FRACTION

sedimented as a 44 S peak of presumably single strands. A comparatively small amount of label was possibly incorporated into an incompletely resolved 26 S form. The synthesis of single-stranded 44 S RNA and of double-stranded 20 S RNA in flavivirus-infected cells has been amply confirmed by others (Trent *et al.*, 1969; Nishimura and Tsukeda, 1971; Zebovitz *et al.*, 1972, 1974; Takeda *et al.*, 1977; Boulton and Westaway, 1977). The 20 S RNase-resistant RNA from JE virus-infected cells was separable by benzoylated DEAE-cellulose chromatography into two peaks (Fukui, 1973). The size of the replicating 44 S RNA measured by gel electrophoresis is the same as that of the virion RNA for several flaviviruses, $4.0–4.2 \times 10^6$ daltons (Boulton and Westaway, 1977; Naeve and Trent, 1978; Wengler *et al.*, 1978).

Wengler *et al.* (1978) showed that West Nile virus-specified 20 S RNA from infected cells was RNase-resistant and was converted to 44 S RNA by denaturation; its hybridization to plus-strand 44 S RNA and fingerprint analyses indicated that it contained plus and minus strands, confirming the data of Stollar *et al.* (1967) on DEN-2 20 S RNA. The West Nile 44 S RNA from infected cells contained the single cap structure m^7GpppAmpN and no poly(A), as in the virion RNA. Analyses of intracellular DEN-2 44 S RNA by Cleaves and Dubin (1979) have confirmed this cap structure. The latter authors note that flavivirus plus-strand RNA exhibits ribose methylation of the penultimate base (a type I cap) at the 5′-terminus, hence differs from alphavirus plus-strand RNA (a type 0 cap) (Dubin *et al.*, 1977), and that the absence of methylation in the N^6-position of the penultimate Am distinguishes both togavirus genera from negative-strand RNA viruses.

The significance of a 26 S RNA species is uncertain. In contrast to the results with DEN-2 virus (Stollar *et al.*, 1967) and with Kunjin virus (Boulton and Westaway, 1977) showing a minor amount of heterodisperse 26 S RNA, less RNase-resistant than 20 S RNA (see Fig. 1), a prominent peak of largely RNase-resistant 26 S RNA from

Fig. 1. Separation and RNase treatment of DEN-2 (A) and Kunjin (B and C) virus-specified RNA from infected cells. Infected cells were treated for 1–2 h with 3 μg actinomycin D per milliliter and then labeled with [^{3}H]uridine for 24–26 h (A), 24–28 h (B), and 18–24 h (C). RNA was extracted and then either centrifuged through 5–20% sucrose gradients (A and B) or analyzed by electrophoresis in 2.5% polyacrylamide gels (C). The RNA was treated with RNase [2 μg/ml for 10 min at 37°C (open circles)] either before electrophoresis (C) or after sedimentation (A and B). The profile of the coelectrophoresed ^{14}C-labeled rRNA markers (28 and 18S) is coplotted (triangles) in (C). [(A) From stollar *et al.* (1967), reproduced with permission); (B) From R. W. Boulton and E. G. Westaway (unpublished results); (C) From Boulton and Westaway (1977), reproduced with permission.]

SLE virus-infected cells was resolved by Trent *et al.* (1969). In JE virus-infected cells synthesis has been reported of partially RNase-resistant 26 S RNA (Nishimura and Tsukeda, 1971; Takeda *et al.*, 1977) and of RNase-sensitive 26 S RNA (Zebovitz *et al.*, 1972, 1974). The reports of RNase-sensitive 26 S RNA need to be viewed with some caution because actinomycin D was added to cells only at the same time as the label or it was added prior to infection at a dose which still permitted significant incorporation of [^{3}H]uridine into host RNA including 28 S rRNA.

The balance of evidence indicates that small, variable amounts of heterodisperse and partially RNase-resistant 26 S RNA are found in flavivirus-infected cells, but it has not been satisfactorily resolved by gel electrophoresis (Boulton and Westaway, 1977; Naeve and Trent, 1978; Wengler *et al.*, 1978). After scrutiny of the publications dealing with flavivirus RNA, Naeve and Trent (1978) noted that the 26 S RNA sedimentation peak was prominent only when a high multiplicity of infection was employed and suggested that 26 S RNA was defective interfering particle RNA. However, the occurrence of 26 S RNA in progeny flavivirus particles released from infected cells has never been reported. The other possibilities are that flavivirus 26 S RNA is produced in small amounts as a normal product of replication, or that it represents a mixture of 44 S RNA and 20 S double-stranded RNA, either complexed or partially degraded, compatible with the single report on its base composition (Trent *et al.*, 1969).

The sedimentation analyses of JE virus-specified intracellular RNA discussed above show also a prominent peak of 8–12 S RNA either RNase-sensitive (Zebovitz *et al.*, 1972, 1974; Takeda *et al.*, 1977) or RNase-resistant (Nishimura and Tsukeda, 1971). It was important to establish whether the 8–12 S RNA or the 26 S RNA of flavivirus-infected cells included RNA of a size and configuration such that it could function as mRNA because of the relevance to replication strategy. The 8–12 S RNA remains near the top of sucrose density gradients and may represent only degraded RNA; the peak of partially RNase-resistant 26 S RNA may mask or overlap with a single-stranded 26 S messenger species. Both these problems are solved by polyacrylamide gel electrophoresis, in which the smaller species of RNA move farthest from the origin and the double-stranded species remain at or near the origin. Under these conditions it was shown (Boulton and Westaway, 1977) that the predominant species of single-stranded RNA in Kunjin virus-infected Vero cells was genome-sized RNA; no single-stranded RNA equivalent to 26 S RNA or smaller (down to the lowest detectable limit of 130,000 molecular weight) was found (Fig.

1). Confirmation came from similar analyses by Naeve and Trent (1978) and Wengler *et al.* (1978) who reported that the only low-molecular-weight RNA species in flavivirus-infected vertebrate cells were single-stranded pieces of 50×10^3 (Uganda S virus-specified) and 65×10^3 and 42×10^3 molecular weight (West Nile virus-specified), respectively. The RNA of 65×10^3 molecular weight was of plus-strand polarity, was not "capped," and contained no poly(A) sequence; it had no mRNA activity in cell-free translation systems. Hence the low-molecular-weight RNA species, including the 8–12 S RNA discussed above, are almost certainly degraded or incomplete and inactive portions of 44 S plus-strand RNA.

In summary, the flavivirus intracellular RNA species larger than 10^5 in molecular weight comprise single-stranded 44 S RNA [apparently identical to the genome in regard to size, presence of capped 5′-termini, and absence of poly(A)], double-stranded 20 S RNA (predominantly RNase-resistant), and variable and often minor amounts of heterodisperse and partially RNase-resistant 26 S RNA. The RNase resistance of 20 S RNA is more than 60% and as much as 90% but cannot be defined precisely because the heterodisperse 26 S RNA trails into the 20 S RNA during sedimentation.

C. Synthesis of Virus-Specified RNA

1. RNA Synthesis during the Latent Period

The long latent period (12 h or more) in flavivirus-infected cells is associated with barely detectable synthesis of virus-specified RNA. However, addition to cells prior to 6 h postinfection of the base analog 6-azauridine, at concentrations well below the toxic dose, inhibited the production of infectious DEN-2 virus, and the effect could be reversed by uracil during the 0–6 h period (Stollar *et al.*, 1966). Similarly, addition of the adenosine analog formycin A to JE virus-infected cells showed that viral RNA synthesis began shortly after 6 h and 6–7 h in advance of release of mature virus (Nishimura and Tsukeda, 1971).

Identification of the initial RNA product(s) has been elusive. Stollar *et al.* (1967) reported that RNA from DEN-2 virus-infected cells labeled for 4–6 or 8–10 h was indistinguishable from uninfected cell RNA. At 10–12 h and later, 44 S RNA and RNase-resistant 20 S RNA were resolved, but only "a small hump" at about 26 S was observed; the base composition of 44 S RNA was equivalent to that of 44 S virion RNA. In JE virus-infected cells during a 30-min pulse early in infection (7.5–8 h) no detectable label was incorporated into 44, 26, or 20 S (Nishimura and Tsukeda, 1971); although these RNA species were labeled during a 6-h chase, this probably represents incorporation of residual [^{3}H]uridine. In Kunjin virus-infected cells, no increase in the rate of actinomycin-resistant RNA synthesis was detected during the latent period, and no viral RNA species were resolved by sedimentation of RNA from cells labeled 6–8 h postinfection (Boulton and Westaway, 1977).

In contrast to the above results, Trent *et al.* (1969) observed at 6–8 h only in SLE virus-infected cells a "spike" of synthesis of RNA which comprised 44 S RNA and largely RNase-resistant 26 and 20 S RNA. The 44 S RNA synthesized subsequently in infected cells had virtually the same base composition as virion 44 S RNA and was infectious, but the ratio of its plaque-forming capacity to [^{3}H]uridine content was very low compared to that of later (maximum) harvests.

In summary, although RNA synthesis commences by about 6 h postinfection (Stollar *et al.*, 1966; Nishimura and Tsukeda, 1971), this early viral RNA (apparently very small amounts of 44 S and of largely RNase-resistant 20–26 S RNA, identified only by Trent *et al.*, 1969) is synthesized at a slow, constant rate. In order to complete the replication cycle, both plus and minus strands of 44 S RNA must be produced during the latent period.

2. *RNA Synthesis after the Latent Period*

After 12 h, the rate of incorporation of [^{3}H]uridine into flavivirus-infected cells increases 4- to 8-fold to reach a maximum rate by 24–30 h; the increase in synthesis or accumulation of 44 S RNA is apparently much greater than that of 20 S RNA (Stollar *et al.*, 1967; Trent *et al.*, 1969; Nishimura and Tsukeda, 1971; Boulton and Westaway, 1977). When corresponding growth curves are also compared, the rise in [^{3}H]uridine incorporation is far outstripped by the 10^3-fold or greater increase in the release of plaque-forming units (PFUs) from cells; the yield of intracellular infectious RNA also rises by more than 100-fold during the same period (Stollar *et al.*, 1967; Trent *et al.*, 1969). Clearly the increased amounts of 44 S RNA synthesized late in infection are readily available for efficient incorporation into virions and are extractable from cells in a form more infectious than when extracted during the latent period. Reports on the base composition of 20 S double-stranded RNA labeled during this period differ and are discussed in Section II,C,3. Conflicting reports on the nature and variable synthesis of 26 S RNA have been summarized previously (Section II,B).

Persistent flavivirus infections have been established in mammalian and in mosquito cell lines (reviewed in Schlesinger, 1977). Definition of the forms of viral RNA produced in these cultures or found in any associated defective interfering particles may assist in further understanding of the relationships discussed below. Although defective interfering particles appeared to be responsible for persistent JE virus infections in two mammalian cell lines (Schmaljohn and Blair, 1977), no new RNA species were seen by sedimentation analysis of JE virus-specific RNA from persistently infected cells or their clones (Schmaljohn and Blair, 1979).

3. *Relationship of 44, 26, and 20 S RNA*

Base analysis of intracellular 44 S RNA at all detectable periods was equivalent to that of the genome (Stollar *et al.*, 1967; Trent *et al.*, 1969). The role of the double-stranded 20 S RNA as a possible replicating form was indicated by base analysis of 20 S RNA labeled with ^{32}P and extracted at various times after infection. Early in infection approximate base pairing in labeled dengue virus 20 S RNA was observed by Stollar *et al.* (1967), indicating synthesis of both positive and negative strands, but after 12 h the labeled base composition differed and was similar to that of virion 44 S RNA (positive strand). However, in SLE virus-infected cells, base analyses of 20 S RNA labeled early in infection (7–10 h) and at least until 16–19 h appeared to comprise equally labeled positive and negative strands (Trent *et al.*, 1969). Nevertheless, an excess of complete positive strands was produced from at least 16 h in all infections, because of the relatively large amounts of 44 S RNA accumulating in infected cells and incorporated into released virions. The results of Stollar *et al.* (1967) indicate that the minus strand (synthesized prior to 12 h postinfection) is stable for at least 20 h and must continue to function as template for transcription into 44 S positive strands. Clearly the 20 S double-stranded RNA is associated with production of 44 S single-stranded RNA which after the latent period is predominantly of positive-strand polarity and functions in polyribosomes as mRNA (Naeve and Trent, 1978; Cleaves and Schlesinger, 1977).

Several reports indicate a possible role for 26 S RNA in the transition from 20 S double-stranded RNA to 44 S single-stranded RNA. After a 10-min pulse with [^{3}H]uridine in JE virus-infected cells pretreated with actinomycin D and glucosamine at noninhibitory concentrations, extraction of a nuclear pellet yielded 20 S RNA, whereas the cytoplasmic fraction contained only 44 S RNA (Takeda *et al.*, 1977, 1978). After a 10-min chase the nuclear pellet yielded only a 26 S RNA peak (the 20 S RNA was then absent), and after a 20-min chase no 26 S RNA remained but 26 S RNA was then present in cytoplasmic fractions; the RNase resistances were not recorded. These and earlier results in SLE virus-infected cells (Trent *et al.*, 1969) suggest a role for 26 S RNA as an intermediate structure during a transition from recently labeled RNase-resistant 20 S RNA to fully RNase-sensitive 44 S RNA. Such a role is in accord with its heterodisperse sedimentation and 30–40% RNase resistance (Fig. 1), but note that flavivirus 26 S RNA sediments more slowly than the 44 S genome, in contrast to the behavior of other partially RNase-resistant structures such as the poliovirus replicative intermediate [range 18–70 S compared to the 35 S genome (Baltimore, 1968)].

In summary, pulse label is incorporated initially into predominantly RNase-resistant 20 S RNA, the incorporation being into full-length positive strands (and negative strands early in infection). In some infections label accumulates more slowly and transiently in heterodisperse and partially RNase-resistant 26 S RNA. Within about 10 min label appears in single strands as 44 S RNA, released from 20 S (and possibly 26 S) RNA.

4. *Structure of 20 S RNA*

The template for transcription of flavivirus RNA resides in the RNase-resistant 20 S RNA, based on (1) its presence, together with 44 S RNA and virus-specified antigens, in an active replication complex isolated by Qureshi and Trent (1972), (2) the presence of full-length positive and negative strands (Stollar *et al.*, 1967) which could function as templates, (3) the immediate incorporation of label which is later transferred to other RNA species (Trent *et al.*, 1969; Takeda *et al.*, 1977, 1978), and (4) evidence that it was the major RNA species labeled during *in vitro* assays with JE virus RNA polymerase (Cardiff *et al.*, 1973a; Zebovitz *et al.*, 1974).

The 20 S flavivirus RNA differs from the replicative intermediate (RI) containing the template for other positive-strand RNA animal viruses in the following respects: (1) The 20 S RNA is as much as 90% RNase-resistant after both short and long pulses of label. In contrast, the RNase resistance of totally labeled (after 40 min) poliovirus RI is 37% (Baltimore, 1968); for Sindbis virus, the RNase resistance of pulse-labeled RNA falls rapidly after 1 min to only 42% after 2 min and 30% after 4 min (Simmons and Strauss, 1972). (2) Flavivirus 20 S RNA sediments as a relatively sharp band compared to poliovirus RI which is heterodisperse, sedimenting at 18–70 S (Baltimore, 1968). For Sindbis virus, the RI structures sediment as sharp, discrete bands only after prior RNase digestion (Simmons and Strauss, 1972). (3) Base pairing of 20 S RNA is observed (by base analysis) without any prior requirement for RNase digestion. This property, and those in (1) and (2) are characteristic of replicative forms, which are derived as "cores" by RNase digestion of RIs, as for Sindbis virus (Simmons and Strauss, 1972), or are considered inactive double-stranded by-products of transcription accumulating late in (for example) poliovirus infection (Baltimore, 1969).

What then is the form of 20 S RNA? Why does it differ from conventional RIs in being apparently so deficient in digestible single-stranded tails? In an analysis of the structure of totally labeled poliovirus RI with 37% RNase resistance, Baltimore (1968) calculated an average value (N) of 6.5 growing strands per RI, based on the formula:

$$\text{Percentage RNase resistance} = [2/(2 + N/2)] \times 100$$

When this formula is applied to flavivirus 20 S RNA, $N = 1$ when RNase resistance is 80%, a commonly observed value ($N < 1$ for greater values). If incorporation of label is asymmetric, as seems to occur late in infection or during very short pulses, $N = 1$ when RNase resistance is 66%, because the formula for asymmetric labeling becomes $100[1/(1 + N/2)]$ [see Baltimore (1968)]. Under the latter conditions, N for poliovirus is 3.5, or 6.5 if allowance is made for slow uptake of [^{3}H]uridine during short pulses (Baltimore, 1969). The mathematical derivation of N was confirmed by electron microscopy of poliovirus RI, which showed, on the average, four to five single-stranded tails attached to the double-stranded core (Savage *et al.*, 1971).

It follows that the structure of flavivirus 20 S RNA is best represented as a single-stranded template equivalent in length to the genome, or 44 S RNA, which is base-paired for most of its length with a single nascent complementary strand which probably collapses onto the template during phenol extraction. The significance of this interpretation will be discussed in Section II,D.

5. *RNA Polymerase and Site(s) of Transcription*

The site of synthesis of flavivirus RNA is clearly located in smooth membranes of density about 1.10 (Boulton and Westaway, 1976), and a growing body of evidence implicates the perinuclear region of infected cells (Takehara, 1971; Brawner *et al.*, 1973; Cardiff *et al.*, 1973b; Zebovitz *et al.*, 1974; Lubiniecki and Henry, 1974; Kos *et al.*, 1975a; Takeda *et al.*, 1977, 1978). Some reports suggested a role of the nucleus per se; those based on inhibition of yield by actinomycin D (Yamazaki, 1968; Zebovitz *et al.*, 1972) can be discounted because of the high doses employed and the lack of appropriate controls for excluding the cytotoxic effects of the drug. Nuclear immunofluorescence, using antiserum which should contain antibodies to all virus-specified proteins including polymerase(s), has not been observed (Bhamarapravati *et al.*, 1964; Southam *et al.*, 1964; Cardiff *et al.*, 1973b; Eldadah and Nathanson, 1967), but Kos *et al.* (1975a) noted that such studies were conducted after the latent period. The yield of JE virus from cells enucleated during the latent period was severely depressed, whereas only a small effect was noted in similar experiments with Sindbis alphavirus (Kos *et al.*, 1975a). However, flavivirus inhibition could result from loss during enucleation of nuclear-associated cytoplasmic membranes (see later). Cleaves and Dubin (1979) interpret the absence of internal m^6Ade in 44 S DEN-2 RNA isolated during the logarithmic phase as evidence that the RNA replication is largely or wholly extranuclear. In summary, a unique role of the cell nucleus in flavivirus replication seems unlikely.

A firm case exists for the involvement of perinuclear membranes in flavivirus replication, based on autoradiography of DEN-2 and JE virus-infected cells labeled after the latent period with [^{3}H]uridine (Lubiniecki and Henry, 1974; Takeda *et al.*, 1978) and on the intense perinuclear localization of fluorescent-labeled antibody specifically to nonstructural antigens beginning at 10 h in dengue virus infections (Cardiff *et al.*, 1973b). Juxtanuclear inclusions were seen by fluorescent antibody and by acridine orange staining in West Nile virus-infected cells (Southam *et al.*, 1964). In addition to the effect of enucleation on yield noted by Kos *et al.* (1975a), the viral RNA polymerase activity of JE virus (Zebovitz *et al.*, 1974) and of SLE virus (Brawner *et al.*, 1973) was closely associated with the nucleus, and in the former case the labeled RNA was removed from the nuclear envelope by mild detergent treatment. After a 10-min pulse of JE virus-infected cells at 15–16 h with [^{3}H]uridine, a large proportion of the incorporation was into 20 S RNA in the nuclear fraction, compared to almost no incorporation into 20 S RNA in the cytoplasmic fraction; however, whereas Takeda *et al.* (1977, 1978) reported that 44 S RNA was concurrently labeled only in cytoplasm, Zebovitz *et al.* (1974) reported that 44 S RNA was labeled only in the nuclear fraction. The conflicting results are difficult to evaluate because of the use of different host cells and cell disruption procedures.

From SLE virus-infected cell extracts treated with EDTA, a detergent-sensitive 250 S replication complex containing viral antigens was separated which retained both RNA polymerase activity and pulse-labeled (15 min) 44 S RNA and 20–26 S partially RNase-resistant RNA (Qureshi and Trent, 1972). The polymerase activity was first detected 6 h after infection and was most active during the rapid phase of viral growth. Similar properties were found in RNA polymerase isolated from cells late in infection with dengue virus or JE virus; the *in vitro* products included 20–26 S heterodisperse RNA (but no 44 S RNA) which for dengue virus was RNase-sensitive and for JE virus was RNase-resistant (Cardiff *et al.*, 1973a; Zebovitz *et al.*, 1974).

In summary, the polypeptide composition of flavivirus RNA polymerase is unknown, but its activity appears to be associated with smooth membranes adjacent to the nuclear membrane.

D. A Suggested Model of Transcription

In Section II,C,4, flavivirus 20 S RNA was represented as comprising a template of 44 S single-stranded RNA which is base-paired for most of its length to a growing complementary strand. A similar structure has been described for the small RNA phages in which commonly only one nascent strand appears to be attached to the RI template; hence in nearly all cases another strand cannot be initiated until the single nascent strand is completed and released (Granboulan and Franklin, 1968; Robertson and Zinder, 1969). An interpretation of the mechanism of flavivirus transcription, compatible with the reported observations, is possible if the RNA phage process is modified to provide for either a delay in release of the polymerase and the newly completed strand after each traverse of the template, or for a rapid transcription rate for most of the template which decreases because of secondary folding as the 5′-end of the template is approached. In either case, complete or almost complete double strands will accumulate (as 20 S RNA). In contrast to RNA phage synthesis (Robertson, 1975), the processes of flavivirus transcription and translation occur at different cell sites (Boulton and Westaway, 1976). The release of 44 S RNA from the 20 S RNA complex may occur in some cases via an intermediate or 26 S RNA which is heterodisperse and partially RNase-resistant (Section II,C,3). The ease of detection and the percentage of RNase resistance of 26 S RNA is probably influenced by whether or not transcription is symmetric (see later).

A slow release of the polymerase from the accumulating completed 20 S double-stranded RNA would provide an opportunity for control of transcription and translation. For this purpose I assume (1) that release or displacement of the polymerase is facilitated by attachment of V2 (core protein) near the 3′-end of each positive strand after it is synthesized, and (2) that V2 attached to the 3′-end of positive-strand RNA (including the infecting genome) is protected from rapid "trimming" to NV1½ (the observed fate of most recently synthesized V2; see Westaway, 1977; Wright and Westaway, 1977), until it is displaced by ribosomes during translation. (Alternatively, the suggested control function of V2 could be performed by one of the nonstructural proteins with an appropriate affinity for the genome.) Attachment of individual proteins specifically to the 3′-end of eukaryotic mRNA is well documented (e.g., Blobel, 1973; Hellerman and Shafritz, 1975). The following elements of control are then possible:

1. Replication from positive strands as templates decreases during the replication cycle because of steric hindrance or blocking by V2 of initiation sites for the polymerase at the 3′-end; hence negative strands will be favored as templates, resulting in asymmetric transcription, and eventually an excess of positive strands will be synthesized for translation and for incorporation into virions.
2. Control as in (1) will not become operative until V2 becomes available in adequate concentration; hence during the latent period both positive and negative strands will continue to be synthesized, but turnover will be slow because of a slow release of polymerase for recycling (limiting amounts of V2). Translation will also be severely restricted during this period, because positive strands (mRNA) tend to be sequestered in base-paired 20 S RNA in the perinuclear region.
3. The attachment of V2 near the 3′-end of positive strands will not interfere with the initiation of translation (nearer the 5′-end), which will be autocatalytic (more V2 product, more positive strands directed from transcription to translation).
4. Positive strands with the core protein V2 attached can initiate the process of assembly of virions.

The model is presented diagrammatically in Fig. 2. It appears to explain many of the observed but puzzling features of flavivirus replication, including the following:

1. The long latent period of 12 h or more.
2. The slow production of detectable amounts of RNA and the even slower (undetectable) synthesis of viral proteins during the latent period.
3. The central role of 20 S RNase-resistant RNA and RNA synthesis and its accumulation.

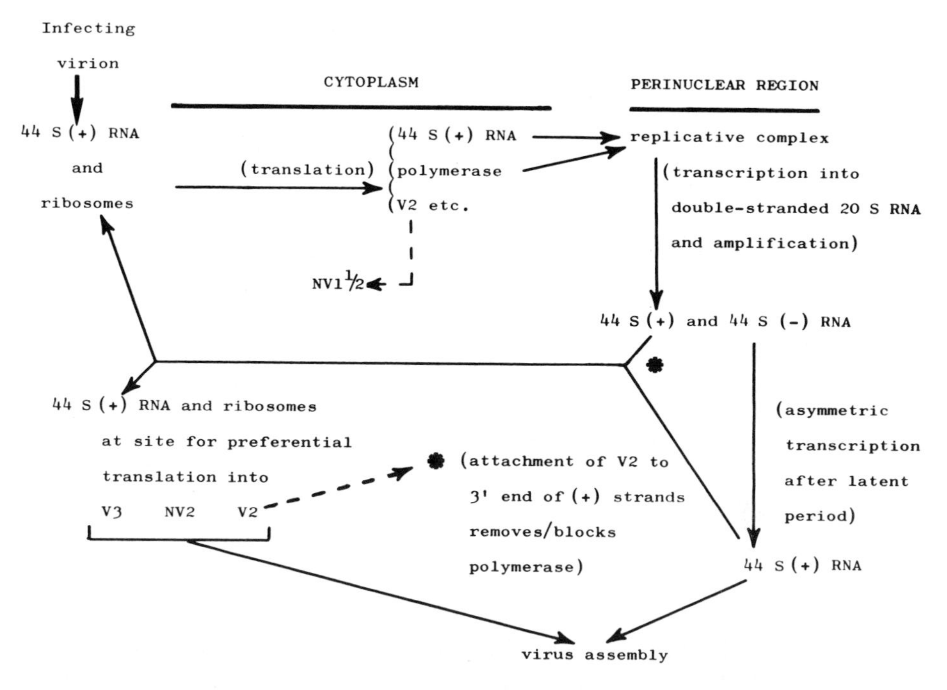

Fig. 2. Model of transcriptional and translational control during the flavivirus replication cycle. It is assumed that only one nascent RNA strand is on each 44 S RNA template and that the polymerase is released slowly from the 20 S RNA complex, after each round of transcription, thus limiting the rate of RNA synthesis. The release of each recently completed 44 S strand from the 20 S RNA complex may occur slowly via an intermediate 26 S RNA which is only partially RNase-resistant (not shown). Escape of some positive 44 S strands to the ribosomes eventually results in the accumulation of V2 which attaches to the 3′-end of positive strands in 20 S RNA complexes, rapidly displacing or blocking initiation by the polymerase on this strand. The movement of released 44 S positive strands to ribosomes is now unrestrained, and the polymerase is diverted to asymmetric transcription from 44 S negative strands. The assumed site of selective translation of virion-associated proteins (V3, NV2, V2) is supported by evidence from puromycin inhibition experiments (Shapiro *et al.*, 1973a; Westaway and Shew, 1977); at this site V2 may escape the posttranslational trimming to NV1½ observed in short-pulse experiments (Westaway, 1977; Wright and Westaway, 1977).

4. The switch to increased synthesis of positive-strand 44 S RNA after the latent period and the ensuing large increase in protein synthesis.

5. The variable amounts (if they have any significance) of labeled partially RNase-resistant 26 S RNA for different flaviviruses, which may be related to the rate of release of polymerase from 20 S RNA, to the rate of RNA synthesis, or to whether transcription is symmetric

(templates fully labeled during long pulses) or asymmetric (only progeny positive strands labeled).

Definitive support or otherwise of the model could be provided by:

1. Electron micrographs of the number of strands in 20 S (and 26 S) RNA and in isolated replication complexes.
2. Inhibition of synthesis of negative strands by adding V2 to an *in vitro* polymerase system.
3. Increased synthesis of negative strands in temperature-shift experiments with *ts* mutants containing a lesion in the core protein cistron.
4. Confirmation of attachment of V2 (or a specific nonstructural protein) to positive strands isolated from polysomes.

The precise role of the perinuclear membranes in transcription also needs to be explored.

III. VIRUS-SPECIFIED PROTEINS

The virions comprise similar molar amounts of envelope glycoprotein V3 (51,000–59,000 daltons), the core protein V2 (about 14,000 daltons), and a "membrane" protein V1 (about 8000 daltons) located between the envelope and core; a slowly sedimenting noninfectious particle (60–80 S and devoid of RNA) released from infected vertebrate cells is a hemagglutinin (SHA) and comprises V3 and V1 in similar proportions and usually the nonstructural glycoprotein NV2 (Stollar, 1969; Shapiro *et al.*, 1971; Trent and Qureshi, 1971; Westaway, 1973, 1975). The latter particle was absent in culture fluids from DEN-2-infected *Aedes albopictus* cells (Sinarachatanant and Olson, 1973).

A. Definition of Nonstructural Proteins and of Glycoproteins

Treatment of flavivirus-infected cells with actinomycin D prior to labeling permits electrophoretic resolution of flavivirus-specified proteins against the high background of continuing host protein synthesis (Westaway and Reedman, 1969; Shapiro *et al.*, 1973a). Additional pretreatment of cells with cycloheximide (Trent and Qureshi, 1971; Shapiro *et al.*, 1971) appears to be unnecessary. The most prominent protein in infected cells is the envelope protein V3, which has the greatest range in molecular weight values, from 51,300 for Kunjin virus to 59,000 for DEN-2 virus, among 11 flaviviruses examined (Shapiro *et al.*, 1971; Trent and Qureshi, 1971; Westaway, 1973; Westaway *et al.*, 1977). Less variation in molecular weight occurs for each of the nonstructural proteins; their representative values for Kunjin virus are as follows: NV5 98,000, NV4 70,500, NV3 44,000, NVX 32,000, NV2½ 21,000, NV2 19,000, and NV1 10,300 (Westaway, 1973)*. Unequivocal identification in different hosts was obtained by difference analysis after coelectrophoresis in sodium dodecyl sulfate (SDS) polyacrylamide gels of labeled infected and uninfected cytoplasm (Fig. 3) and by slab gel electrophoresis (Fig. 4) (Westaway, 1973, 1975; Westaway *et al.*, 1977). The improved resolution in slab gels showed that Kunjin V2 (core protein of virion) was represented in cytoplasm by the slightly faster migrating NV1½ (13,500) (Fig. 4) which is deficient in one of the [^{35}S]methionine-labeled tryptic peptides of V2 (Wright and Westaway, 1977). NV3 and NVX are resolved with difficulty, and their synthesis may vary with different flaviviruses, host cells, and the period postinfection. Comigration with NV3 of a prominent and ubiquitous host protein of similar molecular weight, possibly the highly conserved actin (Pollard and Weihing, 1974), is an irksome coincidence in electropherograms of infected cell extracts from several sources. The minor structural protein V1 is absent in infected cells.

A soluble antigen type-specific by complement fixation and immunodiffusion was separated on DEAE-cellulose columns from cells infected with SLE, JE, DEN-2, and West Nile viruses and appears to be equivalent to NV5 (Qureshi and Trent, 1973a). However, identification of other type-specific and soluble complement-fixing antigens (SCFs) as nonstructural proteins of flaviviruses is less certain and may require comparisons by tryptic peptide mapping (Wright and Westaway, 1977). From supernates or extracts of infected cells, a JE virus SCF of molecular weight 53,000 (slightly smaller than V3) (Eckels *et al.*, 1975) or of molecular weight 40,000 (Rai and Ghosh, 1976) and dengue virus SCFs of average molecular weight 39,000 (Russell *et*

*Agreement has been reached by several flavivirus research groups concerning a proposed change in nomenclature. Structural proteins will be designated E for V3 (envelope), C for V2 (core), and M for V1 (membrane-like). The nonstructural proteins shown to be both unique (unrelated to one another) and stable end products will be designated P, or GP for glycoproteins, followed by the molecular weight × 10^{-3}, the lower case p or gp being used if these criteria have not been met. Thus NV5, NV4, NV3, NVX, NV2½, NV2, and NV1 for Kunjin virus will be specified as P98, P71, gp44, p32, P21, GP19, and P10, respectively. The intracellular equivalents of Kunjin V3 and V2 will be designated P51(E) and P14(C), the established relationships with E and C (see Sections III,A,C) being indicated parenthetically.

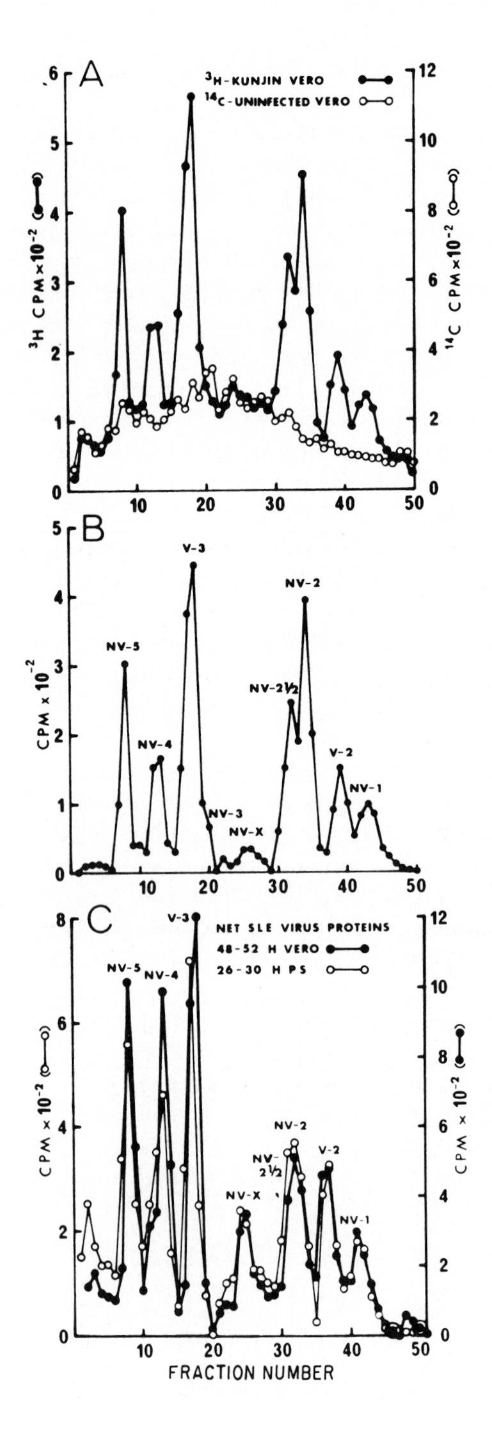
A
^{3}H-KUNJIN VERO
^{14}C-UNINFECTED VERO
^{3}H CPM×10^{-2}
^{14}C CPM×10^{-2}
B
CPM × 10^{-2}
V-3
NV-5
NV-4
NV-3
NV-X
NV-2
NV-2½
V-2
NV-1
C
NET SLE VIRUS PROTEINS
48-52 H VERO
26-30 H PS
CPM × 10^{-2}
FRACTION NUMBER

al., 1970; Cardiff *et al.*, 1970, 1971; McCloud *et al.*, 1971; Brandt *et al.*, 1970; Smith *et al.*, 1970) were isolated. Their relationships to the designated nonstructural proteins or to V3 are unknown. Although dengue SCF contains group-specific and type-specific determinants (Russell *et al.*, 1970), as does V3 (Westaway *et al.*, 1974, 1975; Qureshi and Trent, 1973a; Trent *et al.*, 1976), a relationship of dengue SCF to V3 is probably excluded because (1) anti-DEN-2 SCF does not neutralize infectivity or significantly precipitate virions or SHA (Cardiff *et al.*, 1971), (2) fluorescent antibodies have shown that DEN-2 SCF remains localized in the perinuclear region after the latent period, whereas virion antigens are dispersed throughout the cytoplasm (Cardiff *et al.*, 1973b), and (3) the ultrastructure of SCF particles is similar to that of dengue virus surface subunits which are released during storage, but such disruption produces no increase in the amount of serologically active SCF (Smith *et al.*, 1970).

Stohlman *et al.* (1976) used nonionic detergents to solubilize amino acid-labeled membranes of DEN-2 virus-infected cells and after SDS electrophoresis observed an additional protein designated "Z" migrating between V3 and NV3, the calculated molecular weight being 53,000. The significance of "Z" is difficult to evaluate in the absence of any antigenic data and because uninfected and infected solubilized membrane proteins were not coeletrophoresed; Stohlman *et al.* (1976) believe that "Z" is derived from a larger virus-specified protein (by successive detergent treatments).

Subsequent to the initial report of glycosylation of the envelope protein V3 of DEN-2 virus (Stollar, 1969), the nonstructural proteins NV3 and NV2 were also found to be glycoproteins (Shapiro *et al.*, 1972a, 1973b; Westaway, 1975) (Fig. 4). Very small amounts of labeled glucosamine and mannose were recently shown to be incorporated

---

Fig. 3. Resolution of flavivirus-specified proteins by elimination of labeled host protein background from electropherograms. Cells were treated with 3 μg actinomycin D per milliliter for 4 h prior to labeling for 4 h with [^{3}H]leucine (infected cells) or [^{14}C]leucine (mock-infected cells). Samples of infected and uninfected SDS-dissolved cells were mixed and then electrophoresed in 8% SDS–phosphate polyacrylamide gels. Profiles of "net" virus-specified proteins were obtained by subtraction after an appropriate factor was applied to the ^{14}C-labeled-uninfected cpm so as to match the ^{3}H-labeled cpm in fractions devoid of virus-specified proteins. The numbered proteins are prefixed "V" for structural and "NV" for nonstructural proteins. (A) Kunjin virus-infected and uninfected Vero cells, labeled at 26–30 h. (B) Net Kunjin virus-specified proteins, representing 58% of the total [^{3}H]leucine incorporation in (A). (C) Coincidence of profiles of net SLE virus-specified proteins labeled in Vero cells (48–52 h) and in PS cells (26–30 h), representing 39% of the total [^{3}H]leucine incorporation in both hosts. (From Westaway, 1973, reproduced with permission.)

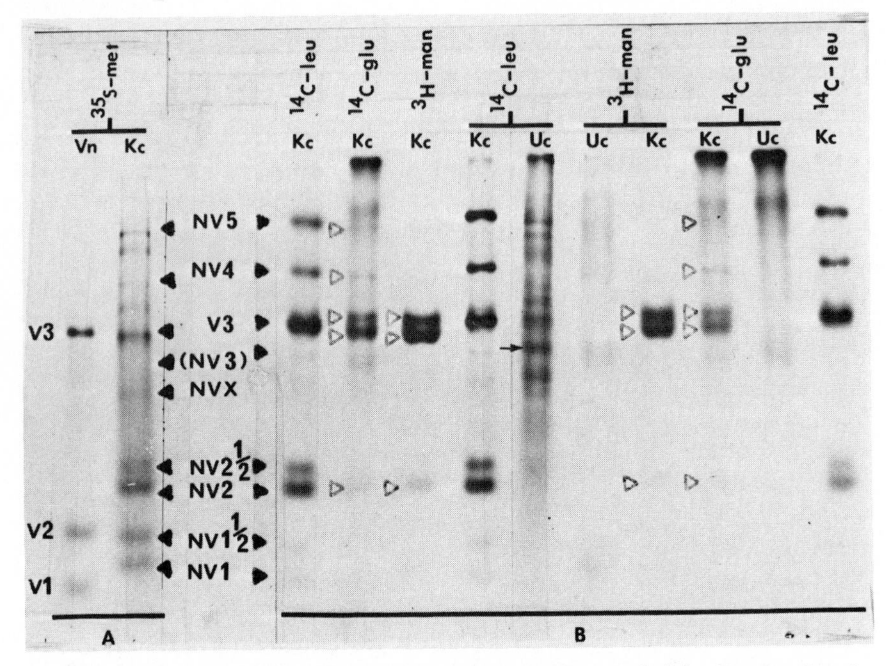

Fig. 4. Identification of proteins and glycoproteins specified by Kunjin virus in Vero cells, analyzed by electrophoresis in 8% SDS-phosphate polyacrylamide slab gels. Kc represents cells labeled at 24 h postinfection for 5 h in amino acids or for 8 h in glucosamine or mannose, after treatment with actinomycin D. Uc represents mock-infected cells. Vn represents a sample of labeled virions purified from infected cell culture fluids. The radioactive substrates added to cells were [^{35}S]methionine (^{35}S-met), [^{3}H]mannose (^{3}H-man), [^{14}C]glucosamine (^{14}C-glu), and [^{14}C]leucine (^{14}C-leu). The intracellular viral proteins labeled in amino acids are indicated by solid pointers and in carbohydrate by open pointers. The bands in the autoradiograph for some of the small proteins have lost definition in the photograph because of the relatively low incorporation of isotope. The host protein which migrates in the expected position of NV3 is marked by an arrow in lane 5 of gel B (refer to the text). (From Westaway and Shew, 1977, reproduced with permission.)

into NV5 and NV4, and this was associated with very small increases in their electrophoretic mobilities (Westaway and Shew, 1977). More prominent electrophoretic microhetereogeneity of NV2 (Shapiro *et al.*, 1973b; Boulton and Westaway, 1976) and of V3 (Westaway and Shew, 1977) was observed when comparisons were made between amino acid-labeled and carbohydrate-labeled infected cytoplasm, and this is discussed in Section III,B. In autoradiograms of slab gels, cytoplasmic V3 labeled in glucosamine or mannose is resolved as a double band (Fig. 4).

In summary, the nine flavivirus-specified proteins found in infected cells are NV5, NV4, V3, NV3, NVX, NV2½, NV2, NV1½ (formerly designated V2), and NV1. Synthesis of NVX and the glycoprotein NV3 is variable; V3 and NV2 are heterogeneous in regard to electrophoretic migration and to extent of glycosylation. The relationship of the SCFs of dengue and of JE viruses to any of these proteins remains unidentified.

B. Proteins in Infected Cell Membranes

Differences in intracellular localization of envelope and core proteins of alphaviruses are related to the site and mode of virus maturation (Bose and Brundige, 1972); similar data for flavivirus proteins could provide positive evidence in this undefined area.

Membrane fractions from several flavivirus–cell systems have been obtained by isopycnic sedimentation of disrupted cytoplasm in discontinuous sucrose gradients, based principally on the methods of Caliguiri and Tamm (1970) as in Fig. 5, and of Bosmann *et al.* (1968). Although the results have not been uniform from one report to another, no preferred membrane site was observed for incorporation of envelope protein V3 or of core protein V2 (Shapiro *et al.*, 1972b; Kos *et al.*, 1975b; Stohlman *et al.*, 1975; Boulton and Westaway, 1976). The common feature of these reports was that all the flavivirus-specified proteins (except V1) were distributed throughout all membrane fractions, and that nearly all the virus-specified protein was associated with detergent-labile membranes. Shapiro *et al.* (1973a) found that of the JE virus-specified polypeptides, none were completely released by treatment with dilute detergent or neutral or alkaline salts; NV5 and NV4 in membranes were selectively degraded by 0.1% trypsin.

Some variation in relative proportions of specific nonstructural flavivirus proteins and in the site of macromolecular synthesis has been noted, e.g., enrichment of NV5 of JE virus (Kos *et al.*, 1975b) and of NV3 of DEN-2 virus (Stohlman *et al.*, 1975) in the less dense (smooth) membrane fractions; other smaller variations concurrently reported do not appear to be significant. Unfortunately in both reports, the conditions employed for disrupting cells were apparently so vigorous that the architecture of the membranes was perturbed sufficiently to cause changes in their buoyant density, resulting in incomplete and abnormal separations. For example, plasma membranes (identified by the specific activity of the marker enzyme 5′-nucleotidase) were concentrated in the smooth membrane fractions (above the 35% sucrose level) of both these reports but were identified by Boulton and Westaway (1976) as intact saclike structures greatly enriched in 5′-

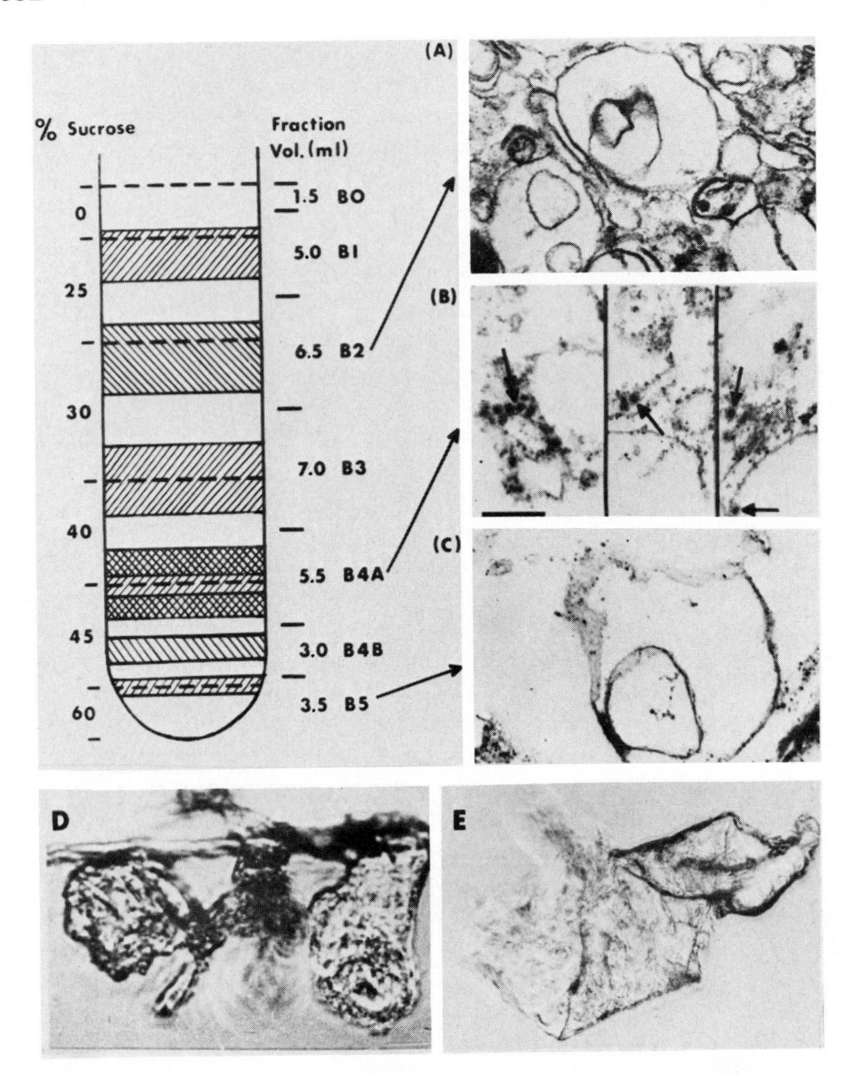

Fig. 5. Schematic representation (top left) of the cell fraction gradient as per the method of Caliguiri and Tamm (1970), and micrographs of the membrane fractions (A–E). Kunjin-infected Vero cells swollen in hypotonic buffer were disrupted gently using only five strokes of a Dounce homogenizer, and the postnuclear fraction (8 ml) was made 30% (w/w) with respect to sucrose and layered in the sucrose gradient above 40% sucrose prior to the addition of less dense layers. Centrifugation at 4°C was for 18 h at 25,000 rpm in an SW25.1 rotor. The electron micrographs show thin sections of the smooth membrane (A), rough membrane (B), and plasma membrane (C) fractions; the bar represents 200 nm. Virions are indicated by arrows. The other fractions (not shown) contained smooth vesicular membranes enriched in the Golgi complex (fraction B1), mixed smooth and rough membranes (fraction B3), and free polyribosomes (fraction

nucleotidase at the 45–60% sucrose interface when much gentler preparative procedures were employed; their identity was confirmed by their recovery using the specific isolation method for plasma membranes of Atkinson and Summers (1971) (Fig. 5) (Boulton and Westaway, 1976). The importance of the mild disruption method used by Atkinson and Summers (1971) for preservation of the integrity of plasma membranes has been ignored in many studies involving membrane fractionation. In Kunjin virus-infected cells the site of RNA synthesis was located in smooth membranes (at the 25–30% sucrose interface) using a 5-min pulse of [^{3}H]uridine, whereas the site of protein synthesis was predominantly in the rough membrane fraction (at the 40–45% sucrose interface) using a 10-min pulse of [^{3}H]leucine (Boulton and Westaway, 1976). In contrast, Stohlman *et al.* (1975) were unable to separate the apparent sites of DEN-2 RNA and protein synthesis, which they reported to be both associated predominantly with the densest membrane fraction (at the 45–60% sucrose interface) after 18-h labeling periods.

Transfer of Kunjin virus-specified proteins after synthesis to plasma membranes occurs within 10 min and is nonselective (Boulton and Westaway, 1976). Surprisingly, the movement of recently synthesized alphavirus proteins to plasma membrane was not detectable after a 20-min pulse, although this was clearly the site of maturation (Richardson and Vance, 1976). The depletion of NV5 in most membrane fractions (Kos *et al.*, 1975b; Stohlman *et al.*, 1975; Boulton and Westaway, 1976) is not surprising, because NV5 is particularly prone to proteolytic degradation after cell disruption in the absence of ionic detergents (Shapiro *et al.*, 1972b; E. G. Westaway and J. L. McKimm, unpublished results). However, some protection of NV5 against degradation occurs in a fraction containing plasma membrane from cells infected with JE virus (Kos *et al.*, 1975b) or with Kunjin virus (Boulton and Westaway, 1976). The incorporation of V3 and the nonstructural proteins in plasma membranes is surprising, because there is no evidence of a role for this site in replication. However, independent confirmation of this incorporation is supplied by studies of immune cytolysis and by immunoperoxidase techniques using electron microscopy (Catanzaro *et al.*, 1974; Brandt and Russell, 1975; Cardiff and Lund, 1976).

B4B). Phase-contrast micrographs at similar magnifications compare the large empty sacs of fraction B5 (D) with plasma membrane isolated specifically by the zonal centrifugation method of Atkinson and Summers (1971) (E). [From Boulton and Westaway (1976, reproduced with permission) and (D and E only) from Boulton (1974).]

The membrane fractionation studies were completed prior to the application (with improved electrophoretic resolution) of slab gel and fluorography techniques, hence no information was obtained on membrane localization of the heterogeneity apparent in V3 and NV2 (see Sections III,C,D). Subsequently, we modified the fractionation method by top-loading the sample on a discontinuous sucrose gradient and using a briefer centrifugation time with a higher *g* value; the membranes in which Kunjin virus proteins were translated and glycosylated were thereby separable from the membrane site of cell protein synthesis (E. G. Westaway and M. L. Ng, submitted for publication).

In summary, virus-specified proteins are incorporated into most cytoplasmic membrane fractions including plasma membrane, and the site of protein synthesis and glycosylation is predominantly the rough endoplasmic reticulum. There is no strong predilection of any protein for a particular membrane, hence no functional role can be inferred for any protein. The wide distribution of virus-specified proteins in membrane fractions distinguishes flaviviruses from alphaviruses.

C. Heterogeneity of V3

Although amino acid-labeled V3 from virions comigrates with amino acid-labeled V3 from infected cells (Westaway and Reedman, 1969; Shapiro *et al.*, 1971; Trent and Qureshi, 1971), recent reports indicate heterogeneity in glycosylated V3 and in the antigenicity of V3. Thus we noted that in Kunjin virus-infected cells, V3 incorporated in cell membranes is labeled in glucosamine and mannose but is deficient in glucosamine in smooth membranes (Boulton and Westaway, 1976); furthermore, V3 is resolved into two bands in autoradiograms of slab gels after electrophoresis of [^{14}C]glucosamine- or [^{3}H]-mannose-labeled V3 from cells (Westaway and Shew, 1977). These two bands usually straddle the position in SDS phosphate gels of V3 from cells labeled in amino acids (Fig. 4). In Laemmli gels, the glycosylated forms of intracellular V3 after pulse-labeling with carbohydrates appear more complex (Wright and Westaway, 1979). Surprisingly, virion V3 labeled either in amino acids or in glucosamine or mannose migrates identically in slab gels with amino acid-labeled cytoplasmic V3 (Westaway and Shew, 1977, and unpublished results). Qureshi and Trent (1973b) solubilized V3 from cells infected with SLE virus and obtained several fractions (by DEAE-cellulose chromatography) of similar antigenicity but which differed markedly in their [^{3}H]glucosamine/[^{14}C]leucine ratios. Although V3 solubilized from JE virions and from cytoplasm of JE and other flavivirus-infected cells cross-reacted broadly in complement fixation (CF) tests (Eckels *et al.*, 1975; Trent *et al.*, 1976), a type-specific CF antigen of slightly lower molecular weight than V3 was purified by isoelectric focusing from JE virus-infected cells (Eckels *et al.*, 1975). Using the ligand concavalin A–Sepharose, Stohlman *et al.* (1976) isolated from DEN-2 virus-infected cells a subpopulation of V3 which still retained hemagglutinating (HA) activity and was type-specific in CF tests but did not block the action of neutralizing antibody. Direct chemical evidence of heterogeneity in V3 was obtained by comparing two-dimensional tryptic peptide maps of [^{35}S]methionine-labeled V3 from infected cells and from virions; 24 peptides in each were identical, but in both cases two peptides mapped differently (Wright *et al.*, 1977).

Clearly V3 in cytoplasm is heterogeneous in regard to carbohydrate content and probably in antigenic specificity. This V3 complex comprises four distinct components which incorporate both glucosamine and mannose (two components), mannose only (in smooth membranes), or neither carbohydrate (apparently the major component). Presumably these variants of V3 are intermediates in essential glycosylation processes prior to the incorporation of V3 into virions. Surprisingly, nearly all the amino acid-labeled V3 in cytoplasm remains unavailable for glycosylation and is stable during long chase periods (Westaway, 1973; Westaway and Shew, 1977).

D. Heterogeneity of NV2

Although designated a nonstructural glycoprotein, NV2 is found in some purified preparations of DEN-2 virus and JE virus (see Section V,C) (Shapiro *et al.*, 1972a, 1973b) and of Kunjin virus (Boulton and Westaway, 1976; Westaway and Shew, 1977). The glucosamine content of "virion" NV2 relative to the amino acid content is similar to the ratio in cytoplasmic NV2 for both DEN-2 virus and JE virus but is much enriched in virion NV2 for Kunjin virus, so much so that it is more prominently labeled than virion V3. For both JE and Kunjin viruses, small, variable differences in electrophoretic mobility are demonstrable for NV2 from virions or from infected cells, and for glucosamine label versus amino acid label in NV2. These differences are seen in some slab gels of Kunjin virus proteins (see Westaway and Shew, 1977, Fig. 2) but often require changes in gel strength for adequate resolution in sliced and counted gels. The possibility that the heterogeneity might be caused by some interconversion with the slightly larger NV2½ (not a glycoprotein) is excluded because the tryptic peptide maps of NV2 and NV2½ are unrelated (Wright and Westaway, 1977).

Like V3, NV2 is clearly a complex of glycosylated and nonglycosylated or only partially glycosylated forms.

IV. PROTEIN SYNTHESIS IN INFECTED CELLS

This continues to be one of the most challenging and difficult areas in flavivirus replication because of the persistence of host protein synthesis and because of the unorthodox nature of results compared to those obtained with other positive-strand RNA animal viruses.

A. Kinetics of Protein Synthesis

In fluorescent antibody studies in DEN-2 virus-infected cells, antibodies to virion proteins and to the nonstructural SCF antigen were localized in the perinuclear region at 10–12 h in 5–10% of cells; by 18 h most cells were affected, and a granular fluorescence associated only with structural antigens was prominent throughout the cytoplasm (Cardiff *et al.*, 1973b). No other evidence of flavivirus-specified protein synthesis during the latent period has been presented. When host protein synthesis was preferentially inhibited by treatment of infected cells with hypertonic salt (Nuss *et al.*, 1975), synthesis of identifiable Kunjin proteins in Vero cells occurred during pulse labeling at 13 h but was not detected earlier (E. G. Westaway, unpublished results). Subsequently, virus-specified protein synthesis increases to a maximum rate at about 20–24 h (Westaway and Reedman, 1969; Shapiro *et al.*, 1971; Trent and Qureshi, 1971).

Infectious center assays show that at 24 h postinfection with Kunjin virus at a multiplicity of infection of 5 PFU/cell, virtually every cell is infected (E. G. Westaway and M. L. Ng, unpublished results). However, host protein synthesis continues, while viral protein synthesis represents a fairly constant proportion of total cell protein synthesis, ranging from less than 20 to nearly 60%, depending on the flavivirus and on the particular host cell (Westaway, 1973, 1975). The proportion and the relative amount of each viral protein synthesized remain constant as total synthesis decreases very late in infection because of cell death. Furthermore, despite the proliferation of host membranes in infected cells (see Section V), the relative amounts of each host protein synthesized throughout the entire period of infection also remain constant and qualitatively equivalent to synthesis in uninfected cells. The evidence is indirect but clear; virtually the same "net" electrophoretic profiles of virus-specified proteins are obtained when ^{14}C-labeled mock-infected cell proteins are coelectrophoresed with ^{3}H-labeled infected cell proteins labeled and harvested during successive periods postinfection in one or more cell lines (Westaway, 1973, 1975).

B. Search for Precursor–Product Relationships

Because of the known strategy of other positive-strand RNA animal viruses, a reasonable expectation was to find some evidence of posttranslational cleavage and of precursor–product relationships among flavivirus-specified proteins. However, no short-lived or additional proteins were observed in Kunjin virus-infected cells labeled in amino acids during pulses as brief as 1 min, in the presence of amino acid analogs, during treatment with inhibitors of proteolytic enzymes, or at elevated (41°C) or low (32°C) temperatures (Westaway, 1973; Westaway and Shew, 1977). The only obvious changes in electrophoretic profiles were delays of about 30 min after very brief pulses in the resolution from each other of NV2 and NV2½, and of 1–4 min in the appearance of NV1 (Westaway and Shew, 1977). Furthermore, no additional virus-specified proteins (possible precursors) either larger or smaller than NV5 were detected under conditions of controlled translation (Westaway, 1977; see also Section IV,D).

No major relationships of the smaller to the larger identified flavivirus-specified proteins were found by two-dimensional tryptic peptide mapping of each of the Kunjin virus-specified proteins labeled in [^{35}S]methionine (Wright *et al.*, 1977; Wright and Westaway, 1977). NV5, NV4, V3, NV2, and V2 were all shown to be unique proteins unrelated to one another; similar results were recently reported for West Nile virus (Wengler *et al.*, 1979). Together these are equivalent to most of the coding capacity of the genome. The distinction between V3 and V2 is of interest because both proteins carry group-reactive antigenic determinants (Trent, 1977). NV1½ is identical with V2 except for a deletion of one labeled tryptic peptide which probably occurs during posttranslational "trimming" (Westaway, 1977; Wengler *et al.*, 1979). NV2½ and NV1 are unrelated by tryptic peptide mapping to all other proteins referred to above, except for a possible relationship to V3 (Wright and Westaway, 1977). It was not possible to obtain sufficient incorporation into NVX, and NV3 could not be resolved or distinguished from a comigrating host protein.

V1 yielded only two methionine-labeled tryptic peptides; similar peptides were identified in digests of NV5, NV4, or V3 (Wright and Westaway, 1977). Surprisingly, the maps show no relationship between V1 and NV2 which was claimed to be the precursor of V1 by Shapiro *et al.* (1972a) based on differences in composition between intracellular JE virions (V3, NV2, V2) and extracellular virions (V3, V2, V1). The heterogeneity in electrophoretic migration and in the extent of glycosylation of NV2, and its apparent structural role in some preparations of virions and SHA, remain an enigma.

Shapiro *et al.* (1972b) showed that limited tryptic digestion of JE virions (and of infected cell membrane fractions) yielded a peptide fragment comigrating with NV3 (molecular weight 45,000); because the fragment from virions must be derived from V3 (molecular weight 53,000), the other product of a single cleavage (molecular weight approximately 8000) would be similar in size to V1 (molecular weight 8700). If only a very small proportion of intracellular V3 suffered this fate, it would be in accord with the very small amount of amino acid or carbohydrate label detectable in NV3 after difference analysis (Westaway, 1973, 1975), with the absence of V1 in cytoplasm (rapidly incorporated into virions), and with the possible relationship of V1 to V3 of Kunjin virus indicated by tryptic peptide analyses (Wright and Westaway, 1977).

The delay in appearance of NV1 during brief pulse-chases, and its slow rate of completion during hypertonic inhibition or pactamycin experiments (Westaway, 1977), suggest that it is derived from near the C-terminus of a larger protein; however, the tryptic peptide maps show a possible relationship only to V3 (Wright and Westaway, 1977), whereas under conditions of puromycin inhibition, synthesis of V3 continued, but that of NV1 was abated (Shapiro *et al.*, 1973a; Westaway and Shew, 1977).

The inability to establish any relationships among the flavivirus-specified proteins in infected cells contrasts with the precursor–product relationships recognized among proteins synthesized in cells infected by picornaviruses (Jacobson *et al.*, 1970) or by alphaviruses (Pfefferkorn and Shapiro, 1974) (see Chapter 13).

C. Indicators for Control of Translation

The size of the flavivirus genome (molecular weight 4.2×10^6) is sufficient to code for all the virus-specified proteins, their total molecular weight being about 370,000 (Boulton and Westaway, 1972; Westaway, 1973). Virtually all the proteins are stable during long labeling and chase periods, but they are not produced in equimolar proportions (Westaway and Reedman, 1969; Shapiro *et al.*, 1971; Westaway and Shew, 1977). Synthesis of NVX appeared to be greatest near the end of the latent period (Westaway, 1973, 1975). These results suggested that some form of translational control existed, and the first direct evidence of control was provided by a selective inhibitory effect of puromycin on synthesis of NV5, NV4, NV2½, and NV1, accompanied by increased synthesis of NV2 and NV1½ or V2 relative to V3 (Shapiro *et al.*, 1973b; Westaway and Shew, 1977) (Fig. 6). Similar results were obtained using cycloheximide as inhibitor during the labeling period; because the mechanism of action of each drug is different, Shapiro *et al.* (1973a) concluded that the synthesis of virus-specified nonstructural, nonglycosylated proteins differed functionally from that of structural or glycosylated proteins.

D. Single versus Multiple Sites of Initiation

1. *Possible Strategies*

The absence of any flavivirus single-stranded RNA smaller than the genome, and the evidence discussed above in Sections IV,B and IV,C, indicate that the translation strategy of flaviviruses is different from that of the other positive-strand RNA animal viruses. The implication is that expression of the flavivirus genome differs fundamentally from that of other viruses in the class IV defined by Baltimore (1971) because of violation of the general principle that in eukaryotic cells initiation of translation and of termination occurs in each case at only a single site on RNA, irrespective of whether the RNA is monocistronic or polycistronic, or host-coded or virus-coded (Baltimore, 1971). Although this useful principle is emphasized in textbooks (Davis *et al.*, 1973; Watson, 1975), it should be noted that in eukaryotic cell-free translation systems, translation of the coat protein cistron of the small RNA phages is initiated and terminated internally on the phage polycistronic messenger (Aviv *et al.*, 1972; Schreier *et al.*, 1973; Morrison and Lodish, 1973).

Despite the lack of evidence for posttranslational cleavage of flavivirus proteins, rapid cleavage proximal to the N-terminal end before completion of a nascent polyprotein would not be detectable in the experiments discussed hitherto, assuming such cleavage was resistant to the various inhibitory conditions employed.

2. *Apparent Gene Sequence during Termination of Translation*

The poliovirus genome was mapped independently by observing the sequence of termination of incorporation of amino acids into trans-

A

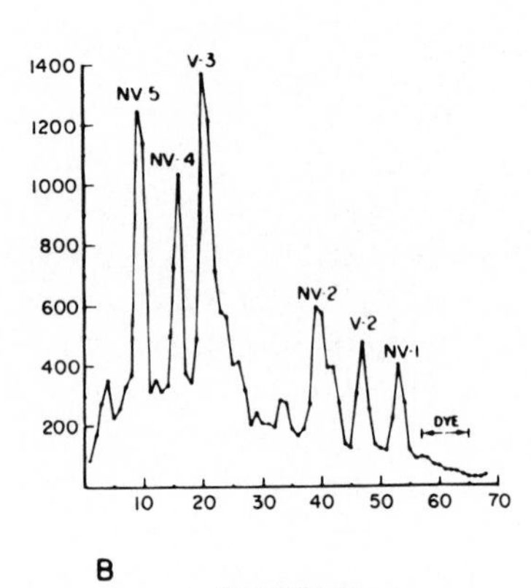

B

PUROMYCIN

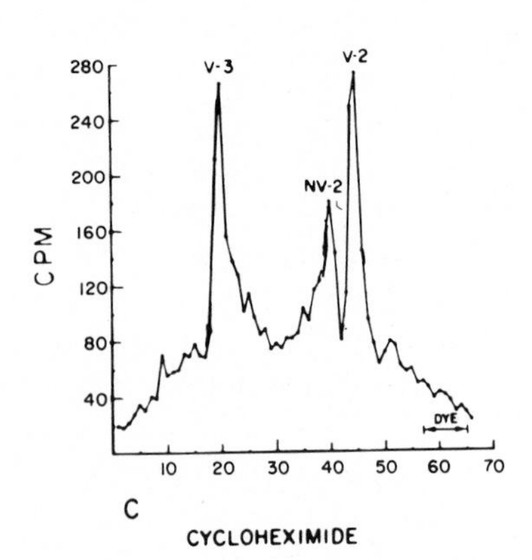

C

CYCLOHEXIMIDE

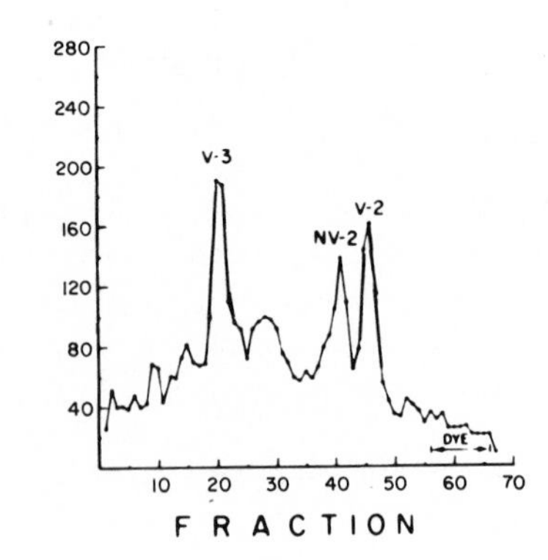

lation products in infected HeLa cells treated with pactamycin (Summers and Maizel, 1971; Taber *et al.*, 1971) or with hypertonic salt (Saborio *et al.*, 1974). Both treatments allow ribosomes to complete their traverse of RNA but prevent reattachment, hence reinitiation of translation. When applied to analysis of termination of incorporation of label into Kunjin virus-specified proteins in Vero cells, both treatments gave essentially the same results (Westaway, 1977). The smaller proteins were in general completed more rapidly, and completion of the largest proteins was in order of increasing molecular weight, namely, V3, NV4, and finally NV5. Within 11–13 min, translation of all virus-specified proteins was completed (Fig. 7). All host proteins were completed in translation in less than 7 min, and this fortuitously improved the resolution in autoradiograms of Kunjin proteins in slab gels. It has been concluded from these experiments that, if translation is initiated at a single site on the viral messenger, the gene sequence from the 5′-end should commence with V2 (NV1½) and terminate in NV4-NV5 at the 3′-end.

3. *Synchronous Reinitiation of Protein Synthesis*

In the poliovirus–HeLa cell system, Saborio *et al.* (1974) showed that reinitiation occurred synchronously after removal of a hypertonic block, and the known gene sequence of the poliovirus genome was thereby confirmed. In similar experiments with the alphavirus Semliki Forest in BHK-21 cells, Clegg (1975) showed that the capsid or C protein was the only protein into which label was incorporated during the first 1 min after reversal of the hypertonic block, and that C was rapidly cleaved from the nascent polyprotein. Synchronous reinitiation of protein synthesis was observed also in Kunjin virus-infected Vero cells, using a hypertonic block of 200 m*M* excess NaCl for 20 min prior to restoration to isotonicity (Westaway, 1977). The validity of these chosen conditions for reinitiation was established by showing (1) that ribosomes were dislodged from polyribosomes after protein synthesis had been terminated, and (2) that upon restoration to isotonic conditions, the order of reinitiation of protein synthesis by poliovirus and by an alphavirus (Sindbis) was exactly that determined by Saborio *et al.* (1974) and by Clegg (1975), respectively. Further-

---

Fig. 6. Effects of inhibitors on protein synthesis in JE virus-infected LLC-MK2 cells. Following addition of actinomycin D at 15 h postinfection, [^{3}H]leucine was added at 26–30 h in the presence of no inhibitor (A), puromycin (80 μg/ml) (B), and (C) (5 μg/ml) cycloheximide. Cell extracts were prepared, and 50 μl of (A) and 80 μl of (B) and (C) were analyzed by electrophoresis. (From Shapiro *et al.*, 1973a, reproduced with permission.)

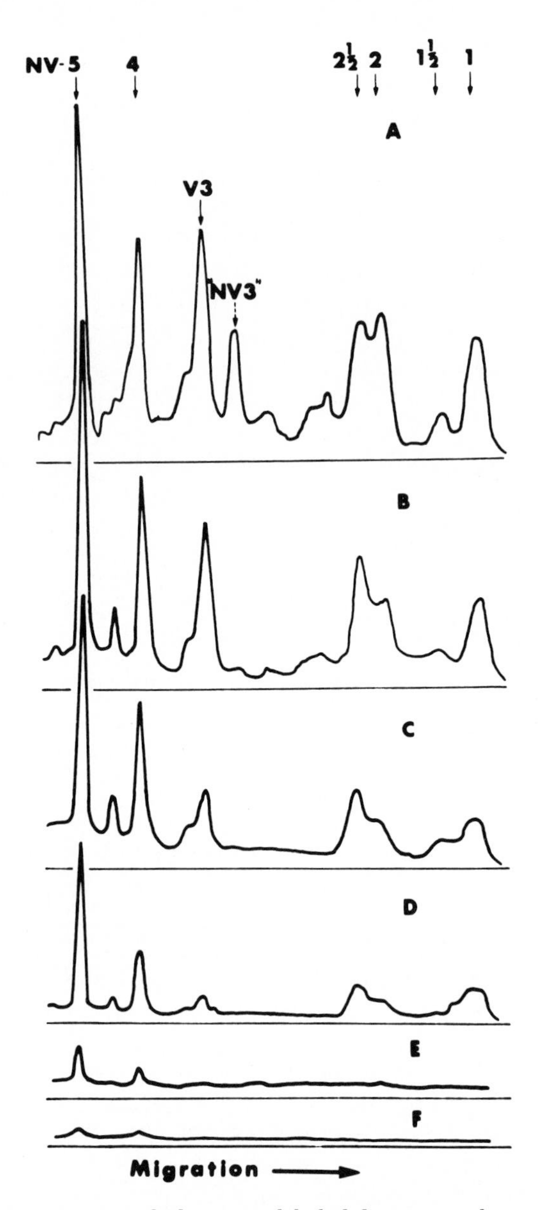

Fig. 7. Kunjin virus-specified proteins labeled during completion of translation in the presence of 200 m*M* excess NaCl and analyzed by slab gel electrophoresis. Infected petri dish Vero cell cultures were treated with hypertonic medium for specified periods, labeled in the same medium plus [^{35}S]methionine (100 μCi/ml) for 4 min, and chased for 30 min with a large excess of nonradioactive methionine. Identical volumes of SDS–cytoplasm were electrophoresed. The densitometer trace in (A) shows the electrophoretic profile of infected cell proteins labeled under isotonic conditions. The periods of treatment with excess NaCl prior to addition of label were (B) 1, (C) 5, (D) 9, (E) 13, and (F) 17 min. See legend for Fig. 4 regarding the designation of "NV3" in (A). (From Westaway, 1977, reproduced with permission.)

more, the total translation times for the poliovirus proteins (12 min) and for the alphavirus structural proteins (about 6 min) were the same as those reported earlier under similar conditions.

In preliminary experiments with Kunjin virus-infected cells reinitiated in translation, all proteins were labeled during a 3-min pulse and their translation was completed within 11 min. When even briefer pulse and chase periods were used (Fig. 8), it was found that (1) no proteins were completed within 3 min after reinitiation, (2) no proteins larger than NV5 were identified, (3) all identifiable proteins were completed in periods proportional to their molecular weight, in total about 300,000, and could be resolved by 9 min after reinitiation, and (4) at least five proteins were labeled within 1 min of reinitiation (Westaway, 1977). The total translation time is thus less than that for the much smaller poliovirus genome (compare 12 min for a product of 210,000 molecular weight), and the initial delay of 3 min before the appearance of any products indicates that all were reinitiated *de novo* rather than representing completion on salt-resistant polyribosomes. Although NV4 and NV5 appeared to map at the 3′-end of the flavivirus messenger in experiments based on order of termination (if a single site of initiation is assumed), no gene sequence could be deduced from these reinitiation experiments.

Initial studies using cell-free systems show that small products are completed first in the translation of tick-borne encephalitis virion RNA (Svitkin *et al.*, 1978) and of West Nile virus 44 S RNA isolated from infected cells (Wengler *et al.*, 1979). Peptide maps of the West Nile polypeptides ranging in size from 11,500 to 90,000 daltons were similar to those of V2 (for smaller products) or of a mixture of V2 and V3 (for larger products). The tick-borne encephalitis polypeptides were selectively diminished in number as the K^+ concentration was increased; at 105 m*M* the two predominant products corresponded in size to V3 (p53) and V2 (p14). Initiation of translation of the nonstructural flavivirus proteins *in vitro* is apparently influenced by ionic conditions and/or by conformational changes induced in the RNA during its isolation. It is clear also from these results that premature termination and read-through occur *in vitro*.

4. *Interpretation*

The most compelling interpretation of all the data obtained with Kunjin virus is that multiple and independent internal initiation of translation occurs on the flavivirus genome functioning as a polycistronic messenger. Flaviviruses are thus the only known positive-strand RNA animal viruses for which available evidence suggests that all the translation products (except possibly one or two very small

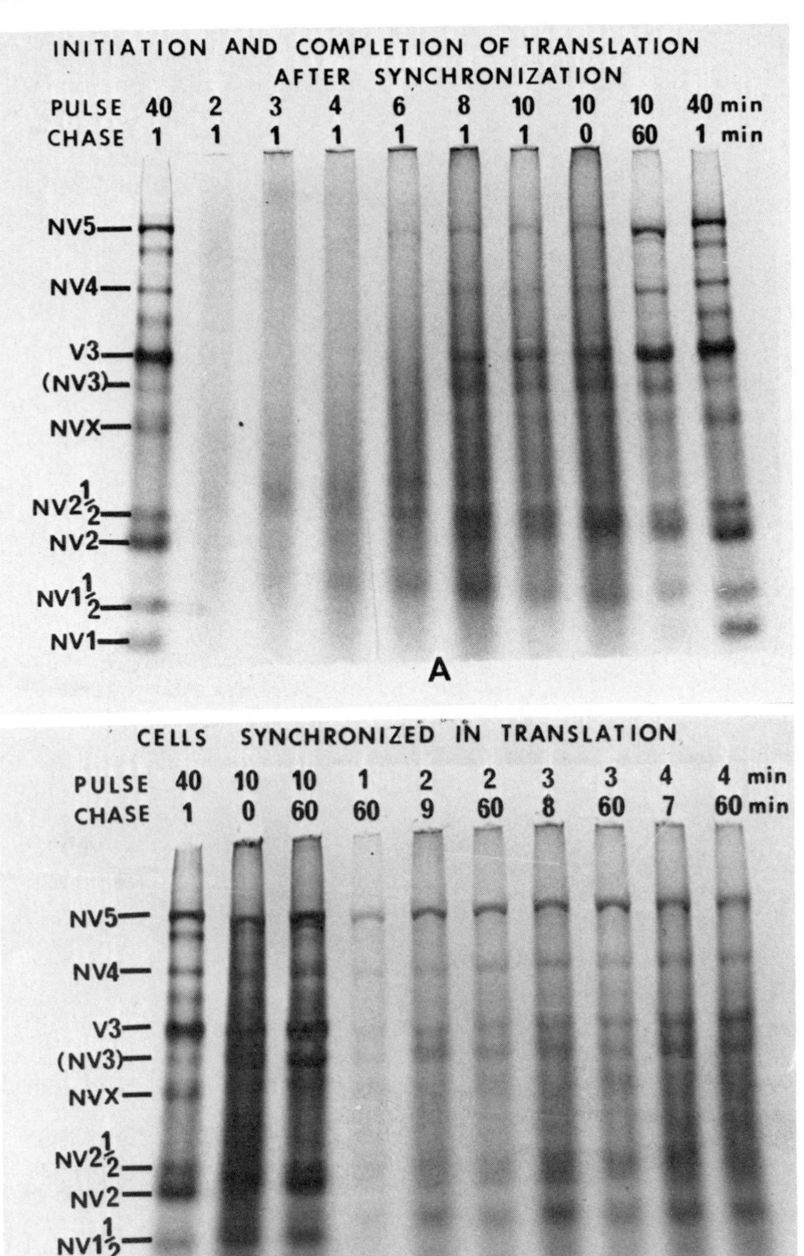
INITIATION AND COMPLETION OF TRANSLATION
AFTER SYNCHRONIZATION
PULSE 40 2 3 4 6 8 10 10 10 40 min
CHASE 1 1 1 1 1 1 1 0 60 1 min
NV5
NV4
V3
(NV3)
NVX
NV2½
NV2
NV1½
NV1
A
CELLS SYNCHRONIZED IN TRANSLATION
PULSE 40 10 10 1 2 2 3 3 4 4 min
CHASE 1 0 60 60 9 60 8 60 7 60 min
NV5
NV4
V3
(NV3)
NVX
NV2½
NV2
NV1½
NV1
B

proteins) appear to be unique, their lack of relationship being confirmed by tryptic peptide mapping (Wright *et al.*, 1977; Wright and Westaway, 1977; Wengler *et al.*, 1979). The presence of only a single m^7G cap residue in each 44 S RNA molecule (Wengler *et al.*, 1978; Cleaves and Dubin, 1979) indicates that, if internal initiation of translation does occur, internal equivalents of m^7G caps (Dubin *et al.*, 1977) are not required.

E. Possible Functions of the Nonstructural Proteins

At present no definite role can be assigned to any of the nonstructural proteins. NV1 and NV3 may be only residues generated during the production of functional products. NV2 must play some role in assembly or maturation of virions, but appears to be neither essential nor inhibitory to infectivity, because extracellular virions have almost the same specific infectivity whether NV2 is incorporated (T forms) or absent (N forms) (Shapiro *et al.*, 1972a) (see Section V,C). NV1½ is clearly derived from V2. The production of NVX is variable, and it must be either very stable or play no role late in infection. The size of NV5 and of NV4, as determined by gel electrophoresis (molecular weights about 98,000 and 70,000, respectively), is conserved for 11 flaviviruses (Westaway *et al.*, 1977), indicating that both proteins may retain an important functional role based on their size and conformation. Because both proteins are also similar in size to the two virus-specified components of the purified alphavirus polymerase with molecular weights of 90,000 and 63,000 (Clewley and Kennedy, 1976), it was suggested that NV5 and NV4 may also be polymerase components (Westaway, 1977).

V. ASSEMBLY AND MATURATION

Late in infection, viruslike particles commence accumulating in cytoplasmic vesicles or as crystalline aggregates, usually membrane-

Fig. 8. Analyses by slab gel electrophoresis of reinitiation and completion of translation of Kunjin virus-specified proteins in infected Vero cells. After treatment with 200 m*M* excess NaCl for 20 min, cells were restored to isotonic conditions and immediately pulse-labeled with [^{35}S]methionine and chased as indicated. Note that in (A) all virus-specified proteins were completed in translation by about 9 min after reinitiation, and that in (B) these proteins were all reinitiated in translation during the first 1 min after restoration to isotonic conditions. To facilitate identification of proteins, all samples in (A) contained similar cpm, and in (B) the samples pulse-labeled for 1, 2, 3, and 4 min contained the same volume of SDS–cytoplasm. (From Westaway, 1977, reproduced with permission.)

bound, while infectious virus continues to be slowly released until cell lysis, which may not occur for several days. The yield of intracellular virus from mechanically disrupted cells is always much less than the yield in the extracellular fluid (Stevens and Schlesinger, 1965; Westaway, 1973). The total yield of infectious virus varies from about 1 PFU/cell to several hundred PFU per cell depending on the flavivirus and the host cell. Factors affecting the release and yield of flaviviruses from cells have been reviewed by Schlesinger (1977).

A. Electron Microscopy of Infected Cells in Mouse Brain

Titers of 10^9–10^{10} LD_{50}/g of brain tissue are commonly obtained by 3–5 days after intracerebral inoculation in 1- or 2-day-old mice. Virions 36–44 nm in diameter, consisting of an electron-dense core about 27 nm in diameter surrounded by a rather uniform lucid envelope, are readily found in thin sections of neuronal cells late in infection by SLE virus (Murphy *et al.*, 1968), Banzi virus (Calberg-Bacq *et al.*, 1975), DEN-2 virus (Sriurairatna *et al.*, 1973), JE virus (Yasuzumi *et al.*, 1964; Yasuzumi and Tsubo, 1965a,b; Oyanagi *et al.*, 1969), yellow fever virus (Blinzinger *et al.*, 1975; David-West *et al.*, 1972), Zika virus (Bell *et al.*, 1971), and Entebbe bat salivary gland virus (Peat and Bell, 1970). Virions are also identifiable in mouse brain cells in the following tick-borne flavivirus infections: Langat virus (Boulton and Webb, 1971; Dalton, 1972), Omsk hemorrhagic fever (Shestopalova *et al.*, 1972), and other viruses of the tick-borne encephalitis complex (Blinzinger *et al.*, 1971; Blinzinger, 1972). In all reports the most dramatic visible changes in cells during infection are produced by proliferation of intracytoplasmic membranes. Many vacuoles appear, and hypertrophy occurs in the endoplasmic reticulum and to a lesser extent in membranes of the Golgi complex. Nearly all the changes to be described are extremely well illustrated in the paper by Murphy *et al.* (1968), which includes numerous electron micrographs at higher magnification than shown in Fig. 9 (see Chapter 8 of this volume).

Because the multiplicity of infection is always perforce much less than 0.1, several cycles of replication ensue before cytopathic signs of infection are detectable. Hypertrophy of membranes is generally visible in a small percentage of neurons at about 40–48 h. Small vesicles appear to be produced in aggregates by budding or invagination followed by release within smooth membranes of the enlarging endoplasmic reticulum (Calberg-Bacq *et al.*, 1975; Bell *et al.*, 1971; Boulton and Webb, 1971; Cardiff *et al.*, 1973b; Dalton, 1972; Peat and Bell,

1970; Sriurairatna *et al.*, 1973; Yasuzumi *et al.*, 1965a). The components of these vesicular aggregates are of varying diameter (40–100 nm), and some may be short, cylindrical, smooth-walled tubules cut transversely (Murphy *et al.*, 1968; Blinzinger, 1972). Many of the small vesicles within the cisternae contain reticular electron-dense centers and are often enclosed together with morphologically mature virions. The appearance of the vesicular aggregates is superficially similar to that of the cytopathic vesicle type I described in alphavirus infections, which is claimed to be the site of alphavirus RNA synthesis (Friedman *et al.*, 1972; Grimley and Friedman, 1970; Grimley *et al.*, 1968). However, in alphavirus infections the small vesicles are uniform in size (60–70 nm diameter) and are arranged peripherally by attachment to the inner lining of the large vesicle.

Within the cisternae of vesicles and membranes of the endoplasmic reticulum, only a few virus particles are found prior to 72 h. Bundles of fibrogranular material with no limiting membrane, similar in appearance to the massed convoluted filaments observed in cells of SLE virus-infected mosquitoes (Whitfield *et al.*, 1973), were noted by Sriurairatna *et al.* (1973) in DEN-2 virus-infected cells at 48 h, but they were hard to find after 72 h; also observed at the same period were numerous parallel filaments (18–25 nm in diameter) close to the nucleus. Degranulation of ribosomes from the endoplasmic reticulum becomes apparent after 72 h, and extensive hypertrophy of the above membrane systems occurs, closely associated with virus development, producing dense regions of overlapping or convoluted microtubules which by 96 h fill large areas of cytoplasm (Fig. 9). Some of these regions exhibit four- or sixfold symmetry (Parker and Stannard, 1967; Murphy *et al.*, 1968) and develop near the nucleus, frequently continuous peripherally with paired parallel membranes of the endoplasmic reticulum. They probably comprise three-dimensional foldings of smooth membranes, producing a meshlike appearance (Murphy *et al.*, 1968; Peat and Bell, 1970; Bell *et al.*, 1971; Calberg-Bacq *et al.*, 1975). The common occurrence of the meshlike structures in relation to virus particles suggests that the structures are interrelated and that their appearance in electron micrographs is influenced by the plane of section. Their precursors may be the transiently observed bundles of fibrogranular material referred to earlier (Sriurairatna *et al.*, 1973). Similar convoluted membrane structures were described in infections by the alphavirus Semliki Forest, but these were not associated with virus particles (Grimley and Friedman, 1970).

Morphologically mature virus particles accumulate in vesicles studded on their outer surface with electron-dense ribosomelike par-

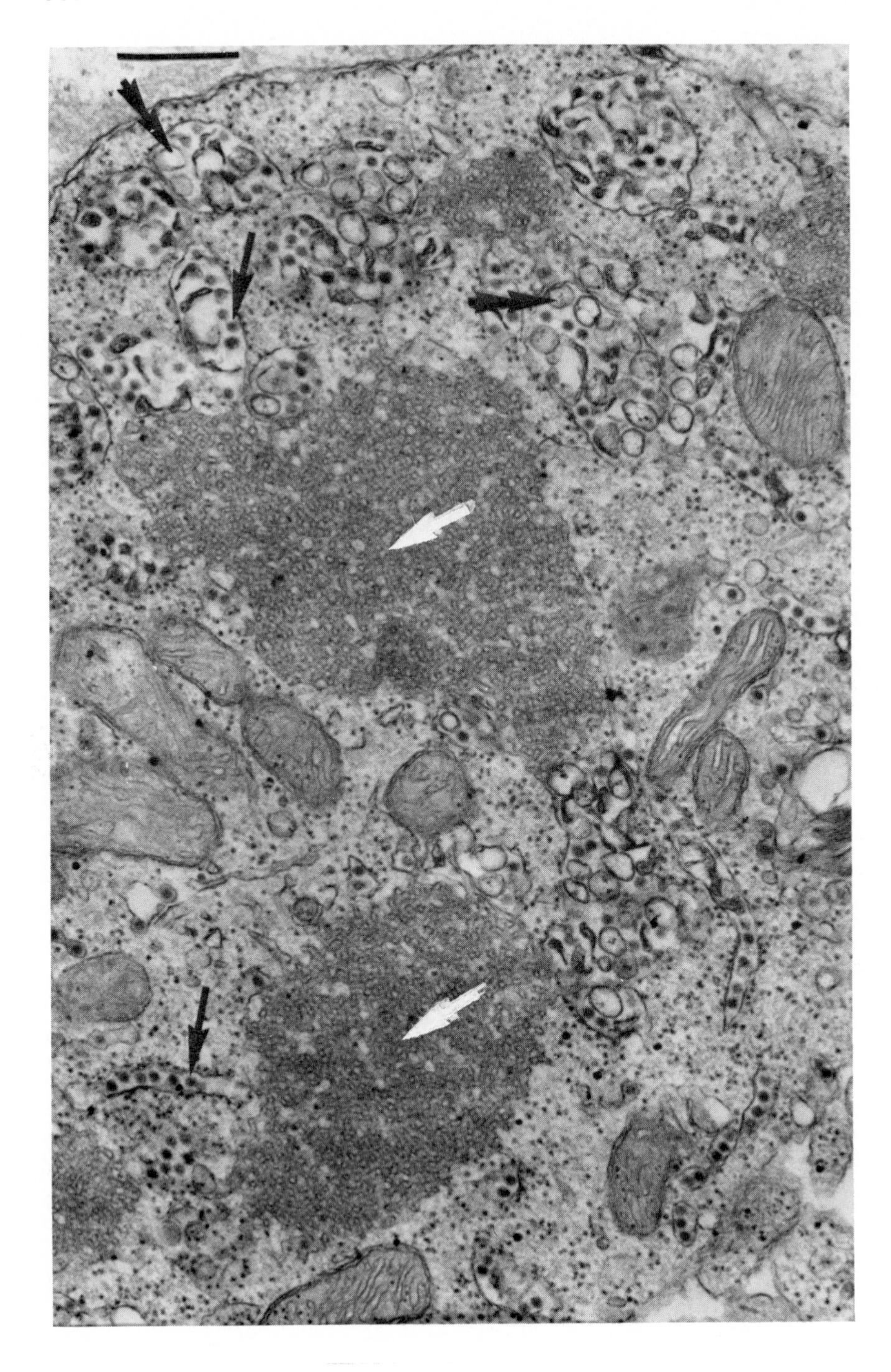

ticles (Murphy *et al.*, 1968; David-West *et al.*, 1972; Calberg-Bacq *et al.*, 1975). Virions also accumulate in vacuoles with smooth membranes or within the Golgi complex (Blinzinger, 1972; Dalton, 1972; Murphy *et al.*, 1968; Oyanagi *et al.*, 1969). Some of these accumulations are enclosed within bundles of smooth membranes or concentric lamellae (Shestopalova *et al.*, 1972; Calberg-Bacq *et al.*, 1975). Single rows of virions enclosed within lamellae of thickened electron-dense smooth membranes appear in many flavivirus infections of mice but receive little or no specific comment (Murphy *et al.*, 1968; Peat and Bell, 1970; Boulton and Webb, 1971; Blinzinger, 1972; Blinzinger *et al.*, 1971; Calberg-Bacq *et al.*, 1975). These unique lamellae were shown by Sriurairatna *et al.* (1973) to extend in a random manner from the perinuclear region, and in some sections they contained "beady rods" 50 nm in diameter the ends of which were connected to or in contact with single rows of virionlike particles (compare Fig. 10). The lamellar membranes bear a precise but distinct relationship to the enclosed particles and rods; their lumen expands peripherally in some cells to contain several rows of virions.

B. Electron Microscopy of Infected Cells in Culture

Because a high multiplicity of infection is possible in monolayer cultures, most cells may be infected synchronously. The more thorough studies in mammalian cell lines are of dengue virus infections by Matsumura *et al.* (1971) and by Demsey *et al.* (1974), and of JE virus infections by Ota (1965). Filshie and Rehacek (1968) studied persistent infections by JE virus and Murray Valley encephalitis (MVE) virus in *Aedes aegypti* cultured mosquito cells. The changes in infected cells are similar in virtually all respects to those described in Section IV,A but are detected earlier. Proliferation of intracytoplasmic membranes occurs, and numerous vesicles are visible by the end of the latent period of about 12 h (Ota, 1965; Matsumura *et al.*, 1971). By 18 h, progeny virions are found in small, crystalline aggregates which become very prominent by 36 h; the aggregates are usually enclosed

---

Fig. 9. Electron micrograph of virus-specific structures seen in a thin section of neuron in baby mouse brain infected 93 h previously with 10^3 LD_{50} of SLE virus. Virions (black arrows), some tadpole-shaped forms, and round membranous vesicles (double arrows), are present within cisternae of endoplasmic reticulum. Randomly packed convoluted membranous masses (white arrows) are associated with cytoplasmic organelles. The top of the figure includes a portion of the cell nucleus. ×35,000. Bar represents 500 nm. (From Murphy *et al.*, 1968, *Lab. Invest.* © 1968 US-Canadian Div. of the IAP.)

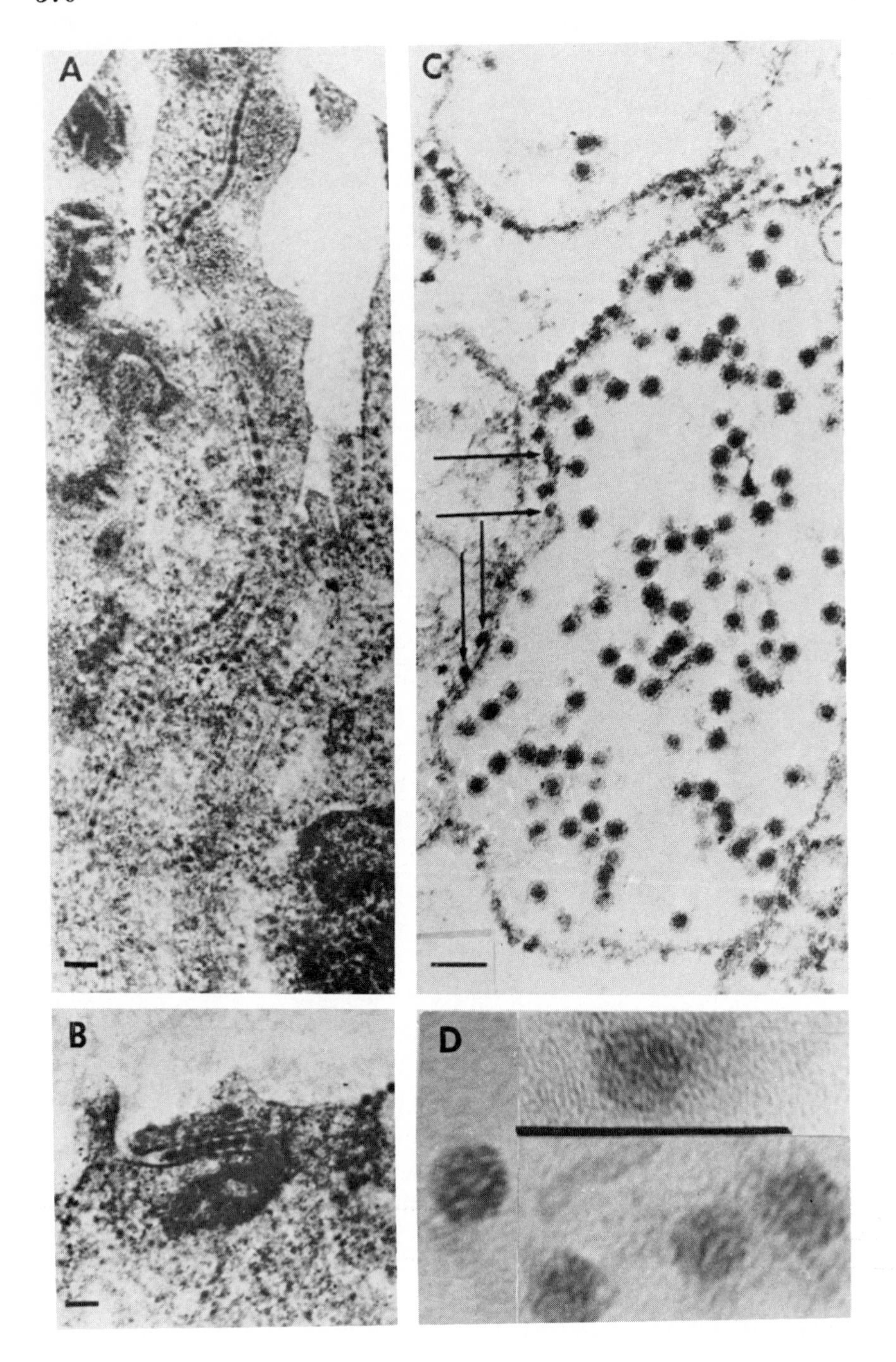
A
B
C
D

in a membrane not always readily detectable (Matsumura *et al.*, 1971) or rarely, as noted in mice, in concentric laminated membranes (Abdelwahab *et al.*, 1964). Single virions which appear to be isolated in cytoplasm are invariably enclosed within a smooth membrane or tube (Ota, 1965; Filshie and Rehacek, 1968). Some of the vesicles containing virus aggregates are apparently ribosome-studded on the exterior surface, but others are completely degranulated (Ota, 1965; Filshie and Rehacek, 1968; Lecatsas and Weiss, 1969; Matsumura *et al.*, 1971).

Meshlike convoluted membranous masses near the nucleus were reported in HEp2 cells infected with West Nile virus (Southam *et al.*, 1964) and in fetal lamb kidney and BHK-21 cells infected with Wesselsbron virus (Parker and Stannard, 1967; Lecatsas and Weiss, 1969). These were associated with virus development and, as in infected mouse brain cells, some membranous masses had elements of fourfold symmetry and others of sixfold symmetry. In DEN-1-infected J-111 cells (human leukocyte cell line) similar "crystalloid" membranes were seen (Matsumura *et al.*, 1977). Broad, parallel filaments appearing by 19 h in cytoplasm may be the precursors of these masses (Lecatsas and Weiss, 1969). Filshie and Rehacek (1968) found that JE virus was more cytolytic than MVE virus in mosquito cells; in the latter infection, the cytoplasm contained large masses of meshlike convoluted tubules of smooth membranes from which radiated lamellae of rough endoplasmic reticulum containing virions; virions accumulated also in vacuoles or in lamellae bound by smooth membranes. Vesicular aggregates were found near the nucleus of dengue virus-infected LLC-MK2 cells; the intravacuolar vesicles were variable in diameter and larger than accompanying virus particles (Cardiff *et al.*, 1973b; Demsey *et al.*, 1974).

Fig. 10. Electron micrographs of thin sections of flavivirus-infected Vero cells, and of virus cores. The bars represent 100 nm. (A) A chain of Kunjin virus particles at 24 h postinfection within a cisterna bounded by thickened electron-dense membranes. Both complete virions and "undifferentiated" beady rods are present. (B) A lamellar cisterna as in (A), showing an exit route for Kunjin virions. (C) Dengue-2 virions at 24 h postinfection within the lumen of a vacuole bounded by a membrane studded with electron-dense ribosomelike particles (arrows). (D) Kunjin virus cores derived by treatment of virions with 0.2% deoxycholate at 4°C to remove the envelope and purified by sedimentation in a sucrose gradient. The cores were identified by the peak of ^{32}P-label with sedimentation coefficient of 130–140 S and then fixed and stained with glutaraldehyde and uranyl acetate, respectively. Note the smooth, spherical shapes compared to the ragged outlines of the ribosomelike particles in (C). [(A) and (B) from Boulton and Westaway (1976, reproduced with permission); (C) from Matsumura *et al.* (1971, reproduced with permission); (D) from Della-Porta (1974).]

In Kunjin virus-infected Vero cells, virions accumulated between 18 and 24 h in vacuoles, but in approximately 10% of cells, virus particles appeared as single chains within cisternae of pairs of thickened electron-dense smooth membranes (Fig. 10). Serial sections showed that these chains represented a cross-section of a lamella, one virion in thickness (Boulton and Westaway, 1976). When infected cell membrane fractions were separated by isopycnic sedimentation (Fig. 5), virions were found associated with the fractions containing rough membranes but were absent from the smooth membrane fraction and from plasma membranes (Boulton and Westaway, 1976). The lamellae were apparently disrupted and not identifiable in membrane fractions. Similar localization of JE virions in separated membrane fractions was reported by Kos *et al.* (1975b).

C. Precursor Particles and Virus Maturation

As already indicated, morphogenesis is associated predominantly with membranes of the endoplasmic reticulum and, to a lesser extent, with those of the Golgi complex. Morphologically mature virions are also occasionally found within distensions of the perinuclear membrane and singly in cytoplasm. Positive morphological evidence of the process of assembly and maturation at any site is lacking.

Most authors urge caution in the interpretation of electron-dense ribosomelike particles arrayed on vesicles or membranes of the endoplasmic reticulum enclosing virions. Matsumura *et al.* (1971) showed that in infected Vero cells the nucleocapsids of Sindbis alphavirus budding through membranes stained uniformly as spherical dense particles, similar to the cores derived from Kunjin virions and which sediment at the same rate as Sindbis virus-derived cores (Fig. 10) (Boulton and Westaway, 1972). In contrast, the particles of similar size arrayed on endoplasmic reticulum membranes in dengue virus-infected Vero cells (Fig. 10) are uneven in shape and staining. If the latter particles function as membrane-bound ribosomes, they are conveniently located for translation of the structural proteins destined for incorporation into virions accumulating within the lumen.

As noted in Section III,B, all the nonstructural proteins, in addition to the envelope protein V3, are incorporated into all cytoplasmic membranes. Hence, if a budding process did occur, the precise site of budding would have to be exquisitely selected. Although some morphologically mature virions accumulate in vacuoles together with vesicular aggregates, no intermediate forms are seen. Virions with the envelope or a pedicle attached to or continuous with plasma membrane or the inner membranous surface of vesicles are found rarely, and these particles could fuse with rather than bud from membranes (Ota, 1965; Filshie and Rehacek, 1968; Murphy *et al.*, 1968; Cardiff *et al.*, 1973b; Matsumura *et al.*, 1977).

Because no direct evidence of budding was obtained from freeze-fracturing of dengue virus-infected cells, Demsey *et al.* (1974) have postulated that envelopes are formed *de novo,* the virions later being totally engulfed, or that the envelope is formed directly by some form of engulfment other than budding. *De novo* formation of the envelope could occur within cisternae by accretion of envelope protein from a protein–lipid domain in modified (thickened) membranes of endoplasmic reticulum (Boulton and Westaway, 1976). Presumably the lipid composition of the flavivirus envelope is host-controlled (Trent, 1973). Interestingly, it has been reported but not yet confirmed that Sindbis (alphavirus) nucleocapsids enclosed within vesicles of infected mosquito cells are apparently enveloped by interaction with the vesicle membranes rather than by budding (Gliedman *et al.*, 1975). Sriurairatna *et al.* (1973) proposed that the final stage of maturation of dengue virus may involve cleavage of particles in beady rods, within degranulated and thickened lamellar membranes. It should be noted that morphologically mature virions are formed within the lumen of membranes of the Golgi complex, which generally appear devoid of ribosomelike or possible precursor particles (Filshie and Rehacek, 1968; Blinzinger, 1972; Whitfield *et al.*, 1973; Calberg-Bacq *et al.*, 1975).

Other reports of possible precursor particles have received little or no support. At an early stage of JE virus infections, the nucleus and cytoplasm of some microglia contained small, dense particles (10–20 nm in diameter) (Yasuzumi *et al.*, 1964); in neurons, uniform (precursor?) particles (20–25 nm) were enclosed in juxtanuclear cytoplasmic inclusions (Yasuzumi and Tsubo, 1965b), and these were "clearly composed of material different from that of mature particles." These findings were not confirmed in similar studies with JE virus by Oyanagi *et al.* (1969), but Murphy *et al.* (1968) referred briefly to similar particles in the nuclei of some SLE virus-infected cells. More recent biochemical and other evidence appears to exclude any involvement of the nucleus in virus development (see Section II,C,5). In a single report (Peat and Bell, 1970), empty, capsidlike structures were observed within dilated "neurotubules" in infected neurons as the first signs of development of Entebbe bat salivary gland virus. However, Blinzinger *et al.* (1971) concluded that similar annular profiles in Zimmern virus-infected neurons were cross sections of tubular structures, helixlike in longitudinal section. Aberrant tailed particles within cisternae containing virions have been reported on only two occasions (Sriurairatna *et al.*, 1973; Murphy *et al.*, 1968).

Biochemical evidence of precursor particles in flavivirus-infected cells is limited and disappointing. Nucleocapsids have not been isolated directly from cells. Shapiro *et al.* (1972a, 1973b) compared the protein composition and specific infectivity (PFU/cpm in the proteins of labeled virions) of extracellular (N forms) and intracellular (I forms) purified virions of JE and of DEN-2 viruses grown in chick cells. The I-form virions comprised V3, V2, and NV2 rather than V1, and their specific infectivity was 0.004 compared to that of N forms; by gel electrophoresis, the purified I forms appeared free of any labeled host protein. Virions also of the same aberrant composition as I forms were harvested from extracellular fluids of infected LLC-MK2 cells maintained in medium containing 6 m*M* Tris (T forms), and NV2 was shown to be a component of the virus envelope. These authors have postulated that I forms "mature" to N forms, gaining increased specific activity, by cleavage of NV2 to V1. However, the assumption of cleavage is now untenable (Wright and Westaway, 1977), and the lower specific infectivity of I forms may be caused by the harsh extraction procedure (two freeze-thaw cycles for cells, homogenization, and sonication) because T forms of the same composition as I forms possess specific activity as high as 0.7 of N forms.

In summary, no convincing evidence of precursor particles or of a budding process has been obtained in flavivirus-infected cells. It seems more likely that assembly occurs rapidly by a condensation or accretion process in cytoplasmic "factories" producing crystalline aggregates, or within the lumen of vesicles or of modified laminated endoplastic reticulum adjacent to or continuous with meshlike masses of convoluted microtubules near the nucleus. Virions accumulate and appear to move toward the cell periphery within distended endoplasmic reticulum or lamellae, in vacuoles, or in vesicles containing vesicular aggregates; the enclosing membranes, if granulated, appear to progress to smooth forms as virions accumulate. Further progress in defining flavivirus morphogenesis requires a thorough concerted effort, combining high-resolution electron microscopy with a variety of cell fractionation techniques and biochemical analyses in at least one defined virus and vertebrate cell system.

D. Exit of Virions from Cells

The membranous structures in which virions accumulate may be interconnected, facilitating the peripheral movement of virions from the juxtanuclear region of maturation. Prior to cell lysis, enlarged vesicles containing virions probably move to the cell perimeter and fuse with the plasma membrane, producing an apparent evagination from which virions can escape (Filshie and Rehacek, 1968; Dalton, 1972). In SLE virus-infected mosquitoes, virus particles are shed into salivary diverticula from the apical end of salivary gland cells by membrane fusion and egestion of endoplasmic reticulum, or by localized breakdown of plasma membrane, without damage to cells (Whitfield *et al.*, 1973). However, the latter method of exit is unique and probably represents evolutionary specialization essential for virus transmission by the vector host. Serial sections through chains of Kunjin virions within lamellae of smooth membranes leading to the plasma membrane show that direct exit from vertebrate cells occurs (Fig. 10). All these methods allow slow and continuous release of progeny virus. From cells undergoing lysis, virus release occurs by breakdown of plasma membrane and release of cytoplasmic contents; thus extracellular virions are commonly found in clusters or still enclosed within vesicles or fragments of endoplasmic reticulum (Murphy *et al.*, 1968; Sriurairatna *et al.*, 1973; Calberg-Bacq *et al.*, 1975).

The yields of infectious virus released into cell culture fluids, and from disrupted cells, are often low compared to the numbers of virions visible in thin sections of some infected cells, but yields of dengue virus PFU and HA may be increased by 6- and 32-fold, respectively, when 40–80 m*M* $MgCl_2$ is added to the culture medium at 10 h postinfection (Matsumura *et al.*, 1972). Schlesinger (1977) comments that such observations lead to the conclusion that the dengue virus masses seen in the cytoplasm of infected Vero and LLC-MK2 cells represent a dead-end accumulation of incompletely processed virions, e.g., deficient in protein V1, as reported by Shapiro *et al.* (1972a). This is supported by an early observation that, after extensive purification, the particle/infectivity ratio of MVE virus extracted from infected mouse brain was approximately 100:1 (Ada *et al.*, 1962). However, as much as 10-fold increases in the infectivity of pools of MVE, JE, and West Nile viruses (Hawkes, 1964; Hawkes and Lafferty, 1967) and 100-fold increases in DEN-2 virus yields (Halstead and O'Rourke, 1977) have been produced by enhancing antibody. It is important therefore to determine the true nature of the masses of virions seen in some infected cell preparations by, for example, labeling cells in amino acids after the latent period and purifying the labeled virions accumulating intracellularly and extracellularly during sequential chase periods. When combined with the radioactivity/infectivity and the particle/infectivity ratios, such data would define the period of intracellular residence of virions prior to release, and the proportion of virions represented by dead-end accumulation. Meanwhile, the true status of intracellular virions and the major routes of exit remain uncertain.

VI. SUMMARY OF FLAVIVIRUS REPLICATION

Serious gaps exist in our knowledge of both the biochemistry and morphogenesis of flavivirus replication. Replication occurs entirely in cytoplasm with a latent period of 12–16 h. Information on adsorption and penetration is meager (see Schlesinger, 1977); subsequently the genome (plus-strand 44 S RNA) is presumably completely uncoated, and an initial round of translation produces the essential but unidentified RNA polymerase. The 44 S RNA is capped but contains no detectable poly(A). Transcription from 44 S RNA occurs on smooth membranes in the perinuclear region, probably in association with developing meshlike masses of convoluted microtubules. Both plus and minus strands are produced in barely detectable amounts until about 12 h; subsequently transcription into plus strands from stable minus-strand templates increases markedly. The very high RNase resistance of the 20 S RNA containing the template indicates that the 20 S RNA includes only one nascent strand and suggests that this strand and the RNA polymerase are released slowly upon completion of transcription. This would account for the long latent period and the unavailability of most progeny 44 S plus strands for translation until a slow leakage to polyribosomes allows development of a control system for asymmetric transcription. A variable amount of a heterodisperse and partially RNase-resistant 26 S RNA may represent an intermediate form during the release of 44 S RNA from the accumulating 20 S RNA. As protein synthesis and virus maturation occur, further hypertrophy of cell membrane systems produces a cytoplasm replete with vesicles, vacuoles, and endoplasmic reticulum with lamellae of paired membranes. These structures are seen with ribosomelike particles on their outer surface, which later become degranulated while mature virions accumulate within their cisternae.

Host protein synthesis is not switched off and may be necessary for generation of proliferating membranes. Translation of viral proteins from the single polycistronic messenger (44 S plus-strand RNA) takes place in rough endoplasmic reticulum, possibly by a process unique for flavivirus-infected eukaryotic cells. Nearly all proteins appear to be initiated internally and independently in translation; no polyprotein precursors are detectable, and proteins with a total molecular weight of at least 300,000 are unrelated by tryptic peptide mapping. The envelope glycoprotein V3 in cytoplasm and the nonstructural glycoprotein NV2 exhibit microheterogeneity in electrophoretic migration. Posttranslational processing of V3 (by glycosylation) and the major core protein V2 (by trimming to NV1½) occurs, and the minor (smallest) protein V1 may be cleaved from a large precursor during maturation. All proteins except V1 are incorporated into most cytoplasmic membrane fractions including plasma membrane. A slowly sedimenting noninfectious particle comprising V3, NV2, and V1 is released from cells, but its morphogenesis and function are unknown.

No nucleocapsids and no particles midway through a budding process have been positively identified. Assembly and maturation probably occur via a condensation process within the cisternae in which virions accumulate. Exit from cells occurs by the release of packets of virions in vesicles during either reverse phagocytosis or cell lysis, and by leakage from cisternae or lamellae opening through the plasma membrane.

The strategy of replication of flaviviruses differs so greatly from that of alphaviruses (see appropriate chapters in this volume) that their joint inclusion as genera in the Togaviridae may require taxonomic reassessment. Despite the inherent difficulties in biochemical studies, flaviviruses offer unique opportunities in analyses of membrane formation and of events of protein translation and control in eukaryotic cells.

ACKNOWLEDGMENT

The work of myself and my colleagues was supported by grants from the National Health and Medical Research Council of Australia.

REFERENCES

Abdelwahab, K. S. E., Almeida, J. D., Doane, F. W., and McLean, D. M. (1964). *Can. Med. Assoc. J.* **90,** 1068–1072.

Ada, G. L., Abbot, A., Anderson, S. G., and Collins, F. D. (1962). *J. Gen. Microbiol.* **29,** 165–170.

Atkinson, P. H., and Summers, D. F. (1971). *J. Biol. Chem.* **246,** 5162–5175.

Aviv, A., Boime, I., Loyd, B., and Leder, P. (1972). *Science* **178,** 1293–1295.

Baltimore, D. (1968). *J. Mol. Biol.* **32,** 359–368.

Baltimore, D. (1969). *In* "The Biochemistry of Viruses" (H. B. Levy, ed.), pp. 101–176. Dekker, New York.

Baltimore, D. (1971). *Bacteriol. Rev.* **35**, 235–241.
Bell, T. M., Field, E. J., and Narang, H. K. (1971). *Arch. Gesamte Virusforsch.* **35**, 183–193.
Bhamarapravati, N., Halstead, S. B., Sookavachana, P., and Boonyapaknavik, V. (1964). *Arch. Pathol.* **77**, 538–543.
Blinzinger, K. (1972). *Ann. Inst. Pasteur, Paris* **123**, 497–519.
Blinzinger, K., Müller, W., and Anzil, A. P. (1971). *Arch. Gesamte Virusforsch.* **35**, 194–202.
Blinzinger, K., Luh, S., and Anzil, A. P. (1975). *Adv. Neurol.* **12**, 459–464.
Blobel, G. (1973). *Proc. Natl. Acad. Sci. U.S.A.* **70**, 924–928.
Bose, H. R., and Brundige, M. A. (1972). *J. Virol.* **9**, 785–791.
Bosmann, H. B., Hagopian, A., and Eylar, E. H. (1968). *Arch. Biochem. Biophys.* **128**, 51–69.
Boulton, P. S., and Webb, H. E. (1971). *Brain* **94**, 411–418.
Boulton, R. W. (1974). Ph.D. Thesis, Monash University, Melbourne, Australia.
Boulton, R. W., and Westaway, E. G. (1972). *Virology* **49**, 283–289.
Boulton, R. W., and Westaway, E. G. (1976). *Virology* **69**, 416–430.
Boulton, R. W., and Westaway, E. G. (1977). *Arch. Virol.* **55**, 201–208.
Brandt, W. E., and Russell, P. K. (1975). *Infect. Immun.* **11**, 330–333.
Brandt, W. E., Chiewsilp, D., Harris, D. L., and Russell, P. K. (1970). *J. Immunol.* **105**, 1565–1568.
Brawner, T. A., Lee, J. C., and Trent, D. W. (1973). *Abstr., Am. Soc. Microbiol., Annu. Meet., 73rd, Miami Beach* p. 202.
Brawner, T. A., Lee, J. C., and Trent, D. W. (1977). *Arch. Virol.* **54**, 147–151.
Calberg-Bacq, C.-M., Rentier-Delrue, F., Osterrieth, P. M., and Duchesne, P. Y. (1975). *J. Ultrastruct. Res.* **53**, 193–203.
Caliguiri, L. A., and Tamm, I. (1970). *Virology* **42**, 100–111.
Cardiff, R. D., and Lund, J. K. (1976). *Infect. Immun.* **13**, 1699–1709.
Cardiff, R. D., McCloud, T. G., Brandt, W. E., and Russell, P. K. (1970). *Virology* **41**, 569–572.
Cardiff, R. D., Brandt, W. E., McCloud, T. G., Shapiro, D., and Russell, P. K. (1971). *J. Virol.* **7**, 15–23.
Cardiff, R. D., Dalrymple, J. M., and Russell, P. K. (1973a). *Arch. Gesamte Virusforsch.* **40**, 392–396.
Cardiff, R. D., Russ, I. B., Brandt, W. E., and Russell, P. K. (1973b). *Infect. Immun.* **7**, 809–816.
"Catalogue of Arthropod-Borne Viruses of the World" (1967). U.S. Public Health Serv. Publ. No. 1760. Washington, D.C.
Catanzaro, P. J., Brandt, W. E., Hogrefe, W. R., and Russell, P. K. (1974). *Infect. Immun.* **10**, 381–388.
Cleaves, G. R., and Dubin, D. T. (1979). *Virology* **96**, 159–165.
Cleaves, G. R., and Schlesinger, R. W. (1977). *Abstr., Am. Soc. Microbiol., Annu. Meet., 77th Honolulu* p. 287.
Clegg, J. C. S. (1975). *Nature (London)* **254**, 454–455.
Clewley, J. P., and Kennedy, S. I. T. (1976). *J. Gen. Virol.* **32**, 395–411.
Dalton, S. (1972). *Ann. Inst. Pasteur, Paris* **123**, 489–496.
David-West, T. S., Labzoffsky, N. A., and Hamvas, J. J. (1972). *Arch. Gesamte Virusforsch.* **36**, 372–379.
Davis, B. D., Dulbecco, R., Eisen, H. N., Ginsberg, H. S., Wood, W. B., Jr., and McCarty, M., eds. (1973). *In* "Microbiology," 2nd ed., pp. 1122–1137. Harper, New York.

Della-Porta, A. J. (1974). Ph.D. Thesis, Monash University, Melbourne, Australia.
Demsey, A., Steere, R. L., Brandt, W. E., and Veltri, B. J. (1974). *J. Ultrastruct. Res.* **46**, 103–116.
Dubin, D. T., Stollar, V., Hsuchen, C.-C., Timko, K., and Guild, G. M. (1977). *Virology* **77**, 457–470.
Eckels, K. H., Hetrick, F. M., and Russell, P. K. (1975). *Infect. Immun.* **11**, 1053–1060.
Eldadah, N., and Nathanson, N. (1967). *Am. J. Epidemiol.* **86**, 776–790.
Filshie, B. K., and Rehacek, J. (1968). *Virology* **34**, 435–443.
Friedman, R. M., Levin, J. G., Grimley, P. M., and Berezesky, I. K. (1972). *J. Virol.* **10**, 504–515.
Fukui, K. (1973). *Kobe J. Med. Sci.* **19**, 23–38.
Gliedman, J. B., Smith, J. F., and Brown, D. T. (1975). *J. Virol.* **16**, 913–926.
Granboulan, N., and Franklin, R. M. (1968). *J. Virol.* **2**, 129–148.
Grimley, P. M., and Friedman, R. M. (1970). *Exp. Mol. Pathol.* **12**, 1–13.
Grimley, P. M., Berezesky, I. K., and Friedman, R. M. (1968). *J. Virol.* **2**, 1326–1338.
Halstead, S. B., and O'Rourke, E. J. (1977). *Nature (London)* **265**, 739–741.
Hawkes, R. A. (1964). *Aust. J. Exp. Biol. Med.* **42**, 465–482.
Hawkes, R. A., and Lafferty, K. J. (1967). *Virology* **33**, 250–261.
Hellerman, J. G., and Shafritz, D. A. (1975). *Proc. Natl. Acad. Sci. U.S.A.* **72**, 1021–1025.
Igarashi, A., Fukunaga, F., and Iukai, K. (1964). *Biken J.* **7**, 111–119.
Jacobson, M. F., Asso, J., and Baltimore, D. (1970). *J. Mol. Biol.* **49**, 657–669.
Kos, K. A., Osborne, B. A., and Goldsby, R. A. (1975a). *J. Virol.* **15**, 913–917.
Kos, K. A., Shapiro, D., Vaituzis, Z., and Russell, P. K. (1975b). *Arch. Virol.* **47**, 217–224.
Lecatsas, G., and Weiss, K. E. (1969). *Arch. Gesamte Virusforsch.* **27**, 332–338.
Lubiniecki, A. S., and Henry, C. J. (1974). *Proc. Soc. Exp. Biol. Med.* **145**, 1165–1169.
McCloud, T. G., Cardiff, R. D., Brandt, W. E., Chiewsilp, D., and Russell, P. K. (1971). *Am. J. Trop. Med. Hyg.* **20**, 964–968.
Matsumura, T., Stollar, V., and Schlesinger, R. W. (1971). *Virology* **46**, 344–355.
Matsumura, T., Stollar, V., and Schlesinger, R. W. (1972). *J. Gen. Virol.* **17**, 343–347.
Matsumura, T., Shiraki, K., Sashikata, T., and Hotta, S. (1977). *Microbiol. Immunol.* **21**, 329–334.
Morrison, T. G., and Lodish, H. F. (1973). *Proc. Natl. Acad. Sci. U.S.A.* **70**, 315–319.
Murphy, F. A., Harrison, A. K., Gary, G. W., Whitfield, S. G., and Forrester, F. T. (1968). *Lab. Invest.* **19**, 652–662.
Naeve, C. W., and Trent, D. W. (1978). *J. Virol.* **25**, 535–545.
Nishimura, C., and Tsukeda, H. (1971). *Jpn. J. Microbiol.* **15**, 309–316.
Nuss, D. L., Oppermann, H., and Koch, G. (1975). *Proc. Natl. Acad. Sci. U.S.A.* **72**, 1258–1262.
Ota, Z. (1965). *Virology* **25**, 372–378.
Oyanagi, S., Ikuta, F., and Ross, E. R. (1969). *Acta Neuropathol.* **13**, 169–181.
Parker, J. R., and Stannard, L. M. (1967). *Arch. Gesamte Virusforsch.* **20**, 469–472.
Peat, A., and Bell, T. M. (1970). *Arch. Gesamte Virusforsch.* **31**, 230–236.
Pfefferkorn, E. R., and Shapiro, D. (1974). *In* "Comprehensive Virology" (H. Fraenkel-Conrat and R. R. Wagner, eds.), Vol. 2, pp. 171–230. Plenum, New York.
Pollard, T. D., and Weihing, R. R. (1974). *Crit. Rev. Biochem.* **2**, 1–65.
Qureshi, A. A., and Trent, D. W. (1972). *J. Virol.* **9**, 565–573.
Qureshi, A. A., and Trent, D. W. (1973a). *Infect. Immun.* **8**, 993–999.
Qureshi, A. A., and Trent, D. W. (1973b). *Infect. Immun.* **7**, 242–248.
Rai, J., and Ghosh, S. N. (1976). *Indian J. Med. Res.* **64**, 981–991.
Richardson, C. D., and Vance, D. E. (1976). *J. Biol. Chem.* **251**, 5544–5550.

Robertson, H. D. (1975). *In* "RNA Phages" (N. D. Zinder, ed.), pp. 113–145. Cold Spring Harbor Lab., Cold Spring Harbor, New York.
Robertson, H. D., and Zinder, N. D. (1969). *J. Biol. Chem.* **244,** 5790–5800.
Russell, P. K., Chiewsilp, D., and Brandt, W. E. (1970). *J. Immunol.* **105,** 838–845.
Saborio, J. L., Pong, S.-S., and Koch, G. (1974). *J. Mol. Biol.* **85,** 195–211.
Savage, T., Granboulan, N., and Girard, M. (1971). *Biochimie* **53,** 533–543.
Schlesinger, R. W. (1977). "Dengue Viruses," Virology Monographs, Vol. 16. Springer-Verlag, Berlin and New York.
Schmaljohn, C., and Blair, C. D. (1977). *J. Virol.* **24,** 580–589.
Schmaljohn, C., and Blair, C. D. (1979). *J. Virol.* **31,** 816–822.
Schreier, M. H., Staehelin, T., Gesteland, R. F., and Spahr, P. F. (1973). *J. Mol. Biol.* **75,** 575–578.
Shapiro, D., Brandt, W. E., Cardiff, R. D., and Russell, P. K. (1971). *Virology* **44,** 108–124.
Shapiro, D., Brandt, W. E., and Russell, P. K. (1972a). *Virology* **50,** 906–911.
Shapiro, D., Kos, K., Brandt, W. E., and Russell, P. K. (1972b). *Virology* **48,** 360–372.
Shapiro, D., Kos, K. A., and Russell, P. K. (1973a). *Virology* **56,** 95–109.
Shapiro, D., Kos, K. A., and Russell, P. K. (1973b). *Virology* **56,** 88–94.
Shestopalova, N. M., Reingold, V. N., Gagarina, A. V., Kornilova, E. A., Popov, G. V., and Chumakov, M. P. (1972). *J. Ultrastruct. Res.* **40,** 458–469.
Simmons, D. T., and Strauss, J. H. (1972). *J. Mol. Biol.* **71,** 615–631.
Sinarachatanant, P., and Olson, L. C. (1973). *J. Virol.* **12,** 275–283.
Smith, T. J., Brandt, W. E., Swanson, J. L., McCown, J. C., and Buescher, E. L. (1970). *J. Virol.* **5,** 524–532.
Southam, C. M., Shipkey, F. H., Babcock, V. I., Bailey, R., and Erlandson, R. A. (1964). *J. Bacteriol.* **88,** 187–199.
Sriurairatna, S., Bhamapravati, N., and Phalavadhtana, O. (1973). *Infect. Immun.* **8,** 1017–1028.
Stevens, T. M., and Schlesinger, R. W. (1965). *Virology* **27,** 103–112.
Stohlman, S. A., Wisseman, C. L., Jr., Eylar, O. R., and Silverman, D. J. (1975). *J. Virol.* **16,** 1017–1026.
Stohlman, S. A., Eylar, O. R., and Wisseman, C. L., Jr. (1976). *J. Virol.* **18,** 132–140.
Stollar, V. (1969). *Virology* **39,** 426–438.
Stollar, V., Stevens, T. M., and Schlesinger, R. W. (1966). *Virology* **30,** 303–312.
Stollar, V., Schlesinger, R. W., and Stevens, T. M. (1967). *Virology* **33,** 650–658.
Strauss, J. H., and Strauss, E. G. (1977). *In* "The Molecular Biology of Animal Viruses" (D. P. Nayak, ed.), Vol. 1, pp. 111–166. Dekker, New York.
Summers, D. F., and Maizel, J. V., Jr. (1971). *Proc. Natl. Acad. Sci. U.S.A.* **68,** 2852–2856.
Svitkin, Y. V., Lyapustin, V. N., Lashkevich, V. A., and Agol, V. I. (1978). *FEBS Lett.* **96,** 211–215.
Taber, R., Rekosh, D., and Baltimore, D. (1971). *J. Virol.* **8,** 395–401.
Takeda, H., Oya, A., Hashimoto, K., and Yamada, M.-A. (1977). *J. Gen. Virol.* **34,** 201–205.
Takeda, H., Oya, A., Hashimoto, K., Yasuda, T., and Yamada, M.-A. (1978). *J. Gen. Virol.* **38,** 281–291.
Takehara, M. (1971). *Arch. Gesamte Virusforsch.* **34,** 266–277.
Theiler, M., and Downs, W. G. (1973). "The Arthropod-Borne Viruses of Vertebrates." Yale Univ. Press, New Haven, Connecticut.
Trent, D. W. (1973). *Abstr., Am. Soc. Microbiol., Annu. Meet., 73rd, Miami Beach,* p. 240.

Trent, D. W. (1977). *J. Virol.* **22**, 608–618.
Trent, D. W., and Qureshi, A. A. (1971). *J. Virol.* **7**, 379–388.
Trent, D. W., Swensen, C. C., and Qureshi, A. A. (1969). *J. Virol.* **3**, 385–394.
Trent, D. W., Harvey, C. L., Qureshi, A., and LeStourgeon, D. (1976). *Infect. Immun.* **13**, 1325–1333.
Watson, J. D. (1975). "Molecular Biology of the Gene," 3rd ed., pp. 455–495. Benjamin, New York.
Wengler, G., Wengler, G., and Gross, H. J. (1978). *Virology* **89**, 423–437.
Wengler, G., Beato, M., and Wengler, G. (1979). *Virology* **96**, 516–529.
Westaway, E. G. (1973). *Virology* **51**, 454–465.
Westaway, E. G. (1975). *J. Gen. Virol.* **27**, 283–292.
Westaway, E. G. (1977). *Virology* **80**, 320–335.
Westaway, E. G., and Reedman, B. M. (1969). *J. Virol.* **4**, 688–693.
Westaway, E. G., and Shew, M. (1977). *Virology* **80**, 309–319.
Westaway, E. G., Della-Porta, A. J., and Reedman, B. M. (1974). *J. Immunol.* **112**, 656–663.
Westaway, E. G., Shew, M., and Della-Porta, A. J. (1975). *Infect. Immun.* **11**, 630–634.
Westaway, E. G., McKimm, J. L., and McLeod, L. G. (1977). *Arch. Virol.* **53**, 305–312.
Whitfield, S. G., Murphy, F. A., and Sudia, W. D. (1973). *Virology* **56**, 70–87.
Wright, P. J., and Westaway, E. G. (1977). *J. Virol.* **24**, 662–672.
Wright, P. J., and Westaway, E. G. (1979). *Abstr. Am. Soc. Microbiol., 79th, Honolulu*, p. 310.
Wright, P. J., Bowden, D. S., and Westaway, E. G. (1977). *J. Virol.* **24**, 651–661.
Yamazaki, S. (1968). *Jpn. J. Microbiol.* **12**, 171–178.
Yasuzumi, G., and Tsubo, I. (1965a). *J. Ultrastruct. Res.* **12**, 304–316.
Yasuzumi, G., and Tsubo, I. (1965b). *J. Ultrastruct. Res.* **12**, 317–327.
Yasuzumi, G., Tsubo, I., Sugihara, R., and Nakai, Y. (1964). *J. Ultrastruct. Res.* **11**, 213–229.
Zebovitz, E., Leong, J. K. L., and Doughty, S. C. (1972). *Arch. Gesamte Virusforsch.* **38**, 319–327.
Zebovitz, E., Leong, J. K. L., and Doughty, S. C. (1974). *Infect. Immun.* **10**, 204–211.

20

Togaviruses in Cultured Arthropod Cells

VICTOR STOLLAR

THE TOGAVIRUSES

ISBN 0-12-625380-3

I. INTRODUCTION

As described in Chapter 2 of this volume, the family Togaviridae has been divided into four genera, *Alphavirus*, *Flavivirus*, *Rubivirus*, and *Pestivirus*. Members of these groups are defined by their physicochemical and antigenic properties. Our concern in this chapter will be only with alpha- and flaviviruses which (with few exceptions) are transmitted in nature by arthropod vectors, mosquitoes in the case of alphaviruses and mosquitoes or ticks in the case of flaviviruses. Because alpha- and flaviviruses have a mandatory replication cycle in an arthropod, they are included in the large complex of viruses known as arboviruses (see Chapters 1, 2, and 6).

The replication of these viruses in intact arthropod vectors is discussed in Chapters 6 and 7. In this chapter, I shall review what is known about the replication of alpha- and flaviviruses in cell cultures derived from arthropods.

Among the viruses which cause disease in humans and domestic animals, arboviruses are unusual in that they can replicate in organisms widely separated phylogenetically (see Chapter 1). Thus, experiments with alpha- and flaviviruses in arthropod and vertebrate cells provide a fascinating study in comparative virology. We may ask, for example, How do different host cells influence the chemical, immunological, and biological properties of a given virus? and How is it that a given virus affects two host cells, for example, a hamster and mosquito cell, so differently—killing the former but not the latter?

In addition to an exercise in comparative virology, the examination of togaviruses in cultured arthropod cells can be expected to provide additional insights into the transmission, maintenance, and persistence in nature of these viruses, among which are many of great medical importance (see Chapter 3).

For further background material, the reader is referred to a volume edited by Weiss (1971) and reviews by Singh (1972) and Mussgay *et al.* (1975).

II. ARTHROPOD CELL LINES USED FOR THE GROWTH OF TOGAVIRUSES

A. Mosquito Cell Lines

At the present time there are a number of cell lines derived from mosquitoes which are suitable for the growth of togaviruses. In general, these lines have been derived from embryonic or larval tissues; their ease of handling resembles that of well-known mammalian cell lines and, at least in some cases, the cells can be grown in medium which, apart from the serum component, is defined.

1. The Aedes albopictus Cell Line of Singh

The *A. albopictus* cell line of Singh is the most widely used line of mosquito cells at the present time. It was derived originally from larvae (Singh, 1967) and grown in a medium devised by Mitsuhashi and Maramorosch (1964), usually referred to as MM medium. As is true for the media of most arthropod cell lines, MM medium is supplemented with fetal calf serum. Two features of this medium are worth noting: First, it is an undefined medium in which amino acids are supplied in the form of lactalbumin hydrolysate and yeast hydrolysate; second, it has an osmotic pressure of 410 mosmol/kg, which is considerably higher than that of serum or of media such as Eagle's medium (290 mosmol/kg), commonly used for the growth of mammalian cells (R. Koo and V. Stollar, unpublished results). Although *A. albopictus* cells are still most commonly grown in MM medium, in several laboratories they have now been adapted to Eagle's medium (Spradling *et al.*, 1975; Igarashi and Stollar, 1976; Sarver and Stollar, 1977). The *A. albopictus* cell line can be grown either as monolayers or in suspension cultures.

Singh also established a line of *Aedes aegypti* cells, but they have not been used as extensively for the study of viruses as the *A. albopictus* cell line.

2. *The Aedes aegypti Cell Line of Peleg*

This cell line was derived from embryonic tissues (Peleg, 1968). It is grown in an undefined medium containing lactalbumin hydrolysate, and the cells are usually grown as monolayers. They have been widely used for the growth of togaviruses and in general give high yields.

3. *Other Mosquito Cell Lines*

Although the "*A. aegypti*" cell line of Grace (1966) was reportedly the first established mosquito cell line, it did not support the growth of togaviruses and on the basis of other information was shown in fact to be a line of moth (*Antheraea eucalypti*) cells (Davey, 1973).

Pudney *et al.* (1978) have described a number of mosquito cell lines derived from various species of *Aedes* and *Anopheles*. Although these lines have not yet been widely circulated, certain of their *Aedes* cell lines do produce high yields of viruses and would be suitable for further study.

A promising cell line which gave high yields of Japanese encephalitis virus (JEV) has been derived from *Culex tritaeniorhynchus* by Hsu *et al.* (1975).

B. Tick Cell Lines

Although much less work has been done with cultured tick cells than with mosquito cells, a number of lines of tick cells have now been established (Varma *et al.*, 1975; Rehacek, 1976; Bhat and Yunker, 1977); and recently Bhat and Yunker (1979) have shown that their RML-14 line from *Dermacentor parumapertus* supports the growth of several tick-borne flaviviruses as well as certain mosquito-borne alpha- and flaviviruses.

III. GROWTH OF VIRUSES IN CELL LINES OF ARTHROPOD ORIGIN

A. Systems Which Produce High Yields of Virus from Mosquito Cells

There have been many reports of the replication of alpha- and flaviviruses in cultured mosquito cells, as well as several describing the replication of flaviviruses in cultured tick cells. I shall not attempt to review or document them all. Instead, the reader is referred to the volume and the reviews mentioned at the beginning of the chapter. The main point to be made is that there are now well-established systems which produce, under conditions of high-multiplicity infection, high yields of alphaviruses and flaviviruses from cultured mosquito cells. In such systems, close to 100% of the cells can be infected, and one-step growth curves can be achieved which, apart from a slightly longer latent period, are similar to those observed with the same virus in vertebrate cells. For example, with Sindbis virus (SV), in *A. albopictus* cells (Singh), we find a latent period of about 5 or 6 h instead of 3 or 4 h as in BHK cells (Stollar *et al.*, 1975); depending on the input multiplicity of virus, the virus yield reaches a maximum between 14 and 24 h after infection.

What appears to have been the first report of the growth of a togavirus *in vitro* in mosquito cells was that of Trager in 1938. He described the growth of Eastern equine encephalitis virus (EEEV) in surviving tissues prepared from various parts of mosquito larvae. Over a period of 28 days, including three passages, the virus replication was equivalent to a 10^5-fold increase. At the end of this time, the virus retained its virulence for mice and was neutralizable by antiserum against EEEV.

Further progress awaited development of the cell lines described in the preceding section. The availability of the mosquito cell lines of Singh and Peleg was quickly followed by the demonstration that they were suitable hosts for the replication of a number of togaviruses.

Table I lists some of the systems shown to produce high yields of togaviruses in cultured mosquito cells. Five alphaviruses, SV, Semliki Forest virus (SFV), EEEV, Venezuelan equine encephalitis virus (VEEV), and Ross River virus (RRV), grew well in the *A. albopictus* line of Singh, and at least three grew well in the *A. aegypti* line of Peleg. Among the flaviviruses, JEV grew extremely well in the *A. albopictus* cell line of Singh, as well as in a more recently developed line from *C. tritaeniorhynchus*. In the case of both SFV and JEV, somewhat higher yields were obtained from Singh's *A. albopictus* line than from Peleg's *A. aegypti* line. A high yield of West Nile virus (WNV) was produced in Peleg's *A. aegypti* cell line, although the time necessary to achieve this yield was longer than in most other systems.

There has been some variation in different systems with respect to the maximum percentage of cells reported to be infected after a culture has been inoculated with one or other of the alphaviruses. For example, when cultures of *A. aegypti* cells were inoculated with SFV, only 8–11% of the cells appeared to be infected (Peleg, 1969b),

TABLE I

Examples of Alpha- and Flaviviruses Which Produce High Yields in Cultured Mosquito Cells

Virus	Cell line	Peak titer	Time peak titer achieved (h)	Reference[a]
Alphaviruses				
SV	*A. albopictus* (Singh)	2–3 × 10^8 PFU/ml	24	(1)
		2–3 × 10^8 PFU/ml	13	(2)
SFV	*A. albopictus* (Singh)	8 × 10^7 PFU/10^6 cells	30	(3)
		2 × 10^9 PFU/10^6 cells	24	(4)
EEEV	*A. albopictus* (Singh)	6–8 × 10^7 PFU/ml	15	(5)
RRV	*A. albopictus* (Singh)	10^7 PFU/10^6 cells	40	(6)
VEEV	*A. albopictus* (Singh)	10^8 PFU/ml	24	(7)
CHIK	*A. albopictus* (Singh)	> 10^8 PFU/ml	48	(12)
SV	*A. aegypti* (Peleg)	2 × 10^8 PFU/ml	72	(8)
SFV	*A. aegypti* (Peleg)	1 × 10^7 LD_{50}/ml	24–48	(9)
EEEV	*A. aegypti* (Peleg)	7 × 10^8 LD_{50}/ml	72	(9)
VEEV	*A. aegypti* (Peleg)	10^9 PFU/ml	48	(7)
Flaviviruses				
JEV	*A. albopictus* (Singh)	1–2 × 10^9 PFU/ml	48–72	(10)
DV	*A. albopictus* (Singh)	10^8 PFU/ml	72–96	(12)
JEV	*A. aegypti* (Peleg)	4 × 10^7 PFU/ml	72	(10)
JEV	*C. tritaeniorhynchus*	6 × 10^8 LD_{50}/ml	96	(11)
WNV	*A. aegypti* (Peleg)	1 × 10^8 LD_{50}/ml	144	(9)

[a] Key to references: (1) Stevens (1970); (2) Stollar *et al.* (1975); (3) Davey and Dalgarno (1974); (4) Luukkonen *et al.* (1976); (5) Stollar and Shenk (1973); (6) Raghow *et al.* (1973); (7) Esparza and Sanchez (1975); (8) Peleg and Stollar (1974); (9) Peleg (1968); (10) Igarashi *et al.* (1973a); (11) Hsu *et al.* (1975); (12) Igarashi (1978).

whereas, when cultures of *A. albopictus* cells were inoculated with SFV, 80% of the cells produced virus (Davey and Dalgarno, 1974). In the SV–*A. albopictus* cell system described by Igarashi *et al.* (1977) virtually all cells contained viral antigen (demonstrated by the indirect fluorescent antibody method) during the acute stage of infection, i.e., 12–24 h after infection.

Although cultured tick cells support the growth of certain flaviviruses, including those naturally transmitted by ticks, the yields of virus have generally been low (Rehacek, 1976). Recently, however, yields of between 10^6 and 10^7 plaque-forming units (PFU)/ml have been achieved with a number of flaviviruses including WNV, central European tick-borne encephalitis virus, Langat virus, and Omsk hemorrhagic fever virus (Bhat and Yunker, 1979).

B. Influence of the Virus Strain and Cell Line on Yield of Virus

As with any *in vitro* virus cell system, the efficiencies of viral replication vary considerably, depending on the strains of virus and cell used.

This is best illustrated by the report of Igarashi and co-workers (1973a) who investigated yields of three different strains of JEV from Singh's *A. albopictus* and *A. aegypti* cells and from Peleg's *A. aegypti* cells and showed that yields of virus were quite different, depending upon which strain or isolate of JEV was used. Of the three mosquito cell lines, Singh's *A. albopictus* line gave the highest yield, followed by Peleg's *A. aegypti* line and then Singh's *A. aegypti* line. The most efficient system (JEV strain JaoH-0566 in Singh's *A. albopictus* cells) produced yields of 10^9 PFU/ml.

These observations were extended by the finding that, among 20 different clones derived from *A. albopictus* cells, the yields of a given virus varied considerably from clone to clone. The viruses used in these experiments were dengue viruses type 1–4 and chikungunya (CHIK) virus (Igarashi, 1978).

These observations emphasize that, in order to find the most productive system, it is necessary to test a variety of virus strains and cell lines. Exactly why or how the virus yield is influenced by the specific strain of virus or cells is unexplained.

C. Infectious RNA

The RNAs of togaviruses are infectious (Pfefferkorn and Shapiro, 1974) in a variety of cells of vertebrate origin. Peleg (1969a) has shown that RNAs extracted from SFV and WNV are also infectious in his *A. aegypti* cell line. In contrast, the RNAs of encephalomyocarditis virus (EMC) virus and poliovirus, although infectious in vertebrate cells, are not infectious in mosquito cells.

D. Related Virus–Cell Systems

Although we are primarily concerned here with togaviruses and mosquito cells, it is worthwhile to mention several observations made in related systems. Sindbis virus, for example, replicates not only in mosquito cells but also in cultured *Drosophila* cells, producing yields of between 10^8 and 10^9 PFU/ml (Bras-Herring, 1975). Vesicular stomatitis virus (VSV) has been reported by a number of investigators to replicate in A. *albopictus* cells with titers ranging from 10^5 or 10^6 PFU/ml (Artsob and Spence, 1974; Schloemer and Wagner, 1975) to 1.2×10^8 PFU/ml (Sarver and Stollar, 1977). This virus can also be grown in *Drosophila* cells, giving rise to persistently infected (PI) cultures (Mudd *et al.*, 1973).

E. Nutritional Requirements for Virus Replication

There have been few, if any, definitive studies of the nutritional requirements of cultured mosquito cells. Igarashi *et al.* (1973b, 1974a,b) have, however, examined amino acid requirements, both for cellular growth and for replication of JEV. As already pointed out, the A. *albopictus* line of Singh can be grown in a medium which, apart from the serum, is completely defined. Usually such a medium contains the "essential" as well as the "nonessential" amino acids of Eagle (1959). Although fetal calf serum is commonly used as a supplement, these cells can be adapted to calf serum (Igarashi *et al.*, 1973b). Igarashi *et al.* (1973b) also demonstrated that, although not all the nonessential amino acids were necessary for cell growth, serine was of critical importance.

In reference to the requirements for virus replication, it was found that proline deficiency specifically reduced the yield of JEV, a flavivirus, from A. *albopictus* cells (Igarashi *et al.*, 1973b, 1974b). Both the amount of intracellular viral antigen and the amount of infectious virus were reduced. In contrast, proline deficiency did not affect the synthesis of SV RNA or host cell RNA. The effects of proline deficiency could be mimicked by treating infected cells with a proline analog, L-azetidine-2-carboxylic acid (Igarashi *et al.*, 1974a). In these experiments also, the yield of JEV was reduced, but yields of SV and CHIK virus were unaffected.

Interesting findings have also been reported with respect to amino acid requirements for SV replication in A. *albopictus* cells. In this case the omission of methionine from the medium completely abolished both the yield of infectious virus and the synthesis of viral RNA (Stollar, 1978). When other amino acids were omitted singly or in combination, there was either relatively little or no effect on virus replication. It was also found that lowering of the methionine concentration in the medium to 20% of the normal level potentiated the inhibitory action of cycloleucine, a compound earlier reported to prevent the biosynthesis of S-adenosylmethionine (Caboche and Bachellerie, 1977). In contrast to the effects just described in A. *albopictus* cells, chick embryo fibroblasts (CEFs) or BHK-21 cells starved for methionine made normal amounts of SV (Stollar, 1978).

F. Morphogenesis of Sindbis Virus in Cultured Mosquito Cells

This will not be discussed in detail here, as it is the subject of Chapter 8 by Murphy and Chapter 17 by Brown. Several points, however, are worthy of mention. Brown and Gliedman (1973) observed morphological variants of SV in infected mosquito cell cultures at late times after infection. These variants had the same general morphological features as standard virus but measured 37 or 26 nm in diameter, as compared to 50 nm for standard virions (these measurements do not include the spikes). By 5 days after infection, these small variants were the predominant virus population in the extracellular medium. It has not been possible so far to separate physically the small variant particles from the standard virus, and nothing further has been reported concerning the formation of these particles.

The morphogenesis of several alphaviruses has been examined in *A. albopictus* cells. From these studies it seems apparent that the process of budding at the plasma membrane is much less prominent than is commonly observed in CEFs or hamster cells. Thus Brown *et al.* (1976) found little or no virus budding at the surface of SV-infected cells. Nor were they able to find significant numbers of nucleocapsids in the cytoplasm. They were able, however, to see large amounts of both nucleocapsids and complete virions along with ribosomes and membrane structures within cytoplasmic vesicular structures. On the basis of their observations, they suggested that the whole process of viral biosynthesis and assembly occurred within cytoplasmic vesicles and that virus was released from the cell through movement of vesicles to the cell surface, followed by reverse phagocytosis. It was also postulated that the sequestration of virus biosynthesis in vesicles is in some way related to the lack of a cytopathic effect (CPE) in mosquito cells.

On the other hand, observations by Raghow *et al.* (1973) and by Stollar *et al.* (1979) indicated that both RRV and SV budded at the cell surface. In the latter case, free cytoplasmic nucleocapsids were also seen. It seems clear therefore that maturation of alphaviruses in cultured mosquito cells may take place either at the plasma membrane or in association with cytoplasmic membranes. The latter process appears to assume increased importance at later times after infection and in persistently infected (PI) cells. Other factors such as strain of virus, type of cell, and composition of media may influence which of these two morphogenetic systems predominates at any given time in a specific system. The evidence indicates, however, that viral assembly in association with intracytoplasmic membranes is of greater relative importance in mosquito cells than in cells of vertebrate origin.

IV. COMPARATIVE ASPECTS OF TOGAVIRUS REPLICATION IN MOSQUITO AND VERTEBRATE CELLS

A. Carbohydrates

Of the major chemical components of togaviruses, it is generally accepted that the RNA and proteins are specified by the viral genome, whereas the lipid and the carbohydrate moieties of the envelope glycoproteins are either host-derived or are specified by host enzymes (see appropriate chapters). Thus, it is the lipid and carbohydrate of togaviruses which should be primarily influenced by the nature of the host cells.

Sialic acid is invariably found in glycoproteins of vertebrate cells, but not in cells of invertebrates, such as insects (Warren, 1963). Stollar *et al.* (1976) showed that cultured mosquito cells were devoid of sialic acid and lacked the enzyme sialyl transferase which transfers the sialic acid moiety onto the terminal position of glycoproteins.

In order to characterize virus grown in different cells, SV was grown in BHK-21 cells (SV_{BHK}), CEFs (SV_{CEF}), and *A. albopictus* cells (SV_{AA}) and labeled with radioactive glucosamine, an efficient precursor of sialic acid. Purified virus was then treated with neuraminidase, either from *Vibrio cholerae* or from *Clostridium perfringens*. Each enzyme was able to release approximately 10% of the radioactivity from SV_{BHK} or SV_{CEF}, but less than 1% from SV_{AA}. It was therefore concluded that SV grown in mosquito cells lacked sialic acid.

Comparable experiments were also carried out with SFV grown in mosquito cells (SFV_{AA}) and hamster cells (SFV_{BHK}) (Luukkonen *et al.*, 1977). Both SFV_{BHK} and SFV_{AA} contained approximately the same amounts of hexosamine, mannose, and fucose. But as was the case with SV, SFV_{BHK} contained sialic acid and SFV_{AA} lacked sialic acid. In addition, galactose which was present in SFV_{BHK} was present only in trace amounts in SFV_{AA}. In these experiments, the various carbohydrates were measured by appropriate colorimetric assays. These results with SV and SFV clearly demonstrate the important influence the host cell exerts over the carbohydrate composition of alphaviruses.

B. Lipids

There have been several reports of lipid analyses of cultured mosquito cells, including the *A. albopictus* line of Singh (Jenkin *et al.*, 1971; Yang *et al.*, 1974; Luukkonen *et al.*, 1973, 1976). Not surprisingly, comparisons with lipids of mammalian cells showed marked differences (Jenkin *et al.*, 1971; Luukkonen *et al.*, 1973).

A detailed study has been reported comparing the phospholipids of SFV_{AA} with those of SFV_{BHK} (Luukkonen *et al.*, 1976). Cells were labeled to equilibrium with inorganic [^{32}P]phosphate. This required between 72 and 120 h. Cells were infected, the virus purified, and the various lipid classes in the virus quantitated on the basis of their radioactivity. Although the total amounts of phospholipid were similar in the two viruses, notable quantitative differences were evident. Thus, in terms of total phospholipid, SFV_{AA} contained, relative to SFV_{BHK}, more phosphatidylethanolamine (62 versus 23%) and ceramide phosphoethanolamine (9.5% versus none), but less phosphatidylcholine (14 versus 42%) and ceramide phosphocholine ($<1\%$ versus 16%). Overall, SFV_{AA} contained predominantly ethanolamine phospholipids, whereas SFV_{BHK} contained mainly choline phospholipids. A further difference was a cholesterol content in SFV_{AA} only 20% of that in SFV_{BHK}.

In both cases, ceramide phospholipids were enriched in the virus relative to the host, whereas other components such as diphosphatidylglycerol and phosphatidylinositol, which were major constituents of the host lipids, were not found in the viral lipids.

Two main conclusions can be drawn from these results. First, as was true for the viral carbohydrates, the host cell exerts a strong influence over the composition of the viral lipids. These effects are especially marked when a virus is grown in cell types which are phylogenetically quite distinct and thus differ themselves in lipid composition. Second, the viral lipids are not a precise reflection of the cellular lipids. Rather, the envelope proteins must play some role in selecting and rejecting certain lipid classes for inclusion in the viral membrane.

A more detailed discussion of the lipids of togaviruses is presented in Chapters 10 and 19.

Moore *et al.* (1976) used the fluorescent probe 1,6-diphenol-1,3,5-hexatriene to measure the microviscosity of the membranes of SV_{BHK} and SV_{AA}; SV_{BHK} was grown at 37°C and SV_{AA} at 22°C. From their observations they concluded that the membrane of SV_{BHK} had a higher microviscosity than that of SV_{AA} and that the microviscosity of both viral membranes was higher than that of intact host cell membranes. If, as suggested, the microviscosity of a membrane is influenced both by its lipid and its protein composition, the difference in microviscosity between SV_{BHK} and SV_{AA} membranes must be due to the difference in lipid composition, since the viral proteins are similar whether the virus is grown in *A. albopictus* or in BHK cells (Luukkonen *et al.*, 1977).

C. RNA and Structural Proteins of the Virion

Careful experiments have not yet been described which compare the RNAs of a given alphavirus grown in mosquito cells and in vertebrate cells. It would be predicted, however, that the size and primary nucleotide sequence would not differ. Both RNAs should also be infectious in either cell type.

The alphavirus genome does undergo at least one major posttranscriptional modification, namely, capping and methylation of the 5′-terminus. For virus grown in CEFs or BHK cells, the cap is a simple one, m^7G^5pppAp, with no methylation of the penultimate base or ribose (Hefti *et al.*, 1976). The enzymology of the capping and methylation of alphavirus RNA is unknown. Thus, it is not certain whether the enzymes involved are virus-specified or host-specified. If the latter, then the nature of the cap found in SV_{AA} RNA might be different from that of virion RNA from virus grown in CEFs or BHK cells.

The polyadenylate sequence found at the 3′-terminus of alphavirus genome RNA is copied from the poly(U) sequence located at the 5′-terminus of the negative strand (Sawicki and Gomatos, 1976). Thus, the length of the poly(A) of the viral genome should be the same, no matter which host cell the virus is grown in. It is possible, however, that in addition to poly(A) synthesis from a poly(U) template, poly(A) sequences might subsequently be added posttranscriptionally by a polyadenylate polymerase. Such a process might vary from cell to cell.

When the structural proteins of SFV_{BHK} and SFV_{AA} were compared by acrylamide gel electrophoresis and staining with Coomassie blue, the relative intensity and appearance of the major viral polypeptides (E1, E2, and C) were similar for the two viruses (Luukkonen *et al.*, 1977). However, both of the envelope proteins (E1 and E2) from SFV_{AA} moved slightly more rapidly than their counterparts from SFV_{BHK}. Similar observations were made for radioactively labeled E1 and E2 in *A. albopictus* and BHK-21 cells infected with SV (Sarver and Stollar, 1978). These differences in the mobility of the envelope proteins likely reflect differences in glycosylation in the two cell types (see above). However, the possibility of different cleavage patterns for precursor peptides in BHK and *A. albopictus* cells has not been ruled out.

D. Infectivity and Antigenicity of Togaviruses Grown in Mosquito Cells

From the preceding section, it is clear that alphaviruses grown in mosquito cells and in vertebrate cells differ with respect to their lipid and carbohydrate content. Because these differences are profound, and because of the charge associated with sialic acid residues, it might be expected that these variations in lipid and carbohydrate composition could influence such viral properties as adsorption, hemagglutination (HA), and antigenicity. Experiments bearing on these questions were described by Stollar *et al.* (1976).

Sindbis virus was grown in CEFs, in BHK cells, and in *A. albopictus* cells. Virus was concentrated and then assayed for infectivity by plaque formation on CEFs, for HA, and for particle count (see Table II). When HA or infectivity was expressed on a per particle basis, there was no significant difference among SV_{BHK}, SV_{CEF}, and SV_{AA}. The PFU/HA ratios were also similar in the three virus preparations. Thus, with respect to HA, and to infectivity as measured on CEFs, the origin of the virus and its variation in biochemical composition had no effect. These results were consistent with an earlier report by Kennedy (1974) that removal of sialic acid from SFV did not affect either its HA or its infectivity.

It would, of course, be highly detrimental for the transmission of alphaviruses if alterations in one host rendered the virus less infective for the alternate host, be it the arthropod or vertebrate host. Thus, it seems clear that, under the selective pressures of evolution, alpha- and flaviviruses have evolved in such a way that, in spite of different host-induced modifications in each of the alternate hosts, these viruses retain their capacity for efficient infection of and replication in both insect and vertebrate cells.

To test whether the host cell affected antigenic properties of the virus, antisera were prepared in rabbits against SV_{AA} and SV_{CEF}. Prior to immunization, each virus was concentrated, and purified by velocity gradient centrifugation followed by centrifugation to equilibrium in sucrose–D_2O. Each antiserum was tested against SV_{CEF} and against SV_{AA} by hemagglutination inhibition (HI), by complement fixation, by

TABLE II

Correlation of Viral Infectivity, Viral Hemagglutinin, and Particle Counts

Origin of virus	Postinfection virus yield tested (h)	PFU/ml ($\times 10^9$)	HAU/ml ($\times 10^4$)[a]	Particles/ml ($\times 10^{11}$)	Particles per PFU	Particles per HAU ($\times 10^7$)
CEF	1–12	18.0	4.1	6.5	36	1.6
BHK	1–12	47.0	16.0	13.0	28	0.8
Aedes albopictus	1–12	6.0	1.0	1.4	23	1.4
Aedes albopictus	24–36	2.6	3.1	2.4	91[b]	0.7
Aedes albopictus	48–60	3.5	0.4	0.8	24	2.2

[a] HAU, Hemagglutination units.
[b] This value may be too high because of a low estimate for the PFU/ml; see Stollar *et al.* (1976).

neutralization, and by immune (complement-dependent) virolysis (Stollar, 1975). By none of these methods was either antiserum able to distinguish between SV_{CEF} and SV_{AA}. The results of the complement fixation experiments are shown in Fig. 1, and the results of the neutralization experiments in Fig. 2. Thus with respect to antisera produced in rabbits, the variation in lipid and carbohydrate composition resulting from growth in different hosts had no effect on the antigenic properties of SV as measured by four different tests.

Antigenic comparisons were also made on type-2 dengue virus (DV) grown in *A. albopictus* cells and in LLC MK2 cells (a monkey kidney cell line) (Sinarachatanant and Olson, 1973). Antiserum was prepared in mice to DV grown in mouse brain (DV_{MB}) and in rabbits against

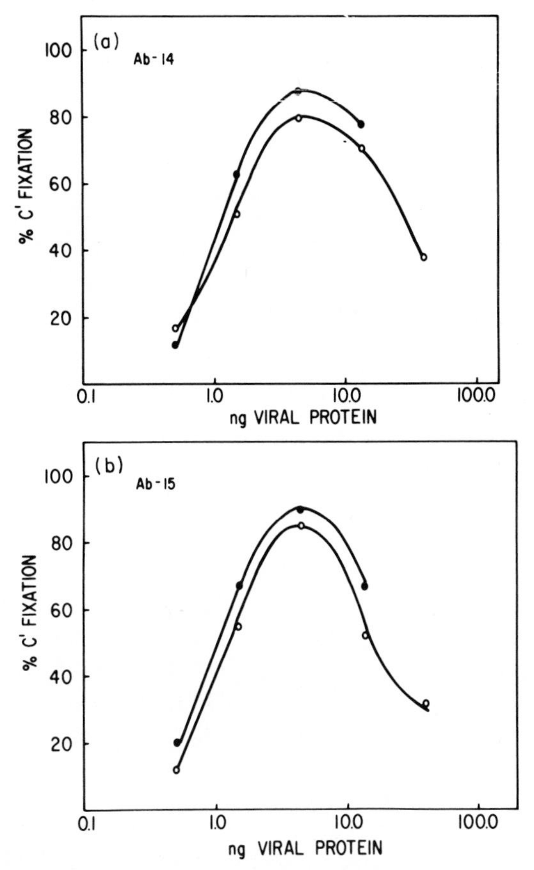

Fig. 1. Microcomplement fixation reactions of SV_{AA} (solid circles) and SV_{CEF} (open circles). (a) With antibodies to SV_{CEF} (Ab-14 final dilution 1:15,000). (b) With antibodies to SV_{AA} (Ab-15, final dilution 1:15,000). (From Stollar *et al.*, 1976.)

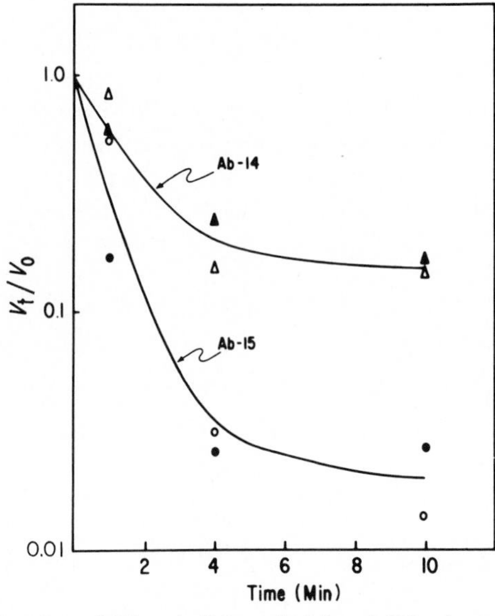

Fig. 2. Neutralization of SV_{CEF} (solid symbols) and SV_{AA} (open symbols) by Ab-14 (anti-SV_{CEF}) and by Ab-15 (anti-SV_{AA}). V_t/V_0 is the ratio of PFU/ml at the indicated times to PFU/ml at zero time. (For further details, see Stollar *et al.*, 1976.)

each of the two host cells LLC MK2 and *A. albopictus*. In HI tests anti-DV serum inhibited HA by $DV_{LLC\ MK2}$ but not by DV_{AA}. Furthermore, HA by DV_{AA} could be inhibited by antiserum to *A. albopictus* cells. The conclusions drawn from these experiments were that the host cells modified the nature of the HA antigens and that DV_{AA} contained antigens in common with *A. albopictus* cells. These conclusions differ from those obtained with SV. Why SV and DV should give different results is not clear. Two points are noteworthy, however. First, the pH optimum for the HA of DV_{AA} was different from that of $DV_{LLC\ MK2}$. Second, Eaton *et al.* (1978) have recently found in the medium of *A. albopictus* cells a hemagglutinin which appears to be of cellular origin. If what appeared to be the DV_{AA} hemagglutinin in the experiments of Sinarachatanant and Olson (1973) were actually an *A. albopictus* hemagglutinin of cellular origin, some of these results might be explained. Furthermore, in contrast to those of the HI experiments, the results of neutralization experiments were much easier to interpret. Thus both $DV_{LLC\ MK2}$ and DV_{AA} were neutralized by anti-DV_{MB} serum, and antiserum to *A. albopictus* cells had little if any effect on the neutralization of DV_{AA}. Therefore, the viral antigens involved in neutralization were similar for both $DV_{LLC\ MK2}$ and DV_{AA}.

E. Viral RNA in Infected Mosquito Cells

When *A. albopictus* cells are infected with SV, the major intracellular viral RNAs seen are the 26 and 42 S single-stranded (ss) and 22 S double-stranded (ds) species. RNase treatment of SV 22 S dsRNA from *A. albopictus* cells permitted the resolution of three replicative form RNAs (Igarashi and Stollar, 1976) similar in size to those described in SV-infected chick cells (Simmons and Strauss, 1972). Both the 26 and the 42 S ssRNAs have been found associated with polysomes and presumably function as messengers (Eaton and Regnery, 1975). All these results suggest that the basic mode of viral RNA transcription is the same in mosquito cells infected with alphaviruses as in infected vertebrate cells. As noted above for the virion RNA, it is possible that when 26 S RNA from an *A. albopictus* cell is examined specific modifications at the 5′-terminus will be found which differ from the capped end group described for SV 26 S RNA from BHK cells (Dubin *et al.*, 1977).

There has been only one report describing viral RNAs made in mosquito cells infected with flaviviruses (Wengler *et al.*, 1978). In Uganda S virus and WNV, no significant differences were found between viral RNAs made in infected hamster (BHK-21) cells and those made in infected *A. albopictus* cells. In both instances, it was observed that (1) The predominant cell-associated viral ssRNA had a sedimentation coefficient of 42 S, was of positive polarity, and thus resembled the RNA extracted from virions; (2) the 42 S ssRNA did not bind to oligo(dT) and, therefore, in all likelihood, lacked a characteristic poly(A) sequence; in the case of 42 S RNA obtained from BHK cells this was confirmed by oligonucleotide fingerprint analysis; (3) infected cells contained one or two low-molecular-weight RNAs ($5–6 \times 10^4$); although the origin and possible role of these RNAs are not known, hybridization experiments showed them to contain viral RNA sequences; a small 6–8 S RNA associated with incomplete virus particles released from dengue-infected KB cells had been noted earlier (Stollar *et al.*, 1966) but was never further characterized.

F. Do Mosquito Cells Produce Interferon?

In suitable cells such as primary CEFs, alphaviruses are excellent inducers of interferon (see, e.g., Atkins *et al.*, 1974). In contrast, only one report has described the production of an interferon-like substance by *A. albopictus* cells (Enzmann, 1973). That this material was interferon-like was based on the observations that (1) actinomycin treatment of A. *albopictus* cells persistently infected with SV led to increased virus production, and (2) medium from PI cultures, treated in a manner similar to that used to prepare interferon from medium of infected vertebrate cells, reduced the yield of SV from mosquito cells. The actinomycin effect could be mediated by mechanisms other than those involving inhibition of interferon production and function. In addition, according to other reports, interferon production by mosquito cells could not be detected after infection with WNV or SV (Murray and Morahan, 1973), after infection with SFV (Peleg, 1969b), or after treatment with poly(I)·poly(C) (Kascsak and Lyons, 1974). Thus, overall, the production of interferon (as defined in vertebrate cell systems) by cultured A. *albopictus* cells does not appear to be an easily reproducible phenomenon. Whether this is generally true for cell lines derived from insects and other invertebrates remains to be seen.

Although the situation with respect to interferon production by mosquito cells is perhaps uncertain, there has been a recent report (Riedel and Brown, 1979) of a novel virus inhibitor produced by SV-infected A. *albopictus* cells. This dialyzable low-molecular-weight material was produced by PI cells but was found in the medium as early as 3 days after infection. It appeared to be a polypeptide, because it was inactivated by protease K as well as by heating. In addition to being cell-specific (it inhibited virus production in A. *albopictus* cells but not in BHK cells), unlike interferon it was also virus-specific [it inhibited the production of SV but not of SFV or WNV (a flavivirus)]. Further characterization of this antiviral agent will clearly be of great interest.

V. CYTOPATHIC EFFECT IN MOSQUITO CELLS INFECTED WITH TOGAVIRUSES

A. Different Response of Cultured Mosquito Cells to Alpha- and Flaviviruses

It has consistently been observed in different laboratories that infections of cultured mosquito cells with alphaviruses are not cytocidal (Peleg, 1968; Stevens, 1970; Davey and Dalgarno, 1974; Esparza and Sanchez, 1975). In contrast, it has been clearly documented that A. *albopictus* cells infected with one of several different flaviviruses (DV, WNV, JEV, or Wesselsbron virus) manifest a marked CPE characterized by extensive cell fusion (Suitor and Paul, 1969; Paul *et al.*, 1969; Djinawi and Olson, 1973). Yet even flavivirus infection does not inevitably produce such effects. For example, high yields of JEV have been obtained from *A. albopictus* cells in the absence of any CPE (A. Igarashi, personal communication). Even when an extensive CPE does follow flavivirus infection, small foci of cells usually survive so that the culture is able to recover (Paul *et al.*, 1969).

A CPE characterized by extensive cell fusion has also been observed after the infection of *A. albopictus* cells with a viral agent (the cell-fusing agent, CFA) which appears to be a togavirus but is neither an alpha- nor a flavivirus (see Section IX; see also Stollar and Thomas, 1975; Igarashi *et al.*, 1976; Chapters 2 and 21 in this volume).

These general observations regarding a CPE following the infection of mosquito cells raise a number of interesting questions related to the fundamental problem of the mechanism of cell killing by viruses. For example, why can alphaviruses which are general extremely cytocidal for vertebrate cells, replicate to high titer in mosquito cells without producing any CPEs? Why do alpha- and flaviviruses differ in their ability to cause CPEs in mosquito cells?

B. Cytopathic Effect-Susceptible and Cytopathic Effect-Resistant Mosquito Cell Clones

The recent finding that cloned variants of *A. albopictus* cells differ markedly in their response to infection with SV (Sarver and Stollar, 1977) should prove extremely useful in studying such questions. In this report it was shown that, although certain clones showed no CPE, others manifested a severe CPE in less than 20 h after infection. It should be pointed out that these clones and the cells from which they were derived were grown in the simpler defined medium described in Section II,A. This medium (E medium) is composed of Eagle's medium plus the nonessential amino acids including glutamine and is supplemented with 10% fetal calf serum. (Most other workers studying togaviruses in mosquito cells have continued to grow their cell cultures in MM medium; see Section II.)

The CPE observed by Sarver and Stollar (1977) had an interesting dependence on temperature (Fig. 3). At 28°C, CPE-susceptible cells (clone LT C-7, for example) tended to assume a more lengthened fibroblastic form and rearranged themselves into a networklike formation, opening up clear spaces on the surface of the petri dish. In contrast, at 34° or 37°C there was clear evidence of cell destruction, with the accumulation of cellular debris. Infected CPE-resistant cells (clone AIS C-3, for example), on the other hand, did not differ in their

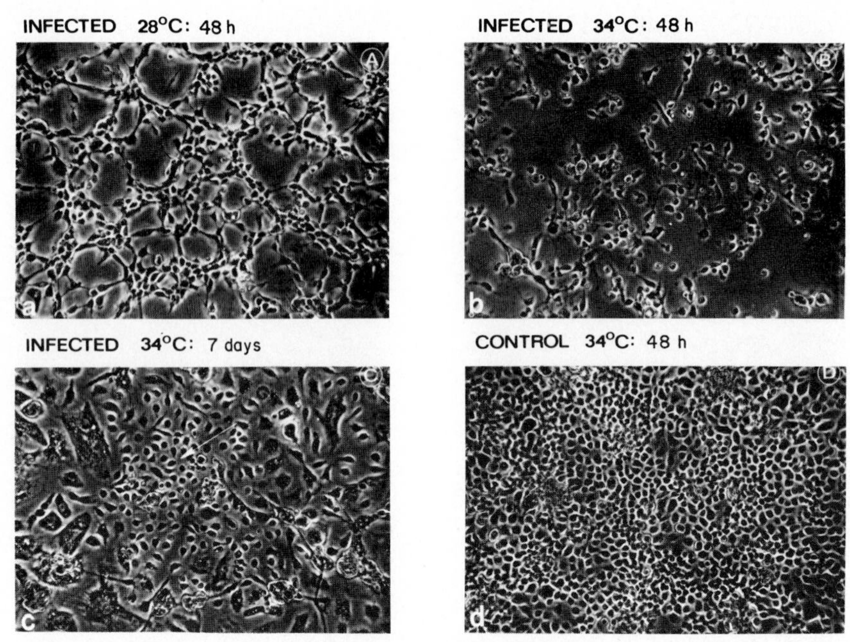

Fig. 3. Light microscopy of mock-infected and SV-infected *A. albopictus* LT C-7 cells (CPE-susceptible). (a) At 48 h after infection at 28°C. (b) At 48 h after infection at 34°C. (c) At 7 days after infection at 34°C. (d) At 48 h after mock infection at 34°C. Phase-contrast. (From Sarver and Stollar, 1977.)

appearance from mock-infected cells at 28°, 34°, or 37°C. The morphological changes described above were apparent by 48 h at 28°C and within 16–18 h at 34°C.

In addition to clones which showed no CPE (0), or extreme CPE (4+) after infection with SV, other clones manifested minimal or intermediate degrees of cell damage (1–2+).

When a CPE-susceptible clone was recloned, 31 of 31 or 100% of the secondary clones were CPE-susceptible. Of 11 clones derived from a CPE-resistant clone (AIS C-3), 8 were CPE-resistant and 3 were CPE-susceptible. Clones derived from one of the secondary resistant clones were all CPE-resistant.

There was no convincing correlation between amounts of virus produced and the occurrence of CPE. LT C-7 cells (CPE-susceptible) produced either similar amounts of virus after infection with SV, or two- to threefold more than AIS C-3 cells (CPE-resistant) (Table III). These differences did not appear significant. With both cell types, virus production was greater at 28°C than at 34°C. The differing re-

TABLE III

Growth of SV, EEEV, and VSV in *Aedes albopictus* (LT C-7) and *Aedes albopictus* (AIS C-3) Cells[a]

Experiment number	Virus	Time after infection (h)	Viral yield (PFU/ml × 10^{-7}) *A. albopictus,* LT C-7 28°C	34°C	*A. albopictus,* AIS C-3 28°C	34°C
1	SV	2	0.15	0.1	0.16	0.14
		10	15.0	29.5	5.1	5.5
		24	150.0	32.5	49.0	22.0
		48	400.0	36.0	190.0	36.0
2	EEEV	24	15.0	4.8	12.0	5.0
		48	16.5	1.3	9.3	1.5

[a] Mosquito cell monolayers were infected with SV or EEEV at an input multiplicity of approximately 100 PFU/ml. Samples of medium were taken at the times indicated and were assayed on primary CEF monolayers. (From Sarver and Stollar, 1977.)

sponse to virus infection was not limited to SV but was also seen after infection with EEEV. Both LT C-7 and AIS C-3 cells yielded identical amounts of extracellular virus after infection with EEEV, although the C-7 cells showed marked CPE, whereas the AIS C-3 cells showed none.

Viral RNA synthesis in the two cell types showed significant quantitative differences and was consistently higher in the CPE-susceptible cells than in the CPE-resistant cells (Fig. 4). Also of considerable interest was the observation that, whereas in AIS C-3 cells the level of viral RNA synthesis—as measured by [^{3}H]uridine incorporation into trichloroacetic acid (TCA)-precipitable material—was the same at 28°, 34°, and 37°C, in LT C-7 cells viral RNA synthesis increased more than twofold at 34°C compared to 28°C and showed a further slight increase at 37°C. Thus, both viral RNA synthesis and CPE showed an important dependence on temperature. The association between increased viral RNA synthesis and the occurrence of CPE does not necessarily indicate a cause–effect relationship. Both phenomena may be due to some other primary effect.

In experiments in which host RNA synthesis was compared in the two cell types, by 8 h after infection, that is, before any obvious morphological changes were evident, there was a marked depression of RNA synthesis in the CPE-susceptible cells relative to that in the

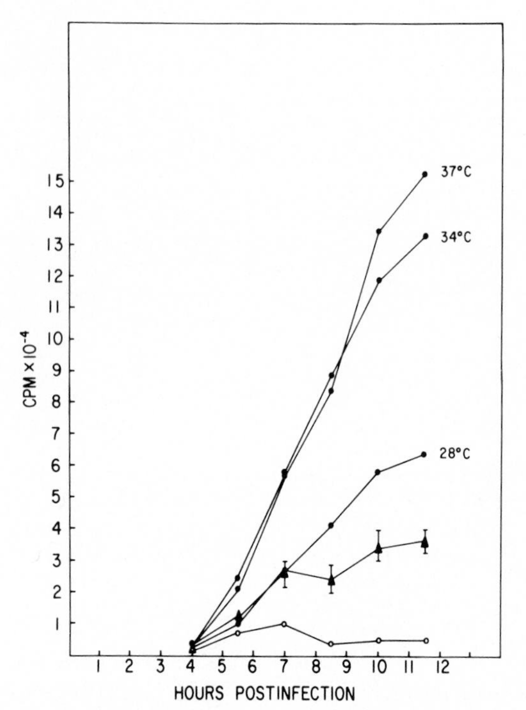

Fig. 4. Virus-directed RNA synthesis in SV-infected *A. albopictus* LT C-7 (CPE-susceptible) (solid circles) and in *A. albopictus* AIS C-3 (CPE-resistant) (triangles) cells at 28°, 34°, and 37°C. Open circles, C-7 and AIS mock-infected cells. Cultures were infected with SV, treated with actinomycin, and then labeled with [^{3}H]uridine. Values represent incorporation into TCA-precipitable material. [^{3}H]Uridine was added at 3 h, and cultures were harvested at various times thereafter. (From Sarver and Stollar, 1977.)

CPE-resistant cells (Fig. 5). In the latter, there was either no depression of host RNA synthesis or a relatively mild decrease from which recovery quickly followed. Similarly, virus infection caused a profound inhibition of host protein synthesis in the CPE-susceptible cells.

Since these cells had been adapted from MM medium to E medium as described above, the effect of maintaining them in MM medium after infection was tested. Although there was a delay of 24–48 h in the appearance of a CPE, the same differential response in terms of occurrence or absence of a CPE in LT C-7 and AIS C-3 cells was seen after infection with SV.

Igarashi (who initially isolated the cell clones described in the paper by Sarver and Stollar, 1977) subsequently isolated a clone (C-

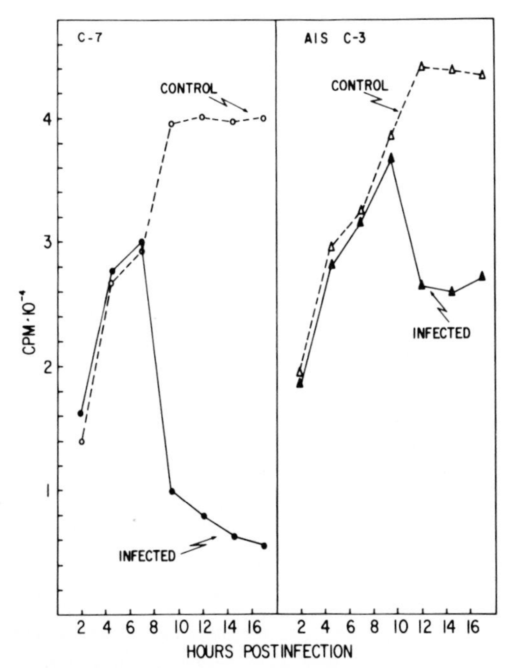

Fig. 5. Host RNA synthesis in *A. albopictus* LT C-7 (CPE-susceptible) and in AIS C-3 (CPE-resistant) cells after infection with SV. Cell cultures were infected with SV and incubated at 34°C. At the indicated times, individual cultures were pulsed with [^{3}H]uridine for 15 min. Cells were then harvested, nuclei collected, and TCA-precipitable radioactivity in the nuclei measured. (For further details, see Sarver and Stollar, 1977.)

6/36) of *A. albopictus* cells which, in contrast to the parental uncloned population, showed a CPE following infection with any of dengue viruses types 1–4 or with CHIK virus (Igarashi, 1978). Since of all the clones tested, clone C-6/36 also produced the highest yields of virus, this work points to an association between the amount of virus produced and the occurrence of a CPE.

C. Effects of Ribavirin (Virazole)

Ribavirin is a nucleoside analog of guanosine, which inhibits a number of RNA and DNA viruses *in vivo* as well as *in vitro*. At a concentration of 150 μg/ml it reduced the yield of SV from *A. albopictus* cells 100-fold and completely protected the cells from CPE (Sarver and Stollar, 1978). At this same concentration, there was, however, little or no significant effect on viral RNA synthesis, on the production of infectious RNA, on nucleocapsid assembly, or on the synthesis and glycosylation of the viral structural proteins. In contrast, there was a marked diminution in the number of cells containing viral antigen (from 90 to 15%, measured by the indirect fluorescent antibody method) when infected mosquito cell cultures were treated with ribavirin. In both *A. albopictus* cells and BHK-21 cells ribavirin markedly depressed both DNA and RNA synthesis as measured by thymidine and uridine incorporation, respectively. Nevertheless, in BHK-21 cells, ribavirin neither inhibited virus production nor prevented CPE. There is still no satisfactory explanation for how ribavirin prevents virus production in mosquito cells, or why it is without effect in BHK-21 cells. These findings suggest that certain host functions, the nature of which is still unclear, are required in the late stages of viral maturation and that these functions are inhibited by ribavirin. (See note added in proof on p. 619.)

VI. MOSQUITO CELLS PERSISTENTLY INFECTED WITH TOGAVIRUSES

A. Properties of Virus from Persistently Infected Cultures

As already noted, with the exception of the system described in the preceding section, infection of *A. albopictus* or *A. aegypti* cells with alphaviruses produces high levels of virus generally without any CPE. Although infection with flaviviruses often causes a CPE in *A. albopictus* cells, the cultures usually recover after a number of days or weeks. Thus, infection with either alpha- or flavivirus readily leads to the production of PI cultures (Rehacek, 1968; Peleg, 1969b; Stollar and Shenk, 1973; Shenk *et al.*, 1974). Little has been reported concerning the properties of cultures persistently infected with flaviviruses, and the rest of this section will therefore be devoted to mosquito cells persistently infected with alphaviruses.

The evolution of mosquito cell cultures persistently infected with SV has been carefully described by Igarashi *et al.* (1977). Replicate cultures of *A. albopictus* cells were infected with SV at an input multiplicity of 5–10 PFU/cell. The cultures were split at weekly intervals, and the titer of infective virus assayed periodically. During the first week, titers of between 10^8 and 10^9 PFU/ml were found. Thereafter, the titer dropped slowly over a period of several weeks until by 5 weeks a plateau of between 10^5 and 10^6 PFU/ml was reached. These levels of virus persisted for many months or as long as such cultures

TABLE IV

Temperature-Sensitive Plaque Formation on BHK Cell Monolayers by Sindbis Virus from Persistently Infected Cultures on *Aedes albopictus*[a]

Incubation temperature (°C)	SV_{PI} (PFU/ml)	EOP[b]	SV_{STD} (PFU/ml)	EOP[b]
28	4.1×10^6	1.0	1.6×10^8	1.0
34	4.4×10^6	1.1	1.7×10^8	1.1
37	8.2×10^4	2.0×10^{-2}	1.7×10^8	1.1
39.5	< 10	$< 2.4 \times 10^{-6}$	1.6×10^8	1.0

[a] Adsorption was at 28°C after which the cultures were incubated at the indicated temperatures for 48 h.

[b] Values represent the titer at the indicated temperature divided by the titer at 28°C.

were maintained. These observations indicate that after the first week, for reasons not yet understood, certain factors act to regulate or control production of the infectious virus. Davey and Dalgarno (1974), studying SFV-infected *A. albopictus* cells, had earlier found a marked shutoff of viral RNA synthesis within 18 h of infection and of virus production within 48 h.

Invariably, the virus produced by *A. albopictus* cells persistently infected with SV (PI SV–*A. albopictus*) eventually becomes small-plaque and predominantly temperature-sensitive (Stollar and Shenk, 1973; Stollar *et al.*, 1974; Shenk *et al.*, 1974). Such virus produces plaques at 28° or 34°C but not at 39.5°C (Table IV). The plaques produced at 34°C by SV_{PI} are easily distinguished on the basis of size from the plaques produced at the same temperature by standard virus (SV_{STD}). As demonstrated in the study of Igarashi *et al.* (1977), the virus population of PI cultures shifted to the *ts* phenotype over a period of several weeks but, by 12 weeks after the initial infection, virus from PI cultures was predominantly temperature-sensitive, and by 20 weeks no plaques could be detected at 39.5°C. Evolution of the persistent infection in replicate cultures was similar, although there were variations with respect to absolute levels of virus, precise size of plaques, and the exact time required for the predominance of *ts* virus. The association of *ts* virus with persistent virus infection is commonly seen not only with togaviruses but also with many systems involving PI mammalian cell cultures maintained at 37°C (Preble and Youngner, 1975).

When 20 *ts* mutants obtained from PI SV–*A. albopictus* were characterized (Shenk *et al.*, 1974) with respect to phenotype at the nonpermissive temperature, 19 out of 20 were found to be RNA^+ and one was RNA^-. All were thermolabile relative to the standard virus. No complementation could be demonstrated between any of these *ts* mutants even when the single RNA^- mutant was tested with several different RNA^+ mutants. Since the *ts* mutants described in this report originated from cultures derived from a single initial infection, it is possible that all or most of the RNA^+ mutants had the same lesion. Alternatively, these mutants may all have contained more than one mutation. Either of these possibilities might explain the failure to demonstrate complementation with these mutants. In later experiments, *ts* mutants were obtained from the 10 replicate PI SV–*A. albopictus* cultures described by Igarashi *et al.* (1977). About 12 weeks after initial infection, two plaque isolates were picked from each replicate PI culture and examined with respect to the ability to synthesize viral RNA at 39.5°C (the nonpermissive temperature). In most cases, the two plaque isolates from a given culture shared the same phenotype with respect to viral RNA synthesis, but overall about 25–30% of these *ts* isolates were RNA^- (V. Stollar, unpublished observations).

Working along similar lines, Maeda *et al.* (1979) have obtained a number of *ts* mutants of WEEV from PI cultures of *A. albopictus* cells. Although *ts* mutants obtained up to 30 days following infection were, for the most part leaky, mutants obtained at later times (80–170 days) were stable and had leak yields of less than 10^{-4}. Characterization of 24 stable mutants revealed 3 RNA^+, 1 RNA^-, and 20 multiple-site mutants. Complementation was demonstrated between certain of these mutants and mutants these workers had earlier obtained by chemical mutagenesis.

It has been observed in certain laboratories that virus from PI mosquito cell cultures shows a reduced virulence for mice. This has been demonstrated for a number of viruses, including at least one flavivirus (Rehacek, 1968; Banerjee and Singh, 1969; Peleg, 1971, 1975). Although in several reports on SFV and SV (Peleg, 1971, 1975; Stollar *et al.*, 1974) there was a correlation between reduced mouse virulence and the small-plaque or *ts* phenotype, in other reports decreased virulence was noted without any mention of *ts* or small-plaque virus (Rehacek, 1968; Banerjee and Singh, 1969).

B. Defective Interfering Particles in Persistently Infected Cultures

The major species of alphavirus RNA seen in infected mosquito cells during the acute stage of the infection are 42 and 26 S ssRNA species and a 22 S dsRNA species (see Section IV,E). In PI SV–*A. albopictus* cultures, however, a smaller dsRNA, about 12 S in size, is also found (Stollar and Shenk, 1973). As shown by Igarashi *et al.* (1977), this species became clearly evident at 10 weeks after infection and was seen consistently thereafter. The relative amounts of the two dsRNA species, 22 and 12 S, showed some variation, but at different times either species could be the predominant one.

As discussed in Chapter 15, a 12 S dsRNA can be seen when vertebrate cells are infected with alphavirus defective interfering (DI) particles. If, as seems likely, the 12 S dsRNA from PI SV–*A. albopictus* cultures is a DI RNA, it would be expected that a ss DI RNA should also be present in these cells. Such was shown to be the case by Eaton (1977) who found a 20 S ssRNA in mosquito cells persistently infected with SV. A 20 S ss DI RNA is of the appropriate size to correspond to a 12 S ds DI RNA (Guild *et al.*, 1977) (see Chapter 15).

The finding of new and smaller viral RNAs, both ss and ds, in PI cultures, suggested that these cultures were producing defective particles, or at least that they supported the intracellular replication of defective viral genomes. This idea has now been strongly reinforced by the demonstration that, when BHK cells were infected with medium from PI SV–*A. albopictus* cultures, the pattern of viral RNA in the BHK cells was one characteristic of infection with DI particles (Eaton, 1977). Thus, not only were defective genomes being replicated in PI mosquito cells, but DI particles must have been present in the extracellular medium.

C. Resistance to Superinfection with Homologous Virus

It is characteristic of PI cultures of many different types that they resist superinfection with the homologous virus, but permit replication of viruses in the same group if they are not very closely related. For example, PI SV–*A. albopictus* and PI SV–*A. aegypti* cultures both restricted superinfecting SV_{STD} but permitted a nearly *normal* yield of another alphavirus, EEEV (Stollar and Shenk, 1973; Peleg and Stollar, 1974). The block to the superinfecting homologous virus appears to be an early one, occurring sometime before replication of its RNA. Thus, the pattern of viral dsRNA in PI SV–*A. albopictus* cells was identical whether they were mock-infected or superinfected with SV_{STD}. As already noted, the PI cultures contained both 22 and 12 S dsRNA. Infection with EEEV on the other hand led to a large increase in synthesis of the 22 S species and, interestingly, also of 12 S ds DI RNA (Stollar and Shenk, 1973).

The means by which PI cultures restrict superinfecting homologous virus is unexplained. There does not, however, appear to be any defect in the adsorption to the cells of superinfecting virus (Igarashi *et al.*, 1977). As noted in Section VII, there is good evidence for both the replication of DI RNAs in PI SV–*A. albopictus* cells and for the release of DI particles into the medium of such cultures; and it is likely that both extracellular DI particles and intracellular defective genomes can interfere efficiently with superinfecting standard virus.

Another means of explaining the restriction of superinfecting homologous virus would be interference by *ts* virus. Temperature-sensitive SV generated in PI mosquito cell cultures, cloned, and then plaque-purified and grown to stocks, either in BHK or in mosquito cells, has been shown to interfere extremely well with SV_{STD} (Stollar *et al.*, 1974; Igarashi *et al.*, 1977).

However, neither DI particles, at least as indicated by the intracellular viral RNA patterns, nor *ts* virus appear until 10–12 weeks after a culture is initially infected. Homologous interference, on the other hand, occurs much earlier (Peleg and Stollar, 1974). It may be necessary, therefore, to seek some other explanation for the resistance of PI cultures to superinfecting homologous virus. It may, for example, be related to the competition for a critical intracellular binding site, or for some specific host protein essential for function of the viral RNA polymerase. It seems safe to state that, in PI cultures, several different mechanisms may operate to prevent the replication of superinfecting homologous virus. One or more of these may be effective at the same time, depending on the age and the state of the PI culture.

D. How Is Virus Maintained in Persistently Infected Cultures?

Although persistent viral infections of mosquito cells are readily established, with or without an initial CPE, the question must also be addressed as to how the infection is maintained. The two alternatives, and neither need operate to the exclusion of the other, are: (1) Virus is passed from an infected parent cell to the daughter cell at the time of cell division, i.e., vertical transmission; and (2) virus released from infected cells continually infects those cells in the culture which at any given time are uninfected, i.e., horizontal transmission. The addition of antiviral serum should markedly diminish the efficiency of the second but should not affect the first mechanism.

When anti-SV serum was added to the medium of PI SV–*A. albopictus* cultures, the cultures were cured of the viral infection, probably within 10 weeks, and clearly by 24 weeks (Igarashi *et al.*, 1977). These observations indicate that, although over the short term the infection can be maintained by vertical transmission, for perpetuation of the long-term infection, horizontal transmission is essential.

To further understand the cellular dynamics of PI cultures, cloning experiments were carried out (Igarashi *et al.*, 1977). Infection of a clonally derived population was indicated by the presence of extracellular virus, by intracellular viral antigen (detected by the fluorescent antibody method), and by resistance to superinfection with SV_{STD}. Uninfected clones were negative with respect to each of these properties. When clones were examined from two different PI cultures (5 or 7 months after they were initially infected) from 42% (16/38) to 58% (14/24) were virus-positive. These values were consistent with what was observed when uncloned PI cultures were tested for viral antigen by the fluorescent antibody method. When four virus-positive clones were recloned, out of 47 clones isolated, 29 were virus-positive and 18 virus-negative. The fact that virus-negative clones were readily isolated upon recloning of virus-positive clones indicates that cells within a PI culture have the ability to cure themselves of virus infection if not exposed to reinfection (isolation of clones was carried out in the presence of antiserum to prevent cross-contamination of clones).

Another indication that cells in a PI culture are able to regulate and control viral replication was obtained from an experiment in which cells from a PI SV–*A. albopictus* culture were subcultured and then monitored daily for the number of cells and the titer of infective virus in the medium (Igarashi *et al.*, 1977). For 2–3 days after subculture the virus titer rose rapidly and then leveled off and fell, indicating a marked shutoff of virus production (Fig. 6). The cell number, on the other hand, continued to increase for 5 or 6 days, or well beyond the time when viral replication ceased. In parallel with the virus titer, viral RNA synthesis also decreased, at a time when the cell number was continuing to increase. This shutoff of viral biosynthesis in PI cultures recalls the early inhibition of viral biosynthesis described by Davey and Dalgarno (1974). The mechanisms by which cells regulate virus replication are unknown; however, it seems that, if such mechanisms were extremely efficient, individual cells would be cured, would become virus-negative, and would remain so until reinfected with virus from neighboring cells. The curing of PI cultures with antivirus serum indicates (1) that given sufficient time and barring continual reinfection, cellular mechanisms are able to cure individual cells of the virus, and (2) that continual reinfection is necessary to maintain a persistent infection.

Of the various types of persistent infection described by Walker (1964), mosquito cells persistently infected with alphaviruses resem-

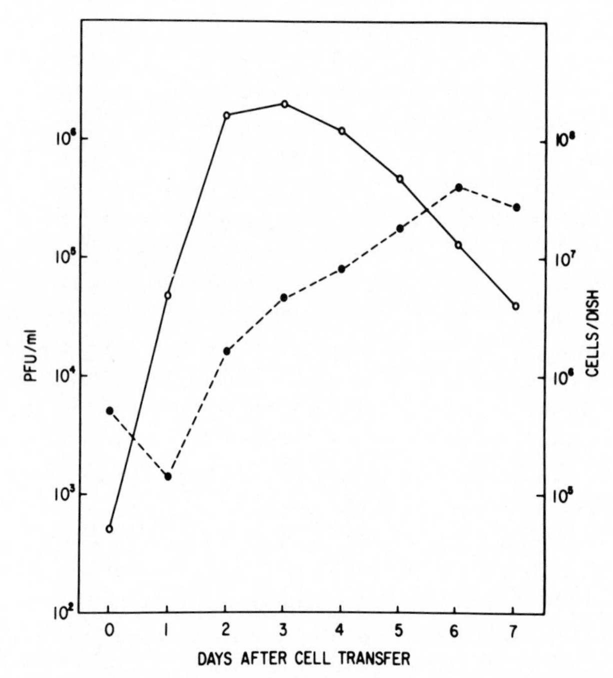

Fig. 6. Cell growth and virus release after subculture of *A. albopictus* cells persistently infected with SV. Replicate cell cultures were prepared from a PI culture 8 months after the initial infection. Each day thereafter for 1 week, one plate was taken to measure the titer of infective virus in the medium (open circles) and the total cell number per plate (solid circles). (From Igarashi *et al.*, 1977.)

ble most closely those classified as "regulated infections." Such systems have the following general properties: (1) Antibody or other extraneous antivirus factors are not needed to maintain equilibrium; (2) the culture is resistant to superinfecting homologous virus; (3) all or a large fraction of the cells are infected; and (4) infected cells divide and grow into infected colonies. As described by Walker, persistent infections of this type are not cured by antiviral serum. This criterion clearly does not apply to PI SV–*A. albopictus* cultures.

VII. MOSQUITO CELLS AND DEFECTIVE INTERFERING PARTICLES OF ALPHAVIRUSES

The subject of DI particles is covered in Chapter 15. Therefore, only a brief summary of those aspects related specifically to mosquito cells will be presented here.

Defective interfering particles of SV or SFV are readily produced upon serial undiluted passage in CEFs or BHK cells. In contrast, Igarashi and Stollar (1976) reported that, when similar procedures were carried out with SV and cultured *A. albopictus* cells, there was little or no evidence, either by interference experiments or by examination of intracellular viral RNA patterns, for the production of DI particles (Igarashi and Stollar, 1976). Furthermore, stocks produced in CEFs or BHK cells which contained DI particles of SV or SFV, when used to infect mosquito cells, behaved as if they did not contain the DI particles (Stollar *et al.*, 1975; Eaton, 1975; Igarashi and Stollar, 1976). Thus, togavirus DI particles produced in vertebrate cells appear to be inert in mosquito cells. More recent work, however, suggests that the question of DI particles in mosquito cells is probably not as clear-cut as the preceding reports seem to indicate. King *et al.* (1979) have observed, for example, that when a serial undiluted passage series of SV in *A. albopictus* was carried out using 48-h harvests (instead of 16-h harvests), there were clear indications that DI particles were produced; and this has since been confirmed in our laboratory (K. Kowal and V. Stollar, unpublished results).

In Section VI,B evidence was presented for the production of SV DI particles by PI SV–*A. albopictus* cultures. It is of considerable interest that such DI particles were able to replicate in Vero cells in the presence of helper virus and that in turn *their progeny* (i.e., from Vero cells) *were able to replicate in A. albopictus cells* (Eaton, 1978). Thus, the inertness in mosquito cells of DI particles produced in vertebrate cells is not mirrored by a reciprocal inertness in vertebrate cells of DI particles produced in PI mosquito cell cultures.

VIII. SINDBIS VIRUS MUTANTS IN CULTURED MOSQUITO CELLS

The production of *ts* and small-plaque viral mutants in PI SV–*A. albopictus* cultures has been described (Stollar and Shenk, 1973; Shenk *et al.*, 1974). A similar phenomenon has been observed in *A. aegypti* cells persistently infected with SV (Peleg and Stollar, 1974; Stollar *et al.*, 1974). When virus from PI SV–*A. albopictus* cultures was cloned and grown into stocks in CEFs, efficiencies of plating (EOP) at 39.5°C relative to 34°C ranged from $<10^{-7}$ to 7×10^{-5} (Shenk *et al.*, 1974). Of 20 clones tested, 19 proved to be RNA-positive; that is, they were able to synthesize significant amounts of viral RNA at the nonpermissive temperature. The *ts* mutant of SV from PI *A. aegypti* cells was also RNA-positive (Stollar *et al.*, 1974). When the clonally derived *ts* stocks from PI SV–A. *albopictus* cultures were serially passaged without dilution at 34°C in BHK cells, the EOP 39.5°C/34°C, rose within four passages to between 2×10^{-4} and 8×10^{-2} (Shenk *et al.*, 1974). However, this did not lead to the production of any large plaques by these passaged virus stocks. Thus the *ts* property and small-plaque production do not seem to be necessarily covariant. No complementation was observed in CEFs between any of the *ts* strains obtained from PI SV–A. *albopictus* cultures.

The ability of the *ts* mutants of SV isolated and characterized by Burge and Pfefferkorn (1966) to complement each other in A. *albopictus* cells has been examined by Renz and Brown (1976). The only complementation observed was that of *ts*13 (group C) with *ts*6 (group F) and with *ts*11 (group B) (see Chapter 14). Although *ts*5 and *ts*2 were also in group C, they were unable to complement either *ts*6 or *ts*11. The low overall efficiency of complementation in mosquito cells relative to what has been observed in CEFs was attributed either to a much lower amount of gene product in mosquito cells compared to that in CEFs, or to compartmentalization within mosquito cells, preventing diffusion of viral gene products.

IX. CONTAMINATING VIRUSES IN CULTURED MOSQUITO CELLS

Insect cell lines like those of vertebrate or mammalian origin can harbor unsuspected viruses. Although viruses could be derived from the original tissues used to prepare the primary cultures and the resulting cell line, it is likely that in certain cases the virus was inadvertently introduced while the cell line was being maintained in the laboratory. In the case of viruses which replicate without a CPE, as is usually true of alphaviruses in mosquito cells, an accidental infection can readily occur without being obvious and, if not specifically looked for, may not be detected.

Based on electron microscopic observation, Hirumi *et al.* (1976) described five distinct types of viral particles in one strain of A. *albopictus* cells. Based solely on their morphological features, these appeared to include a parvovirus, an orbivirus, a picornavirus, and a togavirus, as well as a "bacterial virus."

Cunningham *et al.* (1975) studied a virus contaminant in A. *albopictus* cells, which was also discovered on the basis of electron microscopic observations. They described large cytoplasmic crystalloid arrays of virus particles whose size and shape were compatible with those of an alphavirus. When the cells were sonicated and fractionated on sucrose gradients, various gradient fractions eventually produced plaques on Vero cells, and the progeny of these plaques produced a slowly developing CPE in BHK cells. Serological analysis finally identified the CPE-inducing agent as an alphavirus, specifically chikungunya virus. The ultimate origin of this infection is unknown.

In Australia, Regnery (1977) found that a strain of *A. albopictus* cells he was working with was refractory to SFV. This was ultimately shown to be due to an unsuspected persistent infection of the cells with SFV.

The chance observation that medium from *A. aegypti* (Peleg) cell cultures caused massive fusion of *A. albopictus* (Singh) cells led to the isolation and characterization of yet another agent, this time from *A. aegypti* cells (Stollar and Thomas, 1975; Igarashi *et al.*, 1976). Although this agent, the CFA, morphologically, physically, and chemically closely resembled flaviviruses, it showed no serological relationship to this group. In addition, the structural proteins within the viral particles differed in size and location from those characteristic of flaviviruses (see Chapter 18) but had some resemblance to the pattern described for lactic dehydrogenase virus (Darnell and Plagemann, 1975) (see Chapter 21).

The *A. aegypti* cells from which the CFA was obtained showed no CPE or signs of cell fusion which might indicate the presence of a viral agent. Nor did the CFA cause fusion of or replicate in any of several vertebrate cells tested. Thus, the only indicator system for this agent so far is the *A. albopictus* cell line of Singh.

As is the case when fusion of *A. albopictus* cells is caused by a flavivirus, so also following infection and fusion of cells with CFA the cultures eventually recover and reassume a normal appearance but remain persistently infected and do not fuse when superinfected with the CFA. This system illustrates very well how, in the absence of a suitable indicator system, the detection of a viral contaminant may be very difficult. Taxonomically, the CFA may be a pestivirus, but its origin remains unknown.

The presence of an unknown viral agent in a cell line may interfere with the experimental designs of the investigator by inhibiting the replication of the test virus or by distorting the results of biochemical experiments, especially those involving radioisotopes. It is therefore important when beginning work with a mosquito cell line, as with any cell line, to determine whether it is virus-free, to maintain and grow it so as to minimize the chance of viral contamination, and finally to monitor it closely for accidental infection, especially for viruses being worked with in the same laboratory.

X. MUTANTS OF MOSQUITO CELLS AND THEIR USE FOR THE STUDY OF TOGAVIRUSES

The development, isolation, and characterization of mutants of mammalian cells is proceeding rapidly (Siminovitch, 1976). We have used procedures similar to those employed successfully to generate mammalian cell mutants to isolate a number of drug-resistant strains of *A. albopictus* cells, including strains resistant to bromo-

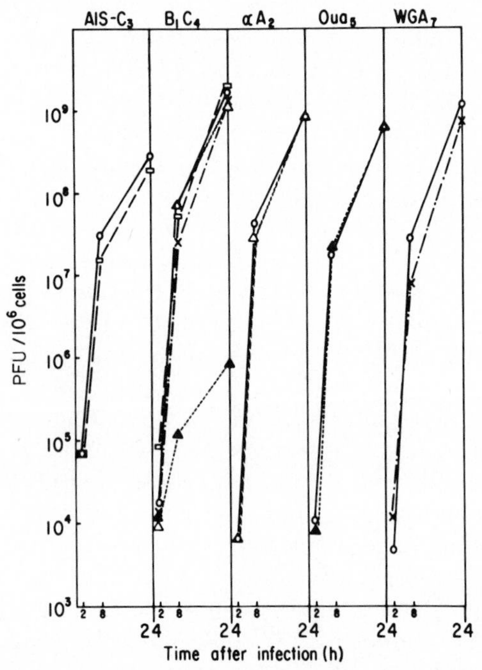

Fig. 7. Yields of SV from mutant *A. albopictus* cells in the presence of bromodeoxyuridine (squares), α-amanitin (open triangles), ouabain (solid triangles), and wheat germ agglutinin (X). Circles, control. AIS C-3 cells (AIS-C_3) are the parent cell strain from which a bromodeoxyuridine-resistant strain (B_1C_4) was developed. The α-amanitin-resistant (αA_2), ouabain-resistant (Oua_5), and wheat germ agglutinin-resistant (WGA_7) strains were all derived from the B_1C_4 cells. Neither α-amanitin, ouabain, nor wheat germ agglutinin affected the yield of SV. Ouabain, however, depressed the yield of SV from ouabain sensitive cells (B_1C_4) but not from ouabain-resistant cells (Oua_5).

deoxyuridine, α-amanitin, or ouabain (Mento and Stollar, 1978a). The replication of SV in cells sensitive or resistant to each of these drugs was examined. Bromodeoxyuridine and α-amanitin had no effect on virus yields, whether tested in cells sensitive or resistant to these drugs (Fig. 7). In cells sensitive to ouabain, however, the virus yield was markedly reduced when ouabain was added. In ouabain-resistant cells, the drug had no effect. These results indicate that ouabain reduces virus yields by an effect on the host cell rather than by a direct effect on the virus. Furthermore, these experiments have shown that some step in virus replication is apparently sensitive

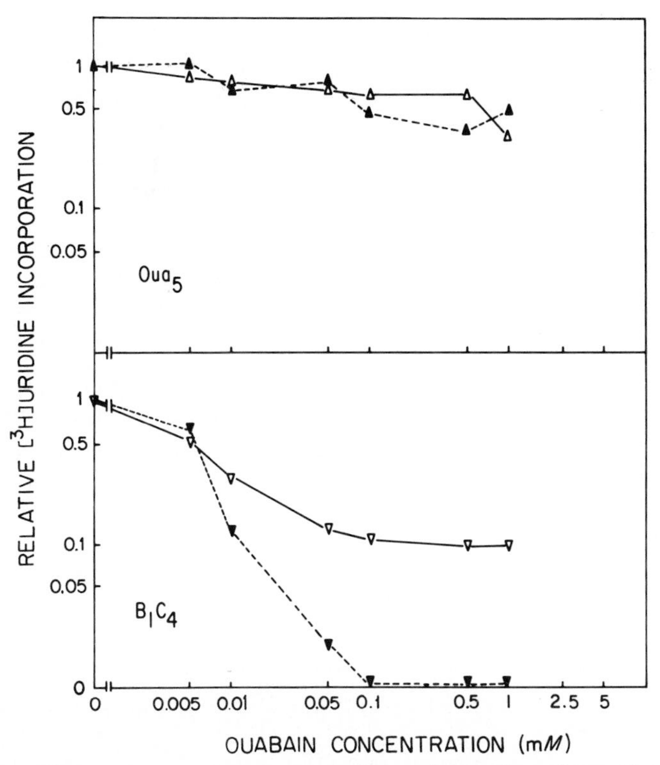

Fig. 8. Relative [^{3}H]uridine incorporation into host and viral RNA by ouabain-sensitive (B_1C_4) and ouabain-resistant (Oua_5) *A. albopictus* cells in the presence of various concentrations of ouabain. Exponentially growing cells were mock-infected (solid lines) or infected with SV (dashed lines) and labeled from 5 to 7 h after infection. Incorporation into host RNA was measured in the mock-infected cells in the absence of actinomycin, and incorporation into viral RNA was measured in the infected cultures in the presence of actinomycin. The values for both host and viral RNA synthesis were expressed relative to values for uninfected and infected cells labeled without ouabain (arbitrarily set as equal to 1.0). (From Mento and Stollar, 1978b.)

to ouabain-induced alterations of the intracellular ionic environment. Although the precise step affected is not known, it has been demonstrated that viral RNA synthesis is markedly reduced by ouabain and to a greater extent than host RNA synthesis (Fig. 8). Reduction of virus yields following the addition of ouabain was seen not only in mosquito cells but also in BHK cells (Mento and Stollar, 1978b).

By using polyethylene glycol to fuse an α-amanitin-resistant cell with an ouabain-resistant cell, a hybrid clone was obtained which was resistant to both drugs. These markers, therefore, behaved codominantly. The yield of SV from the hybrid cells was only slightly reduced by the addition of ouabain. Thus, in this respect also, the hybrid cells closely resembled the ouabain-resistant cells.

It is to be expected that new applications will continue to be found for the use of mutant cells to study togaviruses, specifically to determine which host functions are essential for viral replication. The techniques of somatic cell genetics should also prove useful for the further study of virus-induced CPE in *A. albopictus* cells (see Section V).

XI. CONCLUDING REMARKS

The methodology of working with togaviruses in cultured mosquito cells has now developed to such a stage that the same kinds of genetic, morphological, and biochemical experiments that have been carried out in vertebrate cells can be performed in mosquito cells.

Although there have already been a few investigations pertaining to the comparative biochemistry of viral replication in mosquito and vertebrate cells, much remains to be learned, for example, about the influence of the different host cells on the composition of the virus. Because alpha- and flaviviruses can replicate in cells which are phylogenetically far removed from each other, they provide fascinating systems for experiments in comparative virology at many different levels.

The investigation of togavirus DI particles in mosquito cells has raised interesting questions as to how different cells favor or inhibit generation and replication of defective particles and genomes. The study of PI mosquito cells may yield information applicable to the understanding of various types of persistent infection in higher organisms. To this writer, however, one of the most fascinating aspects of togaviruses and mosquito cells relates to the subject of virus-induced CPE. As pointed out in the body of this chapter, the comparative study of togaviruses, especially alphaviruses in vertebrate and in mosquito cells, and in CPE-resistant and CPE-susceptible cells, provides us with a unique system for examining a central and a fundamentally important problem in animal virology—How do viruses kill cells?

Note Added in Proof. More recent experiments from our laboratory have altered our interpretation of how ribavirin inhibits viral replication. We now know that ribavirin blocks SV replication by inhibiting viral RNA synthesis [Malinoski and Stollar (1980), *Virology* **102,** 473–476]. This work showed that in *A. albopictus* cells Actinomycin completely reversed the action of ribavirin. Thus when viral RNA or protein synthesis was measured in the presence of Actinomycin the inhibition expected with ribavirin was abolished. In the absence of Actinomycin, viral RNA synthesis (as measured by infectivity or by the appearance of viral dsRNA) was reduced in proportion to the reduction in the yield of infectious virus.

REFERENCES

Artsob, H., and Spence, L. (1974). *Acta Virol. (Engl. Ed.)* **18,** 331–340.

Atkins, G. J., Johnston, M. D., Westmacott, L. M., and Burke, D. C. (1974). *J. Gen. Virol.* **25,** 381–390.

Banerjee, K., and Singh, K. R. P. (1969). *Indian J. Med. Res.* **57,** 1003–1005.

Bhat, U. K. M., and Yunker, C. E. (1977). *J. Parasitol.* **63,** 1092–1098.

Bhat, U. K. M., and Yunker, C. E. (1979). *In* "Arctic and Tropical Arboviruses" (E. Kurstak, ed.), Chapter 17, pp. 263–275. Academic Press, New York.

Bras-Herring, F. (1975). *Arch. Virol.* **48,** 121–129.

Brown, D. T., and Gliedman, J. B. (1973). *J. Virol.* **12,** 1534–1539.

Brown, D. T., Smith, J. F., Gliedman, J. B., Riedel, B., Filtzer, D., and Renz, D. (1976). *In* "Invertebrate Tissue Culture: Applications in Medicine, Biology and Agriculture" (E. Kurstak and K. Maramorosch, eds.), Chapter 3, pp. 35–48. Academic Press, New York.

Burge, B. W., and Pfefferkorn, E. R. (1966). *Virology* **30,** 214–223.

Caboche, M., and Bachellerie, J.-P. (1977). *Eur. J. Biochem.* **74,** 19–29.

Cunningham, A., Buckley, S. M., Casals, J., and Webb, S. R. (1975). *J. Gen. Virol.* **27,** 97–100.

Darnell, B. M., and Plagemann, P. G. (1975). *J. Virol.* **16,** 420–433.

Davey, M. W. (1973). Ph.D. Thesis, Australian Natl. Univ., Canberra.

Davey, M. W., and Dalgarno, L. (1974). *J. Gen. Virol.* **24,** 453–463.

Djinawi, N. J., and Olson, L. C. (1973). *Arch. Gesamte Virusforsch.* **43,** 144–151.

Dubin, D. T., Stollar, V., Hsuchen, C.-C., Timko, K., and Guild, G. M. (1977). *Virology* **77,** 457–470.

Eagle, H. (1959). *Science* **130,** 432–437.

Eaton, B. T. (1975). *Virology* **68,** 534–537.

Eaton, B. T. (1977). *Virology* **77,** 843–848.

Eaton, B. T. (1978). *In* "Viruses and Environment" (E. Kurstak and K. Maramorosch, eds.), pp. 181–201. Academic Press, New York.

Eaton, B. T., and Regnery, R. L. (1975). *J. Gen. Virol.* **29,** 35–49.
Eaton, B. T., Ward, R., and Artsob, H. (1978). *Intervirology* **9,** 362–369.
Enzmann, P.-J. (1973). *Arch. Gesamte Virusforsch.* **40,** 382–389.
Esparza, J., and Sanchez, A. (1975). *Arch. Virol.* **49,** 273–280.
Grace, T. D. C. (1966). *Nature (London)* **211,** 366–367.
Guild, G. M., Flores, L., and Stollar, V. (1977). *Virology* **77,** 158–174.
Hefti, E., Bishop, D. H. L., Dubin, D. T., and Stollar, V. (1976). *J. Virol.* **17**, 149–159.
Hirumi, H., Hirumi, K., Speyer, G., Yunker, C. E., Thomas, L. A., Cory, J., and Sweet, B. H. (1976). *In Vitro* **12,** 83–97.
Hsu, S. H., Wang, B. T., Huang, M. H., Wong, W. J., and Cross, J. H. (1975). *Am. J. Trop. Med. Hyg.* **24,** 881–888.
Igarashi, A. (1978). *J. Gen. Virol.* **40,** 531–544.
Igarashi, A., and Stollar, V. (1976). *J. Virol.* **19,** 398–408.
Igarashi, A., Sasao, F., Wungkobkiat, S., and Fukai, K. (1973a). *Biken J.* **16,** 17–23.
Igarashi, A., Sasao, F., and Fukai, K. (1973b). *Biken J.* **16,** 95–101.
Igarashi, A., Sasao, F., and Fukai, K. (1974a). *Biken J.* **17,** 35–37.
Igarashi, A., Sasao, F., and Fukai, K. (1974b). *Biken J.* **17,** 39–49.
Igarashi, A., Harrap, K. A., Casals, J., and Stollar, V. (1976). *Virology* **74,** 174–187.
Igarashi, A., Koo, R., and Stollar, V. (1977). *Virology* **82,** 69–83.
Jenkin, H., Townsend, D., Makino, S., and Yang, T. (1971). *Curr. Top. Microbiol. Immunol.* **55,** 97–102.
Kascsak, R. J., and Lyons, M. J. (1974). *Arch. Gesamte Virusforsch.* **45,** 149–154.
Kennedy, S. I. T. (1974). *J. Gen. Virol.* **23,** 129–143.
King, C.-C., King, M. W., Garry, R. F., Wank, M.-M., Ulug, E. T., and Waite, M. R. F. (1979). *Virology* **96,** 229–238.
Luukkonen, A., Brummer-Korvenkontio, M., and Renkonen, O. (1973). *Biochim. Biophys. Acta* **326,** 256–261.
Luukkonen, A., Kääriäinen, L., and Renkonen, O. (1976). *Biochim. Biophys. Acta* **450,** 109–120.
Luukkonen, A., Von Bonsdorff, C.-H., and Renkonen, O. (1977). *Virology* **78,** 331–335.
Maeda, S., Hashimoto, K., and Simizu, B. (1979). *Virology* **92,** 532–541.
Mento, S. J., and Stollar, V. (1978a). *Somat. Cell Genet.* **4,** 179–191.
Mento, S. J., and Stollar, V. (1978b). *Virology* **87,** 58–65.
Mitsuhashi, J., and Maramorosch, K. (1964). *Contrib. Boyce Thompson Inst.* **22,** 435–460.
Moore, N. F., Barenholz, Y., and Wagner, R. R. (1976). *J. Virol.* **19,** 126–135.
Mudd, J. A., Leavitt, R. W., Kingsbury, D. T., and Holland, J. J. (1973). *J. Gen. Virol.* **20,** 341–351.
Murray, A. M., and Morahan, P. A. (1973). *Proc. Soc. Exp. Biol. Med.* **142,** 11–15.
Mussgay, M., Enzmann, P. J., Weiland, E., and Horzinek, M. C. (1975). *Prog. Med. Virol.* **19,** 258–323.
Paul, S. D., Singh, K. R. P., and Bhat, U. K. M. (1969). *Indian J. Med. Res.* **57,** 339–348.
Peleg, J. (1968). *Virology* **35,** 617–619.
Peleg, J. (1969a). *Nature (London)* **221,** 193–194.
Peleg, J. (1969b). *J. Gen. Virol.* **5,** 463–471.
Peleg, J. (1971). *Curr. Top. Microbiol. Immunol.* **55,** 155–161.
Peleg, J. (1975). *Ann. N. Y. Acad. Sci.* **266,** 204–213.
Peleg, J., and Stollar, V. (1974). *Arch. Gesamte Virusforsch.* **45,** 309–318.
Pfefferkorn, E. R., and Shapiro, D. (1974). *In* "Comprehensive Virology" (H. Fraenkel-Conrat and R. R. Wagner, eds.), Vol. 2, Chapter 4, pp. 171–230. Plenum, New York.

Preble, O. T., and Youngner, J. S. (1975). *J. Infect. Dis.* **131,** 467–473.
Pudney, M., Leake, C. J., and Varma, M. G. R. (1979). *In* "Arctic and Tropical Arboviruses" (E. Kurstak, ed.), Chapter 16, pp. 245–262. Academic Press, New York.
Raghow, R. S., Grace, T. D. C., Filshie, B. K., Bartley, W., and Dalgarno, L. (1973). *J. Gen. Virol.* **21,** 109–122.
Regnery, R. L. (1977). Ph.D. Thesis, Australian Natl. Univ., Canberra.
Rehacek, J. (1968). *Acta Virol. (Engl. Ed.)* **12,** 340–346.
Rehacek, J. (1976). *In* "Invertebrate Tissue Culture: Applications in Medicine, Biology and Agriculture" (E. Kurstak and K. Maramorosch, eds.), Chapter 2, pp. 21–33. Academic Press, New York.
Renz, D., and Brown, D. T. (1976). *J. Virol.* **19,** 775–781.
Riedel, B., and Brown, D. T. (1979). *J. Virol.* **29,** 51–60.
Sarver, N., and Stollar, V. (1977). *Virology* **80,** 390–400.
Sarver, N., and Stollar, V. (1978). *Virology* **91,** 267–282.
Sawicki, D. L., and Gomatos, P. J. (1976). *J. Virol.* **20,** 446–464.
Schloemer, R. H., and Wagner, R. R. (1975). *J. Virol.* **15,** 1029–1032.
Shenk, T. E., Koshelnyk, K. A., and Stollar, V. (1974). *J. Virol.* **13,** 439–447.
Siminovitch, L. (1976). *Cell* **7,** 1–11.
Simmons, D. T., and Strauss, J. H. (1972). *J. Mol. Biol.* **71,** 615–631.
Sinarachatanant, P., and Olson, L. C. (1973). *J. Virol.* **12,** 275–283.
Singh, K. R. P. (1967). *Curr. Sci.* **36,** 506–508.
Singh, K. R. P. (1972). *Adv. Virus Res.* **17,** 187–206.
Spradling, A., Hui, H., and Penman, S. (1975). *Cell* **4,** 131–137.
Stevens, T. M. (1970). *Proc. Soc. Exp. Biol. Med.* **134,** 356–361.
Stollar, V. (1975). *Virology* **66,** 620–624.
Stollar, V. (1978). *Virology* **91,** 504–507.
Stollar, V., and Shenk, T. E. (1973). *J. Virol.* **11,** 592–595.
Stollar, V., and Thomas, V. L. (1975). *Virology* **64,** 367–377.
Stollar, V., Stevens, T. M., and Schlesinger, R. W. (1966). *Virology* **30,** 303–312.
Stollar, V., Peleg, J., and Shenk, T. E. (1974). *Intervirology* **2,** 337–344.
Stollar, V., Shenk, T. E., Koo, R., Igarashi, A., and Schlesinger, R. W. (1975). *Ann. N.Y. Acad. Sci.* **266,** 214–231.
Stollar, V., Stollar, B. D., Koo, R., Harrap, K. A., and Schlesinger, R. W. (1976). *Virology* **69,** 104–115.
Stollar, V., Harrap, K., Thomas, V., and Sarver, N. (1979). *In* "Arctic and Tropical Arboviruses" (E. Krustak, ed.), Chapter 18, pp. 277–296. Academic Press, New York.
Suitor, E. C., Jr., and Paul, F. J. (1969). *Virology* **38,** 482–485.
Trager, W. (1938). *Am. J. Trop. Med.* **18,** 387–393.
Varma, M. G. (1975). *J. Med. Entomol.* **11,** 698–706.
Walker, D. L. (1964). *Prog. Med. Virol.* **6,** 111–148.
Warren, L. (1963). *Comp. Biochem. Physiol.* **10,** 153–171.
Weiss, E., ed. (1971). *Curr. Top. Microbiol. Immunol.* **55.**
Wengler, G., Wengler, G., and Gross, H. J. (1978). *Virology* **89,** 423–437.
Yang, T. K., McMeans, E., Anderson, L. E., and Jenkin, H. M. (1974). *Lipids* **9,** 1009–1013.

21

Non-Arbo Togaviruses

MARGO A. BRINTON

I. INTRODUCTION

In 1970, the arboviruses were subdivided on the basis of virion structure by the Vertebrate Virus Subcommittee of the International Committee on Nomenclature of Viruses. The term "arbovirus" is now applied to any virus which has a biological cycle in both an arthropod and a vertebrate host. Viruses from a number of taxonomic groups are represented among the arboviruses. Certain picorna-, reo-, and rhabdoviruses are arboviruses. Initially, all classified togaviruses were also arboviruses; there was, however, no reason to expect that all viruses possessing the taxonomic characteristics of togaviruses would have or retain biological cycles in both arthropod and vertebrate hosts.

THE TOGAVIRUSES

ISBN 0-12-625380-3

Recently rubella (RU) virus, bovine viral diarrhea virus (BVDV), hog cholera virus (HCV), lactate dehydrogenase virus (LDV), equine arteritis virus (EAV), and simian hemorrhagic fever (SHF) virus have been found to be structurally similar to togaviruses, and none of these viruses are known to replicate in an arthropod host (Horzinek, 1973; Trousdale *et al.*, 1975). A number of bat viruses, such as Rio Bravo virus, Montana *Myotis* leukoencephalitis virus (MMLV), Dakar bat virus, and Entebbe bat virus, and the two rodent viruses, Modoc virus and Cowbone Ridge virus (CRV), also seem to be non-arbo togaviruses (Andrewes and Pereira, 1972). *Aedes aegypti* cell-fusing agent (CFA), which replicates in mosquito cells but has no known vertebrate host, likewise displays structural similarities to togaviruses (Igarashi *et al.*, 1976). It is quite possible that many additional, as yet unidentified, non-arbo togaviruses exist in nature.

Among the non-arbo togaviruses listed above only the four bat viruses and the two rodent viruses show any antigenic similarity to members of either the *Alphavirus* or *Flavivirus* genera of the Togaviridae family. The remainder appear to be representatives of three or more additional genera. Bovine viral diarrhea virus, HCV, and probably also Border disease virus, constitute the *Pestivirus* genus, while RU virus seems to be the only known member of the *Rubivirus* genus. Lactate dehydrogenase virus, EAV, CFA, and possibly SHF virus do not appear to fit into any of the present togavirus genera.

A comprehensive review of the research on each non-arbo togavirus has not been attempted in this chapter. Instead, the intent has been to concentrate on presenting the available physicochemical and morphological data and to give only a brief summary of clinical data.

II. PESTIVIRUSES

A. Bovine Viral Diarrhea Virus

Bovine viral diarrhea virus, also called mucosal disease virus, is the causative agent of mucosal disease in cattle and can induce congenital malformations (Dinter *et al.*, 1962; Ward, 1969).

1. Physicochemical Properties

Bovine viral diarrhea virions are sensitive to ethyl ether, chloroform, and trypsin (Hafez and Liess, 1972b; Hermodsson and Dinter, 1962; Ditchfield and Doane, 1964; Gillespie *et al.*, 1963; Castrucci *et al.*, 1968; Tanaka *et al.*, 1968), and the presence of 1 *M*

$MgCL_2$ was found to increase, rather than decrease, the heat sensitivity of this virus (Hafez and Liess, 1972b). Strains of BVDV are most stable in the pH range from 5.7 to 9.3, with maximum stability at pH 7.4 (Hafez and Liess, 1972b).

Horzinek *et al.* (1971) observed a heterogeneous size distribution of BVDV infectivity after rate zonal centrifugation in a glycerol density gradient. Electron microscopic examination of fractions from such gradients revealed that the faster sedimenting fractions contained virions with swollen envelopes and membrane debris. The buoyant density as determined by infectivity after equilibrium density centrifugation in CsCl ranged from 1.13 to 1.15 g/cm^3 (Fernelius, 1968a; Hafez and Liess, 1972b). However, after isopycnic centrifugation of radioactively labeled BVDV, in sucrose density gradients, the radioactive profile peaked at a density of approximately 1.12 to 1.13 g/cm^3 (Parks *et al.*, 1972; Maess and Reczko, 1970; M. A. Brinton, unpublished data). Electron microscopic and centrifugation data indicate that the BVDV envelope is very fragile and that host cell proteins and membrane fragments are difficult to separate from BVD virions (Horzinek *et al.*, 1971; Pritchett and Zee, 1975; Pritchett *et al.*, 1975).

a. Morphology. Early filtration studies indicated a virus size between 50 and 100 nm (Fernelius, 1968b). Particles of varying diameter were observed by electron microscopy by a number of investigators (Fernelius, 1968b; Hafez *et al.*, 1968; Ritchie and Fernelius, 1969). Maess and Reczko (1970) reported that BVD virions appeared spherical and measured 57 ± 7 nm in diameter with a core that was 24 ± 4 nm in diameter. Horzinek *et al.* (1971) reported a similar particle diameter but found the average diameter of the core to be 28 ± 3 nm. Particles which resemble virions have been observed which have diameters up to 60 nm (Horzinek *et al.*, 1971; Hafez *et al.*, 1968), but the presence of these larger particles appears to depend on the method of preparation and is probably the result of artificial virion distortion. Envelope projections were not visible on the surface of BVD virions. Horzinek *et al.* (1971) observed some particles which seemed to have a "rosary" envelope structure, electron-lucent spherules about 6 nm in diameter which were arranged around the core. The particles with these structures on their surfaces were of a lesser average diameter than the normal virions, and it is possible that the spherules can be visualized only when some surface material normally present on the virions has been removed. This virus has been found not to agglutinate mammalian or avian erythrocytes (Hafez and Liess, 1972a) (for further discussion see Chapter 8).

b. Genome. The RNA of BVDV was demonstrated to be infectious and sensitive to RNase (Diderholm and Dinter, 1966). RNA extracted from isopycnically banded [^{3}H]uridine-labeled BVDV with sodium dodecyl sulfate (SDS) was analyzed by rate zonal sedimentation on sucrose density gradients and by polyacrylamide gel electrophoresis (Pritchett *et al.*, 1975). The major RNA component was found to have a sedimentation coefficient of 38 S. Two minor components were estimated to have *s* values of 31 and 24. The molecular weights for these RNAs were calculated to be 3.22×10^6 for the 38 S molecule, 2.09×10^6 for the 31 S molecule, and 1.22×10^6 for the 24 S molecule. All three RNAs were found to be susceptible to RNase digestion.

Analysis of the RNA molecules obtained from two BVDV strains, one cytopathic and one noncytopathic, revealed that the 38–39 S RNA was the major species of RNA in both strains, but different minor components were observed. A 35 and a 46 S species were seen in extracts of the cytopathic strain virion, and a 27 and a 47 S species in extracts of the noncytopathic strain virions (M. A. Brinton, unpublished data). It is not presently known whether these minor RNA species represent aberrant size genomes, degradation products of the genome RNA, or cellular RNA contaminants. The infectivity of the various size classes has not yet been assessed. However, an infectious RNA has been extracted from BVDV-infected cells (Horzinek, 1976). Infectivity was assayed by plaque formation on calf testicle cells pretreated with hypertonic buffers. The infectious RNA had a sedimentation coefficient of 37–40 S in sucrose gradients made in 0.1 *M* NaCl.

When BVDV RNA was centrifuged through gradients under conditions of low ionic strength (Pritchett *et al.*, 1975), a single broad peak of radioactivity was observed. The sedimentation coefficient of the center of this peak was calculated to be 24 S. It was not determined, however, whether this shift was due to disaggregation or to a conformational change in the BVDV RNA molecules.

c. Structural Proteins. The separation of BVDV structural proteins by polyacrylamide gel electrophoresis has been reported (Frost and Liess, 1973; Pritchett and Zee, 1975). Frost and Liess (1973) found three polypeptide bands in extensively purified BVDV preparations which had been disrupted with either Triton N101 or SDS. Molecular weights were not calculated. The electrophoretic pattern and the number of proteins varied when virus preparations were extracted with Tween 80, ethyl ether, or urea and Nonidet P-40. Pritchett and Zee (1975) found that, even after purification of BVDV by differential and isopycnic centrifugation, residual host proteins were still present in the viral preparations. Virus-specific proteins were distinguished through the use of coelectrophoresis of ^{3}H-labeled virus with ^{14}C-labeled uninfected host cell material. Four unique components were found in virus preparations: viral component Vc1 and Vc3 migrated heterogeneously and had molecular weights ranging from 93,000 to 110,000 and 50,000 to 59,000, respectively. The molecular weight of Vc2 was calculated to be 70,000 and that of Vc4 to be 27,000 (Fig. 1). It has not yet been determined whether any of these proteins are glycosylated or which of them is the core protein.

2. *Cultivation*

BVDV was initially propagated in primary bovine kidney and testis cell cultures (Carbrey *et al.*, 1971; Edward *et al.*, 1967; Gillespie *et al.*, 1960a,b; Gutekunst and Malmquist, 1963). However, because bovine

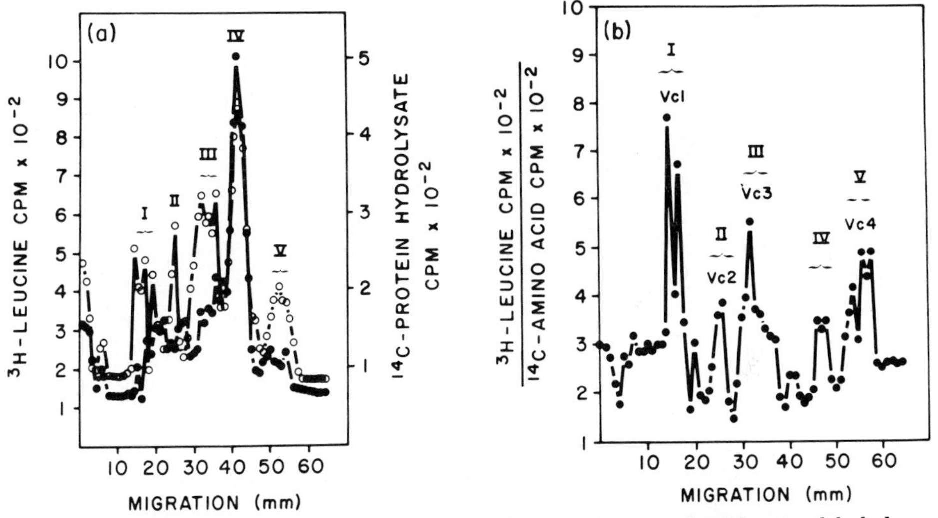

Fig. 1. Structural proteins of BVDV. (a) Coelectrophoresis of [^{3}H]leucine-labeled BVDV (NADL strain) proteins (open circles) and [^{14}C]protein hydrolysate-labeled uninfected host cell proteins on 7.5% polyacrylamide SDS gels (solid circles). Virus propagated in bovine bone marrow cells was isopycnically banded on a sucrose density gradient prior to disruption for electrophoresis. Host material was obtained from the density region of the virus band from a gradient layered with culture fluid from mock-infected cells. Vc, Viral component. Estimated molecular weight: Vc1 = 93,000–110,000; Vc2 = 70,000; Vc3 = 50,000–59,000; Vc4 = 27,000. (Reproduced with permission from Pritchett and Zee, 1975, *Am. J. Vet. Res.* **36**, 1731–1734.) (b) Ratio of [^{3}H]-leucine-labeled BVDV (NADL strain) proteins to [^{14}C]protein hydrolysate-labeled host cell components after coelectrophoresis on 7.5% polyacrylamide gels (solid circles) from (a).

viral diarrhea is widespread in the cattle population and can cross the placental barrier, it proved difficult to prepare consistently primary cultures not already infected with the virus (Carbrey *et al.*, 1971). The virus has subsequently been propagated in cell lines of bovine tubinate cells (McClurkin *et al.*, 1974) and fetal bovine endometrial cells, which are known to be free of BVDV infection.

A large number of BVDV strains exist. Some are cytopathogenic for tissue culture cells, while others are not. However, cytopathogenicity does not seem to be correlated with ability to produce disease or mortality in cattle. For example, the National Animal Disease Laboratory (NADL) strain, which is noncytopathogenic in tissue culture, is fully virulent for neonatal calves (Fernelius *et al.*, 1973). Most of the cytopathogenic BVDV strains have been observed to become noncytopathogenic after passage in pigs or rabbits (Baker *et al.*, 1954; Fernelius *et al.*, 1969b).

The growth of BVDV is highly sensitive to inhibition by proflavine or acriflavine. It has been possible to select acriflavine-resistant mutants, and these mutants are stable through numerous cell culture passages, as well as through passage in calves (Rockborn *et al.*, 1974; Dinter and Diderholm, 1972). This marker is useful in following the yield of virus after *in vivo* infections, since BVDV strains are widely disseminated in cattle populations and an experimental infection or vaccination may coincide with a natural infection (McKercher *et al.*, 1968).

Homologous interference between strains of BVDV has been reported (Gillespie *et al.*, 1962). Cultures became resistant to challenge with a cytopathogenic strain 3 days after infection with a noncytopathogenic strain (McKercher *et al.*, 1968). Heterologous interference has also been reported between BVDV and bovine enteroviruses (Schiff and Storz, 1972). No interference was observed when Newcastle disease virus (NDV) (Diderholm and Dinter, 1966), vesicular stomatitis virus (VSV), or pseudorabies (Torlone *et al.*, 1965) was the superinfecting virus. Bovine viral diarrhea virus is one of the endogenous bovine viruses which sometimes contaminate commercially supplied fetal bovine sera (Kniazeff *et al.*, 1973).

Bovine viral diarrhea virus is sensitive to inhibition by interferon. Although interferon was demonstrated in calf kidney cell cultures persistently infected with BVDV, constant levels of virus were released by such cultures (Diderholm and Dinter, 1966). Interferon production was also detected in pigs infected with BVDV (Welsh and Pfau, 1972). Infection of cells with BVDV appears to suppress the induction of interferon after superinfection with NDV as well as the inhibitory action of added exogenous interferon.

The only indication so far that defective interfering BVDV particles exist is the production of "bull's eye" plaques by a cloned BVDV strain (Diderholm *et al.*, 1974). The formation of similar concentric rings in plaques in other systems has been attributed to the presence of defective interfering particles (Welsh and Pfau, 1972).

3. *Pathogenesis*

The consequences of infection of cattle with BVDV vary from an inapparent infection to a severe fatal disease. Detection of anti-BVDV antibody in much of the world's cattle population suggests that mild infections are common. Infected pregnant cows may have normal pregnancies, have abortions, or bear malformed calves, regardless of whether or not the infection is inapparent or severe (Kahrs, 1971). Classically, BVDV infection is characterized by fever, leukopenia, diarrhea, lacrimation, nasal discharge, and oral ulcerations, as well as lesions in the hooves and lymph nodes. Most clinical disease occurs in cattle between 6 and 24 months of age (Kahrs *et al.*, 1966) and is often fatal. However, occasionally a diseased animal may survive and subsequently develop a chronic debilitating disease (mucosal disease) manifested by weight loss, intermittent diarrhea, inefficient feed utilization, and respiratory symptoms (Kahrs, 1971).

Bovine viral diarrhea virus causes a brief mild disease in young pigs, but infected pigs shed virus in nasal mucus and feces for periods of 2–3 weeks. Swine appear to be an intermediate host for the virus, and natural infections of swine may be the source of naturally occurring variants of BVDV (Fernelius *et al.*, 1969a,b, 1973).

4. *Immunology*

Bovine viral diarrhea virus displays an affinity for cells of the immune system. Necrotic foci in lymph nodes and spleen, destruction of Peyer's patches, and general atrophy of lymphatic tissues are observed in infected animals (Muscoplat *et al.*, 1973). Peripheral lymphocytes from calves infected with BVD virus were found to be unresponsive to stimulation with phytohemagglutinin (PHA). The usual proliferative response induced by the addition of PHA was depressed when BVDV was added at the same time as PHA. Heat-killed or uv-irradiated virus did not cause a depression of mitogen stimulation. Lymphocytes infected with BVDV seem unable to undergo blast transformation in response to mitogenic stimulation and may also be impaired in other immunological functions.

Immune complexes consisting of BVDV, antiviral antibody, and complement have been found in the renal glomeruli of cattle with mucosal disease (Prager and Liess, 1976).

French and Snowdon (1964), reported that cattle clinically affected by a BVDV infection could transmit infection to sheep kept in the same field, and pigs were reported to show evidence of BVDV infection after association with cattle (Castrucci *et al.*, 1974; Stewart *et al.*, 1971). Also, pigs were found to have high anti-BVDV antibody titers after contact with BVDV-vaccinated cattle (Stewart *et al.*, 1971).

5. *Vaccines*

Attenuated strains of BVDV were used for a time to immunize cattle and, although the vaccines were efficacious in preventing subsequent BVD infection, a condition resembling mucosal disease was observed to develop in some cattle shortly after vaccination (Peter *et al.*, 1967). Vaccination of pregnant animals can lead to fetal malformation or death (Brown *et al.*, 1974, 1975; Archbald, 1974).

There appear to be wide serological differences among strains of BVDV. However, vaccines prepared from different strains seem to provide adequate protection against most field strains of the virus.

B. Hog Cholera Virus

Hog cholera virus, also known as swine fever virus, displays a serological cross-reaction with BVDV and with the virus that causes Border disease in sheep (Plant *et al.*, 1973; Huck *et al.*, 1975).

1. *Physicochemical Properties*

Initial studies revealed that HCV is a chloroform- and ether-labile, moderately trypsin-sensitive virus which contains an RNA genome (Dinter, 1963; Loan, 1964; McKissick and Gustafson, 1967).

This virus is fairly stable to pH changes; a pH less than 1.4 or greater than 13 is necessary to inactivate virus within an hour (Slavin, 1938). Virions are quickly inactivated when dried in air but persist for up to several months in pork or garbage (Dunne, 1958; Mateva *et al.*, 1966; Torlone *et al.*, 1965). The buoyant density of hog cholera virions has been determined to be between 1.15 and 1.17 g/cm^3 on CsCl gradients (Horzinek, 1966, 1967; Cunliffe and Rebers, 1968; Ushimi *et al.*, 1969; Dunne, 1958). CsCl gradient centrifugation is known to induce envelope disruption in togaviruses (Aaslestad *et al.*, 1968; Horzinek *et al.*, 1971). Buoyant density determinations from sucrose or glycerol gradient centrifugation have not been reported.

a. Morphology. After concentration on glycerol gradients, spherical HC virions with a diameter of 53 ± 14 nm were observed by electron microscopy. The virion envelope measured 6 nm in width (Horzinek *et al.*, 1971) and, after incubation of virus with 4 *M* urea, spherical cores 27 ± 3 nm across were seen. "Rosary" envelope structures, similar to those described for BVDV, were occasionally observed on the surface of hog cholera virions. Enzmann and Weiland (1978) observed spherical virus particles with a diameter of 42 ± 8 nm, which had projections on their surface about 6–8 nm in length, in virus preparations which had been purified on sucrose gradients. Disruption of virions tends to occur during centrifugation through CsCl gradients (Dunne, 1958; Cunliffe and Rebers, 1968; Ushimi *et al.*, 1969) (see Chapter 8 for further details).

b. Genome. The RNA of HCV was found to have a sedimentation coefficient of 40–45 S (Enzmann and Rehberg, 1977). A molecular weight of 4×10^6 was estimated for this RNA after electrophoresis in polyacrylamide agarose gels (Enzmann and Rehberg, 1977). The genomes of HCV and Sindbis virus appear to differ somewhat in their secondary structure, as indicated by differences in their sedimentation in sucrose gradients and their migration in gels (Enzmann and Rehberg, 1977).

c. Structural Proteins. Three main polypeptides have been resolved by polyacrylamide gel electrophoresis of two strains of HCV grown either in PK-15 or SK cells (Enzmann and Rehberg, 1977). The molecular weights of these three polypeptides were calculated to be 54,000–56,000, 45,000–47,000, and 35,000–37,000 (Fig. 2). Only these three proteins were isolated by radioimmune precipitation (Enzmann and Weiland, 1978).

Occasionally three additional polypeptides were detected by gel electrophoresis. One of these had a molecular weight of 65,000 and may represent an uncleaved precursor. The two other proteins, which had molecular weights ranging from 30,000 to 35,000 and 18,000 to 25,000, may arise from degradation of the structural proteins. Only the 46,000-molecular-weight protein was found to stain positively with the periodic acid Schiff reagent, indicating that it was a glycoprotein (Enzmann and Rehberg, 1977). However, both the 54,000–56,000 and the 45,000–47,000 polypeptides were shown to be glycoproteins by labeling with sodium [^{3}H]borohydride (Enzmann and Weiland, 1978). A glycoprotein of HCV has been shown to bind to lectins, and this

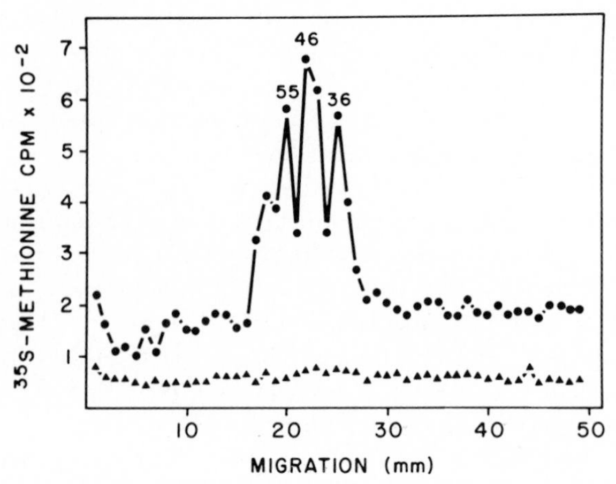

Fig. 2. Structural proteins of HCV. Electrophoresis pattern on 7.5% polyacrylamide SDS gels of [^{35}S]methionine-labeled HCV (strain ALD-970a) propagated in PK-15 cells (circles) and in culture fluid from mock-infected cells (triangles). Estimated molecular weights: 55,000, 46,000, and 36,000. (Reproduced with permission from Enzmann and Rehberg, Z. *Naturforsch. Teil C* **32**, 456–458, 1977.)

ability is being utilized to purify the glycoprotein (Dalsgaard and Overby, 1976). The location of the various structural proteins within the virion has not yet been described, although it is probable that the two glycoproteins are associated with the viral envelope.

3. *Cultivation*

Hog cholera virus replicates in cultures of porcine tissue, such as spleen, kidney, testicle, and leukocytes obtained from peripheral blood. Most HCV strains are noncytopathogenic, but some cytopathic strains have been developed (Bachmann *et al.*, 1967; Gillespie *et al.*, 1960a,b). However, some of these "cytopathic strains" have been found to be contaminated with other porcine viruses (Bodon, 1965; Horzinek *et al.*, 1967b). Noncytopathogenic strains can persist in cell cultures. Persistent infections have been maintained in porcine kidney cells for as long as 75 passages. Virus recovered from persistently infected porcine cultures has sometimes been found to be attenuated (Sato *et al.*, 1964; Loan and Gustafson, 1964; Torlone and Titoli, 1964). Interferon is induced in tissue culture by infection with cytopathic strains of HCV (Lin *et al.*, 1969). However, persistently infected cultures did not appear to produce interferon (Torlone *et al.*, 1965).

The infectivity of HCV is titered by immunofluorescent microplaque techniques (Aynaud, 1968; Danner and Bachmann, 1970; Pirtle and Kniazeff, 1968). Under single growth cycle conditions in pig kidney cells, viral antigen is first detected 4 h after infection.

The noncytopathogenic GPE strain of HCV was found to interfere with the replication of both VSV and Western equine encephalomyelitis virus in pig kidney cells (Shimizu *et al.*, 1970; Fukusho *et al.*, 1976). A reverse plaque assay has been developed which makes use of the interference between HCV and VSV (Fukusho *et al.*, 1976).

4. *Pathogenesis*

Only swine are naturally infected by HCV. This virus seems to be spread by direct contact between infected pigs, or by feeding virus-contaminated garbage to pigs. Shope (1958) suggested that swine lungworm might be a reservoir for HCV. Houseflies, birds, and man may transfer the virus mechanically (Dunne, 1958). Virulent HCV is highly contagious. The disease is characterized by fever, vomiting, eye discharge, diarrhea, generalized hemorrhages, leukopenia, and encephalitis. The virus primarily attacks cells of the reticuloendothelial system (Dunne, 1958; Seifried and Cain, 1932). Infected pigs die either from the virus infection itself or from a secondary infection with bacteria, which causes pneumonia or ulcerative enteritis.

5. *Vaccines*

Because of their potential for spreading hog cholera, live virus vaccines are now banned in the United States. Previously, vaccination was carried out with either inactivated virus, attenuated virus strains, or by simultaneous inoculation of virulent virus and antiserum. The latter two methods led to persistence of infection within herds, while the first gave only temporary immunity. Although vaccination has been responsible for the degree of control of hog cholera obtained in the United States for the last 50 years, its role in the persistence of the disease is also acknowledged (Dunne, 1958). Attenuated vaccine strains of HCV often produce a chronic type of infection characterized by a persistent viremia. Individuals in immune herds were repeatedly found to have detectable virus levels (Carbrey and Kresse, 1967; Dunne *et al.*, 1955; Baker and Sheffy, 1960). Infection of young pigs with attenuated virus strains can lead to runting (Baker and Sheffy, 1960), while inoculation of breeding sows with attenuated HCV can subsequently lead to abortions or malformations of fetuses. The vaccinated breeding sow is probably the most important reservoir of HCV. She does not display disease symptoms, but live piglets born to such a sow are usually weak, runted, and carriers of a weakly virulent HCV which can spread to other susceptible pigs (Huck and Aston, 1964; Cowart and Morehouse, 1967; Schwartz *et al.*, 1967; Dunne and Clark, 1968). All vaccination of swine with live virus was stopped when it was realized that the eradication of hog cholera would be impossible as long as any form of live virus was used for vaccination. Research on HCV has now been restricted to a few government facilities with proper containment areas.

C. Border Disease Virus

There have been numerous reports of Border disease occurring in sheep in the United States, Australia, and New Zealand during the last 20 years. This disease is apparently caused by a virus antigenically similar to BVDV and HCV (Plant *et al.*, 1973).

1. Physicochemical Properties

To date only a few studies of the characteristics of Border disease virus have been carried out. The virus will pass through a membrane filter with a 50-nm pore diameter. It is sensitive to heating at 56°C for 30 min and moderately sensitive to ether (Vantsis *et al.*, 1976).

On sucrose density gradients the peak of infectivity was found to band at a density of 1.115 g/cm^3.

2. Cultivation

Border disease virus was propagated in secondary cultures of fetal lamb kidney (FLK) cells. After about 12 passages in FLK cultures the virus was able to replicate in secondary cultures of calf kidney cells. Border disease virus produced an obvious cytopathic effect (CPE) in FLK monolayers about 7 days after infection. The calf kidney cultures were destroyed more slowly and less completely than the FLK monolayers after infection with the same virus inoculum.

3. Pathogenesis

Lambs infected with Border disease virus have a history of poor growth, and many are born with rhythmic clonic spasm. Their fleece is abnormally pigmented and fuzzy (Osburn *et al.*, 1972, 1973; Manktelow *et al.*, 1969). Inoculation of pregnant ewes with a cell suspension or with cell-free extracts of brain, spleen, or spinal cord from an infected lamb results in an inapparent infection in the ewes, and most of the fetuses that are not aborted are born with Border disease (Jackson *et al.*, 1972; Manktelow *et al.*, 1969; Gardiner *et al.*, 1972). In contrast, inoculation of pregnant ewes with BVDV resulted in abortions, fetal mummification, or lambs born with hydrocephalus and cerebellar hypoplasia (Snowdon *et al.*, 1975). Hydrocephalus is not a symptom of Border disease (Ward, 1971).

4. *Immunology*

Antisera obtained from sheep infected with Border disease virus showed serological cross-reaction with both BVDV and HCV (Osburn *et al.*, 1973; Plant *et al.*, 1973). Ewes injected with Border disease virus showed low titers of neutralizing antibody to the NADL strain of BVDV, and cattle recovering from BVDV infection had low titers of antibody to Border disease virus. In both cases the animals had high titers of neutralizing antibody to the virus with which they were infected. Border disease virus was neutralized to only a limited degree by hyperimmune serum from a BVDV-infected animal.

D. Conclusion

Although limited information is presently available on the physicochemical properties of the pestiviruses, it appears that they do not fit into either the alpha- or the flavivirus group. Antigenically they form a distinct group. In size and morphological characteristics these viruses are similar to flavivirus, but the distribution and molecular weights of their structural proteins appear to resemble more closely those seen in alphaviruses (Enzmann and Weiland, 1978) (see Table I). However, analysis of pestivirus structural proteins is still preliminary, and sufficient data are not available to compare their structural proteins adequately. Buoyant densities determined after centrifugation of pestiviruses through CsCl density gradients may not be reliable, since similar viruses are known to be disrupted in such gradients.

III. RUBIVIRUS

Rubella

Natural infections of RU virus occur only in humans. The disease resulting from RU virus infection has been named German measles.

1. *Physicochemical Properties*

A wide range of buoyant density values has been reported for RU virions. In CsCl density gradients, values of 1.085 (Russell *et al.*,

TABLE I

Physical Parameters of Non-Arbo Togaviruses

Virus	Average virion diameter (nm)	Average core diameter (nm)	Sedimentation coefficient of genome RNA (S)	Molecular weight of structural proteins (× 10^{-3})[k]
Pestiviruses				
Bovine diarrhea virus[a]	57 ± 7	28 ± 3	38	93–110 70 50–59 27
Hog cholera virus[b]	53 ± 14	27 ± 3	40–45	54–56 (G) 45–47 (G) 35–37
Border disease virus[c]	< 50	?	?	?
Rubivirus				
Rubella[d]	58 ± 7	33 ± 1	40	60–63.6(G) 47–56 (G) 29–35 (N)
Other				
Lactate dehydrogenase-elevating virus[e]	55	30–35	48	24–44 (G) 17–18 (M) 13–15(N)
Equine arteritis virus[f]	60 ± 13	35	48	21(28–40) (G) 14–15 (M) 12–13 (N)

Simian hemorrhagic fever virus[g]	40–45	22–25	15–30	Two glycoproteins plus two polypeptides
Cell-fusing agent[h]	45–52	23–28	40–42	49 (G) 16.5 (G) 13 (N)
Alphavirus				
Sindbis[i]	61 ± 2	25 ± 0.4	49	50 (G) 50 (G) 30–34 (N) 10 (G)[l]
Flavivirus				
Dengue[j]	48–50	26	45	53–63 (G) 13.5–17 (N) 8.5–8.7 (M)

[a] Horzinek *et al.* (1971); Pritchett *et al.* (1975); Pritchett and Zee (1975).
[b] Horzinek *et al.* (1971); Enzmann and Rehberg (1977).
[c] Vantsis *et al.* (1976).
[d] Horzinek *et al.* (1971); Hovi and Vaheri (1970a); Vaheri and Hovi (1972).
[e] Horzinek *et al.* (1975); Brinton-Darnell and Plagemann (1975); Darnell and Plagemann (1972); Michaelides and Schlesinger (1973).
[f] Horzinek *et al.* (1971); van der Zeijst *et al.* (1975); Hyllseth (1973); Zeegers *et al.* (1976).
[g] Wood *et al.* (1970); Trousdale *et al.* (1975); Tauraso *et al.* (1968).
[h] Stollar and Thomas (1975); Igarashi *et al.* (1976).
[i] Horzinek and Mussgay (1969); Simmons and Strauss (1972); Strauss *et al.* (1969); Garoff *et al.* (1974).
[j] Murphy *et al.* (1968); Matsumura *et al.* (1971); Igarashi *et al.* (1964); Stevens and Schlesinger (1965); Shapiro *et al.* (1972); Stollar (1969); see Chapter 18.
[k] G, Glycoprotein; N, nucleocapsid protein; M, membrane protein. See Chapter 8 for electron micrographs of many of the non-arbo togaviruses.
[l] Found only in SFV (see Chapter 9).

1967), 1.12 (Thomsen *et al.*, 1968), and 1.32 g/cm^3 (Amstey *et al.*, 1968) have been estimated. The disparity of these values probably reflects the sensitivity of RU virions to degradation in CsCl. In sucrose or potassium citrate gradients, virion densities of 1.18–1.19 g/cm^3 have been obtained (McCombs and Rawls, 1968; Liebhaber and Gross, 1972; Furukawa *et al.*, 1967a,b; Magnusson and Skaaret, 1967; Sedwick and Sokol, 1970). Even in these gradients the density of RU virions was observed to vary depending on the suspending medium used to prepare the gradients. When sucrose gradients were prepared in distilled water, the virion density was 1.16 g/cm^3, whereas in gradients prepared in Tris–EDTA buffer the density was 1.18 g/cm^3. This density shift was reversible and was attributed to the existence of the virus in different states of hydration (McCombs and Rawls, 1968). The sedimentation coefficient of RU virus in sucrose has been estimated to be 342 by Russell *et al.* (1967) and 240 by Thomsen *et al.* (1968). Rubella virions are sensitive to lipid solvents and are quickly inactivated at pH 3 (Chagnon and Laflamme, 1964).

a. Morphology. Rubella virus particles are roughly spherical and have an average diameter of 58 ± 7 nm after purification in sucrose density gradients. The core has an average diameter of 33 ± 1 nm (Horzinek *et al.*, 1971). Ringlike structures, approximately 11 nm in diameter, are sometimes observed on the surface of disintegrating particles. Holmes *et al.* (1969) and Bardeletti *et al.* (1975) observed 5- to 6-nm spikes on the viral envelope. Virion morphology is readily distorted by drying or by ionic strength shifts; if virions are prefixed, however, most are spherical (see Chapter 8 for details).

b. Genome. The genome of RU virus is RNA. Thymidine analogs do not inhibit virus replication (Maassab and Cochran, 1964; Prinzie, 1964; Maes *et al.*, 1966), and the RU genome incorporates radioactive [5-^{3}H]uridine (Brodersen and Thomssen, 1969; Hovi and Vaheri, 1970a). Rubella virus RNA is single-stranded, as indicated by its sensitivity to degradation by pancreatic RNase, its sedimentation characteristics, and its buoyant density (Hovi and Vaheri, 1970a,b). Rubella RNA sedimented at 40 S in a sucrose gradient made with 0.1 *M* saline, at 29 S in 1 m*M* saline, and at 55 S in 0.1 *M* saline and 1 m*M* $MgCl_2$. In comparison, rRNA sedimented at 28, 25, and 38 S, respectively, under the same conditions. The observed effect of low ionic strength on the sedimentation velocity of RU RNA seems to indicate that the coiling of RU RNA is more compact than that of rRNA. Similar effects of ionic strength have been noted with other single-stranded viral RNAs (Pfefferkorn, 1968). Addition of a divalent cation such as Mg^{2+} is known to increase the secondary structure of single-stranded RNAs by metal bonding (Spirin, 1962). In the presence of $MgCl_2$ the increase in the $s_{20,w}$ value of RU RNA was more pronounced than that observed for rRNA (Hovi and Vaheri, 1970a). Based on a sedimentation coefficient of 40 S, the molecular weight of RU RNA has been calculated to be approximately 3×10^6. The buoyant density of the RNA in cesium sulfate is 1.634 g/cm³ (Hovi and Vaheri, 1970a). Infectivity of RU RNA has been demonstrated by Hovi and Vaheri (1970a).

The nonionic detergent Nonidet P-40 has been used to degrade RU virions (Hovi and Vaheri, 1970a). A detergent concentration of 0.015% v/v is sufficient also to dissociate the nucleoprotein core and liberate viral RNA if polyvinyl sulfate (20 μg/ml) is present. Nonidet P-40 extraction of virus in the absence of polyvinyl sulfate results in the liberation of an intact core. The deproteinization of the viral RNA in the presence of polyvinyl sulfate is thought to be due to a competition for core proteins between the polyvinyl sulfate anion and the RNA molecules (Pons *et al.*, 1969).

c. **Structural Proteins.** The proteins of RU virus have been analyzed by polyacrylamide gel electrophoresis. Liebhaber and Gross (1972) reported that gradient-purified RU virions grown in Vero cell cultures contained eight species of viral-specific proteins. These polypeptides were grouped into three main areas after electrophoresis on 5% acrylamide gels. These areas were designated region I, region II, and region III. Within each region the authors observed heterogenicity. The molecular weights of the protein(s) within region I were estimated to be between 29,800 and 32,000, those within region II ranged from 47,000 to 56,000, while region III contained molecules of 60,000 to 63,000. The proteins within region II and region III incorporated fucose and thus were designated glycoproteins. The amount of protein resolved in region II varied from experiment to experiment. In some experiments, small amounts of two proteins with molecular weights greater than 65,000 were observed (Fig. 3b).

Vaheri and Hovi (1972), using 10% acrylamide gels to analyze the proteins of BHK-21/13S-grown RU virus, found three virus-specific proteins. Viral protein VP1 and VP2 incorporated glucosamine and thus were designated glycoproteins. Their average molecular weights were estimated to be 62,500 (VP1) and 47,500 (VP2). VP2 was represented by a broad peak which contained molecules over a molecular weight range of 45,000–50,000 (Fig. 3a). The amount of protein observed in the VP2 region varied from experiment to experiment. In some experiments two shoulders on this peak were observed. Vaheri and Hovi (1972) suggested that the heterogeneous migration of VP2 might be due to the presence of a large amount of carbohydrate. Also, some of the material migrating in the region of VP2 might be host glycoproteins which remain associated with RU virus throughout the purification procedure or which are an intrinsic part of the virion.

Disruption of RU virions with neutral detergents released a core

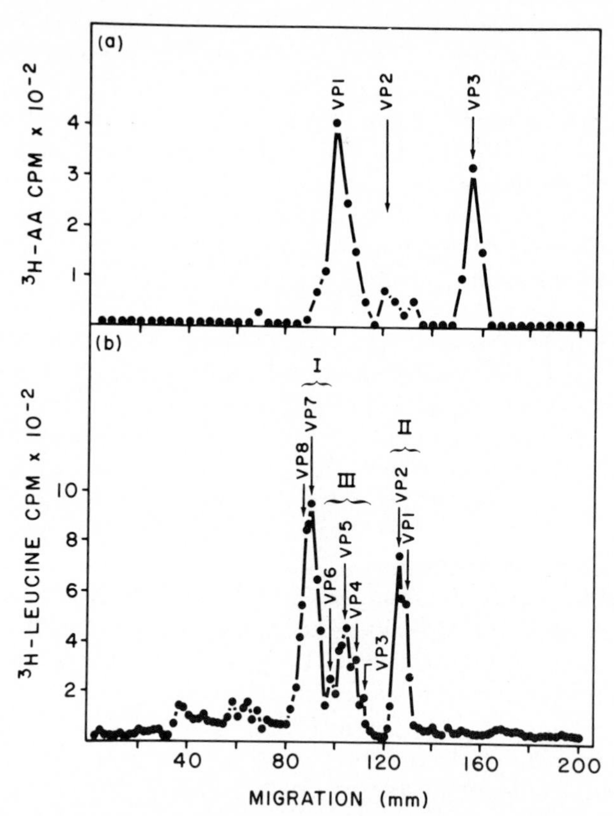

Fig. 3. Structural proteins of RU virus. (a) Polypeptide pattern of [^{3}H]amino acid-labeled RU virus grown in BHK-21/13S cells electrophoresed on 10% polyacrylamide SDS gels. VP, Viral protein. Estimated molecular weights: VP1 = 62,500; VP2 = 47,500; VP3 = 35,000. (Reproduced with permission from Vaheri and Hovi, *J. Virol.* **9**, 10–16, 1972.) (b) Electrophoresis pattern of [^{3}H]leucine-labeled RU virus propagated in Vero cells. Viral polypeptides were carboxymethylated prior to electrophoresis on 5% polyacrylamide SDS gels. Estimated molecular weights: VPI = 29,800; VP2 = 32,000; VP3 = 47,100; VP4 = 51,800; VP5 = 54,000; VP6 = 56,600; VP7 = 60,600; VP8 = 63,600. (Reproduced with permission from Liebhaber and Gross, *Virology* **47**, 684–693, 1972.)

particle which sedimented at 150 S in sucrose density gradients and had a buoyant density of 1.28–1.31 g/cm^3. Only VP3 (35,000) remained associated with the core and was thus the RU virus nucleoprotein. VP3 was found to be relatively rich in arginine. Assuming uniform labeling of all polypeptides, Vaheri and Hovi (1972) calculated the VP1/VP2/VP3 molecular ratio in the virions to be 5:1:5. The three protein peaks reported by Vaheri and Hovi (1972) and the three regions reported by Liebhaber and Gross (1972) contain molecules of approximately the same molecular weight (see Fig. 3).

Bardeletti *et al.* (1975) also analyzed the proteins of BHK-21/13S-grown RU virus. These investigators used 7.5% acrylamide gels and either a high or a low concentration of the denaturing agents, urea and SDS. Comparison of the proteins observed with the two methods revealed that only one of the major components was common to both techniques. Two small polypeptides were only observed when the higher concentration of denaturing agents was used. Virus grown in RK-13 cells gave identical patterns to the BHK-21 virus. Major components with molecular weights of 58,000, 48,000, 31,000, and 16,500 and minor components of 75,000, 40,000, 26,000, and 10,000 were observed under conditions of a high concentration of denaturing agents. In contrast, major components of 63,000, 42,000, and 30,000 and minor components of 88,000, 76,000, 54,000, 36,000, and 24,000 were identified using a low concentration of denaturing agents. Again three main protein peaks were observed.

The composition of the gels used by the three laboratories for electrophoresis of RU viral proteins differed significantly. Gel composition has been shown to influence the apparent molecular weight calculated for viral polypeptides (Allison *et al.*, 1974). Although many of the differences in the polypeptides observed may be attributed to the different methods used for virus purification and to the conditions of electrophoresis, in no system was good resolution of the proteins achieved. It is not yet clear whether the minor components are distinct viral structural proteins, host proteins which remain associated with purified virus, or degradation products of viral structural proteins. In general, it appears that RU virions contain at least two glycoproteins and at least one nucleocapsid protein.

Rubella virions contain 0.03 mg of RNA and 0.245 mg of total lipids (of which 0.169 mg is phospholipids) per milligram of protein (Bardeletti *et al.*, 1975). Rubella virus infection produces a significant increase in the total lipid of BHK-21/13 S cells after infection. Although the total percentage of phospholipids decreased, an increase in the percentage of phosphatidylcholine and another as yet unidentified nonphosphorylated lipid was observed. A significant level of cardiolipin, a phospholipid usually considered specific to the inner mitochondrial membrane, was found in purified RU virus. This may indicate that some virion maturation can occur through mitochondrial membranes (Bardeletti and Gautheron, 1976).

2. *Cultivation*

Rubella virus was initially propagated in primary human (Weller and Neva, 1962) and monkey (Parkman *et al.*, 1962) tissues. Subsequently, it has been grown in a wide variety of cell lines, the most common one being BHK-21 (Beale *et al.*, 1963; McCarthy *et al.*, 1963; Vaheri *et al.*, 1965). Rubella virus can be assayed in tissue culture by plaque assay (Rhim *et al.*, 1967; Vaheri *et al.*, 1967), by immunofluorescence (Hobbins and Smith, 1968), by hemagglutination (Sedwick and Sokol, 1970), or by positive or negative hemadsorption (Schmidt *et al.*, 1969; Rawls *et al.*, 1967). Virus replication occurs only in the cytoplasm (Woods *et al.*, 1966). Studies of RU-infected Vero cells indicate that host cell protein synthesis is inhibited as viral protein synthesis increases. A decrease in the incorporation of labeled amino acids into acid-precipitable material and an increase in single ribosomes was observed 24–36 h after infection (Killian and Dorsett, 1974). Rubella virus matures by budding from cytoplasmic membranes. Mature virus particles are observed to accumulate in cisternae of the endoplasmic reticulum. Late in infection budding is sometimes observed to occur at the plasma membrane (Von Bonsdorff and Vaheri, 1969). Accumulation of cores was not observed in RU-infected cells, either in the cytoplasm or around vacuoles (Holmes *et al.*, 1969).

Rubella virus can interfere with the replication of several other viruses, such as NDV (Parkman *et al.*, 1962; Kleiman and Carver, 1977). One attenuated vaccine strain (RA27/3) and a *ts* mutant of RU virus have been found to be defective in their ability to interfere with NDV (Milfune and Matsuo, 1975).

In cytoplasmic extracts of BHK-21 cells infected with RU virus an infectious virus-specific single-stranded RNA was found which sedimented at 38–40 S. Virus-specific RNAs were also observed to sediment in a heterogeneous pattern in the 18–28 S region of the gradients (Sedwick and Sokol, 1970; Hovi and Vaheri, 1970b). This RNA was partially resistant to RNase treatment, and some of it may represent replicative intermediates. After RNase treatment, a component sedimenting at 18–20 S remained.

Hovi and Vaheri (1970b) reported that, when 0.3 μg/ml of actinomycin D was added to cultures at the time of infection with RU virus, the extracellular virus yield was reduced threefold. However, when actinomycin D (0.3 μg/ml) was added later than 15 h after infection, the production of virus was unaffected. Woods and Robbins (1968) and Sedwick and Sokol (1970) observed the same phenomenon. Sedwick and Sokol (1970) attributed this reduction in virus yield to the toxic effect of actinomycin D on the host cells. Trypan blue staining showed that the number of viable cells in uninfected cultures exposed to actinomycin D (1 or 2 μg/ml) for 30 h was only 20–40% of the untreated control.

3. *Pathogenicity*

Rubella virus usually causes a mild disease (German measles) in humans, which is characterized by a generalized rash, enlargement of lymph nodes, and a slight fever. Complications are rare. However, babies born to mothers infected with RU during the first 4 months of their pregnancy show a high incidence of congenital abnormalities (Hotchin, 1971). Inhibition of mitosis by RU virus infection has been demonstrated in a variety of embryonic tissues (Plotkin *et al.*, 1965; Rawls and Melnick, 1966) and the virus has been shown to be able to exert an inhibitory effect on the growth of human embryonic bone (Heggie, 1976).

Rubella virus has been recovered from infants born with the RU syndrome and also from normal infants whose mothers were infected in the later stages of pregnancy. Virus was obtained from pharyngeal secretions, cerebrospinal fluid, blood, and urine. In some cases, RU virus could be isolated from congenitally infected infants as long as 3 years after birth. Spontaneous carrier cultures can be readily derived from cultured postmortem tissues obtained from infants congenitally infected with RU virus. These chronically infected cells were found to have a reduced growth rate and a shortened life span. This carrier state was not dependent on serum inhibitors or RU antibodies for its maintenance. Anti-RU antibody treatment did not cure the cultures. All cells in the carrier population were found to be capable of producing virus. The persistence of virus in congenitally affected infants may well favor the survival of RU virus in nature (Alford *et al.*, 1964; Menser *et al.*, 1967; Monif *et al.*, 1965). (See Chapter 3.)

4. *Immunology*

Rubella virions contain a hemagglutinin. Agglutination is observed with 1-day-old chick red blood cells (RBCs), adult goose RBCs, and sheep RBCs under conditions similar to those used for togavirus hemagglutination (Stewart *et al.*, 1967; Furukawa *et al.*, 1967a;

Liebhaber, 1970). An early purification method for RU virus utilized binding and subsequent elution from erythrocytes (Holmes and Warburton, 1967). Hemagglutination inhibition tests have been employed to assess the possibility of an antigenic relationship of RU to more than 200 arboviruses, including members of the alpha- and flavivirus groups, as well as bunyaviruses (Mettler *et al.*, 1968). No positive reactions were observed between the RU immune sera and any of the arbovirus antigens, or between any of the arbovirus immune sera and the RU antigen preparations. This extensive study provides the best evidence that RU virus has no antigenic relationship to any known togaviruses.

Rubella virions are sensitive to immune lysis. The genome in immunologically damaged RU virus is susceptible to attack by RNase (Schluederberg *et al.*, 1976). Complement-fixing (CF) antigens have been found in supernatants from RU-infected cell cultures. One CF antigen is associated with purified RU virions, while another is soluble and appears in the culture fluid late in the course of infection (Furukawa *et al.*, 1967b). Bardeletti and Gautheron (1976) have reported that purified RU virus also has a weak neuraminidase activity which is Ca^{2+}-independent.

5. *Vaccines*

Attenuated vaccines have been developed in an effort to prevent RU virus-induced congenital disease. Infection with the vaccine strains induces antibody conversions in over 90% of the recipients. However, the resulting antibody titers are lower and of a more restricted range than those produced by a virulent virus strain (Horstmann *et al.*, 1970; LeBouvier and Plotkin, 1971), and the immunity to reinfection is limited (Abrutyn *et al.*, 1970).

IV. OTHER VIRUSES

A. Lactate Dehydrogenase-Elevating Virus

Lactate dehydrogenase-elevating virus replicates only in mice and primary cultures of mouse tissues. The amount of LDV in the blood during the initial viremia is unusually high; virus titers rise to 10^{10}–10^{11} 50% infectious doses per milliliter (ID_{50}/ml) of plasma by 14–24 h after infection (Riley, 1974; Notkins, 1965a). The virus titer subsequently drops until a stable level of about 10^5 ID_{50}/ml is reached approximately 8 weeks after infection. This level then persists for the lifetime of the mouse (Notkins and Shochat, 1963). Infected mice show no signs of illness and live a normal life span. The infection is characterized by a permanent 10-fold elevation in the plasma levels of lactate dehydrogenase (LDH) (Riley *et al.*, 1960) and of several other, but not all, plasma enzymes (Plagemann *et al.*, 1962). Evidence suggests that plasma LDH levels are elevated because of impaired functioning of the plasma enzyme clearance mechanism (Notkins, 1965a,b; Bailey *et al.*, 1964; Mahy *et al.*, 1965, 1967) and, possibly, also because of an increased influx of LDH into the bloodstreams of infected mice (Riley, 1968; Brinton and Plagemann, 1977).

Research conducted on LDV up to 1975 has been comprehensively reviewed by Rowson and Mahy (1975).

1. *Physicochemical Properties*

In glycerol gradients, LDV has been reported to have a sedimentation coefficient of about 200 S, while its nucleocapsid sediments at 176 S (Horzinek *et al.*, 1975). The density of LDV has been reported to be as low as 1.12 g/cm^3 and as high as 1.20 g/cm^3 (Rowson and Mahy, 1975). However, the most recent reports estimate the density to be about 1.13 to 1.14 g/cm^3 (Michaelides and Schlesinger, 1973; Horzinek *et al.*, 1975; Brinton-Darnell and Plagemann, 1975).

After treatment of LDV with nonionic detergent, a subviral particle was released which sedimented at a density of 1.17 g/cm^3. This was an unusually low density for a particle which appeared to be composed of the viral genome and the smallest polypeptide, VP1. The RNA within these particles was sensitive to digestion by RNase, and the particles were also sensitive to degradation by trypsin. The low density of this supposed LDV core particle was unexpected and presently cannot be explained (Brinton-Darnell and Plagemann, 1975).

Lability of the virion envelope is indicated by the extreme sensitivity of LDV to detergent treatment and by the fact that the virions tend to disintegrate upon preparation for electron microscopy. A brief incubation with 0.01% Nonidet P-40 or Triton X was sufficient to disrupt LDV, while this treatment had no effect on Sindbis virus (Brinton-Darnell and Plagemann, 1975).

a. **Morphology.** The diameter of LDV has been reported to range from 40 to 70 nm (see review in Rowson and Mahy, 1975). Several investigators have observed distorted particle shapes when unfixed LDV preparations are stained with phosphotungstic acid (PTA). This is especially true after virions have been purified on sucrose gradients (Brinton-Darnell and Plagemann, 1975; Horzinek *et al.*, 1975). The use of prefixed preparations has revealed that the LDV is spherical. Horzinek *et al.* (1975) reported that the diameter of the virion varied between 50 and 80 nm depending on the methods of fixation and staining employed. Brinton-Darnell and Plagemann (1975) observed virions with an average diameter of about 55 nm. The diameter of the nucleocapsid of LDV is approximately 30–35 nm (Horzinek *et al.*, 1975; Brinton-Darnell and Plagemann, 1975).

The surface of the LDV membrane appears to be free of projections (Horzinek *et al.*, 1975). Small, hollow spherules about 8–14 nm in diameter were observed on the surfaces of disintegrating virions (Brinton-Darnell and Plagemann, 1975) (see Chapter 8).

b. Genome. The genome of LDV is a single-stranded infectious 48 S RNA molecule with a molecular weight of approximately 5–6 × 10^6 (Notkins and Scheele, 1963; Darnell and Plagemann, 1972; Brinton-Darnell and Plagemann, 1975; Niwa *et al.*, 1973). Estimates of the size of the LDV RNA as determined by radiation target size (Rowson *et al.*, 1968), zonal sedimentation (Darnell and Plagemann, 1972), and gel electrophoresis (Niwa *et al.*, 1973) agree quite well.

c. Structural Proteins. Virions of LDV appear to have at least three structural proteins. The largest of these, VP3, is the only glycoprotein, while VP1 is the nucleocapsid protein. Both VP2, which is nonglycosylated, and VP3 are removed by detergent treatment. The molecular weights of the LDV proteins have been estimated by two laboratories. Using 15% acrylamide gels, Michaelides and Schlesinger (1973) calculated the molecular weight of VP1 to be 13,000, that of VP2 to 17,000, and that of VP3 to be 28,000. Brinton-Darnell and Plagemann (1975) estimated molecular weights of 15,000, 18,000, and 24,000–44,000 for VP1, VP2, and VP3, respectively, on 7.5% acrylamide gels (Fig. 4). The molecular weight of VP3 could not be estimated with accuracy because it always appeared as a heterogeneous peak. In contrast, the glycoprotein of the flavivirus, West Nile virus, migrated as a homogeneous band under similar conditions of preparation and analysis (Brinton-Darnell and Plagemann, 1975).

The molar ratio of the three proteins within LDV virions has been estimated to be 3–5:1:1 by Michaelides and Schlesinger (1973) and 2–3:1:1 by Brinton-Darnell and Plagemann (1975).

The carbohydrate moiety of the LDV glycoprotein (VP3) appears to be located on the outer surface of the viral envelope, even though no surface spikes are observed in the electron microscope, since LDV can be precipitated by concanavalin A.

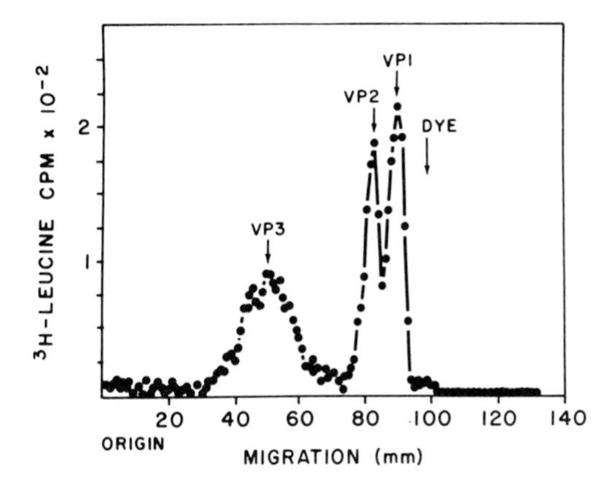

Fig. 4. Structural proteins of LDV. Electrophoresis pattern of [^{3}H]leucine-labeled LDV propagated in primary mouse macrophages. Gels were 7.5% polyacrylamide and contained SDS. VP, Viral protein. Estimated molecular weights: VP1 = 15,000; VP2 = 18,000; VP3 = 24,000–44,000. (Reproduced with permission from Brinton-Darnell and Plagemann, *J. Virol.* **16**, 420–433, 1975.)

2. *Cultivation*

Lactate dehydrogenase virus replication has been detected only in primary mouse cell cultures which contain phagocytic cells of the reticuloendothelial system (DuBuy and Johnson, 1968; Snodgrass *et al.*, 1972; Brinton-Darnell *et al.*, 1975). The restricted host range of LDV appears not to be due to its inability to adsorb to cells other than macrophages or to cells from other species. Even LDV pseudotypes (LDV nucleocapsids surrounded by Sindbis envelopes) were incapable of inducing a productive infection in mouse L cells or in chicken embryo fibroblasts (Lagwinska *et al.*, 1975).

Electron microscopic observation of thin sections of infected macrophages showed that LDV matured by budding from the cytoplasm into intracytoplasmic vesicles. Free cores were not observed around vesicles or in the cytoplasm (Brinton-Darnell *et al.*, 1975).

Studies on the synthesis of LDV RNA in macrophage cultures in the presence of actinomycin D showed that 48 S viral RNA could first be detected intracellularly between 4 and 5 h after infection (Brinton-Darnell *et al.*, 1975). The time period required between the completion of a viral genome and its appearance in mature extracellular virions is about 1.5 h. Heterogeneous RNA sedimenting between 10 and 40 S was detected in infected cells by 3 h after infection. This RNA was found to be partially resistant to RNase treatment. After

RNase treatment, the predominant species of remaining RNA sedimented at approximately 27 S. Although this heterogeneous RNA may represent degradation products of viral RNA, the partial RNase resistance indicates that viral replicative intermediates may be present in this region of the gradient.

The synthesis of intracellular viral RNA was completely inhibited by the addition of cycloheximide to the culture medium 0.5–1 h after infection of cells, and no virion-associated RNA polymerase activity was detectable in preparations of partially purified virions.

Yamazaki and Notkins (1973) reported that the 24-h yield of infectious LDV from primary mouse embryo cultures was reduced by 75–90% when actinomycin D was added to the culture media during the first 9 h of infection. The rapid toxic effect of actinomycin D on macrophage function and integrity indicates that the observed reduction in virus replication may not be specific, but a reflection of a general cytotoxicity (Brinton and Plagemann, 1980).

Infection by LDV causes no detectable CPE, and the virions do not agglutinate RBCs (Rowson and Mahy, 1975). The only assay method currently available utilizes the increase in plasma LDH levels in infected mice as an indication of infection (Riley *et al.*, 1960; Plagemann *et al.*, 1962). The ability of LDV to induce interferon in macrophage cultures may provide another method for assaying LDV in the future (Lagwinska *et al.*, 1975).

3. *Immunology and Pathology*

Infection of mice with LDV has been reported to cause subtle alterations in the immune response, such as stimulation of the production of humoral antibody (Notkins *et al.*, 1966b; Mergenhagen *et al.*, 1967) and depression of cell-mediated immunity (Howard *et al.*, 1969; Snodgrass *et al.*, 1972). Several investigators have reported suppression of the growth of certain mouse tumors by LDV infection (Notkins, 1965a; Riley, 1966, 1968; Bailey *et al.*, 1965; Gregory *et al.*, 1965; Brinton-Darnell and Brand, 1977), but LDV-induced tumor enhancement and normal tumor growth have also been observed (Plagemann and Swim, 1966; Michaelides and Schlesinger, 1974). The timing of the LDV infection and the inoculation of tumor cells have been shown to be critical in determining the effect of LDV on the growth of the tumor (Michaelides and Schlesinger, 1974). During the acute phase of LDV infection, a depression of cell-mediated immunity was observed, while during the early phase of the chronic stage of infection a transient enhancement of cell-mediated immunity occurred. Degeneration of T-cell-dependent areas was observed during the first 4 days after infection (Snodgrass *et al.*, 1972; Michaelides and Schlesinger, 1974). By the seventh day, however, repopulation of lymphocytes had begun and a transient overpopulation seemed to occur (Michaelides and Schlesinger, 1974).

Mice display a normal antibody response to LDV beginning 7 days after infection (Notkins and Shochat, 1963; Porter *et al.*, 1969; Evans and Riley, 1968). During the lifelong viremia, virus is present as an infectious complex with antibody (Notkins *et al.*, 1966a, 1968). Although significant numbers of immune complex deposits are found in glomeruli of infected mice, only mild lesions are observed and immune complex glomerulonephritis does not develop (Porter and Porter, 1971). Recently it has been shown that C58 mice, and to a lesser degree AKR mice, which have been immunosuppressed by the aging process, by irradiation, or by cyclophosphamide treatment, develop a polioencephalitis after infection with LDV (Martinez *et al.*, 1980).

B. Equine Arteritis Virus

Equine arteritis virus received its name from the observation that infection of horses with this virus is characterized by medial necrosis of small muscular arteries (Jones *et al.*, 1957).

1. *Physicochemical Properties*

The sedimentation coefficient of EAV has been calculated to be approximately 224 S ± 8 (van der Zeijst *et al.*, 1975), while EAV nucleocapsids sediment at 158 S (Zeegers *et al.*, 1976). Previous estimates of the buoyant density of EAV ranged between 1.17 and 1.18 g/cm^3 (Hyllseth, 1970). More recently, van der Zeijst *et al.* (1975) reported that, although they observed a small peak of infectivity and radioactivity in sucrose gradients at a density of 1.17 g/cm^3, the main peak of virus banded at a density of 1.155 g/cm^3. In all cases, a sharp, symmetrical peak of radioactivity and infectivity was observed.

a. Morphology. Equine arteritis virions have been found to have an average diameter of 60 ± 13 nm, with an isometric core structure measuring 35 nm across (Horzinek *et al.*, 1971). Virus particles are roughly spherical. No prominent projections were observed on their surfaces, but ringlike structures, 12–14 nm in diameter, were observed on the surface of some virions and attached to some detergent-released cores (Horzinek *et al.*, 1971; Hyllseth, 1973) (see Chapter 8).

b. Genome. Radwan *et al.* (1973) have reported that [^{3}H]uridine is incorporated into the genome of EAV, indicating that the nucleic acid of this virus is an RNA molecule. Further studies by van der Zeijst *et al.* (1975) have confirmed this conclusion. The EAV genome is an infectious, single-stranded RNA with a molecular weight of 4×10^6. The molecular weight was calculated both from acrylamide–agarose gel migration data and sucrose gradient sedimentation data. A pronounced degree of secondary structure was indicated by variation in the sedimentation coefficient in sucrose gradients of different ionic strength and in formaldehyde sucrose gradients. An $s_{20,w}$ value of 48 S was calculated after sedimentation through isokinetic sucrose gradients made with 0.1 *M* saline. The single-stranded nature of the RNA indicated by its RNase sensitivity was confirmed by its density of 1.65 g/cm^3, which was measured after centrifugation in Cs_2SO_4 gradients. Heating in the presence of formaldehyde, followed by centrifugation in formaldehyde-containing gradients, demonstrated that the EAV genome was a single continuous strand of RNA. Semliki Forest virus (SFV) RNA was used in the studies described above as a control togavirus RNA. RNA from EAV and SFV showed identical sedimentation coefficients in the various types of gradients employed, suggesting that they were molecules of the same molecular weight with comparable amounts of secondary structure.

c. Structural Proteins. The analysis of the proteins of EAV by polyacrylamide gel electrophoresis has so far been reported by two laboratories (Hyllseth, 1973; Zeegers *et al.*, 1976). Radioactively labeled EAV grown in BHK-21 cells and then concentrated and purified by a five-step process, including pelleting, rate zonal centrifugation, and two sequential isopycnic centrifugations, yielded nine virus-associated proteins (Hyllseth, 1973). Two of these proteins, VP7 (15,000 daltons) and VP8 (13,000 daltons), contained a major proportion of the label. Disruption of virions by detergent treatment revealed that only VP8 remained associated with the core, and thus it was designated the nucleoprotein of EAV. Six of the proteins—VP1 (72,000), VP2 (55,000), VP3 (40,000), VP4 (36,000), VP5 (32,000), and VP6 (28,000)—incorporated [^{14}C]glucosamine and thus are glycoproteins (Fig. 5). Some variation in VP3, VP4, VP5, and VP6 was observed in different runs.

Subsequently, Zeegers *et al.* (1976) compared EAV-associated proteins in virus pools grown in three different types of cells, Vero, BHK-21, and RK-13. Only three proteins were common to the virus preparations from all three cell lines. These proteins were designated

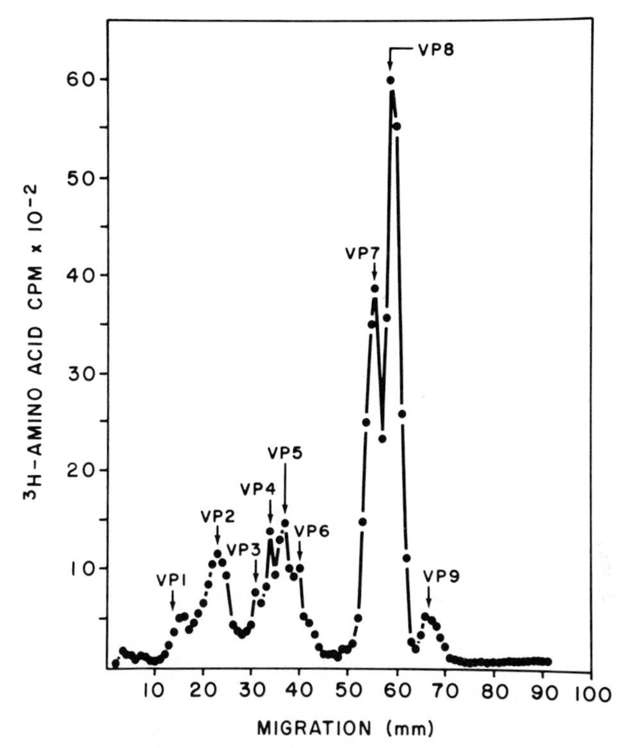

Fig. 5. Structural proteins of EAV. Polypeptide pattern of BHK-21/13S-grown [^{3}H]-leucine-labeled EAV (Bucyrus strain) electrophoresed on 10% polyacrylamide SDS-containing gels. VP, Viral protein. Estimated molecular weights: VP1 = 72,000; VP2 = 55,000; VP3 = 40,000; VP4 = 36,000; VP5 = 32,000; VP6 = 28,000; VP7 = 15,000; VP8 = 13,000; VP9 = 10,500. (Reproduced with permission from Hyllseth, *Arch. Gesamte Virusforsch.* **40**, 177–188, 1973.)

VP1, VP2, and VP3, and their molecular weights were calculated to be 12,000, 14,000, and 21,000, respectively. When comparable labeling efficiencies were assumed for each, the VP1/VP2/VP3 molar ratio was estimated to be 27:5:1. VP1 was found to be the core protein. This protein is phosphorylated, and its molecular weight is close to that calculated by Hyllseth (1973) for the nucleocapsid protein VP8 (13,000). The 14,000 (VP2) protein of Zeegers *et al.* (1976) seems very similar to the 15,000 (VP7) of Hyllseth (1973). This protein is removed with the envelope by detergent treatment but does not incorporate glucosamine. Zeegers *et al.* (1976) found that only one of the glycoproteins, VP3 (21,000), was present in the virus preparations from all three cell systems. However, the amount of VP3 seems to be too low for it to be the only glycoprotein present in the virion. The glycoproteins VP3, VP4, VP5, and VP6 observed by Hyllseth (1973) were not well resolved and migrated as a "complex" over the molecular weight range of 28,000–40,000. In light of the data reported by Zeegers *et al.* (1976), it seems that some of the glycoproteins observed by Hyllseth (1973) might be cellular contaminants. However, it is interesting that these contaminants consistently remain associated with EAV even after a number of sequential gradient centrifugations. The acrylamide gel pattern of EAV proteins is strikingly similar to that of LDV proteins (compare Figs. 4 and 5).

2. *Cultivation*

Replication of EAV occurs in primary cultures of horse kidney, rabbit kidney, and hamster kidney cells and also in such cell lines as BHK-21, RK-13, and Vero (Konishi *et al.*, 1975; Zeegers *et al.*, 1976). It produces plaques in BHK-21 cells (Hyllseth, 1969). In Vero cells, intracellular viral RNA synthesis begins at about 4 h after infection, and progeny virus starts to be released between 8 and 10 h after infection. Immunofluorescent studies revealed that viral antigen first appeared as a small number of granules in the perinuclear region and increased during the course of the infection filling the cytoplasm (Inoue *et al.*, 1975; Breese and McCollum, 1971).

Equine arteritis virus particles mature by budding into cisternae of the endoplasmic reticulum (Magnusson *et al.*, 1970).

3. *Pathogenesis*

Equine arteritis virus is extremely contagious for horses under natural conditions. It usually infects young animals and probably enters the body through the respiratory tract. The disease it causes is characterized by fever, conjunctivitis, rhinitis, edema of the legs, enteritis, colitis, and necrosis of small arteries. This virus is one cause of abortions in mares; it spreads from the infected mare to the fetus. The mare has an inapparent infection, while the fetus develops a fatal infection. Young animals with fatal cases develop bronchopneumonia.

C. Simian Hemorrhagic Fever Virus

To date, SHF virus has not been found to cross-react serologically with any of the arboviruses, not even with those known to cause a hemorrhagic disease (Casals, 1971) (cf. Chapter 3).

1. *Physicochemical Properties*

Virions are enveloped and pH sensitive. The buoyant density of SHF virus has been estimated to be 1.19–1.20 g/cm^3 in sucrose density gradients and 1.18 g/cm^3 in potassium tartrate gradients (Wood *et al.*, 1970; Trousdale *et al.*, 1975).

a. Morphology. Virions were observed to be spherical. Their outer diameter is 45–50 nm, with an average diameter of 48 nm. The nucleocapsids have an average diameter of 25 nm (Wood *et al.*, 1970; Trousdale *et al.*, 1975).

b. Genome. That the genome of SHF virus is an RNA molecule was first indicated by studies with metabolic inhibitors (Tauraso *et al.*, 1968; Wood *et al.*, 1970). RNA from SHF virus is sensitive to degradation by RNase (Trousdale *et al.*, 1975). When purified, ^{32}P-labeled virions were extracted with phenol–SDS and the nucleic acid analyzed by polyacrylamide gel electrophoresis, a broad peak of radioactivity estimated to be 15–30 S was observed. Semliki Forest virus 43 S RNA was used as a marker. The broadness of the SHF RNA peak suggests that partial degradation of the RNA may have occurred during preparation.

c. Structural Proteins. Preliminary characterization of SHF proteins has been reported by Trousdale *et al.* (1975). Four peaks of radioactivity were observed after gel electrophoresis of purified SHF virus, which had been doubly labeled with ^{14}C-labeled amino acids and [^{3}H]glucosamine (Fig. 6). Structural protein (SP) peak 1 was found to incorporate glucosamine. A small amount of glucosamine also appeared to be incorporated into SP4. Peak SP2 was present in a low concentration as compared to the other three proteins. No molecular-weight estimates were reported, nor was any indication given of whether SP3 or SP4 represented the SHF virus nucleoprotein.

2. *Cultivation*

The host range of SHF virus is limited. Only simian species and two tissue culture lines derived from simians are susceptible to SHF virus infection. The virus was first grown successfully in the embryonic rhesus monkey cell line MA-104. After six serial passages in MA-104 cells, the virus was adapted to grow in an African green monkey kidney cell line, BSC01 (Wood *et al.*, 1970).

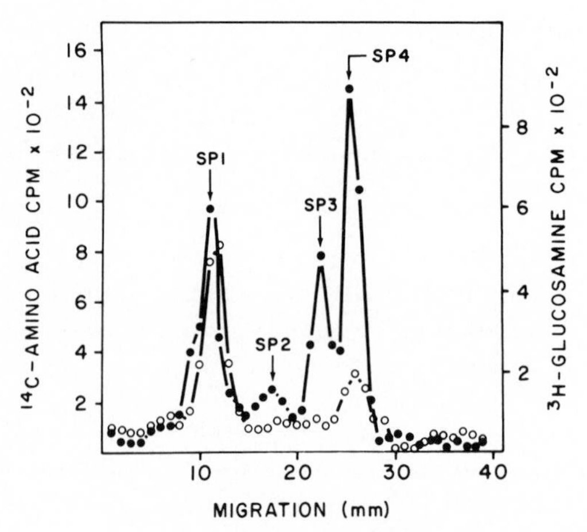

Fig. 6. Structural proteins of SHF. Electrophoresis pattern of SHF virus propagated in MA-104 cells and doubly labeled with ^{14}C-labeled amino acids (solid circles) and [^{3}H]glucosamine (open circles). Gels were 10% polyacrylamide and contained SDS. SP, Structural protein. Molecular weights not estimated. (Reproduced with permission from Trousdale *et al.*, *Proc. Soc. Exp. Biol. Med.* **150**, 701–711, 1975.)

Infected cells were reported to contain vacuoles with accumulations of enveloped virions. Corelike particles were observed budding from the cytoplasm into the vacuoles. The virions contain a high percentage of phosphatidylcholine, which also suggests that the viral envelope originates in the internal membranes of the host cell (Renkonen *et al.*, 1971). Infected cells contain proliferated cytoplasmic membranes which appear as broad, undulating lamellae. The surfaces of these lamellae are sometimes surrounded with electron-dense particles, the largest of which are about the size of viral cores (Wood *et al.*, 1970).

3. *Pathogenesis*

The disease picture observed among the various species of monkeys is very interesting. Rhesus monkeys (*Macaca mulatta*) and all other macaques develop a rapid, fatal hemorrhagic disease after SHF virus infection, while patas monkeys (*Erythrocebus patas*) develop a chronic infection and show no signs of disease (London, 1977). The hemorrhagic disease is characterized by fever, mild facial edema, dehydration, hemorrhages of the skin, bloody diarrhea, and death. There is also a generalized destruction of lymphoid tissues (Allen *et al.*, 1968; Abildgaard *et al.*, 1975). The infection spreads rapidly among rhesus monkeys by aerosol and by direct contact.

Chronically infected patas monkeys have very high levels of virus in their blood and tissues. A concentration as high as 10^{12} ID_{50}/ml of blood has been observed. Baboons (*Papio papio*) and African green monkeys (*Arcopithecus aethiops*) were also found to be able to support latent SHF infections. Transmission of infection from chronically infected monkeys to susceptible rhesus monkeys does not occur by contact or by aerosol. The epizootic occurring in the NIH rhesus colony in 1972 was inadvertently initiated by mechanical transfer of blood from a chronically infected patas monkey to a rhesus monkey (London, 1977).

D. Cell-Fusing Agent

The CFA appears to be a virus which replicates only in mosquito cells. Attempts to transmit this virus to a vertebrate host have not been successful.

1. *Physicochemical Properties*

In sucrose–D_2O density gradients the CFA banded at a density of 1.195 g/cm^3. Its sedimentation coefficient was estimated to be approximately 205–210 S on linear sucrose gradients (Igarashi *et al.*, 1976).

a. Morphology. The virus is spherical with an outer diameter between 45 and 52 nm. The core of the virus measures 23–28 nm. Spikes were resolved on the surfaces of the virions (Stollar and Thomas, 1975) (see Chapter 8).

b. Genome. The CFA contains a genome of single-stranded RNA with an *s* value of about 40–42 S. Analysis of infected cell extracts indicated that a 40–42 S single-stranded RNA and a 22 S double-stranded RNA were the predominant species found. A single-stranded peak at 26 S was not observed (Igarashi *et al.*, 1976).

c. Structural Proteins. Virions contain three structural proteins, VP1, VP2, and VP3, with molecular weights of 13,000, 16,500, and 49,000, respectively (Igarashi *et al.*, 1976). VP2 and VP3 are glycoproteins associated with the viral envelope, while VP1 is the nucleocapsid protein (Fig. 7).

2. *Cultivation*

This virus was isolated from the medium of an *A. aegypti* cell line (Peleg, 1968). Infection of *Aedes albopictus* cells with the CFA leads to the formation of syncytia which appear approximately 48–72 h after infection. However, under conditions of low Ca^{2+} and high cell den-

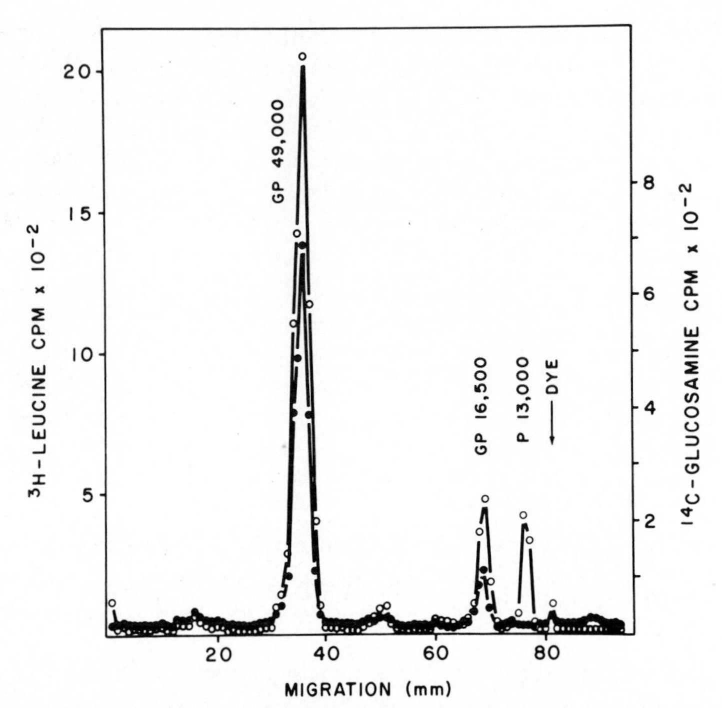

Fig. 7. Structural proteins of the CFA. Electrophoresis pattern of the CFA grown in *A. albopictus* cells labeled with either [^{3}H]leucine (open circles) or with [^{14}C]-glucosamine (solid circles) on a discontinuous SDS polyacrylamide gel composed of a 10% resolving gel and a 5% stacking gel. GP, Glycoprotein; P, polypeptide. Estimated molecular weights given in figure. (Reproduced with permission from Igarashi *et al.*, *Virology* **74**, 174–187, 1976.)

sity, cell fusion is inhibited, while virus yields are unaffected (Igarashi *et al.*, 1976).

The virus matures by budding through intracytoplasmic membranes. Areas in the cytoplasm were seen where cores seemed to be assembled. Cores were also observed to accumulate along membranes of the endoplasmic reticulum. Virions bud through these membranes into the cisternal space (Igarashi *et al.*, 1976).

3. *Immunology*

The CFA showed no serological cross-reaction with a number of flaviviruses, even though in its size and morphology it appears to be similar to flaviviruses.

The presence of a hemagglutinin on the virion could not be demonstrated (Igarashi *et al.*, 1976).

V. NON-ARBO FLAVIVIRUSES

A number of viruses have been isolated from bats and rodents which are antigenically related to flaviviruses but which do not appear to have a biological cycle in arthropods.

Rio Bravo virus or bat salivary gland virus (BSGV) has been isolated from bats in California, Texas, and Mexico (Burns and Farinacci, 1956; Baer and Woodall, 1966). Infection rates in nine groups of 50–125 adult bats (*Tadarida brasiliensis mexicana*) varied from 0 to 7%. No evidence of disease has been found in bats infected with BSGV. Virus has been demonstrated to be present only in saliva and salivary glands (Baer and Woodall, 1966). The virus causes paralysis and death in newborn mice regardless of the route of inoculation and in adult mice after intracerebral inoculation. The virus does not kill adult mice after intraperitoneal injection, but it can be recovered from the salivary glands, kidneys, and mammary glands of these mice. Monkeys and man can be infected with BSGV, and a brief febrile disease is observed (Sulkin *et al.*, 1962). Bat salivary gland virus failed to replicate in three genera of mosquitoes experimentally exposed to it (Sulkin *et al.*, 1962).

Bat salivary gland virus grows in HeLa, hamster kidney, chick embryo, human diploid, and calf kidney cultures. The virus can be titered by plaque assay in Vero and LLC-MK2 cells (Stim, 1969), has a hemagglutinin, and is rapidly inactivated by low pH (Wilhelm and Gerone, 1970).

Four further bat viruses, Dakar bat virus (Brès and Chambon, 1964), Entebbe bat virus (Lumsden *et al.*, 1961), MMLV, and Phnom-Penh bat virus, have been isolated. All four show a serological similarity to flaviviruses. These viruses cross-react with each other but are not identical. Filtration studies on MMLV indicate a virion size between 30 and 100 nm (Bell and Thomas, 1964). These non-arbo flaviviruses do contain a hemagglutinin (Berge, 1975).

The transmission of MMLV to a mouse by a bat bite, the isolation of virus from bat saliva, and the chronic nature of the infection in bats suggest that arthropod parasites are not essential to the maintenance of MMLV in nature (Williams *et al.*, 1964). Phnom-Penh bat virus was not isolated from any of a large number of mosquitoes and other arthropods caught in the region of Cambodia where the virus was isolated from bats. Also, in the laboratory, *A. aegypti* and *Culex pipiens* which fed on infected suckling mice showed a titer of 10^8 LD_{50} soon after feeding, but no virus could be isolated from these mosquitoes 5, 10, or 15 days later (Salaün *et al.*, 1975; Berge, 1975).

The bat viruses produce a fatal disease in mice after an experimental infection, and they can be cultivated in a number of cell lines such as LLC-MK2, Vero, and BHK-21 (Berge, 1975).

Modoc virus was recovered from the mammary glands and from throat swabs of mice (*Peromyscus manicatus*) in California. This virus cross-reacts serologically with flaviviruses and with Rio Bravo bat virus. It has not yet been tested against the other bat viruses (Casals, 1960; Johnson, 1970). Modoc virus replicates in a number of laboratory cell lines and produces plaques in Vero and LLC-MK2 cells (Stim, 1969).

Cowbone Ridge virus was isolated from a cotton rat (*Sigmodon hispidus*) in Florida. This virus showed serological relatedness to a number of flaviviruses, Modoc virus, and Rio Bravo virus. The diameter of this virus has been estimated to be about 38 nm by electron microscopy of thin sections of infected mouse brains (Calisher *et al.*, 1969). Virion accumulation was observed in the lumens of the endoplasmic reticulum of infected cells. CRV grows in a variety of cell lines and plaques in L cells. Only rodents have so far been found to be susceptible to infection with CRV.

Apoi virus, Jutiapa virus, and Saboya virus isolated from rodents, Sokuluk virus, Carey Island virus isolated from bats, and Israel turkey meningoencephalitis virus may also be non-arbo flaviviruses. They are newly identified members of the flavivirus group, but it is not presently known whether any of these viruses multiply in nature in an arthropod vector (Berge, 1975) (see Chapter 2).

VI. DISCUSSION

The viruses discussed in this chapter have been designated togaviruses on the basis of their small size, spherical shape, envelope, and nucleic acid (Fenner, 1977; Horzinek, 1973). Members of the *Alphavirus* and *Flavivirus* genera are antigenically related to the other members of their particular genus. Alphaviruses bud preferentially from the marginal membrane of the infected cell, whereas flavivirus assembly seems to be associated with internal membranes. Naked nucleocapsids are observed to accumulate around vacuoles in the cytoplasm of alphavirus-infected cells, while in flavivirus-infected cells no such accumulation occurs. Alphaviruses have one or more envelope glycoprotein(s) and a single nonglycosylated nucleocapsid protein. Flaviviruses contain a single envelope glycoprotein, a nonglycosylated membrane-associated protein, and a nucleocapsid protein. Alphaviruses have an average virion diameter of 55–60 nm, while flaviviruses have an average diameter of about 48–50 nm. Alpha- and flaviviruses contain a hemagglutinin and show positive agglutination with erythrocytes from a number of animal species over a narrow pH range. Treatment of these viruses with proteases decreases their hemagglutinating activity and produces virions which appear to have smooth surfaces in the electron microscope (Compans, 1971). Non-arbo togaviruses, LDV, EAV, BVDV, and HCV, have a smooth surface and do not show any hemagglutinating activity.

With the exception of non-arbo flaviviruses, non-arbo togaviruses show no antigenic cross-reaction with any of the members of either the *Alphavirus* or the *Flavivirus* genera. However, the three pestiviruses are antigenically related to each other. Rubella virus, LDV, SHFV, EAV, CFA, and the pestiviruses each share some characteristics with flaviviruses and some with alphaviruses, but none of them fit into either of these genera (see Table I). Unfortunately, study of the physicochemical parameters of most non-arbo togaviruses is still incomplete. In part, this has been due to the technical difficulties encountered during attempts to analyze the characteristics of these viruses. In view of the relatively small number of currently known togaviruses which are neither alpha- nor flaviviruses, it seems premature to speculate on the number of genera that will eventually constitute the togavirus family.

REFERENCES

Aaslestad, H. G., Hoffman, E. J., and Brown, A. (1968). *J. Virol.* **2**, 972–978.

Abildgaard, C., Harrison, J., Espana, C., Spangler, W., and Gribble, D. (1975). *Am. J. Trop. Med. Hyg.* **24**, 537–544.

Abrutyn, E., Herrmann, K. L., Karchmer, A. W., Friedman, J. P., Page, E., and Witte, J. J. (1970). *Am. J. Dis. Child.* **120**, 129–133.

Alford, C. A., Neva, F. A., and Weller, T. H. (1964). *N. Engl. J. Med.* **271**, 1275–1281.

Allen, A. M., Palmer, A. E., Tauraso, N. M., and Shelokov, A. (1968). *Am. J. Trop. Med. Hyg.* **17**, 413–421.

Allison, J. H., Agrawal, H. C., and Moore, B. W. (1974). *Anal. Biochem.* **58**, 592–601.

Amstey, M. S., Hobbins, T. E., and Parkman, P. D. (1968). *Proc. Soc. Exp. Biol. Med.* **127**, 1231–1236.

Andrewes, C., and Pereira, H. G. (1972). "Viruses of Vertebrates," pp. 260–277. Williams & Wilkins, Baltimore, Maryland.

Archbald, L. F. (1974). *Vet. Med. Small Anim. Clin.* Dec., pp. 1540–1541.

Aynaud, J. M. (1968). *Rech. Vet.* **1**, 25–36.

Bachmann, P. A., Sheffy, B. E., and Siegl, G. (1967). *Arch. Gesamte Virusforsch.* **22**, 467–471.

Baer, G. M., and Woodall, D. F. (1966). *Am. J. Trop. Med. Hyg.* **15**, 769–771.

Bailey, J. M., Clough, J., and Stearman, M. (1964). *Proc. Soc. Exp. Biol. Med.* **117,** 350–354.
Bailey, J. M., Clough, J., and Lohaus, A. (1965). *Proc. Soc. Exp. Biol. Med.* **119,** 1200–1204.
Baker, J. A., and Sheffy, B. E. (1960). *Proc. Soc. Exp. Biol. Med.* **105,** 675–678.
Baker, J. A., Yorke, C. J., Gillespie, J. H., and Mitchell, G. B. (1954). *Am. J. Vet. Res.* **15,** 525–531.
Bardeletti, G., and Gautheron, D. C. (1976). *Arch. Virol.* **52,** 19–27.
Bardeletti, G., Kessler, N., and Aymard-Henry, M. (1975). *Arch. Virol.* **49,** 175–186.
Beale, A. J., Christofinis, G. C., and Furminger, I. G. S. (1963). *Lancet* **ii,** 640–641.
Bell, J. F., and Thomas, L. A. (1964). *Am. J. Trop. Med. Hyg.* **13,** 607–612.
Berge, T. O., ed. (1975). "International Catalogue of Arboviruses Including Certain Other Viruses of Vertebrates," DHEW Publ. No. 11 (CDC) 75–8301. U.S. Gov. Print. Off., Washington, D.C.
Bodon, L. (1965). *Acta Vet. Hung.* **15,** 471–472.
Breese, S. S., Jr., and McCollum, W. H. (1971). *Arch. Gesamte Virusforsch.* **35,** 290–295.
Brès, P., and Chambon, L. (1964). *Ann. Inst. Pasteur, Paris* **107,** 34–43.
Brinton, M. A., and Plagemann, P. G. W. (1977). *Abstr., Am. Soc. Microbiol., Annu. Meet., New Orleans* p. 327.
Brinton, M. A., and Plagemann, P. G. W. (1980). *Intervirology* **12,** 349–356.
Brinton-Darnell, M., and Brand, I. (1977). *J. Natl. Cancer Inst.* **59,** 1027–1029.
Brinton-Darnell, M., and Plagemann, P. G. W. (1975). *J. Virol.* **16,** 420–433.
Brinton-Darnell, M., Collins, J. K., and Plagemann, P. G. W. (1975). *Virology* **65,** 187–195.
Brodersen, M., and Thomssen, R. (1969). *Arch. Gesamte Virusforsch.* **26,** 118–126.
Brown, T. T., DeLahunta, A., Bistner, S. I., Scott, F. W., and McEntee, K. (1974). *Vet. Pathol.* **11,** 486–505.
Brown, T. T., Bistner, S. I., DeLahunta, A., Scott, F. W., and McEntee, K. (1975). *Vet. Pathol.* **12,** 394–404.
Burns, K. F., and Farinacci, C. J. (1956). *Science* **123,** 227–228.
Calisher, C. H., Davie, J., Coleman, P. H., Lord, R. D., and Work, T. H. (1969). *Am. J. Epidemiol.* **89,** 211–216.
Carbrey, E. A., and Kresse, J. I. (1967). *Proc. U.S. Livestock Sanit. Assoc., Annu. Meet., 71st, Phoenix, Ariz.* p. 335.
Carbrey, E. A., Brown, L. N., Chow, T. L., Kahrs, R. F., McKercher, D. G., Smithies, L. K., and Tamoglia, T. W. (1971). *Proc. U.S. Anim. Health Assoc.* **75,** 629–648.
Casals, J. (1960). *Can. Med. Assoc. J.* **82,** 355–358.
Casals, J. (1971). "Marburg Virus Disease," pp. 98–104. Springer-Verlag, Berlin and New York.
Castrucci, G., Cilli, V., and Gagliardi, G. (1968). *Arch. Gesamte Virusforsch.* **24,** 48–64.
Castrucci, G., Titoli, F., Ranucci, S., Castro Portugal, F. L., Cilli, V., and Pedini, B. (1974). *Boll. Ist. Sieroter. Milan.* **53,** 585–591.
Chagnon, A., and Laflamme, P. (1964). *Can. J. Microbiol.* **10,** 501–502.
Compans, R. W. (1971). *Nature (London), New Biol.* **229,** 114–116.
Cowart, W. O., and Morehouse, L. G. (1967). *J. Am. Vet. Med. Assoc.* **151,** 1788–1794.
Cunliffe, H. R., and Rebers, P. A. (1968). *Can. J. Comp. Med. Vet. Sci.* **32,** 409–411.
Dalsgaard, K., and Overby, E. (1976). *Acta Vet. Scand.* **17,** 465–474.
Danner, K., and Bachmann, P. A. (1970). *Zentralbl. Veterinaermed.* **17,** 353–362.
Darnell, M. A., and Plagemann, P. G. W. (1972). *J. Virol.* **10,** 1082–1085.
Diderholm, H., and Dinter, Z. (1966). *Proc. Soc. Exp. Biol. Med.* **121,** 976–980.

Diderholm, H., Klingeborn, B., and Dinter, Z. (1974). *Arch. Gesamte Virusforsch.* **45**, 169–172.
Dinter, Z. (1963). *Zentralbl. Bakteriol., Parasitenkd., Infektionskr. Hyg., Abt. 1: Orig.* **188**, 475–486.
Dinter, Z., and Diderholm, H. (1972). *Arch. Gesamte Virusforsch.* **38**, 274–277.
Dinter, Z., Hansen, H. J., and Ronéus, O. (1962). *Zentralbl. Veterinaermed.* **9**, 739–747.
Ditchfield, J., and Doane, F. W. (1964). *Can. J. Comp. Med.* **28**, 148–152.
DuBuy, H. G., and Johnson, M. L. (1968). *Proc. Soc. Exp. Biol. Med.* **128**, 1210–1214.
Dunne, H. W. (1958). "Diseases of Swine," 4th ed., pp. 177–239. Iowa State Coll. Press, Ames.
Dunne, H. W., and Clark, C. D. (1968). *Am. J. Vet. Res.* **29**, 787–797.
Dunne, H. W., Reich, C. V., Hokanson, J. F., and Lindstrom, E. S. (1955). *Proc. Am. Vet. Med. Assoc., Minneapolis, Minn., Annu. Meet., 92nd,* p. 148.
Edward, A. G., Calhoon, J. R., and Mills, G. D. (1967). *Lab. Anim. Care* **17**, 102–107.
Enzmann, P. J., and Rehberg, H. (1977). *Z. Naturforsch. Teil C* **32**, 456–458.
Enzmann, P. J., and Weiland, F. (1978). *Arch. Virol.* **57**, 339–348.
Evans, R., and Riley, V. (1968). *J. Gen. Virol.* **3**, 449–452.
Fenner, F. (1977). *Intervirology* **7**, 44–47.
Fernelius, A. L. (1968a). *Arch. Gesamte Virusforsch.* **25**, 211–218.
Fernelius, A. L. (1968b). *Arch. Gesamte Virusforsch.* **25**, 219–226.
Fernelius, A. L., Lambert, G., and Hemness, G. J. (1969a). *Am. J. Vet. Res.* **30**, 1561–1572.
Fernelius, A. L., Lambert, G., and Packer, R. A. (1969b). *Am. J. Vet. Res.* **30**, 1541–1550.
Fernelius, A. L., Amtower, W. C., Lambert, G., McClurkin, A. W., and Matthews, P. J. (1973). *Can. J. Comp. Med.* **37**, 13–20.
French, E. L., and Snowdon, W. A. (1964). *Aust. Vet. J.* **40**, 99–105.
Frost, J. W., and Liess, B. (1973). *Arch. Gesamte Virusforsch.* **42**, 297–299.
Fukusho, A., Ogawa, N., Yamamoto, H., Sawada, M., and Sazawa, H. (1976). *Infect. Immun.* **14**, 332–336.
Furukawa, T., Plotkin, S., Sedwick, D., and Profeta, M. (1967a). *Nature (London)* **215**, 172–173.
Furukawa, T., Vaheri, A., and Plotkin, S. A. (1967b). *Proc. Soc. Exp. Biol. Med.* **125**, 1098–1102.
Gardiner, A. C., Barlow, R. M., Rennie, J. C., and Keir, W. A. (1972). *J. Comp. Pathol.* **82**, 159–163.
Garoff, H., Simons, K., and Renkonen, O. (1974). *Virology* **61**, 493–504.
Gillespie, J. H., Baker, J. A., and McEntee, K. (1960a). *Cornell Vet.* **50**, 73–79.
Gillespie, J. H., Sheffy, B. E., and Baker, J. A. (1960b). *Proc. Soc. Exp. Biol. Med.* **105**, 679–681.
Gillespie, J. H., Madin, S. H., and Darby, N. B. (1962). *Proc. Soc. Exp. Biol. Med.* **110**, 248–250.
Gillespie, J. H., Madin, S. H., and Darby, N. B. (1963). *Cornell Vet.* **53**, 276–282.
Gregory, K. F., Ng, C. W., and Blizzard, L. E. (1965). *Proc. Can. Soc. Microbiol.* No. 52.
Gutekunst, D. E., and Malmquist, W. A. (1963). *Can. J. Comp. Med.* **27**, 121–123.
Hafez, S. M., and Liess, B. (1972a). *Acta Virol. (Engl. Ed.)* **16**, 388–398.
Hafez, S. M., and Liess, B. (1972b). *Acta Virol. (Engl. Ed.)* **16**, 399–408.
Hafez, S. M., Petzoldt, K., and Reczko, E. (1968). *Acta Virol. (Engl. Ed.)* **12**, 471–473.
Heggie, A. D. (1976). *Teratology* **15**, 47–56.
Hermodsson, S., and Dinter, Z. (1962). *Nature (London)* **194**, 893–894.
Hobbins, T. E., and Smith, K. O. (1968). *Proc. Soc. Exp. Biol. Med.* **129**, 407–412.

Holmes, I. H., and Warburton, M. F. (1967). *Lancet* **ii,** 1233–1236.
Holmes, I. H., Wark, M. C., and Warburton, M. F. (1969). *Virology* **37,** 15–25.
Horstmann, D. M., Liebhaber, H., LeBouvier, G. L., Rosenberg, D. A., and Halstead, S. B. (1970). *N. Engl. J. Med.* **283,** 771–778.
Horzinek, M. (1966). *J. Bacteriol.* **92,** 1723–1726.
Horzinek, M. (1967). *Arch. Gesamte Virusforsch.* **21,** 447–453.
Horzinek, M. C. (1973). *J. Gen. Virol.* **20,** 87–103.
Horzinek, M. C. (1976). *In* "Commission of European Communities, Agricultural Research Seminar on Studies on Virus Replication," pp. 9–29. Dir. Gen. Sci. Tech. Inf. Inf. Manage., Luxembourg.
Horzinek, M., and Mussgay, M. (1969). *J. Virol.* **4,** 514–520.
Horzinek, M., Reczko, E., and Petzoldt, K. (1967). *Arch. Gesamte Virusforsch.* **21,** 475–478.
Horzinek, M., Maess, J., and Laufs, R. (1971). *Arch. Virusforsch.* **33,** 306–318.
Horzinek, M. C., van Wielink, P. S., and Ellens, D. J. (1975). *J. Gen. Virol.* **26,** 217–226.
Hotchin, J. (1971). *Monogr. Virol.* **3,** 96–102.
Hovi, T., and Vaheri, A. (1970a). *Virology* **42,** 1–8.
Hovi, T., and Vaheri, A. (1970b). *J. Gen. Virol.* **6,** 77–83.
Howard, R. J., Notkins, A. L., and Mergenhagen, S. E. (1969). *Nature (London)* **221,** 873–874.
Huck, R. A., and Aston, F. W. (1964). *Vet. Res.* **76,** 1151–1154.
Huck, R. A., Evans, D. H., Woods, D. G., King, A. A., Stuart, P., and Shaw, I. G. (1975). *Br. Vet. J.* **131,** 427–435.
Hyllseth, B. (1969). *Arch. Gesamte Virusforsch.* **28,** 26–33.
Hyllseth, B. (1970). *Arch. Gesamte Virusforsch.* **30,** 97–104.
Hyllseth, B. (1973). *Arch. Gesamte Virusforsch.* **40,** 177–188.
Igarashi, A., Fukunaga, T., and Fukai, K. (1964). *Biken J.* **7,** 111–119.
Igarashi, A., Harrap, K. A., Casals, J., and Stollar, V. (1976). *Virology* **74,** 174–187.
Inoue, T., Yanagawa, R., Shinagawa, M., and Akiyama, Y. (1975). *Jpn. J. Vet. Sci.* **37,** 569–575.
Jackson, T. A., Osburn, B. I., and Crenshaw, G. L. (1972). *Vet. Rec.* **91,** 223–224.
Johnson, H. N. (1970). *Am. J. Trop. Med. Hyg.* **19,** 537–539.
Jones, T. C., Doll, E. R., and Bryans, J. C. (1957). *Cornell Vet.* **47,** 52–68.
Kahrs, R. F. (1971). *J. Am. Vet. Med. Assoc.* **159,** 1383–1386.
Kahrs, R. F., Baker, J. A., and Robson, D. S. (1966). *Proc. U.S. Livestock Sanit. Assoc.* **70,** 145–153.
Killian, C. S., and Dorsett, P. H. (1974). *Abstr., Am. Soc. Microbiol., Annu. Meet., Chicago* p. 261.
Kleiman, M. B., and Carver, D. H. (1977). *J. Gen. Virol.* **36,** 335–340.
Kniazeff, A. J., Wopschall, L. J., Hopps, H. E., and Morris, C. S. (1973). *In Vitro* **11,** 400–403.
Konishi, S., Akashi, H., Sentsui, H., and Ogata, M. (1975). *Jpn. J. Vet. Sci.* **37,** 259–267.
Lagwinska, E., Stewart, C. C., Adles, C., and Schlesinger, S. (1975). *Virology* **65,** 204–214.
LeBouvier, G. L., and Plotkin, S. (1971). *J. Infect. Dis.* **123,** 220–223.
Liebhaber, H. (1970). *J. Immunol.* **104,** 818–825.
Liebhaber, H., and Gross, P. A. (1972). *Virology* **47,** 684–693.
Lin, T. C., Kang, B. J., Shimizu, Y., Kumagai, T., and Sasahara, J. (1969). *Natl. Inst. Anim. Health Q.* **9,** 10–19.
Loan, R. W. (1964). *Am. J. Vet. Res.* **25,** 1366–1370.

Loan, R. W., and Gustafson, D. P. (1964). *Am. J. Vet. Res.* **25**, 1120–1122.
London, W. T. (1977). *Nature (London)* **268**, 344–345.
Lumsden, W. H. R., Williams, M. C., and Mason, P. J. (1961). *Ann. Trop. Med. Parasitol.* **55**, 389–397.
Maassab, H. F., and Cochran, K. W. (1964). *Proc. Soc. Exp. Biol. Med.* **117**, 410–413.
McCarthy, K., Taylor-Robinson, C. H., and Pillinger, S. E. (1963). *Lancet* **ii**, 593–598.
McClurkin, A. W., Pirtle, E. C., Coria, M. F., and Smith, R. L. (1974). *Arch. Gesamte Virusforsch.* **45**, 285–289.
McCombs, R. M., and Rawls, W. E. (1968). *J. Virol.* **2**, 409–414.
McKercher, D. G., Saito, J. K., Crenshaw, G. L., and Bushnell, R. B. (1968). *J. Am. Vet. Med. Assoc.* **152**, 1621–1624.
McKissick, G. E., and Gustafson, D. P. (1967). *Am. J. Vet. Res.* **28**, 909–914.
Maes, R., Vaheri, A., Sedwick, D., and Plotkin, S. (1966). *Nature (London)* **210**, 384–385.
Maess, J., and Reczko, E. (1970). *Arch. Gesamte Virusforsch.* **30**, 39–46.
Magnusson, P., and Skaaret, P. (1967). *Arch. Gesamte Virusforsch.* **20**, 374–382.
Magnusson, P., Hyllseth, B., and Marusyk, H. (1970). *Arch. Gesamte Virusforsch.* **30**, 105–112.
Mahy, B. W., Rowson, K. E., Parr, C. W., and Salaman, M. H. (1965). *J. Exp. Med.* **122**, 967–981.
Martinez, D., Brinton, M. A., Tachovsky, T. G., and Phelps, A. H. (1980). *Infect. Immun.* **27**, 979–987.
Mahy, B. W., Rowson, K. E., and Parr, C. W. (1967). *J. Exp. Med.* **125**, 277–288.
Manktelow, B. W., Porter, W. L., and Lewis, K. H. C. (1969). *N.Z. Vet. J.* **17**, 245–248.
Mateva, V., Milanov, M., and Chilev, D. (1966). *Vet. Sbir., Sofia* 63, 9. *Vet. Bull. (London)* **36**, 797 (Abstr.).
Matsumura, T., Stollar, V., and Schlesinger, R. W. (1971). *Virology* **46**, 344–355.
Menser, M. A., Harley, J. D., Hertzberg, R., Dorman, D. C., and Murphy, A. M. (1967). *Lancet* **ii**, 387–388.
Mergenhagen, S. E., Notkins, A. L., and Dougherty, S. F. (1967). *J. Immunol.* **99**, 576–581.
Mettler, N. E., Petrelli, R. L., and Casals, J. (1968). *Virology* **36**, 503–504.
Michaelides, M. C., and Schlesinger, S. (1973). *Virology* **55**, 211–217.
Michaelides, M. C., and Schlesinger, S. (1974). *J. Immunol.* **112**, 1560–1564.
Mifune, K., and Matsuo, S. (1975). *Virology* **63**, 278–281.
Monif, G. R. G., Avery, G. B., Korones, S. B., and Sever, J. L. (1965). *Lancet* **i**, 723.
Murphy, F. A., Harrison, A. K., Gary, G. W., Jr., Whitfield, S. G., and Forrester, F. T. (1968). *Lab. Invest.* **19**, 652–662.
Muscoplat, C. C., Johnson, D. W., and Stevens, J. B. (1973). *Am. J. Vet. Res.* **34**, 753–755.
Niwa, A., Yamazaki, S., Bader, J., and Notkins, A. L. (1973). *J. Virol.* **12**, 401–404.
Notkins, A. L. (1965a). *Bacteriol. Rev.* **29**, 143–160.
Notkins, A. L. (1965b). *Fed. Proc., Fed. Am. Soc. Exp. Biol.* **24**, 378.
Notkins, A. L., and Scheele, C. (1963). *Virology* **20**, 640–642.
Notkins, A. L., and Shochat, S. J. (1963). *J. Exp. Med.* **117**, 735–747.
Notkins, A. L., Mahar, S., Scheele, C., and Goffman, J. (1966a). *J. Exp. Med.* **124**, 81–97.
Notkins, A. L., Mergenhagen, S. E., Rizzo, A. A., Scheele, C., and Waldmann, T. A. (1966b). *J. Exp. Med.* **123**, 347–364.
Notkins, A. L., Mage, M., Ashe, W. K., and Mahar, S. J. (1968). *J. Immunol.* **100**, 314–320.
Osburn, B. I., Crenshaw, G. L., and Jackson, T. A. (1972). *J. Am. Vet. Med. Assoc.* **160**, 442–445.

Osburn, B. I., Clarke, G. L., Stewart, W. C., and Sawyer, M. (1973). *J. Am. Vet. Med. Assoc.* **163**, 1165–1167.
Parkman, D., Buescher, E. L., and Arnstein, E. S. (1962). *Proc. Soc. Exp. Biol. Med.* **111**, 225–230.
Parks, J. B., Pritchett, R. F., and Zee, Y. C. (1972). *Proc. Soc. Exp. Biol. Med.* **140**, 595–598.
Peleg, J. (1968). *Virology* **35**, 617–619.
Peter, C. P., Tyler, D. E., and Ramsey, F. K. (1967). *J. Am. Vet. Med. Assoc.* **150**, 46–52.
Pfefferkorn, E. R. (1968). *In* "Molecular Basis of Virology" (H. Fraenkel-Conrat, ed.), pp. 332–350. Reinhold, New York.
Pirtle, E. C., and Kniazeff, A. J. (1968). *Am. J. Vet. Res.* **29**, 1033–1040.
Plagemann, P. G. W., and Swim, H. E. (1966). *Proc. Soc. Exp. Biol. Med.* **121**, 1142–1146.
Plagemann, P. G. W., Watanabe, M., and Swim, H. E. (1962). *Proc. Soc. Exp. Biol. Med.* **111**, 749–754.
Plant, J. W., Littlejohns, I. R., Gardiner, A. C., Vantsis, J. T., and Huck, R. A. (1973). *Vet. Rec.* **92**, 455.
Plotkin, S. A., Boué, A., and Boué, J. G. (1965). *Am. J. Epidemiol.* **81**, 71–85.
Pons, M. W., Schulze, I. T., and Hirst, C. K. (1969). *Virology* **39**, 153–161.
Porter, D. D., and Porter, H. G. (1971). *J. Immunol.* **106**, 1264–1266.
Porter, D. D., Porter, H. G., and Deerhake, B. B. (1969). *J. Immunol.* **102**, 431–436.
Prager, D., and Liess, B. (1976). *Zentralbl. Veterinaermed.* **23**, 458–469.
Prinzie, A. (1964). *Proc. Int. Symp. Stand. Vaccines Measles Serol. Rubella, Lyons* pp. 122–125.
Pritchett, R. F., and Zee, Y. C. (1975). *Am. J. Vet. Res.* **36**, 1731–1734.
Pritchett, R., Manning, J. S., and Zee, Y. C. (1975). *J. Virol.* **15**, 1342–1347.
Radwan, A. I., Burger, D., and Davis, W. C. (1973). *Virology* **53**, 372–378.
Rawls, W. E., and Melnick, J. L. (1966). *J. Exp. Med.* **123**, 795–816.
Rawls, W. E., Desmyter, J., and Melnick, J. L. (1967). *Proc. Soc. Exp. Biol. Med.* **124**, 167–172.
Renkonen, O., Kaarainen, L., Simons, K., and Gahmberg, C. G. (1971). *Virology* **46**, 318–326.
Rhim, J. S., Schell, K., and Huebner, R. J. (1967). *Proc. Soc. Exp. Biol. Med.* **125**, 1271–1274.
Riley, V. (1966). *Science* **153**, 1657–1658.
Riley, V. (1968). *Methods Cancer Res.* **4**, 493–618.
Riley, V. (1974). *Prog. Med. Virol.* **18**, 198–213.
Riley, V., Lilly, F., Huerto, E., and Bardell, D. (1960). *Science* **132**, 545–547.
Ritchie, A. E., and Fernelius, A. L. (1969). *Arch. Gesamte Virusforsch.* **28**, 369–389.
Rockborn, G., Diderholm, H., and Dinter, Z. (1974). *Arch. Gesamte Virusforsch.* **45**, 128–134.
Rowson, K. E. K., and Mahy, B. W. J. (1975). "Lactic Dehydrogenase Virus," Virology Monographs. 13. Springer-Verlag, Berlin and New York.
Rowson, K. E. K., Parr, I. B., and Alper, T. (1968). *Virology* **36**, 157–159.
Russell, B., Selzer, G., and Goetze, H. (1967). *J. Gen. Virol.* **1**, 305–310.
Salaün, J. J., Klein, J. M., and Mebrard, G. (1975). International Catalogue of Arboviruses Including Certain Other Viruses of Vertebrates. DHEW, Pub. No. 11, CDC75-8301, p. 570. U.S. Gov. Printing Office, Washington, D.C.
Sato, V., Nishimura, Y., Hanaki, T., and Norbuto, K. (1964). *Arch. Gesamte Virusforsch.* **14**, 394–403.

Schiff, L. J., and Storz, J. (1972). *Arch. Gesamte Virusforsch.* **36,** 218–225.
Schluederberg, A., Ajello, C., and Evans, B. (1976). *Infect. Immun.* **14,** 1097–1102.
Schmidt, N. J., Lennette, E. H., and Dennis, J. (1969). *Proc. Soc. Exp. Biol. Med.* **132,** 128–133.
Schwartz, W. L., Solorzano, R. F., Hamlin, H. H., and Thigpen, J. E. (1967). *J. Am. Vet. Med. Assoc.* **150,** 192–195.
Sedwick, W. D., and Sokol, F. (1970). *J. Virol.* **5,** 478–489.
Seifried, O., and Cain, C. B. (1932). *J. Exp. Med.* **56,** 351–362.
Shapiro, D., Trent, D., Brandt, W. E., and Russell, P. K. (1972). *Infect. Immun.* **6,** 206–209.
Shimizu, Y., Furuuchi, S., Kumagai, T., and Sasahara, J. (1970). *Am. J. Vet. Res.* **31,** 1787–1794.
Shope, R. E. (1958). *J. Exp. Med.* **107,** 609–622.
Simmons, D. T., and Strauss, J. H. (1972). *J. Mol. Biol.* **71,** 599–613.
Slavin, G. (1938). *J. Comp. Pathol. Ther.* **51,** 213–224.
Snodgrass, M. J., Lowry, D. S., and Hanna, M. G., Jr. (1972). *J. Immunol.* **108,** 877–892.
Snowdon, W. A., Parsonson, I. M., and Brown, M. L. (1975). *J. Comp. Pathol.* **85,** 241–251.
Spirin, A. S. (1962). *Prog. Nucleic Acid Res. Mol. Biol.* **1,** 301.
Stevens, T. M., and Schlesinger, R. W. (1965). *Virology* **27,** 103–112.
Stewart, G. L., Parkman, P. D., Hopps, H. E., Douglas, R. D., Hamilton, J. P., and Mayer, H. M., Jr. (1967). *N. Engl. J. Med.* **276,** 554–557.
Stewart, W. C., Carbrey, E. A., Jenney, E. W., Brown, C. L., and Kresse, J. I. (1971). *J. Am. Vet. Med. Assoc.* **159,** 1556–1563.
Stim, T. B. (1969). *J. Gen. Virol.* **5,** 329–338.
Stollar, V. (1969). *Virology* **39,** 426–438.
Stollar, V., and Thomas, V. L. (1975). *Virology* **69,** 367–377.
Strauss, J. H., Burge, B. W., and Darnell, J. E., Jr. (1969). *Virology* **37,** 367–376.
Sulkin, S. E., Burns, K. F., Shelton, D. F., and Wallis, C. (1962). *Tex. Rep. Biol. Med.* **20,** 113–127.
Tanaka, Y., Unaba, Y., Omori, T., and Matumoto, M. (1968). *Jpn. J. Microbiol.* **12,** 201–210.
Tauraso, N. M., Shelokov, A., Palmer, A. E., and Allen, A. M. (1968). *Am. J. Trop. Med. Hyg.* **17,** 422–431.
Thomsen, R., Laufs, R., and Miller, J. (1968). *Arch. Gesamte Virusforsch.* **23,** 332–345.
Torlone, V., and Titoli, F. (1964). *Vet. Bull.* (*London*) **35,** 630. (Abstr.)
Torlone, V., Titoli, F., and Gialleti, L. (1965). *Life Sci.* **4,** 1707–1713.
Trousdale, M. D., Trent, D. W., and Shelokov, A. (1975). *Proc. Soc. Exp. Biol. Med.* **150,** 707–711.
Ushimi, C., Tajima, M., Tanaka, S., Nakajima, H., Shimizu, Y., and Furuuchi, S. (1969). *Natl. Inst. Anim. Health Q.* **9,** 28–34.
Vaheri, A., and Hovi, T. (1972). *J. Virol.* **9,** 10–16.
Vaheri, A., Sedwick, W. D., Plotkin, S. A., and Maes, R. (1965). *Virology* **27,** 239–241.
Vaheri, A., Sedwick, W. D., and Plotkin, S. A. (1967). *Proc. Soc. Exp. Biol. Med.* **125,** 1086–1092.
van der Zeijst, B. A. M., Horzinek, M. C., and Moennig, V. (1975). *Virology* **68,** 418–425.
Vantsis, J. T., Barlow, R. M., Fraser, J., Rennie, J. C., and Mould, D. L. (1976). *J. Comp. Pathol.* **86,** 111–120.
Von Bonsdorff, C. H., and Vaheri, A. (1969). *J. Gen. Virol.* **5,** 47–51.
Ward, G. W. (1969). *Cornell Vet.* **59,** 570–576.

Ward, G. W. (1971). *Cornell Vet.* **61**, 179–185.
Weller, T. H., and Neva, F. (1962). *Proc. Soc. Exp. Biol. Med.* **111**, 215–225.
Welsh, R. M., and Pfau, C. J. (1972). *J. Gen. Virol.* **14**, 177–187.
Wilhelm, A. R., and Gerone, P. J. (1970). *Appl. Microbiol.* **20**, 612–615.
Williams, M. C., Simpson, D. I. H., and Shepherd, R. C. (1964). *Nature (London)* **203**, 670.
Wood, O., Tauraso, N., and Liebhaber, H. (1970). *J. Gen. Virol.* **7**, 129–136.
Woods, W. A., and Robbins, F. C. (1968). *J. Gen. Virol.* **3**, 43–49.
Woods, W. A., Johnson, R. T., Hostetler, D. D., Lepow, M. L., and Robbins, F. C. (1966). *J. Immunol.* **96**, 253–260.
Yamazaki, S., and Notkins, A. L. (1973). *J. Virol.* **11**, 473–478.
Zeegers, J. J. W., Van der Zeijst, B. A. M., and Horzinek, M. C. (1976). *Virology* **73**, 200–205.

Index

A

L

M

N

O

S